GEOTECHNICAL ENGINEERING: PRINCIPLES AND PRACTICES

Second Edition

GEOTECHNICAL ENGINEERING: PRINCIPLES AND PRACTICES

Second Edition

Donald P. Coduto
California State Polytechnic University, Pomona
Man-chu Ronald Yeung
California State Polytechnic University, Pomona
William A. Kitch
California State Polytechnic University, Pomona

PEARSON

Upper Saddle River Boston Columbus San Francisco New York
Indianapolis London Toronto Sydney Singapore Tokyo Montreal
Dubai Madrid Hong Kong Mexico City Munich Paris Amsterdam Cape Town

Vice President and Editorial Director, ECS: Marcia J. Horton
Executive Editor: Holly Stark
Editorial Assistant: Keri Rand
Vice President, Production: Vince O'Brien
Marketing Manager: Tim Galligan
Marketing Assistant: Mack Patterson
Senior Managing Editor: Scott Disanno
Production Project Manager: Clare Romeo
Senior Operations Specialist: Alan Fischer
Operations Specialist: Lisa McDowell
Art Director: Jayne Conte
Cover Designer: Suzanne Behnke
Cover Illustration/Photo(s): David Nunuk/nunukphotos.com
Manager, Rights and Permissions: Zina Arabia
Manager, Visual Research: Beth Brenzel
Image Permission Coordinator: Debbie Latronica
Manager, Cover Visual Research & Permissions: Karen Sanatar
Composition: Laserwords Private Limited, Chennai, India
Full-Service Project Management: Gowri Vasanthkumar, Laserwords
Printer/Binder: LSC Communications
Typeface: 10/12 Times Ten

Credits and acknowledgments borrowed from other sources and reproduced, with permission, in this textbook appear on appropriate page within text.

About the cover photo

The Tower of Pisa began to tilt during its construction in the 12th century. By the time construction was completed in 1370, the tower was tilting approximately 2° to the south. The tilt of the tower gradually increased over the centuries until it reached 5½° in 1990, by which time the top of the tower was approximately 4½ meters out of plumb. The tower was closed to the public in 1990 for fear that it might collapse, and a major international effort was undertaken to restore it to a safe condition. The cover photo shows the technique used to stabilize the tower. It involved carefully removing soil from under the foundation on the north side of the tower with the drilling equipment shown in the photo. This allowed the tower to settle back to the north and reduced the tilt by ½°, moving the top of the tower back to the north by ½ meter. The tower is now back to the position it was in during the 1830s and has not moved significantly since being stabilized. The tower was reopened to the public in 2001 (Burland et al. 2003 and 2009).

Library of Congress Cataloging-in-Publication Data

Coduto, Donald P.
Geotechnical engineering: principles and practices / Donald P. Coduto, Man-chu Ronald Yeung, William A. Kitch. -- 2nd ed.
p. cm.
Includes bibliographical references and index.
ISBN 0-13-236868-4
1. Engineering geology. I. Yeung, Man-chu Ronald, II. Kitch, William A. III. Title.
TA705.C62 2011
624.1'51--dc22

2010000164

Prentice Hall
is an imprint of

www.pearsonhighered.com

ISBN-13: 978-0-13-236868-1
ISBN-10: 0-13-236868-4

3 2019

Contents

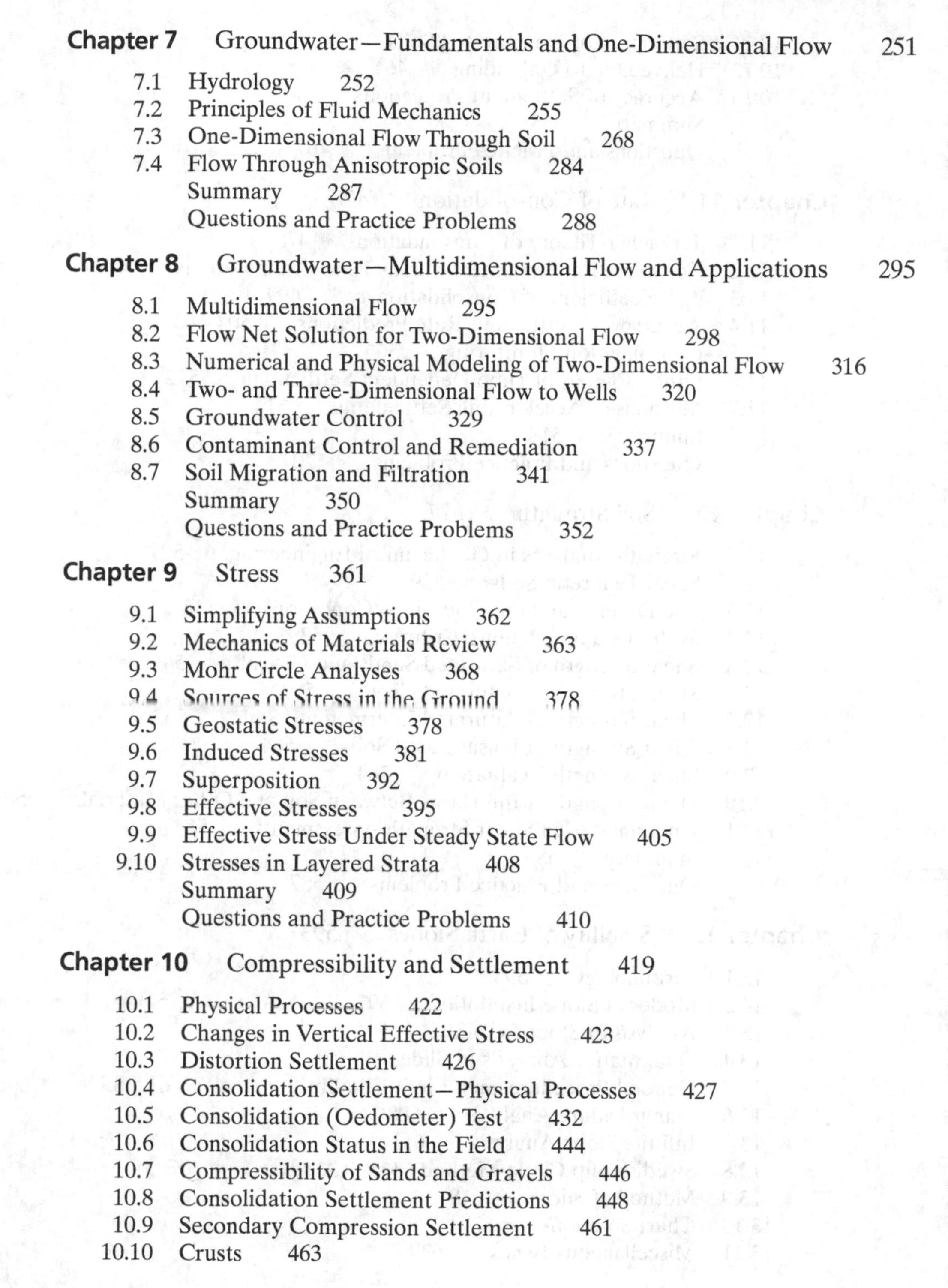

Preface

This second edition of *Geotechnical Engineering: Principles and Practices* reflects our experiences using the first edition with our own students over the past 12 years, as well as constructive suggestions we received from faculty, students, and practicing engineers. Some topics have been condensed or eliminated, while others have been expanded, clarified, or added. The primary target audience remains the same: undergraduate civil engineering students in their first geotechnical engineering course.

WHAT IS NEW IN THIS EDITION

This edition benefits from the addition of two new co-authors: Man-chu Ronald Yeung and William A. Kitch. The three of us are colleagues at Cal Poly Pomona. Each brings unique experiences and skills to the manuscript, and we believe the final product has benefitted from this collaboration.

We have strengthened the coverage of "principles" to facilitate a better understanding of fundamental geotechnical engineering concepts and to provide a firmer foundation for more advanced studies. These discussions include extensive narratives intended to impart an understanding of the underlying physical processes, not just an ability to do computations. The book also retains a "practices" component, which introduces students to the practical application of these principles to real engineering problems. Although this book is not intended to be a comprehensive treatise on geotechnical engineering practice, we have found that a mix of theory and application facilitates the most effective learning.

Every chapter has some updates, some of which are quite extensive. The most noteworthy improvements include the following:

- The chapter on soil strength has been updated to provide more in-depth coverage of the underlying physical behavior.
- The chapter on slope stability has been enhanced.
- The coverage of structural foundations has been expanded to two chapters.
- A new appendix, *Finite Difference Solutions to Flow Problems*, has been added.

- Stand-alone chapters on geoenvironmental engineering, dams and levees, soil improvement, and geotechnical earthquake engineering have been eliminated. Key points from these chapters have been condensed and moved to other locations.
- Many of the homework problems have been updated and new homework problems have been added.

The solutions manual as well as PowerPoint figures of all images and tables from this book can be downloaded electronically from our Instructor's Resource Center located at www.pearsonhighered.com/Coduto <http://www.pearsonhighered.com/Coduto>. The material available through the Instructor Resource Center is provided solely for the use of instructors in teaching their courses and assessing student learning. All requests for instructor access are verified against our customer database and/or through contacting the requestor's institution. Contact your local sales representative for additional assistance or support.

Acknowledgments

This new edition would not have been possible without the help and support of many people. First and foremost, we thank Christopher Sandoval, who is one of our alumni and a practicing geotechnical engineer. Chris handled a great deal of the organizational and editorial work, allowing us to focus our energies on the detailed contents of the manuscript. We also thank Holly Stark and Clare Romeo at Pearson/Prentice-Hall, and Gowri Vasanthkumar at Laserwords, who provided professional advice and support throughout the writing and editing process. Finally, we thank our families for their patience and understanding.

Donald P. Coduto
Man-chu Ronald Yeung
William A. Kitch

Claremont, California
January 2010

About the Authors

Donald P. Coduto is currently a professor of geotechnical engineering and chair of the Civil Engineering Department at the California State Polytechnic University, Pomona. He earned a B.S. in Civil Engineering from the California State Polytechnic University, Pomona, an M.S. in Geotechnical Engineering from the University of California, Berkeley, and an MBA from the Claremont Graduate University. He is an ASCE Fellow, a licensed Civil Engineer and a licensed Geotechnical Engineer, and has worked on a variety of geotechnical projects for both private and public sector clients.

Man-chu Ronald Yeung is currently a professor of civil engineering at the California State Polytechnic University, Pomona. He received a B.S. in Civil Engineering in 1986, an M.S. in Geotechnical Engineering in 1987, and a Ph.D. in Civil Engineering in 1991, all from the University of California, Berkeley. Before joining Cal Poly Pomona in 2005, Dr. Yeung had worked for several consulting firms and taught at several universities including Montana Tech, San Jose State University, and The University of Hong Kong. He is currently a member of the Editorial Board of the ASCE Journal of Geotechnical and Geoenvironmental Engineering, a member of the ASCE Rock Mechanics Committee, and the Treasurer of the Geotechnical Engineering Technical Group of the ASCE Los Angeles Section. He has been a registered Civil Engineer in California since 1994.

William A. Kitch is currently an associate professor of civil engineering at the California State Polytechnic University, Pomona. He received his B.S. in Civil Engineering in 1982 and his M.S. in Civil Engineering in 1983, both from the University of Illinois, Urbana-Champaign. He earned his Ph.D. in Civil Engineering in 1991 from the University of Texas at Austin. He is a retired Lt Col in the US Air Force and had over 23 years of practicing engineering experience in both the private and public sectors. He is a registered Civil Engineer in California and Colorado.

Notation and Units of Measurement

There is no universally accepted notation in geotechnical engineering. However, the notation used in this book, as described in the following table, is generally consistent with popular usage.

Symbol	Description	Typical Units		Defined on Page
		English	SI	
A	Cross-sectional area	ft^2	m^2	258
A	Base area of foundation	ft^2	m^2	674
A	Percentage of soil passing #200 sieve	percent	percent	345
A_f	Cross-sectional area at failure	in^2	mm^2	571
A_0	Initial cross-sectional area	in^2	mm^2	571
a	Cross-sectional area of standpipe	ft^2	cm^2	279
a	Length of a cell in flow net	ft	m	301
a_y	Yield acceleration	ft/s^2	m/s^2	635
B	Width of loaded area (such as a footing)	ft	m	384
b	Width of a cell in flow net	ft	m	301
b	Unit length	ft	m	608
C	Hazen's coefficient	—	$cm/s/mm^2$	281
C_A	Aging factor	Unitless	Unitless	130
C_B	SPT borehole diameter correction	Unitless	Unitless	93
C_c	Coefficient of curvature	Unitless	Unitless	145
C_c	Compression index	Unitless	Unitless	441
C_{OCR}	Overconsolidation correction factor	Unitless	Unitless	130
C_P	Grain size correction factor	Unitless	Unitless	130
C_R	SPT rod length correction	Unitless	Unitless	93
C_R	Relative compaction	Percent	Percent	222
C_r	Recompression index	Unitless	Unitless	441
C_S	SPT sampler correction	Unitless	Unitless	93
C_u	Coefficient of uniformity	Unitless	Unitless	145
C_α	Secondary compression index	Unitless	Unitless	461
C_1	Correction factor of footing depth	Unitless	Unitless	694

Symbol	Description	Typical Units		Defined on Page
		English	SI	
C_2	Correction factor of creep	Unitless	Unitless	694
C_1	Correction factor of footing shape	Unitless	Unitless	694
c_T	Total cohesion	lb/ft^2	kPa	545
c'	Effective cohesion	lb/ft^2	kPa	533
c_v	Coefficient of consolidation	ft^2/day	m^2/day	479
D	Particle diameter	in	mm	144
D	Depth to failure surface	ft	m	611
D	Depth of footing	ft	m	657
D_{li}	The size of openings in the larger of two sieves	—	cm	282
D_r	Relative density	Percent	Percent	129
D_{si}	The size of openings in the smaller of two sieves	—	cm	282
D_w	Depth from ground surface to groundwater table	ft	m	256
D_{10}	Particle size at which 10% is finer (comparable definition for D values with other subscripts)	—	mm	145
d	Diameter of capillary rise tube	in	mm	267
d	Diameter of vane	in	mm	580
d	Moment arm	ft	m	615
d_{85}	Particle size at which 85% of the soil to be filtered is finer	—	mm	345
E	Modulus of elasticity	lb/ft^2	kPa	366
E	Normal side force	lb	kN	611
E_D	DMT modulus	lb/ft^2	kPa	104
E_m	SPT hammer efficiency	Unitless	Unitless	93
E_s	Equivalent modulus of soil layer	lb/ft^2	kPa	694
e	Void ratio	Unitless	Unitless	128
e	Base of natural logarithms	2.7183	2.7183	
e_{max}	Maximum void ratio	Unitless	Unitless	129
e_{min}	Minimum void ratio	Unitless	Unitless	129
e_p	Void ratio at end of primary consolidation	Unitless	Unitless	461
e_0	Initial void ratio	Unitless	Unitless	436
F	Fines content (% passing #200 sieve)	Percent	Percent	168
F	Factor of safety	Unitless	Unitless	543
f_i	Fraction of soil between two sieve size	Unitless	Unitless	282
f_{sc}	CPT cone side friction	T/ft^2	MPa or kg/cm^2	100
f_{sc1}	CPT cone side friction corrected for overburden stress	T/ft^2	MPa or kg/cm^2	100
G	Shear modulus	lb/ft^2	kPa	366
G_h	Equivalent fluid density	lb/ft^3	kN/m^3	742
G_L	Specific gravity of soil–water mixture	Unitless	Unitless	144
G_s	Specific gravity of solids	Unitless	Unitless	127
g	Acceleration due to gravity	ft/s^2	m/s^2	127
H	Thickness of soil strata or soil layer	ft	m	285

Symbol	Description	Typical Units		Defined on Page
		English	SI	
H	Height of wall	ft	m	722
H_a	Saturated thickness of aquifer	ft	m	327
H_{dr}	Maximum drainage distance	ft	m	481
H_{fill}	Thickness of proposed fill	ft	m	424
h	Total head	ft	m	259
h_c	Height of capillary rise	ft	m	266
h_p	Pressure head	ft	m	259
h_v	Velocity head	ft	m	259
h_w	Total head inside well casing during pumping	ft	m	320
h_z	Elevation head	ft	m	259
h_0	Total head in aquifer before pumping	ft	m	320
h_1	Total head in farthest observation well	ft	m	327
h_2	Total head in nearest observation well	ft	m	327
I_D	DMT material index	Unitless	Unitless	104
I_L	Liquidity index	Unitless	Unitless	156
I_P	Plasticity index	Unitless	Unitless	155
I_σ	Influence factor for vertical stress	Unitless	Unitless	383
I_ε	Strain influence factor	Unitless	Unitless	694
$I_{\varepsilon p}$	Peak strain influence factor	Unitless	Unitless	694
i	Hydraulic gradient	Unitless	Unitless	261
j	Seepage force per unit volume of soil	lb/ft^3	kN/m^3	406
K	Coefficient of lateral earth pressure	Unitless	Unitless	398
K_a	Coefficient of active earth pressure	Unitless	Unitless	724
K_D	DMT horizontal stress index	Unitless	Unitless	104
K_p	Coefficient of passive earth pressure	Unitless	Unitless	726
K_0	Coefficient of lateral earth pressure at rest	Unitless	Unitless	721
k	Hydraulic conductivity	ft/s	cm/s	271
k_{eq}	Equivalent hydraulic conductivity	ft/s	cm/s	307
k_n	Hydraulic conductivity normal to fabric	ft/s	cm/s	348
k_x	Horizontal hydraulic conductivity	ft/s	cm/s	284
k_z	Vertical hydraulic conductivity	ft/s	cm/s	284
L	Length perpendicular to cross-section	ft	m	300
L	Length of loaded area (such as a footing)	ft	m	384
LL	Liquid limit (see w_L)	Unitless	Unitless	153
l	Distance the water travels	ft	m	261
l	Length along shear surface	ft	m	608
M	Mass	lb_m	kg	124
M_c	Mass of can	lb_m	kg	125
M_s	Mass of solids	lb_m	kg	124
M_w	Mass of water	lb_m	g	124
M_1	Mass of moist sample and can	lb_m	g	125
M_2	Mass of dry sample and can	lb_m	g	125
m	Slice width	ft	m	624
N	SPT blow count recorded in field	Blows/ft	Blows/300 mm	90
N	Normal force	lb	kN	611

Symbol	Description	Typical Units		Defined on Page
		English	SI	
N_a	Total normal force between wall and soil under the active condition	lb	kN	732
N_c, N_q, N_γ	Bearing capacity factors	Unitless	Unitless	678
N_D	Number of equipotential drops	Unitless	Unitless	302
N_F	Number of flow tubes	Unitless	Unitless	302
N_F	Factor in Cousins' charts	Unitless	Unitless	628
N_p	Total normal force between wall and soil under the passive condition	lb	kN	735
N_{60}	SPT blow count corrected for field procedures	Blows/ft	Blows/ 300 mm	93
$N_{1,60}$	SPT blow count corrected for field procedures and overburden stress	Blows/ft	Blows/ 300 mm	94
n	Porosity	Percent	Percent	129
n_a	Air porosity	Percent	Percent	129
n_e	Effective porosity	Percent	Percent	275
n_w	Water porosity	Percent	Percent	129
O_{95}	Equivalent opening size of geotextile	—	mm	347
OCR	Overconsolidation ratio	Unitless	Unitless	446
P	Point load	lb	kN	381
P_f	Normal load at failure	lb	kN	573
P_a	Total resultant force between wall and soil under the active condition	lb	kN	732
P_p	Total resultant force between wall and soil under the passive condition	lb	kN	734
P_0	Resultant horizontal force between wall and soil under the at-rest condition	lb	kN	723
PI	Plasticity index (see I_P)	Unitless	Unitless	155
PL	Plastic limit (see w_P)	Unitless	Unitless	154
p_a	Atmospheric pressure	lb/ft^2	kPa	695
p_a	Resultant active pressure	lb/ft^2	kPa	729
p_p	Resultant passive pressure	lb/ft^2	kPa	731
Q	Flow rate	ft^3/s	m^3/s	258
Q_c	Compressibility factor	Unitless	Unitless	131
q	Bearing pressure (or gross bearing pressure)	lb/ft^2	kPa	383
q_a	Allowable bearing capacity	lb/ft^2	kPa	682
q_c	CPT cone resistance	T/ft^2	MPa or kg/cm^2	96
q_{c1}	CPT cone resistance corrected for overburden stress	T/ft^2	MPa or kg/cm^2	98
q_u	Unconfined compressive strength	lb/ft^2	kPa	573
q_{ult}	Ultimate bearing capacity	lb/ft^2	kPa	677
R	Distance from load to point	ft	m	381
R	Radius of slip circle	ft	m	613
R_f	Friction ratio (cone penetration test)	Percent	Percent	97
r	Horizontal component of distance from load to point	ft	m	381

Symbol	Description	Typical Units		Defined on Page
		English	SI	
r_w	Radius of well casing	ft	m	320
r'_w	Equivalent radius of well casing	ft	m	328
r_0	Radius of influence	ft	m	320
r_1	Radius from pumped well to farthest observation well	ft	m	327
r_2	Radius from pumped well to nearest observation well	ft	m	327
S	Number of stories in a building	Unitless	Unitless	81
S	Degree of saturation	Percent	Percent	125
S	Shear side force	lb	kN	611
S_0	Specific surface of soil particle, per unit volume	—	1/cm	281
S_F	Shape factor	Unitless	Unitless	282
S_t	Sensitivity	Unitless	Unitless	560
s	Shear strength	lb/ft^2	kPa	530
s_u	Undrained shear strength	lb/ft^2	kPa	556
T	Pore shape factor	Unitless	Unitless	281
T	Shear force	lb	kN	609
T	Thickness of footing	ft	m	657
T_f	Torque at failure	in-lb	N-m	580
T_v	Time factor	Unitless	Unitless	485
t	Time	s	s	130
t	Thickness of geotextile	in	mm	348
t_{adj}	Adjusted time (settlement computations)	years	yr	496
t_c	Duration of construction period	years	yr	496
t_p	Time required to complete primary consolidation	years	yr	462
t_{90}	Time to complete 90% of primary consolidation	years	yr	501
U	Degree of consolidation	Percent	Percent	491
u	Pore water pressure	lb/ft^2	kPa	264
u_e	Excess pore water pressure	lb/ft^2	kPa	429
u_f	Pore water pressure at failure	lb/ft^2	kPa	577
$u_{e,normal}$	Excess pore water pressure due to induced mean normal stress	lb/ft^2	kPa	553
$u_{e,shear}$	Excess pore water pressure due to induced deviator stress	lb/ft^2	kPa	553
u_h	Hydrostatic pore water pressure	lb/ft^2	kPa	264
V	Volume	ft^3	m^3	124
V_a	Volume of air	ft^3	m^3	124
V_a	Total shear force between wall and soil under the active condition	lb	kN	732
V_p	Total shear force between wall and soil under the passive condition	lb	kN	735
V_{cone}	Volume of sand cone below valve	ft^3	m^3	231
V_f	Volume of fill	yd^3	m^3	239
V_m	Volume of Proctor mold	ft^3	m^3	216
V_s	Volume of solids	ft^3	m^3	124

Symbol	Description	Typical Units		Defined on Page
		English	SI	
V_v	Volume of voids	ft^3	m^3	124
V_w	Volume of water	ft^3	m^3	124
v	Velocity	ft/s	m/s	144
v_a	Apparent velocity	ft/s	m/s	274
v_s	Seepage velocity	ft/s	m/s	275
W	Weight	lb	kN	124
W_f	Weight of foundation	lb	kN	674
W_m	Weight of Proctor mold	lb	kN	216
W_{ms}	Weight of Proctor mold + soil	lb	kN	216
W_s	Weight of solids	lb	kN	124
W_w	Weight of water	lb	kN	124
W_1	Initial weight of sand cone apparatus	lb	kN	230
W_2	Final weight of sand cone apparatus	lb	kN	230
w	Moisture content	Percent	Percent	124
w_L	Liquid limit	Percent	Percent	153
w_o	Optimum moisture content	Percent	Percent	214
w_P	Plastic limit	Percent	Percent	154
x	Horizontal distance or coordinate	ft	m	255
x_f	Horizontal distance from load	ft	m	381
y	Horizontal distance or coordinate	ft	m	255
y_f	Horizontal distance from load	ft	m	381
z	Depth below ground surface	ft	m	255
z_{dr}	Vertical distance from point to nearest drainage boundary	ft	m	485
z_f	Depth below loaded area	ft	m	381
z_w	Depth below groundwater table	ft	m	256
α	Horizontal angle between strike and the vertical plane on which an apparent dip is to be computed	deg	deg	41
α	Inclination of shear surface	deg	deg	608
α	Inclination of back surface of wall from vertical	deg	deg	739
β	Inclination of ground surface from horizontal	deg	deg	627
γ	Shear strain	Unitless	Unitless	366
γ	Unit weight	lb/ft^3	kN/m^3	126
γ'	Effective unit weight	lb/ft^3	kN/m^3	681
γ_b	Buoyant unit weight	lb/ft^3	kN/m^3	127
γ_c	Unit weight of concrete	lb/ft^3	kN/m^3	674
γ_d	Dry unit weight	lb/ft^3	kN/m^3	126
γ_{fill}	Unit weight of fill	lb/ft^3	kN/m^3	424
$\gamma_{d,c}$	Average dry unit weight in cut area	lb/ft^3	kN/m^3	239
$\gamma_{d,f}$	Average dry unit weight in fill area	lb/ft^3	kN/m^3	239
$\gamma_{d,max}$	Maximum dry unit weight	lb/ft^3	kN/m^3	214
γ_{sand}	Unit weight of sand in sand cone test	lb/ft^3	kN/m^3	231
γ_w	Unit weight of water	lb/ft^3	kN/m^3	126

Symbol	Description	Typical Units		Defined on Page
		English	SI	
Δe	Change in void ratio	Unitless	Unitless	436
Δh	Head loss	ft	m	260
ΔV	Change in volume during grading	yd^3	m^3	239
$\Delta\sigma_z$	Induced vertical total stress	lb/ft^2	kPa	381
δ	Dip	deg	deg	41
δ	Total settlement	in	mm	423
δ_a	Apparent dip	deg	deg	41
δ_a	Allowable total settlement	in	mm	685
δ_c	Consolidation settlement	in	mm	422
$\delta_{c,ult}$	Ultimate consolidation settlement	in	mm	450
δ_D	Differential settlement	in	mm	685
δ_{Da}	Allowable differential settlement	in	mm	685
δ_d	Distortion settlement	in	mm	422
δ_s	Secondary compression settlement	in	mm	422
ε	Normal strain	Unitless	Unitless	365
ε_f	Axial strain at failure	Unitless	Unitless	571
ε_z	Vertical normal strain	Unitless	Unitless	434
$\varepsilon_{\parallel}$	Normal strain parallel to load	Unitless	Unitless	366
$\varepsilon_{\perp}$	Normal strain perpendicular to load	Unitless	Unitless	366
η	Dynamic viscosity of soil–water mixture	—	Poise	144
θ	Angle (stress analysis)	deg	deg	320
θ	Angle (slope stability analysis)	deg	deg	614
θ_1	Angle between the plane of interest and the major principal stress plane	deg	deg	372
λ	Vane shear correction factor	Unitless	Unitless	582
$\lambda_{c\phi}$	Factor in slope stability computations	Unitless	Unitless	627
μ	Viscosity of pore fluid	—	Pa-s	281
μ	Coefficient of friction	Unitless	Unitless	530
ν	Poisson's ratio	Unitless	Unitless	366
ρ	Density	lb_m/ft^3	kg/m^3	127
ρ_d	Dry density	lb_m/ft^3	kg/m^3	127
ρ_w	Density of water	lb_m/ft^3	kg/m^3	127
σ	Normal stress	lb/ft^2	kPa	363
σ	Normal pressure acting on a wall	lb/ft^2	kPa	732
σ'	Effective stress	lb/ft^2	kPa	397
σ'_c	Preconsolidation stress	lb/ft^2	kPa	438
σ'_D	Effective stress at depth *D* below the ground surface	lb/ft^2	kPa	677
σ_d	Deviator stress	lb/in^2	kPa	376
σ_m	Mean normal stress	lb/in^2	kPa	376
σ'_m	Overconsolidation margin	lb/ft^2	kPa	446
σ_x	Horizontal total stress	lb/ft^2	kPa	364
σ'_x	Horizontal effective stress	lb/ft^2	kPa	398
σ_y	Horizontal total stress	lb/ft^2	kPa	364
σ'_y	Horizontal effective stress	lb/ft^2	kPa	398

Symbol	Description	Typical Units English	Typical Units SI	Defined on Page
σ_z	Vertical total stress	lb/ft^2	kPa	364
σ'_z	Vertical effective stress	lb/ft^2	kPa	397
σ_{df}	Deviator stress at failure	lb/ft^2	kPa	576
σ'_{zp}	Initial vertical effective stress at depth of the peak strain influence factor	lb/ft^2	kPa	694
σ'_{zf}	Final vertical effective stress	lb/ft^2	kPa	423
σ'_{z0}	Initial vertical effective stress	lb/ft^2	kPa	423
σ_1	Major principal stress	lb/ft^2	kPa	371
σ_2	Intermediate principal stress	lb/ft^2	kPa	371
σ_3	Minor principal stress	lb/ft^2	kPa	371
σ_{1f}	Total major principal stress at failure	lb/ft^2	kPa	576
σ_{3f}	Total minor principal stress at failure	lb/ft^2	kPa	576
σ'_{1f}	Effective major principal stress at failure	lb/ft^2	kPa	576
σ'_{3f}	Effective minor principal stress at failure	lb/ft^2	kPa	576
τ	Shear stress	lb/ft^2	kPa	364
τ	Shear stress acting a wall	lb/ft^2	kPa	732
τ_e	Equilibrium shear stress	lb/ft^2	kPa	605
τ_{max}	Maximum shear stress	lb/ft^2	kPa	373
τ_{zx}	Shear acting on the z plane in the x direction	lb/ft^2	kPa	371
Φ	Potential function	ft^2/s	m^2/s	298
ϕ'	Effective friction angle	deg	deg	530
ϕ_T	Total friction angle	deg	deg	545
ϕ_w	Wall–soil interface friction angle	deg	deg	739
Ψ	Flow function (or stream function)	ft^2/s	m^2/s	299
ψ	Permittivity of geotextile filter	s^{-1}	s^{-1}	348
ψ	Factor in Bishop's Equation	Unitless	Unitless	625

CHAPTER 1

Introduction to Geotechnical Engineering

Virtually every structure is supported by soil or rock. Those that aren't either fly, float, or fall over.

Richard L. Handy (1995)

Geotechnical engineering is the branch of civil engineering that deals with soil, rock, and underground water, and their relation to the design, construction, and operation of engineering projects. This discipline is also called *soils engineering* or *ground engineering*. Nearly, all civil engineering projects must be supported by the ground, and thus require at least some geotechnical engineering. Figure 1.1 illustrates geotechnical aspects of some typical civil engineering projects.

Typical issues addressed by geotechnical engineers include the following:

- What are the soils and rocks in the subsurface at a construction site?
- Can the soils and rocks beneath a construction site safely support the proposed project?
- What groundwater conditions currently exist, how might they change in the future, and what impact do they have on the project?
- What will be the impact of any planned excavation, grading, or filling?
- Are the natural or proposed earth slopes stable? If not, what must we do to stabilize them?
- What kinds of foundations are necessary to support planned structures, and how should we design them?
- If the project requires retaining walls, what kind would be best and how should we design them?

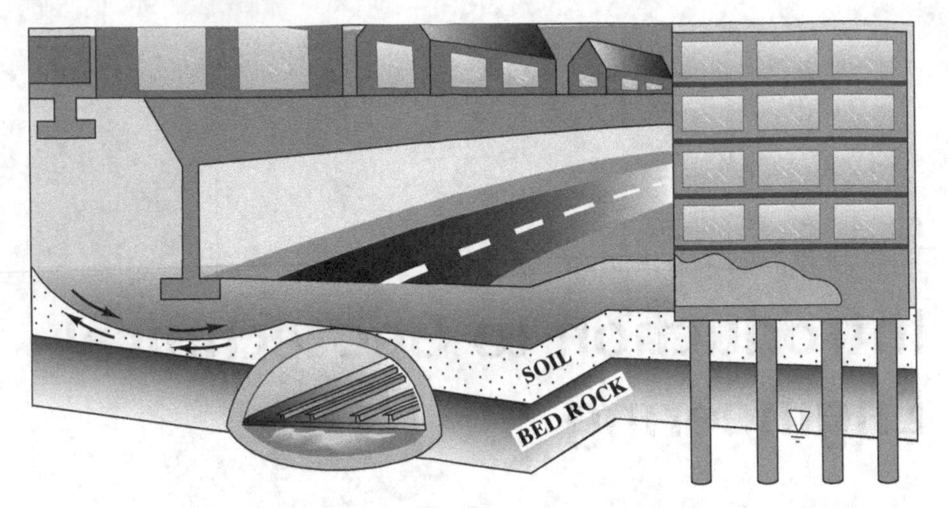

FIGURE 1.1 This figure illustrates geotechnical aspects of some typical civil engineering projects including deep foundations for high rise building, design of a tunnel in soil and rock, shallow foundations for homes, stability of earth slopes, and design of retaining walls.

- If the project requires a tunnel or underground opening, how should it be excavated and supported?
- How will the site respond to potential earthquakes?
- Has the ground become contaminated with chemical or biological materials? Do these materials represent a health or safety hazard? If so, what must we do to rectify the problem?

Sometimes these issues are simple and straightforward and require very little geotechnical engineering. However, in other cases, they are very complex and require extensive exploration, testing, and analysis. At difficult sites, geotechnical concerns may even control the project's technical and economic feasibility.

Geotechnical engineering is closely related to *engineering geology*, which is a branch of geology, as shown in Figure 1.2. Individuals from both disciplines often work together, making contributions from their own expertise to solve practical problems. The discipline combining both geotechnical engineering and engineering geology is sometimes called *geotechnics*.

1.1 GEOTECHNICAL ENGINEERING DESIGN PROCESS

Geotechnical engineers usually begin a geotechnical design project by assessing the underground conditions and the engineering properties of the various strata at the project site. We call this process site exploration and characterization, which is arguably the most important task because it provides input data for all subsequent tasks. Geologists and engineering geologists have a large role to play during this task.

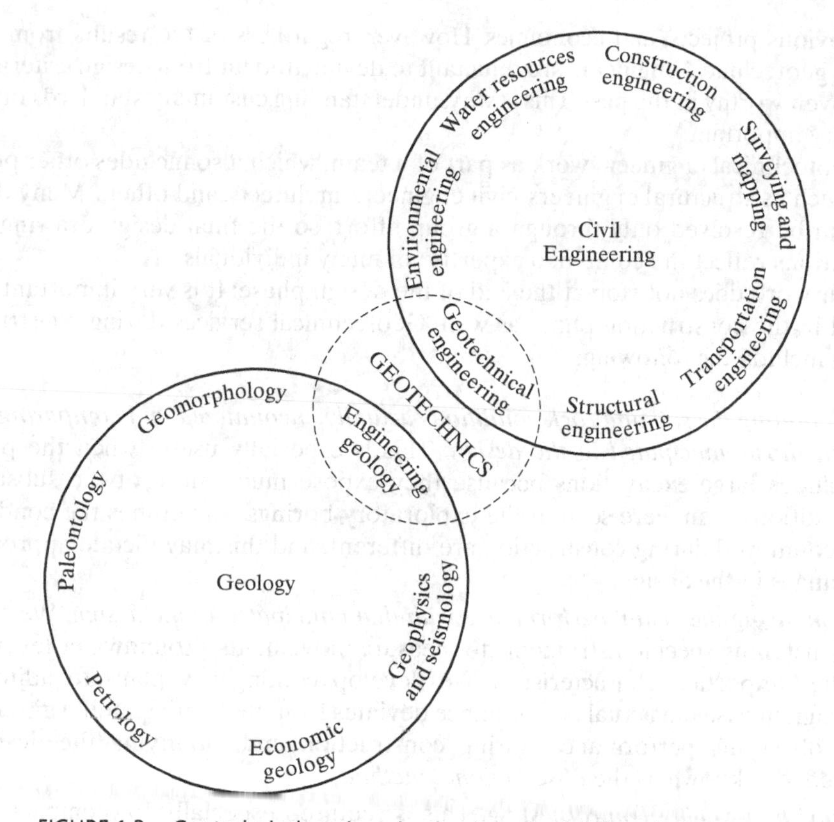

FIGURE 1.2 Geotechnical engineering is a branch of civil engineering, whereas engineering geology is a branch of geology. These two disciplines are closely related, and the discipline combining the two is sometimes called *geotechnics*. Note: This illustration is not a complete listing of the branches of either discipline.

This task usually involves drilling vertical holes called *exploratory borings* into the ground, obtaining soil and rock samples, and testing these samples in a laboratory. It may also involve conducting tests in situ (in-place).

The next step is to perform engineering analyses based on the information gained from the site exploration and characterization program. The tools we use to perform these analyses are collectively known as *soil mechanics* and *rock mechanics;* they include empirical, analytical, and numerical methods. Thus, soil mechanics and rock mechanics are to geotechnical engineering what structural mechanics is to structural engineering. In these fields, "mechanics" refers to the empirical, analytical, and numerical tools, whereas "engineering" is a broader term that includes the entire design and construction process. This book covers only topics in soil mechanics and their application to geotechnical engineering.

We then use the analysis results to develop geotechnical input for design purposes. The design process also requires the application of engineering judgment, experience from previous projects, and economics. However, regardless of the results from these analyses, geotechnical engineers are reluctant to deviate too far from design criteria that have proven worthy in the past. This is why, understanding customary standards of practice is very important.

Geotechnical engineers work as part of a team, which also includes other professionals such as structural engineers, civil engineers, architects, and others. Many design issues can be resolved only through a group effort, so the final design drawings and specifications reflect the combined expertise of many individuals.

Our work does not stop at the end of the design phase: It is very important to be involved in the construction phase as well. Geotechnical services during construction typically include the following:

- *Examining the soil and rock conditions actually encountered and comparing them with those anticipated in the design.* This is especially useful when the project includes large excavations because they expose much more of the subsurface conditions than were seen in the exploratory borings. Sometimes the conditions encountered during construction are different, and this may dictate appropriate changes in the design.
- *Comparing the actual performance with that anticipated in the design.* We do this by installing special instruments to measure movements, groundwater levels, and other important characteristics. We develop contingency plans to adjust the design in case the actual performance deviates from that anticipated. This process of observing performance during construction and modifying the design as needed is known as the *observational method.*
- *Providing quality control testing.* This is required especially for compacted fills and structural foundations.

Occasionally geotechnical services continue beyond the end of construction. For example, sites prone to long-term settlements may require monitoring for months or years after construction. Postconstruction activities also can include investigations of facilities that have not performed satisfactorily and development of remedial measures.

1.2 HISTORICAL DEVELOPMENT

Although this book primarily focuses on current methods of evaluating soil and rock for engineering purposes, it also occasionally discusses the historical development of geotechnical engineering. Part of the reason for these historical vignettes is to gain an appreciation for our heritage as engineers. Another, perhaps more important reason is to help us understand how technical advances have occurred in the past, because this guides us in developing future advancements.

Early Methods (Prior to 1850)

People have been building structures, dams, roadways, aqueducts, and other projects for thousands of years. However, until recently, these projects did not include any rational engineering assessment of the underlying soil or rock. Early construction was based on common sense, experience, intuition, and rules of thumb, and builders passed this collective wisdom orally from generation to generation, often through trade guilds. Early scientists were concerned with more lofty matters, and generally considered the study of soil and rock beneath their dignity.

Sometimes builders used crude tests to assess the soil conditions. For example, the Italian architect Palladio (1508–1580) wrote that firm ground could be confirmed "if the ground does not resound or tremble if something heavy is dropped. In order to ascertain this, one can observe whether some drum skins placed on the ground vibrate and give off a weak sound or whether the water in a vessel placed on the ground gets into motion" (Flodin and Broms, 1981). The primary objective of such assessments seems to have been the identification and subsequent avoidance of sites with poor soil conditions.

These design methods were usually satisfactory so long as the construction projects were modest in scope, similar to previous projects (and thus tied to experience) and built away from obviously poor sites. Using these methods, the ancient builders sometimes accomplished amazing feats of construction, some of which still exist. For example, some dams in India have been in service for more than 2000 years. Unfortunately, the ancient builders also experienced some dramatic failures.

During the Middle Ages, builders began constructing larger and more sophisticated structures such as the cathedrals and related buildings in Europe. These projects pressed beyond the limits of experience, so the old rules of thumb did not always apply and unfortunate failures sometimes occurred. The failures produced new rules of thumb that guided subsequent projects. These trial-and-error methods of developing design criteria continued through the Renaissance and into the beginning of the Industrial Revolution, but it became increasingly evident that they were very tedious and expensive ways to learn.

The Leaning Tower of Pisa, shown in Figure 1.3, is the most famous example of soil-related problems from this era. Construction began in AD 1173 and continued off-and-on for nearly 200 years. The tower began to tilt during construction, so the builders attempted to compensate by providing a slight taper to the upper stories. The movement continued after construction, and by 1982, the top of the 58.4 m (192 ft) tall structure was 5.6 m (18.4 ft) off plumb.

Modern investigations of the subsurface conditions have found a weak clay stratum about 11 m below the ground surface. This clay is very compressible, and has settled under the concentrated weight of the tower. The south side has settled more than the north, which has caused the tower to tilt. *Foundation Design: Principles and Practices* (Coduto, 2001), the companion volume to this book, describes the tower in more detail.

Slowly, builders began to apply the scientific method to various aspects of construction. These efforts attempted to determine why the things we build behave the way they do, express this behavior mathematically, and develop analysis and design methods based on these understandings. Initially, these efforts focused almost exclusively on structural issues. Leonardo da Vinci (1452–1519), the artist/scientist, was one of the few to briefly study the behavior of soils. He observed the angle of repose in sands, proposed test methods to determine the bearing capacity of soils, and speculated

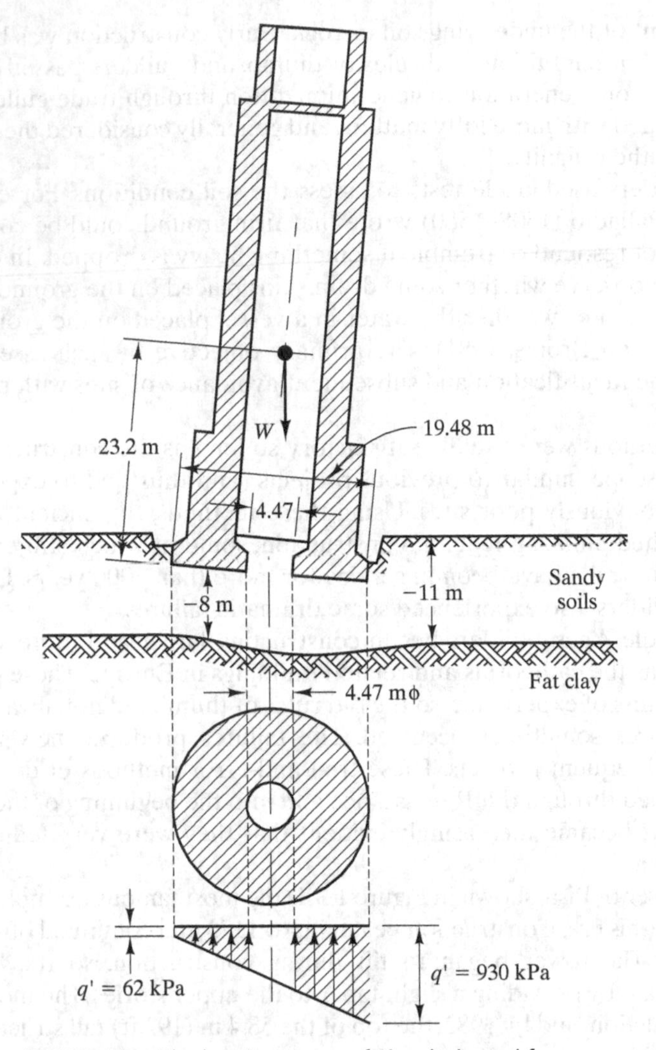

FIGURE 1.3 The leaning tower of Pisa. (Adapted from Terzaghi 1934a.)

on the processes of groundwater hydrology. However, daVinci's ideas on soils do not appear to have extended beyond the pages of his notebooks, and had no impact on design or construction methods.

Engineers and scientists began to address the engineering behavior of soil more seriously during the seventeenth and eighteenth centuries. Most of this early work focused on the analysis and design of retaining walls (Skempton, 1979). It was generally dictated by military needs, and was mostly performed by individuals associated with the army, especially in France. Henri Gautier, B.F. Belidor, Charles Augustin Coulomb, and others developed methods of predicting the forces imparted by soil onto retaining walls, which led to more rational design methods. Coulomb's work, which he published in 1776, is often considered the first example of rational soil mechanics, and still forms the basis for computation of earth pressures acting on walls. Unfortunately, much of this work extended well beyond the ability of eighteenth century builders and designers to measure relevant engineering properties in soil, and thus was difficult to apply to practical problems.

Some scientific investigations of soil behavior continued during the early nineteenth century, including studies of the stability of earth slopes and other topics. However, this work had limited usefulness, was not widely disseminated, and had very little impact on the vast majority of construction projects.

Late Nineteenth Century Developments

The last half of the nineteenth century was a period of rapid industrialization, which produced tremendous growth in both the amount and scope of construction projects. Railroad lines were expanding, urban areas were growing (with the resulting need for infrastructure construction), ports were being enlarged, and larger buildings were being built. Iron and steel had become common civil engineering materials, and reinforced concrete was beginning to appear. These projects drove advances in structural engineering, hydraulic engineering, and other fields. Some of these advancements later became useful to geotechnical engineers. For example, Henri Darcy's work on flow through sand filters for water purification purposes later became the basis for analyses of groundwater flow, and various developments in mechanics of materials, such as that of Otto Mohr, later would be applied to soil. However, progress in geotechnical engineering still lagged behind.

When working at sites with potentially problematic soils, especially soft clays, some engineers drove steel rods into the ground to roughly assess the soil conditions. These tests were called *soundings*. Sometimes engineers used exploratory borings to obtain soil samples, although the information gained from them was purely qualitative. Some advances in analysis and design methods were also developed during this period, including empirical formulas for determining the load capacity of pile foundations. However, we had not yet developed important unifying concepts and did not understand how soils behave, so efforts at assessing subsurface conditions were of limited value.

The increasing size of civil engineering projects, especially after 1880, raised more concerns about the consequences of failure, yet overly conservative designs were too expensive. The time was ripe for geotechnical engineering to emerge as a clearly defined discipline within civil engineering, for inventing better techniques of assessing soil and rock conditions, and for developing sound methods of integrating them into civil engineering practice. Some of the American pioneers in foundation engineering active during this period included:

- *Frederick Baumann (1826–1921)*, a prominent architect in Chicago who published a pamphlet in 1873 that recommended each building column should have an individual footing with a base area proportional to the column load. This was the first written recognition of this relationship in the United States.
- *William Sooy Smith (1830–1916)*, an engineering graduate of both Ohio University and West Point who cofounded the engineering firm Parkinson & Smith in 1857.

During the civil war, he served in the Union army rising to the rank of Brigadier General. After the war, he returned to civil engineering. He was the first engineer to use pneumatic caisson foundations and pioneered the use of deep foundations in Chicago.

- *Daniel Moran (1864–1937)*, a foundation engineer and builder who worked on the foundations for many large buildings in New York's financial district, major bridges across the country, and other monumental projects. He pioneered new methods of constructing bridge foundations.
- *Lazarus White (1874–1953)*, a foundation engineer and builder who developed new methods of design and construction, underpinning, and other advances. He was also a cofounder of the firm Spencer, White, and Prentis.

Geotechnical Engineering in Sweden—Early Twentieth Century

The first large-scale attempts at geotechnical engineering occurred in Sweden during the early decades of the twentieth century (Bjerrum and Flodin, 1960). The Swedes even introduced the word "geotechnical" (in Swedish, Geotekniska) during this period.

Sweden was a likely place for these early developments because extremely poor soil conditions underlie much of the country. Soft, weak clays are present beneath the most populated areas, and they are the source of many problems including excessive settlement and catastrophic landslides. Many of these clays are very *sensitive*, which means they lose strength when disturbed, and thus are prone to dramatic failures.

Old place names that translate to "Earth Fall," "Land Fall," and "Clay Fall," illustrate the long history of landslides in this region. In Norway, which has similar soil problems, landslides killed an average of 17 persons per year between 1871 and 1940 (Flodin and Broms, 1981). These problems became much worse when construction projects created cuts and fills that further destabilized marginal ground.

The city of Göteborg, Sweden suffered from extensive soil-related problems during the early development of its port facilities. These projects required dredging soils, building quays (facilities for docking ships), and other works on the soft clays that existed in the harbor. This was very difficult, and several major landslides occurred during and after construction of these facilities. Port engineers began to assess these landslides and develop methods of safely building port facilities.

Meanwhile, the Swedish State Railways needed to make cuts and fills to provide alignments for new tracks, and these steepened slopes prompted more landslides. One of the more disastrous failures occurred in 1913 when 185 m (600 ft) of track slipped into Lake Aspen. This event prompted the formation of the Geotekniska Kommission (Geotechnical Commission) of the Swedish State Railways to study the problem and to develop solutions. They intended the new term *geotechnical* to reflect the commission's reliance on both geology and civil engineering.

The most prominent engineer in this effort was Wolmar Fellenius (1876–1957) (Figure 1.4). A graduate of the Royal Institute of Technology in Stockholm, Fellenius had become familiar with soil problems when he served as the port engineer in Göteborg. When the commission was formed, he was a professor of hydraulics at the Royal Institute, a position he held until his retirement in 1942. Fellenius became the

FIGURE 1.4 Wolmar Fellenius was the port engineer in Göteborg, Sweden, a professor at the Royal Institute of Technology in Stockholm, and the chairman of the Geotechnical Commissions of the Swedish State Railways. He and his colleagues conducted the first large-scale geotechnical studies. In the process, they developed many of the exploration, sampling, testing and analysis techniques that are used today. (Photograph courtesy of Professor Bengt Fellenius.)

chair of the commission, and John Olsson, also a civil engineer, did much of the day-to-day work.

The commission had to begin by developing new methods of drilling and sampling soils, which was very difficult in the soft clays. They were the first to develop methods of obtaining undisturbed samples of these soils. Then they developed laboratory test equipment, studied the behavior of these soils, and produced new methods of analysis and design. They investigated more than 300 sites and collected 20,000 soil samples. This work was truly a pioneering effort, and is a testimony to the resourcefulness and insights of these men. Their soil mechanics laboratory, established in 1914, appears to have been the first of its kind in the world.

More railroad failures occurred while the commission's work was in progress, most notably, the 1918 landslide at Vita Sikudden, which killed 41 people. These failures further emphasized the importance of their work.

The commission's final report, completed in 1922, was the world's first comprehensive geotechnical report. It presented their methods of investigation and analysis and contained recommendations on how to avoid future landslides. Afterward, many committee members continued to develop new test equipment and refine their analysis and design methods.

These early developments in Sweden represented the first significant efforts at geotechnical engineering, and they influenced subsequent construction in Scandinavia. However, the rest of the world had little or no knowledge of this work until much later. The task of promoting geotechnical engineering on a widespread international level required more people to spread the message, and one of them soon became recognized as a leader in this effort: Karl Terzaghi (Figure 1.5).

Karl Terzaghi

Karl Terzaghi (1883–1963) has often been called *the father of soil mechanics*. Although he was only one of many people who ushered in the new profession we now call *geotechnical engineering*, his influence and early leadership were especially noteworthy. Terzaghi, more than any other, set the tone and direction of the profession and promoted it as a legitimate branch of civil engineering.

Terzaghi was born in Prague, which was then part of Austria. His initial academic work was in mechanical engineering, and he earned an undergraduate degree in that subject. However, he found that it was not to his liking, so his first engineering job was with a civil engineering firm in Vienna that specialized in reinforced concrete. He worked at construction sites in many European locations, which also gave him opportunities to pursue one of his favorite subjects: geology. He later earned a doctorate based on his work in reinforced concrete design.

Throughout this period, Terzaghi became increasingly interested in the ignorance of civil engineers in matters relating to earthwork and foundation design. Although structural design had already reached a high level of sophistication, the design of earthwork and foundations was based on unreliable empirical rules. He felt that this topic needed a more scientific approach and decided to focus his attention on developing rational design methods.

FIGURE 1.5 Karl Terzaghi in 1951. (Photograph courtesy of Margaret Terzaghi-Howe.)

In 1916, he accepted a teaching position at the Imperial School of Engineers in Istanbul (then known as *Constantinople*), and later moved to Robert College, also in Istanbul, where he began research on the behavior of soils, including studies of piping failures in sands beneath dams and settlement in clays. The work on clays eventually led to his theory of consolidation, which we will study in Chapters 10 and 11. This theory, which has since been verified, is considered one of the most significant milestones in civil engineering.

If we wish to define a certain time as the "birth" of geotechnical engineering as a widely recognized discipline, it would be the year 1925, for that was when Terzaghi published the first comprehensive book on the subject. He gave it the title *Erdbaumechanik auf Bodenphysikalischer Grundlage* (German for *The Mechanics of Earth Construction Based on Soil Physics*; Terzaghi, 1925a) and published it in Vienna (Figure 1.6). *Erdbaumechanik* addressed various aspects of what we would now call geotechnical engineering, and did so from a rational perspective that recognized the importance of field observations.

In 1925, Terzaghi also accepted a visiting lectureship position at the Massachusetts Institute of Technology, where he soon became recognized as the leader of a new branch of civil engineering. The same year, he published a series of English articles in the American journal *Engineering News Record* (Terzaghi, 1925b) and a paper in the *Journal of the Boston Society of Civil Engineers* (Terzaghi, 1925c). He also expanded

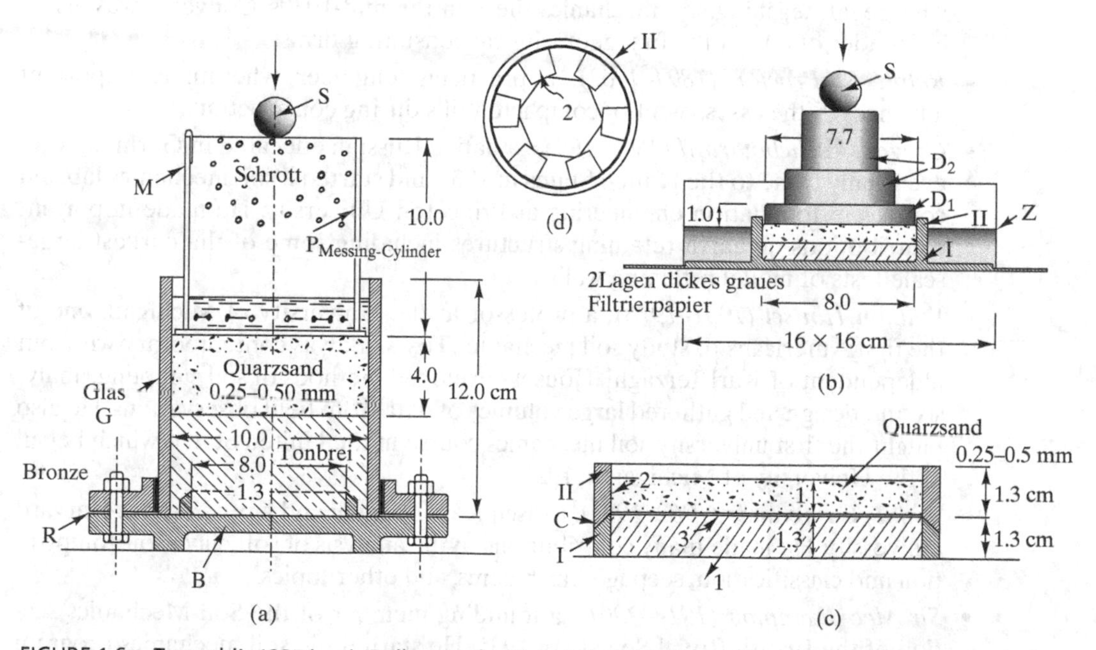

FIGURE 1.6 Terzaghi's 1925 book *Erdbaumechanik* included this illustration of a consolidometer, which is a laboratory device for measuring the settlement of soils. Terzaghi used devices like this to develop his theory of consolidation, which we will discuss in Chapters 10 and 11.

his research interests to include frost heave, pavement design, and other topics, along with continuing his interests in foundations and dams.

In 1929, he returned to Vienna and began serving as a professor at the Technical University. During the next several years, he continued his research activities, along with an active speaking and consulting schedule that brought him to many places in Europe, Asia, Africa, and North America. This work generated extensive interest in soil mechanics. Then, in 1939, following Hitler's Anschluss, he returned to the United States and accepted a professorship at Harvard University. He continued teaching, consulting, and lecturing around the world, but Harvard remained his home for the rest of his life.

Terzaghi had a remarkable ability to develop rational and practical solutions to real engineering problems from a jumble of what had previously been a maze of incoherent facts and observations. He came onto the engineering scene at the right time and with the necessary skills, rising from obscurity to lead the establishment of geotechnical engineering as a rational and legitimate branch of civil engineering.

Additional Twentieth Century Developments

Other prominent engineers also made important contributions to the fledgling profession of geotechnical engineering during the 1920s and 1930s. Some of them were from academic circles, whereas others were practicing engineers and/or contractors. Those who had the most impact on engineering practice in the United States included the following:

- *Fred Converse (1892–1987)*, a professor at the California Institute of Technology who began teaching soil mechanics there in the mid-1930s. Converse was also a cofounder of one of the first geotechnical consulting firms.
- *Ralph R. Proctor (1894–1962)*, a practicing engineer who made important advances in the assessment of compacted fills during construction.
- *Gregory Tschebotarioff (1899–1985)*, a native Russian educated in Germany who eventually came to the United States in 1937 and started a soil mechanics lab and courses in foundation engineering at Princeton University. He made important contributions on earth retaining structures including some of the earliest large-scale tests of model retaining walls.
- *William Housel (1901–1978)*, a professor at the University of Michigan, one of the first Americans to study soil mechanics. His work was contemporary with, but independent of, Karl Terzaghi. Housel developed methods of soil sampling, analysis, and design and gathered large volumes of data from field observations. He also taught the first university soil mechanics course in the United States, which began at the University of Michigan in 1927.
- *Arthur Casagrande (1902–1981)*, a disciple of Terzaghi and a professor at Harvard University. He made many contributions to the analysis of soft clays, soil composition and classification, seepage, earth dams, and other topics.
- *Sir Alec Skempton (1914–2001)*, a founding member of the Soil Mechanics section of the British Royal Society in 1935. He started the soil mechanics program at Imperial College in 1947. Skempton made indispensable contributions to the understanding of effective stress and pore pressures in clays as well as bearing capacity and slope stability analysis.

We reached another important milestone in 1936 when the First International Conference on Soil Mechanics and Foundation Engineering met in Cambridge, Massachusetts. It was the first significant professional conference devoted exclusively to this topic, and its published proceedings represented more technical material on this subject than all of the material published before the conference. The International Society of Soil Mechanics and Foundation Engineering was founded during that conference. Both the society and its conferences continue to be an important means of disseminating knowledge.

In spite of these advancements, Terzaghi was presenting lectures with titles like "Soil Mechanics—A New Chapter in Engineering Science" as late as 1939 to professional audiences who apparently had very little familiarity with the subject (Terzaghi, 1939). This increased awareness of the usefulness of geotechnical engineering, along with the massive construction projects of the 1950s and 1960s, finally established geotechnical engineering as a routine part of nearly all significant civil engineering projects.

Ralph Peck

Ralph Peck (1912–2008), perhaps the most influential of Terzaghi's protégées, began his geotechnical studies as a laboratory assistant to Casagrande at Harvard in 1938 (Figure 1.7). At this time, the city of Chicago was undertaking a major subway project, so the local chapter of the American Society of Civil Engineers invited Terzaghi to speak on tunneling. His chosen topic, "The Danger of Tunneling Beneath Large Cities Founded on Soft Clays," so alarmed the audience that he was soon retained as

FIGURE 1.7 Ralph Peck in 1999. (Photograph courtesy Norwegian Geotechnical Institute.)

a consultant to the city. Terzaghi insisted that the city establish a laboratory to test soils and monitor the tunneling operations (DiBiagio and Flaate, 2000), and chose Peck to manage the laboratory and observe the construction. This was the beginning of several decades of collaboration between these two men.

The knowledge gained by Terzaghi and Peck during the Chicago subway project led to major breakthroughs in the areas of strength of soft clays, earth pressures on braced excavations and tunnel linings, and settlement due to tunneling. Their collaboration also generated the book *Soil Mechanics in Engineering Practice*, a landmark text that is still in print today (in its third edition) more than 60 years after its first publication.

Peck's chief legacy to the geotechnical engineering discipline was as an advocate for the role of judgment in geotechnical engineering practice. "Theory and calculation are no substitute for judgment, but are the basis for sounder judgment," Peck once said (DiBiagio and Flaate, 2000). He had a firm grasp of the uncertainties involved in geotechnical design and knew that engineering judgment was essential in managing these uncertainties. Peck developed the *observational method* discussed earlier in this chapter, and applied this disciplined approach to solving engineering problems (Peck, 1969). Peck attributed this approach to Terzaghi, but it was Peck who defined and promulgated it.

H. Bolton Seed

H. Bolton Seed (1922–1989) was another influential protégée of Terzaghi. In the late 1950s and 1960s, dramatic foundation failures were caused by a number of earthquakes including earthquakes in Jaltipan, Mexico (1959), Cañete, Chile (1960), Niigata, Japan (1964), and Anchorage, Alaska (1964). The study of soil behavior during earthquakes spawned the discipline of geotechnical earthquake engineering. While several engineers had prominent roles in the development of this area of geotechnical engineering, Seed can rightly be called the *father of geotechnical earthquake engineering*. His research on the phenomenon of soil liquefaction was pioneering, and his methods for analyzing the safety of earth dams during earthquakes have been adopted throughout the world.

1.3 MODERN GEOTECHNICAL ENGINEERING

During the last century we have made substantial progress in advancing geotechnical engineering. Our abilities to assess subsurface conditions, predict soil behavior, and accommodate this behavior using appropriate designs are now far better than before. A large number of consulting firms specialize in geotechnical engineering, many government agencies have geotechnical engineering departments, and nearly all civil engineers have occasion to work with geotechnical engineers on their projects. About 15% of American Society of Civil Engineers members now identify geotechnical engineering as their primary or secondary area of interest. Geotechnical engineers also have expanded into new areas, most notably *geoenvironmental engineering*, which deals with underground environmental problems.

As geotechnical engineering matured, it also developed a "personality" that is slightly different from other civil engineering disciplines. These personality traits include the following:

- We work with soil and rock, which are natural geologic materials. As such, their engineering properties are more complex and difficult to characterize than those of manufactured materials such as concrete or steel. Soil and rock properties also vary significantly from one project site to another, and even at different locations within a single site. Therefore, we devote a significant part of our work and budget to site characterization. Unlike structural engineers, who can simply look up material properties in a book, geotechnical engineers must measure the properties of the materials encountered at a site, by sampling material at the site and testing it in a laboratory. To accomplish these tasks, geotechnical engineers and their staffs spend a great deal of time in the field and laboratory.
- Practical economic constraints limit the number of exploratory borings we can drill and the number of laboratory tests we can perform. As a result, we have direct knowledge of only a very small portion of the soil or rock beneath a project site. This introduces many potential sources of error: What are the subsurface conditions between and beyond the borings? Are the samples truly representative of the geologic materials underlying the site? How much sample disturbance has occurred during drilling, recovery, transport, and laboratory testing? What effect does this sample disturbance have on the measured engineering properties?
- Because of the potentially large errors in our site characterization programs, we use a large measure of engineering judgment when bringing the laboratory and field data into our analyses. In addition, our ability to perform quantitative analyses far exceeds the accuracy of the data on which they are based. Therefore, the results of these analyses are usually not very precise. As a result, we typically use larger factors of safety and more conservative designs.
- We rely more heavily on "engineering judgment," which is a combination of experience, subjectivity, reliance on precedent, and other factors.
- We have a more extensive involvement during construction and frequently revise our design recommendations when conditions encountered during construction are different from those anticipated. This is the essence of Peck's observational method.

Geotechnical engineers also spend a great deal of time interacting with others, including general civil engineers, structural engineers, architects, building officials, geologists, contractors, attorneys, and owners. As a result, good written and oral communication skills are very important.

Geotechnical engineers continue to face new technical challenges. The high cost of real estate, especially in urban areas, often dictates the need to build on sites with poor soil conditions—sites we would have rejected in the past. These difficult sites pose special problems, and have resulted in the development of new construction materials and techniques, such as ground improvement methods. However, the construction industry, which includes all branches of civil engineering, has also become very competitive, and is largely driven by the marketplace. Clients demand high-quality services and expect to receive them quickly and inexpensively. Thus, it has become very important to work efficiently. It has also enhanced the demand for innovative construction methods, such as mechanically stabilized earth walls, that are more cost-effective than previous solutions.

1.4 ACCURACY OF GEOTECHNICAL ENGINEERING ANALYSES

Although the many advances in geotechnical engineering over the last century have greatly improved our ability to predict the behavior of soil and rock, we still need to maintain a healthy sense of skepticism. Most of our analyses are handicapped by the uncertainties introduced by the site exploration and characterization program. In addition, our mathematical models of soil and rock behavior are only approximate, and often do not explicitly consider important factors. Simply because an equation is available to describe a certain process does not mean that we can expect it to give us precise results!

One of the most common mistakes among students studying geotechnical engineering, and even among some practicing engineers, is to overestimate the accuracy of geotechnical analyses. The widespread availability of digital computers and the related software has made this problem even worse, because our ability to perform analyses has far surpassed the technical and economic realities of obtaining the underlying soil and rock data. This apparent computational process often leads to overconfidence, and ultimately may result in construction failures.

Most of the example problems in this book have been solved to a precision of three significant figures. This has been done for clarity and to avoid excessive round-off errors. However, few, if any, geotechnical analyses are really this accurate. In reality, the actual behavior often varies from the predicted behavior by 50% or more. Therefore, it is best to perform most geotechnical analyses to no more than two or three significant figures, and recognize that the true precision is really much less.

There are occasions when more precise analyses are useful, especially when conducting "what-if" studies or when actual performance data is available from the field. More precise analyses also may be appropriate for very sophisticated projects that have a correspondingly intense site exploration and characterization programs. However, it is very important to avoid placing too much confidence in the results, for it is very easy to perform analyses to much greater levels of precision than are justified by the data.

1.5 A PICTORIAL OVERVIEW OF GEOTECHNICAL ENGINEERING

Virtually all civil engineering projects require at least some geotechnical engineering. This section shows the role of geotechnical engineering and geotechnical engineers at work through photographs.

Geotechnical Aspects of Structures

Civil engineering structures come in many sizes and shapes. They are constructed of many different materials. One thing they all have in common is that they are tied to the earth in some manner. Figures 1.8–1.11 illustrate geotechnical aspects of various structures.

FIGURE 1.8 Buildings—the Sears Tower in Chicago is one of the tallest buildings in the world. It needs massive foundations to transmit the structural loads into the ground. The design of these foundations depends on the nature of the underlying soils. Geotechnical engineers are responsible for assessing these soil conditions and developing suitable foundation designs.

FIGURE 1.9 Bridges—the foundation for the south pier of the Golden Gate Bridge in San Francisco had to be built in the open sea. It extends down to bedrock, some 30 m (100 ft) below the water level and 12 m (40 ft) below the channel bottom. This was especially difficult to build because of the tremendous tidal currents at this site.

FIGURE 1.10 Dams—Oroville Dam in California is one of the largest earth dams in the world. It is made of 61,000,000 m^3 (80,000,000 yd^3) of compacted soil. The design and construction of such dams require extensive geotechnical engineering.

FIGURE 1.11 Tunnels—the Ted Williams Tunnel is part of the Central Artery Project in Boston. This prefabricated tunnel section was floated to the job site, and then sunk into a prepared trench in the bottom of the bay. Its integrity depends on proper support from the underlying soils.

Geotechnical Failures

Geotechnical engineers try to avoid failures like those shown in Figures 1.12–1.15. The study of such failures has greatly advanced our understanding of geotechniques and improved geotechnical engineering practices.

FIGURE 1.12 This house was built near the top of a slope and had a beautiful view of the Pacific Ocean. Unfortunately, a landslide occurred during a wet winter, undermining the house and causing part of its floor to fall away.

FIGURE 1.13 Teton Dam in Idaho failed in 1976, only a few months after the embankment had been completed and the reservoir began to be filled. This failure killed 11–14 people and caused about $400 million of property damage. (Courtesy of the Bureau of Reclamation.)

FIGURE 1.14 The 1964 Niigata Earthquake in Japan caused extensive liquefaction in this port city. These apartment buildings rotated when the underlying soils liquefied. (Courtesy of Earthquake Engineering Research Center Library, Berkeley, California.)

FIGURE 1.15 The approach fill to this highway bridge has settled because the underlying soils are soft clays and silts. However, the bridge has not settled because it is supported on piles. Although this "failure" is not as dramatic as the others, it is a source of additional maintenance costs, and can be a safety hazard to motorists and pedestrians.

Determining Geologic Material Properties

We use a variety of techniques to assess the subsurface conditions and the properties of soil and rock in Figures 1.16–1.19.

FIGURE 1.16 Performing a field reconnaissance. This is the top of a recent landslide, and the man in the photograph is examining the soil and rock exposed in the scarp.

FIGURE 1.17 Drilling exploratory borings to obtain soil and rock samples. This rig drills holes up to 30 m (100 ft) deep

FIGURE 1.18 Testing samples in a soil mechanics laboratory. These tests help us determine the engineering properties of the soil or rock.

FIGURE 1.19 Monitoring geotechnical instruments. These instruments measure groundwater levels and pressures, soil movements, and other similar attributes.

Geotechnical Construction

Geotechnical engineers also are actively involved in construction. Examples of geotechnical construction are shown in Figures 1.20–1.23.

FIGURE 1.20 This rig is drilling a hole in the ground that will be filled with reinforced concrete to form a drilled shaft foundation.

FIGURE 1.21 This 11 m (35 ft) deep excavation extends 10 m (30 ft) below the groundwater table. In addition, a river is present just beyond the excavation on the left side of the photograph. Therefore, it was necessary to first install an extensive dewatering system to draw down the groundwater table.

FIGURE 1.22 This rig is installing a series of wick drains, which help accelerate the settlements that will occur as a result of a proposed fill.

FIGURE 1.23 The fill for this highway near Fort St. John, British Columbia is being reinforced with geogrids, thus allowing the side slopes to be steeper than would be possible with an unreinforced fill.

CHAPTER 2

Engineering Geology

And so geology, once considered mostly a descriptive and historical science, has in recent years taken on the aspect of an applied science. Instead of being largely speculative as perhaps it used to be, geology has become factual, quantitative, and immensely practical. It became so first in mining as an aid in the search for metals; then in the recovery of fuels and the search for oil; and now in engineering in the search for more perfect adjustment of man's structures to nature's limitations and for greater safety in public works.

Charles P. Berkey, Pioneer Engineering Geologist, 1939

Geology is the science of rocks, minerals, soils, and subsurface water, including the study of their formation, structure, and behavior. As the above quotation indicates, geology was once confined to purely academic studies, but it has since expanded into a practical science as well. *Engineering geology* is the branch that deals with the application of geologic principles to engineering works.

Unlike geotechnical engineers, whose training is in civil engineering, engineering geologists have a background in geology. Their work includes mapping, describing, and characterizing the rocks and soils at a construction site; assessing stability issues, such as landslides; and appraising local seismicity and earthquake potentials. In other words, engineering geologists provide crucial information about the site that geotechnical engineers will use in analysis and design. This is why you will find in many projects geologists or engineering geologists working closely with geotechnical engineers as a team. Therefore, it is important for the geologist to have some understanding of engineering, and the engineer to have some understanding of geology. Some individuals have even acquired full professional credentials in both fields.

This chapter explores fundamental principles of geology and their application to geotechnical engineering, with extra emphasis on the geological origin of soils. These principles are important to geotechnical engineers because they help us understand the nature of the subsurface conditions, design effective site exploration and characterization programs, and interpret data obtained from such programs.

2.1 THE GEOLOGIC CYCLE

The earth's crust, where civil engineering projects are built, is constantly being shaped and reshaped by geologic processes powered by the sun and by tectonic forces originated from within the earth. However, the geologic processes acting on the earth's crust are extremely slow by human standards. Even during an entire lifetime, one can expect to directly observe only a minutely small amount of progress in these processes. Therefore, geologists must rely primarily on observations of the earth as it presently exists (i.e., on the *results* of these processes) to develop their theories.

Geologic theories are organized around a framework known as the *geologic cycle*. This cycle, shown in Figure 2.1, includes many processes acting simultaneously. The most important of these begins with molten magma from within the earth crystallizing into rock, then continues with the rock being broken down into soil, and

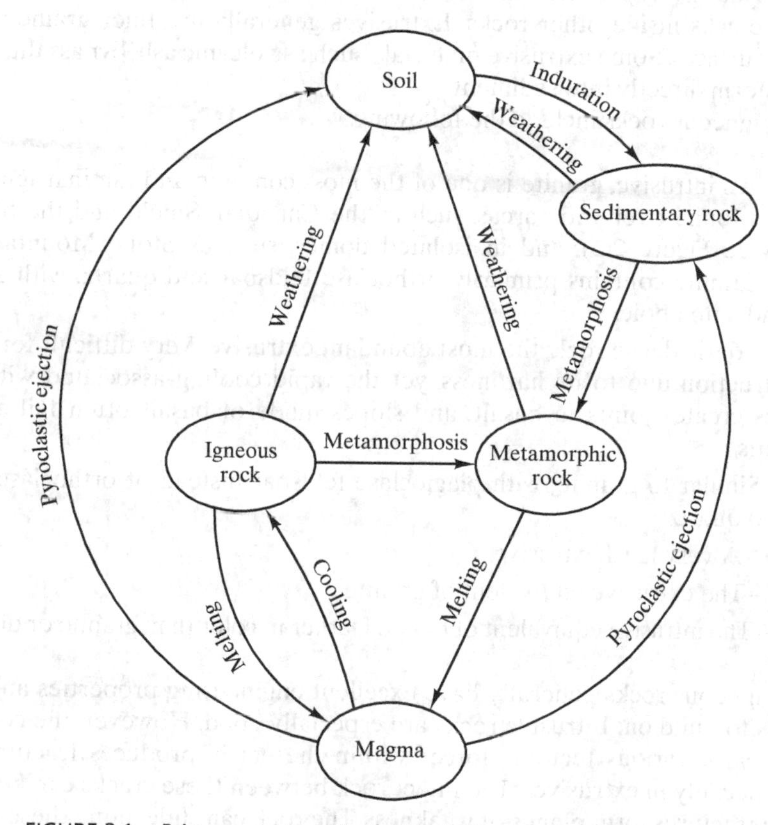

FIGURE 2.1 Primary processes in the geologic cycle.

that soil then being converted back into rock. This process repeats itself over and over again through geologic time.

2.2 ROCKS

In geology, rocks are classified by how they are formed. Under such a genetic classification, the three major rock types are igneous, sedimentary, and metamorphic rocks, as discussed next.

Igneous Rocks

The geologic cycle begins with *magma*, a molten rock deep inside the earth. This magma cools as it moves upward toward the ground surface, forming *igneous rocks*. There are two primary types of igneous rocks: *intrusives* (also called *plutonic rocks*) form below the ground surface from the relatively slow cooling of magma, whereas *extrusives* (also called *volcanic rocks*) result from the relatively rapid cooling of molten lava at the ground surface, such as through a volcano. Intrusives include both large bodies of rock (known as *plutons*) and smaller sheet-like bodies (known as *sills* and *dikes*) that fill cracks inside other rocks. Extrusives generally are finer grained and have smoother surfaces. Some extrusive materials, such as volcanic ash, bypass the rock stage and transform directly into sediment.

Common igneous rocks include the following.

Granite—An intrusive, granite is one of the most common and familiar igneous rocks. It is found over wide areas, such as the Canadian Shield and the Sierra Nevada (see Figure 2.2), and in isolated domes, such as Stone Mountain in Georgia. Granite contains primarily orthoclase feldspar and quartz, with some biotite and amphibole.

Basalt—A dark, dense rock; the most abundant extrusive. Very difficult for tunnel construction due to its hardness, yet the rapid cooling associated with all extrusives creates joints in basalt, and slopes made of basalt often fail along these joints.

Diorite—Similar to granite, with plagioclase feldspar instead of orthoclase and little or no quartz.

Andesite—A very hard extrusive.

Rhyolite—The extrusive equivalent of granite.

Gabbro—The intrusive equivalent of basalt. Darker in color than granite or diorite.

Unweathered igneous rocks generally have excellent engineering properties and are good materials to build on. Intrusive rocks are especially good. However, the cooling process, along with various tectonic forces within the earth, produces fractures in these rocks, especially in extrusives. The intact rock between these cracks can be very strong, but the fractures form planes of weakness. The rock can slide along these weak planes, potentially causing instability in the rock mass. The engineering properties of

FIGURE 2.2 Half Dome in Yosemite National Park. The near vertical face was carved by glaciers. This rock, often classified as granite, is more accurately called *granodiorite*—a material halfway between granite and diorite.

weathered igneous rocks are less desirable because the rock is changing into a more soil-like material.

Sedimentary Rocks

Soil deposits can be transformed back into rock through the hardening process called *induration* or *lithification*, thus forming the second major category of rocks: *sedimentary rocks*. There are two types of such rocks: clastic rocks and carbonates.

Clastic Rocks Clastic rocks form when deep soil deposits become hardened as a result of pressure from overlying strata and cementation through precipitation of water-soluble

TABLE 2.1 Common Clastic Sedimentary Rocks[a]

Texture and Average Particle Size	Composition	Rock Name
Coarse-grained, gravel-size (>2 mm)	Rounded fragments of any rock type; quartz, quartzite, chert dominant	Conglomerate
	Angular fragments of any rock type; quartz, quartzite, chert dominant	Breccia
Medium-grained, sand-size (0.06–2 mm)	Quartz with minor accessory minerals	Sandstone
	Quartz with at least 25% feldspar	Arkose
	Quartz, rock fragments, and considerable clay	Graywacke
Fine-grained, silt-size (0.002–0.06 mm)	Quartz and clay minerals	Siltstone
Very fine-grained, clay-size (<0.002 mm)	Quartz and clay minerals	Claystone and shale

[a]Adapted from Hamblin and Howard, 1975.

minerals such as calcium carbonate or iron oxide. Because of their mode of deposition, many clastic rocks are *layered* or *stratified*, which makes them quite different from *massive* formations. The interfaces between these layers are called *bedding planes*. Table 2.1 presents a list of common clastic rocks. Shale and sandstone are the most common.

Often, various types of clastic rocks are *interbedded*. For example, a sequence might contain a 1-m thick bed of sandstone, then 5 m of siltstone, 0.5 m of claystone, and so on.

Most *conglomerate, breccia, sandstone*, and *arkose* rocks generally have favorable engineering properties. Those cemented with silica or iron oxide are especially durable, but may be difficult to excavate. However, some are only weakly indurated, often cemented only with clay or other water-soluble minerals. These may behave much like a soil, and may be much easier to excavate.

Fine-grained and very fine-grained clastic rocks are more common, and much more problematic. Sometimes the term *mudstone* is used to collectively describe these rocks, but they are more precisely described as *siltstone* (when the rock is derived from silt), *claystone* (when derived from clay and slightly to mildly indurated), or *shale* (when derived from clay and well-indurated). Nearly all these have distinct bedding planes, as shown in Figure 2.3, and are subject to opening and shearing along these planes. All except shale are usually easy to excavate with conventional earthmoving equipment.

Some fine-grained and very fine-grained clastic rocks are also subject to *slaking*, which is a deterioration after excavation and exposure to the atmosphere and wetting-and-drying cycles. Rocks that exhibit strong slaking will rapidly degenerate to soil, and thus can create problems for engineering structures built on them.

Carbonates A different type of sedimentary rock forms when organic materials accumulate and become indurated. Because of their organic origin, they are called *carbonates*. Common carbonate rocks include the following:

Limestone—The most common type of carbonate rock, limestone is composed primarily of calcite ($CaCO_3$). Most limestones formed from the accumulation of marine organisms on the bottom of the ocean, and usually extend over large areas.

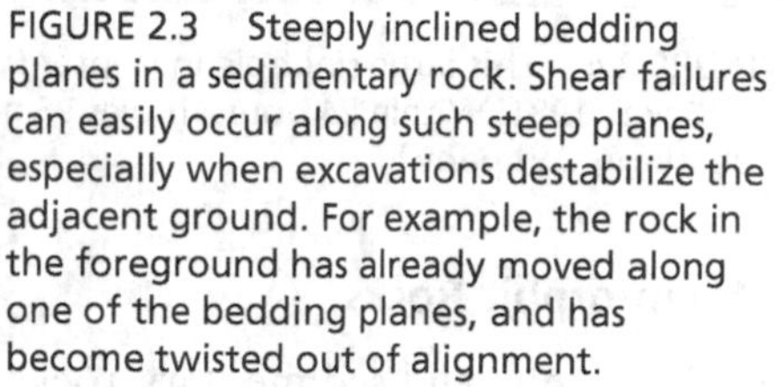

FIGURE 2.3 Steeply inclined bedding planes in a sedimentary rock. Shear failures can easily occur along such steep planes, especially when excavations destabilize the adjacent ground. For example, the rock in the foreground has already moved along one of the bedding planes, and has become twisted out of alignment.

Some of these deposits were later uplifted by tectonic forces in the earth and now exist below land areas. For example, much of Florida is underlain by limestone.

Chalk—Similar to limestone, but much softer and more porous.

Dolomite—Similar to limestone, except based on the mineral dolomite instead of calcite.

Some carbonate rocks also have bedding, but it is usually less distinct than in clastic rocks.

Carbonate rocks, especially limestone, can be dissolved by long exposure to water, especially if it contains a mild solution of carbonic acid. Groundwater often gains small quantities of this acid through exposure to carbon dioxide in the ground. This process often produces *karst topography*, which exposes very ragged rock at the ground surface and many underground caves and passageways. In such a topography, streams sometimes "mysteriously" disappear into the ground, only to reappear elsewhere.

Sometimes the rock is covered with soil, so the surface expressions of karst topography may be hidden. Nevertheless, the underground caverns remain, and sometimes the ground above caves into them. This creates a *sinkhole*, such as the one in Figure 2.4. This caving process can be triggered by the lowering of the groundwater table, which often occurs when wells are installed for water supply purposes.

In areas underlain by carbonate rock, especially limestone, geotechnical engineers are concerned about the formation of sinkholes beneath large and important structures. We use exploratory borings, geophysical methods, and other techniques (see Chapter 3) to locate hidden underground caverns, then either avoid building above these features, or fill them with grout.

FIGURE 2.4 This large sinkhole in Winter Park, Florida, suddenly appeared on May 8, 1981. Within 24 hours, it was 75 m (250 ft) in diameter. (GeoPhoto Publishing Company.)

Metamorphic Rocks

Both igneous and sedimentary rocks can be subjected to intense heat and pressure while deep in the earth's crust. These conditions produce more dramatic changes in the minerals within the rock, thus forming the third type of rock—*metamorphic rocks*. The metamorphic processes generally improve the engineering behavior of these rocks by increasing their hardness and strength. Nevertheless, some metamorphic rocks can still be problematic.

Some metamorphic rocks are *foliated*, which means they have oriented grains similar to bedding planes in sedimentary rocks. These *foliations* are important because the shear strength is less for shear stresses acting parallel to the foliations. Other metamorphic rocks are *nonfoliated* and have no such orientations.

Common metamorphic rocks include the following:

Foliated rocks:

- *Slate*—Derived principally from shale; dense; can be readily split into thin sheets parallel to the foliation (such sheets are used to make chalkboards).
- *Schist*—A strongly foliated rock with a large mica content; this type of foliation is called *schistosity;* prone to sliding along foliation planes.
- *Gneiss*—Pronounced "nice"; derived from granite and similar rocks; contains banded foliations.

Nonfoliated rocks:

- *Quartzite*—Composed principally or entirely of quartz; derived from sandstone; very strong and hard.
- *Marble*—Derived from limestone or dolomite; used for decorative purposes and for statues.

Unweathered, nonfoliated rocks generally provide excellent support for engineering works, and are similar to intrusive igneous rocks in their quality. However, some foliated rocks are prone to slippage along the foliation planes. Schist is the most notable in this regard because of its strong foliation and the presence of mica. The 1928 failure of St. Francis Dam in California (Rogers, 1995) has been partially attributed to shearing in schist, and the 1959 failure of Malpasset Dam in France (Goodman, 1993) to shearing in a schistose gneiss.

Metamorphic rocks are also subject to weathering, thus forming weathered rock, residual soils, and transported soils and beginning the geologic cycle anew.

2.3 ROCK-FORMING MINERALS

Minerals are naturally formed elements or compounds with specific structures and chemical compositions. As the basic constituents of rocks, minerals control much of rock behavior. Some minerals are very strong and resistant to deterioration, and produce rocks with similar properties, while others are much softer and produce weaker rocks.

More than 2000 different minerals are present in the earth's crust. They can be identified by their physical and chemical properties, by standardized tests, or by examination under a microscope. Only a few of them occur in large quantities, and they form the material for most rocks. The most common minerals are as follows.

Feldspar—This is the most abundant mineral and an important component of many kinds of rocks. *Orthoclase* feldspars contain potassium ($KAlSi_3O_8$) and usually range from white to pink. *Plagioclase* feldspars contain sodium ($NaAlSi_3O_8$), calcium ($CaAl_2Si_2O_8$), or both, and range from white to gray to black. Feldspars have a moderate hardness.

Quartz—Also very common, quartz is another major ingredient in many kinds of rocks. It is a silicate (SiO_2) and usually has a translucent to milky white color, as shown in Figure 2.5. Quartz is harder than most minerals, and thus is very resistant to weathering. *Chert* is a type of quartz sometimes found in some sedimentary rocks. It can cause problems when used as a concrete aggregate.

Ferromagnesian minerals—A class of minerals, all of which contain both iron and magnesium. This class includes pyroxene, amphibole, hornblende, and olivine. These minerals have a dark color and a moderate hardness.

Iron oxides—Another class of minerals, all of which contain iron (Fe_2O_3), includes limonite and magnetite. Although less common, these minerals give a distinctive rusty color to some rocks and soils, and can act as cementing agents.

Calcite—A mineral made of calcium carbonate ($CaCO_3$), and is usually white, pink, or gray. It is soluble in water, and thus can be transported by groundwater into cracks in rocks where it precipitates out of solution. It can also precipitate in soil, becoming a cementing agent. Calcite is much softer than quartz or feldspar, and effervesces vigorously when come into contact with dilute hydrochloric acid.

FIGURE 2.5 A large quartz crystal. Quartz crystals in rocks are normally much smaller.

Dolomite—Similar to calcite, with magnesium added, it shows less vigorous reaction with dilute hydrochloric acid.

Mica—Translucent thin sheets or flakes. *Muscovite* has silvery flakes; while *biotite* is dark gray or black. These sheets have a very low coefficient of friction, which can produce shear failures in certain rocks, such as schist.

Gypsum—A very soft mineral often occurring as a precipitate in sedimentary rocks. It is colorless to white and has economic value when found in thick deposits. For example, it is used to make drywall. Gypsum is water soluble, and thus can dissolve under the action of groundwater, which can lead to other problems.

When a rock breaks down into soil through *weathering*, as discussed later in the chapter, many of the minerals in the rock remain in their original forms. For example, many sand grains are made of quartz, and thus reflect its engineering properties. Other minerals undergo chemical and physical changes and take on new properties. For example, feldspar often experiences such changes, and forms clay minerals (discussed in Chapter 4). Soil can also acquire other materials, including organic matter, man-made materials, and water.

2.4 STRUCTURAL GEOLOGY

Structural geology is the study of the three-dimensional distribution of rock formations and the orientations of weakness planes they contain. Under certain conditions, these weakness planes will control rock behavior. Therefore, structural geology is an important part of engineering geology because it gives us important insights into rock behavior, and engineering geologists routinely develop detailed geologic maps that describe rock structures.

Bedding Planes and Foliation

All sedimentary rocks formed in horizontal or near-horizontal layers, and these layers often reflect alternating cycles of deposition. This process produces parallel bedding planes as shown in Figure 2.3. The shear strength along these weakness planes is typically much less than across them, a condition we call *anisotropic strength*. When these rocks were uplifted by tectonic forces in the earth, the bedding planes could be rotated to a different angle, as shown in Figure 2.6. Because the rock shears much more easily along the bedding planes, their orientation is important. Many landslides have occurred on slopes with unfavorable bedding orientations. Therefore, engineering geologists and geotechnical engineers are very careful to compare the attitudes of bedding planes with the orientations of proposed slopes.

Some metamorphic rocks have similar planes of weakness called foliation. This characteristic is called *schistosity*, and like bedding planes, the foliation can also be mapped.

Folds

Tectonic forces also distort rock masses. When horizontal compressive forces are present, the rock distorts into a wavy pattern called *folds* as shown in Figure 2.7. Sometimes these folds are gradual; other times they are very abrupt. When folds are oriented concave downward, they are called *anticlines*, and when concave upward they are called *synclines*.

Fractures

Fractures are cracks in a rock mass. Their orientation is very important because the shear strength along these fractures is less than that of the intact rock mass, so they form potential failure surfaces. There are three types of fractures: joints, shears, and faults.

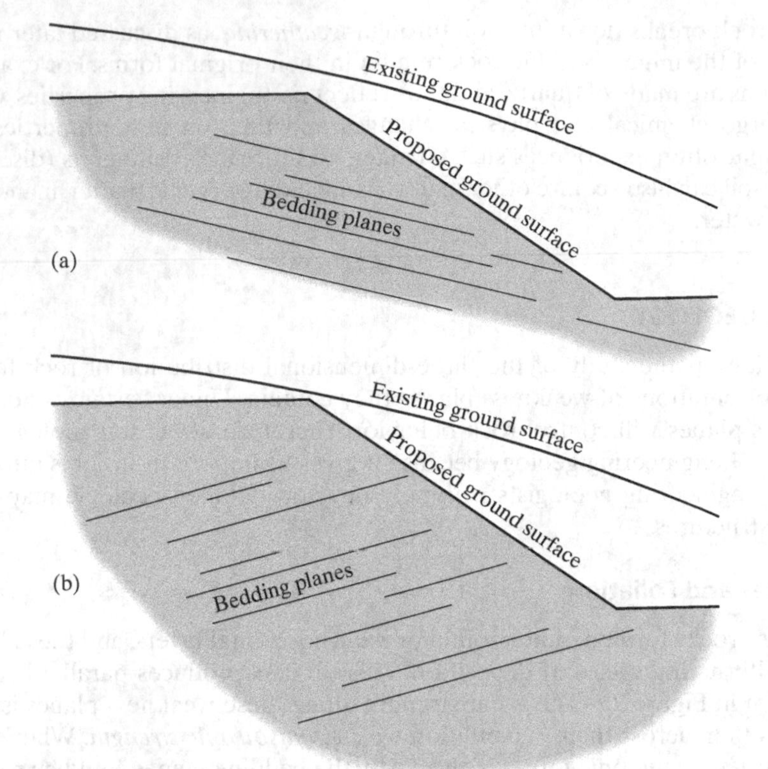

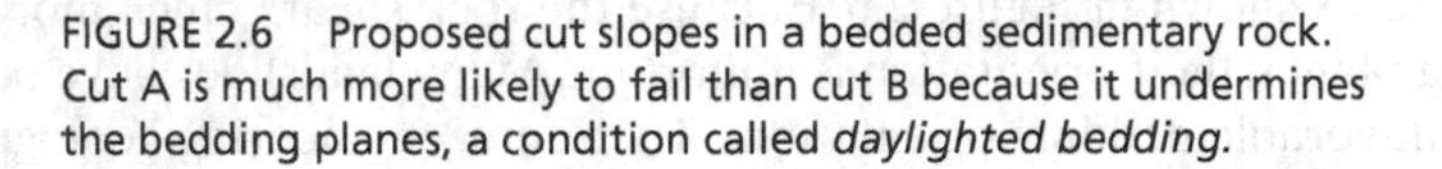
FIGURE 2.6 Proposed cut slopes in a bedded sedimentary rock. Cut A is much more likely to fail than cut B because it undermines the bedding planes, a condition called *daylighted bedding*.

Joints are relatively minor tensile fractures that have experienced no or minimal shear movements. They can be the result of cooling (in the case of igneous rocks), tensile tectonic stresses, or tensile stresses from lateral movements of adjacent rocks. Joints usually occur at fairly regular spacings, and a group of such joints is called a *joint set*.

Shears are fractures that have experienced a small shear displacement, perhaps a few centimeters. They are caused by various stresses in the ground, and do not appear in sets as joints do. Zones in the rock mass that contain numerous shears, or shear zones, are often conduits of groundwater.

Faults are similar to shears, except they have experienced much greater shear displacements. Although there is no standard for distinguishing the two, many geologists would use the term *fault* when the shear displacement exceeds about 1 m. Such movements are sometimes associated with earthquakes. Based on the theory of plate tectonics, the earth's crust is made of many distinct crustal plates, at the boundaries of which are many major active faults. One example is the San Andreas Fault in California. Due

FIGURE 2.7 Folds in a sedimentary rock. (GeoPhoto Publishing Company.)

to the spreading or colliding at the plate boundaries driven by tectonic forces, strains can build up in the rocks. When the strain-induced stresses are too high for the rocks to take, the rock ruptures causing an earthquake, releasing the stored strain energy as seismic waves propagate from the point of rupture (called *the focus of the earthquake*) to the ground surface.

Faults are classified according to their geometry and direction of movement, as shown in Figure 2.8. *Dip-slip faults* are those whose movement is primarily along the dip. It is a *normal fault* if the overhanging block is moving downward, or a *reverse fault* if it is moving upward. A reverse fault with a very small dip angle is called a *thrust fault*. Conversely, *strike-slip faults* are those whose movement is primarily along the strike. They can be either *right-lateral* or *left-lateral* depending on the relative motion of the two sides. Some faults experience both dip-slip and strike-slip movements. The *fault trace* is the intersection of the fault and the ground surface.

The term *discontinuity* is often used in this context to include bedding planes, foliations, joints, shears, faults, and all other similar defects in rocks. Because the orientations of these features often control rock mass behavior, extensive analytical methods have been developed to systematically evaluate discontinuity data gathered in the field (Priest, 1993).

Strike and Dip

When developing geologic maps, we are interested in both the presence of certain geologic structures and their orientations in space. For example, a rock mass may be unstable if it has joints oriented in a certain direction, but much more stable if they are oriented in a different direction. For similar reasons, we are also interested in the orientations of faults, bedding planes, and other geologic structures.

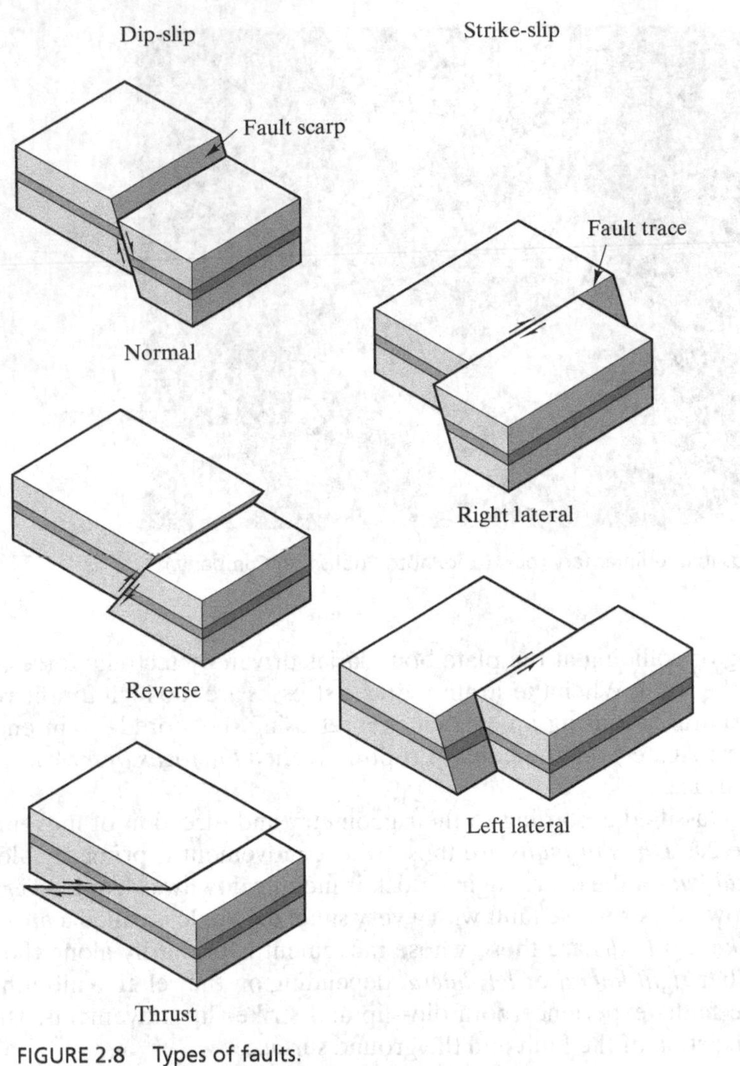

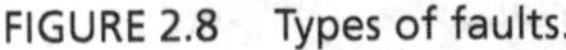
FIGURE 2.8 Types of faults.

Many of these structures are roughly planar, at least for short distances, and therefore may be described by defining the orientation of this plane in space. We express this orientation using the *strike* and *dip*, as shown in Figure 2.9.

The strike of a plane is the compass direction of the intersection of the plane and the horizontal, and is expressed as a bearing from true north. For example, if a fault has a strike of N30°W, then the intersection of the fault plane with a horizontal plane traces a line oriented 30° west of true north. The dip is the angle between the geologic surface and the horizontal, and is measured in a vertical plane oriented perpendicular to the strike. The dip also needs a direction. For example, a fault with a N30°W strike may have a dip of 20° northeasterly. When expressed together, these data are called an *attitude*, and may be written in a condensed form as N30°W; 20°NE. Although the

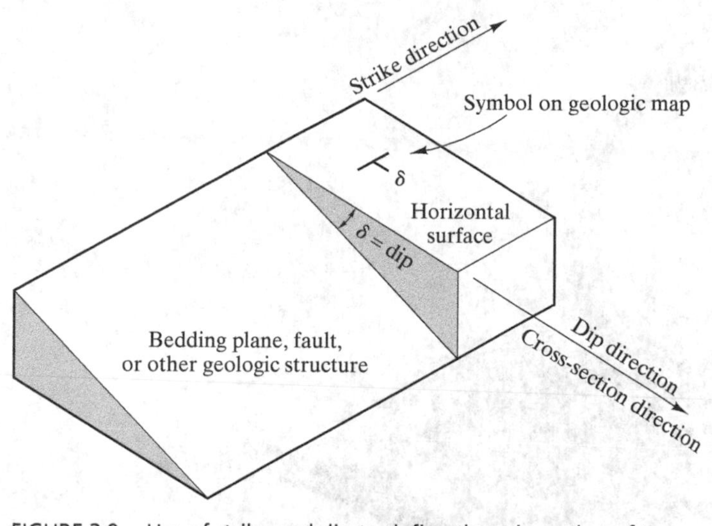

FIGURE 2.9 Use of strike and dip to define the orientation of a geologic structure. (Adapted from *Engineering Geology* by Richard E. Goodman, Copyright ©1993. Reprinted by permission of John Wiley & Sons.)

strike direction given is "exact," the dip direction given can be approximate. In this case, there are only two possibilities for the dip direction, NE or SW, so the purpose of the given dip direction is simply to distinguish between these two possibilities. The "exact" dip direction is 90° from the strike.

Attitudes are usually measured in the field using a *Brunton compass*, as shown in Figure 2.10. This device includes both a compass and a level, and thus can be used to measure both strikes and dips. The measured attitudes are then recorded graphically on geologic maps using the symbol shown in Figure 2.11. This symbol may be modified to indicate the type of structure being identified.

Sometimes we need to know the dip as it would appear in a vertical plane other than the one perpendicular to the strike. Figure 2.12 shows such a plane. For example, we may have drawn a cross-section that is oriented perpendicular to a slope, but at some angle other than 90° from the strike, and need to know the dip angle as it appears in that cross-section. This dip is called the *apparent dip* and may be computed using:

$$\tan \delta_a = \tan \delta \sin \alpha \tag{2.1}$$

where:

δ_a = apparent dip

δ = dip

α = horizontal angle between strike and the vertical plane on which the apparent dip is to be computed

FIGURE 2.10 A Brunton compass is used to measure bedrock attitudes and other geologic features in the field.

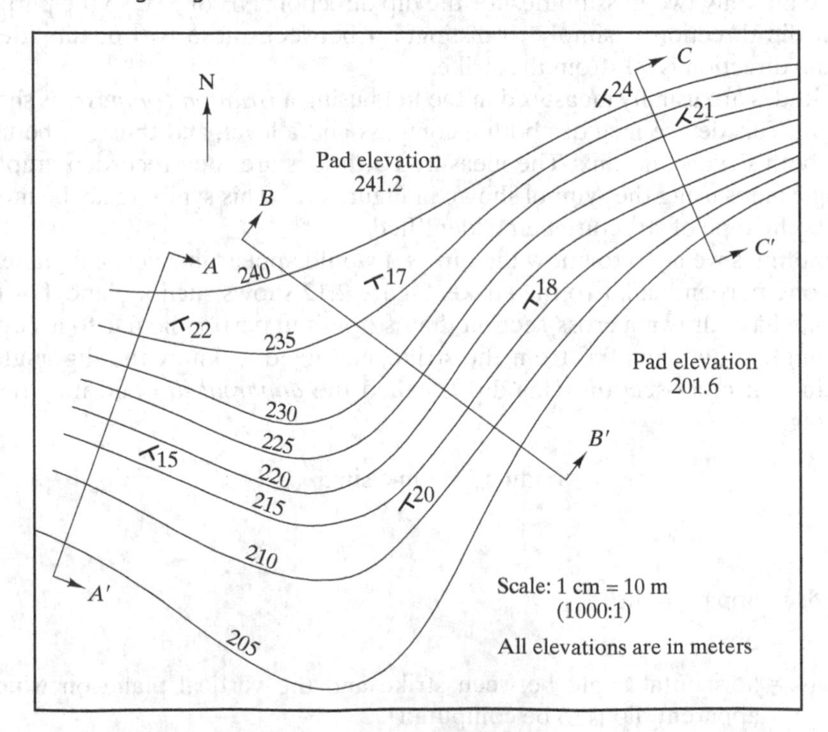

FIGURE 2.11 Geologic map showing bedrock attitudes. In this case, the attitudes represent the bedding planes in a sedimentary rock.

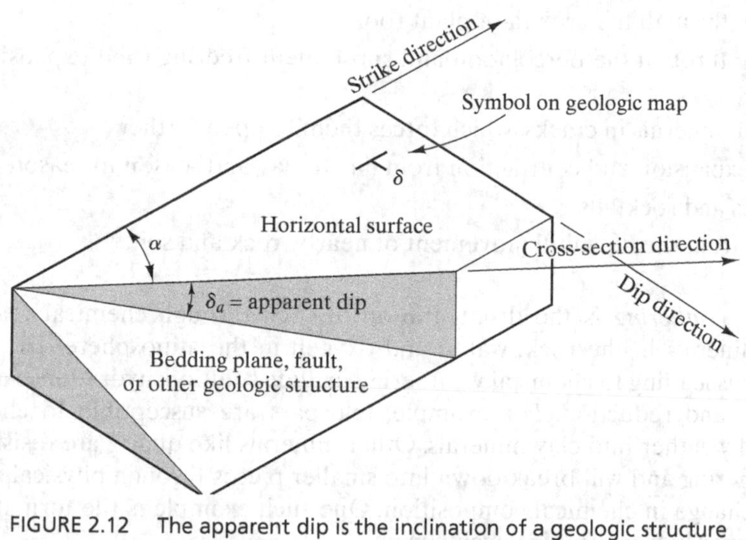

FIGURE 2.12 The apparent dip is the inclination of a geologic structure as seen on any vertical plane. It is always less than or equal to the true dip, which is determined on the vertical plane perpendicular to the strike. (Adapted from *Engineering Geology* by Richard E. Goodman, Copyright ©1993. Reprinted by permission of John Wiley & Sons.)

Example 2.1

Compute the apparent dip of the bedding planes as they would appear in the central portion of Section *B-B′* in Figure 2.11.

Solution:

Base analysis on the 17° measured attitude. The angle between its strike and Section *B-B′* is 65°. Therefore, using Equation 2.1:

$$\begin{aligned} \tan \delta_a &= \tan \delta \sin \alpha \\ &= \tan 17° \sin 65° \\ \delta_a &= 15° \end{aligned}$$

Thus, the bedding plane will appear to be flatter than it really is.

2.5 WEATHERING

Rocks exposed to the atmosphere are immediately subjected to physical (or mechanical), chemical, and biological breakdown through weathering. *Physical* or *mechanical weathering* is the disintegration of rocks into smaller particles through physical processes, including:

- The erosive action of water, ice, and wind
- Opening of cracks as a result of unloading due to the erosion of overlying soil and rock

- Loosening through the growth of plant roots
- Loosening through the percolation and subsequent freezing (and expansion) of water
- Growth of minerals in cracks, which forces them to open further
- Thermal expansion and contraction from day to day and season to season
- Landslides and rockfalls
- Abrasion from the downhill movement of nearby rock and soil.

Chemical weathering is the disintegration of rock through chemical reactions between the minerals in the rock, water, and oxygen in the atmosphere. The major types of reactions leading to chemical weathering include solution, hydration, carbonation, oxidation, and reduction. For example, feldspars are susceptible to chemical weathering and weather into clay minerals. Other minerals like quartz are resistant to chemical weathering and will break down into smaller pieces through physical weathering with no change in chemical composition. One such example is the formation of quartz sand primarily from physical weathering.

Biological weathering is the disintegration of rocks into smaller particles caused by biological activities that produce organic acids.

The rock passes through various stages of weathering, eventually being broken down into small particles, the material we call soil. These soil particles may remain in place, forming a *residual soil*, or they may be transported away from their parent rock through processes discussed later in this chapter, thus forming a *transported soil*. Figure 2.13 shows an accumulation of fallen rock fragments called *talus* at the base of a rock slope, which is the beginning of one process of soil transport.

FIGURE 2.13 Talus accumulation at the base of a rock slope in eastern Washington state.

Weathering processes continue even after the rock becomes a soil. As soils become older, they change due to continued weathering. The rate of change depends on many factors, including:

- The general climate, especially precipitation and temperature (note that climates in the past were often quite different from those today)
- The physical and chemical makeup of the soil
- The elevation and slope of the ground surface
- The depth to the groundwater table
- The type and extent of flora and fauna
- The presence of microorganisms
- The drainage characteristics of the soil.

2.6 SOIL FORMATION, TRANSPORT, AND DEPOSITION

Geologists classify soils into two major categories: *residual soils* and *transported soils*. Different types of soils under these categories are described next. This discussion focuses on the inorganic components within a soil. Organic soils and their origins are discussed in Chapter 4. Also note that geotechnical engineers use different soil classification systems (discussed in Chapter 5) for engineering purposes.

Residual Soils

When the rock weathering process is faster than the transport processes induced by water, wind, and gravity, much of the resulting soil remains in place. It is known as a residual soil, and typically retains many of the characteristics of the parent rock. The transition with depth from soil to weathered rock to fresh rock is typically gradual with no distinct boundaries.

In tropical regions, residual soil layers can be very thick, sometimes extending for hundreds of meters before reaching unweathered bedrock. Cooler and more arid regions normally have much thinner layers, and often no residual soil at all.

The soil type depends on the character of the parent rock. For example, *decomposed granite* (or simply "DG") is a sandy residual soil derived from the weathering of granitic rocks. DG is commonly used in construction projects as a high-quality fill material. Shales, which are sedimentary rocks that consist largely of clay minerals, weather to form clayey residual soils.

Saprolite is a general term for residual soils that are not extensively weathered and still retain much of the structure of the parent rock. Some have used the term rotten rock to describe saprolite. They typically include small concretions (harder, less weathered fragments) surrounded by more weathered material. Extensive saprolite deposits exist in the Piedmont area of the eastern United States (the zone between the Appalachian Mountains and the coastal plain) (Smith, 1987).

Laterite is a residual soil found in tropical regions. This type of soil is cemented with iron oxides, which gives it a high dry strength.

The engineering properties of residual soils range from poor to good, and generally improve with depth.

Transported Soils

Transported soils are formed by the deposition of sediments that have been transported from their places of origin by various agents. Different transporting agents lead to different soil types, as discussed next.

Glacial Soils Much of the earth's land area was once covered with huge masses of ice called *glaciers*. In North America, glaciers once extended as far south as the Ohio River, as shown in Figure 2.14. In Europe, glaciers once existed as far south as Germany. Many of these areas are now heavily populated, so the geologic remains of glaciation have much practical significance.

Glaciers had a dramatic effect on the landscape and created a category of soils called *glacial soils*. Glacial ice was not stationary; it moved along the ground, often grinding down some areas and filling in others. In some locations, glaciers reamed out valleys, leaving long lakes, such as the Finger Lakes of upstate New York. The Great Lakes also have been attributed to glacial action. Figure 2.15 shows how the moving ice strips away weathered rock, leaving a hard, unweathered surface in its wake.

Glaciers grind down the rock and soil, and transport these materials over long distances, even hundreds of kilometers, so the resulting deposits often contain a mixture of materials from many different sources. These deposits can also have a wide

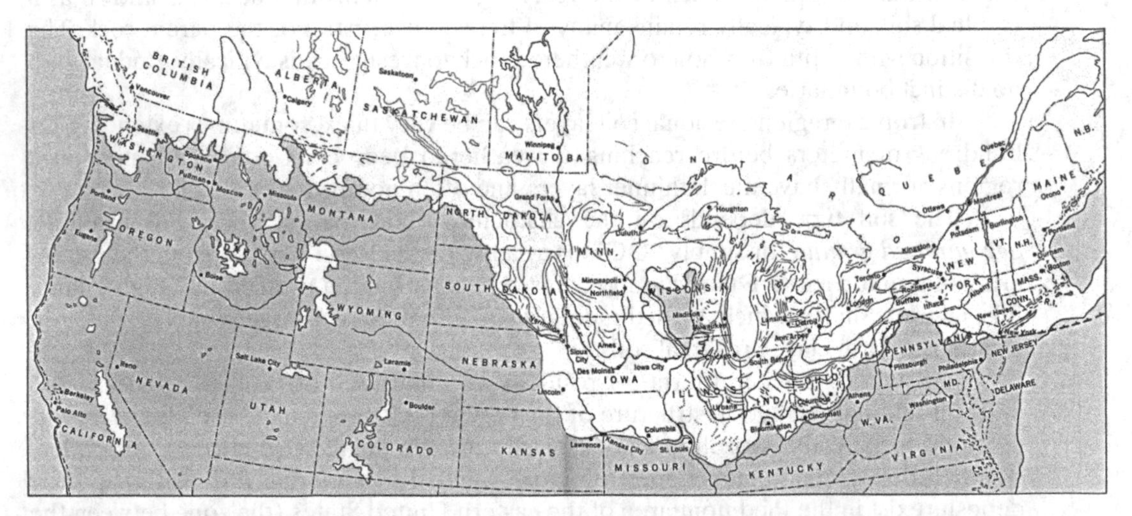

FIGURE 2.14 Southern extent of glaciation in North America during the various ice ages. The white areas were once covered with glaciers, and the heavy lines in these areas indicate locations of major moraines. (Adapted from *Engineering Geology* by Richard E. Goodman, Copyright ©1993. Reprinted by permission of John Wiley & Sons.)

FIGURE 2.15 Effects of glaciation on metamorphic rock in Manitoba. The striations, gouging, and polishing of the rock surface are all due to the moving ice. (Geological Survey of Canada.)

range of hardness and particle size, and are among the most complex and heterogeneous of all soils. The term *drift* encompasses all glacial soils, which then can be divided into three categories: till, glaciofluvial soils, and glaciolacustrine soils.

Till is the soil deposited directly by the glacier. It typically contains a wide variety of particle sizes, ranging from clay to gravel. Soil that was bulldozed by the glacier then deposited in ridges or mounds is called *ablation till*, as shown in Figure 2.16. These ridges and mounds are called *moraines*, and are loose and easy to excavate. In contrast, soil caught beneath the glacier, called *lodgement till*, has been heavily consolidated under the weight of the ice. Because of these heavy consolidation pressures and the wide range of particle sizes, lodgement till has a very high unit weight and is often nearly as strong as concrete. Lodgement till is sometimes called *hardpan*. It provides excellent support for structural foundations, but is very difficult to excavate.

Geotechnical site assessments need to carefully distinguish between ablation till and lodgement till. Both engineers and contractors need to be aware of the difference and plan accordingly. For example, construction of the St. Lawrence Seaway along the US–Canada border during the 1950s encountered extensive deposits of lodgement till that caused significant problems and delays. This problem was especially acute on the Cornwall Canal section of the seaway, causing one contractor to go bankrupt, another to default, and a third to file a $5.5 million claim on a $6.5 million contract (Legget and Hatheway, 1988).

FIGURE 2.16 The glacier in the background, which is part of the Athabasca Glacier in Alberta, is retreating and has left these moraines in its wake. The horizontal mounds of soil in the foreground are terminal moraines, and the ridges at the base of the mountain along the sides of the glacier are lateral moraines. Notice the wide range of particles sizes in these moraines.

When the glaciers melted, they generated large quantities of runoff. This water eroded much of the till and deposited it downstream, forming *glaciofluvial soils* (or *outwash*). Because of the sorting action of the water, these deposits are generally more uniform than till, and many of them are excellent sources of sand and gravel for use as concrete aggregates.

The fine-grained portions of the till often remained suspended in the runoff water until reaching a lake or the ocean, where they finally settled to the bottom. These are called *glaciolacustrine soils* and *glaciomarine soils*. Sometimes silts and clays were deposited in alternating layers according to the seasons, thus forming a banded soil called *varved clay*. The individual layers in varved clays are typically only a few millimeters thick, and are often separated by organic strata. These soils are soft and compressible, and thus are especially prone to problems with shear failure and excessive settlement.

Glaciomarine soils that formed in seawater are especially problematic because they have a high *sensitivity* (they lose shear strength when disturbed, as discussed in Chapter 12), and thus are prone to disastrous landslides. Such deposits are found in the Ottawa and St. Lawrence river valleys in eastern Canada (known as Champlain, Laurentian or Leda clays) and in southern Scandinavia. Figure 2.17 shows a flowslide in Leda clay adjacent to the South Nation River near Ottawa, Ontario. It resulted in the loss of 50 acres of farmland (Sowers, 1992).

FIGURE 2.17 The 1971 South Nation River flowslide near Ottawa, Ontario. This failure occurred in a soft marine soil called *Leda Clay*. (Geological Survey of Canada.)

Soils in the Chicago area are good examples of glacial deposits, and are typical of conditions in the Great Lakes region (Chung and Finno, 1992). The bedrock in this area consists of a marine dolomite that was overridden by successive advances and retreats of continental glaciers. At times this area was under ancient Lake Chicago, which varied in elevation from 18 m above to 30 m below the present level of Lake Michigan. These glaciers left both lodgement till and moraines, glaciolacustrine clays (deposited in the ancient lake), and glaciofluvial deposits in the riverbottoms, as shown in Figure 2.18.

Alluvial Soils *Alluvial soils* (also known as *fluvial soils* or *alluvium*) are those transported to their present position by rivers and streams. These soils are very common, and a very large number of engineering structures are built on them. Alluvium often contains extensive groundwater aquifers, so it is also important in the development of water supply wells and in geoenvironmental engineering.

When the river or stream is flowing rapidly, the silts and clays remain in suspension and are carried downstream; only sands, gravels, and boulders are deposited. However, when the water flows more slowly, more of the finer soils are also deposited. Rivers flow rapidly during periods of heavy rainfall or snowmelt, and slowly during periods of drought, so alluvial soils often contain alternating horizontal layers of different soil types.

The water also slows when the stream reaches the foot of a canyon, and tends to deposit much of its soil load there. This process forms *alluvial fans*, as shown in

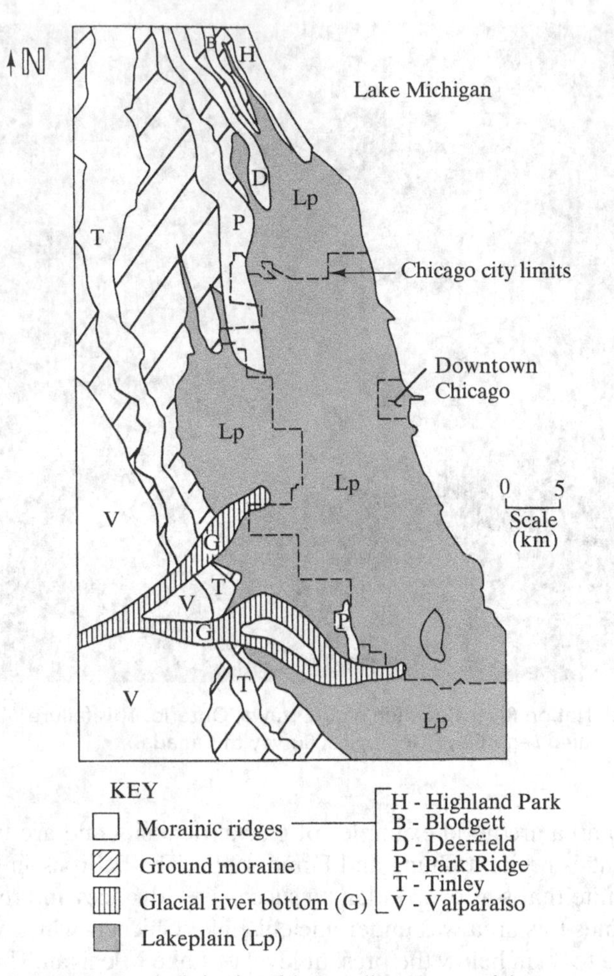

FIGURE 2.18 Soils in the Chicago, Illinois, area. (Chung and Finno, 1992.)

Figure 2.19, which are one of the most obvious alluvial soils. They are especially common in arid areas.

Large boulders are sometimes carried by water, especially in steep terrain, and deposited in the upper reaches of alluvial deposits as shown in Figure 2.20. Sometimes such boulders are subsequently covered with finer soils and become obscured. However, they can cause extensive problems when engineers attempt to drill exploratory borings or contractors try to make excavations or drive pile foundations.

Rivers in relatively flat terrain move much more slowly and often change course, creating complex alluvial deposits. Some of these are called *braided stream deposits* and *meander belt deposits*, as shown in Figure 2.21. In addition, the deposition

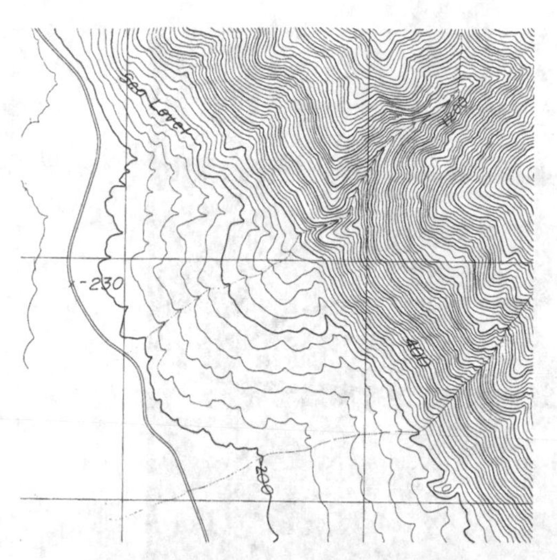

FIGURE 2.19 Topographic map of an alluvial fan in Death Valley, California. Soils eroded in the mountains are deposited at the foot of the canyon, thus forming a fan-shaped alluvial deposit. (USGS Gold Valley quadrangle map.)

FIGURE 2.20 Most alluvial soils consist of gravel, sand, silt, and clay. However, cobbles and boulders can also be present, especially along the base of mountains. For example, these large boulders were carried here by water and thus are an alluvial soil. They are located near top of an alluvial fan that spreads out from a steep canyon.

characteristics at a given location can change with time, so one type of alluvial soil is often underlain by other types.

In arid areas, evaporation draws most of the water out of the soil, leaving any dissolved chemicals behind. The resulting deposits of calcium carbonate, calcium sulfate, and other substances often act as cementing agents, converting the alluvial soil into a very hard material called *caliche*. These deposits are common in the southwestern states, and can be very troublesome to contractors who need to excavate through them.

FIGURE 2.21 The meanders in this river are forming a broad deposit of alluvial soils. (GeoPhoto Publishing Company.)

Most alluvial soils have moderately good engineering properties, and typically provide fair to good support for buildings and other structures.

Lacustrine and Marine Soils *Lacustrine soils* are those deposited beneath lakes. These deposits may still be underwater, or may now be exposed due to the lowering of the lake water level, such as the glaciolacustrine soils in Chicago (Figure 2.18). Most lacustrine soils are primarily silts and clays. Their suitability for foundation support ranges from poor to average.

Marine soils are also deposited underwater, except that they form in the ocean. *Deltas* are a special type of marine deposit formed where rivers meet larger bodies of water, and gradually build up to the water surface. Examples include the Mississippi River Delta and the Nile River Delta. This mode of deposition creates a very flat terrain, so the water flows very slowly. The resulting soil deposits are primarily silts and clays, and are very soft. Because of their deposition mode, most lacustrine and marine soils are very uniform and consistent. Thus, although their engineering properties are often poor, they may be more predictable than other more erratic soils.

Some sands also accumulate as marine deposits, especially in areas where rivers discharge into the sea at a steeper gradient. This sand is moved and sorted by the waves and currents, and some of it is deposited back on shore as *beach sands*. These sands are typically very poorly graded (i.e., they have a narrow range of particle sizes), have well-rounded particles, and are very loose. Beach deposits typically move parallel to the shoreline, and this movement can be interrupted by the construction of jetties and other harbor improvements. As a result, sand can accumulate on one side of the jetty, and be almost nonexistent on the other side. Changes in sea level elevations can leave beach deposits oriented along previous shorelines.

Deeper marine deposits are more uniform and often contain organic material from marine organisms. Those that have a large organic content are called *oozes*, one of the most descriptive of all soil names. The construction of offshore oil drilling platforms requires exploration and assessment of these soils.

Some lacustrine and marine soils have been covered with fill. This is especially common in urban areas adjacent to bays, such as Boston and San Francisco. The demand for real estate in these areas often leads to reclaiming such land, as shown in Figure 2.22. However, this reclaimed land is often a difficult place to build upon, because the underlying lacustrine and marine deposits are weak and compressible. Sometimes these soils have special names, such as Boston Blue Clay and San Francisco Bay Mud.

Aeolian Soils *Aeolian soils* (also known as *eolian soils*) are those deposited by wind. This mode of transport generally produces very poorly-graded soils (i.e., a narrow range of particle sizes) because of the strong sorting power of wind. These soils are also usually very loose, and thus have only fair engineering properties.

There are three primary modes of wind-induced soil transport (see Figure 2.23): *Suspension* occurs when wind lifts individual silt particles to high altitudes and transports them for great distances. This process can create large dust storms, such as

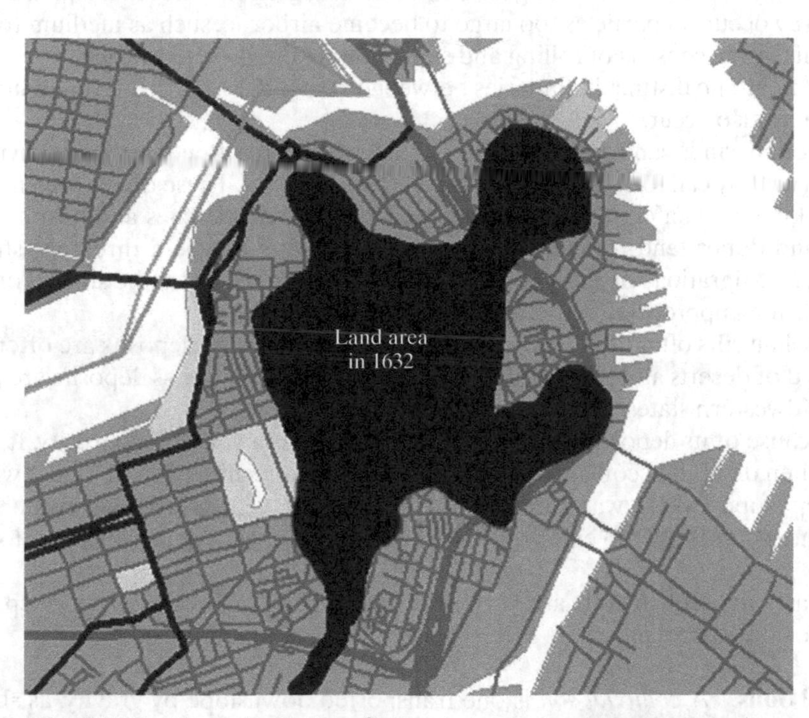

FIGURE 2.22 When the Puritans first settled in Boston, Massachusetts, the land area was as shown by the black zone in this map. It was connected to the mainland via a narrow isthmus. Since then, the city has been extended by placing fill in the adjacent water, thus forming the shoreline as it now exists.

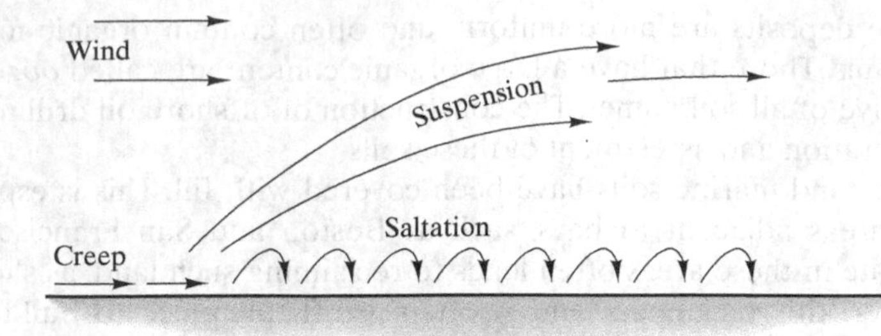

FIGURE 2.23 Modes of aeolian transport.

those that occurred in Oklahoma and surrounding states during the "dust bowl" drought of the 1930s.

Saltation (from the Latin *saltatio*—to dance) is the intermediate process where soil particles become temporarily airborne, then fall back to earth. Upon landing, the particle bounces or dislodges another particle, thus initiating another flight. This motion occurs in fine sands, and typical bounce distances are on the order of 4 m. Particles moving by saltation do not gain much altitude, generally not more than 1 m.

Creep occurs in particles too large to become airborne, such as medium to coarse sands. This mode consists of rolling and sliding along the ground surface.

There are no distinct boundaries between these processes, so intermediate modes of transport also occur.

Aeolian sands can form horizontal strata, which are often interbedded with alluvial soils, or they can form irregular hills called *sand dunes*. These dunes are among the most striking aeolian deposits, and are found along some beaches and in some desert areas. Sand dunes tend to migrate downwind, and thus can be a threat, as shown in Figure 2.24. Migrations of 3 m/yr are not unusual, but this rate can be slowed or halted by establishing appropriate vegetation on the dune.

Aeolian silts often form deep deposits called *loess*. Such deposits are often found downwind of deserts and glacial outwash deposits. Extensive loess deposits are present in the Midwestern states.

Because of its deposition mode, loess typically has a very high porosity. It is fairly strong when dry, but becomes weak when wetted. As a result, it can be stable when cut to a steep slope (where water infiltration is minimal), yet unstable when the slope is flatter and water is able to enter the soil. Figure 2.25 shows a near-vertical cut slope in loess.

Nearly all aeolian soils are very prone to erosion, and often have deep gullies. Good erosion control measures are especially important in these soils.

Colluvial Soils A *colluvial soil* is one transported downslope by gravity, as shown in Figure 2.26. There are two types of downslope movement—slow and rapid. Both types occur only on or near sloping ground.

Slow movement, which is typically on the order of millimeters per year, is called *creep*. It occurs because of gravity-induced downslope shear stresses, the expansion

FIGURE 2.24 This sand dune near the beach in Marina, California, is slowly migrating to the right and has partially buried the fence.

FIGURE 2.25 The slope in the center of this photograph is a cut made in a loess deposit near the Mississippi River in Tennessee. Notice how it is stable in spite of being near-vertical.

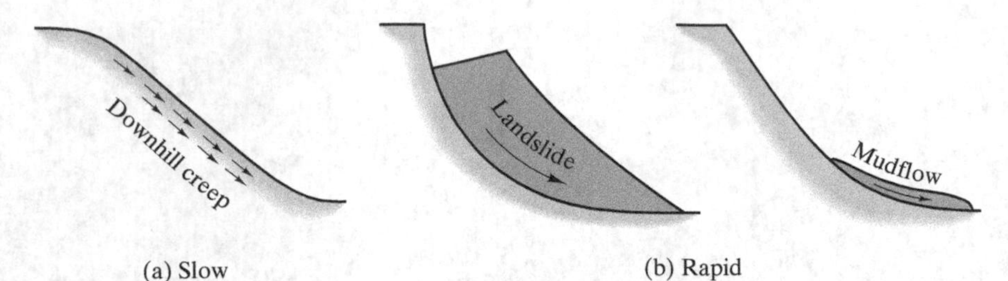

FIGURE 2.26 Colluvial soils: (a) slowly formed by creep; (b) rapidly formed by landslides or mudflows.

and contraction of clays, frost action, and other processes. Creep typically extends to depths of 0.3 to 3 m, with the greatest displacements occurring at the ground surface. In spite of the name, this process is entirely different from the "creep" process in aeolian soils.

Such slow movements might first appear to be inconsequential, but in time they can produce significant distortions in structures founded on such soils. Foundations that extend through creeping soils to firm ground below may be subjected to significant downslope forces from these soils, and need to be designed accordingly. In addition, the engineering properties of the soil deteriorate as it moves downhill, thus producing a material that is inferior to the parent soils.

Rapid downslope movements, such as landslides or mudflows, are more dramatic events which we will discuss in Chapter 13. Although these rapid movements can occur in any type of soil, the product is considered to be a colluvial soil.

Although colluvial soils occur naturally, construction activities sometimes accelerate their formation. For example, making an excavation at the toe of a slope may change a slow creep into a landslide.

2.7 ROCK AND SOIL AS GEOMATERIALS

Geotechnical engineers may encounter in practice a wide variety of *geomaterials*, ranging from very soft soils like the Young San Francisco Bay Mud to massive hard rocks like the Sierra Nevada granodiorite that has been called *the best rock on earth*. There are countless different types of geomaterials between these two extremes, including intermediate materials we call hard soils or soft rocks. A geotechnical engineer should be prepared to handle the behavior of any geomaterials encountered, be they rock, soil, or in-between materials.

Soil and Rock Definitions

Both geologists and engineers frequently divide geomaterials into two broad categories: *rock* and *soil*. Although this may seem to be a simple distinction, in reality it is not and has often been a source of confusion. To a geologist, rock is "any naturally formed aggregate or mass of mineral matter, whether or not coherent, constituting an essential and appreciable part of the earth's crust" (American Geological Institute, 1976). This definition focuses on the modes of origin and structure of the material. On the other hand, engineers (and contractors) sometimes consider rock to be a "hard, durable material that cannot be excavated without blasting," a definition based on strength and durability.

Unfortunately, these two definitions sometimes produce conflicting classifications, especially in intermediate materials. For example, some materials that are defined as rock in terms of their geologic origin are soft enough to be easily excavated with the same equipment used for soil. They may even look like soil. A soft claystone is a good example. Conversely, some cemented soils, such as caliche, are "hard as rock" and very difficult to excavate. This difficulty in classifying some materials has often led to construction lawsuits, because contractors are typically paid more to excavate "rock." It can also be a problem when piles are to be driven to "rock."

Therefore, it is important for both engineering geologists and geotechnical engineers to properly communicate the nature of geomaterials (rock vs. soil) to other members of the design and construction teams. Our thought processes tend to use the geologists' definitions because they help us interpret the subsurface conditions, but contractors and other engineers usually interpret our comments in light of the engineers' definitions.

Sometimes this difficulty can be overcome by using the terms *hard rock* and *soft rock*, where the latter is capable of being excavated by conventional earthmoving equipment. However, this definition can also lead to confusion and is not entirely satisfactory. In Chapter 6, we will discuss more specific classification methods to be used in excavation specifications

Another aspect of dividing geomaterials into rock and soil is that this distinction often determines the kinds of subsurface data we need to acquire, the tests we will perform, and the analyses we will conduct. This is because there are important differences between these two materials, including the following (Goodman, 1990):

- Rocks are generally cemented; soils are rarely cemented
- Rocks usually have much lower porosity than soils
- Rocks can be found in states of decay with greatly altered properties and attributes; effects of weathering on soils are more subtle and generally less variable
- Rock masses are often discontinuous; soil masses usually can be represented as continuous
- Rocks have more complex and generally unknowable stress histories. In many rock masses, the least principal stress is vertical; in most soils the greatest principal stress is vertical.

Geomaterial Structure and Scale Effects

Broadly speaking, soil and rock can be analyzed either as continuous materials or discontinuous materials. In a continuum model, we ignore the behavior of and interaction between individual discrete components of the material (soil particles or rock blocks)

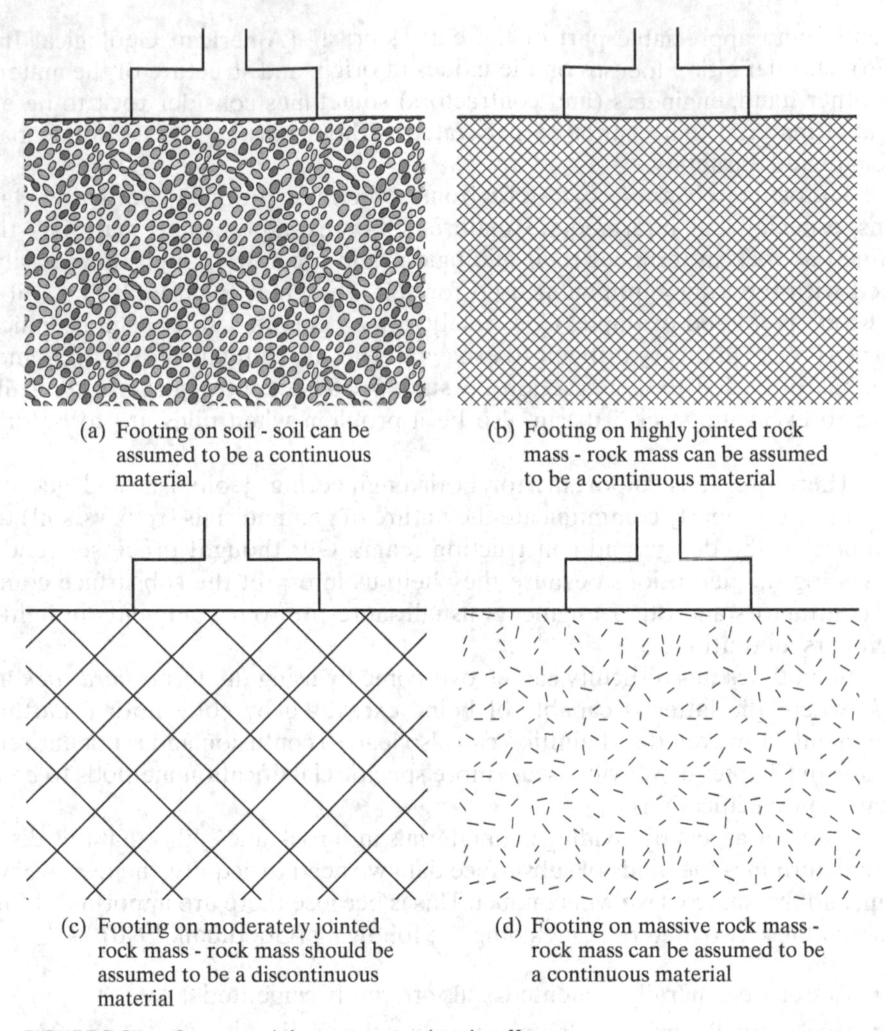

(a) Footing on soil - soil can be assumed to be a continuous material

(b) Footing on highly jointed rock mass - rock mass can be assumed to be a continuous material

(c) Footing on moderately jointed rock mass - rock mass should be assumed to be a discontinuous material

(d) Footing on massive rock mass - rock mass can be assumed to be a continuous material

FIGURE 2.27 Geomaterial structure and scale effects.

and treat the material as an equivalent continuous material. In a discontinuum model, we model explicitly the discrete components (soil particles or rock blocks) and their interaction. Generally, soil mechanics has used continuum-based models, and rock mechanics has used both continuum- and discontinuum-based models.

Given a problem at hand, the geotechnical engineer must determine the appropriate analysis techniques to be used for the problem. Figure 2.27 schematically shows how the scale of an engineering structure relative to that of some feature of the geomaterial involved would determine what assumption on the geomaterial would be appropriate in the analysis. Figures 2.27(a) and (b) show a footing having a dimension that is much larger than that of a soil particle and the joint spacings of a highly jointed rock mass, respectively; therefore, it would be appropriate to assume the soil or the rock mass as a continuous material in the analysis of the footing. In these cases, soil mechanics and continuum-based rock mechanics techniques would be applicable.

Figure 2.27(c) shows a footing having a dimension that is on the same order of magnitude as that of the joint spacing of a moderately jointed rock mass; therefore, it would be appropriate to assume the rock mass as a truly discontinuous material in the analysis of the footing. In this case, soil mechanics or continuum-based rock mechanics techniques would not be applicable and discontinuum-based rock mechanics techniques would be applicable.

Figure 2.27(d) shows a footing founded on a massive rock mass that is virtually jointless; therefore, it would be appropriate to assume the rock mass as a continuous material in the analysis of the footing. In this case, continuum-based rock mechanics and solid mechanics techniques would be applicable.

SUMMARY

Major Points

1. Engineering geology is a profession closely related to geotechnical engineering. It deals with the application of geologic principles to engineering works, and is especially useful at sites where rock is at or near the ground surface.
2. It is important for geologists to have some understanding of engineering, and for engineers to have some understanding of geology.
3. The earth's crust is always changing through many processes that can be organized into a framework called the geologic cycle. These processes can be very slow, and we must understand them to properly interpret geologic profiles.
4. There are three major rock types: igneous, sedimentary, and metamorphic.
5. Minerals are naturally formed elements or compounds with specific structures and chemical compositions. They are the basic constituents of rocks and soils.
6. Properly identifying the configuration and orientation of rock formations is at least as important as identifying the rock types contained in them. This study is called structural geology.
7. Soils are formed through several different geologic processes. Understanding these processes gives us insight into the engineering behavior of these soils.
8. Rocks disintegrate into smaller particles through weathering, eventually becoming soils.
9. There are three major types of weathering: physical (or mechanical), chemical, and biological.
10. Geologists classify soils into two broad categories: residual soils and transported soils.
11. Geotechnical engineers may encounter in practice a wide variety of geomaterials, ranging from soft clays to massive hard rocks.

12. Earth materials may be divided into two broad categories, rock and soil. Unfortunately, everyone does not agree on how to distinguish between the two, especially in intermediate materials.
13. Whether continuum- or discontinuum-based soil mechanics or rock mechanics techniques would be applicable to a geotechnical design problem would depend on the scale of the engineering structure relative to that of some feature of the geomaterial involved.

Vocabulary

ablation till
aeolian soils
alluvial fan
alluvial soil
alluvium
andesite
anisotropic strength
anticline
apparent dip
arkose
attitude
basalt
bedding planes
biological weathering
breccia
Brunton compass
calcite
carbonates
chalk
chemical weathering
clastic rocks
claystone
colluvial soil
conglomerate
continuum
creep
decomposed granite
diorite
dip
dip-slip fault
discontinuity
discontinuum
dolomite
drift
engineering geology
extrusives
fault
fault trace
feldspar
ferromagnesian minerals
folds
foliations
fracture
gabbro
geologic cycle
geology
geomaterial
glacial soil
glaciofluvial soil
glaciolacustrine soil
glaciomarine soil
gneiss
granodiorite
granite
graywacke
gypsum
hardpan
igneous rocks
intrusives
iron oxides
joint
karst topography
lacustrine soils
laterite
left-lateral fault
limestone
lodgement till
loess
magma
marble
marine soils
mechanical weathering
metamorphic rocks
mica
minerals
moraine
mudstone
normal fault
physical weathering
quartz
quartzite
residual soil
reverse fault
rhyolite
right-lateral fault
rock
rock mechanics
saltation
sand dune
sandstone
saprolite
scale effects
schist
sedimentary rocks
shale
shear zone
siltstone
sinkhole
slate
soil
soil mechanics
strike
strike-slip fault
structural geology
suspension

syncline
thrust fault
till
transported soil
varved clay
weathering

QUESTIONS AND PRACTICE PROBLEMS

Section 2.1 The Geologic Cycle

2.1 Describe the interrelationships in the geologic cycle among magma, different types of rocks, and soil.

Section 2.2 Rocks

2.2 Which would probably provide better support for a large, heavy building, a diorite or a shale? Why?

2.3 Would a tunnel excavated in a granite require more or less support than a tunnel excavated in a mudstone? Why?

2.4 Fossils are imprints in rock of ancient plants and animals. What type of rock might contain fossils? What type would never contain fossils? Explain.

2.5 What type of rock is most likely to develop sinkholes? Why?

2.6 In general, how does the age of a rock affect its engineering characteristics?

Section 2.3 Rock-Forming Minerals

2.7 Name four common minerals. For each mineral named, state one characteristic or property of the mineral and describe how this characteristic or property may be reflected in the property of a rock that contains the mineral.

Section 2.4 Structural Geology

2.8 Define "bedding planes" and explain why it is important to assess their orientations as a part of slope stability analyses.

2.9 The bedding planes in a certain sedimentary rock have an average strike of N43°E and an average dip of 38°SE, as shown by the attitude in Figure 2.28. A 15-m tall, east-west striking

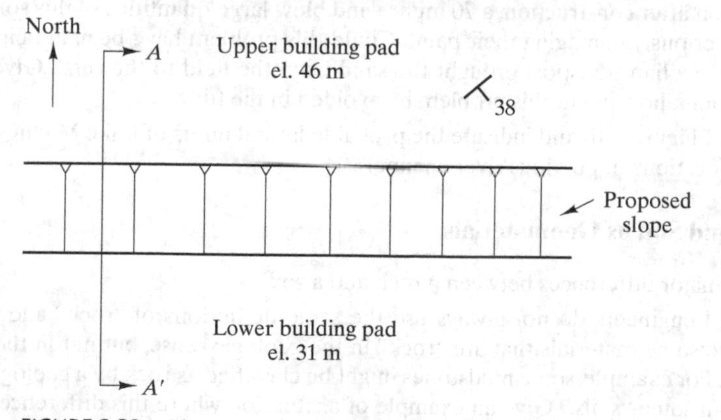

FIGURE 2.28 Plan view of proposed slope for Problem 2.9. el. = elevation.

cut slope inclined at 34° with the horizontal is to be made in this rock. The ground surface above and below this proposed slope will be nearly level. Compute the apparent dip of the bedding planes as they will appear in cross-section *A*-*A*′, then draw this cross-section. Your drawing should show the ground surface and the bedding planes. Do these bedding planes pose a potential slope stability problem? Explain.

2.10 Draw cross-section *A*-*A*′ in Figure 2.11 and compute the apparent dip of the bedding planes as they would appear in this cross-section. There are two nearby attitudes, so compute the apparent dip for each. Then, sketch in the bedding planes on the cross-section. Do these bedding planes pose a potential stability problem? Why or why not?

2.11 Draw cross-section *C*-*C*′ in Figure 2.11 and compute the apparent dip of the bedding planes as they would appear in this cross-section. There are two nearby attitudes, so compute the apparent dip for each. Then, sketch in the bedding planes on the cross-section. Do these bedding planes pose a potential stability problem? Why or why not?

Section 2.5 Weathering

2.12 Name the three major types of weathering and define each.

2.13 Discuss how weathering affects the engineering properties, for example, density and strength of a rock.

Section 2.6 Soil Formation, Transport, and Deposition

2.14 Define residual soils. Name one type of residual soil and describe its typical engineering characteristics.

2.15 Define transported soils. Name the major types of transported soils and describe how each is formed.

2.16 Explain the difference between ablation till and lodgement till. Which would provide better support for heavy civil engineering projects? Why?

2.17 Which would probably provide better support for a proposed structure, an alluvial sand or an aeolian sand? Why?

2.18 A new car dealership has recently been built in an area known for occasional strong winds. Unfortunately, an open field of fine sandy soil exists immediately upwind of the dealership. Soon after construction, a 70 mi/hr wind blew large quantities of this soil onto the new cars, seriously damaging their paint. Could this problem have been anticipated? What mode of aeolian transport brought the sand from the field to the cars? Given the current conditions, how might this problem be avoided in the future?

2.19 Make a copy of Figure 2.18 and indicate the probable lateral limits of Lake Michigan and the probable locations of previous river channels.

Section 2.7 Rock and Soil as Geomaterials

2.20 What are the major differences between a rock and a soil?

2.21 Geologists and engineers do not always use the same definitions of "rock" and "soil." Thus, there are some materials that are "rock" in the geologic sense, but not in the engineering sense. For example, some mudstones might be classified as rock by a geologist, yet be weaker than some "soils." Give an example of a situation where this difference could cause problems in the design or construction of a civil engineering project.

Comprehensive

2.22 A proposed construction site is underlain by a sedimentary rock containing rounded gravel-size particles and sand. The gravel-size particles represent about 75 percent of the total mass. What is the name of this rock? Would you expect it to provide good support for the proposed structural foundations? Would it be difficult to excavate?

2.23 As the glaciers in North America melted, the runoff formed a large lake in what is now southern Manitoba, eastern North Dakota, and western Minnesota. Called Lake Agassiz, it was larger than all of the current Great Lakes combined. The present Lake Winnipeg is a remnant of this ancient lake. The City of Winnipeg is located on the ancient lakebed. What kinds of soils would you expect to find beneath the city, and what would be their likely geologic origin? What would be the typical engineering characteristics of these soils?

2.24 Would you expect to find till in Houston, Texas? Why or why not?

2.25 A heavy structure is to be built on a site adjacent to the Hudson River near Albany, NY. This area was once covered with glaciers that left deposits of lodgement till and glaciofluvial soils. Since then, the river has deposited alluvial soils over the glacial deposits. The design engineer wishes to support the structure on pile foundations extending to the lodgement till, and you are planning a series of exploratory borings to determine the depth to these strata. What characteristics would you expect in the lodgement till, i.e., how would you recognize it?

2.26 New Orleans, Louisiana is located near the mouth of the Mississippi River. What geological process has been the dominant source of the soils beneath this city? What engineering characteristics would you expect from these soils, i.e., overall quality, uniform or erratic, etc.? Explain.

2.27 A project is to be built on a moderately sloping site immediately below the mouth of a canyon near Phoenix, Arizona. Using the geologic terms described in Section 2.6, what type of soil is most likely to be found? Why? What engineering characteristics would you expect from these soils? Explain.

2.28 A varved clay deposit has been progressively buried by other deposits and eventually has been lithified into a sedimentary rock. What type of rock is it? Would you expect its bedding planes to be distinct or vague? What engineering characteristics would you expect from this rock? Explain.

CHAPTER 3

Site Exploration and Characterization

The process of exploring to characterize or define small scale properties of substrata at construction sites is unique to geotechnical engineering. In other engineering disciplines, material properties are specified during design, or before construction or manufacture, and then controlled to meet the specification. Unfortunately, subsurface properties cannot be specified; they must be deduced through exploration.

Charles H. Dowding (1979)

Most engineers work with manufactured products that have very consistent and predictable engineering properties. For example, when a structural engineer designs a W18×55 beam to be made of A36 structural steel, he or she can be confident that the yield strength will be 36 k/in.^{2}, the modulus of elasticity will be 29×10^3 k/in.^{2}, the moment of inertia will be 891 in.^{4}, and so on. There is no need to test A36 steel every time someone wants to design a beam; we simply specify what is to be used and the contractor is obligated to supply it.

Geotechnical engineers do not have this luxury. We work with soils and rocks, which are natural materials whose engineering properties can vary dramatically from one place to another. For example, one site may be underlain by strong, hard deposits, such as lodgement till, and can safely support heavy loads, while another may be underlain by soft, weak deposits, such as varved clay, and thus requires careful design and construction techniques to support even nominal loads. In addition, we need to work with whatever soils and/or rocks that are present at our site. Thus, instead of specifying required properties, our task is to determine the existing properties at our site. This process is called *site characterization.*

This distinction is no small matter, because the site characterization efforts typically represent a very large share of the geotechnical engineering budget. We often spend more time and money exploring the subsurface conditions and defining the engineering characteristics of the materials encountered than we do performing our analyses and developing our designs.

The objectives of a site exploration and characterization program include the following:

- Determining the locations and thicknesses of soil and rock strata
- Determining the location of the groundwater table, along with other important groundwater-related issues
- Recovering samples for testing and evaluation
- Conducting tests, either in the field or in the laboratory, to measure relevant engineering properties
- Defining special problems and concerns.

Unfortunately, most of what we want to know is hidden underground and thus very difficult to discern. We can explore the subsurface conditions using borings and other techniques, and recover samples for testing and evaluation, but even the most thorough exploration program encounters only a small fraction of the soils and rocks below the site. We do not know what soil conditions exist between borings, and must rely on interpolation based on a knowledge of geology and soil deposition processes. In addition, we never can be completely sure if our samples are truly representative, or if we have missed some important underground feature. These uncertainties represent the single largest source of problems for geotechnical engineers. We overcome them using a combination of techniques, including the following:

- Recognizing the uncertainties and applying appropriate conservatism and factors of safety to our analyses and designs
- Using a knowledge of the local geology to interpret the available subsurface information
- Observing and monitoring conditions during construction, and being prepared to modify the design based on newly acquired information
- Acknowledging that 100% reliability is not attainable, and accepting some risk of failure due to unforeseen conditions.

3.1 PROJECT ASSESSMENT

Before planning a site exploration and characterization program, the geotechnical engineer must gather certain information on the proposed development. This information would include such matters as:

- The types, locations, and approximate dimensions of the proposed improvements (i.e., a nine-story building is to be built here, a parking lot there, and an access road to connect the project with the main highway over there)
- The type of construction, structural loads, and allowable settlements

- The existing topography and any proposed grading
- The presence of previous development on the site, if any.

All these factors have an impact on the methods and thoroughness of the program. For example, a proposed nuclear power plant to be built on a difficult site would require very extensive site exploration and characterization, while a one-story wood frame building on a good site may require only minimal effort.

3.2 LITERATURE SEARCH

The first step in gathering information on a site often consists of reviewing published sources. Sometimes these efforts reveal the results of extensive work already performed on the site, and very little additional exploration may be necessary. More often, literature searches provide only a general understanding of the local soils and rocks.

Sources of relevant literature include the following:

- *Geologic maps*, which are representations of the soil and rock types exposed at the ground surface, and usually show the extent of various geologic formations, alignments of faults, major landslides, and other geologic features. They also may include cross-sections showing subsurface conditions. Published maps usually cover areas much larger than a single project site. Scales of about 1:24,000 (the same as for a U.S. Geological Survey (USGS) 7.5 minute quad map) are common.

  Studying the local geology helps alert us to potential problems at the site and helps us interpret the data gathered from our surface and subsurface exploration programs, which are virtually always limited to the site under consideration. If bedrock is exposed at our site, geologic maps help us identify the formation to which it belongs, and thus assist in the identification of potential problems. For example, the Bearpaw Formation in Montana, Saskatchewan, and the surrounding area is a shale with large quantities of montmorillonite, a highly expansive clay mineral. Structures built on this formation often have problems when it becomes wet and swells, so the identification of its presence on a site signals the need for special precautions.

  Sometimes our site characterization efforts involve developing new large-scale geologic maps that describe our site in more detail.

- *Soil survey reports*, which contain maps of the near-surface soil conditions. These maps are developed primarily for agricultural purposes, but can provide useful information for engineers. In the United States, soil survey reports are produced primarily by the Natural Resources Conservation Service (formerly known as the Soil Conservation Service).

  Typical soil surveys encompass areas about the size of a county, with mapping scales of about 1:15,000 to 1:24,000. The soil survey maps are accompanied by reports that include limited test data, along with qualitative evaluations of each soil series. Although these surveys alone are usually not sufficiently detailed to allow the development of detailed geotechnical designs, they do identify the general surface soil conditions in the area, and can be helpful in planning more detailed site-specific investigations.

- *Geotechnical investigation reports* from other nearby projects, or even previous projects on our site, are often available, especially in urban areas. These reports can be very valuable because they normally include borings, soil tests, and other relevant data.
- *Historic groundwater data* is sometimes available from maps or reports. This data may be used to predict the worst-case groundwater conditions that might occur during the life of a project.
- *Seismic hazard maps* produced by the USGS and the California Geological Survey (CGS) may be used to help assess the potential effects of earthquakes on a project.

3.3 REMOTE SENSING

Remote sensing is the process of detecting features on the earth's surface from some remote location, such as an aircraft or spacecraft. This can be done using aerial photographs, radar, and other types of sensors. For geotechnical engineers, aerial photographs are the most useful remote sensing tool.

Conventional Aerial Photographs

Aerial photographs or simply *airphotos* are taken from airplanes using special cameras. Some of these are *oblique*, which means they view the landscape at some angle, while others are *vertical*, or looking straight down. The latter are more common, and generally more useful. Figure 3.1 shows a vertical airphoto. Both black-and-white and color photos are available, usually on 9 in. × 9 in. (229 mm × 229 mm) negatives. Color photographs are more useful because they reveal geologic information in more detail.

Sometimes overlapping vertical airphotos are used to form a *stereo pair*. When viewed through a stereoscope, as shown in Figure 3.2, these photos present a three-dimensional image of the ground.

The scale of airphotos generally is between 1:3,000 and 1:40,000, which allows us to identify important geologic features, such as landslides, faults, and erosional features, and helps us understand site topography and drainage patterns. This technique is especially useful at sites where observations from the ground are blocked by forests. Viewing old airphotos also helps determine the site history, including previous buildings, old cuts and fills, and so on.

Infrared Aerial Photographs

It also is possible to take aerial photographs using a special film that is sensitive to both the visible and infrared spectra. The colors are shifted from that of normal color photographs (yellow objects appear green, etc.) and reflected infrared light is

FIGURE 3.1 Vertical aerial photograph. The ocean is at the bottom of the photo. The land area includes a major highway, residential areas, and agricultural areas. The dark vertical line near the center of the photograph is trees along a creek. (Pacific Western, Inc., Santa Barbara, CA.)

FIGURE 3.2 The first author using a stereoscope to view aerial photographs in three dimensions.

shown as red. This is valuable because vegetation reflects infrared, and thus is easily discernible.

Healthy, vigorous vegetation reflects the most, and is bright red in the photographs. This normally indicates the presence of water, and thus can be used to locate springs and seepage zones. This technique is especially useful in certain slope stability studies, because water from these springs and seeps may cause future landslides, or may be part of the explanation of a past landslide.

3.4 FIELD RECONNAISSANCE AND SURFACE EXPLORATION

The *field reconnaissance* consists of "walking the site" and visually assessing the local conditions. It includes obtaining answers to such questions as the following:

- Is there any evidence of previous development on the site?
- Is there any evidence of previous grading on the site?
- Is there evidence of landslides or other stability problems?
- Are nearby structures performing satisfactorily?
- What are the surface drainage conditions?
- What types of soil and/or rock are exposed at the ground surface?
- Will access problems limit the types of subsurface exploration techniques that can be used?
- Might the proposed construction affect existing improvements? For example, a fragile old building adjacent to the site might be damaged by vibrations from pile driving.
- Do any offsite conditions affect the proposed development? For example, potential flooding, mudflows, or rockfalls from offsite might affect the property.

This work also includes marking the locations of proposed exploratory borings and trenches. When rock is exposed, the field reconnaissance often will include geologic mapping.

Depending on the site conditions, a field reconnaissance also might include detailed mapping of the surface conditions. For example, if peat bogs (depressions filled with highly organic soils) are present, their lateral extent must be carefully recorded. If rock is exposed at or near the ground surface, geologic mapping by an engineering geologist may be required.

Besides manual mapping by an engineering geologist, there are emerging alternative mapping techniques that can provide a large amount of data in a short period of time. Two such techniques are *laser scanning* and *photogrammetry*. These techniques can be applied to map even areas that are inaccessible as long as these areas can be scanned or photographed from a distance.

3.5 SUBSURFACE EXPLORATION

Although information on the soil and rock conditions exposed at the ground surface is very valuable, geotechnical engineers also need to evaluate the subsurface (underground) conditions. The geophysical methods described later in this chapter can provide some insight, but we primarily rely on soil and rock samples obtained by drilling vertical holes known as *borings*, or by digging exploratory trenches or pits. These subsurface exploration activities usually are the heart of a site characterization program, and typically are the most expensive part because they require the mobilization of both equipment and labor.

Exploratory Borings

The most common method of exploring the subsurface conditions is to drill a series of vertical holes in the ground. These are known as *borings* or *exploratory borings* and are typically 75 to 600 mm (3–24 in.) in diameter and 2 to 30 m (7–100 ft) deep.

Small, shallow borings can be made with lightweight hand-operated augers, as shown in Figure 3.3. This equipment is inexpensive and portable, but limited in its capabilities and generally suitable only for very small projects with boring depths less than about 4 m (13 ft). Some additional capacity can be gained by using portable power-operated equipment, but such equipment is still too limited for most projects.

Geotechnical engineers usually use much heavier equipment powered by larger engines. Sometimes it is mounted on skids or small roll-in units, as shown in Figure 3.4, but most often it is truck-mounted as shown in Figure 3.5. These *truck-mounted drill rigs* perform at least 90% of geotechnical drilling, and can drill to depths of 30 m (100 ft) with little difficulty. Some truck-mounted rigs can drill to 60 m (200 ft) or even more, but such capabilities are rarely needed.

Drilling Methods

Different methods are available to advance the boring, depending on the anticipated soil and rock conditions.

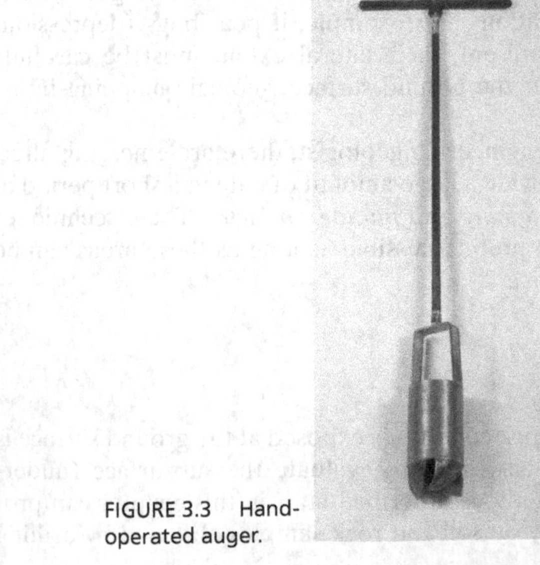

FIGURE 3.3 Hand-operated auger.

FIGURE 3.4 Limited access rig that can be moved through narrow openings. It is connected to a truck-mounted hydraulic pump via the hoses in the foreground.

Drilling in Firm and Dense Soils The simplest drilling methods use a *flight auger* or a *bucket auger*, as shown in Figures 3.6 and 3.7, respectively, to produce an open hole. With either type, the auger is lowered into the hole and rotated to dig into the soil. Then, it is removed, the soil is discharged onto the ground, and the process is repeated. The hole is free of equipment between these cycles, which allows the driller to insert sampling equipment at desired depths and obtain undisturbed samples.

These methods are comparatively inexpensive, so long as they are used in suitable conditions, such as firm and dense soils or soft rocks. However, they can meet *refusal* (the inability to progress further) when they encounter hard boulders or hard bedrock. This is especially likely when the boring diameter is small, since even large

FIGURE 3.5 Truck-mounted drill rig.

FIGURE 3.6 A crew drilling an exploratory boring using a truck-mounted flight auger. These augers typically have outside diameters of 75–200 mm (3–8 in.). (Foremost Mobile Drilling Co.)

FIGURE 3.7 Using a bucket auger. The bucket has just come out of the hole and will be tilted back by the drill rig. The driller's helper will then pull the rope, which will open the bottom and release the soil. Most bucket augers have diameters between 300 and 900 mm (12–36 in.).

cobbles might block the drilling. Sometimes this problem can be overcome by using a larger diameter auger (i.e., one that is larger than the cobbles and boulders). Alternatively, some rigs can switch to a coring mode and continue as described below. Otherwise, it becomes necessary to use some other type of drilling method.

Drilling in Soils Prone to Caving or Squeezing Open hole methods encounter problems in soils prone to *caving* (i.e., the sides of the boring falling in) or *squeezing* (the soil moving inwards, reducing the boring diameter). Caving is most likely in loose sands and gravels, especially below the groundwater table, while squeezing is likely in soft saturated

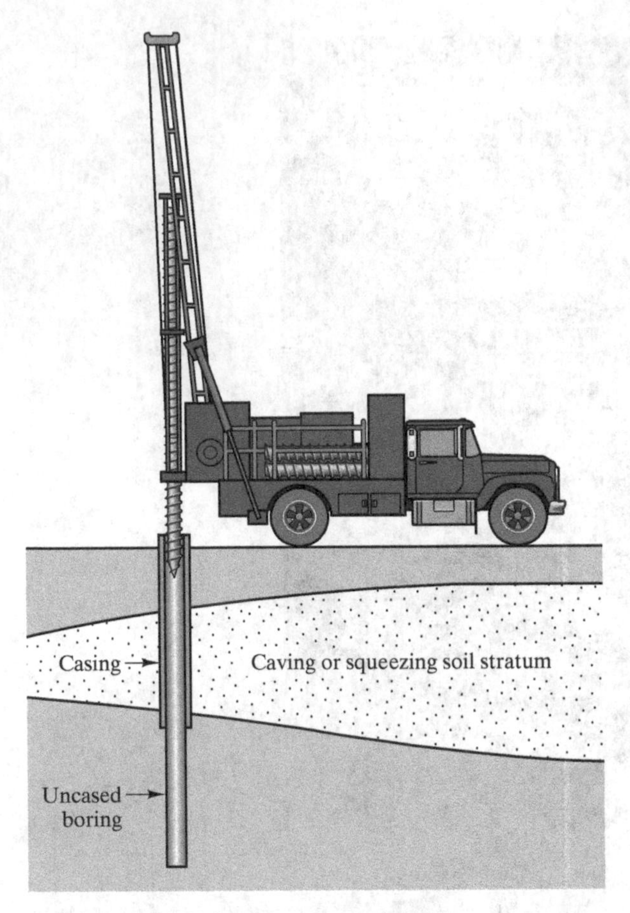

FIGURE 3.8 Use of casing to prevent caving and squeezing.

silts and clays. In such cases, it becomes necessary to provide some type of lateral support inside the hole during drilling.

One method of supporting the hole is to install *casing* (see Figure 3.8), which is a temporary lining made of a steel pipe. This method is especially useful if only the upper soils are prone to caving, because the casing does not need to extend for the entire depth of the boring.

Another, more common method is to use a *hollow stem auger*, as shown in Figure 3.9. Each auger section has a pipe core known as a *stem*, with a temporary plug on the bottom of the first section. The driller screws these augers into the ground, adding sections as needed. Unlike conventional augers, it is not necessary to remove them to obtain samples. Instead, the driller removes the temporary plug and inserts the sampler through the stem and into the soils below the bottom auger section, as shown in Figure 3.10. Then the sample is recovered, the plug is replaced, and drilling continues to the next sample depth. When the boring is completed, the augers are removed. Hollow stem drill rigs with 200 mm (8 in.) diameter augers are very common, and are often used even when caving is not a problem.

FIGURE 3.9 Use of a hollow stem auger.

The third method is to fill the boring with *drilling mud* or slurry, which is a mixture of bentonite or attapulgite clay and water. This material provides a hydrostatic pressure on the walls of the boring, as shown in Figure 3.11, thus preventing caving or squeezing. These borings are usually advanced using the *rotary wash method*, which flushes the drill cuttings up to the ground surface by circulating the mud with a pump. When samples are needed, the drilling tools are removed from the hole and the sampling tools are lowered through the mud to the bottom. Special drilling tools may be added if the boring reaches hard soils or rock.

Coring Drilling through rock, especially hard rock, requires different methods and equipment. Engineers usually use *coring*, which simultaneously advances the hole and obtains nearly continuous undisturbed samples. This is fundamentally a different

FIGURE 3.10 Lowering a soil sampler through a hollow stem auger.

method that consists of grinding away an annular zone with a rotary diamond drill bit, leaving a cylindrical core, which is captured by a *core barrel* and removed from the ground. The cuttings are removed by circulating drilling fluid, water, or air. Figure 3.12 shows a core sampler partway through a *core run*, which is the segment sampled during one stroke of the sampler. Coring also can be done in hard soil.

After each core run, the sample is brought to the ground surface and placed in a wooden *core box* for examination and storage, as shown in Figure 3.13. This permits detailed logging of the hole and provides high-quality samples for laboratory testing. Most cores are 48 or 54 mm in diameter.

Coring logs often record the *rock quality designation* (RQD), which is the percentage of core in pieces 100 mm or longer. It is a useful measure of rock fracturing, and thus an indicator of stability. RQD values greater than 90% typically indicate excellent rock, while values less than 50% indicate poor or very poor rock.

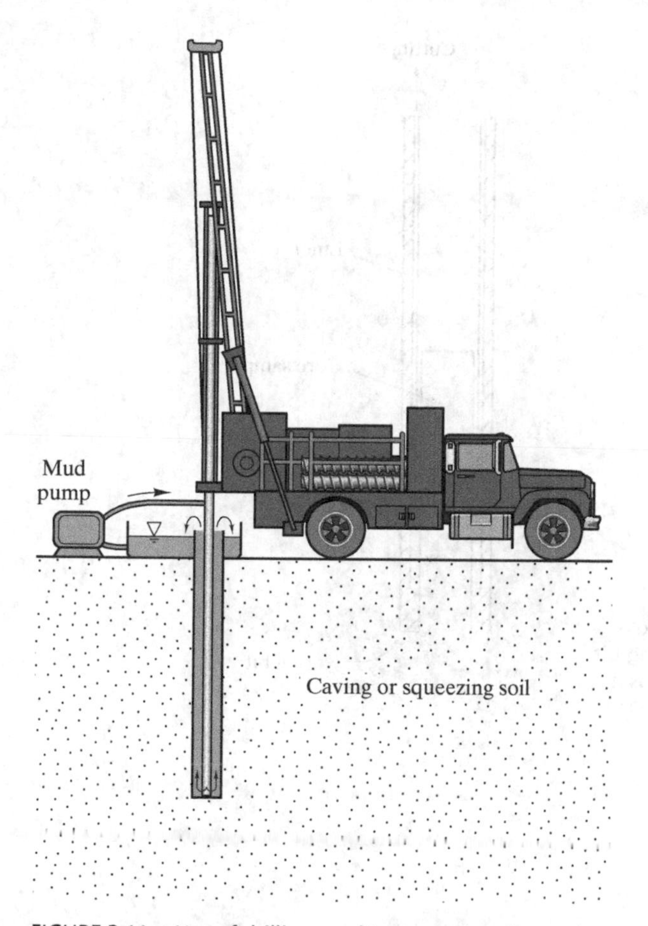

FIGURE 3.11 Use of drilling mud to prevent caving and squeezing. The mud provides a hydrostatic pressure to the sides of the boring, thus keeping the adjacent soils in place.

The *core recovery*, which is the total sample length recovered from each core run divided by the run length, also should be recorded. Often some of the sample is "lost," especially in weak or friable rocks.

Unfortunately, the weakest and most fractured zones, which are the most important zones to identify, are those most likely to be lost during coring. Also, most coring does not retain the in situ orientation of the sample, so information on the directions of joints, bedding planes, etc. is lost. Both of these problems can be at least partially overcome by using downhole cameras that take photographs or video recordings of the hole after the core has been removed.

Boring Logs The conditions encountered in an exploratory boring are recorded on a *boring log*, such as the one shown in Figure 3.14. The vertical position on these logs represents depth, and the various columns describe certain characteristics of the soils and rocks encountered. These logs also indicate the sample locations and might include

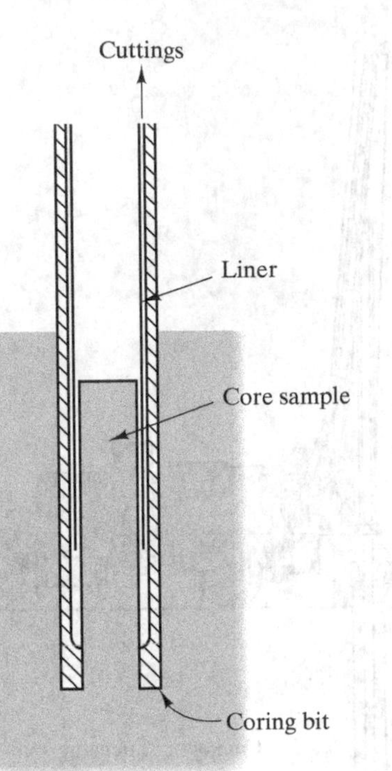

FIGURE 3.12 A sample being recovered by coring, the coring bit cuts an annular-shaped hole as it penetrates into the ground.

FIGURE 3.13 Core samples in a core box.

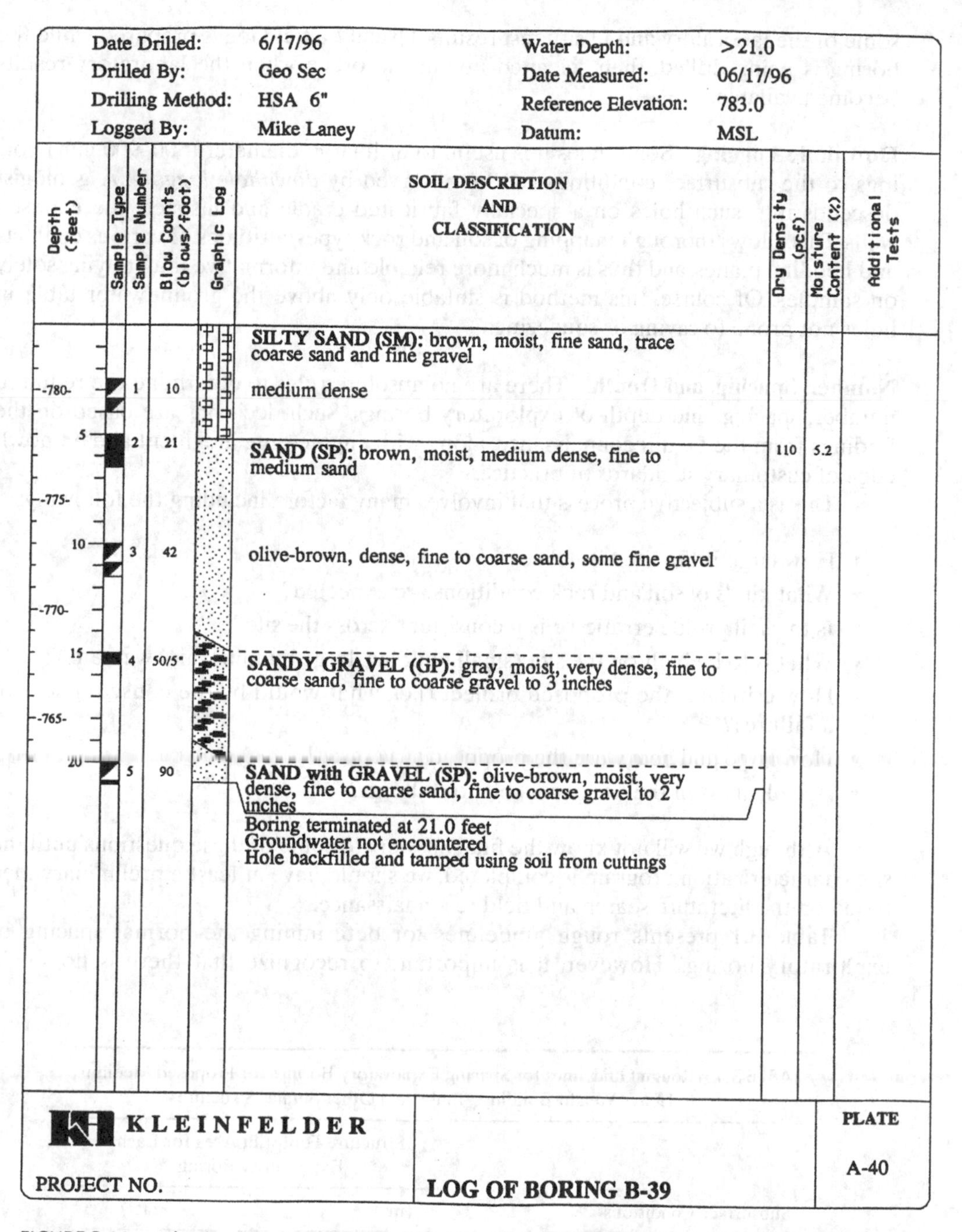

FIGURE 3.14 A boring log. Samples 2 and 4 were obtained using a heavy-wall sampler, and the corresponding blow counts are the number of hammer blows required to drive the sampler. Samples 1, 3, and 5 were obtained from standard penetration tests, and the corresponding blow counts are the N_{60} values, as discussed later in this chapter. (Kleinfelder, Inc.)

some of the laboratory and in situ test results. Usually, a field log is prepared while the boring is being drilled, then "cleaned up" in the office when the laboratory results become available.

Downhole Logging Sometimes it is useful to drill large-diameter (500–900 mm) borings so the subsurface conditions can be observed by *downhole logging*. A geologist descends into such holes on a specially fabricated cradle and inspects the exposed walls. This allows thorough mapping of soil and rock types, attitudes of various contacts and bedding planes, and thus is much more reliable and informative than relying solely on samples. Of course, this method is suitable only above the groundwater table in holes not prone to caving or squeezing.

Number, Spacing, and Depth There are no absolute rules to determine the required number, spacing, and depth of exploratory borings. Such decisions are based on the findings from the field reconnaissance, along with engineering judgment and a knowledge of customary standards of practice.

This is a subjective process that involves many factors, including the following:

- How large is the site?
- What kinds of soil and rock conditions are expected?
- Is the soil profile erratic, or is it consistent across the site?
- What is to be built on the site (small building, large building, highway, etc.)?
- How critical is the proposed project (i.e., what would be the consequences of a failure)?
- How large and heavy are the proposed structures?
- Are all areas of the site accessible to drill rigs?

Although we will not know the final answers to some of these questions until the site characterization program is completed, we should have at least a preliminary idea based on the literature search and field reconnaissance.

Table 3.1 presents rough guidelines for determining the normal spacing of exploratory borings. However, it is important to recognize that there is no single

TABLE 3.1 Rough Guidelines for Spacing Exploratory Borings for Proposed Medium to Heavy Weight Buildings, Tanks, and Other Similar Structures

	Structure Footprint Area for Each Exploratory Boring	
Subsurface Conditions	(m^2)	(ft^2)
Poor quality and/or erratic	100–300	1000–3000
Average	200–400	2000–4000
High quality and uniform	300–1000	3000–10000

TABLE 3.2 Rough Guidelines for Depths of Exploratory Borings for Buildings on Shallow Foundations[a]

Subsurface Conditions	Minimum Depth of Borings (S = number of stories; D = anticipated depth of foundation) (m)	(ft)
Poor	$6S^{0.7} + D$	$20S^{0.7} + D$
Average	$5S^{0.7} + D$	$15S^{0.7} + D$
Good	$3S^{0.7} + D$	$10S^{0.7} + D$

[a]Adapted from Sowers, 1979.

"correct" solution for the required number and depth of borings, and these guidelines must be tempered with appropriate engineering judgment.

Borings for buildings and other structures on shallow foundations generally should extend at least to the depths described in Table 3.2. If fill is present, the borings must extend through it and into the natural ground below, and if soft soils are present, the borings should extend through them and into firmer soils below. For heavy structures, at least some of the borings should be carried down to bedrock, if possible, but certainly well below the depth of any proposed deep foundations.

On large projects, the drilling program might be divided into two phases: a preliminary phase to determine the general soil profile and a final phase based on the results of the preliminary borings.

Example 3.1

A three-story steel frame office building is to be built on a site where the soils are expected to be of average quality and average uniformity. The building will have a 30 m × 40 m footprint and is expected to be supported on spread footing foundations located about 1 m below the ground surface. The site appears to be in its natural condition, with no evidence of previous grading. Bedrock is several hundred feet below the ground surface. Determine the required number and depth of the borings.

Solution:

As per Table 3.1, one boring will be needed for every 200–400 m^2 of footprint area. Since the total footprint area is $30 \times 40 = 1200$ m^2, use four borings.

As per Table 3.2, the minimum depth is $5S^{0.7} + D = 5(3)^{0.7} + 1 = 12$ m. However, it would be good to drill at least one of the borings to a greater depth.

Exploration plan:

3 borings to 12 m
1 boring to 16 m

Exploratory Trenches and Pits

Sometimes it is only necessary to explore the upper 3 m (10 ft) of soil. This might be the case for lightweight projects on sites where the soil conditions are known to be good, or on sites with old shallow fills of questionable quality. Additional shallow investigations also might be necessary to supplement a program of exploratory borings.

In such cases, geotechnical engineers often dig *exploratory trenches* (also known as *test pits*) using a backhoe, as shown in Figure 3.15. These techniques provide more information than a boring of comparable depth (because more of the soil is exposed), and often are less expensive. The log from a typical exploratory trench is shown in Figure 3.16.

Two special precautions are in order when using exploratory trenches. First, these trenches must be adequately shored or laid back to a sufficiently flat slope before anyone enters them. Many individuals (including one of the first author's former colleagues) have been killed by neglecting to enforce this basic safety measure. Second, these trenches must be properly backfilled to avoid creating an artificial soft zone that might affect future construction.

3.6 SOIL AND ROCK SAMPLING

The primary purpose of drilling exploratory borings and digging exploratory trenches is to obtain representative soil and rock samples. We use these samples to determine

FIGURE 3.15 This exploratory trench was dug by the backhoe in the background, and has been stabilized using aluminum-hydraulic shoring. An engineering geologist is logging the soil conditions in one wall of the trench. In this case, the purpose of the trench is to locate a fault.

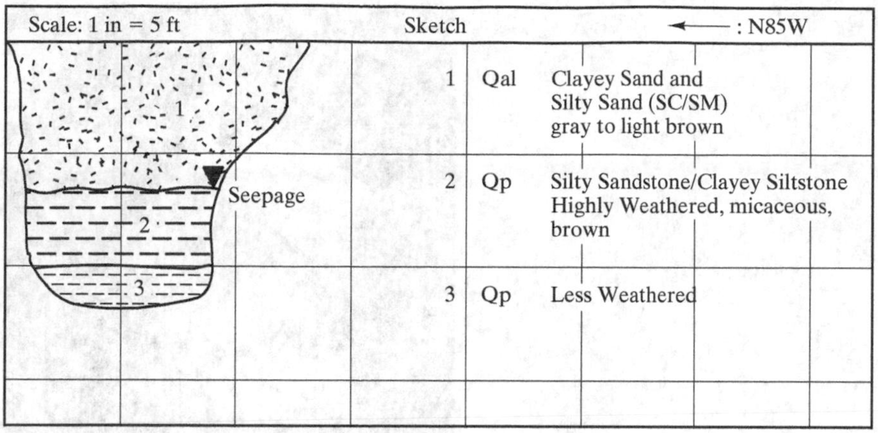

Scale: 1 in = 5 ft	Sketch		←: N85W
1	1	Qal	Clayey Sand and Silty Sand (SC/SM) gray to light brown
2 Seepage	2	Qp	Silty Sandstone/Clayey Siltstone Highly Weathered, micaceous, brown
3	3	Qp	Less Weathered

FIGURE 3.16 Log from an exploratory trench. (Courtesy of Converse Consultants.)

the subsurface profile and to measure the engineering properties of the subsurface materials. There are two categories of samples: disturbed and undisturbed, which are discussed in the following.

Disturbed Samples

A *disturbed sample* (also called a *bulk sample*) is one obtained with no attempt to retain the in-place structure of the soil or rock. The driller might obtain such a sample by removing cuttings from the bottom of a flight auger and placing them in a bag. Disturbed samples, such as the one in Figure 3.17, are suitable for many purposes, such as classification and compaction tests.

Undisturbed Samples

The greater challenge in soil sampling is to obtain *undisturbed samples*, which are necessary for many soil tests. Except for coring, which recovers undisturbed samples as the hole is advanced, drilling operations must stop periodically to permit insertion of special sampling tools into the hole as shown in Figure 3.10.

In a truly undisturbed soil sample, the soil is recovered completely intact and its in-place structure and stresses are not modified in any way. Unfortunately, the following problems make it impossible to obtain such samples:

- Shearing and compression that occurs during the process of inserting the sampling tool or sampler; disturbance due to this mechanism may be quantified by the *area ratio* of the sampler, which is the ratio of the annular cross-sectional area of the sampler tube to the circular cross-sectional area of the sampler itself.
- Release of in situ stresses as the sample is removed from the ground.

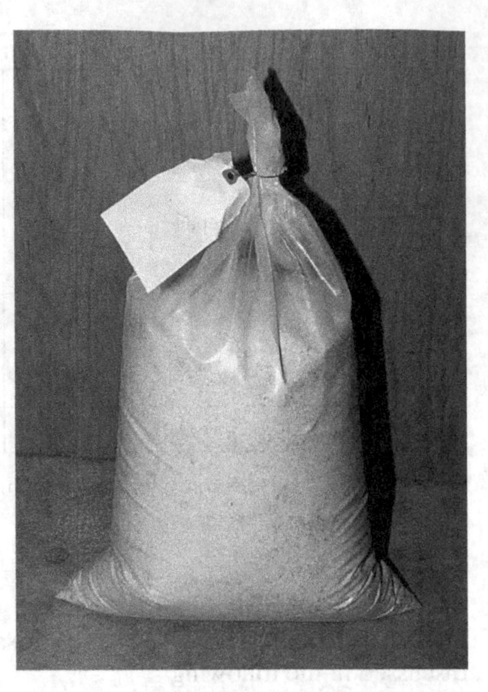

FIGURE 3.17 A typical disturbed sample stored in a plastic bag. The label attached to the bag identifies the sample.

- Possible drying and desiccation
- Vibrations during recovery and transport

Additional disturbances can occur in the laboratory as the sample is removed from its container. Thus, many engineers prefer to use the term *relatively undisturbed* to describe their samples. Sands are especially prone to disturbance during sampling. Nevertheless, geotechnical engineers have developed various methods of obtaining high-quality samples of most soils.

Shelby Tube Samplers In the mid-1930s, Mr. H.A. Mohr developed the *Shelby tube sampler*, shown in Figure 3.18(a), which soon became the most common soil sampling tool (Hvorslev, 1949). It also is known as a *thin-wall sampler* ("Shelby tubing" is a trade name for the seamless steel tubing from which the sampler is manufactured). Figure 3.18(b) shows a Shelby tube sampler attached to a standard head assembly. Most Shelby tube samplers have a 3.00 in. (76.2 mm) outside diameter and 1/16 in. (1.6 mm) wall thickness.

The head assembly is attached to a series of drilling rods, lowered to the bottom of the boring, then smoothly pressed into the natural ground below. This smooth pressing is accomplished by attaching a hydraulic cylinder to the top of the rods and using the drill rig as a reaction. Sometimes it is necessary to pound the sampler in by striking the rods with a 63.5 kg (140 lb) hammer, as shown in Figure 3.25, but this method can produce significantly more sample disturbance. The sampler is then pulled out of the ground with the soil retained inside, capped, and brought to the laboratory.

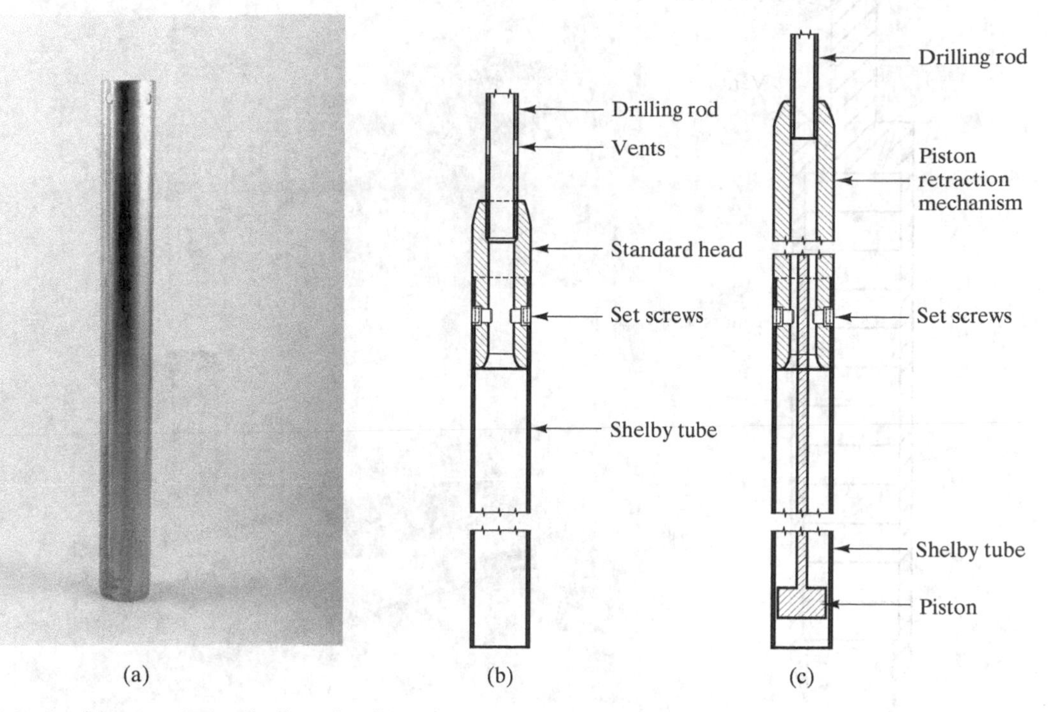

FIGURE 3.18 (a) A 3 × 36 in. Shelby tube; (b) a Shelby tube attached to a standard head with four screws; (c) a Shelby tube attached to a piston sampler.

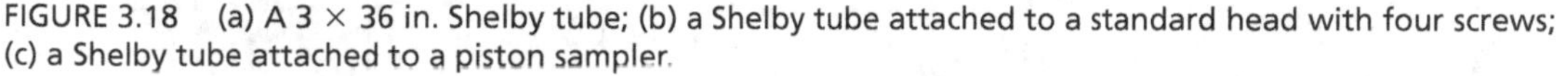

The standard head assembly has vents to allow water and air trapped above the sample to escape as it is inserted into the ground. However, some backpressure remains, and it can compress the sample. This problem is avoided in the *piston sampler* in Figure 3.18(c) in which a piston is placed inside the Shelby tube sampler. The piston is initially at the bottom of the tube, and remains at a constant elevation as the tube is advanced, thus shielding the soil from the backpressure.

Heavy-Wall Samplers Although Shelby tube samplers generally provide very good results in soft soils, they are difficult to use in hard soils. The tube may bend or collapse due to the heavy loads required to press or drive it into such soils, or it may become jammed into the ground and impossible to retrieve. The usual solution is to use a sampler with heavier walls as shown in Figure 3.19. Although these heavy walls induce more disturbance, they also provide sufficient strength and durability to survive hard soil conditions. These *heavy-wall samplers* are almost always pounded into the bottom of the boring.

Heavy-wall samplers usually contain brass or stainless steel liners as shown in Figures 3.19 and 3.20, and these liners contain the soil sample. After being extracted from the boring, the sampler is opened and the soil and liners are removed and placed in a protective cylinder for transport to the laboratory and storage.

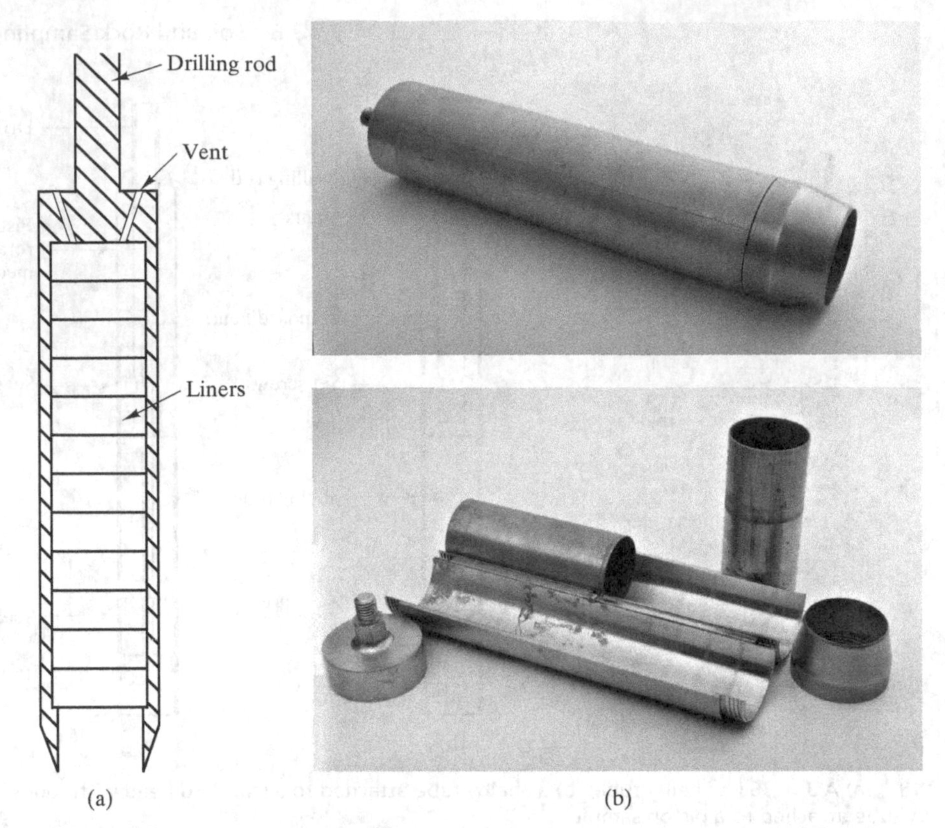

FIGURE 3.19 A heavy-wall sampler. (a) Cross-section showing liners, many different sizes are used; (b) the sampler can be opened to retrieve the liners, and thus is often called *a split barrel sampler*. The liners are available in difference lengths.

FIGURE 3.20 Soil samples from heavy-wall samplers are contained in liners and stored in plastic tubes. In this case, each liner has an outside diameter of 2.5 in., a height of 1.0 in., and is made of brass.

3.7 GROUNDWATER EXPLORATION AND MONITORING

The presence of water in soil pores or rock fissures has a very significant impact on the engineering behavior of the soil or rock, so site characterization programs also need to assess groundwater conditions. When drilling a boring or excavating a trench, we may observe small seeps, with moisture trickling into the hole. These may be due to small nonuniformities in the soil conditions that have trapped water at a certain level. Larger zones of trapped water are known as *perched groundwater*. If we continue drilling to a great enough depth, the *groundwater table* is eventually encountered, which is the level to which water fills an open boring. Soils below the groundwater table are said to be *saturated*, which means all of their voids are filled with water.

Sometimes the water quickly flows into the hole, reaching equilibrium in an hour or less. In these cases, the groundwater table can be located in the open hole before it is backfilled. However, water in silty and clayey soils may require many hours or even days to reach equilibrium, and leaving the hole open that long may pose safety problems. In addition, the groundwater table often changes with time, and we may wish to monitor these changes. The solution to both problems is to install an observation well in the boring as shown in Figure 3.21. It consists of a slotted plastic pipe backfilled with pervious soils (or even the drill cuttings) and sealed with an impervious cap. Groundwater is able to flow freely into or out of this pipe, so the water level inside is the groundwater table. The depth to this water can be measured using the electronic probe shown in Figure 3.22.

We will discuss groundwater in much more detail in Chapters 7 and 8.

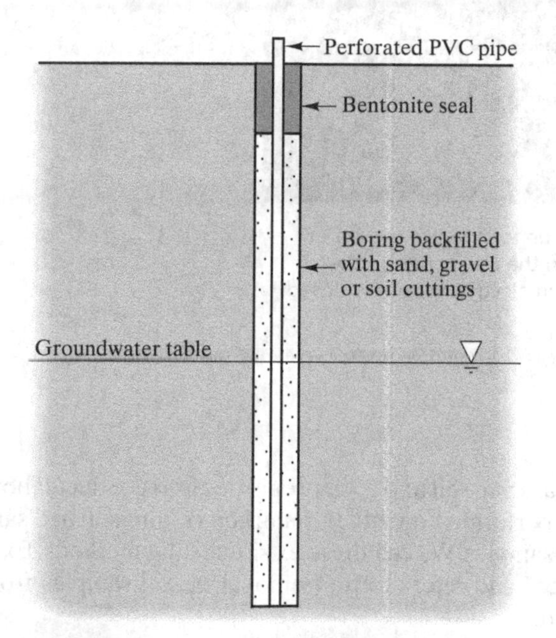

FIGURE 3.21 An observation well.

FIGURE 3.22 Using an electronic probe to measure the water level inside an observation well. When the electrodes on the bottom of the tape touch the water, an electrical circuit is closed and a buzzer sounds inside the reel.

3.8 EX SITU TESTING

The most common method of measuring soil and rock properties is to conduct laboratory tests. Some of these tests may be performed on either disturbed or undisturbed samples, while others require undisturbed samples. We call these ex situ testing methods. Ex situ is Latin for "out of its original place" and refers to the removal of soil samples from the ground and testing them elsewhere.

We will discuss various laboratory tests throughout this book. For clarity, these discussions are in the chapters related to the engineering properties being measured:

Moisture content test	Chapter 4
Unit weight test	Chapter 4
Specific gravity test	Chapter 4
Relative density test	Chapter 4
Sieve analysis	Chapter 4
Hydrometer analysis	Chapter 4
Atterberg limits tests	Chapter 4
Proctor compaction test	Chapter 6
Hydraulic conductivity test	Chapter 7
Consolidation test	Chapter 10
Direct shear test	Chapter 12
Triaxial compression test	Chapter 12
Unconfined compression test	Chapter 12
Ring shear test	Chapter 12

There also are many other laboratory tests we will not cover in this book (see Bardet, 1997).

3.9 IN SITU TESTING

The primary alternative to laboratory testing is to conduct in situ (Latin for in-place) tests. These consist of bringing special equipment to the field, inserting it into the ground, and testing the soil or rock while it is still underground. Such methods are especially useful in soils that are difficult to sample, such as clean sands. The additional data obtained give us more insight into the soil variability beneath a proposed construction site.

Geotechnical engineers often use the raw data obtained from in situ tests as general indicators of soil properties. For example, some in situ tests involve pounding or pressing something into the ground. If this is difficult to do, the soil must be strong; if it is easy to do, the soil must be weak. We also have developed empirical correlations between in situ test results and specific engineering properties. We will discuss some of these correlations in the following chapters:

Relative density	Chapter 4
Consistency	Chapter 5
Shear strength	Chapter 12
Settlement of foundations	Chapter 15

The discussions in this chapter are limited to describing the test procedures, adjusting the results, performing basic interpretations, and discussing the advantages and disadvantages of each test method.

Standard Penetration Test

One of the oldest and most common in situ tests is the *standard penetration test* (*SPT*). It was developed in the late 1920s and has been used extensively in North and South America, the United Kingdom, Japan, and elsewhere. Because of this long record of experience, the SPT is well-established in engineering practice. It is performed inside an exploratory boring using inexpensive and readily available equipment, and thus adds little cost to a site characterization program.

Although the SPT also is plagued by many problems that affect its accuracy and reproducibility, it probably will continue to be used for the foreseeable future, primarily because of its low cost. However, it is partially being replaced by other test methods, especially on larger and more critical projects.

Test Procedure The test procedure was not standardized until 1958 when ASTM standard D1586 first appeared. It is essentially as follows[1]:

1. Drill a 60–200 mm (2.5–8 in.) diameter exploratory boring to the depth of the first test.
2. Insert the SPT sampler (also known as a *split-spoon sampler*) into the boring. The shape and dimensions of this sampler are shown in Figure 3.23. It is connected via steel rods to a 63.5 kg (140 lb) hammer, as shown in Figure 3.24.
3. Using either a rope and cathead arrangement or an automatic tripping mechanism, raise the hammer a distance of 760 mm (30 in.) and allow it to fall. This energy drives the sampler into the bottom of the boring. Repeat this process until the sampler has penetrated a distance of 460 mm (18 in.), recording the number of hammer blows required for each 150 mm (6 in.) interval. Stop the test if more than 50 blows are required for any of the intervals, or if more than 100 total blows are required. Either of these events is known as *refusal* and is so noted on the boring log.
4. Compute the N-value by summing the blow counts for the last 300 mm (12 in.) of penetration. The blow count for the first 150 mm (6 in.) is retained for reference purposes, but not used to compute N because the bottom of the boring is likely to be disturbed by the drilling process and may be covered with loose soil left in the boring. Note that the N-value is the same regardless of whether the engineer is using English or SI units.
5. Extract the SPT sampler, then remove and save the soil sample.
6. Drill the boring to the depth of the next test and repeat steps 2 through 6 as required.

Thus, N-values may be obtained at intervals no closer than 500 mm (20 in.). Typically these tests are performed at 1.5–5 m (5–15 ft) intervals.

[1]See the ASTM D1586 standard for the complete procedure.

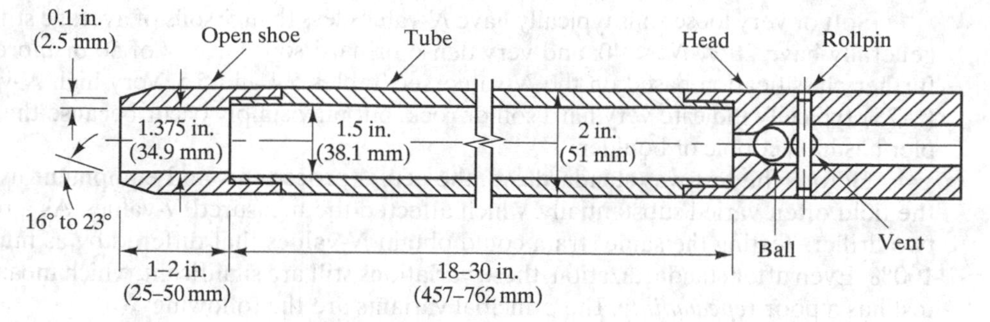

FIGURE 3.23 The SPT sampler. (Adapted from ASTM D1586; Copyright ASTM, used with permission.)

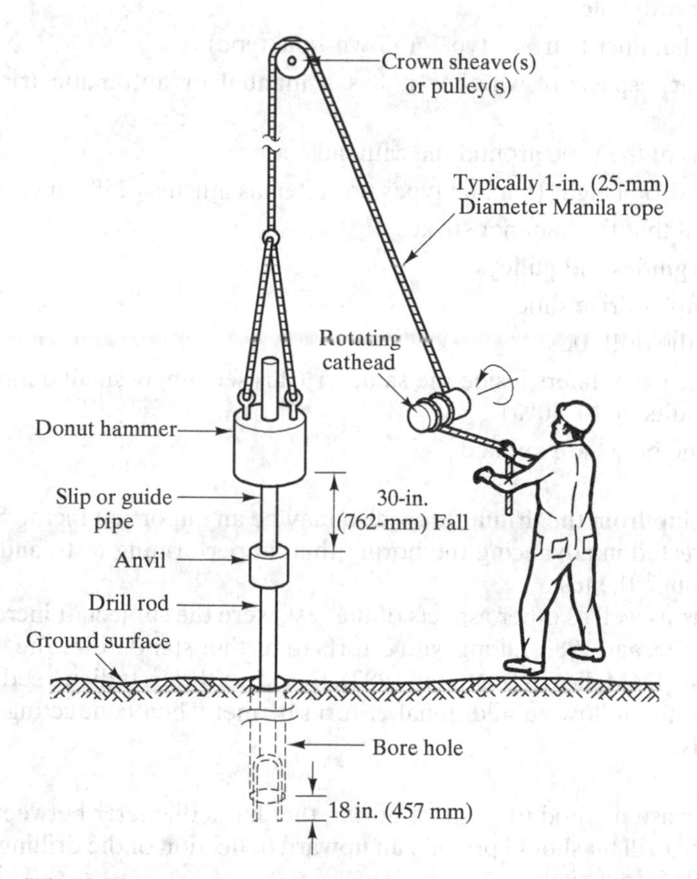

FIGURE 3.24 The SPT sampler in place in the boring with hammer, rope, and cathead. (Adapted from Kovacs et al., 1981.)

Soft or very loose soils typically have N-values less than 5; soils of average stiffness generally have $20 < N < 40$; and very dense or hard soils have N of 50 or more. For further classification based on the N-value, see Tables 5.4 and 5.5. Very high N-values (>75) typically indicate very hard soil or rock, but may simply occur because the sampler has hit a cobble or boulder.

Before the test was standardized, the actual procedures and equipment used in the field often varied substantially, which affected the measured N-values. As a result, two drillers testing the same strata could obtain N-values that differed by as much as 100%. Even after standardization, these variations still are significant, which means the test has a poor *repeatability*. The principal variants are the following:

- Method of drilling
- Cleanliness at the bottom of the hole (lack of loose dirt) before the test
- Presence or lack of drilling mud
- Diameter of the drill hole
- Location of the hammer (surface type or down-hole type)
- Type of hammer, especially whether it has a manual or automatic tripping mechanism
- Number of turns of the rope around the cathead
- Actual hammer drop height (manual types are often as much as 25% in error)
- Mass of the anvil that the hammer strikes
- Friction in rope guides and pulleys
- Wear in the sampler drive shoe
- Straightness of the drill rods
- Presence or absence of liners inside the sampler (this seemingly small detail can alter the test results by 10–30%)
- Rate at which the blows are applied.

Poor workmanship from the drilling crew also may be an important factor. Some crews are more interested in advancing the boring than in performing tests, and thus may tend to rush through the test.

These variations, as well as other aspects of the test, were the subject of increased scrutiny during the 1970s and 1980s, along with efforts to further standardize the "standard" penetration test (DeMello, 1971; Nixon, 1982). Based on these studies, Seed et al. (1985) recommended the following additional criteria be met when conducting standard penetration tests:

- Use the rotary wash method to create a boring that has a diameter between 200 and 250 mm. The drill bit should provide an upward deflection of the drilling mud (tricone or baffled drag bit).
- If the sampler is made to accommodate liners, then these liners should be used so the inside diameter is 35 mm.

- Use A or AW size drill rods for depths less than 15 m, and N or NW size for greater depths.
- Use a hammer that has an efficiency of 60%.
- Apply the hammer blows at a rate of 30 to 40 per minute.

Fortunately, automatic hammers are becoming more popular. They are much more consistent than hand-operated hammers, and thus improve the reliability of the test.

In spite of these disadvantages, the SPT does have at least three important advantages over other in situ test methods: First, it obtains a sample of the soil being tested. This permits direct soil classification. Most of the other methods do not include sample recovery, so soil classification must be based on conventional sampling from nearby borings and on correlations between the test results and soil type. Second, it is very fast and inexpensive because it is performed in borings that would have been drilled anyway. Finally, nearly all drill rigs used for soil exploration are equipped to perform this test, whereas other in situ tests require specialized equipment that may not be readily available.

Corrections to Test Results We can improve the raw SPT data by applying certain correction factors, thus significantly improving its repeatability. The variations in testing procedures may be at least partially compensated by converting the N recorded in the field to N_{60} as follows (Skempton, 1986):

$$N_{60} = \frac{E_m C_B C_S C_R N}{0.60} \tag{3.1}$$

where:

N_{60} = SPT N-value corrected for field procedures

E_m = hammer efficiency (from Table 3.3)

C_B = borehole diameter correction (from Table 3.4)

C_S = sampler correction (from Table 3.4)

C_R = rod length correction (from Table 3.4)

N = SPT N-value recorded in the field

Many different hammer designs are in common use, none of which is 100% efficient. Some common hammer designs are shown in Figure 3.25, and typical hammer efficiencies are listed in Table 3.3. Many of the SPT-based design correlations were developed using hammers that had an efficiency of about 60%, so Equation 3.1 corrects the results from other hammers to that which would have been obtained if a 60% efficient hammer was used.

The SPT data also may be adjusted using an *overburden correction* that compensates for depth effects. Tests performed near the bottom of uniform soil deposits have higher N-values than those performed near the top, so the overburden correction adjusts the measured N-values to what they would have been if the vertical effective

TABLE 3.3 SPT Hammer Efficiencies[a]

Country	Hammer Type	Hammer Release Mechanism	Hammer Efficiency E_m
Argentina	Donut	Cathead	0.45
Brazil	Pin Weight	Hand Dropped	0.72
China	Automatic	Trip	0.60
	Donut	Hand dropped	0.55
	Donut	Cathead	0.50
Colombia	Donut	Cathead	0.50
Japan	Donut	Tombi trigger	0.78–0.85
	Donut	Cathead 2 turns + special release	0.65–0.67
UK	Automatic	Trip	0.73
USA	Safety	2 turns on cathead	0.55–0.60
	Donut	2 turns on cathead	0.45
Venezuela	Donut	Cathead	0.43

[a]Adapted from Clayton, 1990.

TABLE 3.4 Borehole, Sampler, and Rod Correction Factors[a]

Factor	Equipment Variables	Value
Borehole diameter factor, C_B	65–115 mm (2.5–4.5 in.)	1.00
	150 mm (6 in.)	1.05
	200 mm (8 in.)	1.15
Sampling method factor, C_S	Standard sampler	1.00
	Sampler without liner (not recommended)	1.20
Rod length factor, C_R	3–4 m (10–13 ft)	0.75
	4–6 m (13–20 ft)	0.85
	6–10 m (20–30 ft)	0.95
	>10 m (>30 ft)	1.00

[a]Adapted from Skempton, 1986.

stress, σ'_z, was 100 kPa(2000 lb/ft^2). Chapter 9 will discuss σ'_z and how to compute it, but for now think of it as a compressive stress produced by the weight of the overlying soil. Until then, the value of σ'_z will be given in any problem statements.

The corrected value, $N_{1,60}$, is (Liao and Whitman, 1986)

$$N_{1,60} = N_{60}\sqrt{\frac{2000\ \text{lb/ft}^2}{\sigma'_z}} \quad \text{(3.2–English)}$$

$$N_{1,60} = N_{60}\sqrt{\frac{100\ \text{kPa}}{\sigma'_z}} \quad \text{(3.2–SI)}$$

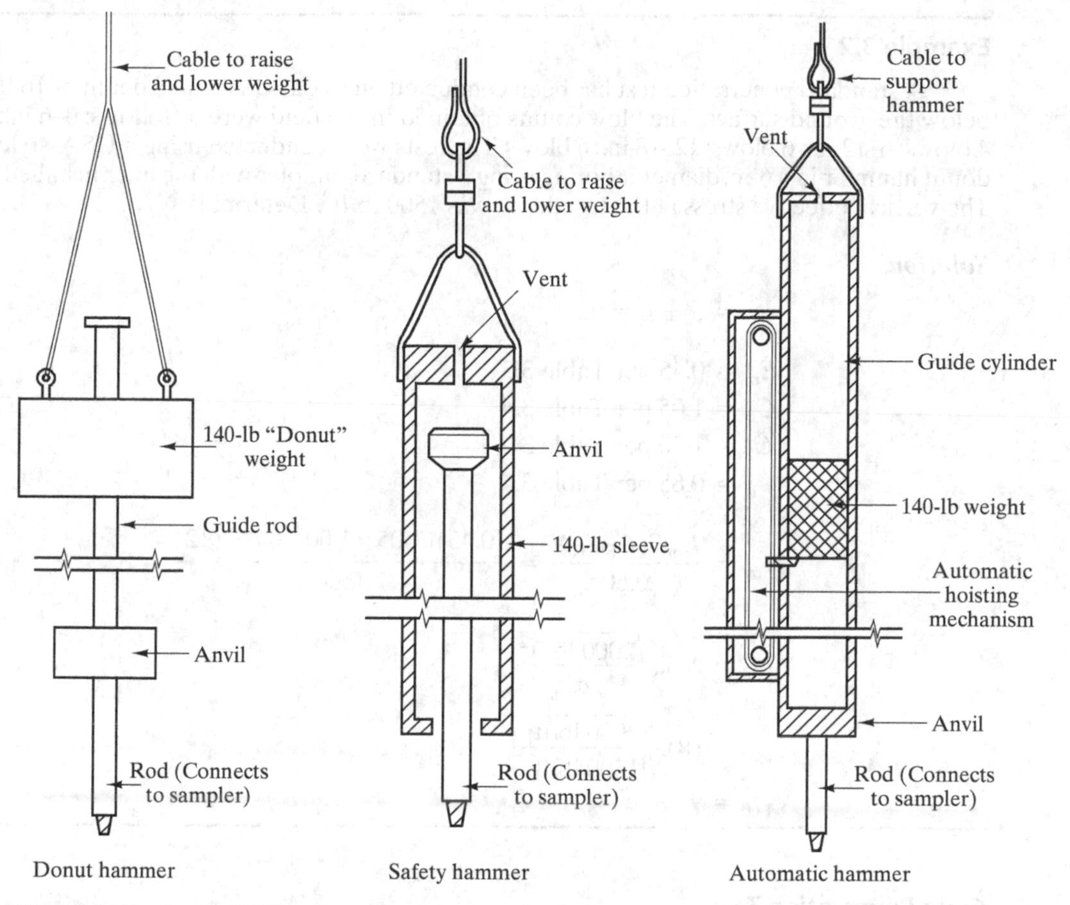

FIGURE 3.25 Types of SPT hammers.

where:

$N_{1,60}$ = SPT N-value corrected for field procedures and overburden stress
σ'_z = vertical effective stress at the test location (kPa or lb/ft^2), as defined in Chapter 9
N_{60} = SPT N-value corrected for field procedures

The use of SPT correction factors is often a confusing issue. Corrections for field procedures (Equation 3.1) are always appropriate, but the overburden correction may or may not be appropriate depending on the procedures used by those who developed the analysis method under consideration. We will identify the proper value by using the appropriate subscripts.

Example 3.2

A standard penetration test has been conducted on a coarse sand at a depth of 16 ft below the ground surface. The blow counts obtained in the field were as follows: 0–6 in.: 4 blows; 6–12 in.: 6 blows; 12–18 in.: 6 blows. The tests were conducted using a USA-style donut hammer in a 6-in. diameter boring using a standard sampler with the liner installed. The vertical effective stress at the test depth was 1500 lb/ft^2. Determine $N_{1,60}$.

Solution:

$$N = 6 + 6 = 12$$

$$\begin{aligned} E_m &= 0.45 \text{ per Table 3.3} \\ C_B &= 1.05 \text{ per Table 3.4} \\ C_S &= 1.00 \text{ per Table 3.4} \\ C_R &= 0.85 \text{ per Table 3.4} \end{aligned}$$

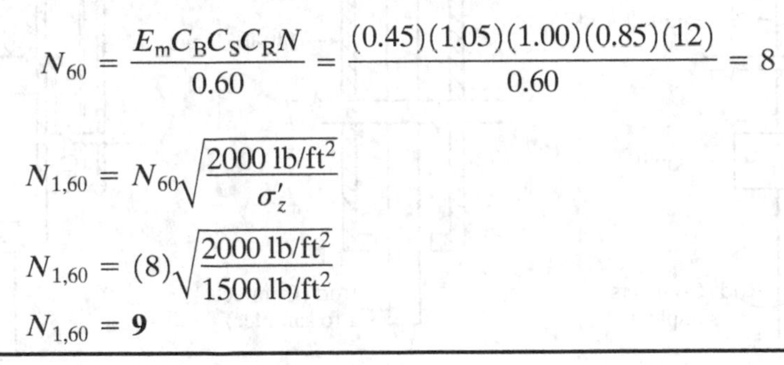

$$N_{60} = \frac{E_m C_B C_S C_R N}{0.60} = \frac{(0.45)(1.05)(1.00)(0.85)(12)}{0.60} = 8$$

$$N_{1,60} = N_{60}\sqrt{\frac{2000 \text{ lb/ft}^2}{\sigma'_z}}$$

$$N_{1,60} = (8)\sqrt{\frac{2000 \text{ lb/ft}^2}{1500 \text{ lb/ft}^2}}$$

$$N_{1,60} = \mathbf{9}$$

Cone Penetration Test

The *cone penetration test* (CPT) [ASTM D3441] is another common in situ test (Schmertmann, 1978; De Ruiter, 1981; Meigh, 1987; Robertson and Campanella, 1989; Briaud and Miran, 1991). Most of the early development of this test occurred in western Europe in the 1930s and again in the 1950s. Further development has occurred in recent decades in both Europe and North America. Although many different styles and configurations have been used, the current standard grew out of work performed in the Netherlands, so it is sometimes called a *Dutch cone*. The CPT has been used extensively in Europe for many years and is becoming increasingly popular in North America and elsewhere.

There are two major types of cones: the original *mechanical cone* and the *electric cone*, as shown in Figure 3.26. In current geotechnical practice, the mechanical cone is no longer used and has been replaced by the electric cone as the standard. Each type of cone has two parts, a 35.7 mm diameter cone-shaped tip with a 60° apex angle and a 35.7 mm diameter × 133.7 mm long cylindrical sleeve. A hydraulic ram pushes this assembly into the ground and instruments measure the resistance to penetration. The *cone resistance*, q_c, is the total force acting on the cone divided by its projected area (10 cm^2); the *cone side friction*, f_{sc}, is the total frictional force acting on the friction

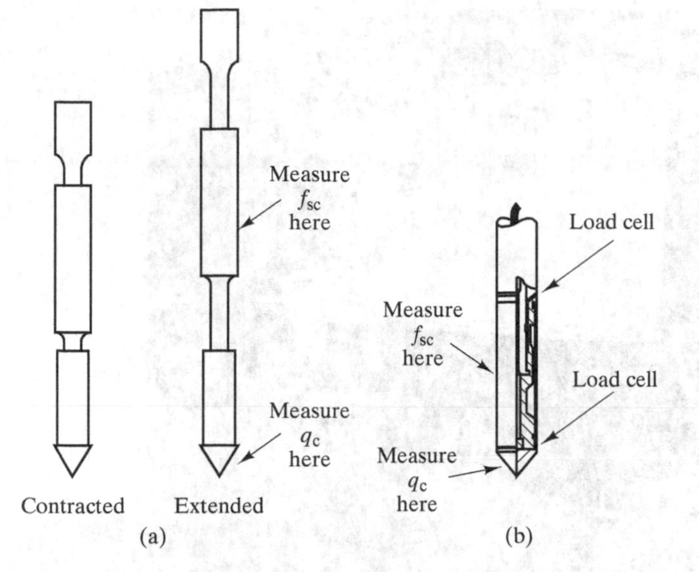

FIGURE 3.26 Types of cones: (a) mechanical cone (also known as *a Begemann cone*); (b) electric cone (also known as *a Fugro cone*).

sleeve divided by its surface area (150 cm^2). It is common to express the side friction in terms of the *friction ratio*, R_f:

$$R_f = \frac{f_{sc}}{q_c} \times 100\% \tag{3.3}$$

The operation of the two types of cones differs in that the mechanical cone is advanced in stages and measures q_c and f_{sc} at intervals of about 20 cm, whereas the electric cone includes built-in strain gages and is able to measure q_c and f_{sc} continuously with depth. In either case, the CPT defines the soil profile with much greater resolution than does the SPT.

CPT rigs are often mounted in large three-axle trucks such as the one in Figure 3.27. These are typically capable of producing maximum thrusts of 100–200 kN (10–20 tons). Smaller, trailer-mounted or truck-mounted rigs also are available.

The CPT has been the subject of extensive research and development (Robertson and Campanella, 1983) and thus is becoming increasingly useful to the practicing engineer. Some of this research effort has been conducted using cones equipped with pore pressure transducers in order to measure the excess pore water pressures that develop while conducting the test. These are known as *piezocones*, and the enhanced procedure is known as a *CPTU test*. These devices are especially useful in saturated clays.

A typical plot of CPT results is shown in Figure 3.28.

The CPT is an especially useful way to evaluate soil profiles. Since it retrieves data continuously with depth (with electric cones), the CPT is able to detect fine changes in the stratigraphy. Therefore, engineers often use the CPT in the first phase of subsurface investigation, saving boring and sampling for the second phase.

FIGURE 3.27 A truck-mounted CPT rig. A hydraulic ram located inside the truck pushes the cone into the ground, using the weight of the truck as reaction.

It also is much less prone to error due to differences in equipment and technique, and thus is more repeatable and reliable than the SPT.

Although the CPT has many advantages over the SPT, there are at least three important disadvantages:

- No soil sample is recovered during the test, so there is no opportunity to inspect the soils tested.
- The test is unreliable or unusable in soils with cementation or significant gravel content.
- Although the cost per foot of penetration is less than that for borings, it is necessary to mobilize a special rig to perform the CPT.

Overburden Correction Most analysis methods use the CPT results directly from the field, but some require the use of an overburden correction factor. This factor is identical to the one applied to SPT results:

$$q_{c1} = q_c \sqrt{\frac{2000 \text{ lb/ft}^2}{\sigma'_z}} \quad \text{(3.4–English)}$$

$$q_{c1} = q_c \sqrt{\frac{100 \text{ kPa}}{\sigma'_z}} \quad \text{(3.4–SI)}$$

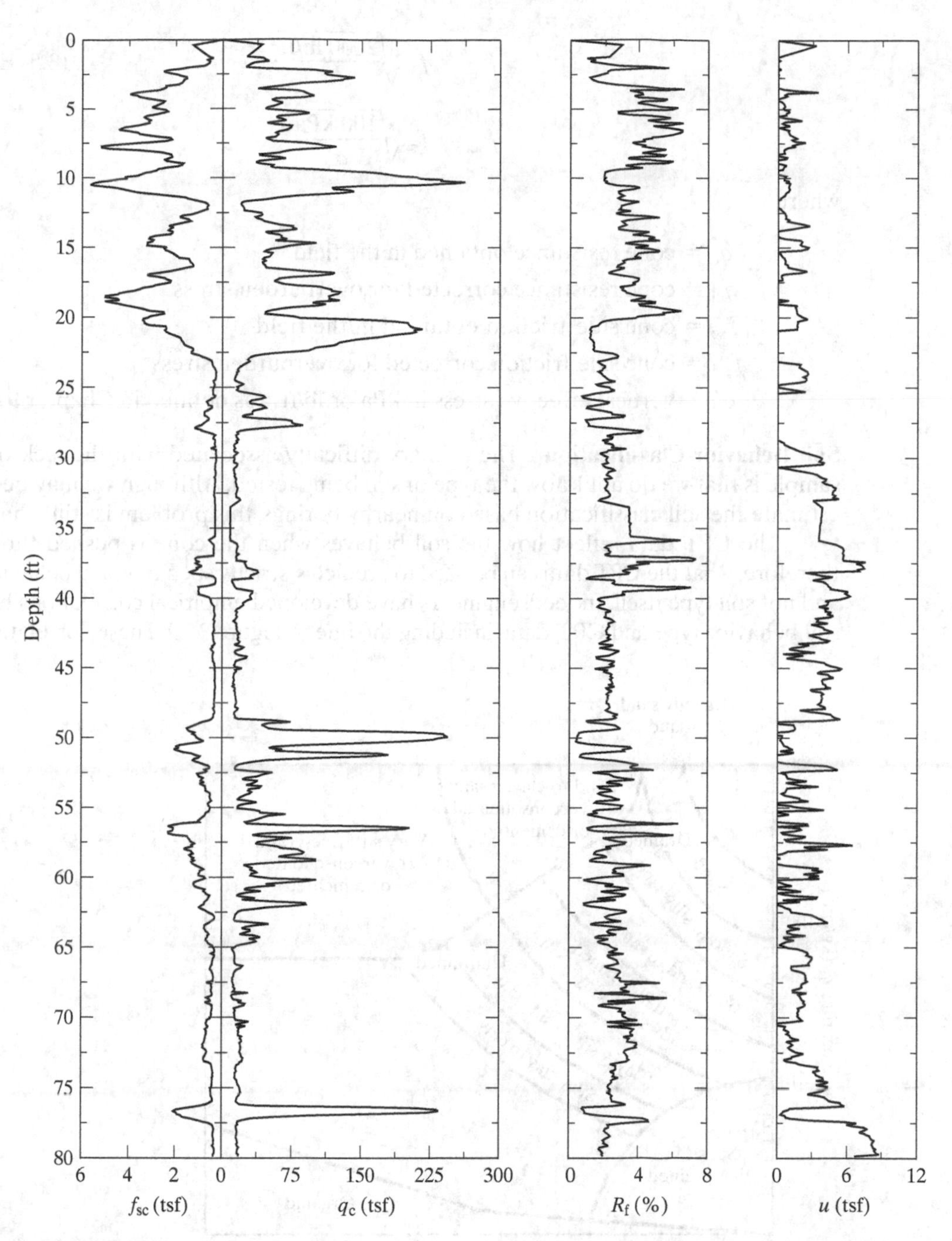

FIGURE 3.28 Sample CPT results were obtained from a peizocone, and thus include a plot of pore water pressure, u versus depth. All stresses and pressures are expressed in tons per square foot (tsf). For practical purposes, 1 tsf = 1 kg/cm^2. (Alta Geo Cone Penetrometer Testing Services, Sandy, Utah.)

$$f_{scl} = f_{sc}\sqrt{\frac{2000 \text{ lb/ft}^2}{\sigma'_z}} \tag{3.5–English}$$

$$f_{scl} = f_{sc}\sqrt{\frac{100 \text{ kPa}}{\sigma'_z}} \tag{3.5–SI}$$

where:

q_c = cone resistance obtained in the field

q_{cl} = cone resistance corrected for overburden stress

f_{sc} = cone side friction obtained in the field

f_{scl} = cone side friction corrected for overburden stress

σ'_z = vertical effective stress in kPa or lb/ft^2 (as defined in Chapter 9)

Soil Behavior Classification The primary difficulty associated with the lack of a soil sample is that we do not know the type of soil being tested. Although we may be able to estimate the soil classification based on nearby borings, this problem is still a handicap.

The CPT data reflect how the soil behaves when the cone is pushed through it; therefore, what the CPT data can be used to predict is, strictly speaking, soil behavior type, and not soil type itself. Indeed, engineers have developed empirical correlations between soil behavior type and CPT data, including the one in Figure 3.29. These correlations can

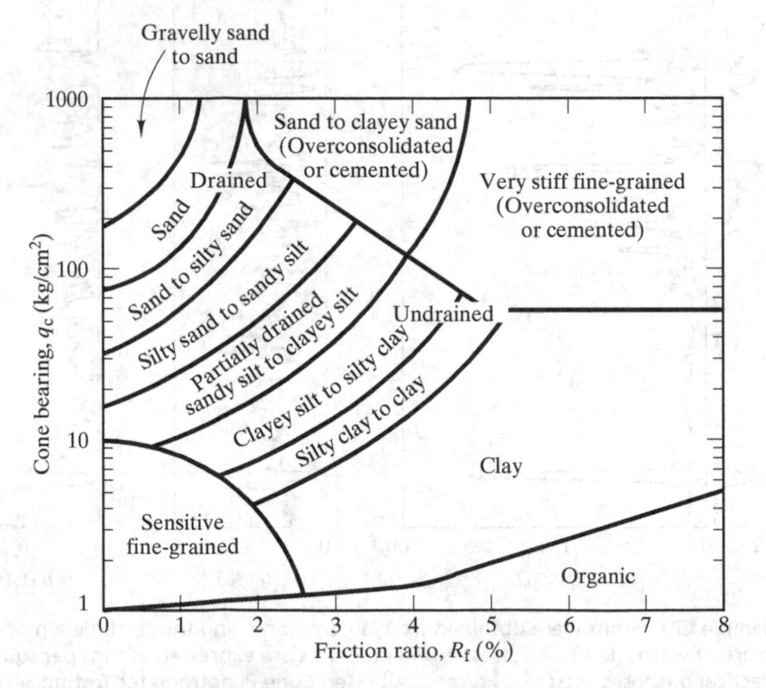

FIGURE 3.29 Soil behavior type chart based on CPT results. (Adapted from Robertson and Campanella, 1983.)

be programmed into a computer, and often are printed along with the test results. However, they need to be used with caution, and are not nearly as precise as a visual classification of real soil samples.

Correlation with SPT Geotechnical engineers also have developed empirical correlations between CPT and SPT data. The one shown in Figure 3.30 presents the q_c/N_{60} ratio as a function of the mean particle size, D_{50} (as defined in Chapter 4).

Pressuremeter Test

In 1954, a young French engineering student named Louis Ménard began to develop a new type of in situ test: the pressuremeter test (PMT). Although Kögler had done some limited work on a similar test some 20 years earlier, it was Ménard who made it a practical reality.

The pressuremeter is a cylindrical balloon that is inserted into the ground and inflated, as shown in Figures 3.31 and 3.32. Measurements of volume and pressure can be used to evaluate the in situ stress, and the compressibility and strength of the adjacent soil, and thus the behavior of a foundation (Baguelin et al., 1978; Briaud, 1992).

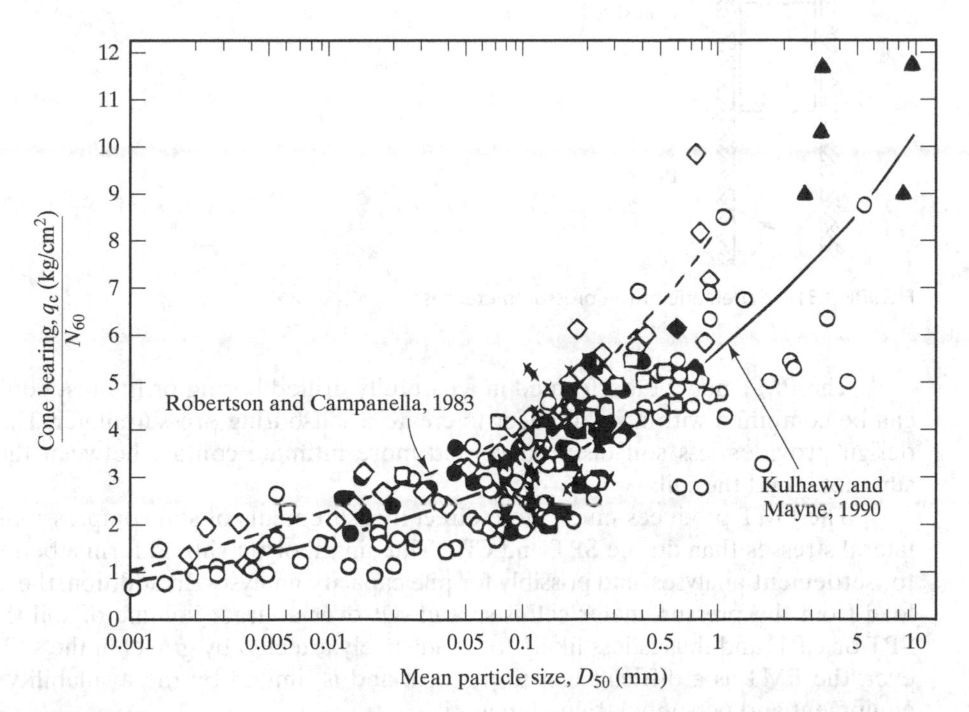

FIGURE 3.30 Correlation between the q_c/N_{60} ratio and the mean particle size, D_{50} (as defined in Chapter 4). (Adapted from Kulhawy and Mayne, 1990, Copyright Electric Power Research Institute, used with permission.)

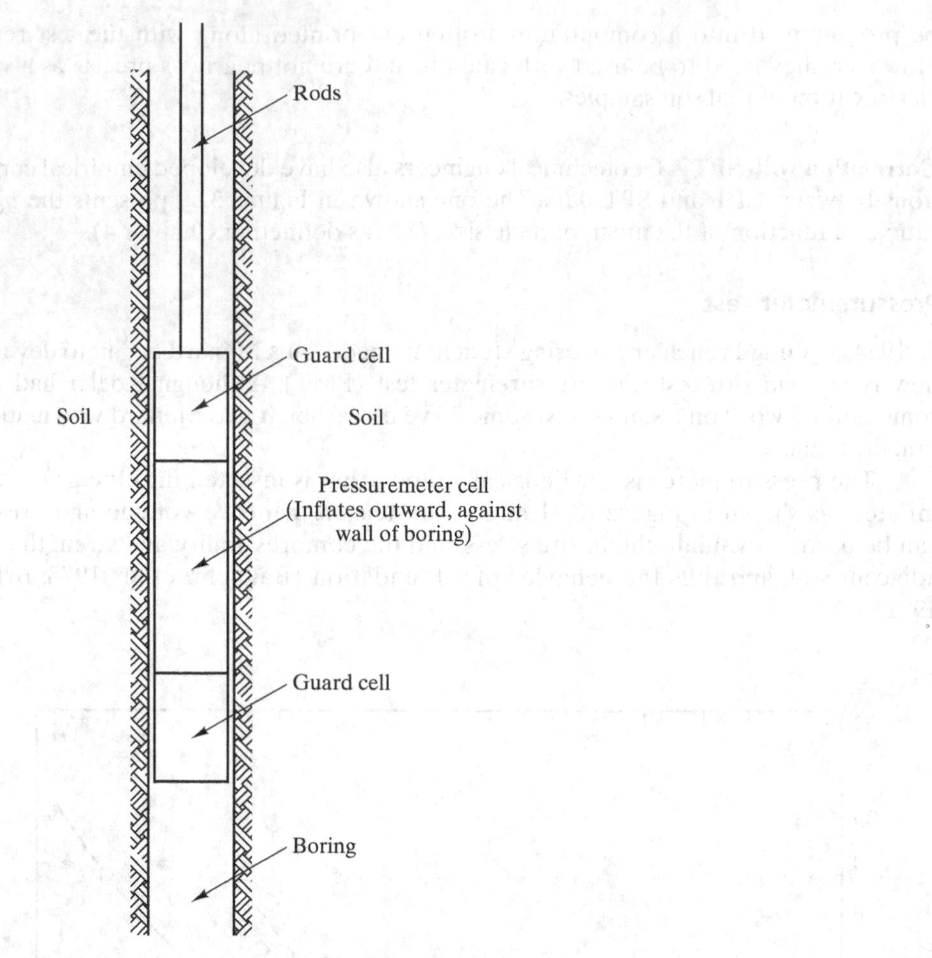

FIGURE 3.31 Schematic of the pressuremeter test.

The PMT may be performed in a carefully drilled boring or the test equipment can be combined with a small auger to create a self-boring pressuremeter. The latter design provides less soil disturbance and more intimate contact between the pressuremeter and the soil.

The PMT produces much more direct measurements of soil compressibility and lateral stresses than do the SPT and CPT. Thus, in theory, it should form a better basis for settlement analyses, and possibly for pile capacity analyses. In addition, the applied load from the pressuremeter cell is spread out over a larger volume of soil than the SPT or CPT, and thus is less likely to be adversely affected by gravel in the soil. However, the PMT is a difficult test to perform and is limited by the availability of the equipment and personnel trained to use it.

Although the PMT is widely used in France and Germany, it is used only occasionally in other parts of the world. However, it may become more popular in the future.

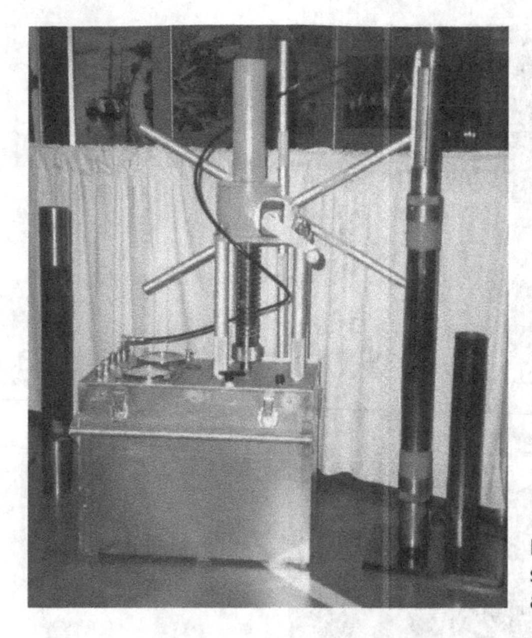

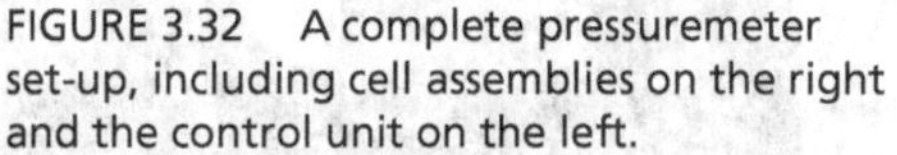

FIGURE 3.32 A complete pressuremeter set-up, including cell assemblies on the right and the control unit on the left.

Dilatometer Test

The dilatometer (Marchetti, 1980; Schmertmann, 1986b, 1988a, 1988b), which is one of the newer in situ test devices, was developed during the late 1970s in Italy by Silvano Marchetti. It is also known as a *flat dilatometer* or a *Marchetti dilatometer* and consists of a 95 mm wide, 15 mm thick metal blade with a thin, flat, circular, steel membrane on one side, as shown in Figure 3.33.

The dilatometer test (DMT) is conducted as follows (Schmertmann, 1986a):

1. Press the dilatometer into the soil to the desired depth using a CPT rig or some other suitable device.
2. Apply nitrogen gas pressure to the membrane to press it outward. Record the pressure required to move the center of the membrane 0.05 mm into the soil (the *A* pressure) and that required to move its center 1.10 mm into the soil (the *B* pressure).
3. Depressurize the membrane and record the pressure acting on the membrane when it returns to its original position. This is the *C* pressure and is a measure of the pore water pressure in the soil.
4. Advance the dilatometer 150 to 300 mm deeper into the ground and repeat the test. Continue until reaching the desired depth.

Each of these test sequences typically requires one to two minutes to complete, so a typical *sounding* (a complete series of DMT tests between the ground surface and the desired depth) may require about 2 hours. In contrast, a comparable CPT sounding might be completed in about 30 minutes.

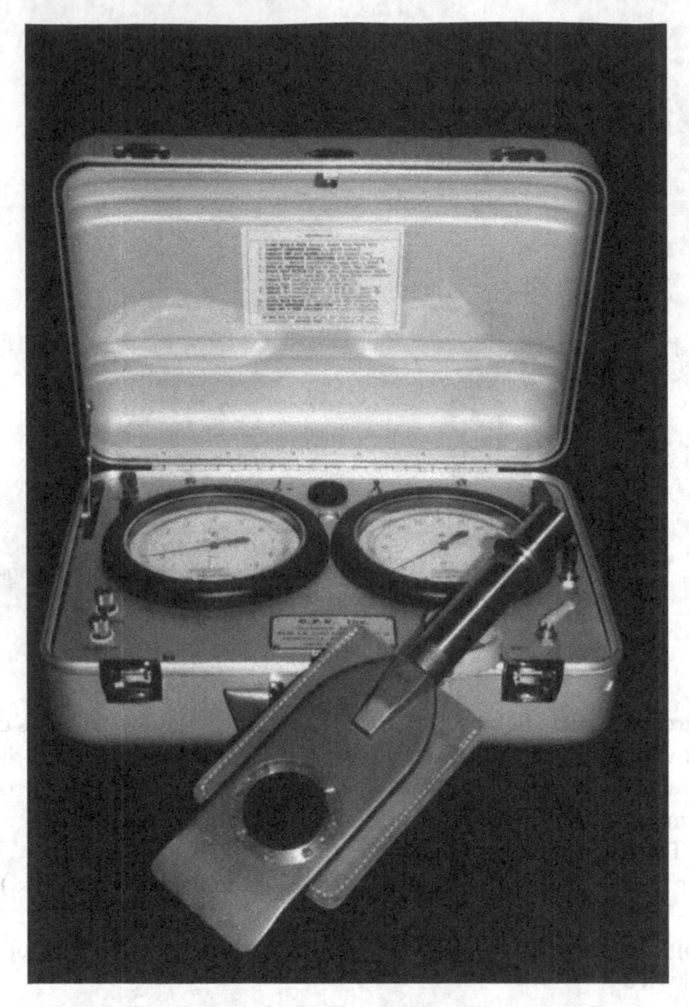

FIGURE 3.33 The Marchetti dilatometer with its control unit. (Courtesy GPE, Inc., Gainesville, FL.)

The primary benefit of the DMT is that it measures the lateral stress condition and compressibility of the soil. These are determined from the A, B, and C pressures and certain equipment calibration factors and expressed as the *DMT indices*, as follows:

I_D = material index (a normalized modulus)
K_D = horizontal stress index (a normalized lateral stress)
E_D = dilatometer modulus (theoretical elastic modulus)

Researchers have developed correlations between these indices and soil classification as well as certain engineering properties (Schmertmann, 1988b; Kulhawy and Mayne, 1990).

The CPT and DMT are complementary tests (Schmertmann, 1988b). The cone is a good way to evaluate soil strength, whereas the dilatometer assesses compressibility and in situ stresses. These three kinds of information form the basis for most foundation engineering analyses. In addition, the dilatometer blade is most easily pressed into the ground using a conventional CPT rig, so it is a simple matter to conduct both CPT and DMT tests while mobilizing only a minimum of equipment.

The dilatometer test is a relative newcomer, and thus has not yet become a common engineering tool. Engineers have had only limited experience with it and the analysis and design methods based on DMT results are not yet well developed. However, its relatively low cost, versatility, and compatibility with the CPT suggest that it may enjoy widespread use in the future. It has very good repeatability, and can be used in soft to moderately stiff soils (i.e., those with $N \leq 40$), and provides more direct measurements of stress–strain properties.

Becker Penetration Test

Soils that contain a large percentage of gravel and those that contain cobbles or boulders create problems for most in situ test methods. Often, the in situ test device is not able to penetrate through such soils (it meets refusal) or the results are not representative because the particles are about the same size as the test device. Frequently, even conventional drilling equipment cannot penetrate through these soils.

One method of penetrating through these very coarse-grained soils is to use a *Becker hammer drill*. This device, developed in Canada, uses a small diesel pile-driving hammer and percussion action to drive a 135 to 230 mm (5.5–9.0 in.) diameter double-wall steel casing into the ground. The cuttings are sent to the top by blowing air through the casing. This technique has been used successfully on very dense and coarse soils.

The Becker hammer drill also can be used to assess the penetration resistance of these soils using the *Becker penetration test*, in which the hammer blow-count is monitored. The number of blows required to advance the casing 1 ft (300 mm) is the Becker blow-count, N_B. Several correlations are available to convert it to an equivalent SPT N-value (Harder and Seed, 1986). One of these correlation methods also considers the bounce chamber pressure in the diesel hammer.

Other In Situ Tests

Many other in situ tests are available, some of which are discussed in other parts of this book. These include the field density test (Chapter 6), the hydraulic conductivity test (Chapter 8), and the vane shear test (Chapter 12).

Comparison of In Situ Test Methods

Each of the in situ test methods has its strengths and weaknesses. Table 3.5 compares some of the important attributes of the tests described in this chapter.

TABLE 3.5 Assessment of In Situ Test Methods[a]

	Standard Penetration Test	Cone Penetration Test	Pressuremeter Test	Dilatometer Test	Becker Penetration Test
Simplicity and durability of apparatus	Simple; rugged	Complex; rugged	Complex; delicate	Complex; moderately rugged	Simple; rugged
Ease of testing	Easy	Easy	Complex	Easy	Easy
Continuous profile or point values	Point	Continuous	Point	Point	Continuous
Basis for interpretation	Empirical	Empirical; theory	Empirical; theory	Empirical; theory	Empirical
Suitable soils	All except gravels	All except gravels	All	All except gravels	Sands through boulders
Equipment availability and use in practice	Universally available: used routinely	Generally available; used routinely	Difficult to locate; used on special projects	Difficult to locate; used on special projects	Difficult to locate; used on special projects
Potential for future development	Limited	Great	Great	Great	Uncertain

[a]Adapted from Mitchell, 1978; used with permission of ASCE.

3.10 GEOPHYSICAL EXPLORATION

Geophysics is the use of various principles of physics to discern geologic profiles and to measure certain properties of the ground. Many such techniques are available, such as the following:

- Inducing seismic waves and measuring their propagation
- Assessing natural gravitational and magnetic fields and their variation across the surface of the earth
- Passing electrical currents through the ground and measuring their propagation
- Sending radar waves or other kinds of radiation into the ground and recording its transmission, absorption, or reflection

Geophysical exploration methods were originally developed by mining and petroleum geologists searching for valuable minerals and oil deep below the earth's surface. They were able to distinguish between different geologic strata by observing their physical properties, and thus locate the most promising sites for mining and oil drilling (Dobrin, 1988).

Later, geotechnical engineers and engineering geologists began to use some of these methods to assist in site characterization studies. Although geophysics is not nearly as precise as drilling borings and obtaining samples, it has the benefit of covering large areas at a small cost, and sometimes can locate features that might be missed by conventional borings. Geophysical methods also can be used as a first step in the exploration process, thus guiding the placement and depth of exploratory borings.

The applicability of geophysical methods for geotechnical site characterization is fairly limited, and only a small fraction of practical projects can benefit from them. However, in certain circumstances, they can be very valuable as a supplement to, but not a substitute for, exploratory borings.

Seismic Geophysical Tests

Seismic geophysical tests are common in situ tests used to define stratigraphy and to measure dynamic soil properties. In these tests, a source generates seismic waves that propagate through the soil, and the waves are measured by distant receivers. Depending on the set-up of the test and the locations of the source and receivers, there are different types of tests, including the *seismic refraction test*, *seismic cross-hole test*, and the *seismic down-hole test.*

The seismic refraction test is commonly used by geotechnical engineers and engineering geologists. It consists of sending seismic waves into the ground and measuring their arrival at various points on the ground surface. The seismic waves can be generated at the source location by striking the ground with a heavy object, such as a sledge hammer, or by detonating a small explosive. The resulting seismic waves are measured by receivers called *geophones* aligned in an array and connected to a *seismograph*, as shown in Figure 3.34.

The wave velocity depends on the physical properties of the soil or rock, most notably its density, modulus of elasticity, porosity, and the frequency of discontinuities in it. Dense, massive rocks, such as granite, have high wave velocities; softer rocks, such as limestone are intermediate; and soils have very low velocities. Seismic refraction uses these differences to discern the subsurface profile.

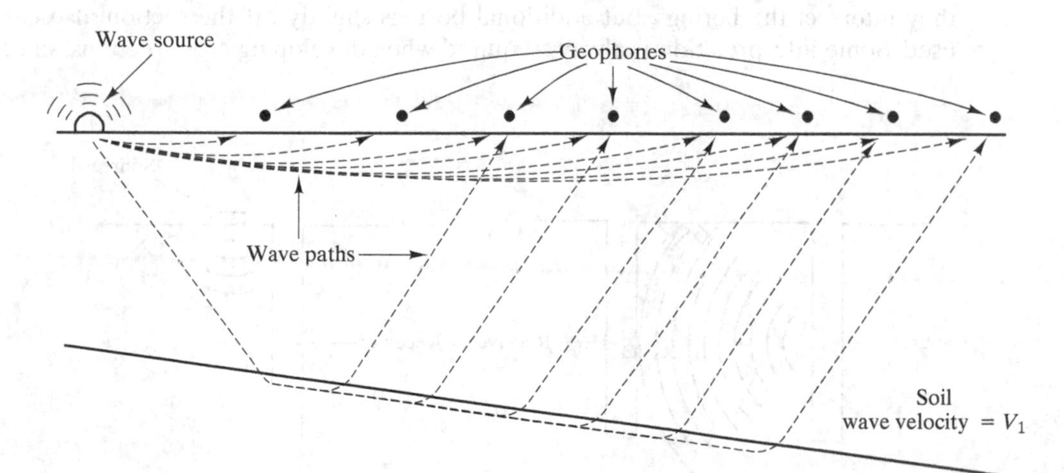

FIGURE 3.34 Use of seismic refraction to measure the depth to a hard layer, such as bedrock.

In Figure 3.34, various strata of soil are underlain by bedrock, and we wish to find the depth to this rock. To obtain this information, we discharge a small explosive and record the wave arrivals at each of the geophones. Some waves travel through the soil and arrive at the various geophones. Simultaneously, other waves travel down to the rock, horizontally through it, and back up to the ground surface. Although this path is longer, the wave velocity through the rock is much higher, so at some distance from the wave source, the latter arrives before the former. By comparing the wave arrival times at the geophones, we can determine the depth to bedrock.

Seismic cross-hole and seismic down-hole tests are depicted schematically in Figure 3.35. In these tests, the wave velocity can be calculated by dividing the distance between the source and receiver by the corresponding travel time measured by the receiver.

Wave velocity data also can be used to evaluate the rippability (ease of excavation) of various soil and rock strata, as discussed in Chapter 6. Finally, geotechnical engineers use this and other dynamic data to analyze the propagation of earthquake motions through the ground, which is part of the process of predicting ground motions due to future earthquakes.

3.11 SYNTHESIS AND INTERPRETATION

Cross-Sections

Site characterization programs often generate large amounts of information that can be difficult to sort through and synthesize. In addition, this data is spread throughout three dimensions, so visualization can be difficult.

One useful method of compiling subsurface data is to draw vertical *cross-sections* across the site, as shown in Figure 3.36. These sections are most easily developed when they intersect the borings, but additional borings slightly off the section also can be used. Some interpretation is always required when developing cross-sections, since we

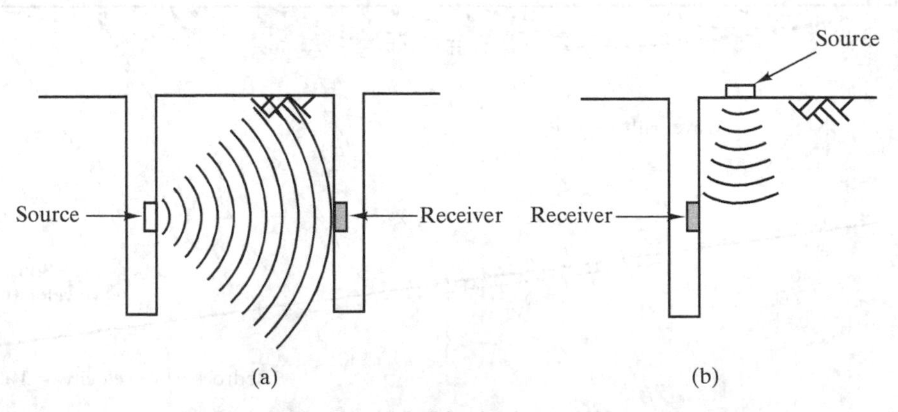

FIGURE 3.35 (a) Seismic cross-hole test; (b) seismic down-hole test.

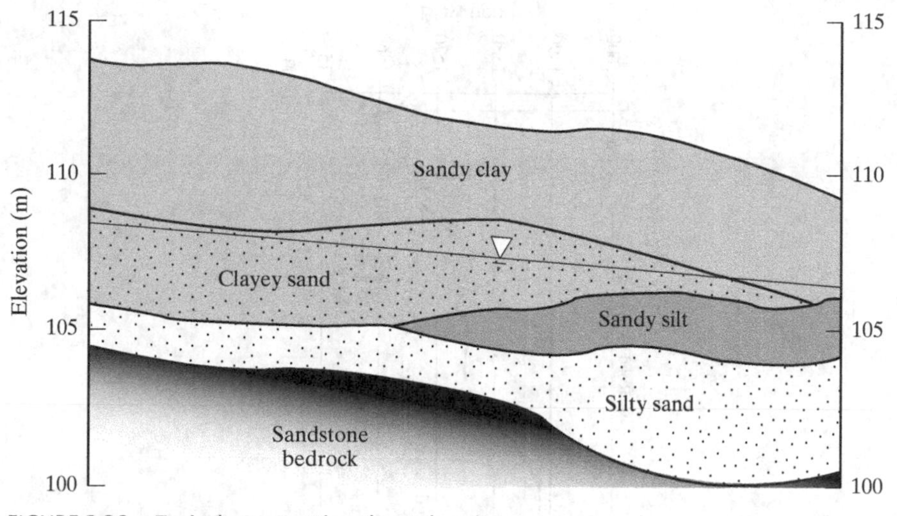

FIGURE 3.36 Typical cross-section through a site.

do not know what conditions exist between the borings. Two perpendicular sections can help geotechnical engineers visualize the site in three dimensions.

Sometimes cross-sections show only the subsurface conditions actually encountered in the exploratory borings, as in Figure 3.37. This method leaves the interpretation to the reader.

One-Dimensional Design Profiles

Although cross-sections are important tools for understanding subsurface variations across a site, many geotechnical analyses are based on one-dimensional profiles. For example, the settlement analyses we will conduct in Chapters 10 and 11 are one-dimensional and compute the settlement at a point on the ground surface due to compression of the soils immediately below that point. If we need to know the settlement at other points, the analysis needs to be repeated as necessary.

A one-dimensional design profile is similar to a boring log in that it describes subsurface conditions as a function of depth, as shown in Figure 3.38. However, the profile used for design probably will be a compilation of several borings and not exactly like any one of them. If the subsurface conditions are fairly uniform across the site (at least by geotechnical engineering standards), then we often use a single representative profile for design.

The development of these design profiles requires a great deal of engineering judgment along with interpolation and extrapolation of the data. It is important to have a feel for the approximate magnitude of the many uncertainties in this process and reflect them in an appropriate degree of conservatism. This judgment comes primarily with experience combined with a thorough understanding of the field and laboratory methodologies.

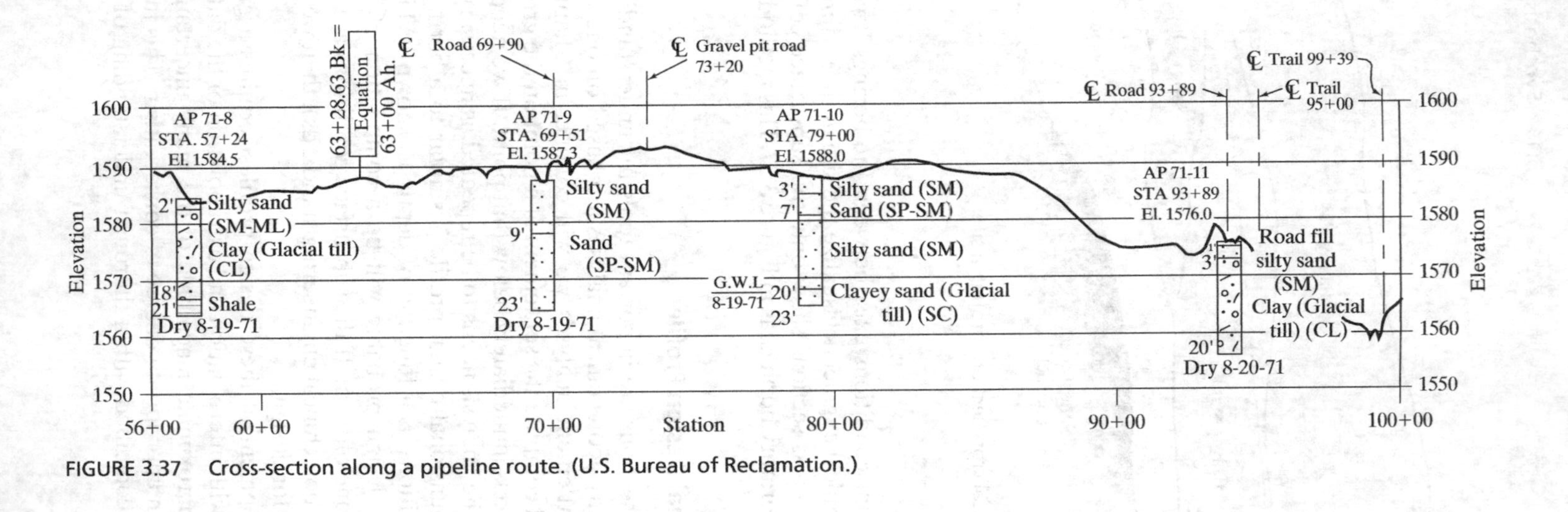

FIGURE 3.37 Cross-section along a pipeline route. (U.S. Bureau of Reclamation.)

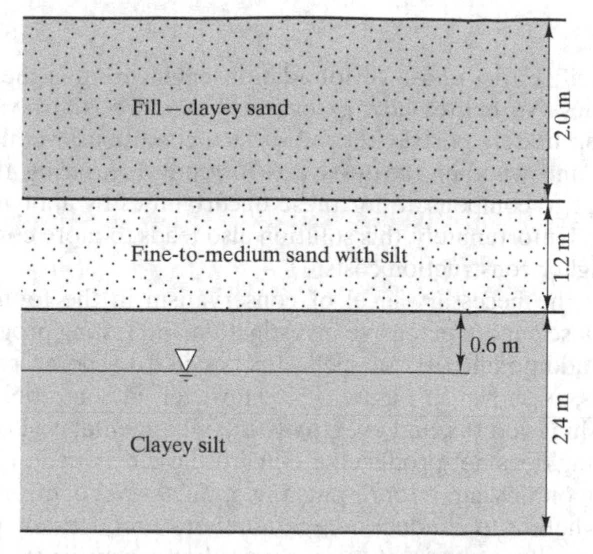

FIGURE 3.38 A typical one-dimensional design soil profile.

Geotechnical Investigation Reports

The final results of a site characterization program are usually presented in a *geotechnical investigation report* that includes copies of all boring logs, laboratory test results, cross-sections, etc., along with the engineer's interpretations. These reports are virtually always prepared in the context of a specific project, and thus include geotechnical recommendations for design of foundations, slopes, retaining walls, and other features. For example, a report for a proposed building might have an outline similar to the following:

Scope and Purpose
Proposed Development
Field Exploration
Groundwater Monitoring
Laboratory Testing
Analysis of Subsurface Conditions
Design Recommendations
  Grading
  Foundations
  Retaining walls
  Pavements
Closure
Appendix A—Boring Logs
Appendix B—Laboratory Test Results
Appendix C—Recommended Construction Specifications

3.12 ECONOMICS

The site investigation and soil testing phase of foundation engineering is the single largest source of uncertainties. No matter how extensive it is, there is always some doubt whether the borings accurately portray the subsurface conditions, whether the samples are representative, and whether the tests are correctly measuring the soil properties. Engineers attempt to compensate for these uncertainties by applying factors of safety in our analyses. Unfortunately, this solution also leads to more conservative designs and therefore higher construction costs.

In an effort to reduce the necessary level of conservatism in the foundation design, the engineer may choose a more extensive investigation and testing program to better define the soils. The additional costs of such efforts will, to a point, result in decreased construction costs, as shown in Figure 3.39. However, at some point, this becomes a matter of diminishing returns, and eventually the incremental cost of additional investigation and testing does not produce an equal or larger reduction in construction costs. The minimum on this curve represents the optimal level of effort.

We also must decide whether to conduct a large number of moderately precise tests (such as the SPT) or a smaller number of more precise but expensive tests (such

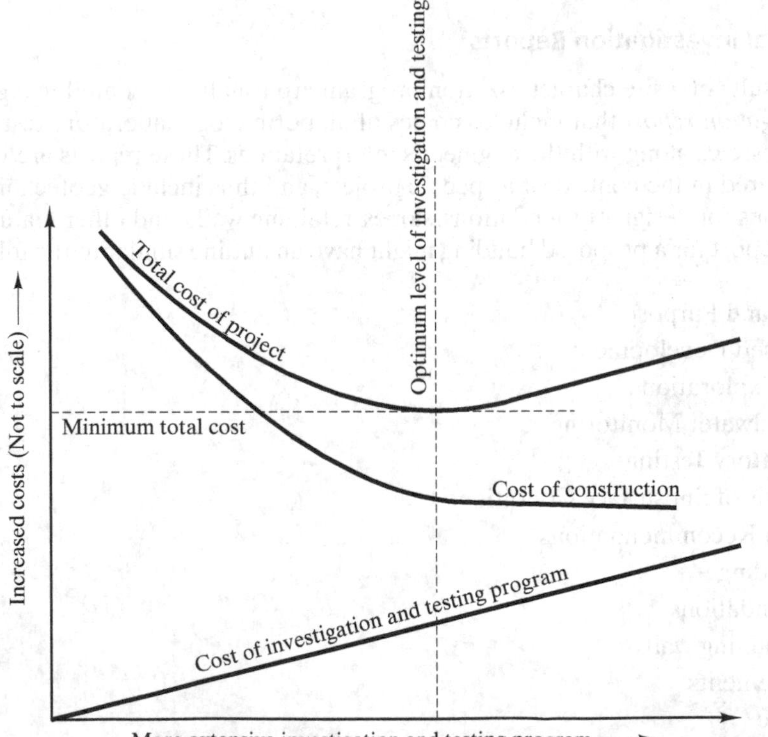

FIGURE 3.39 Cost-effectiveness of more extensive site characterization programs.

as the PMT). Handy (1980) suggested the most cost-effective test is the one with a variability consistent with the variability of the soil profile. Thus, a few precise tests might be appropriate in a uniform soil deposit, but more data points, even if they are less precise, are more valuable in an erratic deposit.

3.13 GEOTECHNICAL MONITORING DURING CONSTRUCTION

One of the most important ways geotechnical engineers have to deal with the many uncertainties of site characterization is continued monitoring of subsurface conditions during construction. Often, new information becomes evident during construction, especially if the construction involves making excavations. For example, if a highway cut is to be made into a hillside, geotechnical engineers and engineering geologists base its design on exploratory borings as discussed in this chapter. Then, when the cut is actually made, we examine the newly exposed ground and compare it to the anticipated conditions. If new conditions are found, then the design may need to be changed accordingly.

Another way of dealing with these uncertainties is to conduct full-scale tests in the field. For example, we have methods of predicting the load-bearing capacity of pile foundations, but often conduct full-scale load tests on piles to verify the computed capacity. Such tests involve installing a real pile at the project site and loading it. These tests might be performed before construction, or more often at the beginning of construction. If the load test indicates capacities significantly different than those anticipated, then we modify the design, perhaps by adding new piles.

A third way is to install *geotechnical instrumentation* into the ground. These are devices specifically designed to measure certain attributes of soil or rock. For example, an *inclinometer* is a geotechnical instrument that measures horizontal movements in the ground. We could install one or more inclinometers in a slow-moving landslide and use the resulting data to help assess the depth and direction of movement, and to judge the effectiveness of stabilization measures.

These techniques of continuing the design process through the construction period are known as the *observational method* (Peck, 1969), and form an important part of geotechnical engineering practice. They also represent a significant difference between geotechnical engineering and structural engineering. Structural engineers rarely, if ever, need to use these methods because they work with materials that are much more predictable and thus do not need such verification.

SUMMARY

Major Points

1. An important difference between geotechnical practice and that of most other branches of engineering is that we must work with natural materials, not manufactured products. These materials, soil and rock, vary significantly from place to

place, so each building site requires a site exploration and characterization program to define the subsurface profile and relevant engineering properties. We then design our project based on the results of this program.

2. The initial stages of a site exploration and characterization program typically consist of gathering published data, reviewing air photos (if applicable), and conducting a field reconnaissance. These tasks are in preparation for the subsurface exploration.
3. Exploration of the subsurface conditions is generally the most important, and most expensive, component. This is usually accomplished by drilling exploratory borings and obtaining disturbed and undisturbed samples. Exploratory trenches and pits also can be useful at some sites. A wide variety of methods and tools is available, and the proper choice depends on the site conditions, cost, and other factors.
4. Geotechnical engineers divide soil and rock tests into two broad categories: ex situ and in situ. Ex situ tests are conducted in the laboratory, and thus are subject to sample disturbance problems. In situ tests are conducted in the field, but suffer from less control and less precision. Often we use both methods to take advantage of the strengths of each.
5. In some cases, geophysical methods, such as seismic refraction, can be used to assess subsurface conditions. However, these methods supplement conventional testing, but do not replace it.
6. Once the site characterization data have been collected, geotechnical engineers synthesize them and present the findings and recommendations in a geotechnical investigation report.
7. In well-managed projects, site characterization continues through construction, since further data often become available and may dictate changes in the design.

Vocabulary

aerial photographs
Becker penetration test
boring log
bucket auger
caving
cone penetration test (CPT)
coring
dilatometer test (DMT)
disturbed sample
downhole logging
drill rig
drilling mud
exploratory borings
exploratory trenches
ex situ testing
field reconnaissance
flight auger
geophysical exploration
geotechnical investigation report
heavy-wall sampler
hollow stem auger
in situ testing
laser scanning
overburden correction
photogrammetry
pressuremeter test (PMT)
remote sensing
rock quality designation (RQD)
sample disturbance
seismic cross-hole test
seismic down-hole test
seismic hazard map
seismic refraction test
Shelby tube sampler
site characterization
site exploration
soil behavior classification
sounding
squeezing
standard penetration test (SPT)
subsurface exploration
undisturbed sample

QUESTIONS AND PRACTICE PROBLEMS

Introduction

3.1 Define site exploration and site characterization.

3.2 List in chronological order the steps involved in performing a site exploration and characterization program.

Section 3.1 Project Assessment

3.3 What information should be gathered for the planning of a site exploration and characterization program?

Section 3.2 Literature Search

3.4 What information can be obtained from geologic maps?

Section 3.5 Subsurface Exploration

3.5 Describe the following drilling methods and give advantages and disadvantages of each: solid stem auger, hollow stem auger, rotary wash, and coring.

3.6 Describe the advantages and disadvantages of using an exploratory trench in site exploration.

Section 3.6 Soil and Rock Sampling

3.7 Describe the concept of sample disturbance and explain the relationship between sampler type and disturbance.

3.8 Describe the Shelby tube sampler and the heavy-wall sampler and state the advantages and disadvantages of each.

3.9 A one-story, 50-m wide × 90-m long manufacturing building is to be built on a site underlain by medium dense to dense silty sand with occasional gravel. This soil probably has better-than-average engineering properties and average uniformity. There are no indications of previous grading or fill at this site, and the groundwater table is believed to be about 30 m below the ground surface. We anticipate supporting this building on spread footing foundations located about 0.5 m below the ground surface. There are no accessibility problems at this site.

(a) How many exploratory borings will be required, and to what depths should they be drilled?

(b) What type of drilling and sampling equipment would you recommend for this project?

3.10 A one-story, 20-m wide × 50-m long concrete tilt-up office building is to be built at a site near wetlands. Previous exploratory borings at nearby sites encountered about 1 m of moderately stiff clayey fill underlain by about 4 m of very soft organic silts and clays, then 15 m of progressively stiffer sandy clays and clayey sands. Limestone bedrock is located about 20 m below the ground surface. The groundwater table is thought to be at a depth of about 0.5 m. Because of the soft soils, we will probably need to support this building on deep foundations that extend at least into the stiffer soils, and possibly to bedrock. There are no accessibility problems at this site.

(a) How many exploratory borings will be required, and to what depth should they be drilled?

(b) What type of drilling and sampling equipment would you recommend for this project, and what kinds of problems should the field crew be prepared to solve?

3.11 A 10-story steel-frame office building with a 200 ft × 200 ft footprint is to be built on a site underlain by alluvial sands and silts. These soils are fairly uniform and probably have good engineering properties. The building will have one 12-ft deep basement and will probably be supported on either a mat foundation[2] located 5 ft below the bottom of the basement, or a deep foundation extending about 60 ft below the bottom of the basement. The groundwater table is about 30 ft below the ground surface and bedrock is several 100 ft below the ground surface. There are no accessibility problems at this site.

(a) How many exploratory borings will be required, and to what depth should they be drilled?

(b) What type of drilling and sampling equipment would you recommend for this project?

3.12 A small commercial development consisting of a one-story supermarket and a one-story retail store building is to be built on the site shown in Figure 3.40. The proposed spread footing foundations will be located at a depth of 2 ft below the ground surface. The site has never been developed before, but a study of old aerial photographs indicates a fill was placed in the northeast section. This fill appears to be up to 5-ft thick, probably was not compacted, and most likely will need to be removed during construction. However, we may be able to reuse this material as fill, so long as it does not contain trash or other deleterious substances. The remainder of the soils are probably stiff clayey silts and sandy silts. The groundwater table is believed to be about 20 ft below the ground surface. There are no accessibility problems at this site.

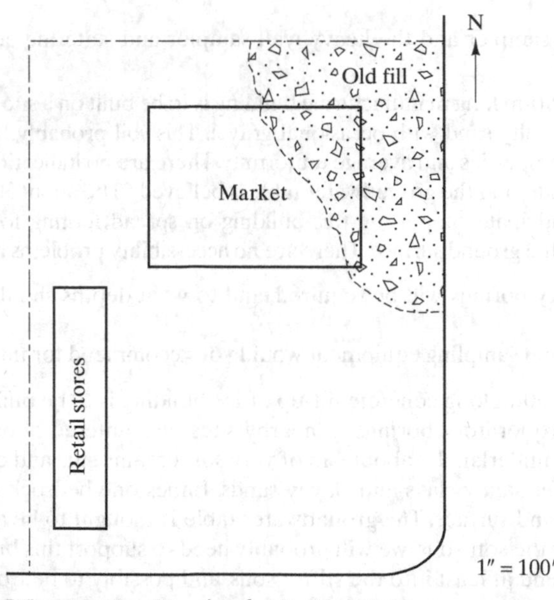

FIGURE 3.40 Site plan for Problem 3.12.

[2]A mat foundation is a type of shallow foundation that encompasses the entire footprint of the building. See Chapter 14 for more details.

Develop a subsurface exploration program and present it as a 250–350 word memo to your field crew instructing them what to do. Be sure to include a copy of the site plan marked with the proposed location of each activity.

Section 3.7 Groundwater Exploration and Monitoring

3.13 Describe how to install an observation well and what it measures.

Section 3.9 In Situ Testing

3.14 The vertical effective stress at Sample 1 in Figure 3.14 is 405 lb/ft^2. Compute $N_{1,60}$.

3.15 The vertical effective stress at Sample 3 in Figure 3.14 is 1270 lb/ft^2. Compute $N_{1,60}$.

3.16 A standard penetration test has been performed at a depth of 6.5 m in a medium sand using a standard sampler and a USA-style donut hammer. The N-value recorded in the field was 16. The boring diameter was about 100 mm, and the vertical effective stress at the test location was 85 kPa. Compute $N_{1,60}$.

3.17 Using Figure 3.29, classify the soils between depths of 23 and 48 ft in the CPT results presented in Figure 3.28. Why are there spikes in the q_c, f_{sc}, and R_f curves between these depths?

3.18 Using Figure 3.29, classify the soils between depths of 66 and 80 ft in the CPT results presented in Figure 3.28. Why are there spikes in the q_c, f_{sc}, and R_f curves between depths of 76 and 78 ft?

3.19 A cone penetration test on a sandy soil with mean particle size of 0.5 mm produced a q_c of 80 kg/cm^2. Estimate the equivalent SPT N_{60}-value.

3.20 Compare the standard penetration test and the Becker penetration test.

3.21 Describe how to perform a pressuremeter test and what it measures.

3.22 Describe how to perform a dilatometer test and what it measures.

Comprehensive

3.23 An engineer is planning to use a 24-in. diameter bucket auger similar to the one in Figure 3.7 to drill several exploratory borings at a site adjacent to a lake. The underlying soils are probably soft clays and silts with N-values of less than 5. Is this a wise choice? Why or why not?

3.24 What type of soil sampling equipment would be most appropriate for the soils described in Problem 3.23? Why?

3.25 A large compacted fill is to be placed on a site underlain by a 15-m thick layer of saturated clay. The weight of this fill will cause the clay layer to consolidate, which will result in large settlements at the ground surface. Since these settlements would have an adverse effect on buildings and other improvements planned for this site, a settlement rate analysis, similar to those we will discuss in Chapter 11, is to be performed to estimate the time required for a certain percentage of the settlement to be completed.

A series of exploratory borings have already been drilled at this site, samples have been recovered, and laboratory tests have been performed to evaluate the consolidation properties of the clay. However, to complete the settlement rate analysis, we need to know if thin horizontal sand seams are present in the clay, and the approximate spacing between these seams. If they exist at all, these seams are probably less than 100 mm thick. Although some of the undisturbed samples contained sand seams, more information is needed.

What kind of additional exploration would you do to determine whether or not more sand seams are present? Be sure to consider both technical feasibility and cost, and explain the reason for your choice.

3.26 A level building pad is to be built at the site shown in Figure 3.41 by cutting and filling as shown. The final pad elevation is to be 215 ft. Then, a three-story steel-frame office building is to be built.

Five exploratory borings have been drilled to determine the subsurface conditions. The logs from these borings were summarized as follows:

Boring 1

Groundwater table depth = 44 ft

Depth (ft)	Soil or Rock Conditions
0–18	Sandy clay
18–35	Clayey sand
35–52	Silty sand
52–55	Sandstone bedrock

Boring 2

Groundwater table depth = 31 ft

Depth (ft)	Soil or Rock Conditions
0–28	Clayey sand
28–36	Silty sand
36–39	Sandstone bedrock

Boring 3

Groundwater table depth = 41 ft

Depth (ft)	Soil or Rock Conditions
0–34	Clayey sand
34–48	Silty sand
48–52	Sandstone bedrock

Boring 4

Groundwater table depth = 40 ft

Depth (ft)	Soil or Rock Conditions
0–33	Clayey sand
33–45	Silty sand
45–47	Sandstone bedrock

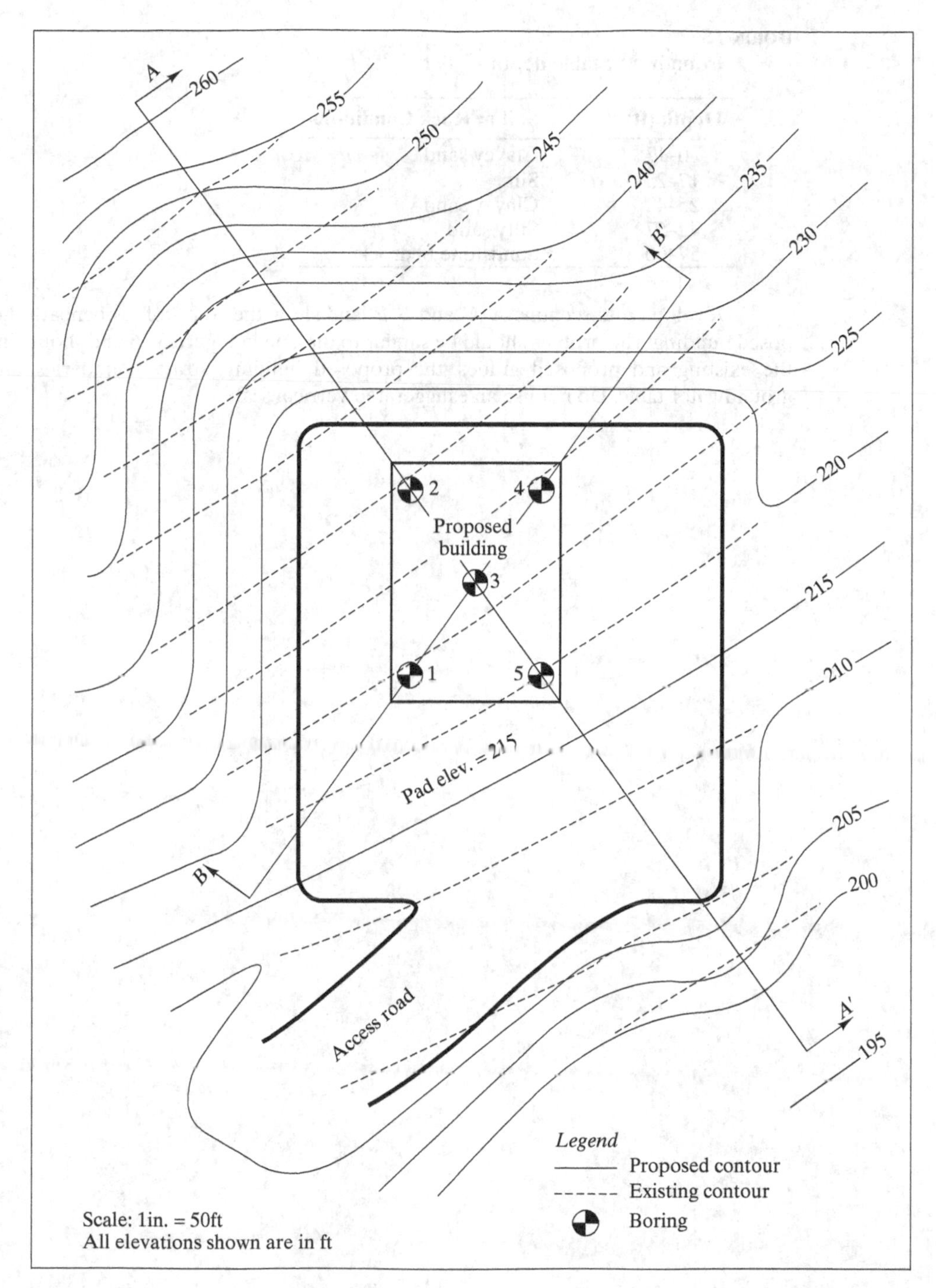

FIGURE 3.41 Site plan for Problem 3.26.

Boring 5

Groundwater table depth = 49 ft

Depth (ft)	Soil or Rock Conditions
0–17	Clayey sand
17–25	Silt
25–42	Clayey sand
42–57	Silty sand
57–60	Sandstone bedrock

Develop cross-sections *A-A′* and *B-B′* and show the soil profiles beneath the proposed building. The profiles should be similar to the one in Figure 3.36 and should include the existing and proposed grades, the proposed building, strata boundaries, and the groundwater table. Do not use an exaggerated vertical scale.

CHAPTER 4

Soil Composition

... in engineering practice, difficulties with soils are almost exclusively due not to the soils themselves but to the water contained in their voids. On a planet without any water there would be no need for soil mechanics.

Karl Terzaghi, 1939

Once the soil and rock samples have been brought to the laboratory, we normally conduct appropriate tests on these samples to develop data for our analyses. Some of these tests measure familiar engineering properties, such as shear strength, while others focus on the sample's composition and structure.

The composition of soils and rocks is quite different from that of other civil engineering materials, such as steel, concrete, or wood. These differences include the following:

- Soils and rocks are *natural materials*, not manufactured products. As such, their engineering properties can vary significantly from place to place and even across a single building site. Although wood also is a natural product, it is sorted and graded before being used in construction. In general, we cannot sort soils or rocks, and must accommodate whatever is present on our site.
- Soil is a *particulate material* that consists of individual particles. It is not a continuous mass. Some rock strata, especially certain sedimentary rocks, also can be treated as particulates. Even a rock that appears to be a continuous mass virtually always contains cracks and fissures that make it discontinuous.
- A soil can contain *all three phases of matter* (solid, liquid, and gas) simultaneously, and these three phases can be present in varying proportions. Rock also can contain all three phases, although the liquid and gaseous phases may be confined mainly to the cracks and fissures.

This chapter discusses the methods we use to assess the composition of a soil and the parameters we use to describe this composition.

FIGURE 4.1 Microphotograph of a medium sand at 5 × magnification.

4.1 SOIL AS A PARTICULATE MATERIAL

Most civil engineering materials consist of a continuous mass held together with molecular bonds, and the mechanical properties of such materials depend on their chemical makeup and on the nature of these bonds. For example, the shear strength of steel depends on the strength of the molecular bonds, and shear failure requires breaking them. In contrast, soil is a particulate material that consists of individual particles assembled together as shown, for example, in Figure 4.1. Its engineering properties depend largely on the interaction between these particles, and only secondarily on their internal properties. This is especially true in gravels, sands, and silts. For example, it has been observed that when a soil fails, it almost always fails by shearing along some failure surface in the soil, with the particles rolling over and sliding past each other along the failure surface, and not by the particles breaking internally. Breakage of individual particles is typically minimal. Thus, the shear strength of a soil depends on such factors as the coefficient of friction between the particles, the tightness of packing, and so on, rather than the chemical bonds inside the particles. The topic of soil strength is discussed in more detail in Chapter 12.

Clays also have a particulate structure, but the nature of the particles is quite different, as discussed later in this chapter. In clays, there is much more interaction between the particles and the pore water, so their behavior is more complex than that of other soils.

4.2 THE THREE PHASES

Soils also are different from most civil engineering materials in that a soil can simultaneously contain solid, liquid, and gaseous phases. The liquid and gaseous phases are contained in the *voids* or *pores* between the solid particles. In addition, the three phases in a soil often interact, and these interactions have important effects on the soil's behavior.

The solid phase is always present in soil and usually consists of particles derived from the weathering of rocks, as discussed in Chapter 2. It also can include organic material.

The liquid phase is usually present in a soil and most often consists of water. However, it also can include other liquids, such as the following:

- Gasoline or other chemicals that have leaked out of underground tanks or pipelines, or infiltrated from the ground surface. The cleanup of such contaminants can be a significant problem and has been the object of many "superfund" sites identified by the Environmental Protection Agency.
- Leachate that has escaped from a sanitary landfill. This leachate can contaminate groundwater and make it unfit for drinking.
- Sea water moving inland through the soil, which often occurs when wells are installed near the ocean to pump out fresh water for municipal purposes.
- Petroleum that occurs naturally and seeps through the soil.

Although these constituents usually represent a small fraction of the liquids in a soil, they can be very important, especially if the groundwater is to be pumped and used for domestic purposes, but becomes contaminated.

If the liquid phase does not completely fill the voids, the remaining space is occupied by the gaseous phase. It is usually air, but can include other gasses, such as the following:

- Methane (CH_4) and carbon dioxide (CO_2) from the decomposition of organic material. These gasses are often found near sanitary landfills and near highly organic natural soils.
- Petroleum vapors from contaminated soils.
- Hydrogen sulfide (H_2S) from the decomposition of sulfur-bearing materials. It is sometimes found near sanitary landfills, sewage facilities, and coal deposits.

Gasses other than air can be important because they sometimes pose safety hazards to workers in utility vaults, small excavations, mines, and other underground areas. Some, such as hydrogen sulfide, are poisonous, while others, such as methane, are flammable. For example, methane gas permeating up from the ground became trapped in a small room of a department store in Los Angeles. In 1985, this gas was accidently ignited, creating an explosion that blew out the windows and collapsed part of the roof. Twenty-three people required hospital treatment as a result of the blast (Hamilton and Meehan, 1992).

Even gasses that are neither toxic nor flammable can be dangerous because they displace oxygen, thus suffocating workers. This problem has caused injury and death, but can be avoided by using special detection equipment and providing adequate ventilation.

Clearly, components other than water and air can be very important. However, they generally represent only a small portion of the soil weight and volume. Therefore, for the purposes of this chapter, we will assume that the liquid and gaseous phases will consist only of water and air, respectively, and will refer to them as "pore water" and "pore air."

4.3 WEIGHT–VOLUME RELATIONSHIPS

It is helpful to identify the relative proportions of solids, water, and air in a soil, because these proportions have a significant effect on the soil's behavior. Therefore, geotechnical engineers have developed methods to quantify the relative weights and volumes of these components.

Phase Diagrams

Phase diagrams, such as those shown in Figure 4.2, indicate the relative proportions of solids, water, and air in a soil. The weights or masses of the three components are depicted on the left side of the diagram, while their volumes on the right side. Defined also in Figure 4.2 are the symbols for the weights, masses and volumes.

Definitions of Weight–Volume Parameters

Geotechnical engineers have defined several *weight–volume parameters* based on the weights or masses and volumes of the three phases shown in the phase diagrams. These parameters, also called *index properties*, give important information on the composition of a particular soil, and form part of the basic language of soil mechanics. While these parameters may not be direct measures of soil properties or soil behavior, they are often correlated empirically with soil properties and soil behavior.

Moisture Content and Degree of Saturation One of the most common soil parameters is the *moisture content*, w (also known as the *water content*). It is a measure of how much water there is in a soil and is defined as the ratio of the weight or mass of water to the weight or mass of solids:

$$w = \frac{W_w}{W_s} \times 100\% \tag{4.1}$$

$$w = \frac{M_w}{M_s} \times 100\% \tag{4.2}$$

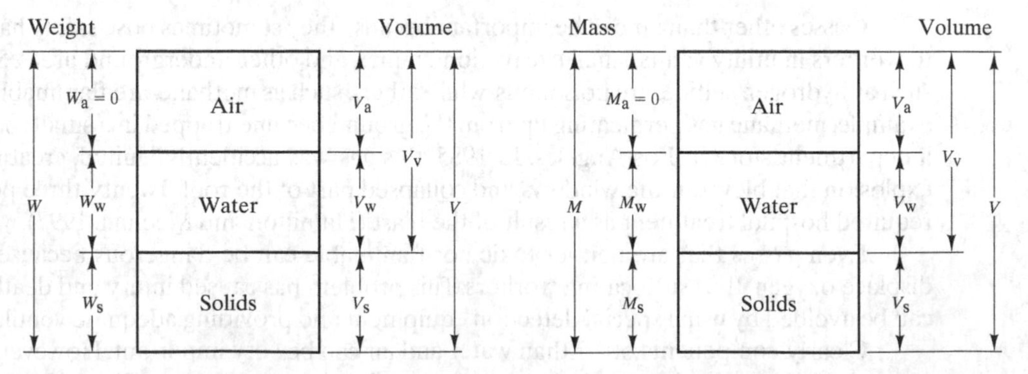

FIGURE 4.2 Phase diagrams describe the relative proportions of solids, water, and air in a soil. One side presents weights or masses, while the other presents volumes.

A small w indicates a dry soil, while a large w indicates a wet one. Values in the field are usually between 3% and 70%, but values greater than 100% are sometimes found in soft soils below the groundwater table, which simply means such soils have in them more water than solids. Note that the moisture content should, by normal practice, be given as a percentage.

Because of the definitions in Equations 4.1 and 4.2, with the denominators being W_s and M_s, and not, as one might expect, W and M, respectively, the moisture content can actually be larger than 100%. Equations 4.1 and 4.2 are good formulas to memorize, but don't make the common mistake of using the wrong denominators.

The moisture content can be easily measured in the laboratory by conducting a *moisture content test* (ASTM D2216) as follows:

1. Obtain a small can to hold the soil sample and find the can's mass, M_c.
2. Place a representative sample of the soil into the can and find the total mass, M_1.
3. Place the can of soil into an oven with a constant temperature of $110 \pm 5°C$ and leave it there until the soil is completely dry. This usually requires 12 to 16 hours.
4. Determine the mass of the dry sample and can, M_2.
5. Compute the moisture content using this equation:

$$w = \frac{M_1 - M_2}{M_2 - M_c} \times 100\% \tag{4.3}$$

A similar parameter is the *degree of saturation*, S, which is the percentage of the voids that are filled with water:

$$S = \frac{V_w}{V_v} \times 100\% \tag{4.4}$$

This is similar to moisture content in that both are equal to zero when there is no water. However, S has a maximum value of 100%, which occurs when all of the voids are filled with water. We use the term *saturated* to describe this condition. Soils below the groundwater table are generally saturated. Values of S above the groundwater table are usually between 5% and 100%, although values approaching zero can be found in very arid areas. Note that the degree of saturation, like the moisture content, should, by normal practice, given as a percentage.

Capillary forces, discussed in Chapter 7, can draw water upward from the groundwater table, often producing soils with $S = 100\%$ well above the groundwater table. Therefore, we should not use saturation computations to determine the location of the groundwater table. Instead, we may use observation wells as discussed in Section 3.7.

Example 4.1

A moisture content test was performed on a soil sample obtained from the field. A small amount of this soil sample was put in a steel can and weighed. The mass of the can plus the soil in it was 18.68 g. After oven-drying, the mass of the can and dried soil was 16.22 g. The mass of the can was 8.30 g. Compute the moisture content of the soil.

Solution:

The moisture content can be computed using Equation 4.3:

$$w = \frac{M_w}{M_s} \times 100\% = \frac{18.68\text{ g} - 16.22\text{ g}}{16.22\text{ g} - 8.30\text{ g}} \times 100\% = \mathbf{31.1\%}$$

Unit Weight and Density The parameter *unit weight*, γ, is defined by:

$$\gamma = \frac{W}{V} \tag{4.5}$$

The unit weight of undisturbed soil samples can easily be determined in the laboratory by measuring their physical dimensions and weighing them. This method produces reliable assessments of γ for many soils. However, it is affected by sample disturbance, especially in sandy and gravelly soils. Sometimes unit weight measurements are made on supposedly "undisturbed" samples that in reality have significant disturbance. Such measurements are very misleading, so it is best not to even attempt unit weight measurements on poor quality samples. Table 4.1 presents typical ranges of γ for various soils.

Two variations of unit weight also are commonly used, the *dry unit weight*, γ_d, and the *unit weight of water*, γ_w:

$$\gamma_d = \frac{W_s}{V} \tag{4.6}$$

$$\gamma_w = \frac{W_w}{V_w} \tag{4.7}$$

TABLE 4.1 Typical Unit Weights

Soil Type and Unified Soil Classification (See Section 5.3)	Typical Unit Weight, γ			
	Above Groundwater Table		Below Groundwater Table	
	(lb/ft³)	(kN/m³)	(lb/ft³)	(kN/m³)
GP—Poorly-graded gravel	110–130	17.5–20.5	125–140	19.5–22.0
GW—Well-graded gravel	110–140	17.5–22.0	125–150	19.5–23.5
GM—Silty gravel	100–130	16.0–20.5	125–140	19.5–22.0
GC—Clayey gravel	100–130	16.0–20.5	125–140	19.5–22.0
SP—Poorly-graded sand	95–125	15.0–19.5	120–135	19.0–21.0
SW—Well-graded sand	95–135	15.0–21.0	120–145	19.0–23.0
SM—Silty sand	80–135	12.5–21.0	110–140	17.5–22.0
SC—Clayey sand	85–130	13.5–20.5	110–135	17.5–21.0
ML—Low plasticity silt	75–110	11.5–17.5	80–130	12.5–20.5
MH—High plasticity silt	75–110	11.5–17.5	75–130	11.5–20.5
CL—Low plasticity clay	80–110	12.5–17.5	75–130	11.5–20.5
CH—High plasticity clay	80–110	12.5–17.5	70–125	11.0–19.5

Normally, $\gamma_w = 9.81$ kN/m^3 or 62.4 lb/ft^3 for fresh water and $\gamma_w = 10.1$ kN/m^3 or 64.0 lb/ft^3 for sea water. In practice, lb/ft^3 is often written as pcf.

For soils below the groundwater table, some computations use the *buoyant unit weight*, γ_b, also called the *submerged unit weight*:

$$\gamma_b = \gamma - \gamma_w \tag{4.8}$$

We also can define similar parameters based on mass instead of weight. These become the *density*, ρ, *dry density*, ρ_d, and *density of water*, ρ_w:

$$\rho = \frac{M}{V} \tag{4.9}$$

$$\rho_d = \frac{M_s}{V} \tag{4.10}$$

$$\rho_w = \frac{M_w}{V_w} \tag{4.11}$$

Design values of $\rho_w = 1000$ kg/m^3 or 1.94 slug/ft^3 for fresh water.

Based on $F = Ma$, the unit weight and density are related by

$$\gamma = \rho g \tag{4.12}$$

where:

g = acceleration due to gravity = 9.81 m/s^2 or 32.2 ft/s^2

For most geotechnical computations, unit weight is more useful than density because we use it to compute stresses due to the weight of the soil. A notable exception is dynamic analyses that need to consider inertial effects that are best presented in terms of mass.

Unfortunately, geotechnical engineers are often careless in our use of these terms. Many of us use the term *density* when we really mean *unit weight*. This usage is technically incorrect, even though it is common in conversation, reports, and even in technical literature, and can be confusing. As a general rule, one can assume the speaker or writer really means unit weight unless the discussion involves a soil dynamics problem. In this book, we will avoid such confusion by always using the proper terms.

Specific Gravity of Solids The specific gravity of any material is the ratio of its density to that of water. In the case of soils, we compute it for the solid phase only, and define the *specific gravity of solids*, G_s:

$$G_s = \frac{M_s}{V_s \rho_w} \tag{4.13}$$

$$G_s = \frac{W_s}{V_s \gamma_w} \tag{4.14}$$

TABLE 4.2 Specific Gravity of Selected Non-Clay Minerals

Mineral	G_s
Quartz	2.65
Feldspar	2.54–2.76
Hornblende	3.00–3.50
Mica	2.76–3.20
Calcite	2.71
Hematite	5.20
Limonite	3.6–4.0
Gypsum	2.32
Talc	2.70–2.80
Olivene	3.27–4.50

TABLE 4.3 Specific Gravity of Selected Clay Minerals

Mineral	G_s
Kaolinite	2.62–2.66
Montmorillonite	2.75–2.78
Illite	2.60–2.86
Chlorite	2.60–2.96

This is quite different from the specific gravity of the entire soil mass, which would include the solids, water, and air. Therefore, do not make the common mistake of computing G_s as γ/γ_w.

Tables 4.2 and 4.3 list G_s values for common soil minerals. Although a standard laboratory test is available to measure G_s (ASTM D854), it is usually not needed because nearly all real soils have $2.60 < G_s < 2.80$, which is a very narrow range. The additional precision obtained by performing a test is generally not worth the expense. For most practical problems, it is sufficient to estimate G_s from the following list:

- Clean, light colored sand of quartz and feldspar 2.65
- Dark-colored sand 2.70
- Sand–silt–clay mixtures 2.70
- Clay 2.72

Nevertheless, some unusual soils have G_s values well outside these limits. For example, the olivene sands in Hawaii have G_s values as high as 4.50, while organic soils have low G_s values, sometimes less than 2.0.

Void Ratio and Porosity The relative volumes of voids and solids in a soil may be quantified using the parameter *void ratio*, *e*:

$$e = \frac{V_v}{V_s} \tag{4.15}$$

Thus, densely packed soils have low void ratios. Typical values in the field range from 0.1 to 2.5. Note that the void ratio should, by normal practice, be given as a numerical value.

The *porosity*, n, is a similar parameter:

$$n = \frac{V_v}{V} \times 100\% \tag{4.16}$$

It typically is between 9% and 70% and is related to e as follows:

$$n = \frac{e}{1 + e} \times 100\% \tag{4.17}$$

Note that the porosity should, by normal practice, given as a percentage. Sometimes geotechnical engineers divide the porosity into two parts: the *water porosity*, n_w, and the *air porosity*, n_a:

$$n_w = \frac{V_w}{V} \times 100\% \tag{4.18}$$

$$n_a = \frac{V_a}{V} \times 100\% \tag{4.19}$$

Relative Density The *relative density* is a special weight–volume parameter used in sandy and gravelly soils. It is defined as:

$$D_r = \frac{e_{max} - e}{e_{max} - e_{min}} \times 100\% \tag{4.20}$$

where:

D_r = relative density
e = void ratio
e_{min} = minimum void ratio
e_{max} = maximum void ratio

The values of e_{min} and e_{max} represent the soil in very dense and very loose conditions, respectively, and are determined by standard laboratory tests (ASTM D4253 and D4254, respectively). Thus, loose soils have low values of D_r, while dense soils have high values. In theory, the lowest possible value of D_r is 0% and the highest possible value is 100%. Thus, D_r is often more useful than e because we can easily compare the field value to the lowest and highest possible values. In other words, the D_r of a soil tells us how dense the soil is relative to how dense it could possibly be. Table 4.4 presents a classification of soil consistency based on its relative density. Note that the relative density should, by normal practice, given as a percentage.

Another way to determine D_r is by using empirical correlations with in situ test data. These methods are often preferred, because it is so difficult to obtain sufficiently undisturbed samples to obtain a reliable value of e, especially in sandy soils. In addition, in situ tests, especially the cone penetration test (CPT), give a better representation of the variability of the soil strata, because coarse-grained soils inevitably have some zones that are looser and denser than average.

TABLE 4.4 Consistency of Coarse-Grained Soils at Various Relative Densities[a]

Relative Density, D_r (%)	Classification
0–15	Very loose
15–35	Loose
35–65	Medium dense[b]
65–85	Dense
85–100	Very dense

[a]Lambe and Whitman, 1969; adapted by permission of John Wiley and Sons, Inc. Other classification systems have been proposed by others that use these terms, but with different values for the corresponding relative densities.
[b]Lambe and Whitman used the term *medium*, but *medium dense* is probably better because "medium" is also used to describe particle size, e.g., medium sand in Table 4.6.

Equations 4.21 and 4.25 present empirical correlations between D_r and the standard penetration test (SPT) and CPT results (Kulhawy and Mayne, 1990). These correlations were developed by analyzing empirical data. The CPT correlation is probably more reliable than the SPT correlation.

Equations 4.21 and 4.25 include several parameters that have not yet been defined. The *vertical effective stress*, σ'_z, is a measure of the compressive stress in the ground at the depth where the test data is being evaluated; the *overconsolidation ratio* (OCR) is a measure of the stress history of the soil; and D_{50} is the particle size where 50% of the soil particles is finer than this size. All of these will be defined and discussed in more detail later in this book. In the meantime, the required values will simply be given in any problem statement.

The relative density can be calculated from SPT data using the following empirical formulas:

$$D_r = \sqrt{\frac{N_{1,60}}{C_p C_A C_{OCR}}} \times 100\% \tag{4.21}$$

$$C_P = 60 + 25\log D_{50} \tag{4.22}$$

$$C_A = 1.2 + 0.05\log\left(\frac{t}{100}\right) \tag{4.23}$$

$$C_{OCR} = \text{OCR}^{0.18} \tag{4.24}$$

where:

$N_{1,60}$ = corrected SPT N-value, as defined in Chapter 3
C_P = grain size correction factor
C_A = aging correction factor
C_{OCR} = overconsolidation correction factor
D_{50} = particle size, in mm, where 50% of the soil particles is finer than this size, as defined in Section 4.4

t = age of soil (time since deposition in years). If no age information data is available, use $t = 100$ years.

OCR = overconsolidation ratio, as defined in Chapter 10. If no information is available to assess the OCR, use a value of 2.

With CPT data, the relative density can be calculated using the following empirical formulas:

$$D_r = \sqrt{\left(\frac{q_c}{315\, Q_c\, \text{OCR}^{0.18}}\right)\sqrt{\frac{2000\ \text{lb/ft}^2}{\sigma'_z}}} \times 100\% \qquad \text{(4.25–English)}$$

$$D_r = \sqrt{\left(\frac{q_c}{315\, Q_c\, \text{OCR}^{0.18}}\right)\sqrt{\frac{100\ \text{kPa}}{\sigma'_z}}} \times 100\% \qquad \text{(4.25–SI)}$$

where:

q_c = cone resistance (kg/cm² or ton/ft²), as defined in Chapter 3

Q_c = compressibility factor
= 0.91 for highly compressible sands
= 1.00 for moderately compressible sands
= 1.09 for slightly compressible sands

For this formula, a sand with a high fines content or a high mica content is "highly compressible," whereas a pure quartz sand is "slightly compressible."

σ'_z = vertical effective stress (lb/ft²; kPa), as defined in Chapter 9

Many people confuse relative density with relative compaction. The latter is defined in Chapter 6. Although the names are similar, and they measure similar properties, these two parameters are numerically different. In addition, some people in other professions use the term *relative density* to describe what we call specific gravity. Geotechnical engineers should never use the term in this way.

Table 4.5 presents typical values of e_{min} and e_{max} for various sandy soils. These are not intended to be used in lieu of laboratory or in situ tests, but could be used to check test results or for preliminary analyses.

TABLE 4.5 Typical Values of e_{min} and e_{max} [a]

Soil Description	e_{min} (dense)	e_{max} (loose)
Equal spheres (theoretical values)	0.35	0.92
Clean, poorly-graded medium sand (Ottawa, Illinois)	0.50	0.80
Clean, fine-to-medium sand	0.40	1.0
Uniform inorganic silt	0.40	1.1
Silty sand	0.30	0.90
Clean fine-to-coarse sand	0.20	0.95
Micaceous sand	0.40	1.2
Silty sand and gravel	0.14	0.85

[a] Hough, 1969; adapted by permission of John Wiley and Sons, Inc.

Derived Equations

By combining the definitions just given and the phase diagrams in Figure 4.2, we can derive a series of new weight–volume equations. For example:

Let $V_s = 1$, as shown in Figure 4.3.

$$e = \frac{V_v}{V_s} \quad \rightarrow \quad \text{gives } V_v = e$$

$$S = \frac{V_w}{V_v} \times 100\% \quad \rightarrow \quad \text{gives } V_w = Se$$

$$\gamma_w = \frac{W_w}{V_w} \quad \rightarrow \quad \text{gives } W_w = Se\gamma_w$$

$$G_s = \frac{W_s}{V_s \gamma_w} \quad \rightarrow \quad \text{gives } W_s = G_s\gamma_w\,(1)$$

$$w = \frac{W_w}{W_s} \times 100\% = \frac{Se\gamma_w}{G_s\gamma_w} \times 100\% = \frac{Se}{G_s} \times 100\%$$

Rewriting this equation gives:

$$S = \frac{wG_s}{e} \times 100\% \tag{4.26}$$

Derivations based on any other assumed value of V_s would produce the same equation.

Using similar derivations, we also can develop the following equations:

$$\gamma_d = \frac{\gamma}{1 + w} \tag{4.27}$$

$$\gamma_d = \frac{G_s\gamma_w}{1 + wG_s/S} \tag{4.28}$$

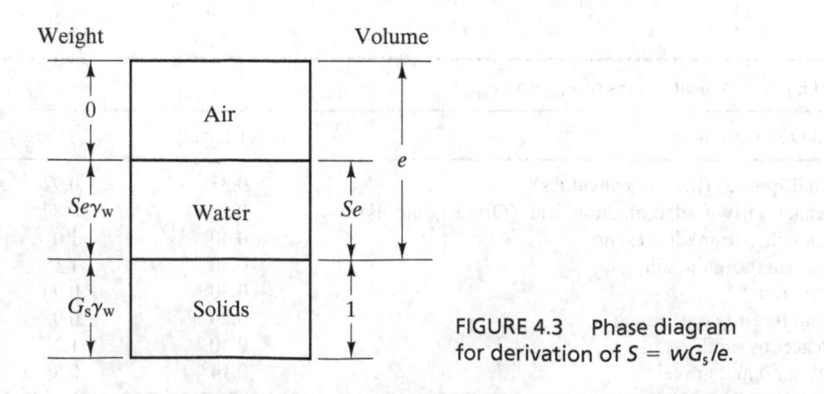

FIGURE 4.3 Phase diagram for derivation of $S = wG_s/e$.

$$W_s = \frac{W}{1 + w} \tag{4.29}$$

$$M_s = \frac{M}{1 + w} \tag{4.30}$$

$$e = \frac{G_s \gamma_w}{\gamma_d} - 1 \tag{4.31}$$

$$w = S\left[\frac{\gamma_w}{\gamma_d} - \frac{1}{G_s}\right] \times 100\% \tag{4.32}$$

$$S = \frac{w}{\dfrac{\gamma_w}{\gamma_d} - \dfrac{1}{G_s}} \times 100\% \tag{4.33}$$

For all of these equations, parameters normally expressed as a percentage must be inserted in decimal form. For example, a degree of saturation of 45% in Equation 4.28 would be expressed as 0.45 not 45.

Solving Weight–Volume Problems

We often encounter problems where one or more of the weight–volume parameters is known and others need to be determined. For example, some parameters, such as moisture content, can be measured in the laboratory, while others, such as void ratio, cannot. Therefore, we need to have a means of computing them.

Often we can perform these computations using derived equations such as Equations 4.26 through 4.33. However, if this method does not work, we can go back to fundamentals and solve the problem using a phase diagram as follows:

1. **Draw** a phase diagram and annotate all of the given weights and volumes. Remember to set $W_a = 0$. If the soil is saturated, we also can set $V_a = 0$.
2. Sometimes no weights or volumes are given in the problem statement, or the only weights or volumes given are equal to zero (i.e., no air or no water). If this is the case, the given data are applicable to the entire soil stratum, regardless of the sample size. However, the phase diagram analysis method requires that we specify a certain quantity of soil, and will not work unless we do so. Therefore, we must assume a quantity. Any one weight or volume may be assumed (but only one). Usually we assume $V = 1\ \text{m}^3$ or $V = 1\ \text{ft}^3$.
3. Using Equations 4.1 through 4.33 and obvious addition and subtraction from the phase diagram (e.g., $V_a = V - V_s - V_w$), determine all the remaining weights and volumes.
4. Compute the values of the required parameters using the weights and volumes of the three phases and Equations 4.1 through 4.20.

Example 4.2

A density test was performed on a soil sample obtained from the field. The soil sample was obtained from a cylindrical sampler similar to those shown in Figures 3.19 and 3.20. The sample was contained in 4 brass rings, each of which had an inside diameter of 2.43 in. and was 1.0 in. high. The net weight of the soil in the four rings was 1.35 lb. Compute the unit weight of the soil.

Solution:

The total volume of the soil V is equal to the inside volume of four brass rings:

$$V = \frac{1}{4}\pi(2.43 \text{ in.})^2(4 \text{ in.}) \times \left(\frac{1 \text{ ft}^3}{1728 \text{ in.}^3}\right) = 0.0107 \text{ ft}^3$$

Therefore, the unit weight of the soil is given by:

$$\gamma = \frac{W}{V} = \frac{1.35 \text{ lb}}{0.0107 \text{ ft}^3} = \mathbf{126\ lb/ft^3}$$

Example 4.3

A 27.50 lb soil sample has a volume of 0.220 ft^3, a moisture content of 10.2%, and a specific gravity of solids of 2.65. Compute the unit weight, dry unit weight, degree of saturation, void ratio, and porosity.

Solution using fundamental and derived equations:

$$\gamma = \frac{W}{V} = \frac{27.50 \text{ lb}}{0.220 \text{ ft}^3} = \mathbf{125\ lb/ft^3}$$

$$\gamma_d = \frac{\gamma}{1 + w} = \frac{125.0 \text{ lb/ft}^3}{1 + 0.102} = \mathbf{113\ lb/ft^3}$$

$$S = \frac{w}{\dfrac{\gamma_w}{\gamma_d} - \dfrac{1}{G_s}} \times 100\% = \frac{0.102}{\dfrac{62.4 \text{ lb/ft}^3}{113.4 \text{ lb/ft}^3} - \dfrac{1}{2.65}} \times 100\% = \mathbf{59\%}$$

$$e = \frac{G_s\gamma_w}{\gamma_d} - 1 = \frac{(2.65)(62.4 \text{ lb/ft}^3)}{113.4 \text{ lb/ft}^3} - 1 = \mathbf{0.458}$$

$$n = \frac{e}{1 + e} \times 100\% = \frac{0.4582}{1 + 0.4582} \times 100\% = \mathbf{31.4\%}$$

Solution using a phase diagram:

Although the fundamental and derived equations were sufficient to solve this problem, and would be the easiest method, we also will illustrate a solution using a phase diagram.

Step 1: Draw and annotate a phase diagram (see Figure 4.4).

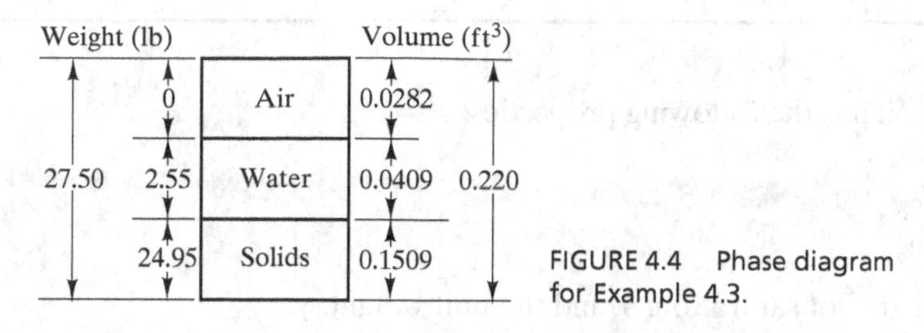

FIGURE 4.4 Phase diagram for Example 4.3.

Step 2: Assume one weight or volume.

Not applicable to this problem, because weights and volumes have been given.

Step 3: Determine weights and volumes in phase diagram.

$$W_s = \frac{W}{1 + w} = \frac{27.50 \text{ lb}}{1 + 0.102} = 24.95 \text{ lb}$$

$$W_w = W - W_s = 27.50 \text{ lb} - 24.95 \text{ lb} = 2.55 \text{ lb}$$

$$G_s = \frac{W_s}{V_s \gamma_w}$$

$$2.65 = \frac{24.95 \text{ lb}}{V_s (62.4 \text{ lb/ft}^3)} \quad \rightarrow \quad V_s = 0.1509 \text{ ft}^3$$

$$\gamma_w = \frac{W_w}{V_w}$$

$$62.4 \text{ lb/ft}^3 = \frac{2.55 \text{ lb}}{V_w} \quad \rightarrow \quad V_w = 0.0409 \text{ ft}^3$$

$$V_a = V - V_w - V_s$$

$$= 0.220 \text{ ft}^3 - 0.0409 \text{ ft}^3 - 0.1509 \text{ ft}^3 = 0.0282 \text{ ft}^3$$

Step 4: Compute parameters.

$$\gamma = \frac{W}{V} = \frac{27.50 \text{ lb}}{0.220 \text{ ft}^3} = \mathbf{125 \text{ lb/ft}^3}$$

$$n = \frac{V_v}{V} \times 100\% = \frac{0.0282 \text{ ft}^3 + 0.0409 \text{ ft}^3}{0.220 \text{ ft}^3} \times 100\% = \mathbf{31.4\%}$$

$$e = \frac{V_v}{V_s} = \frac{0.0409 \text{ ft}^3 + 0.0282 \text{ ft}^3}{0.1509 \text{ ft}^3} = \mathbf{0.458}$$

$$\gamma_d = \frac{W_s}{V} = \frac{24.95 \text{ lb}}{0.220 \text{ ft}^3} = \mathbf{113 \text{ lb/ft}^3}$$

$$S = \frac{V_w}{V_v} \times 100\% = \frac{0.0409 \text{ ft}^3}{0.0409 \text{ ft}^3 + 0.0282 \text{ ft}^3} \times 100\% = \mathbf{59\%}$$

Example 4.4

A certain soil has the following properties:

$G_s = 2.71$
$n = 41.9\%$
$w = 21.3\%$

Find the degree of saturation, S, and the unit weight, γ.

Solution:

Although this problem could be solved using Equations 4.17 and 4.26–4.28, we will use the phase diagram method to illustrate Step 2 of the procedure described above.

Step 1: Draw and annotate phase diagram (see Figure 4.5).

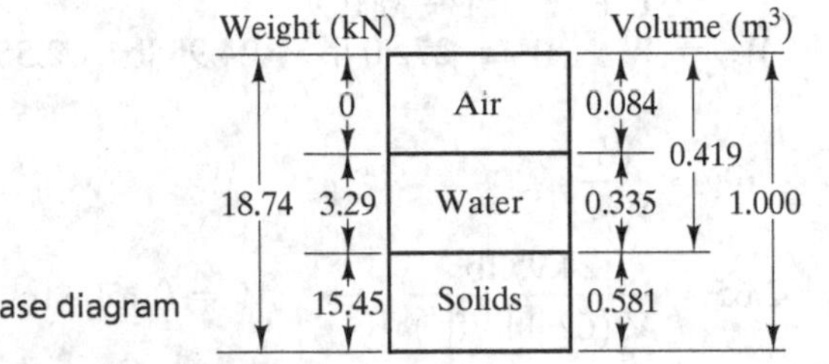

FIGURE 4.5 Phase diagram for Example 4.4.

Step 2: Assume one weight or volume.

No weights or volumes were stated, so all of the given data applies to the entire soil strata. Therefore, we will develop a phase diagram for an assumed total volume V of 1 m^3.

Step 3: Determine weights and volumes in phase diagram.

$$n = \frac{V_v}{V} \times 100\% \quad \rightarrow \quad 0.419 = \frac{V_v}{1} \rightarrow V_v = 0.419 \text{ m}^3$$

$$V_s = V - V_v = 1.000 \text{ m}^3 - 0.419 \text{ m}^3 = 0.581 \text{ m}^3$$

$$G_s = \frac{W_s}{V_s \gamma_w} \quad \rightarrow \quad 2.71 = \frac{W_s}{(0.581 \text{ m}^3)(9.81 \text{ kN/m}^3)} \rightarrow W_s = 15.45 \text{ kN}$$

$$w = \frac{W_w}{W_s} \times 100\% \rightarrow 0.213 = \frac{W_w}{15.45 \text{ kN}} \rightarrow W_w = 3.29 \text{ kN}$$

$$W = W_s + W_w = 15.45 \text{ kN} + 3.29 \text{ kN} = 18.74 \text{ kN}$$

$$\gamma_w = \frac{W_w}{V_w} \quad \rightarrow \quad 9.81 \text{ kN/m}^3 = \frac{3.29 \text{ kN}}{V_w} \rightarrow V_w = 0.335 \text{ m}^3$$

$$V_a = V_v - V_w = 0.419 \text{ m}^3 - 0.335 \text{ m}^3 = 0.084 \text{ m}^3$$

Step 4: Compute parameters.

$$S = \frac{V_w}{V_v} \times 100\% = \frac{0.335\ \text{m}^3}{0.335\ \text{m}^3 + 0.084\ \text{m}^3} \times 100\% = \mathbf{80.0\%}$$

$$\gamma = \frac{W}{V} = \frac{18.74\ \text{kN}}{1\ \text{m}^3} = \mathbf{18.7\ kN/m^3}$$

Example 4.5

The standard method of measuring the specific gravity of solids (ASTM D854) uses a calibrated glass flask known as a *pycnometer*, as shown in Figure 4.6. The pycnometer is first filled with water and set on a balance to find its mass. Then, it is refilled with a known mass of dry soil plus water so the total volume is the same as before. Again, its mass is determined. From this data, we can compute G_s.

Using this technique on a certain soil sample, we have obtained the following data:

Mass of soil = 81.80 g
Moisture content of soil = 11.2%
Mass of pycnometer + water = 327.12 g
Mass of pycnometer + soil + water = 373.18 g
Volume of pycnometer = 250.00 ml

Compute G_s for this soil.

FIGURE 4.6 Use of a pycnometer to measure G_s in the laboratory.

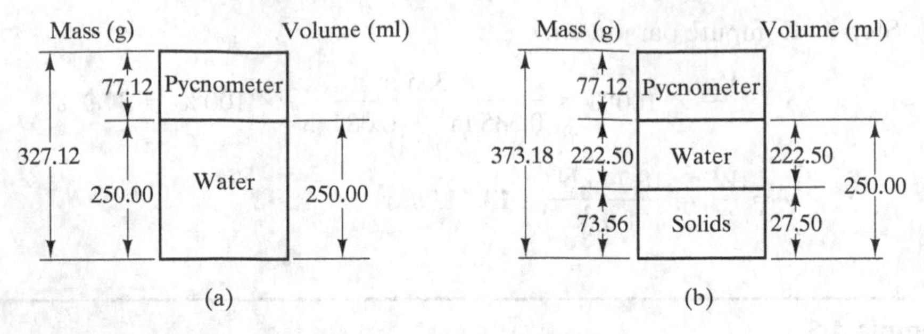

FIGURE 4.7 Modified phase diagrams for Example 4.5.

Solution:

Using the phase diagrams shown in Figure 4.7:

$$(M_w)_A = V\rho_w = (250.00 \text{ ml})(1 \text{ g/ml}) = 250.00 \text{ g}$$

$$M_P = M_A - (M_w)_A = 327.12 \text{ g} - 250.00 \text{ g} = 77.12 \text{ g}$$

$$(M_s)_B = \frac{M}{1 + w} = \frac{81.8 \text{ g}}{1 + 0.112} = 73.56 \text{ g}$$

$$(M_w)_B = 373.18 \text{ g} - 77.12 \text{ g} - 73.56 \text{ g} = 222.50 \text{ g}$$

$$(V_w)_B = \frac{(M_w)_B}{\rho_w} = \frac{222.50 \text{ g}}{1 \text{ g/ml}} = 222.50 \text{ ml}$$

$$(V_s)_B = 250.00 \text{ ml} - 222.50 \text{ ml} = 27.50 \text{ ml}$$

$$G_s = \frac{M_s}{V_s\rho_w} = \frac{73.56 \text{ g}}{(27.50 \text{ ml})(1 \text{ g/ml})} = \mathbf{2.67}$$

Note: A real laboratory specific gravity test also considers various correction factors to account for the density of water as a function of temperature and other variables. For simplicity, we have not considered these factors here.

4.4 PARTICLE SIZE AND SHAPE

The individual solid particles in a soil can have different sizes and shapes, and these characteristics also have a significant effect on its engineering behavior. Therefore, geotechnical engineers often assess the distribution of particle sizes in a soil and the shapes of the particles in the soil.

Particle Size Classification

Several systems have been developed to classify a soil particle based on its size. We will examine only one: the ASTM system, as described in Table 4.6. According to this system, particles are classified according to their ability to pass through a series of standard sieves. A sieve consists of a carefully manufactured mesh of wires with a

TABLE 4.6 ASTM Particle Size Classification (Per ASTM D2487)

Sieve Size		Particle Size			
Passes	Retained on	(inch)	(mm)	Soil Classification	
	12 in.	> 12	> 300	Boulder	Rock Fragments
12 in.	3 in.	3–12	75–300	Cobble	
3 in.	3/4 in.	0.75–3	19.0–75	Coarse gravel	Soil
3/4 in.	#4	0.19–0.75	4.75–19.0	Fine gravel	
#4	#10	0.079–0.19	2.00–4.75	Coarse sand	
#10	#40	0.017–0.079	0.425–2.00	Medium sand	
#40	#200	0.003–0.017	0.075–0.425	Fine sand	
#200		< 0.003	< 0.075	Fines (silt + clay)	

FIGURE 4.8 An 8-in. (200 mm) diameter sieve used for soil testing. This one is a 1-in. sieve. Notice how the smaller pieces of gravel have passed through, while the larger pieces have not.

specified opening size, as shown in Figure 4.8. Particles larger than 3 in. (76.2 mm), or more precisely, particles that cannot go through the 3 in. sieve, are known as *rock fragments*. Smaller particles are defined as *soil.*

Figures 4.9 and 4.10 show photographs of samples from each category. Natural deposits often include a mixture of both rock fragments and soil.

Laboratory Tests

Although the distribution of particle sizes can often be estimated by eye, two laboratory tests are commonly used to provide more precise assessments: the *sieve analysis* and the *hydrometer analysis*. In this context, the word "analysis" means a laboratory test, not a series of computations.

Boulder Cobble

FIGURE 4.9 Boulders are particles larger than 12 in. in size; cobbles are particles between 3 and 12 in. in size.

Sieve Analysis A sieve analysis is a laboratory test that measures the particle size distribution of a soil by passing the soil through a series of sieves. The larger sieves are identified by their opening sizes. For example, a 3/4-in. sieve will barely pass a 3/4-in. diameter sphere. Smaller sieves are numbered, with the number indicating the openings per inch. For example, a #8 sieve has 8 openings per inch or 64 per square inch. However, the size of each opening is less than 1/8 in. because of the widths of the wires. Table 4.7 presents opening sizes for standard sieves used in North America.

Most people can just barely see 0.1 mm diameter objects without using a magnifying glass, and this nearly equals the #200 sieve size. This size also represents the border between gritty and smooth textures. Thus, referring back to Table 4.6, sands and silts could be visually distinguished by looking at the particles (if you can see them, it is sand) and by feeling for grittiness (if it feels gritty, it is sand). Both of these evaluations are best done with wetted soil samples.

The sieve analysis (ASTM D422) consists of preparing a soil sample with known weight of solids, W_s, and passing it through the stack of sieves as shown in Figure 4.11. The sieves are arranged in order with the coarsest one on top of the stack and decreasing sieve sizes down the stack. A pan is placed below the finest sieve to collect particles passing through it. The weight retained on each sieve is then expressed as a percentage of the total weight.

The percentage passing the #200 sieve, sometimes called the *<200 (less than 200) portion*, *−200 (minus 200) portion* or the *fines content*, is especially noteworthy. Soils

FIGURE 4.10 Full-scale photos of coarse and fine gravel; coarse, medium, and fine sand.

TABLE 4.7 ASTM Standard Sieves (Per ASTM D422 and E100)

Sieve Identification	Opening Size (inch)	Opening Size (mm)
3 in.	3.0	75.0
2 in.	2.0	50.0
$1\frac{1}{2}$ in.	1.5	37.5
1 in.	1.0	25.0
3/4 in.	0.75	19.0
3/8 in.	0.374	9.5
#4	0.187	4.75
#8	0.0929	2.36
#10	0.0787	2.00
#16	0.0465	1.18
#20	0.0335	0.850
#30	0.0236	0.600
#40	0.0167	0.425
#50	0.0118	0.300
#60	0.00984	0.250
#100	0.00591	0.150
#140	0.00417	0.106
#200	0.00295	0.075

that have less than about 5% passing the #200 sieve are called *clean*, while those that have more are called *dirty*. For example, a *clean sand* is primarily sand, with less than 5% fines (silt or clay particles).

Hydrometer Analysis Although the sieve analysis works very well for particles larger than the #200 sieve (sand, gravel, and coarser particles) and it determines the total amount of fines, it does not give the distribution of finer particles (silt and clay particles). The smallest clay particles are only about 1×10^{-4} mm in size, which is about the same size as a smoke particle. Practically speaking, it would be very difficult to separate the silt and clay particles into different size ranges by sieving because it would require very fine sieves that are almost impossible to manufacture. Therefore, we need to use another technique that gives us the particle size distribution indirectly—the hydrometer analysis. This procedure (ASTM D422) consists of placing a soil sample with a known W_s into a 1000 ml graduated cylinder and filling it with water. The laboratory technician vigorously shakes the cylinder to place the soil particles in suspension. When the suspension reaches a uniform density, the cylinder is placed upright on a table as shown in Figure 4.12.

Once the cylinder has been set upright, the soil particles begin to settle to the bottom. We describe this downward motion using Stoke's Law:

$$v = \frac{D^2 \gamma_w (G_s - G_L)}{18\eta} \tag{4.34}$$

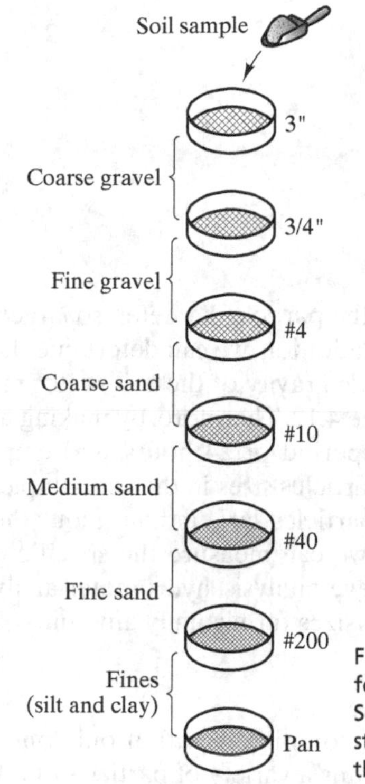

FIGURE 4.11 A series of sieves for conducting a sieve analysis. Some of the sieve sizes used in a standard test are different than those shown here.

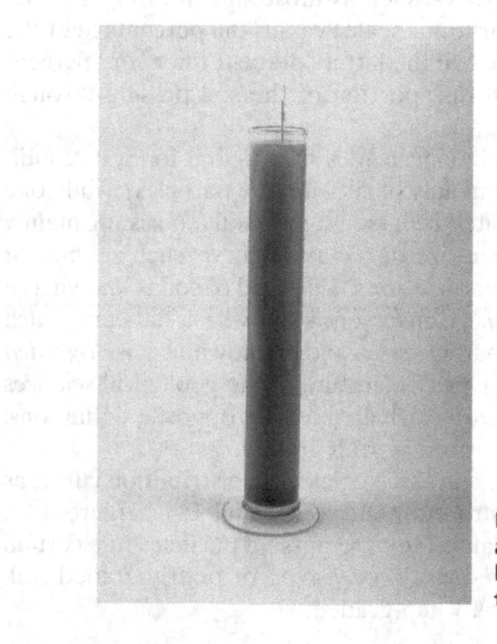

FIGURE 4.12 Equipment used in a hydrometer analysis. Part of the hydrometer is protruding from the top of the fluid.

where:

v = velocity of settling soil particle
D = particle diameter
γ_w = unit weight of water
G_s = specific gravity of solid particles
G_L = specific gravity of soil-water mixture
η = dynamic viscosity of soil-water mixture

The velocity is proportional to the square of the particle diameter, so larger particles settle much more quickly than smaller ones. In addition, we can determine the mass of solids still in suspension by measuring the specific gravity of the soil–water mixture, which is done using the hydrometer shown in Figure 4.12. Therefore, by making a series of specific gravity measurements, usually over a period of 24 hours, and employing Stoke's law, we can determine the distribution of particles sizes in the soil sample.

The hydrometer analysis is unsuitable for particles larger than about the #100 sieve size because they settle more quickly than we can measure the specific gravity using a hydrometer. However, by performing a sieve analysis, hydrometer analysis, or both, we can determine the distribution of particle sizes for virtually any soil.

Particle Size Distribution Curves

A real soil rarely contains only particles that fall completely within only one of the categories listed in Table 4.6. It almost always contains a variety of particles of different sizes mixed together. Therefore, we need to have an effective means of presenting the distribution of particle sizes in a soil. This method is to use the *particle size distribution curve*, also called the *grain size distribution curve*, such as those shown in Figure 4.13. These are plots of the particle size (on a logarithmic scale) versus the percentage of the particles by weight smaller than that size. We call the latter "percent finer" or "percent passing," which generates mental images of that portion of the soil passing through a sieve.

A curve on the left side of the diagram in Figure 4.13, such as that for Soil A, indicates a primarily fine-grained soil (consisting mainly of silt and clay particles), while one on the right side, such as that for Soil B, indicates a coarse-grained soil (consisting mainly of sand and gravel particles). A steep particle size distribution curve, such as that for Soil C, reflects a soil with a narrow range of particle sizes. This kind of soil is known as a *poorly-graded soil* (or a *uniformly graded soil*). Conversely, a soil with a flat curve, such as that for Soil D, contains a wide range of particle sizes and is known as a *well-graded soil*. Figure 4.14 shows photographs of both types. Literature in the geological sciences often uses the terms *well-sorted* and *poorly-sorted*, which have the opposite definitions: Well-sorted = poorly-graded, while poorly-sorted = well-graded.

For some soils, there is a nearly flat zone in the particle size distribution curve, as shown for example by the particle size distribution curve for Soil E in Figure 4.13. These soils are described as *gap-graded* because they are missing particles in a certain size range. Gap-graded soils are sometimes considered a type of poorly-graded soil. Aggregates used to make concrete are typically gap-graded.

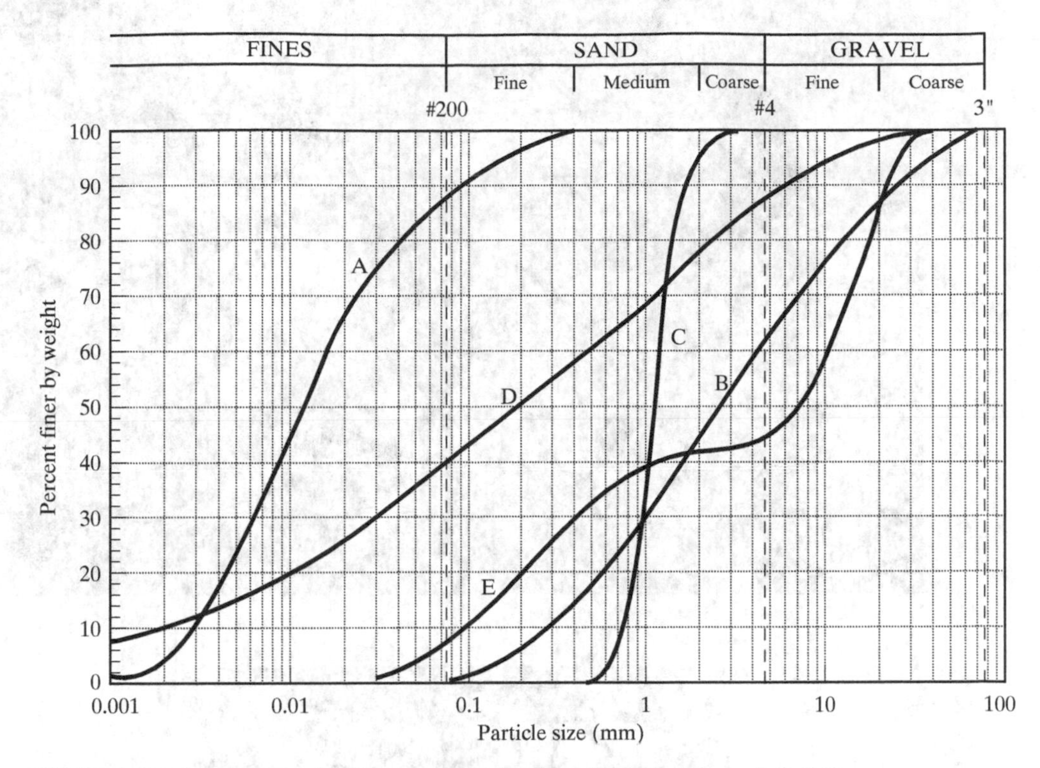

FIGURE 4.13 Particle size distribution curves for five soils (Soil A through Soil E).

We can determine the percentage of each type of soil particle by weight by comparing the percents passing the appropriate sieve sizes as listed in Table 4.6. For example, to determine the amount of sand in a soil, subtract the percent passing the #200 sieve from the percent passing the #4 sieve.

The particle sizes that correspond to certain percent-passing values for a given soil are known as the *D-sizes*. For example, D_{10} is the particle size that corresponds to 10% passing. In other words, 10% of the soil particles is finer than D_{10}. Two additional parameters, the *coefficient of uniformity*, C_u, and the *coefficient of curvature*, C_c, are based on the *D*-sizes:

$$C_u = \frac{D_{60}}{D_{10}} \tag{4.35}$$

$$C_c = \frac{(D_{30})^2}{D_{10}D_{60}} \tag{4.36}$$

Steep curves, which reflect poorly-graded soils, have low values of C_u, while flat curves (well-graded soils) have high values. Soils with smooth curves have C_c values between about 1 and 3, while irregular curves have higher or lower values. For example, most gap-graded soils have C_c values outside this range.

(a)

(b)

FIGURE 4.14 (a) A well-graded soil has a wide range of particles sizes, in this case ranging from fine sand to coarse gravel. (b) A poorly-graded soil has a narrow range of particle sizes. This particular soil is a poorly-graded gravel that is commercially produced in rock crushing plants. It is called *pea gravel*, even though the particles are slightly larger than most peas. Both photographs are full-scale. A soil technician who once worked for the first author apparently misunderstood the origin of the name pea gravel, because his notes always referred to it as *pee gravel!*

Example 4.6

Determine the following for Soil D in Figure 4.13:

- Percent gravel, sand, and fines
- C_u and C_c

Solution:

Percent finer data from particle size distribution curve:

#200—40%

#4—88%

3 in.—100%

Percentage of each type of soil particle:

$$\text{Gravel} = 3\text{ in.} - \#4 = 100\% - 88\% = \mathbf{12\%}$$
$$\text{Sand} = \#4 - \#200 = 88\% - 40\% = \mathbf{48\%}$$
$$\text{Fines} = \#200 = \mathbf{40\%}$$

D-sizes from particle size distribution curve:

$$D_{10} = 0.0019\text{ mm}$$
$$D_{30} = 0.030\text{ mm}$$
$$D_{60} = 0.49\text{ mm}$$

C_u and C_c:

$$C_u = \frac{D_{60}}{D_{10}} = \frac{0.49\text{ mm}}{0.0019\text{ mm}} = \mathbf{260}$$

$$C_c = \frac{D_{30}^2}{D_{10}D_{60}} = \frac{(0.030\text{ mm})^2}{(0.0019\text{ mm})(0.49\text{ mm})} = \mathbf{1.0}$$

Note: This is an exceptionally large value of C_u, which reflects the very flat particle size distribution curve. Most C_u values for real soils are less than 20.

Particle Shape

The shape of silt, sand, and gravel particles varies from very angular to well rounded, as shown in Figure 4.15 (Youd, 1973). A soil containing angular particles is most often found near the parent rock from which it is derived, while a soil containing rounded particles is most often found farther away as the particles have experienced more abrasion during the transportation process.

The particle shape has some effect on soil properties. For example, everything else being equal, a soil with angular particles has a greater shear strength than one with rounded ones because it is more difficult to make angular particles slide or roll past one another. This is one of the reasons aggregate base material used beneath highway

FIGURE 4.15 Classification of particle shape for silts, sands, and gravels.

pavements is often made of rocks that have been passed through a rock crusher to create a very angular gravel. Clay particles have an entirely different shape, and are discussed later in the next section.

Some nonclay particles are much flatter than any of the samples shown here. One example is mica, which is plate-shaped. Although mica never represents a large portion of the total weight, even a small amount can affect a soil's behavior. Sands that include mica are known as *micaceous sands*.

4.5 CLAY SOILS

Soils that consist of silt, sand, or gravel particles are primarily the result of physical and mild chemical weathering processes and retain much of the chemical structure of their parent rocks. However, this is not the case with clay soils because they have experienced extensive chemical weathering and have been changed into a new material quite different from the parent rocks. As a result, the engineering properties and behavior of clays also are quite different from those of coarser soils.

Formation and Structure of Clay Minerals

Several different chemical weathering processes form *clay minerals*, which are the materials from which clay particles are made. We will examine one of these processes as an example. This process changes orthoclase feldspar, a mineral found in granite and many other rocks, into a clay mineral called *kaolinite*. The process begins when water acquires carbon dioxide as it falls through the atmosphere and seeps through soil, thus forming a weak carbonic acid solution. This acid reacts with orthoclase feldspar according to the following chemical formula (Goodman, 1993):

$$4KAlSi_3O_8 + 2H_2CO_3 + 2H_2O \rightarrow 2K_2CO_3 + Al_4(OH)_8Si_4O_{10} + 8SiO_2$$

$$\text{orthoclase} + \text{carbonic acid} + \text{water} \rightarrow \text{potassium carbonate} + \text{kaolinite} + \text{silica}$$

Finally, the potassium carbonate and silica are carried off in solution by groundwater, ultimately to be deposited elsewhere, leaving kaolinite clay where orthoclase feldspar once existed.

These various chemical weathering processes form sheet-like chemical structures. There are two types of sheets: *tetrahedral* or *silica* sheets consist of silicon and oxygen atoms; *octahedral* or *alumina* sheets have aluminum atoms and hydroxyls (OH). Sometimes octahedral sheets have magnesium atoms instead of aluminum, thus forming *magnesia* sheets. These sheets then combine in various ways to form dozens of different clay minerals, each with its own chemistry and structure. The three most common ones are as follows:

Kaolinite: Kaolinite consists of alternating silica and alumina sheets, as shown in Figure 4.16(a). These sheets are held together with strong chemical bonds, so kaolinite is a very stable clay. Unlike most other clay minerals, kaolinite does not expand appreciably when wetted, so it is used to make pottery. It also is an important ingredient in paper, paint, and other products, including pharmaceuticals (i.e., kaopectate).

Montmorillonite: Belonging to the *smectite* group of clay minerals, montmorillonite has layers made of two silica sheets and one alumina sheet, as shown in Figure 4.16(b). The bonding between these layers is very weak, so large quantities of water can easily enter and separate them, thus causing the clay to swell. This property can be very troublesome or very useful, depending on the situation. Problems with soil expansion include extensive distortions in structures, highways, and other civil engineering projects. However, this expansive behavior and the low permeability of montmorillonite can be useful for sealing borings or providing groundwater barriers. *Bentonite*, an expansive clay consisting primarily of montmorillonite, is commercially mined and sold for such purposes.

Illite: Illite has layers similar to those in montmorillonite, but contains potassium ions between the layers, as shown in Figure 4.16(c). The chemical bonds in this structure are stronger than those in montmorillonite but weaker than those in kaolinite, so illite expands slightly when wetted. Glacial clays in the Great Lakes region consist primarily of illite.

Other common clay minerals include *vermiculite*, *chlorite*, and *attapulgite*.

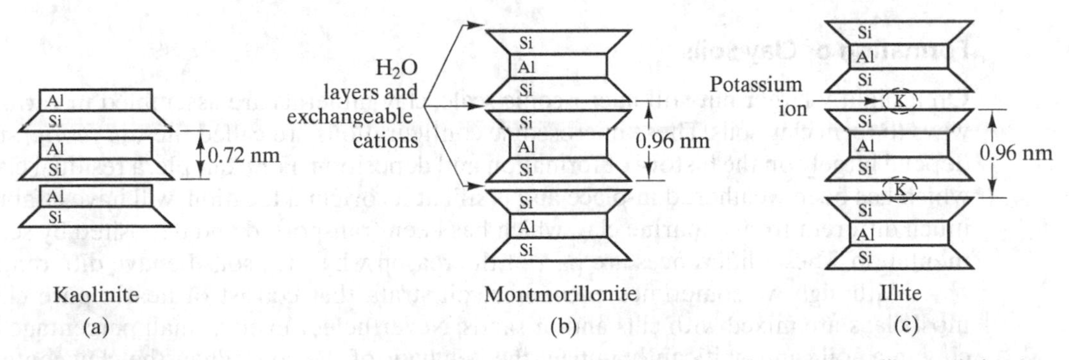

FIGURE 4.16 Structures of common clay minerals: (a) kaolinite consists of alternating silica and alumina sheets; (b) montmorillonite consists of alumina sheets sandwiched between silica sheets, with water and exchangeable cations possibly present between the silica sheets; (c) illite is similar to montmorillonite, but contains potassium ions between the silica sheets. (Adapted from Holtz and Kovacs, 1981.)

Individual clay particles are extremely small (less than 2 μm (2×10^{-6} m) in size). They cannot be seen with optical microscopes, and require an electron microscope for scientific study (see Figure 4.17). These images show how the shape of clay particles is platy or flaky and is substantially different from the equidimensional or bulky shapes of coarser particles.

Properties of Clays

Because of the small size and platy shape of clay particles, the surface area to mass ratio of clay particles is much greater than that of coarser particles. This ratio is known as the *specific surface*. For example, montmorillonite has a specific surface up to about 800 m^2/g, which means that 3.5 g of this clay has a surface area equal to that of a football field.

The extremely large specific surface of clays provides more contact area between particles, and thus more opportunity for various interparticle forces to develop. These surface forces dominate clay behavior. In the presence of water, naturally negatively charged clay particles *adsorb* water molecules and ions. The water molecules are said to be adsorbed on particle surfaces because they are strongly attracted to the particle surfaces. Because of the large surface area that provides more places for water molecules to attach to, clays can have a great affinity for water and can adsorb a great amount of water. Some clays can easily absorb several times their dry weight in water. Montmorillonite clays have the greatest specific surface, so it is no surprise that they have the greatest affinity for water, leading to their swelling behavior. The interactions among the adsorbed water, ions present, and the clay minerals are quite complex, but the net effect is that the engineering properties vary as the moisture content varies. For example, the shear strength of a given clay at a moisture content of 50% will be less than at a moisture content of 10%.

This behavior is quite different from the behavior of sands, because a sand has a specific surface that is much smaller than that of a clay and the sand particles are more inert. Other than changes in pressure within the pore water, which affect all soils and are discussed later in this book, variations in moisture content have very little effect on the behavior of sands.

Formation of Clay Soils

On a slightly larger but still microscopic scale, clay minerals are assembled in various ways to form clay soils. These microscopic configurations are called the *soil fabric*, and depend largely on the history of formation and deposition. For example, a residual clay, which has been weathered in-place and is still at its original location, will have a fabric much different from a marine clay, which has been transported and deposited by sedimentation. These differences are part of the reason why such soils behave differently.

Although we sometimes encounter soil strata that consist of nearly pure clay, most clays are mixed with silts and/or sands. Nevertheless, even a small percentage of clay in a soil can significantly impact the behavior of the soil. When the clay content exceeds about 50%, the sand and silt particles are essentially floating in the clay, and have very little effect on the engineering properties of the soil.

Kaolinite

Montmorillonite

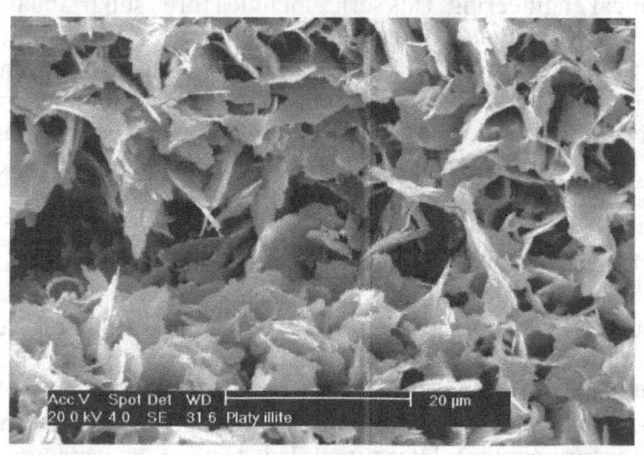

Illite

FIGURE 4.17 Electron microscope images of kaolinite, montmorillonite, and illite clays. Note the platy or flaky particle shapes, which are quite different from the equidimensional or bulky shapes of coarser particles. (Images courtesy of Clay Mineral Society.)

4.6 PLASTICITY AND THE ATTERBERG LIMITS

From the previous discussion, it is clear that silt and clay particles are two very different kinds of soil particles. Yet, the classification system described in Table 4.6 used the term *fines* to describe all particles that pass through a #200 sieve. It makes no attempt to distinguish between silts and clays. Some classification systems draw the line between them based on particle size as determined from a hydrometer test, typically at 0.001 to 0.005 mm. Although such systems can be useful, they also can be misleading because the biggest difference between silt and clay is not their particle sizes, but their physical and chemical structures, as discussed earlier.

It would be impractical to use electron microscopes or other sophisticated equipment to distinguish between clays and silts on a routine basis. Instead, we distinguish them by assessing a property called *plasticity*, which can be determined much more easily and inexpensively. In this context, the term *plasticity* describes the response of a soil to changes in moisture content. When the consistency of a soil changes from hard and rigid to soft and pliable by having water added to it, the soil is said to be exhibiting plasticity. Clays can be very plastic and silts only slightly plastic, whereas clean sands and gravels do not exhibit any plasticity at all. This assessment can be made using visual–manual procedures, and with experience, one can distinguish between clays and silts simply with the hands and a water bottle. More formal assessments of plasticity are performed using the Atterberg limits.

The Atterberg Limits

In 1911, the Swedish soil scientist Albert Atterberg (1846–1916) developed a series of tests to evaluate the relationship between moisture content and soil consistency (Atterberg, 1911; Blackall, 1952). Then, in the 1930s, Karl Terzaghi and Arthur Casagrande adapted these tests for civil engineering purposes, and they soon became a routine part of geotechnical engineering. This series includes three separate laboratory tests: the *liquid limit test*, *plastic limit test*, and the *shrinkage limit test*. Together they are known as the *Atterberg limit tests* (ASTM D4318 for liquid and plastic limits and D427 for shrinkage limit). The liquid limit and plastic limit tests are routinely performed in many soil mechanics laboratories and will be described below. However, the shrinkage limit test is less useful, and is rarely performed by geotechnical engineers, and therefore, will not be described here.

The Liquid Limit Test There are two major versions of the liquid limit test. One version (ASTM D4318), which is popular in the United States, uses the *liquid limit device*, the so-called Casagrande cup, as shown in Figure 4.18. In this test, a soil sample is placed in the cup, and a groove is cut in the soil using a standard tool. The cup is then repeatedly dropped onto a hard rubber pad, and the number of drops required to close the bottom of the groove along a longitudinal distance of 1/2 in. is recorded. The soil is then removed and its moisture content determined. This test is then repeated at various moisture contents, producing a semilog plot of the number of drops in log scale versus the moisture content. By definition, the soil is said to be at its liquid limit when exactly 25 drops are required to close the bottom of the groove along a distance of

FIGURE 4.18 This device is used to perform the liquid limit test. The soil pat is in place and has been grooved with the grooving tool shown on the right. The next step is to begin turning the crank, which will repeatedly drop the cup.

1/2 in. The corresponding moisture content is determined from the semilog plot of the test data. The liquid limit, *LL* or w_L, is this moisture content rounded to the nearest percent but expressed without the percentage sign. For example, if the corresponding moisture content is 45%, the liquid limit is 45.

The other version of the liquid limit test that is popular in Europe and Asia uses a *fall cone* to penetrate a soil sample and correlates the penetration to the liquid limit. There are two major variations of the *fall cone method*, one adopted in the British Standards and one in the Swedish Standards; only the British Standard (BS 1377) fall cone method will be described here. The British Standard (BS 1377) fall cone method requires a standard cone that weighs 0.78 N (80 g) and has an apex angle of 30°, as shown in Figure 4.19. A soil is at its liquid limit if the fall cone penetrates the soil by 20 mm in 5 s when allowed to fall freely from a position where the tip of the cone is just touching the top surface of the soil. As in the ASTM D4318 test, the moisture content corresponding to the liquid limit may be obtained by performing several fall cone tests on soils at different moisture contents to obtain a semilog plot of the cone penetration in log scale versus the moisture content. The moisture content corresponding to a 20-mm cone penetration will give the liquid limit, which again is this moisture content rounded to the nearest percent but expressed without the percentage sign.

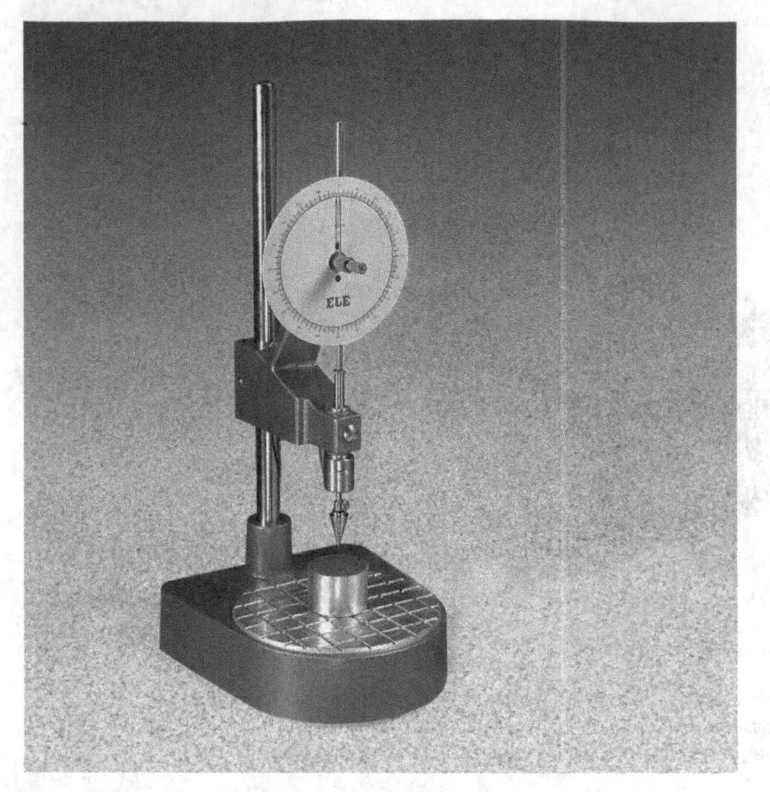

FIGURE 4.19 The British Standard (BS 1377) fall cone apparatus. (Courtesy of ELE International.)

The Plastic Limit Test The plastic limit test procedure involves carefully rolling the soil sample into threads, as shown in Figure 4.20. As this rolling process continues, the thread becomes drier and thinner and eventually breaks. If the soil is too dry, it breaks at a large diameter. If it is too wet, it breaks at a much smaller diameter. By definition, the soil is at its plastic limit when it breaks at a diameter of 1/8 in. (3 mm). If this happens, the plastic limit, PL or w_P, is the corresponding moisture content rounded to the nearest percent but expressed without the percentage sign.

Consistency and Plasticity Assessments Based on Atterberg Limits

The Atterberg limits help engineers assess the plasticity of a fine-grained soil and its consistency at various moisture contents. Figure 4.21 shows the changes in these characteristics with changes in moisture content.

When a fine-grained soil is at a moisture content higher than that corresponding to its liquid limit, it is said to be in a *liquid state*. In this state, the soil can flow and deform easily.

When a fine-grained soil is at a moisture content between moisture contents corresponding to its liquid and plastic limits, it is said to be in a *plastic state*. In this state, the

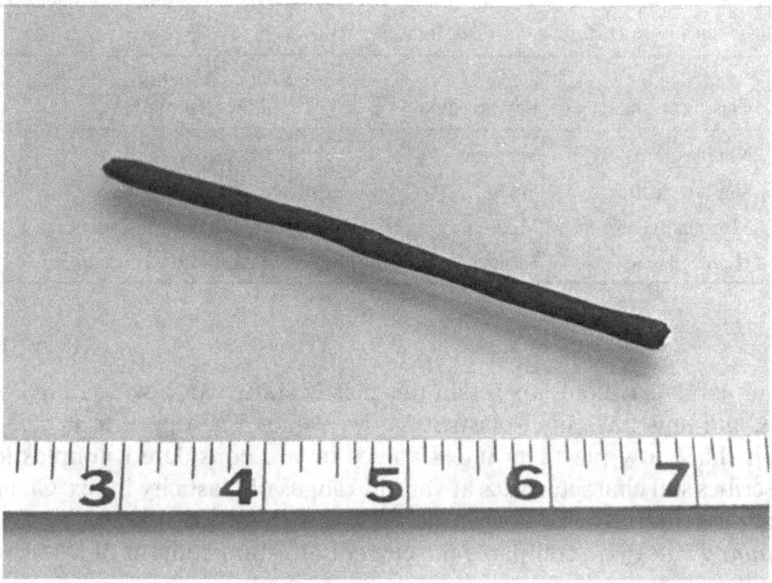

FIGURE 4.20 A plastic limit test is being performed on this soil sample. It has been rolled until it broke, which occurred at a diameter of 1/8 in. (see scale). Thus, by definition, this soil is at the plastic limit. The value of w_L will be determined by performing a moisture content test. The scale is in inches.

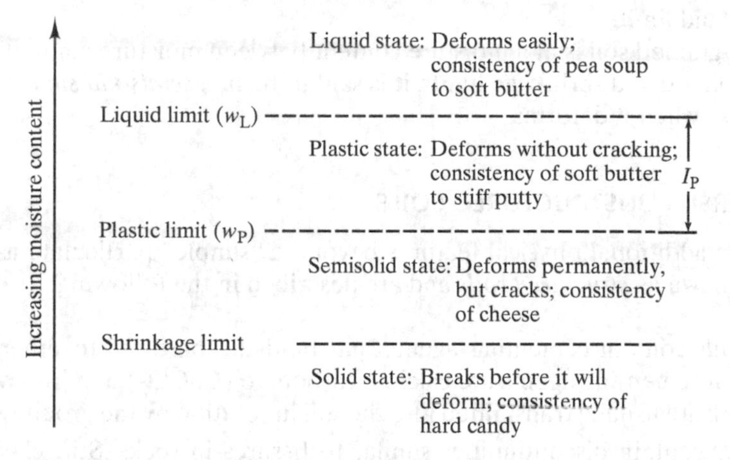

FIGURE 4.21 Consistency of fine-grained soils at different moisture contents (Sowers, 1979).

soil can be easily molded without cracking or breaking. The children's toy *play-dough* has a similar consistency. This property relates to the amount and type of clay in the soil. The *plasticity index*, PI or I_P, is a measure of the range of moisture contents that encompasses the plastic state:

$$I_P = w_L - w_P \tag{4.37}$$

TABLE 4.8 Characteristics of Soils With Different Plasticity Indices[a]

Plasticity Index, I_P	Classification	Dry Strength	Visual–Manual Identification of Dry Sample	
0–3	Nonplastic	Very low	Falls apart easily	Increasing clay content ⇓
3–15	Slightly plastic	Slight	Easily crushed with fingers	
15–30	Medium plastic	Medium	Difficult to crush with fingers	
> 30	Highly plastic	High	Impossible to crush with fingers	

[a]Sowers, 1979.

Soils with large clay contents retain this plastic state over a wide range of moisture contents, and thus have high plasticity index values. The opposite is true of silty soils: silty soils have low plasticity index values or are considered nonplastic (NP). Table 4.8 describes soil characteristics at various ranges of plasticity index. Clean sands and gravels are considered to be nonplastic.

The *liquidity index*, I_L, compares the current moisture content of a soil, w, to its Atterberg limits:

$$I_L = \frac{w - w_P}{I_P} \tag{4.38}$$

Thus, a liquidity index of 0 means the soil is currently at its plastic limit, and 1 means it is at its liquid limit.

When a fine-grained soil is at a moisture content between moisture contents corresponding to its plastic and shrinkage limits, it is said to be in a *semisolid state*. In this state, the soil cracks when it deforms.

4.7 STRUCTURED VERSUS UNSTRUCTURED SOILS

Many soils contain additional physical features beyond a "simple" particulate assemblage. These are known as *structured soils* and are described in the following:

- *Cemented soils* contain cementing agents that bind the particles together. The most common cementing agents are calcium carbonate ($CaCO_3$) and iron oxides (Fe_2O_3). Both are usually transmitted into the soil in solution by the groundwater.
- *Fissured soils* contain discontinuities similar to fissures in rocks. Stiff clays are especially likely to contain fissures.
- *Sensitive clays* are those with a flocculated structure of clay particles that resemble a house of cards. These soils are very sensitive to disturbance, which destroys this delicate structure. The landslide in Figure 2.17 occurred in a sensitive clay.

Unstructured soils are those that do not contain such special features. Most geotechnical analyses are based on unstructured soils, and thus often need to be modified when working with structured soils.

4.8 ORGANIC SOILS

Technically, any material that contains carbon is "organic." However, engineers and geologists use a more narrow definition when we apply the term to soils. An *organic soil* is one that contains a significant amount of organic material recently derived from plants or animals. It needs to be fresh enough to still be in the process of decomposition, and thus retains a distinctive texture, color, and odor.

Some soils contain carbon, but are not recently derived from plants or animals and thus are not considered organic in this context. For example, some sands contain calcium carbonate (calcite), which arrived as a chemical precipitate.

The identification of organic soils is very important, because they are much weaker and more compressible than inorganic soils, and thus do not provide suitable support for most engineering projects. If such soils are present, we usually try to avoid them, excavate them, or drive piles through them to reach more suitable deposits.

The term *peat* refers to a highly organic soil derived primarily from plant materials. It has a dark brown to black color, a spongy consistency, and an organic odor. Usually plant fibers are visible, but in the advanced stages of decomposition, they may not be evident. Peat often occurs in *bogs*, which are pits filled with organic material. Bogs are typically covered with a live growth of moss. Thoroughly decomposed peat is sometimes called *muck*, although this term also is used to describe waste soil or rock, such as the cuttings from tunnels. Peat can be subjected to the same processes that form sedimentary and metamorphic rocks, thus forming *bituminous coal* and *anthracite coal*, respectively.

Swamps are larger than bogs and may contain a wider variety of materials. They are typically fed by slow streams or lakes. The Everglades in Florida is a noteworthy example.

Although many organic soil deposits create obvious topographic features such as swamps and bogs, others are buried underground, having been covered with inorganic alluvial soils. These often are difficult to detect and can be the source of large differential settlements. For example, such buried deposits are present near the coast in Orange County, California, and have been the source of unexpected settlement problems.

Organic deposits also may be mixed with inorganic soils, especially silts and clays, producing soils that are not as bad as peat, but worse than inorganic deposits.

SUMMARY

Major Points

1. Soil is a particulate material, so its engineering properties depend primarily on the interaction between these particles. This is especially true of gravels, sands, and silts. Clays also are particulates, but their behavior is much more complex because of the interaction between the particles and the pore water.
2. Soil can include all three phases of matter simultaneously, and their relative proportions are important. Geotechnical engineers have developed a series of weight–volume parameters to describe these proportions.

3. The distribution of particle sizes in a soil also is important, and this distribution can be determined by performing a sieve analysis and/or a hydrometer analysis. The results are presented as a particle size distribution curve.
4. The solid particles also have various shapes, which impact their behavior.
5. Clays are formed by chemical weathering processes. The individual particles are much smaller than sands or silts, and their engineering behavior is much more dependent on the moisture content.
6. Clays and silts are often distinguished from each other by assessing their plasticity, which reflects their affinity for water. The Atterberg limit tests, especially the plastic limit and liquid limit tests, help us do this.
7. Structured soils are those with special features, such as cementation, fissures, or flocculated structures. They behave differently from unstructured soils, which do not contain these features.
8. Organic soils are those with a significant quantity of organic matter. Their engineering properties are much worse than those of inorganic soils.

Vocabulary

Atterberg limits
bog
boulder
buoyant unit weight
cemented soil
clay
cobble
consistency
degree of saturation
density
dry density
dry unit weight
fall cone method
fines
fissured soil
gap-graded soil
grain size distribution curve
gravel
hydrometer analysis
illite
kaolinite
liquid limit
liquidity index
moisture content
montmorillonite
muck
organic soil
particle shape
particle size distribution curve
particulate material
peat
phase diagram
plastic limit
plasticity
plasticity index
poorly-graded soil
porosity
pycnometer
relative density
sand
sensitive clay
shrinkage limit
sieve analysis
silt
specific gravity of solids
structured soil
submerged unit weight
unit weight
unstructured soil
void ratio
water content
weight–volume parameter
well-graded soil

QUESTIONS AND PRACTICE PROBLEMS

Section 4.3 Weight–Volume Relationships

4.1 A cube of soil measures 1.5 ft on each side and weighs 375 lb. It's moisture content is 26.0% and the specific gravity of solids is 2.72. Compute the void ratio, porosity, degree of saturation, unit weight and dry unit weight of this soil.

4.2 An undisturbed block sample of clay weighs 101.4 kg and has dimensions of 0.4 m × 0.4 m. × 0.4 m. Its moisture content is 25.0%. Assuming a reasonable value of the specific gravity of solids, compute the unit weight, dry unit weight, void ratio, porosity and degree of saturation of the clay.

4.3 A sample of soil is compacted into a 9.44×10^{-4} m^3 laboratory mold. The mass of the compacted soil is 1.91 kg, and its moisture content is 14.5%. Using a specific gravity of solids of 2.66, compute the degree of saturation, density (kg/m^3), unit weight (kN/m^3) and dry unit weight of this compacted soil.

4.4 A sample of soil was compacted into a 1/30 ft^3 laboratory mold. The weight of the compacted soil was 4.1 lb and its moisture content 13.1%. Using a specific gravity of solids of 2.70, compute the unit weight, dry unit weight and degree of saturation of this compacted soil. This compacted soil sample was then submerged in water. After 2 weeks, it was found that the sample had swelled and that its total volume had increased by 5%. Compute the new unit weight and moisture content of the soil sample after 2 weeks of submersion in water.

4.5 The moisture content of a saturate soil is 36.0%. Assuming the specific gravity of soils is 2.68, compute the void ratio, porosity and unit weight (lb/ft^3 or kN/m^3) of this soil.

4.6 A soil sample obtained from below the groundwater table has a moisture content of 23.5% and a specific gravity of solids of 2.72. Compute its unit weight, dry unit weight, buoyant unit weight, void ratio, porosity and degree of saturation.

4.7 A sample of clay was obtained from a point below the groundwater table. A moisture content test on this sample produced the following data:

Mass of can = 10.88 g
Mass of can + moist soil = 116.02 g
Mass of can + dry soil = 85.34 g

(a) Compute the moisture content.
(b) Assume a reasonable value for G_s, and then compute the void ratio, unit weight, dry unit weight and buoyant unit weight.

4.8 An undisturbed cylindrical soil sample is 60 mm in diameter and 152 mm long. It has a mass of 816 g. After finding the mass of the entire sample, a small portion was removed and a moisture content test was performed on it. The results of this test on the sub-sample were:

Mass of can = 22.01 g
Mass of can + moist soil = 124.97 g
Mass of can + dry soil = 112.72 g
Using $G_s = 2.70$, compute w, γ, γ_d, e, and S.

4.9 An undisturbed cylindrical soil sample is 2.4 in. in diameter and 6 in. long. It has a weight of 1.95 lb. After finding the weight of the entire sample, a small portion was removed and a moisture content test was performed on it. The results of this test on the sub-sample were:

Mass of can = 20.50 g
Mass of can + moist soil = 110.46 g
Mass of can + dry soil = 96.81 g
Using $G_s = 2.66$, compute w, γ, γ_d, e, and S.

4.10 A strata of clean, light-colored quartz sand located below the groundwater table has a moisture content of 25.6%. The minimum and maximum void ratios of this soil are 0.380 and 1.109, respectively. Select an appropriate value of G_s for this soil, compute its relative density, and determine its consistency using Table 4.4.

4.11 A contractor needs 214 yd^3 of aggregate base material for a highway construction project. It will be compacted to a dry unit weight of 130 lb/ft^3. This material is available in a stockpile at a local material supply yard, but is sold by the ton, not by the cubic yard. The moisture content of the stockpile is 7.0%.

- **(a)** How many tons of aggregate base material should the contractor purchase to have exactly the correct volume of compacted material?
- **(b)** The contractor purchased the material per the computation in part (a), and it exactly met the needs at the project site. An intense rainstorm occurred the following week, which delayed further construction and raised the moisture content of the stockpile to 19.0%. Now, the contractor needs to prepare another identical section of aggregate base and is ordering the same number of tons as before. How many cubic yards of compacted aggregate base will be produced from this second shipment? How will it compare with the first shipment? Explain.

4.12 A cone penetration test has been conducted, and has measured a cone resistance of 85 kg/cm^2 at a depth of 10 m. The vertical effective stress at this depth is 150 kPa, and the overconsolidation ratio is 2. The soils at this depth are quartz sands. Compute the relative density, and classify the soil using Table 4.4.

4.13 A cone penetration test has been conducted, and has measured a cone resistance of 110 ton/ft^2 at a depth of 20 ft. The vertical effective stress at this depth is 2500 lb/ft^2, and the overconsolidation ratio is 2.5. The soils at this depth are clayey sands. Estimate the relative density, and classify the soil using Table 4.4.

4.14 A standard penetration test has been conducted at a depth of 15 ft in an 8-in. diameter exploratory boring using a USA-style safety hammer and a standard sampler. This test produced an uncorrected N-value of 12. The soil inside the sampler was a fine-to-medium sand with $D_{50} = 0.6$ mm. The vertical effective stress at this depth is 1100 lb/ft^2. Adjust the N-value as described in Chapter 3, and then compute the relative density and classify the soil using Table 4.4.

Section 4.4 Particle Size and Shape

4.15 Determine the percent gravel, percent sand, and percent fines for this soil in Figure 4.13:

- **(a)** Soil A.
- **(b)** Soil B.
- **(c)** Soil C.
- **(d)** Soil D.
- **(e)** Soil E.

4.16 Determine C_u and C_c for this soil in Figure 4.13:

- **(a)** Soil A.
- **(b)** Soil B.
- **(c)** Soil C.
- **(d)** Soil D.
- **(e)** Soil E.

4.17 Plot the particle size distribution curve for each of the three soils the data for which are given below. All three curves should be on the same semilogarithmic diagram.

	% Passing by Weight[a]		
Sieve Number	**Lagoon Clay Beufort, SC**	**Beach Sand Daytona Beach, FL**	**Weathered Tuff Central America**
3/4 in.	100	100	100
½ in.	100	100	98
#4	100	100	95
#10	100	100	93
#20	100	100	88
#40	100	98	82
#60	100	90	75
#100	95	10	72
#200	80	2	68
Particle diameter from hydrometer analysis (mm)			
0.045	61		66
0.010	42		33
0.005	37		21
0.001	27		10

[a]Data from Sowers (1979).

4.18 Determine C_u and C_c for each of the three soils in Problem 4.17. Which of these soils is most well-graded? Why?

4.19 The American Association of State Highway and Transportation Officials (AASHTO) has defined grading requirements for soils to be used as base courses under pavements (AASHTO Designation M 147). The grading requirements for Class C base material are as follows:

Sieve Designation	Percent Passing by Weight
1 in.	100
3/8 in.	50–85
# 4	35–65
# 10	25–50
# 40	15–30
# 200	5–15

Imported soils available from nearby borrow sites are being considered, and they have particle size distributions as described by the curves for Soils A, B, C, D and E in Figure 4.13. Detemine if this soil in Figure 4.13 satisfies the AASHTO particle size requirements for Class C base material:

(a) Soil A.
(b) Soil B.
(c) Soil C.

(d) Soil D.
(e) Soil E.

Section 4.5 Clay Soils

4.20 Borings for observation wells, such as the one shown in Figure 3.21, are normally sealed with an impervious cap near the ground surface. This cap prevents significant quantities of surface water from seeping into the well. For convenience, manufacturers supply a pelletized clay that has been dried and formed into 10-mm diameter balls. The driller then pours these balls into the boring and adds water. As the clay absorbs the water, it expands and seals the boring. What type of clay would be most appropriate for this purpose? Why? Would other clays produce less satisfactory results? Why?

4.21 Compute the specific surface (expressed in m^2/g) for a typical fine sand. State any assumptions, and compare your computed value with that quoted for montmorillonite clay in Section 4.5. Discuss the significance of these two numbers.

Section 4.6 Plasticity and the Atterberg Limits

4.22 Describe how the state and consistency of a very dry clay change as water is added to it, including its state and consistency at the liquid limit and the plastic limit.

4.23 A liquid limit test has been performed on a soil using the Casagrande cup. In this test, exactly 25 drops of the cup were required to close the bottom of the standard groove along a longitudinal distance of 1/2 in. A moisture content test was then performed on the tested soil and yielded the following results:

Mass of soil + can before placing in oven	52.20 g
Mass of soil + can after removal from oven	41.52 g
Mass of can	22.40 g

On the basis of this test only, compute the liquid limit of the soil.

4.24 A plastic limit test has been performed on a soil. Threads of the soil that broke at a diameter of 1/8 in. were collected for a moisture content test, which yielded the following results:

Mass of soil + can before placing in oven	41.30 g
Mass of soil + can after removal from oven	37.40 g
Mass of can	22.20 g

Compute the plastic limit of the soil.

4.25 A soil has a liquid limit of 61 and a plastic limit of 30. A moisture content test performed on an undisturbed sample of this soil yielded the following results:

Mass of soil + can before placing in oven	96.20 g
Mass of soil + can after removal from oven	71.90 g
Mass of can	20.80 g

Compute the following:

(a) The plasticity index.
(b) The moisture content.
(c) The liquidity index.

Present a qualitative description of this soil at its in situ moisture content.

4.26 A soil has $w_P = 30$ and $w_L = 80$. Compute its plasticity index, and then describe the probable clay content (i.e., small, moderate, or high).

4.27 A soil has $w_P = 22$ and $w_L = 49$. What moisture content corresponds to a liquidity index of 0.5?

Comprehensive

4.28 A sand with G_s = 2.66 and e = 0.60 is completely dry. It then becomes wetted by a rising groundwater table. Compute the unit weight (lb/ft^3 or kN/m^3) under the following conditions:

(a) When the sand is completely dry.
(b) When the sand is 40% saturated (S = 40%).
(c) When the sand is completely saturated.

4.29 A soil initially has a degree of saturation of 95% and a unit weight of 129 lb/ft^3. It is then placed in an oven and dried. After removal from the oven, its unit weight is 109 lb/ft^3. Compute the void ratio, porosity, initial moisture content, and specific gravity of the soil, assuming its volume did not change during the drying process.

4.30 A 1.20-m thick strata of sand has a void ratio of 1.81. A contractor passes a vibratory roller over this strata, which densifies it and reduces its void ratio to 1.23. Compute its new thickness.

4.31 A 412-g sample of silty sand with a moisture content of 11.2% has been placed on a #200 sieve. The sample was then "washed" on the sieve, forcing the minus #200 particles to pass through. The soil that remained on the sieve was then oven dried and found to have a mass of 195 g. By visual inspection, it is obvious that all of this soil is smaller than the #4 sieve. Compute the percent sand in the original sample.

4.32 A standard penetration test has been performed on a soil, producing $N_{1,60}$ = 19. A sieve analysis was then performed on the sample obtained from the SPT sampler, producing curve C in Figure 4.13. Assuming this soil is about 150 years old and has OCR = 1.8, compute its relative density and determine its consistency.

4.33 Develop a formula for relative density as a function of γ_d, $\gamma_{d\text{-hi}}$, and $\gamma_{d\text{-lo}}$, where γ_d is the dry unit weight in the field, $\gamma_{d\text{-hi}}$ is the dry unit weight that corresponds to e_{min} and $\gamma_{d\text{-lo}}$ is the dry unit weight that corresponds to e_{max}.

4.34 All masses for the specific gravity test described in Example 4.5 were determined using a balance with a precision of ±0.01 g. The volume of the pycnometer is accurate to within ±0.5%, and the moisture content measurement is accurate to within ±2.0% (i.e., the real w could be as low as 11.0% or as high as 11.4%). Assuming all measurement errors are random, determine the precision of the computed G_s value.

Hint: All errors are random, so the worst case would be if none of the errors were compensating (i.e., each measurement had the maximum possible error, and each contributed to making the computed G_s farther from its true value). Therefore, compute the highest possible value of G_s that is consistent with the stated uncertainties, then compare it with the G_s obtained in Example 4.5.

4.35 A 10,000 ft^3 mass of saturated clay had a void ratio of 0.962 and a specific gravity of solids of 2.71. A fill was then placed over this clay, causing it to compress. This compression is called consolidation, a topic we will discuss in Chapters 10 and 11. During this process, some of the water was squeezed out of the voids. However, the volume of the solids remained unchanged. After the consolidation was complete, the void ratio had become 0.758.

(a) Compute the initial and final moisture content of the clay.
(b) Compute the new volume of the clay.
(c) Compute the volume of water squeezed out of the clay.

4.36 What are the three most common clay minerals? Which one usually causes the most problems for geotechnical engineers? Why?

CHAPTER 5

Soil Classification

Chinese legends record a classification of soils according to color and structure which was made by the engineer Yu during the reign of Emperor Yao, about 4000 years ago.

This is the earliest known soil classification system (Thorp, 1936)

Thus far we have studied several ways of categorizing soil, such as by its geologic origin, mineralogy, particle size, and plasticity index. Each of these methods is useful in the proper context, but more comprehensive systems also are needed to better classify soils for engineering purposes. These systems need to focus on the characteristics that affect their engineering behavior, and must be standardized so that everyone "speaks the same language." This way, engineers can communicate using terms that clearly describe the soil, while still being concise.

Many such soil classification systems have been developed, usually based on the particle size distribution and the Atterberg limits. They often are supplemented by nonstandardized classifications of other properties such as consistency and cementation.

5.1 USDA SOIL CLASSIFICATION SYSTEM

The U.S. Department of Agriculture (USDA) soil classification system (Soil Survey Staff, 1975) is used in soil survey reports and other agricultural documents. Although this system differs from those used by geotechnical engineers, we must understand it to interpret their reports, as discussed in Section 3.2.

The first step in using this system is to determine the percentage by dry weight of each constituent, as follows:

- Coarse fragments: >2.0 mm
- Sand: 0.05–2.0 mm

- Silt: 0.002–0.05 mm
- Clay: <0.002 mm

Note that these divisions do not correspond to those defined by ASTM (Table 4.6), and that clays and silts are distinguished by particle size, not Atterberg limits. Since this system is based entirely on particle size, it is a type of *textural classification system*.

Next, find the total weight of sand + silt + clay (do not include coarse fragments in the total) and convert each weight to a percentage of this total. Then, using the triangle in Figure 5.1, find the soil classification.

If coarse fragments constitute more than 15–20% of the entire sample, then add one of the following terms to the classification obtained from the triangle:

Channery: Fragments of thin, flat sandstone, limestone, or schist up to 6 in. along the longer axis.

Cherty: Angular fragments that are less than 3 in. in diameter, at least 75% of which are chert (a sedimentary rock that occurs as nodules in limestone and shale).

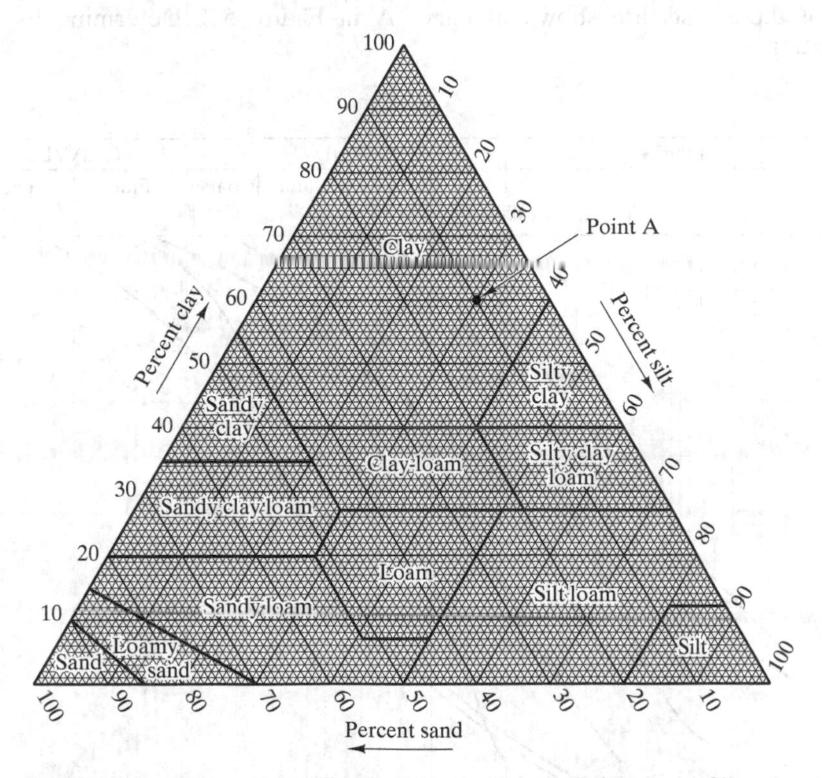

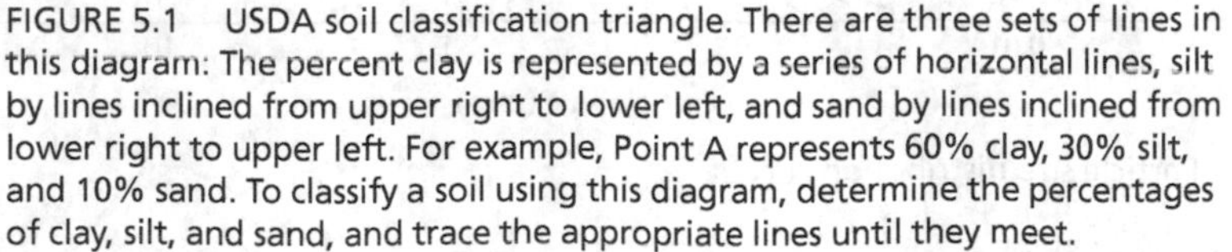

FIGURE 5.1 USDA soil classification triangle. There are three sets of lines in this diagram: The percent clay is represented by a series of horizontal lines, silt by lines inclined from upper right to lower left, and sand by lines inclined from lower right to upper left. For example, Point A represents 60% clay, 30% silt, and 10% sand. To classify a soil using this diagram, determine the percentages of clay, silt, and sand, and trace the appropriate lines until they meet.

Cobbly: Rounded or partially rounded fragments of rock ranging from 3 to 10 in. in diameter.

Flaggy: Relatively thin fragments 6–15 in. long of sandstone, limestone, slate, shale, or schist.

Gravelly: Rounded or angular fragments, not prominently flattened, up to 3 in. in diameter, with less than 75% chert.

Shaly: Flattened fragments of shale less than 6 in. along the longer axis.

Slaty: Fragments of slate less than 6 in. along the longer axis.

Stony: Rock fragments larger than 10 in. in diameter if rounded, or longer than 15 in. along the longer axis if flat.

Example 5.1

Sieve and hydrometer analyses have been performed on a soil sample, and the results of these tests are shown as curve A in Figure 5.2. Determine its USDA classification.

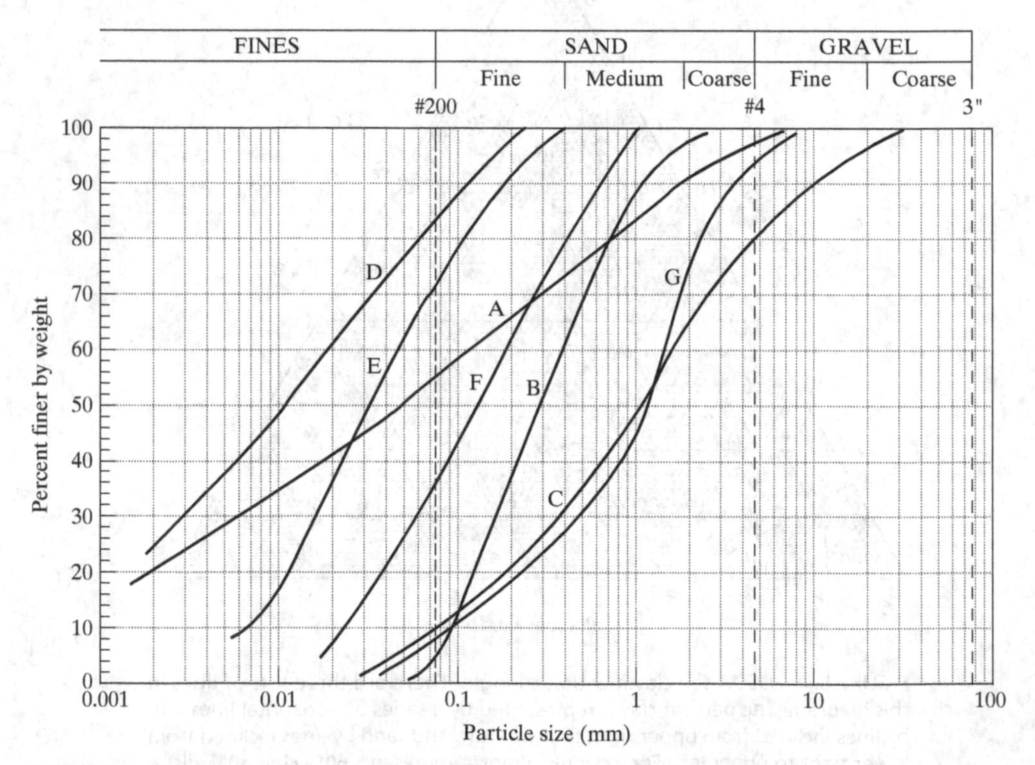

FIGURE 5.2 Particle size distribution curves.

Solution:

From the particle size distribution curve:

Particle size (mm)	Finer (%)
2.0	92
0.05	50
0.002	20

Coarse fragments (%) = 100 − 92 = 8
Sand (%) = 92 − 50 = 42
Silt (%) = 50 − 20 = 30
Clay (%) = 20

Adjust percentages as a total of sand + silt + clay

Sand = 42(100/92) = 46%
Silt = 30(100/92) = 33%
Clay = 20(100/92) = 22%

Using Figure 5.1, this soil plots as a **loam**.

5.2 AASHTO SOIL CLASSIFICATION SYSTEM

Terzaghi and Hogentogler developed one of the first engineering soil classification systems in 1928. It was intended specifically for use in highway construction, and still survives as the American Association of State Highway and Transportation Officials (AASHTO) system (AASHTO, 1993). It rates soils according to their suitability for support of roadway pavements and continues to be widely used on such projects.

The AASHTO system uses both particle size distribution and Atterberg limits data to assign a *group classification* and a *group index* to the soil. The group classification ranges from A-1 (best soils) to A-8 (worst soils). Group index values near 0 indicate good soils, whereas values of 20 or more indicate very poor soils. However, it is important to remember that a soil that is "good" for use as a highway subgrade might be "very poor" for some other purpose.

This system considers only that portion of the soil that passes through a 3-in. sieve. If any plus 3-in. material is present, its percentage by weight should be recorded and noted with the classification.

The AASTHO classification system shown in Table 5.1 can be used to determine the group classification. To use the Table, begin on the left side with A-1-a soils and check each of the criteria. If all have been met, then this is the group classification. If any criterion is not met, step to the right and repeat the process, continuing until reaching the first column for

TABLE 5.1 AASHTO Soil Classification System

General Classification	Granular Materials (35% or less passing No. 200 sieve)[c]							Silt-Clay Materials (more than 35% passing No. 200 sieve)				Highly Organic
Group Classification	A-1			A-2							A-7	
	A-1-a	A-1-b	A-3	A-2-4	A-2-5	A-2-6	A-2-7	A-4	A-5	A-6	A-7-5 A-7-6	A-8
Sieve analysis percent passing:												
#10	≤50											
#40	≤30	≤50	≥51									
#200	≤15	≤25	≤10	≤35	≤35	≤35	≤35	≥36	≥36	≥36	≥36	
Characteristics of fraction passing #40:											b	
Liquid limit				≤40	≥41	≤40	≥41	≤40	≥41	≤40	≥41	
Plasticity index	≤6		NP[a]	≤10	≤10	≥11	≥11	≤10	≤10	≥11	≥11	
Usual types of significant constituent materials	Stone fragments; gravel and sand		Fine sand	Silty or clayey gravel and sand				Silty soils		Clayey soils		Peat or muck
General rating as subgrade	Excellent to good							Fair to poor				Unsuitable

[a]NP indicates the soil is non-plastic (i.e., it has no clay).
[b]The plasticity index of A-7-5 soils is ≤ liquid limit −30. For A-7-6 soils, it is > than the liquid limit −30.
[c]The placement of A-3 before A-2 is necessary for the "left-to-right elimination process" and does not indicate superiority of A-3 over A-2.

which all the criteria have been satisfied. Do not begin at the middle of the chart. Alternatively you may use the flow chart in Figure 5.3 to determine the group classification.

In addition to determining the group classification, you must also compute the group index using Equation 5.1:

$$\text{Group index} = (F - 35)[0.2 + 0.005(w_L - 40)] + 0.01(F - 15)(I_P - 10) \quad (5.1)$$

where:

F = fines content (percent passing #200 sieve), expressed as a whole number without the percent sign

w_L = liquid limit

I_p = plasticity index

When evaluating the group index for A-2-6 or A-2-7 soils, use only the second term in Equation 5.1. For all soils, express the group index as a whole number. Computed group index values of less than zero should be reported as zero.

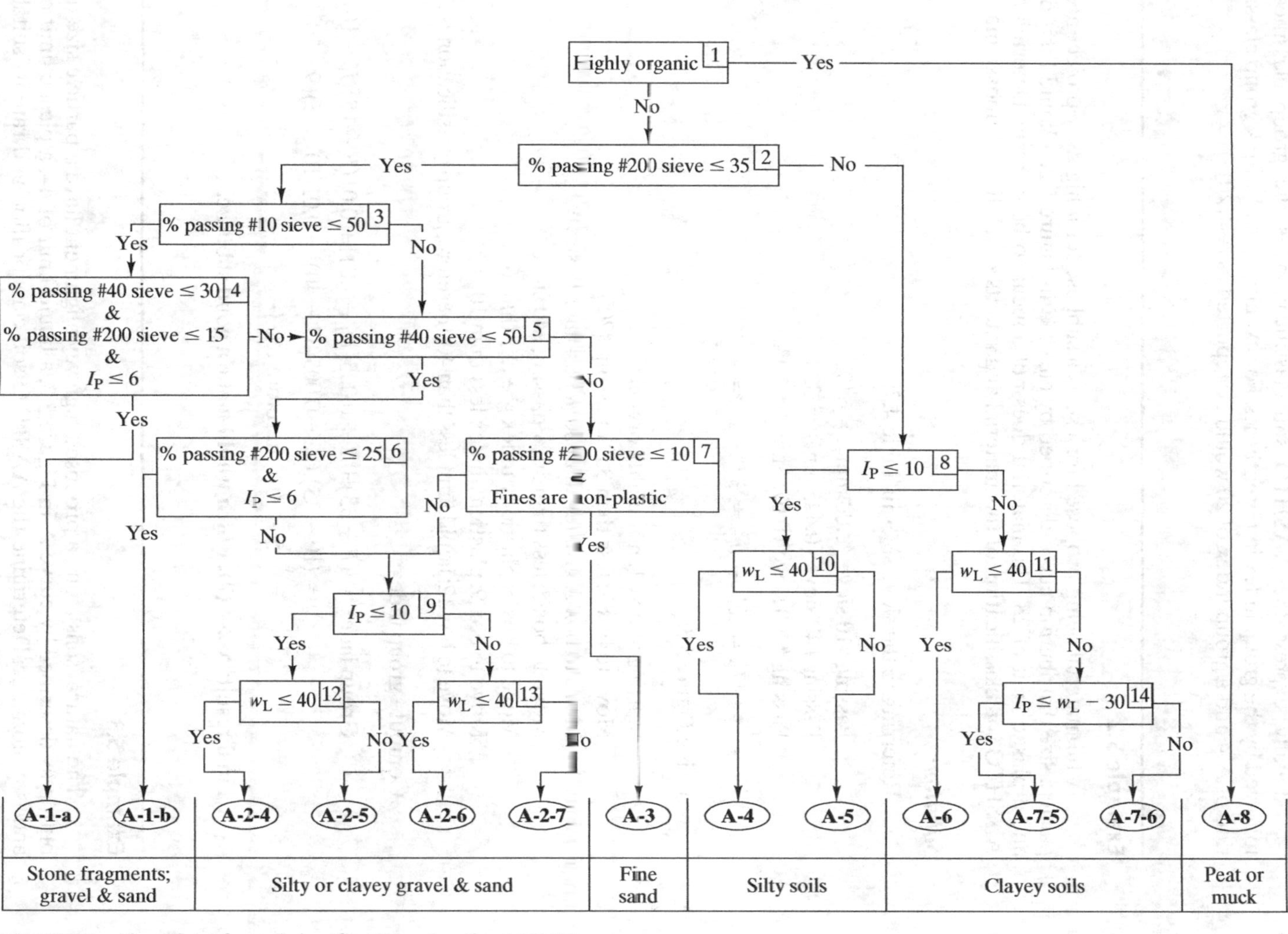

FIGURE 5.3 Flow chart for soil classification using the AASHTO system.

Finally, express the AASHTO soil classification as the group classification followed by the group index in parentheses. For example, a soil with a group classification of A-4 and a group index of 20 would be reported as A-4(20).

Example 5.2

A fill material being proposed for use as a subbase for a highway project has the particle size distribution curve described by curve C in Figure 5.2, a liquid limit of 37, and a plastic limit of 28. The material does not appear to be organic. Determine the AASHTO soil classification of this material and rate its suitability as subbase material.

Solution:

Referring to the sieve sizes in Table 4.7:

Passing #10 sieve (2.00 mm) = 64%
Passing #40 sieve (0.425 mm) = 33%
Passing #200 sieve (0.075 mm) = 10%

$$I_P = w_L - w_P = 37 - 28 = 9$$

Using Figure 5.3:

Starting at box 1: not highly organic
Move to box 2: less than 35% passing #200
Move to box 3: greater than 50% passing #10 sieve
Move to box 5: less than 50% passing #40 sieve
Move to box 6: plasticity index greater than 6
Move to box 9: plasticity index less than 10
Move to box 12: liquid limit less than 40; therefore, group classification is A-2-4

Compute group index:

$$\begin{aligned} \text{Group index} &= (F-35)[0.2+0.005(w_L-40)]+0.01(F-15)(I_P-10) \\ &= (10-35)[0.2+0.005(37-40)]+0.01(10-15)(9-10) \\ &= -5 < 0 \quad \text{therefore use 0} \end{aligned}$$

Final result: **A-2-4 (0), which would make a good subbase.**

Example 5.3

The natural soils along a proposed highway alignment have a particle size distribution as described by curve A in Figure 5.2, a liquid limit of 44, a plastic limit of 21, and is nonorganic. Determine the AASHTO soil classification and rate its suitability for pavement support.

Solution:

Referring to the sieve sizes in Table 4.7:

Passing #10 sieve (2.00 mm) = 92%
Passing #40 sieve (0.425 mm) = 74%
Passing #200 sieve (0.075 mm) = 54%

$$I_P = w_L - w_P = 44 - 21 = 23$$

Using Figure 5.3:

Starting at box 1: not organic
Move to box 2: greater than 35% passing #200
Move to box 8: plasticity index greater than 10
Move to box 11: liquid limit greater than 40
Move to box 14: plasticity index (23) greater than liquid limit minus 30 (14)
Therefore, group classification is A-7-6

Compute group index:

$$\begin{aligned}\text{Group index} &= (F - 35)[0.2 + 0.005(w_L - 40)] + 0.01(F - 15)(I_P - 10) \\ &= (54 - 35)[0.2 + 0.005(44 - 40)] + 0.01(54 - 15)(23 - 10) \\ &= 9\end{aligned}$$

Final result: **A-7-6 (9), which would make a poor subgrade.**

5.3 UNIFIED SOIL CLASSIFICATION SYSTEM (USCS)

Arthur Casagrande developed a new engineering soil classification system for the U.S. Army during World War II (Casagrande, 1948). Since then, it has been updated and is now standardized in ASTM D2487 as the *Unified Soil Classification System* (USCS). Unlike the AASHTO system, the USCS is not limited to any particular kind of project; it is an all-purpose system. The USCS does not attempt to rank soils from good to bad, as does the AASHTO system, because specific soils may be excellent for some purposes and poor for others. Nevertheless, because of its flexibility and ease of use, this has become the most common soil classification system among geotechnical engineers in North America.

In its original form, the classification consisted only of a two- or four-letter *group symbol*. Later, the system was enhanced by the addition of several *group names* for each group symbol. For example, a typical USCS classification would be:

SM—Silty sand with gravel

where "SM" is the group symbol and "Silty sand with gravel" is the group name.

The position of a soil type in the group name indicates its relative importance, as follows:

Noun = Primary component
Adjective = Secondary component (or further explanation of primary component)
"with ..." = Tertiary component

For example, a *clayey sand with gravel* has sand as the most important component, clay as the second most important, and gravel as the third most important. If very little of a soil type is present, then it is not included in the group name at all. For example, a *clayey sand* is similar to the soil just described, except it has less than 15% gravel.

Initial Classification

To use the USCS, begin with an initial classification as follows:

1. Determine if the soil is *highly organic*. Such soils have the following characteristics:
   - Composed primarily of organic material
   - Dark brown, dark gray, or black color
   - Organic odor, especially when wet
   - Soft consistency.

   In addition, fibrous material (remnants of stems, leaves, roots, etc) is often evident.

   If the soil does not have these characteristics (and the vast majority do not), then go to Step 2. However, if it does, then classify it as follows:

Group symbol	Pt
Group name	Peat

   This completes the unified classification for highly organic soils. These soils are very problematic because of their high compressibility and low strength, so the group symbol Pt on a boring log is a red flag to geotechnical engineers.

2. Conduct a sieve analysis to determine the particle size distribution curve. For an informal classification, a particle size distribution curve based on a visual inspection may suffice.

3. On the basis of the particle size distribution curve, determine the percent by weight passing the 3-in., #4, and #200 sieves, then compute the percentages by weight of gravel, sand, and fines using the definitions in Table 4.6.

4. If 100% of the sample passes the 3-in. sieve, go to Step 5. If not, base the classification on the part that passes this sieve (usually called the *minus 3-in. fraction*). To do so, adjust the percentages of gravel, sand, and fines using a procedure similar to that in Example 5.1. Then, perform the classification based on these modified percentages, and note the percentage of cobbles and/or boulders and the maximum particle size with the final classification. For example, if 20% of the soil is cobbles, some as large as 8 in., the USCS classification (after going through the rest of the procedure) might be:

   SW—Well-graded sand with gravel and 20% cobbles, max 8 in.

5. If 5% or more of the soil passes the #200 sieve, then conduct Atterberg limits tests to determine the liquid and plastic limits. Note that the Atterberg limits tests are conducted on the portion of the soil passing the #40 sieve (the −40 portion).
6. If the soil is fine-grained (i.e., ≥50% passes the #200 sieve), follow the directions for fine-grained soils. If the soil is coarse-grained (i.e., <50% passes the #200 sieve), follow the directions for coarse-grained soils.

Classification of Fine-Grained Soils

Fine-grained soils are those that have at least 50% passing the #200 sieve. Thus, these soils are primarily silt and/or clay. Most sand particles that may be present will be floating in a silt or clay matrix, and thus have little impact on the engineering properties. In order to easily distinguish between clays and silts, Casagrande developed the *plasticity chart* in Figure 5.4 that uses the Atterberg limits as the basis for classifying these soils. This chart uses the Atterberg limits to distinguish between clays and silts. Although most fine-grained soils contain both clay and silt, and possibly sand and gravel as well, those that plot above the *A-line* are classified as clays, whereas those below this line are silts.

We use the plasticity chart to determine the group symbol for fine-grained soils. It usually consists of two letters, which are interpreted as follows:

First Letter		**Second Letter**	
M	Predominantly silt	L	Low plasticity
C	Predominantly clay	H	High plasticity
O	Organic		

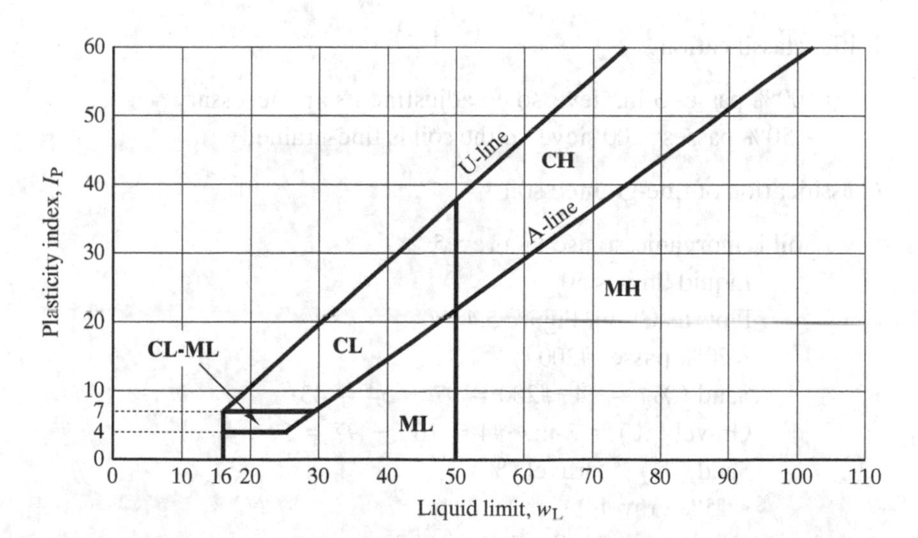

FIGURE 5.4 Plasticity chart (ASTM D2487). The "A-line" separates silts from clays, whereas the "U-line" represents the upper limit of recorded test results. Data that plot above the U-line are probably in error. Note that the vertical axis is the plasticity index, not the plastic limit. Soils identified as "nonplastic" (NP) are classified as ML.

CL soils are known as *lean clays*, whereas CH soils are *fat clays*. The corresponding terms for ML and MH soils are *silt* and *elastic silt*, respectively, even though the stress–strain behavior of MH soils is no more elastic than any other soil.

In this context, an *organic* soil is one that has a noteworthy percentage of organic matter, yet consists primarily of inorganic material. This differs from a *highly organic* soil, as described earlier (group symbol Pt), which contains much more organic material. With experience, one can usually determine whether a fine-grained soil is inorganic (M or C) or organic (O) by visual inspection. Alternatively, we could perform two liquid limit tests, one on an unmodified sample from the field and another on a sample that is first oven-dried. The drying process alters any organics that might be present, and thus changes the liquid limit. If the liquid limit after oven-drying is less than 75% of the original value, then the soil is considered to be organic. If not, then it is inorganic. Do not use the Atterberg limits from the oven dried sample to classify the soil. The tests on the oven dried sample are used only to determine whether or not the soil classifies as organic.

To classify inorganic fine-grained soils, use the flow chart in Figure 5.5 and the plasticity chart in Figure 5.4. For organic fine-grained soils, use Figures 5.4 and 5.6.

Example 5.4

Classify the inorganic soil from Example 5.3 using the USCS.

Solution:

Initial classification

100% passes 3-in. sieve, so no adjustments are necessary
≥50% passes #200 sieve, so the soil is fine-grained

Classification of fine-grained soil

Soil is inorganic, so use Figure 5.5
Liquid limit <50
Plots as CL on Figure 5.4
<70% passes #200
Sand (%) = #4−#200 = 97 − 54 = 43
Gravel (%) = 3-in.−#4 = 100 − 97 = 3
Sand (%) > gravel (%)
<15% gravel
Group name = Sandy lean clay

Final result: **CL—Sandy lean clay.**

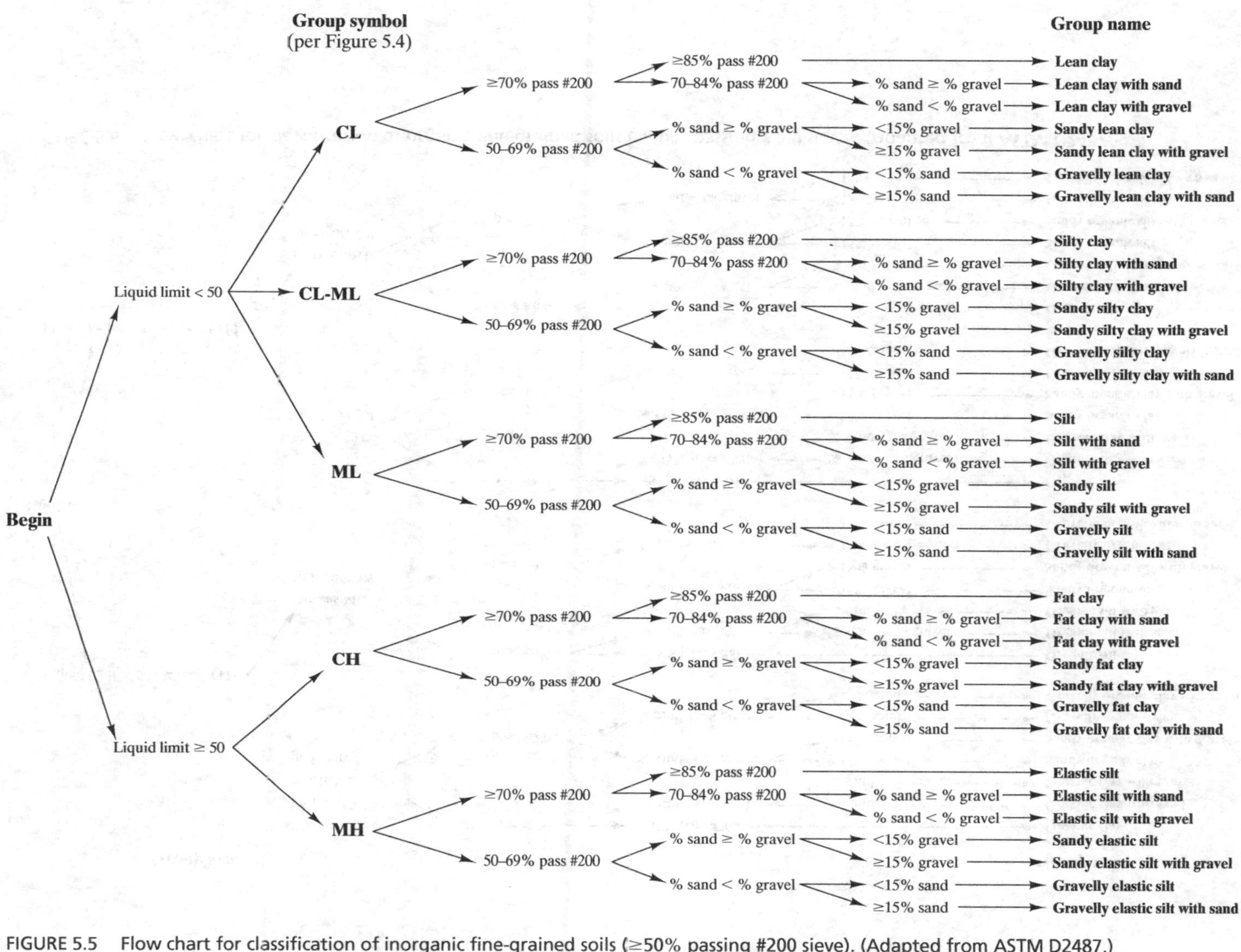

FIGURE 5.5 Flow chart for classification of inorganic fine-grained soils (≥50% passing #200 sieve). (Adapted from ASTM D2487.)

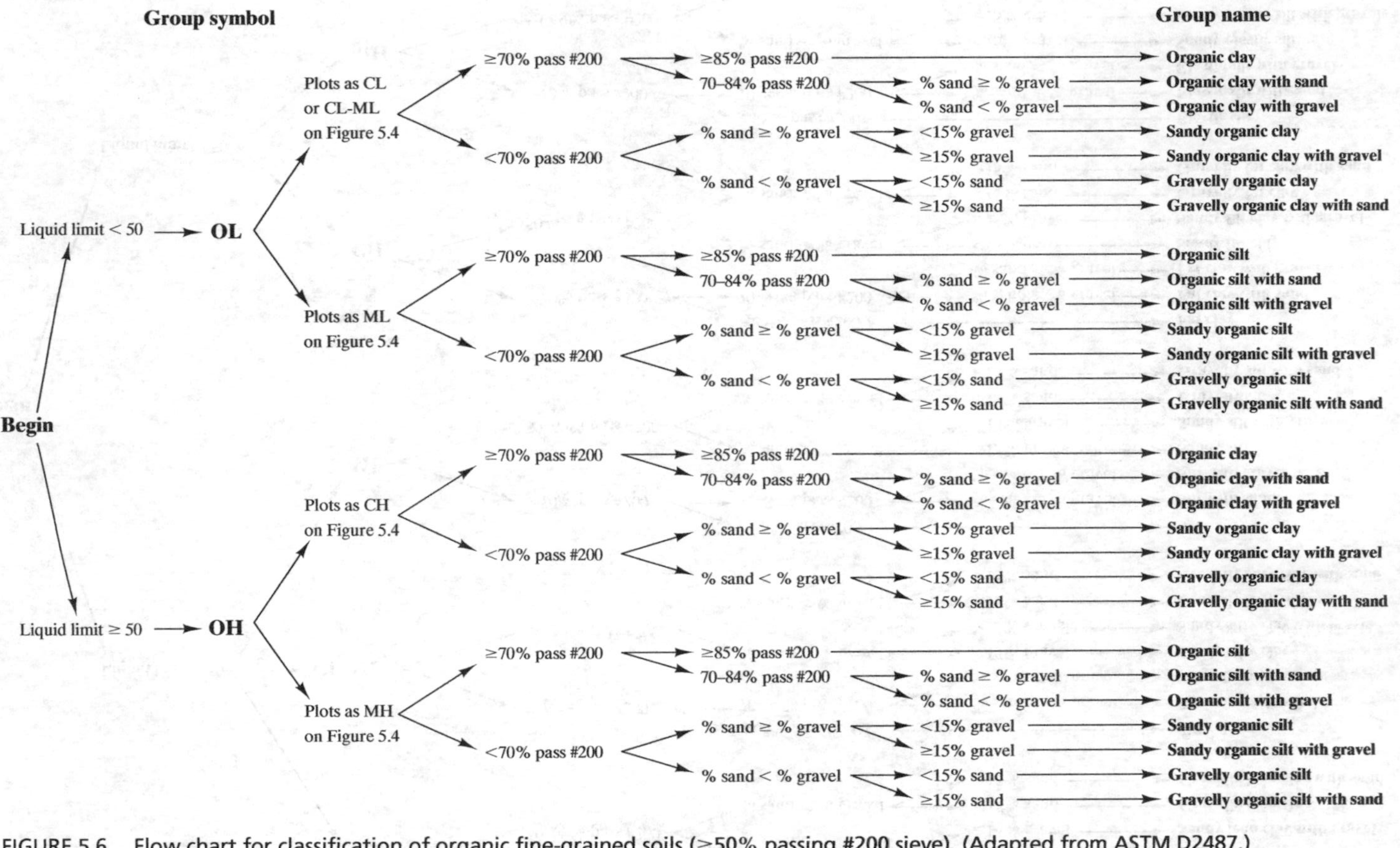

FIGURE 5.6 Flow chart for classification of organic fine-grained soils (≥50% passing #200 sieve). (Adapted from ASTM D2487.)

Classification of Coarse-Grained Soils

Coarse-grained soils are those that have less than 50% passing the #200 sieve. Thus, these soils are primarily sand and/or gravel. The letters forming the group symbols for coarse-grained soils are defined below:

First Letter		**Second Letter**	
S	Predominantly sand	P	Poorly-graded
G	Predominantly gravel	W	Well-graded
		M	Silty
		C	Clayey

As discussed in Chapter 4, poorly-graded soils are those with a narrow range of particle sizes (i.e., a steep particle size distribution curve), whereas well-graded soils have a wide range of particle sizes (i.e., a flatter particle size distribution curve). In this context, silty (M) or clayey (C) indicates a large percentage of silt or clay in a coarse-grained soil.

To classify coarse-grained soils, use the flow chart in Figure 5.7. By inspection of this chart, we see that both sands and gravels are divided into three categories depending on the percentage of fines (percentage passing the #200 sieve):

If <5% fines	Use two-letter group symbol to describe gradation (well- or poorly-graded)
If 5–12% fines	Use four-letter group symbol to describe both gradation and type of fines
If >12% fines	Use two-letter group symbol to describe type of fines (silt or clay)

Note that a coarse-grained soil with less than 5% fines will be classified as either a well- or poorly-graded sand or gravel (e.g., GP or SW) without reference to the type of fine-grained material present. However if a coarse-grained soil contains as little as 5% fines, the presence of the fines will be reflected in the classification. A soil with a great deal of fines (more than 12%) will be classified as a clayey or silty sand or gravel (e.g., GC or SM). A soil with an intermediate amount of fines (5–12%) will carry a dual classification (e.g., GP-GM or SW-SC). The USCS classification reflects these thresholds because as little as 5% fines can have an impact on soil behavior, and 12% fines can have a significant impact on soil behavior.

Example 5.5

Determine the unified soil classification for the inorganic soil B in Figure 5.2.

Solution:

Initial classification

100% passes 3-in. sieve, so no adjustments are necessary
<50% passes #200 sieve, so soil is coarse-grained

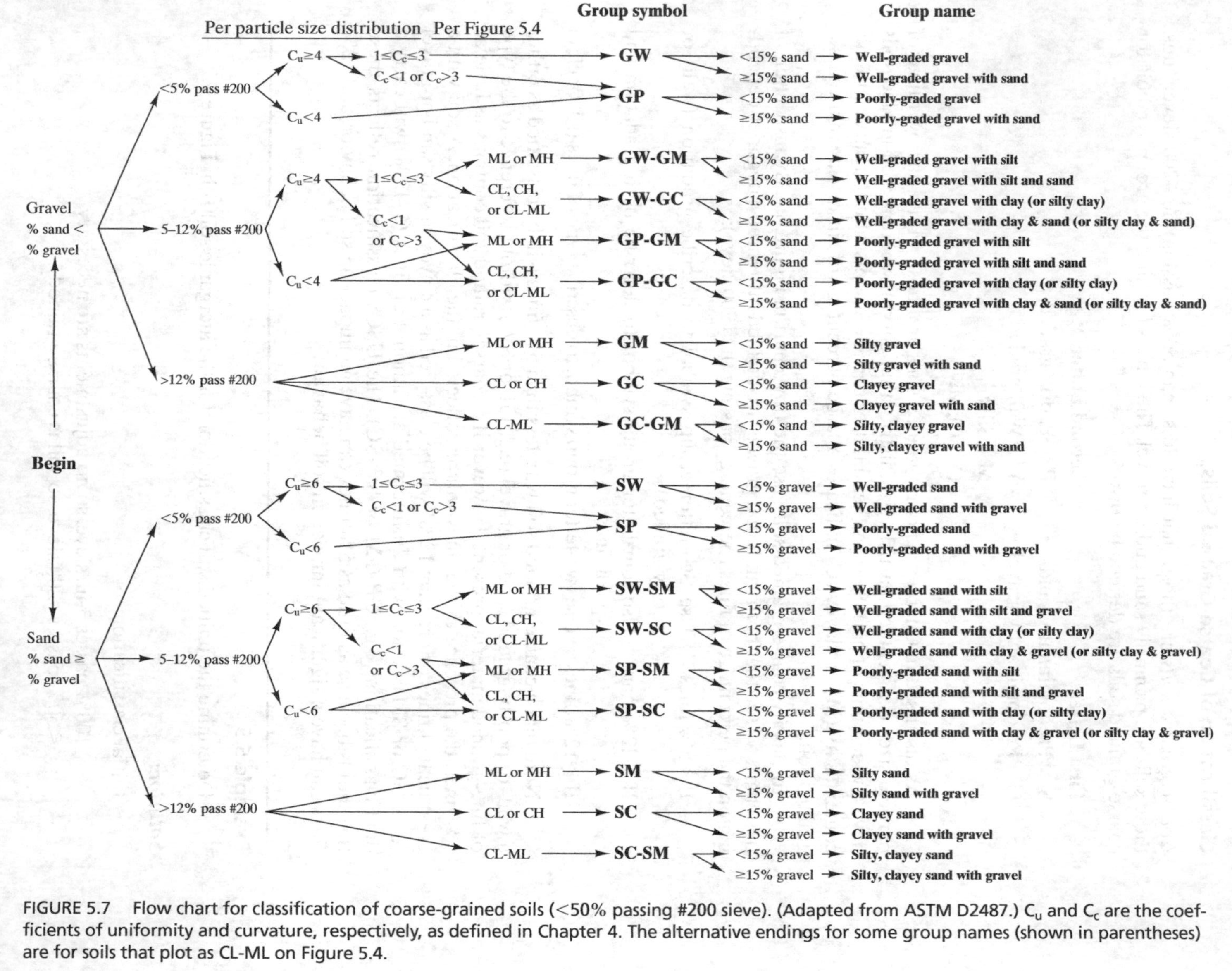

FIGURE 5.7 Flow chart for classification of coarse-grained soils (<50% passing #200 sieve). (Adapted from ASTM D2487.) C_u and C_c are the coefficients of uniformity and curvature, respectively, as defined in Chapter 4. The alternative endings for some group names (shown in parentheses) are for soils that plot as CL-ML on Figure 5.4.

Classification of coarse-grained soils

% gravel = 3-in.−#4 = 100% − 100% = 0%
% sand = #4−#200 = 100% − 4% = 96%
% fines = #200 = 4%
$D_{10} = 0.10$ mm
$D_{30} = 0.17$ mm
$D_{60} = 0.40$ mm
$C_u = D_{60}/D_{10} = 0.40/0.10 = 4.0$
$C_c = D_{30}^2/(D_{10}D_{60}) = 0.17^2/[(0.10)(0.40)] = 0.72$

Using Figure 5.7

% sand > % gravel
<5% fines
$C_u < 6$ and $C_c < 1$, so group symbol is SP
<15% gravel

Final results: **SP—Poorly-graded sand.**

Example 5.6

The inorganic soil C in Figure 5.2 has a liquid limit of 30 and a plastic limit of 25. Determine its unified soil classification.

Solution:

Initial classification

100% passes 3-in. sieve, so no adjustments are necessary
<50% passes #200 sieve, so soil is coarse-grained

Classification of coarse-grained soils

% gravel = 3-in.−#4 = 100% − 80% = 20%
% sand = #4−#200 = 80% − 10% = 70%
% fines = #200 = 10%
$D_{10} = 0.075$ mm
$D_{30} = 0.39$ mm
$D_{60} = 1.7$ mm
$C_u = D_{60}/D_{10} = 1.7/0.075 = 23$
$C_c = D_{30}^2/(D_{10}D_{60}) = 0.39^2/[(0.075)(1.7)] = 1.2$

Using Figure 5.7

% sand > % gravel
5–12% fines
$C_u > 6$ and $1 \leq C_c \leq 3$ (i.e., soil is well-graded)
$I_P = w_L - w_P = 30 - 25 = 5$
Plots as ML on Figure 5.4, so group symbol is SW-SM
>15% gravel

Final result: **SW-SM—Well-graded sand with silt and gravel.**

Classification of Borderline Soils

Sometimes a soil classification is very close to the dividing line between two different group symbols. In such cases, it is acceptable to use both symbols in the classification, with the "correct" symbol first, followed by the "almost correct" symbol. For example, a sand–clay combination with slightly less than 50% fines could be identified as SC/CL.

Soil Assessment Based on USCS Classification

Geotechnical engineers have many ways to assess the suitability of a soil for particular purposes. For example, if a soil is being considered for use as an "impervious" cap over a sanitary landfill, we would perform hydraulic conductivity tests as described in Chapter 7 to determine how easily water flows through it. Soils that restrict the flow of water are best for landfill caps. However, before we perform these specialized tests, geotechnical engineers assess a soil based on its classification. For example, an SW soil would pass water very easily, and thus would be rejected for the landfill cap even without a hydraulic conductivity test. Table 5.2 presents general soil properties based on the unified group symbol, and may be used to assist in such assessments.

5.4 VISUAL–MANUAL SOIL CLASSIFICATION

Complete classification of soils using either the USCS or AASHTO method requires sieve, hydrometer, and Atterberg limits laboratory tests that can take several days. It is not cost effective to perform these tests on every sample. Additionally, engineers often need to rapidly classify soils in the field. Visual-manual classification is a way to quickly classify soils without laboratory testing. The following methods outline the basics of visual-manual classification. A complete visual-manual classification procedure is provided in Appendix C can be viewed online at www.pearsonhighered.com/Coduto.

Distinguishing Fine Sand from Silt and Clay

The #4 sieve that separates gravel from sand has an opening size of 4.75 mm or 3/16 in. It is relatively easy to distinguish sand from gravel by simple visual observation. However,

TABLE 5.2 Assessment of Soil Properties Based on Group Symbol[a]

Group Symbol	Compaction Characteristics	Compressibility and Expansion	Drainage and Hydraulic Conductivity	Value as a Fill Material	Value as a Pavement Subgrade When Not Subject to Frost	Value as a Base Course for Pavement
GW	Good	Almost none	Good drainage; pervious	Very stable	Excellent	Good
GP	Good	Almost none	Good drainage; pervious	Reasonably stable	Excellent to good	Poor to fair
GM	Good	Slight	Poor drainage; semipervious	Reasonably stable	Excellent to good	Fair to poor
GC	Good to fair	Slight	Poor drainage; semipervious	Reasonably stable	Good	Good to fair; not suitable if subject to frost
SW	Good	Almost none	Good drainage; pervious	Very stable	Good	Fair to poor
SP	Good	Almost none	Good drainage; pervious	Reasonably stable when dense	Good to fair	Poor
SM	Good	Slight	Poor drainage; impervious	Reasonably stable when dense	Good to fair	Poor
SC	Good to fair	Slight to medium	Poor drainage; impervious	Reasonably stable	Good to fair	Fair to poor; not suitable if subject to frost
ML	Good to poor	Slight to medium	Poor drainage; impervious	Fair stability, good compaction required	Fair to poor	Not suitable
CL	Good to fair	Medium	No drainage; impervious	Good stability	Fair to poor	Not suitable
OL	Fair to poor	Medium to high	Poor drainage; impervious	Unstable; should not be used	Poor, not suitable	Not suitable
MH	Fair to poor	High	Poor drainage; impervious	Fair to poor stability, good compaction required	Poor	Not suitable
CH	Fair to poor	Very high	No drainage; impervious	Fair stability, expands, weakens, shrinks, cracks	Poor to very poor	Not suitable
OH	Fair to poor	High	No drainage; impervious	Unstable; should not be used	Very poor	Not suitable
Pt	Not suitable	Very high	Fair to poor drainage	Should not be used	Not suitable	Not suitable

[a]Adapted from Sowers, 1979.

distinguishing fine sand from silt or clay can be more difficult. The following methods can assist in this process:

- The #200 sieve approximately corresponds to the smallest particles one can see with the unaided eye. Thus, individual fine sand particles can be distinguished, but individual silt particles cannot. In addition, particles larger than the #200 sieve have a gritty texture, whereas those smaller are pasty.
- Clay and silt particles often clump together, and may look like sand. These clumps will soften and separate into smaller particles when wetted. Therefore, when in doubt, be sure to wet the soil before classifying it.

Distinguishing Clays from Silts

The key to distinguishing clays (CL and CH) from silts (ML and MH) is to observe how a soil responds to changes in water content. Clays are relatively strong when dry and can absorb a significant amount of water and remain in a moldable, plastic state. Silts have a lower dry strength than clays and absorb less water. The following methods can assist in distinguishing between clay and silt.

- Attempt to break a dried clump of the soil about the size of a small marble with your fingers. Clays will be difficult or impossible to break, whereas silts will break or crumble easily.
- Add water to a sample of soil until it is moldable, then try and roll the soil into a thread approximately 5 mm in diameter. Clays can easily be rolled into a thread, whereas silts will crumble and fall apart while being rolled.

Cementing agents, such as calcium carbonate, are sometimes present in sandy or silty soils. These agents can give the soil a high dry strength, even if no clay is present. Again, the key is to wet the sample. Cemented soils will retain their dry strength, whereas clayey soils will soften when wetted.

5.5 SUPPLEMENTAL SOIL CLASSIFICATIONS

Regardless of the textural classification system being used (AASHTO, USCS, etc.), geotechnical engineers often add certain supplementary classifications. Some of these have been standardized in ASTM D2488, but informal classifications are used at least as often as the standard terms. Either way, they provide important information on the in situ soil conditions.

Moisture

Knowledge of the moisture condition of a soil can be very useful. Therefore, *moisture content tests*, as discussed in Chapter 4, are among the most common soil tests. In addition, engineers often give a qualitative assessment of soil moisture using descriptors such as those in Table 5.3.

TABLE 5.3 Moisture Classification

Classification	Description
Dry	Dusty; dry to the touch
Slightly moist	Some moisture, but still has a dry appearance
Moist	Damp; but no visible water
Very moist	Enough moisture to wet the hands
Wet	Saturated; visible free water

Color

The soil color can vary as its moisture content changes, so it is a less reliable classification. Nevertheless, it is useful as a common supplementary soil classification. Although standardized color description systems, such as the Munsell Color Charts, are available, they express colors using an awkward notation (i.e., 10 YR 5/3). Therefore, most engineers just use common color names, such as brown, tan, and gray-brown. Sometimes individual firms or agencies standardize these names, but there is no widely accepted standard.

Consistency

The term *consistency* of a soil is a very useful supplementary classification, but it has a slightly different meaning for fine versus coarse-grained soils. For fine-grained soils, it describes the plastic state of the soil from very soft to very hard. For coarse-grained soils it describes the soil's relative density from very loose to very dense. The behavior of a given soil can change significantly with its consistency. For example, a hard CH soil is quite different from a very soft CH soil; therefore, this supplemental classification is commonly used. Consistency depends on the soil type, moisture content, unit weight, and other factors, and may change in the field with time, especially if the soil becomes wet. Tables 5.4 and 5.5 present classifications of soil consistency, along with qualitative and quantitative descriptions.

When classifying the consistency of coarse-grained soils based on standard penetration test data, it is especially important to apply the overburden correction described in Section 3.9, thus obtaining $N_{1,60}$.

Cementation

Some soils are cemented with certain chemicals such as calcium carbonate ($CaCO_3$) or iron oxide (Fe_2O_3). The presence of calcium carbonate can be determined by applying a small amount of hydrochloric acid (HCl) and noting the reaction. A bubbling action indicates the presence of $CaCO_3$. A lack of bubbling indicates some other cementing agent. Iron oxide gives the soil a red-orange tint, similar to rusty steel. Table 5.6 presents definitions of soil cementation terms.

TABLE 5.4 Consistency Classification for Fine-Grained Soils[a]

Classification	Description	SPT N_{60} value[b]	Undrained Shear Strength, s_u[c] (kPa)	(lb/ft^2)
Very soft	Thumb penetrates easily; extrudes between fingers when squeezed	<2	<12	<250
Soft	Thumb will penetrate soil about 25 mm; molds with light finger pressure	2–4	12–25	250–500
Medium	Thumb will penetrate about 6 mm with moderate effort; molds with strong finger pressure	4–8	25–50	500–1000
Stiff	Thumb indents easily; will penetrate 12 mm with great effort	8–15	50–100	1000–2000
Hard	Thumb will not indent soil; thumbnail readily indents it	15–30	100–200	2000–4000
Very hard	Thumbnail will not indent soil or will indent it only with difficulty	>30	>200	>4000

[a]Adapted from Terzaghi, Peck, and Mesri, 1996 and U.S. Navy, 1982; by permission of John Wiley & Sons, Inc.
[b]The *N*-value is defined in Chapter 4.
[c]The undrained shear strength is defined in Chapter 12, and is half of the unconfined compressive strength.

TABLE 5.5 Consistency Classification for Coarse-Grained Soils[a]

Classification	Description	SPT $N_{1,60}$ value[b]	D_r Relative Density[c]
Very loose	Easy to penetrate with a 12-mm diameter rod pushed by hand	<4	0–15
Loose	Difficult to penetrate with a 12-mm diameter rod pushed by hand	4–10	15–35
Medium dense	Easy to penetrate 300 mm with a 12-mm diameter rod driven with a 2.3-kg (5-lb) hammer	10–17	35–65
Dense	Difficult to penetrate 300 mm with a 12-mm diameter rod driven with a 2.3-kg (5-lb) hammer	17–32	65–85
Very dense	Penetrate only about 150 mm with a 12-mm diameter rod driven with a 2.3-kg (5-lb) hammer	>32	85–100

[a]Adapted from U.S. Navy, 1982, and Lambe and Whitman, 1969; by permission of John Wiley & Sons, Inc.
[b]These values are for sandy soils. If some fine gravel is present, use two-thirds of the field value. If significant quantities of coarse gravel are present, do not use this table.
[c]If CPT data is available, use Equation 4.25 to compute D_r.

TABLE 5.6 Classification of Soil Cementation[a]

Classification	Description
Weak	Crumbles or breaks with handling or little finger pressure
Moderate	Crumbles or breaks with considerable finger pressure
Strong	Will not crumble or break with finger pressure

[a]After ASTM D2488.

Structure

Soil particles can be assembled into many different structures. Often, these need to be noted in the classification of undisturbed samples as follows (ASTM D2488):

- *Stratified*—Alternating layers of varying material or color with layers at least 6 mm thick
- *Laminated*—Alternating layers of varying material or color with the layers less than 6 mm thick
- *Fissured*—Breaks along definite planes of fracture with little resistance to fracturing
- *Slickensided*—Fracture planes appear polished or glossy, sometimes striated (linear markings showing evidence of past movement)
- *Blocky*—Cohesive soil that can be broken down into small angular lumps that resist further breakdown
- *Lensed*—Inclusions of small pockets of different soils, such as small lenses of sand scattered throughout a mass of clay
- *Homogeneous*—Same color and appearance throughout.

5.6 APPLICABILITY AND LIMITATIONS

Standardized soil classification systems are very valuable tools that help geotechnical engineers identify soils and make preliminary assessments of their engineering behavior. However, they also have limitations. Casagrande (1948) said:

> "It is not possible to classify all soils into a relatively small number of groups such that the relation of each soil to the many divergent problems of applied soil mechanics will be adequately presented."

Do not expect too much from a soil classification system. Identifying the proper classification is a good start, but we still need to use other test results, an understanding of soil behavior, engineering judgment, and experience.

SUMMARY

Major Points

1. Standardized soil classification systems are an important part of the language of geotechnical engineering. They help us identify soils and communicate important characteristics to other engineers. In addition, a classification helps us develop preliminary assessments of a soil's behavior.
2. The USDA soil classification system is designed to classify soils according to their suitability for agricultural use. This system classifies soils based solely on their particle size distribution and uses different definitions for sand, silt, and clay than those defined by ASTM. It is important to geotechnical engineers because of the large number of USDA soil survey reports that use this system.
3. The AASHTO soil classification system is designed to classify soils according to their suitability as highway subgrades and ranks soils from good to bad on this basis. This system uses both particle size distribution and Atterberg limits data to classify soils.
4. The USCS is an all-purpose system and therefore, does not rate soils from good to bad. Instead, it provides classifications that allow geotechnical engineers to determine a soil's general suitability for many different applications. This system uses both particle size distribution and Atterberg limits data to classify soils.
5. Formal classification using USCS requires laboratory testing to measure the particle size distribution and Atterberg limits. However, expedient visual-manual classification is often used in the field to provide expedient classification without laboratory testing. When done properly by an experience person, this method can provide accurate classifications.
6. Supplemental soil classifications assist in further describing important characteristics not addressed by the USDA, AASHTO, or USCS systems and are commonly used. Geotechnical engineers must be familiar with supplemental classifications, particularly those for moisture, consistency, and cementation.

Vocabulary

AASHTO soil classification system
blocky
cementation
coarse-grained soil
consistency
fine-grained soil
fines
fissured
group classification
group index
group name
group symbol
highly organic soil
homogeneous
laminated
lensed
organic soil
plasticity chart
slickensided
stratified
structure
Unified Soil Classification System (USCS)
USDA soil classification system

QUESTIONS AND PRACTICE PROBLEMS

5.1 What kinds of soils are contained in *loam* as the term is used in the USDA classification system?

5.2 The USDA classification system pays very little attention to gravel sized portion of a soil. Why?

5.3 What is the purpose of the group index in the AASHTO classification system?

5.4 You have been asked to determine if the soil at a given site is suitable for use as a pavement subbase material. What laboratory tests would you recommend performing on the soil samples taken from the site?

5.5 Is it ever necessary to conduct a hydrometer analysis when classifying soils according to the USCS, or is a sieve analysis always sufficient?

5.6 Explain the difference between a silty sand and a sandy silt in the USCS.

5.7 How would you determine if a soil was a silty sand or a sandy silt if:

(a) Laboratory test equipment was available?
(b) Laboratory test equipment was not available?

Data for Problems 5.8–5.11

As part of a site investigation program for a residential development project you retrieved four different soil samples. None of the soils appear to be organic. At your laboratory you performed particle size analyses and Atterberg limits tests on the samples. The particle size distribution data for the four soil samples are shown in Figure 5.2 as curves D, E, F, and G. From your Atterberg limit tests on the −40 portion of these samples you obtained the following data

Soil Sample Identification	Liquid Limit	Plastic Limit
D	60	33
E	40	30
F	30	26
G	nonplastic	

5.8 Determine the USDA classification for soils D through G using the data given above.

5.9 Determine the AASHTO group classification and group index for soils D through G using the data given above.

5.10 Which of the soils D through G would make the best highway subgrade according to the AASHTO classification system?

5.11 Determine the USCS group symbol and group name for soils D through G using the data given above.

Data for Problems 5.12–5.15

You have collected soil samples as part of the investigation for a dam project. The particle size distribution data for the four soil samples are shown in Figure 5.8 as curves H, I, J, and L. Soil H has a strong organic smell but contains no fibrous material. The other soils are nonorganic.

From your Atterberg limits tests on the −40 portion of these samples, you have obtained the following data

Soil Sample Identification	Liquid Limit	Plastic Limit
H	60	26
I	nonplastic	
J	46	34
L	37	18

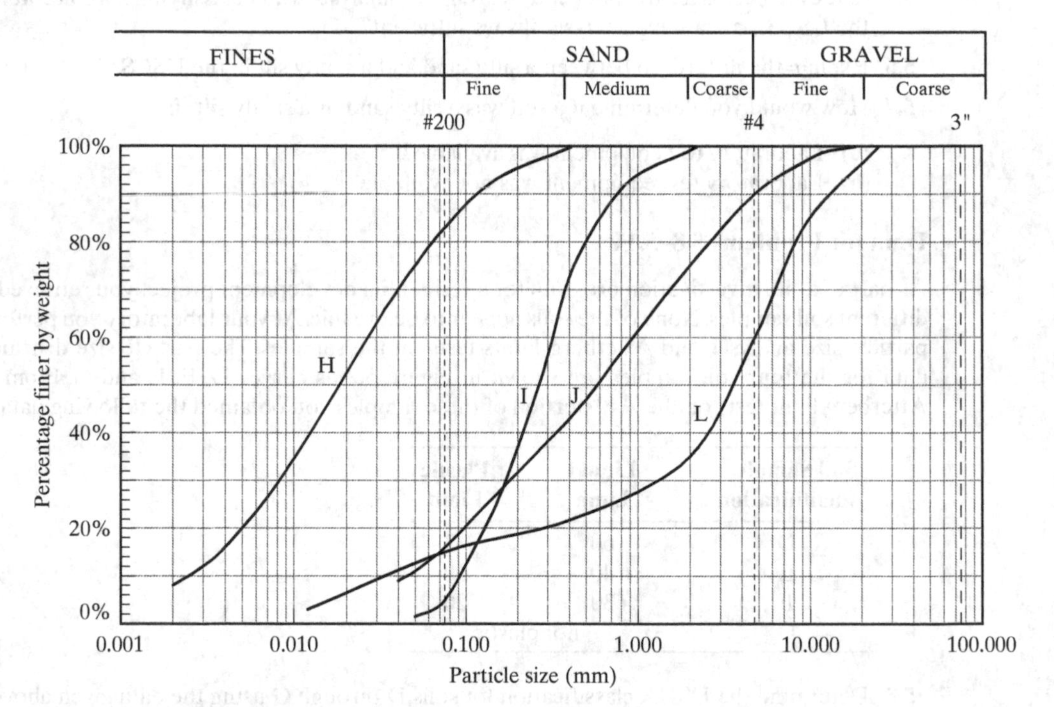

FIGURE 5.8 Particle size distribution curves for Problems 5.12 through 5.15.

5.12 Determine the USDA classification for soils H through L using the data given above. The gravel portion of soil L consists predominantly of well-rounded granitic material.

5.13 Determine the AASHTO group classification and group index for soils H through L using the data given above.

5.14 Which of the soils H through L would make the best highway subgrade according to the AASHTO classification system?

5.15 Determine the USCS group symbol and group name for soils H through L using the data given above.

5.16 You have performed a hydrometer test on a soil and measured the particle size distribution shown in Figure 4.13 as Soil A. The soil has an organic smell but does not contain any plant fibers. The Atterberg limits tests on the −40 portion of the soil measured a plastic limit of 21 and a liquid limit of 40. Determine its unified group symbol and group name.

5.17 You have performed a sieve analysis on a soil sample and measured the particle size distribution shown in Figure 4.13 as Soil C. The fine portion of the soil was clearly nonplastic so you did not measure the Atterberg limits. Determine the unified group symbol and group name for the soil.

5.18 A clean well-graded sand would rate very high in the AASHTO soil classification system, and thus be considered a good soil for use as a highway subgrade. However, it would be a very poor choice for certain other applications. Give one example of a situation where this soil would not be a good choice.

5.19 A sanitary landfill project needs an import soil for use as an impervious cap over the refuse. The following soils are available for this purpose:

Soil 1: GC
Soil 2: SC
Soil 3: ML
Soil 4: CL

The design engineer desires a soil with a high clay content because it will help keep water out of the landfill. Which of the available soils appears to be most promising? Why?

5.20 Which of the following USCS soils would probably receive the best rating for use as a highway subgrade per the AASHTO classification: CL, SM, ML, or SW? Why?

5.21 You are drilling and sampling at a site in an arid region. You encounter a fine-grained soil that is dry and hard. You are able to break chunks of the soil with your fingers but only with difficulty. You add water to the soil until it is in a plastic state and then form a ribbon out of the plastic soil by squeezing it between your thumb and forefinger. It is easy to form a long thin ribbon out of the moistened soil. Is this soil a silt or a clay? Explain.

5.22 You are examining a soil sample retrieved during drilling. The sample has little or no plasticity. The particles are very uniform and about the size of the period at the end of this sentence. What is the most likely USCS classification for this soil?

CHAPTER 6

Excavation, Grading, and Compacted Fill

The wise architect should wash the excavations with the five products of the cow.

The excavation should be made at night and the bricks should be laid in the daytime.

The chief architect should distinguish the two varieties of bricks, namely stony brick and pure brick, and their three genders, and should fix the male bricks in the temples of male deities.

Manasara, a sixth-century architect (Acharya, 1980)

Most civil engineering projects include some *earthwork*, which is the process of changing the configuration of the ground surface. When we remove soil or rock, we make a *cut* or *excavation*; when we add soil or rock, we create a *fill* or *embankment*. These changes make the site more suitable for the proposed development.

For example, virtually all highway and railroad projects require earthwork to create smooth grades and alignments and to provide proper surface drainage. The value of cuts and fills in highway projects is especially evident in mountainous areas where older two-lane roads were built using as little earthwork as possible, resulting in steep grades and sharp turns. When a modern multi-lane interstate highway is built through the same mountains, engineers use more generous cuts and fills, thus producing straighter and smoother alignments and gentler grades. Figure 6.1 shows a deep cut made during construction of such a highway.

Earthwork also is important in building construction, especially in hilly terrain. Cuts and fills are used to create level pads for the buildings, parking lots, and related areas, as shown in Figure 6.2, thus using the land area to better advantage. Even building

FIGURE 6.1 This four-lane highway was created by making a cut on the left side of the photo and in the background, and a fill in the foreground.

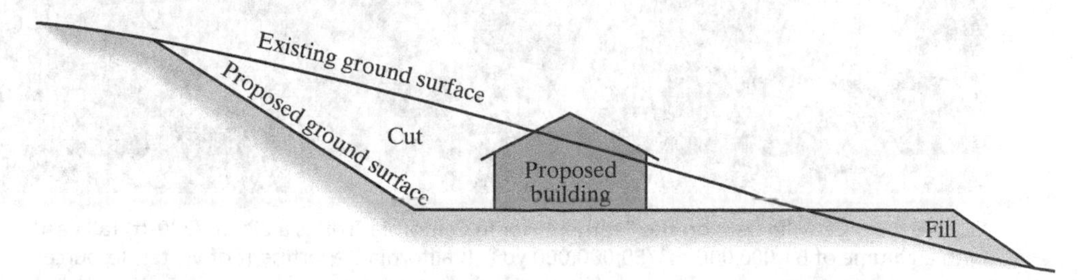

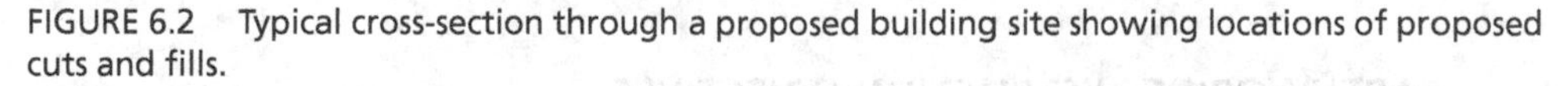

FIGURE 6.2 Typical cross-section through a proposed building site showing locations of proposed cuts and fills.

projects in areas with naturally level terrain often require minor earthwork to provide proper surface drainage.

Large residential developments in hilly areas often include extensive earthwork, sometimes millions of cubic meters for a single project that produces hundreds of residential lots. Such projects provide level building pads and smooth street alignments, and allow more houses to be built in a given area. They also provide a safer development, because a properly graded tract is less prone to landslides, excessive settlement, flooding, and other problems.

Finally, the most impressive earthwork projects are earth dams, such as the one in Figure 6.3. They require very large volumes of carefully placed fill, yet are much less expensive than concrete dams because such fills can be placed very economically. When properly designed and constructed, earth dams also are very safe.

FIGURE 6.3 Oroville Dam on the Feather River in California. This is a 225-m (740-ft) tall earth dam with a volume of 61,000,000 m^3 (80,000,000 yd^3). (California Department of Water Resources.)

6.1 EARTHWORK CONSTRUCTION OBJECTIVES

The most fundamental objective of earthwork construction is to change the ground surface from some initial configuration, typically described by a topographic map, to some final configuration, as described on a new topographic map known as a *grading plan*. These changes are often necessary to properly accommodate the proposed construction, and to maintain proper surface drainage, which is important to the long-term performance of cuts and fills.

Another important requirement is that earthwork must not create slope stability problems (i.e., landslides). This is especially important in hilly and mountainous areas, since the level building pads or road alignments created by the earthwork are possible only because nearby areas are made steeper and thus less stable.

Compacted fills have additional requirements such as the following:

1. Fills must have sufficient shear strength to support both their own weight and external loads, such as foundations or vehicles. Lack of sufficient strength can produce landslides, bearing capacity failures, and rutting of pavements.

2. Fills must be sufficiently stiff to avoid excessive settlement. Soft fills permit foundations to settle excessively, thus damaging buildings, and often produce undesirable changes in surface drainage patterns.
3. Fills must continue to satisfy the above two requirements, even if they become wet.
4. Some fills, such as the core of earth dams or liners for sanitary landfills, must have a sufficiently low hydraulic conductivity to restrict the flow of water. Others, such as aggregate base material below pavements, must have a high hydraulic conductivity to drain water away from critical areas.
5. In areas prone to frost heave (a heaving that occurs when the ground freezes), it is sometimes desirable to have fills made of soils that are not frost-susceptible.

Fills for residential, commercial, and industrial projects built before the 1960s often did not fully satisfy these requirements. As a result, these buildings often experienced excessive settlements, and slope failures such as landslides were common. The term *fill* became associated with shoddy construction. However, not all fills are inherently bad—only poorly constructed fills.

During the 1960s, building codes began to include stricter requirements for fills, and required builders to hire geotechnical engineers. These changes produced significantly better fills, and the problems were significantly reduced. Today, fills built in accordance with accepted standards of practice are at least as good as cuts.

The new grading techniques included many changes, most notably stricter requirements for soil compaction. This means all fills must be packed tightly using heavy equipment.

6.2 CONSTRUCTION METHODS AND EQUIPMENT

Engineers need to have a good understanding of construction methods and equipment because the way something is built often has a significant impact on how it must be designed. This is especially true for geotechnical engineers, because we are usually involved in both the design and construction phases of a project.

Historic Methods

Builders have been making cuts and fills for thousands of years, often with primitive tools and large numbers of slaves and animals. A number of impressive projects were built long before the advent of modern construction equipment by hauling soil in baskets and carts. Among the most noteworthy were the earth dams built for irrigation purposes in India and Sri Lanka. For example, Kalaweva reservoir in Sri Lanka was created by a 19-km long, 21-m tall earth dam built in AD 459. Another even larger dam would have created a huge reservoir. However, this dam was never used because the canal that was to feed the reservoir was built running uphill, suggesting that the builders' expertise in earthwork had far surpassed their skills in surveying (Schuyler, 1905).

The first significant advancements came in the nineteenth century with the introduction of steam power. This led to mechanized earthmoving equipment and began the era of efficient large-scale earthmoving.

Large steam-powered equipment was used to build the Panama Canal, which was the largest earthmoving project of its day. The Culebra Cut (now known as the *Gaillard Cut*), shown in Figure 6.4, was the most difficult part of the construction because of its height and the very difficult geologic conditions. The cut passes through a formation known as the Cucaracha (Cockroach) Shale, a weak sedimentary rock. Unfortunately, the first efforts to build the canal were orchestrated by the arrogant Ferdinand de Lesseps, who despised engineers and began construction without the benefit of a geologic study (Kerisel, 1987). This led to a misguided attempt to build a sea-level canal, which would have required a 109-m cut at Culebra. Massive landslides (created by the construction activities), yellow fever, and financial insolvency eventually halted the first attempt to build the canal.

When construction resumed after geologic studies, locks had been added to raise the canal 25 m above sea level. Even so, huge landslides continued to be a problem, resulting in enormous excavation quantities. The Culebra Cut alone required 75,000,000 m^3(98,000,000 yd^3), a massive quantity even by today's standards. The canal was completed in 1914, and stands as both a great civil engineering achievement and a dramatic case study demonstrating the importance of understanding geology.

FIGURE 6.4 Construction of the Culebra Cut for the Panama Canal, 1907. The steam-powered excavators placed loads of soil and rock into railroad cars that hauled them away. (Library of Congress.)

Hydraulic Fills

During the early decades of the twentieth century, many large earthwork projects were built using a technique called *hydraulic filling*. This method consisted of mixing the soil with large quantities of water, conveying the mixture to the construction site through pipes and flumes, then depositing it at the desired locations. The soil settled in place and the excess water was directed away. No compaction equipment was used. Figure 6.5 shows a hydraulic fill dam under construction.

This technique was popular between 1900 and 1940, especially for earthfill dams, because earthmoving equipment was too small and underpowered for such large projects. Unfortunately, the quality of such fills was poor and they often experienced large settlements and landslides.

The last significant hydraulic fill built in the United States was at Ft. Peck Dam in Montana. A very large, 3,800,000 m^3(5,000,000 yd^3) landslide occurred during construction in 1938 and was blamed on the poor quality of the hydraulic fill. The advent of modern earthmoving equipment was already underway, making hydraulic fills obsolete.

FIGURE 6.5 Construction of San Pablo Dam in California using hydraulic fill techniques, circa 1918. Two fills are being placed simultaneously, one on each side of the photograph, to form the shells. The finer soils flow to the center and form the impervious core. (U.S. Bureau of Reclamation.)

Another serious problem with hydraulic fills became evident in 1971 when the Lower San Fernando Dam near Los Angeles failed during an earthquake measuring 6.4 on the Richter scale. This failure was due to liquefaction of the hydraulic fill soils. As a result, several hydraulic fill dams have been rebuilt or replaced to avoid similar failures.

Modern Earthmoving

The twentieth century brought further advances in earthmoving equipment, especially during the period 1920–1965, so today we can move large quantities of earth at a very low cost. To illustrate these improvements, consider the 1914 excavation costs for the Panama Canal, which were about $0.79/yd^3 (Church, 1981). If we adjust this figure for inflation, it translates to about $17/yd^3 considering the value of the dollar in 2010. However, the 2010 cost of similar work using modern equipment is less than one-quarter of that adjusted cost.

Equipment A thorough discussion of modern earthmoving equipment is well beyond the scope of this book. However, we will cover some of the principal machines and their applications.

The key development in modern earthmoving equipment was the introduction of the *tractor* or *crawler* shown in Figure 6.6. It converts engine power into traction, and thus moves both itself and other equipment. The first tractors were developed for agricultural and military purposes during the early twentieth century. They were mounted on tracks to allow mobility over very rough terrain. Modern track-mounted equipment, such as the one shown in Figure 6.6(a) is very powerful and mobile, but operates at slow speeds, generally no more than 11 km/hr (7 mi/hr). Wheel-mounted tractors, such as the one shown in Figure 6.6(b) also are available, and have the advantage of greater operating speeds, often in excess of 50 km/hr (30 mi/hr). However, wheel-mounted equipment has less traction and is not as well suited for rough terrain. Thus, both types have a role in modern earthmoving.

(a) (b)

FIGURE 6.6 Tractors: (a) track-mounted or crawler type; (b) wheel-mounted type. (Caterpillar, Inc.)

Various kinds of equipment can be attached to tractors to produce productive work. One of the most common accessories is the *bulldozer*, which is a movable steel blade attached to the front of a tractor. For example, the tractor in Figure 6.6(a) is equipped with a bulldozer, and could be used for cutting, moving, spreading, mixing, and other operations.

Another common attachment is a *loader*, as shown in Figure 6.7. It consists of a bucket attached to the front of a tractor and can be used to pick up, transport, and deposit soil. Loaders also may be used for light excavation.

A *hoe* attachment is a bucket used for digging pits or trenches. A special tractor with a loader in the front and a hoe in the back, as shown in Figure 6.8, is called a *backhoe*, and is very commonly seen on construction sites. An *excavator*, shown in Figure 6.9, is larger and mounted on a special chassis instead of on a tractor.

Often, tractors are described by the names of their attachments. For example, the term *bulldozer* (or simply '*dozer*') often refers to a tractor with a bulldozer attachment.

Conventional Earthwork We will use the term *conventional earthwork* to describe the excavation, transport, placement, and compaction of soil or soft rock in areas where equipment can move freely. This process may be divided into several distinct steps, each requiring appropriate equipment and techniques.

Clearing and Grubbing The first step in most earthwork projects is to remove vegetation, trash, debris, and other undesirable materials from the areas to be cut or filled. Stumps, roots, buried objects, and contaminated soils also need to be removed.

FIGURE 6.7 Wheel-mounted loader removing soil from a stockpile. (Caterpillar Inc.)

FIGURE 6.8 A backhoe is a tractor with a loader on the front and a hoe on the back. They are commonly used to dig trenches, as shown here. (Caterpillar Inc.)

FIGURE 6.9 An excavator is a large hoe mounted on the special rotating chassis. This excavator is digging a hole on the right side of the picture and dumping the spoils in a pile on the left side. Excavators also can dump directly into trucks. (Caterpillar Inc.)

FIGURE 6.10 Clearing and grubbing in a former orange grove. The trees are being removed and hauled to the dumpster seen in the background.

Most of these materials would have a detrimental effect on the fill, and must be hauled off the site. The above-ground portion of this work, as shown in Figure 6.10, is called *clearing* and the underground portion is called *grubbing*.

The time and money required for clearing and grubbing varies from site to site. For example, very little work may be required in arid areas because little vegetation is present, while substantial effort is often required in tropical areas because of their dense forests with thick underbrush. In forests of marketable timber, the value of the trees removed during clearing may generate a net profit.

Vegetation may remain in areas that will not be cut or filled, and grading plans often are designed to save desirable vegetation, especially large trees. These areas often require special protection so they are not damaged by the construction activities.

Sometimes clearing and grubbing is accompanied by *stripping*, which consists of removing and storing the topsoil. Such soils are valuable because they contain nutrients for plants. Once the grading is completed, these soils are returned to the top of the graded surface in areas to be landscaped.

Limited quantities of inorganic debris, such as chunks of concrete, bricks, or asphalt pavement, do not need to be hauled away and may be incorporated into the fill so long as they are no larger than about 300 mm (12 in.) and are mixed with a sufficient quantity of soil. Larger debris is called *oversize*, and must be spread out in the fill or placed with other special techniques to avoid the creation of undesirable voids. However, oversized objects should not be placed in fills that are to be penetrated with pile foundations, nor in the upper 3 m (10 ft), as they would cause problems with utility excavations.

Excavation Excavation, which is the removing of soil or rock, can occur at many different locations. Usually, most of the excavation occurs in areas where the proposed ground surface is lower than the existing ground surface. Normally, the excavated materials are then used to make fills at other portions of the project site. Sometimes additional soil is needed, so it becomes necessary to obtain it by excavating at offsite *borrow pits*, which are places where soil is removed to be used as import (the term *borrow* is a misnomer, since we have no intention of bringing the soil back!). Finally, areas to be filled are often prepared by first excavating loose upper soils, thus exposing firm ground on which to place the fill.

Contractors have various kinds of earthmoving equipment that may be used to excavate soils. Loaders are useful when the excavated soil is to be immediately loaded into a truck or conveyor belt, but are generally not used to transport the soil for long distances. *Scrapers* (sometimes called *pans*), such as the one shown in Figure 6.11, are much more efficient for most moderate- to large-size projects because they can load, transport, and unload. Figure 6.12 shows the internal operation of a scraper.

Sometimes the ground is too hard to be excavated with a scraper or loader. This problem often can be overcome by first loosening it with a *ripper* attached to a tractor, as shown in Figure 6.13. This device consists of one or more teeth that are pressed into the ground, and then pulled through to loosen it. The ripping operation can then be followed by excavation using excavating equipment.

When the ground is so hard that even rippers do not work, it may become necessary to use *blasting*. This consists of drilling strategically placed holes into the ground and packing them with an explosive. Then, the explosives are detonated, thus loosening the ground and allowing it to be excavated. Special chemicals also are available for use

FIGURE 6.11 A scraper transporting a load of soil across a construction site. This scraper is equipped with an elevating mechanism at the front of the bowl to assist in loading soil.

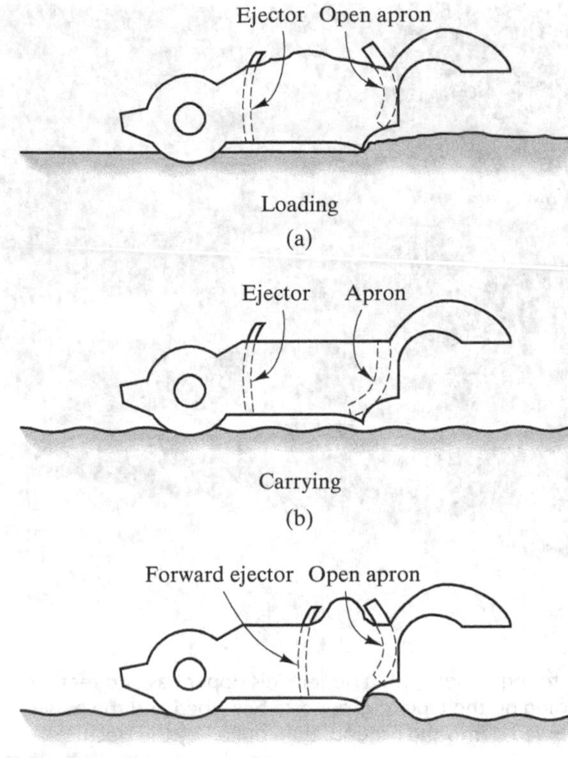

FIGURE 6.12 Operation of a scraper: (a) To load the bowl, the operator opens the apron, moves the ejector plate to the rear, and lowers the front so it digs 100–150 mm (4–6 in.) into the ground. As the scraper moves forward, it fills with soil. In harder ground, it may be necessary to push the scraper with a bulldozer during the loading phase. (b) To transport soil, the apron is closed and the bowl is lifted. Scrapers usually transport soil at speeds of about 32 km/hr (20 mi/hr). (c) Upon reaching the area to be filled, the operator again opens the apron, but lowers the bowl only slightly, leaving it 100–150 mm (4–6 in.) above the ground surface. The ejector is moved forward, pushing the soil out and under the bowl, thus depositing a uniform thickness of soil. (Wood, 1977.)

in areas where the noise and vibration from explosives are unacceptable. The contractor pours these liquid chemicals and a catalyst into the holes. The catalyst causes the chemical to solidify and expand, thus fracturing the rock.

Often the *excavatability* (ease of excavation) or *rippability* (ease of ripping) at a site is evident from a visual inspection and the required equipment and techniques may be selected accordingly. At questionable sites, measurements of the seismic wave velocity

FIGURE 6.13 A track-mounted tractor equipped with a ripper. This ripper has two teeth; others have as many as four or five, depending on the type of ground to be ripped and the power of the tractor. The teeth are hydraulically lowered into the ground, then pulled by the tractor.

from a seismic refraction survey (see Chapter 3) can assist in selecting the proper equipment. Generally, soil and rock with velocities less than about 500 m/s (1600 ft/s) can be excavated without ripping. Higher velocities can be assessed using Figure 6.14.

Rippability also depends on other factors not reflected in the seismic wave velocity, such as the size and spacing of joints, ripper tooth penetration, presence of boulders, and operator technique.

Because the excavatability of soil and rock can vary widely, even on a single job site, it becomes an important issue in computing payment to the contractor. If the excavation quantities are *unclassified*, the contractor receives the same unit price for all materials, while *classified* excavation sets different prices depending on the ease of excavation. Typically, classified excavation is divided into two categories: *common*, which includes soil, and *rock*. Rock excavation would have a much higher unit price. Unfortunately, there are many intermediate materials that could arguably fit into either classification, so this often becomes a point of contention between engineers and contractors, and has been the source of many lawsuits. To avoid such problems, specifications for classified excavation need to clearly define each category.

Transport and Placement Although bulldozers and wheel loaders can transport soils for short distances (i.e., less than 100–150 m), they become very uneconomical with longer hauls. For these projects, it becomes necessary to use other equipment.

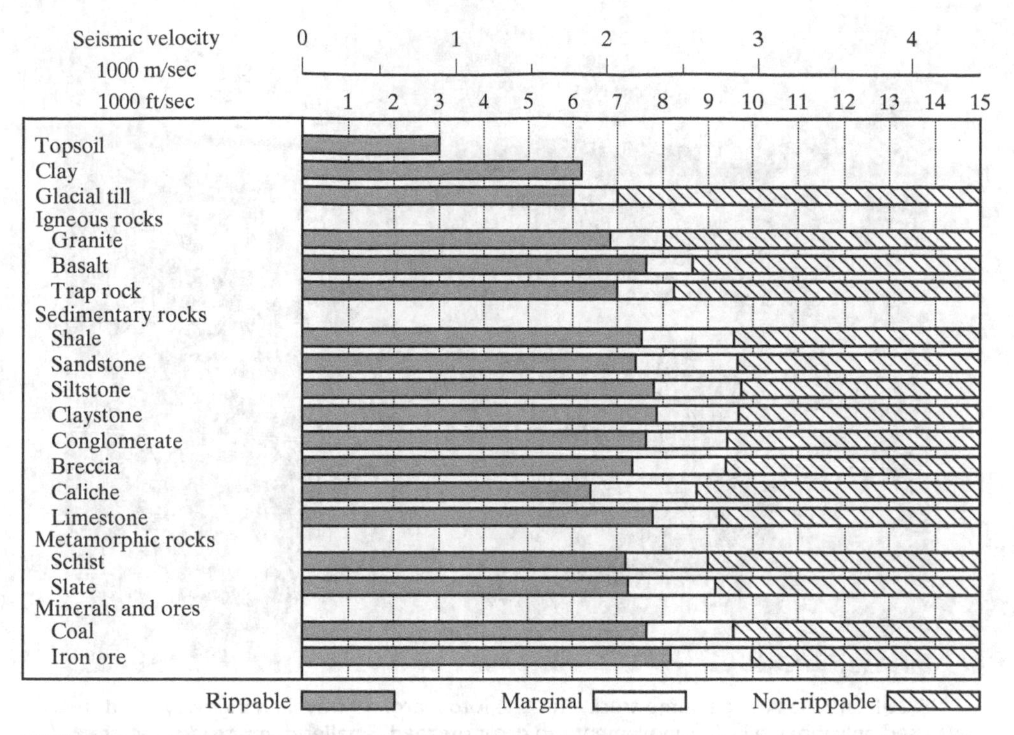

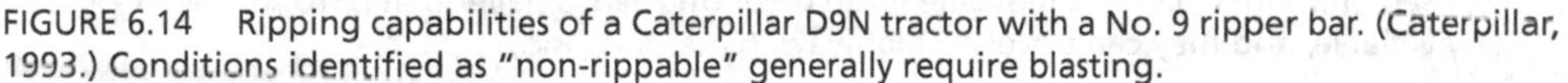

FIGURE 6.14 Ripping capabilities of a Caterpillar D9N tractor with a No. 9 ripper bar. (Caterpillar, 1993.) Conditions identified as "non-rippable" generally require blasting.

Scrapers are very efficient at moderate-length hauls, and many earthwork projects fall into this category. They can excavate, transport, and place the soil, as described above, but cannot be used to haul over public highways.

Dump trucks, like the one shown in Figure 6.15, can be used instead of scrapers, especially when the soil is being excavated by loaders. Most dump trucks can travel over public highways, and they move faster than scrapers. However, this method requires more equipment and more operators.

When soil needs to be hauled for longer distances over the highway, contractors often use *wagons*. These are towed by semi tractors and are most cost-effective when the haul distance is long. Upon arriving at the fill site, the wagons are self-unloading, either through the bottom or the back. Large off-road wagons, such as the one in Figure 6.16, also are available.

When large quantities of soil need to be transported to a confined area, such as with earth dams, a system of *conveyor belts* sometimes is an efficient method. One such system is shown in Figure 6.17. Bull Shoals Dam in Arkansas was built using an exceptionally long series of conveyor belts that stretched 7 mi (11 km) from a borrow pit to the dam.

The selection of equipment to transport soils depends on many factors, most notably the haul distance as described in Table 6.1.

FIGURE 6.15 A large dump truck. This one is too large to travel on highways, and is used only when all of its movement can occur off road. Smaller dump trucks also are available, and they can travel on highways. (Caterpillar Inc.)

FIGURE 6.16 A very large off-road wagon. Upon reaching its destination, this wagon will unload the soil through doors on the bottom. Smaller versions of this design can be towed by semi tractors over the highway.

FIGURE 6.17 Use of conveyor belts to transport soil to the Seven Oaks Dam in California. These belts move soil about 3 km (2 mi) from the borrow site to the dam and then drop it at a convenient location. The loader places it in large dump trucks that haul it a short distance to the construction site.

TABLE 6.1 Economical Haul Distances[a]

Equipment	Economical Haul Distance (m)	Economical Haul Distance (ft)
Bulldozer	<100	<300
Wheel loader	50–150	150–500
Scraper	300–2,500	1,000–8,000
Dump truck	350–6,500	1,100–21,000
Conveyor belt	30–11,000	100–36,000
Wagon	>3,000	>10,000

[a]Adapted from Caterpillar, 1993.

Once the soil arrives at the area to be filled, it must be laid out in thin horizontal *lifts*, typically about 200 mm (8 in.) thick. Each lift must be moisture-conditioned and compacted before the next lift is placed. Thus, the fill is constructed one lift at a time.

Moisture Conditioning The soil must be at the proper moisture content before it is compacted, and will not compact well if it is either too wet or too dry. Usually the moisture content of the soil to be placed is not correct and needs to be adjusted accordingly. If the soil is too dry, this is usually done by spraying it with a water truck, as shown in Figure 6.18. A bulldozer or other equipment is then used to mix the soil so the water is uniformly distributed.

FIGURE 6.18 A water truck spraying a new lift of soil for a fill and preparing it to be compacted.

The grading contractor has a much more difficult problem when the soil is too wet. Mixing it with dryer soil is difficult, especially with clays, because it is very hard to achieve thorough mixing. The result often consists of alternating clumps of wet and dry soil instead of a uniform mixture. The most common technique is to spread the wet soil over a large area and allow the sun to dry it. This works well so long as a rainstorm does not occur.

Compaction The next step is to compact the lift. *Compaction* is the use of equipment to compress soil into a smaller volume, thus increasing its dry unit weight and improving its engineering properties. Figure 6.19 shows the typical changes in the volumes of the three phases in the soil from the time when the soil was in its natural condition to when it is compacted into a fill. Because the solids and water are virtually incompressible and because typically the water in the soil cannot move out of the soil during compaction, compaction produces a reduction in the volume of air, with the volumes of solids and water remaining unchanged during compaction.

Although many early fills were built without any special effort to compact them, some engineers recognized the importance of compaction as early as the nineteenth century. Animals were used as compaction "equipment" on some projects, including a team of 115 goats used to compact an earth dam near Santa Fe, New Mexico, in 1893 (Johnson and Sallberg, 1960). Heavy rollers also began to be used. Initially these rollers were pulled by teams of horses, but by 1920 the horses had been replaced with tractors. Further developments during the twentieth century enhanced the capabilities of compaction equipment. Today, a wide variety of effective compaction equipment is available.

All of the equipment that drives over a fill, from pickup trucks to loaded scrapers, contributes to its compaction. However, we generally cannot rely only on this incidental compaction because of the following reasons:

- Some construction equipment is intentionally designed to have low contact pressures between the tires or tracks and the soil. This allows them to travel more quickly and easily through soft ground. For example, a Caterpillar 973 track

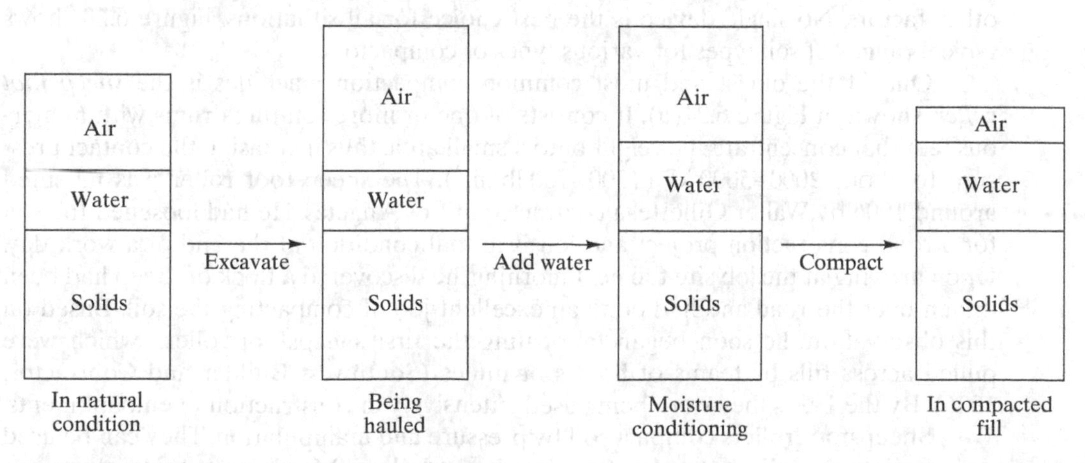

FIGURE 6.19 Phase diagrams showing the changes in soil as it moves from it natural location to a compacted fill. Note that the volume of solids does not change during the process.

loader has a contact pressure of only 83 kPa (12 lb/in.^{2}). Such pressures are too low to produce the required compaction in normal-thickness lifts. Other equipment has much larger contact pressures and does provide good compaction.

- Incidental traffic usually follows common routes, so their compactive effort is not uniformly distributed across the fill. Thus, some areas may receive sufficient compaction, while others receive virtually none.

Therefore, it is usually necessary to use specialized compaction equipment specifically designed for this task. Such equipment is much more efficient and effective. All compaction equipment uses one or more of the following four methods (Spann, 1986):

- *Pressure*: The contact pressure between the equipment and the ground is probably the most important factor in the resulting compaction of the underlying soils. A typical sheepsfoot roller has a contact pressure of about 3500 kPa (500 lb/in.^{2}), which is far greater than the track-mounted equipment described earlier.
- *Impact:* Some equipment imparts a series of blows to the soil, such as by dropping a weight. This adds a dynamic component to the compactive effort.
- *Vibration*: Vibratory compaction equipment utilizes eccentric weights or some other device to induce strong vibrations into the soil, which can enhance its compaction. These vibrations typically have a frequency of 1000–3500 cycles per minute.
- *Manipulation*: Compaction equipment that imparts some shearing forces to the soil can also contribute to better compaction. This action is called *kneading* or *manipulation*. However, excessive manipulation, such as in an overly wet fill, can be detrimental. When such fills are simply being moved around with no compaction occurring, we have a condition called *pumping*.

The proper selection of compaction equipment and methods depends on the type of soil, the size of the project, compaction requirements, required production rate, and other factors. No single device is the best choice for all situations. Figure 6.20 shows typical ranges of soil types for various types of compactors.

One of the oldest and most common compaction machines is the *sheepsfoot roller*, shown in Figure 6.21(a). It consists of one or more rotating drums with numerous feet that concentrate its weight onto a small area, thus increasing the contact pressure to about 2000–5000 kPa (300–700 lb/in.^{2}). The sheepsfoot roller was invented around 1906 by Walter Gillette, a contractor in Los Angeles. He had loosened the soil for a road construction project and left it in that condition at the end of a work day. Upon arriving at the job site the next morning, he discovered a flock of sheep had been driven over the road and had done an excellent job of compacting the soil. Based on this observation, he soon began fabricating the first sheepsfoot rollers, which were pulled across fills by teams of horses or mules (Southwest Builder and Contractor, 1936). By the 1930s they were being used extensively in construction of embankments.

Sheepsfoot rollers compact soil by pressure and manipulation. They can be used on a variety of soils, but work best in silts and clays. Most sheepsfoot rollers can accommodate soil lifts with loose thicknesses of about 200 mm (8 in.).

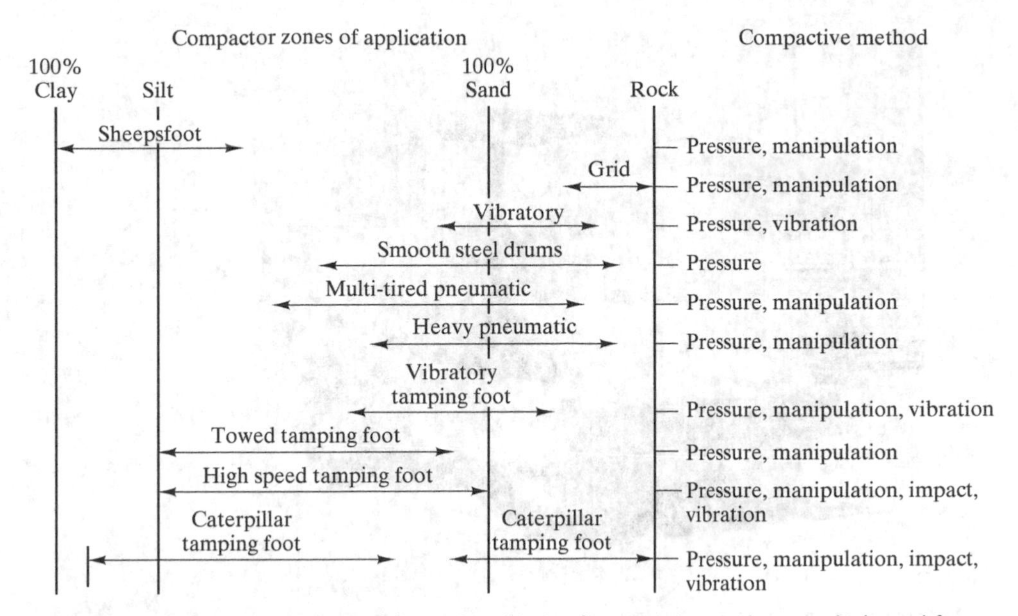

FIGURE 6.20 Soil types best suited for various kinds of compaction equipment. (Adapted from Caterpillar, 1993.)

FIGURE 6.21 Footed rollers: (a) towed sheepsfoot roller; (b) self-propelled tamping foot roller. (Caterpillar Inc.)

Tamping foot rollers, such as the one shown in Figure 6.21(b), are very similar to sheepsfoot rollers, except they use larger feet with a correspondingly smaller contact pressure. They can be operated at a faster speed, but do not compact to as great a depth.

Pneumatic rollers (also known as *rubber-tire rollers*) are heavy units resting on several tires. The contact pressure is typically about 600 kPa (85 lb/in.^{2}). Each tire is able to move up and down independently, so this device is good at finding small soft

FIGURE 6.22 Self-propelled vibratory roller.

spots that rigid compaction equipment, such as sheepsfoot rollers, can miss. The tires also provide a kneading action that enhances compaction. These rollers can compact lifts with loose thicknesses of 250–300 mm (10–12 in.).

Vibratory rollers, such as the one in Figure 6.22, are similar to sheepsfoot or tamping foot rollers, with the addition of a vibrating mechanism. Thus, they use pressure, manipulation, and vibration to compact the soil. Vibration is especially effective in sandy and gravelly soils. The heaviest of these rollers can accommodate loose lift thicknesses of up to 1 m (3 ft), and they provide some compactive effort to depths of about 2 m (7 ft).

Smooth steel-wheel rollers (do not call them "steam rollers"!), such as the one in Figure 6.23, leave a smooth compacted soil surface. The nonvibratory types are not well suited for compacting soil because the contact pressure is much lower than that of sheepsfoot rollers. However, they may be used to *proof roll* a subgrade just before paving (i.e., a final rolling to confirm compaction of the uppermost soils), and to compact the aggregate base course and asphalt pavement.

Fine Grading Once the last lift has been placed and the fill is approximately at the final elevation, we say the *rough grading* is complete. Then the contractor begins *fine grading*, which consists of careful trimming and filling to produce the desired configuration. As the name implies, fine grading is more precise and therefore requires different equipment.

On highway subgrades and large building and parking lot pads, a *motor grader* (or *blade*) is often used, as shown in Figure 6.24. At more confined sites, a small loader equipped with a fine-grading accessory called a *gannon* might be used.

FIGURE 6.23 A smooth steel-wheel roller preparing to compact an asphalt pavement. (Caterpillar Inc.)

FIGURE 6.24 A motor grader fine grading a highway subgrade. (Caterpillar Inc.)

Fine grading slopes is more difficult, but it is nevertheless important both for aesthetics and to avoid future surficial stability problems. One method is to intentionally overfill the slope, then trim them back using a dozer equipped with a cutting arm called a *slope board*.

Small Backfills Another type of earthwork, quite different from conventional grading, is the backfilling of small confined areas, such as behind retaining walls and inside small excavations. These areas are not large enough to accommodate the equipment described earlier, so it becomes necessary to use smaller equipment and more hand labor. Figure 6.25 shows examples of portable compaction equipment that may be used in confined spaces.

Utility Trenches Many civil engineering projects include installation of underground utilities, such as water pipes, electrical lines, sewer pipes, and so on. These are installed by digging a long trench, placing the pipe or conduit, and backfilling. The compaction of these backfills is important, especially because they often are beneath roadways. Unfortunately, many utility trenches are not properly compacted, and they eventually settle. This produces a condition that is both unsightly and hazardous.

Compaction can be difficult because of the narrow width, and because it must be accomplished without breaking the pipe. Often, the zone immediately around the pipe is filled with clean sand that is compacted by *jetting*, which consists simply of injecting water into the sand and relying on gravity and lubrication to compact the soil. The remainder of the trench is backfilled in lifts and compacted using a variety of equipment, such as the hoe-mounted sheepsfoot shown in Figure 6.26.

FIGURE 6.25 Using portable compaction equipment to compact soils in areas that are too confined to accommodate large equipment. (Wacker Corporation.)

FIGURE 6.26 A hoe-mounted sheepsfoot roller compacting a utility trench backfill.

6.3 SOIL COMPACTION CONCEPTS

Soil compaction is one of the most important aspects of earthwork construction, and proper compaction often is the key difference between a poor fill and an excellent fill. Compaction leads to improved engineering properties of a fill, including the following:

- Increased shear strength, which reduces the potential for slope stability problems, such as landslides, and enhances the fill's capacity for supporting loads, such as those from foundations.
- Decreased compressibility, which reduces the potential for excessive settlement.
- Decreased hydraulic conductivity, which inhibits the flow of water through the soil. This may be desirable or undesirable, depending on the situation.
- Decreased void ratio, which reduces the amount of water that can be held in the soil, and thus helps maintain the desirable strength properties.
- Increased erosion resistance, which helps maintain the ground surface in a serviceable condition.

However, soil compaction comes at some cost, so engineers need to have a method of determining how much compaction is needed to produce the desired quality, while not incurring unnecessary costs associated with overcompaction. We also need to know what construction methods will produce the greatest compaction at the lowest cost. Thus, it is necessary to understand the physics of soil compaction and use this knowledge to develop appropriate design and quality control procedures.

The work imparted to compact a soil, which is an indicator of the associated cost, is expressed in terms of the *compactive effort*, which is the work performed per unit volume of soil. In the field, this compactive effort is related to the weight of the compaction equipment, the number of passes of the equipment over each lift (or layer) of soil to be compacted, the lift thickness, and other factors. In the laboratory, the compactive effort can be measured in terms of the weight and number of blows from the compaction hammer and other factors.

How compacted a soil is, or the *degree of compaction*, depends on various factors including:

- The type of soil being compacted.
- The method of compaction.
- The compactive effort.
- The moisture content of the soil being compacted (the *as-compacted moisture content*).

To measure the degree of compaction or fill quality, geotechnical engineers use the dry unit weight, γ_d, with higher values indicating better quality. Although we are really much more interested in strength, compressibility, and other characteristics, the dry unit weight is a convenient proxy and is easier to measure.

Compaction Curves

One of the key considerations in the design and construction of compacted fills is the effect of the as-compacted moisture content on γ_d. To assess this characteristic for the specific soil to be used, we conduct laboratory tests using a constant compactive effort to obtain a *compaction curve*, which is a plot of the as-compacted γ_d versus the as-compacted moisture content, w.

Figure 6.27 presents a typical family of compaction curves for the same soil, each of which corresponds to a different compactive effort. Each curve peaks at a certain moisture content, known as the *optimum moisture content*, w_o, corresponding to the *maximum dry unit weight*, $\gamma_{d,max}$. This is the point at which we obtain the densest fill for a given compactive effort. Note how an increase in compactive effort produces a smaller w_o and a larger $\gamma_{d,max}$.

The mechanics behind the shape of a compaction curve are very complex (Hilf, 1991). In a relatively dry soil, we achieve better compaction by first adding water to raise its moisture content to near-optimum. This water provides lubrication, softens clay bonds, and reduces surface tension forces within the soil. However, if the soil is too wet, there is little or no air left in the voids, and it thus becomes very difficult or impossible to compact. Such soils need to be dried before they are compacted. Based on empirical data, the peak of the compaction curve occurs roughly at a degree of saturation, S, of about 80%, as shown in Figure 6.27.

Although most soils have compaction curves similar to those in Figure 6.27, clean sands and gravels (SP, GP, and some SW and GW soils) often do not. They typically

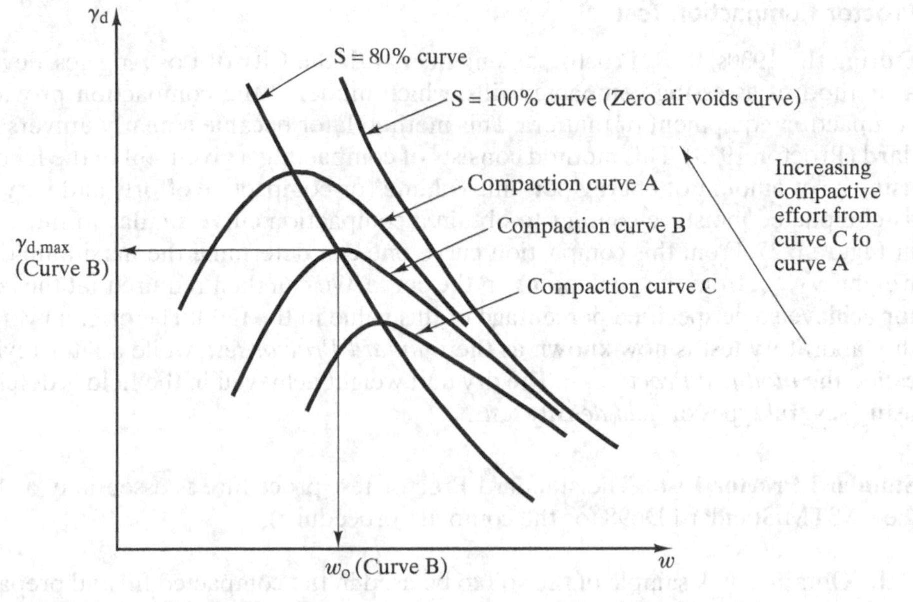

FIGURE 6.27 Typical compaction curves for a given soil showing relationships among dry unit weight, as-compacted moisture content, and compactive effort.

have much flatter compaction curves (i.e., they are less sensitive to moisture content than clays) and these curves sometimes have two small peaks.

6.4 SOIL COMPACTION STANDARDS AND SPECIFICATIONS

Earthwork construction projects require detailed specifications that address a variety of issues, including compaction. The geotechnical engineer develops these specifications, and the contractor is then obligated to satisfy them.

Virtually all compaction specifications include the criterion of achieving a minimum dry unit weight, γ_d, and occasionally, the moisture content is also specified to be within a certain range. Then, during construction, the geotechnical engineer usually has a staff of field engineers and technicians who measure the as-compacted dry unit weight and as-compacted moisture content achieved in the field to verify the contractor's compliance with these specifications. It is important to recognize that the combination of the as-compacted dry unit weight and as-compacted moisture content is an indicator of quality, is easy to measure, and correlates empirically with desirable engineering properties. In other words, we want favorable strength, compressibility, hydraulic conductivity, void ratio, and erosion resistance properties, and know empirically that we have attained them when the dry unit weight criterion and the moisture content criterion, if necessary, have been met. Therefore, the question now becomes what minimum dry unit weight and what moisture content range, if necessary, should be specified. We will answer these questions with the help of two standardized laboratory compaction tests and some empirical observations on the relationship between as-compacted moisture content and soil fabric.

Proctor Compaction Test

During the 1930s, R. R. Proctor, an engineer with the City of Los Angeles, developed a method of assessing compacted fills, which modeled the compaction provided by compaction equipment of that era. This method later became a nearly universal standard (Proctor, 1933). This method consists of compacting a given soil in the laboratory using a set amount of energy per unit volume (or compactive effort) and varying the as-compacted moisture content to obtain a compaction curve similar to those shown in Figure 6.27. From this compaction curve one can determine the maximum dry unit weight, $\gamma_{d,max}$, from the peak point on the curve. We can then require that the contractor achieve some specified percentage of this value in the field. The original version of the laboratory test is now known as the *standard Proctor test*, while a later revision is called the *modified Proctor test*. The dry unit weight achieved in the field is determined using several types of *field density tests*.

Standard Proctor Test The standard Proctor test procedure is essentially as follows (see ASTM Standard D698 for the complete procedure):

1. Obtain a bulk sample of the soil to be used in the compacted fill and prepare it in a specified way.
2. Place some of the prepared soil into a standard 1/30 ft^3 (9.44×10^{-4} m^3) cylindrical steel mold until it is about 40% full. This mold is shown in Figure 6.28.
3. Compact the soil by applying 25 blows from a special 5.5-lb (2.49-kg) hammer that drops from a height of 12 in. (305 mm).
4. Place a second layer of the prepared soil into the mold until it is about 75% full and compact it using 25 blows from the standard hammer.
5. Place the third layer of the prepared soil into the mold and compact it in the same fashion. Thus, we have applied a total of 75 hammer blows, giving a constant compactive effort.
6. Trim the sample so that its volume is exactly 1/30 ft^3, then weigh it. The unit weight, γ, is thus:

$$\gamma = \frac{W_{ms} - W_m}{V_m} \tag{6.1}$$

where:

W_{ms} = weight of mold + soil

W_m = weight of mold

V_m = volume of mold = 1/30 ft^3 = 9.44×10^{-4} m^3

When using English units, γ is expressed in lb/ft^3. With SI units, use kN/m^3, which requires converting kg to kN (1 kN = 102.0 kg).

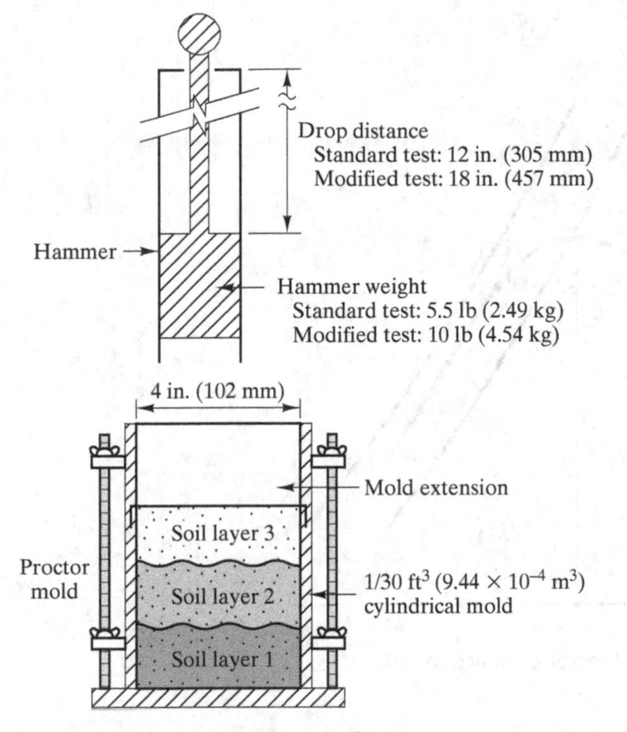

FIGURE 6.28 Mold and hammer for a Proctor compaction test. The standard test uses three layers, as shown, while the modified test uses five.

7. Perform a moisture content test on a representative portion of the compacted sample, then compute the dry unit weight, γ_d, using Equation 4.27.
8. Repeat steps 2–7 three or four times, each with the soil at a different moisture content.

The test results are then plotted on a γ_d versus w diagram as shown in Figure 6.29. Before drawing a compaction curve, it is good practice to first draw two other curves: One representing the variation of dry unit weight with moisture content when $S = 100\%$ (sometimes called the *zero air voids curve*), and the other representing the same variation when $S = 80\%$, as shown in Figures 6.27 and 6.29. These two curves can be developed using Equation 4.32, and are intended to help us draw the compaction curves. The $S = 100\%$ curve represents an upper limit for the compaction data, for it is impossible to have $S > 100\%$. Based on empirical data, the $S = 80\%$ curve should go nearly through the peaks of the compaction curves. In addition, the right limb of a compaction curve should be slightly to the left of and roughly parallel to the $S = 100\%$ curve, because compaction does not remove all of the air. Example 6.1 illustrates this data reduction process.

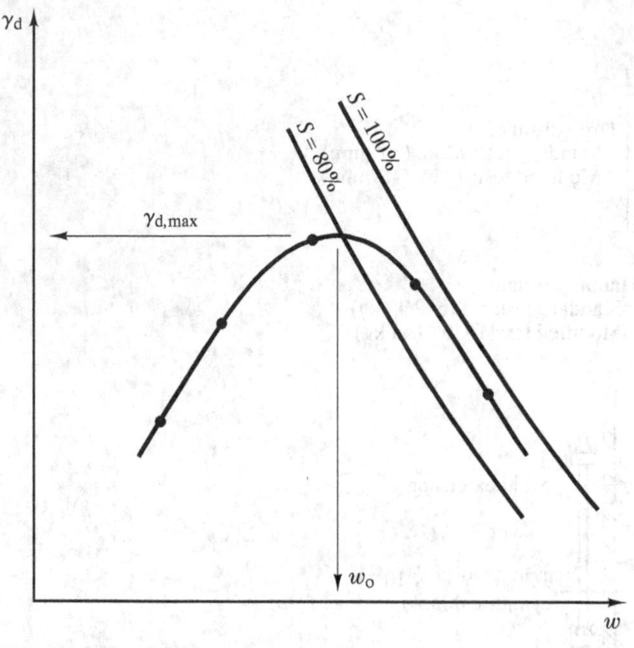

FIGURE 6.29 Results from a Proctor compaction test.

Example 6.1

A series of standard Proctor compaction tests has been performed on a soil sample. The test results are as follows:

	Data Point No.	1	2	3	4	5
	Mass of compacted soil + mold (kg)	3.762	3.921	4.034	4.091	4.040
Moisture Content	Mass of can (g)	20.11	21.24	19.81	20.30	20.99
	Mass of can + wet soil (g)	240.85	227.03	263.45	267.01	240.29
	Mass of can + dry soil (g)	231.32	212.65	241.14	238.81	209.33

The mass of the compaction mold was 2.031 kg. The moisture content tests were performed on small portions of the compacted soil samples.

a. Compute γ_d and w for each data point and plot these results.

b. Develop and plot curves for $S = 80\%$ and $S = 100\%$ using a specific gravity of 2.69.

c. Using the data from parts a and b, draw the Proctor compaction curve and determine $\gamma_{d,max}$ and w_o.

Solution:

a. Computations for data point no. 1:
Using Equations 4.12 and 6.1,

$$\gamma = \frac{W_{ms} - W_m}{V_m}$$

$$= \frac{(M_{ms} - M_m)g}{V_m}$$

$$= \frac{(3.762 \text{ kg} - 2.031 \text{ kg})(9.81 \text{ m/s}^2)}{9.44 \times 10^{-4} \text{ m}^3}\left(\frac{1 \text{ kN}}{1000 \text{ N}}\right)$$

$$= 17.99 \text{ kN/m}^3$$

Using Equation 4.3,

$$w = \frac{M_1 - M_2}{M_2 - M_c} \times 100\%$$

$$= \frac{240.85 \text{ g} - 231.32 \text{ g}}{231.32 \text{ g} - 20.11 \text{ g}} \times 100\%$$

$$= 4.5\%$$

Using Equation 4.27,

$$\gamma_d = \frac{\gamma}{1 + w} = \frac{17.99 \text{ kN/m}^3}{1 + 0.045} = 17.22 \text{ kN/m}^3$$

Thus, this data point plots at $w = 4.5\%$, $\gamma_d = 17.22 \text{ kN/m}^3$. Following the same procedure for the other data points gives:

Data Point No.	**1**	**2**	**3**	**4**	**5**
γ (kN/m^3)	17.99	19.64	20.81	21.41	20.88
w	4.5%	7.5%	10.1%	12.9%	16.4%
γ_d (kN/m^3)	17.22	18.27	18.90	18.96	17.94

b. Using Equation 4.32, select values of γ_d in the range of test data from part a and compute the corresponding values of w for $S = 80\%$ and 100%:

	***w*(%)**	
γ_d (kN/m^3)	**@ *S* = 80%**	**@ *S* = 100%**
16.00	19.3	24.1
18.00	13.9	17.3
20.00	9.5	11.9

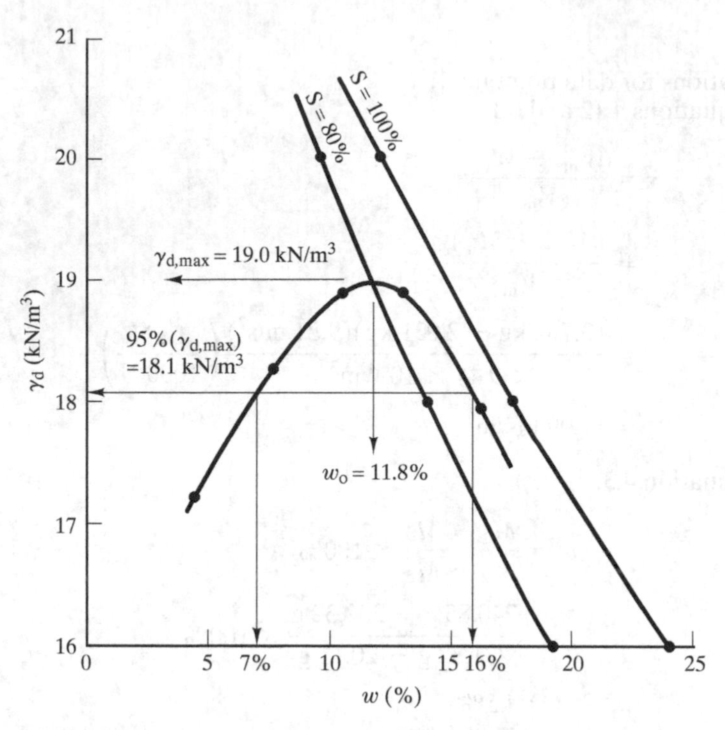

FIGURE 6.30 Laboratory test results for Example 6.1.

c. Plot the data points from part a and the curves from part b, and then draw the compaction curve as shown in Figure 6.30.

The final results are:

$$\gamma_{d,max} = \mathbf{19.0\ kN/m^3}$$
$$w_o = \mathbf{11.8\%}$$

Modified Proctor Test During the 1940s and 1950s truck and aircraft traffic increased significantly in both weight and frequency. Geotechnical engineers found that fills compacted using the standard Proctor test were no longer providing adequate support for modern trucks and aircrafts. At the same time, heavier and more efficient earthmoving and compaction equipment was being developed. The U.S. Army Corps of Engineers addressed these issues by developing the modified Proctor test, which used a higher compactive effort and thereby produced higher values of $\gamma_{d,max}$. The higher compactive effort better modeled the newer construction equipment that was producing fills to support the increased truck and aircraft traffic. The principal differences between the standard and modified tests are shown in Table 6.2 and Figure 6.31. The modified Proctor test was later adopted by the American Association of State Highway and Transportation Officials (AASHTO

TABLE 6.2 Principal Differences between Standard and Modified Proctor Compaction Tests

	Standard Proctor Test	Modified Proctor Test
Standards	ASTM D698 and AASHTO T-99	ASTM D1557 and AASHTO T-180
Hammer weight	5.5 lb (2.49 kg)	10.0 lb (4.54 kg)
Hammer drop height	12 in. (305 mm)	18 in. (457 mm)
Number of soil layers	3	5
Number of hammer blows per layer	25	25
Energy imparted from hammer per unit volume	12,400 ft-lb/ft³ (600 kN-m/m³)	56,000 ft-lb/ft³ (2,700 kN-m/m³)

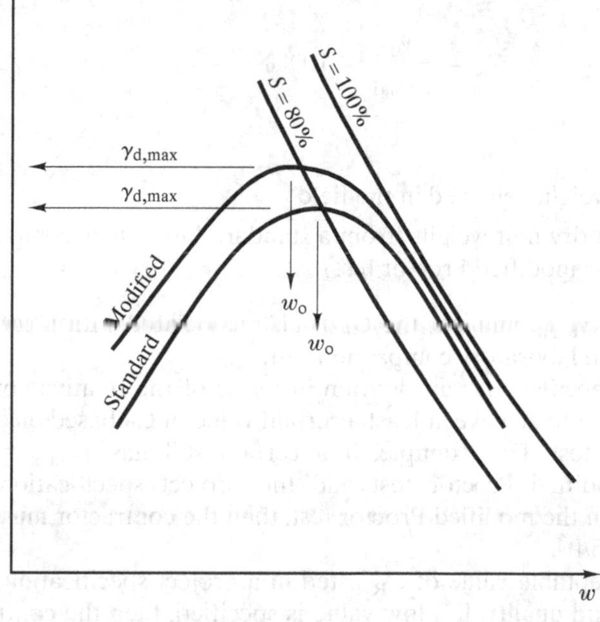

FIGURE 6.31 Comparison of standard Proctor test and modified Proctor test results on the same soil.

Standard T-180) and ASTM (Standard D1557), and is now the most commonly used test standard.

Figure 6.31 shows how the term *maximum* dry unit weight is misleading, because the standard and modified Proctor tests have two different "maximums." For each test, $\gamma_{d,max}$ is the greatest dry unit weight achieved for that level of compactive effort (i.e., for so many blows of a certain hammer per unit volume). A higher compactive effort in the modified Proctor test gives a compaction curve that is above and to the left of the standard Proctor compaction curve. Therefore, the $\gamma_{d,max}$ is higher from a modified Proctor test than from a standard Proctor test. Figure 6.31 also shows how the optimum moisture content, w_o, for the modified test is slightly less than that for the standard test. Because the standard and modified Proctor tests give different compaction curves, it is good practice to state the relevant test method whenever we refer to the maximum dry unit weight or optimum moisture content, even though the current geotechnical practice uses the modified Proctor test almost exclusively.

Relative Compaction Once the maximum dry unit weight has been established, based on a standard laboratory compaction test such as the modified Proctor test, for the soil being used in the compacted fill, we can express the degree of compaction achieved in the field by using the *relative compaction*, C_R:

$$C_R = \frac{\gamma_d}{\gamma_{d,max}} \times 100\% \tag{6.2}$$

where:

γ_d = dry unit weight achieved in the field

$\gamma_{d,max}$ = maximum dry unit weight (from a standard laboratory compaction test such as the modified Proctor test)

Note that like the $\gamma_{d,max}$ and w_o, the C_R is also associated with a certain compactive effort or standard laboratory compaction test.

Most earthwork specifications are written in terms of the relative compaction, and require the contractor to achieve at least a certain value of C_R based on a standard laboratory compaction test. For example, if a certain soil has $\gamma_{d,max} = 120 \text{ lb/ft}^3$ obtained from the modified Proctor test and the project specifications require $C_R \geq 90\%$ based also on the modified Proctor test, then the contractor must compact the soil until $\gamma_d \geq 108 \text{ lb/ft}^3$.

The minimum acceptable value of C_R listed in a project specification is a compromise between cost and quality. If a low value is specified, then the contractor can easily achieve the required compaction and, presumably, will perform the work for a low price. Unfortunately, the quality will be low. Conversely, a high specified value is more difficult to achieve and will cost more, but will produce a high-quality fill. Table 6.3 presents typical requirements based on the modified Proctor test, developed from empirical data.

The γ_d achieved in the field will generally be less than $\gamma_{d,max}$, as measured by the modified Proctor test, even if the moisture content is exactly equal to the corresponding w_o because the contractor usually applies less compactive effort (i.e., so many passes of a certain kind of roller) than used in the modified Proctor test. This is acceptable so long as the project specifications permit relative compactions of less than 100%, which they usually do (see Table 6.3). However, sometimes the contractor

TABLE 6.3 Typical Compaction Specifications

Type of Project	Minimum Required Relative Compaction Based on Modified Proctor Test (%)
Fills to support buildings or roadways	90
Upper 150 mm (6 in.) of subgrade below roadways	95
Aggregate base material below roadways	95
Earth dams	95

applies a large effort and exceeds $\gamma_{d,max}$, thus producing a relative compaction of more than 100%.

The Proctor test is commonly used to specify compaction as described above, but it may not be appropriate for very clean sands and gravels (SP, SW, GP and GW with little or no fines) due to the following reasons:

- These soils may give relatively flat compaction curves or curves with two maximum points, as mentioned in Section 6.3.
- These soils are usually compacted in the field saturated with pressure and vibration, whereas the Proctor test simulates compaction at various moisture contents through pressure, impact and manipulation.

In practice, quality assurance procedures for the compaction of clean sands and gravels often rely heavily on the use of specified construction methods and equipment, and much less on achieving a specified relative compaction.

In Chapter 4, we defined the relative density, D_r, which is based on the minimum and maximum void ratios (Equation 4.20). This parameter is conceptually similar to C_R in that both reflect soil compaction and both are expressed as a percentage, but they are not numerically equal. In addition, some engineers and contractors incorrectly use the term *relative density* when they are really describing relative compaction. This is often a source of confusion, so it is important to be careful when using these terms.

Compaction Methods and Soil Fabric

When a clay is compacted using a certain compactive effort at different moisture contents, not only will the dry unit weight vary with the as-compacted moisture content, the resulting fabric of the compacted soil will also depend on the as-compacted moisture content, as shown in Figure 6.32 (Lambe, 1958). When compacted dry of optimum, clays tend to have a *flocculated fabric* consisting of platy particles oriented randomly, as shown schematically in Figure 6.32. Conversely, clays compacted wet of optimum tend to have a more *oriented* or *dispersed fabric*, in which platy particles are aligned parallel to one another.

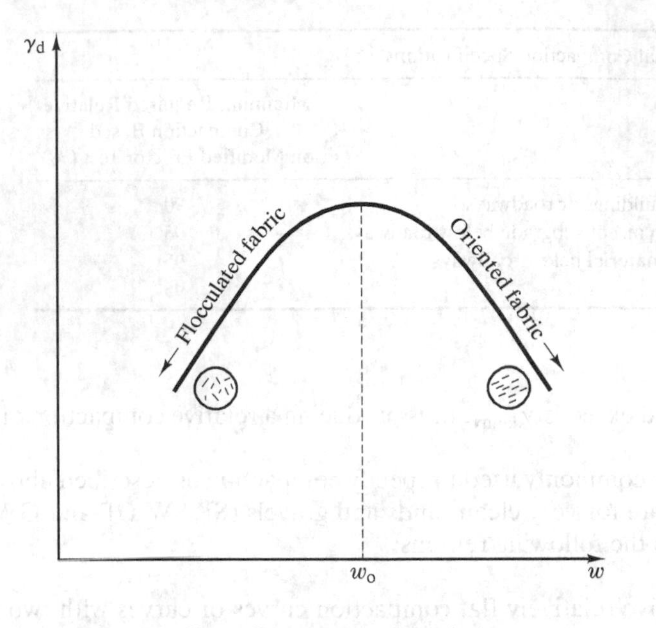

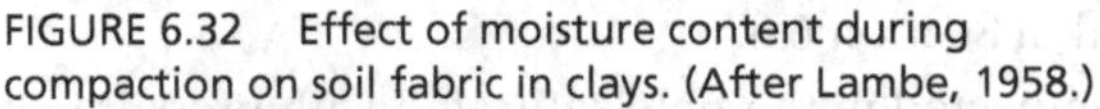
FIGURE 6.32 Effect of moisture content during compaction on soil fabric in clays. (After Lambe, 1958.)

The difference in soil fabric leads to differences in various soil properties. For example, a clay compacted dry of optimum will have a higher *hydraulic conductivity* (i.e., they pass water more easily) than the same clay compacted wet of optimum, even though γ_d is the same for both, as shown by empirical data in Figure 6.33 (Mitchell et al., 1965). Similarly, a clay compacted dry of optimum will have a greater shear strength than the same clay compacted wet of optimum, even though γ_d is the same for both, as shown by test results in Figure 6.34 (Seed and Chan, 1959).

The fabric in clay fills also depends on the compaction method. For example, pressure compaction produces a different fabric than manipulation compaction, even though both may produce the same γ_d. These effects are especially pronounced when the soil is compacted wet of optimum.

Although these fabric effects are generally not considered in routine compacted fill projects, they can become important in critical projects, such as earth dams, landfill liners and very deep fills. For example, clays used to build the core of an earth dam should be placed wet of optimum to obtain a lower hydraulic conductivity. Such projects may include performing a series of laboratory tests on soils compacted using various moisture contents and methods, and developing compaction specifications based on the results of these tests. For example, Daniel and Benson (1990) showed that for the compacted soil data obtained by Mitchell et al. (1965) in Figure 6.33, the zone of as-compacted moisture content and dry unit weight that would result in a hydraulic conductivity less than 1×10^{-7} cm/s is shown in Figure 6.35. Therefore, to satisfy this

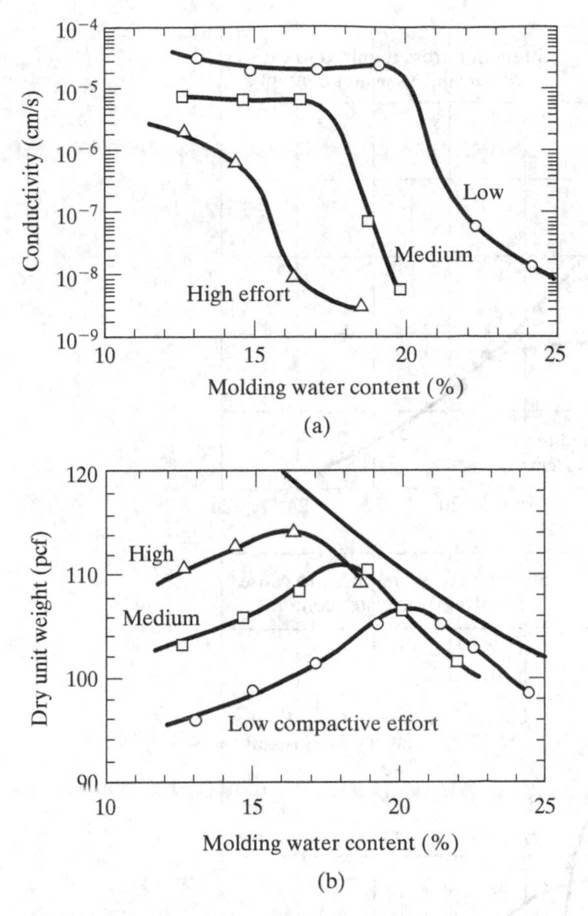

FIGURE 6.33 Empirical data for silty clay compacted by impact compaction: (a) plots of hydraulic conductivity versus molding (as-compacted) water (moisture) content; (b) compaction curves. (Mitchell et al., 1965.)

hydraulic conductivity requirement, a combination of as-compacted moisture content and compactive effort must be selected to give a resulting combination of as-compacted moisture content and dry unit weight that falls within the acceptable zone in Figure 6.35. We also should note that requirements on other soil properties, such as *shear strength*, may produce different acceptable zones on the compaction plot. The intersection of all acceptable zones corresponding to all relevant soil properties would govern the compaction specifications and operation. For example, Daniel and Benson (1990) showed schematically in Figure 6.36 how to obtain an acceptable zone on the compaction plot satisfying both hydraulic conductivity and shear strength requirements for a clay landfill liner.

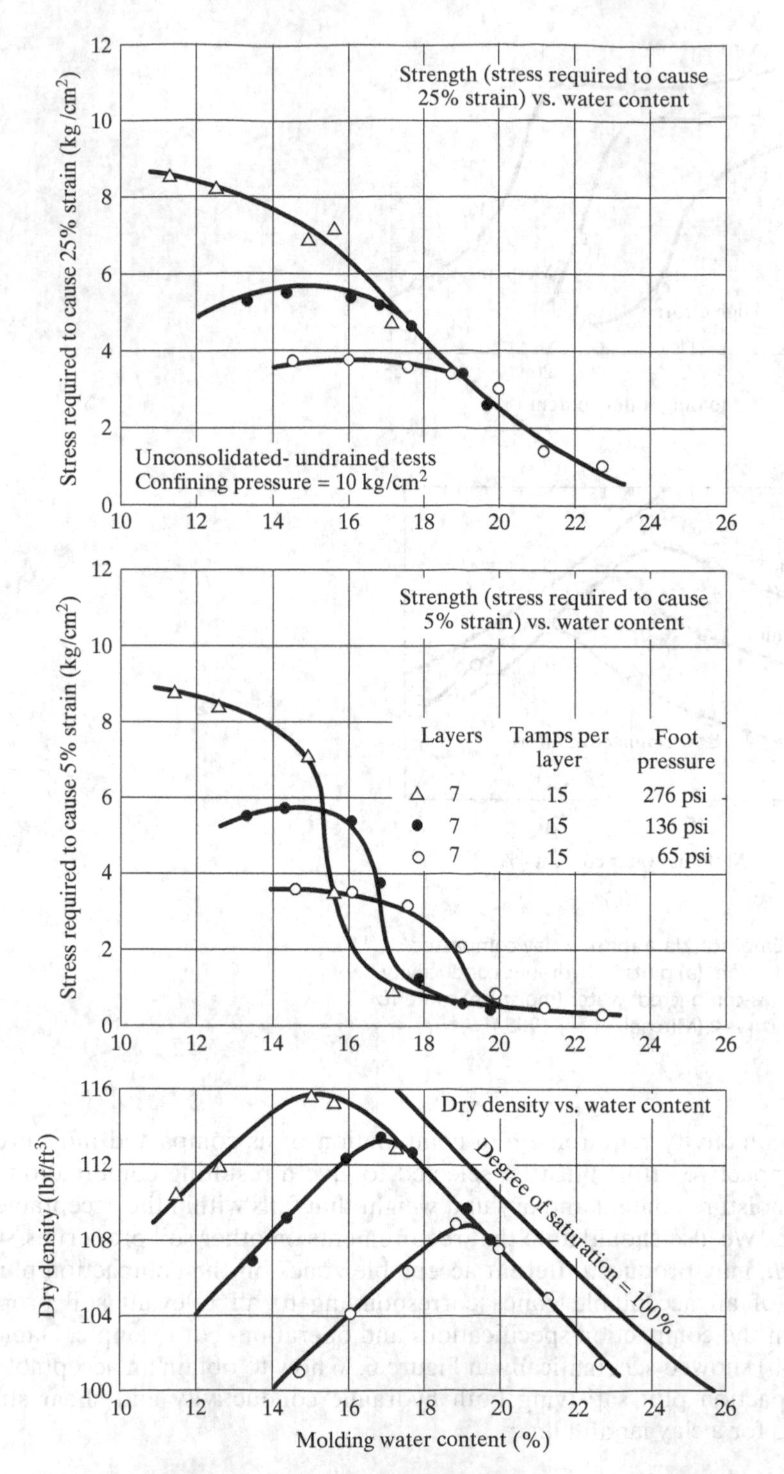

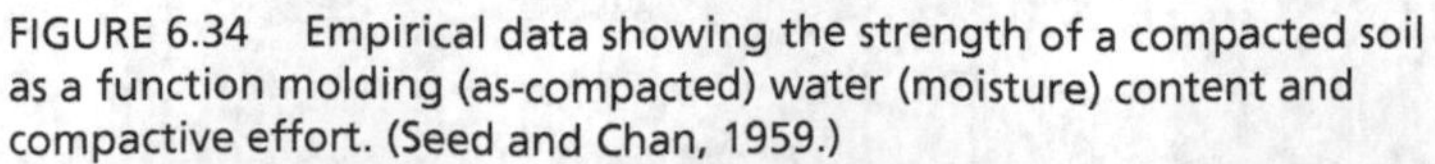

FIGURE 6.34 Empirical data showing the strength of a compacted soil as a function molding (as-compacted) water (moisture) content and compactive effort. (Seed and Chan, 1959.)

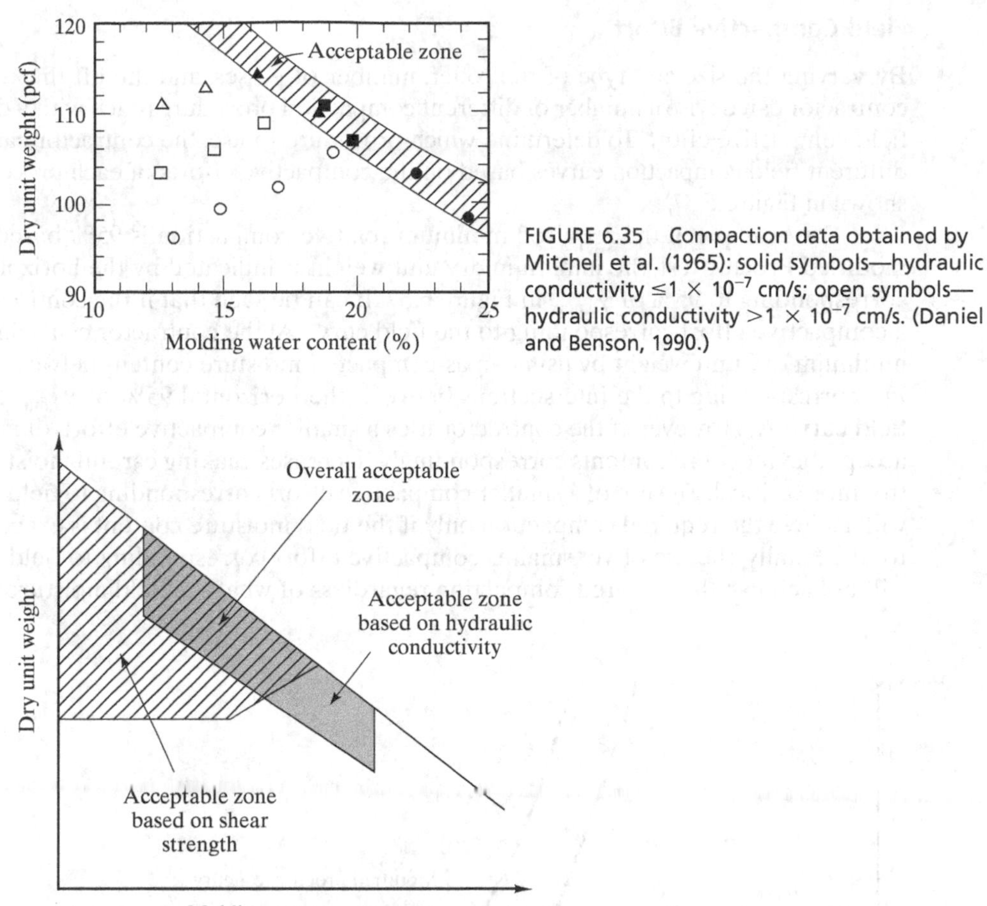

FIGURE 6.35 Compaction data obtained by Mitchell et al. (1965): solid symbols—hydraulic conductivity $\leq 1 \times 10^{-7}$ cm/s; open symbols—hydraulic conductivity $>1 \times 10^{-7}$ cm/s. (Daniel and Benson, 1990.)

FIGURE 6.36 Acceptable zone on the compaction plot satisfying both hydraulic conductivity and shear strength requirements for a clay landfill liner. (Daniel and Benson, 1990.)

6.5 FIELD CONSIDERATIONS AND MONITORING

To ensure that earthwork in the field meets the requirements of the design engineer, grading specifications should dictate the minimum acceptable relative compaction, and in certain cases, may include restrictions on the as-compacted moisture content, limits on the lift thickness, and other concerns. However, we normally do not specify how the contractor satisfies these requirements. The contractor should be free to select the means and methods to satisfy the specified compaction requirements. In general, the contractor has control over the type and size of compaction equipment, number of passes, lift thickness, and the as-compacted water content. The contractor will manipulate these variables to achieve the required compaction in the most efficient and cost effective manner.

Field Compactive Effort

By varying the size and type of the roller, number of passes, and the lift thickness, the contractor can devise a number of different compaction procedures each with a different field compactive effort. To determine which procedure to use, the contractor can create different field compaction curves based on the compactive efforts of each procedure, as shown in Figure 6.37.

For example, if the required minimum relative compaction is 95% based on the modified Proctor test, the minimum dry unit weight is indicated by the horizontal line corresponding to 95% of $\gamma_{d,max}$ in Figure 6.37. It can be seen that if the contractor uses a compactive effort corresponding to the field curve A, the contractor can achieve the minimum dry unit weight by using an as-compacted moisture content between w_a and w_c, corresponding to the intersections between the horizontal 95% of $\gamma_{d,max}$ line and field curve A. However, if the contractor uses a smaller compactive effort, the range of acceptable moisture contents correspondingly decreases, making careful moisture control more critical. The use of a smaller compactive effort corresponding to field curve B will achieve the required compaction only if the field moisture content is exactly equal to w_b. Finally, the use of yet smaller compactive effort corresponding to field curve C will not achieve the required compaction regardless of what the field moisture content

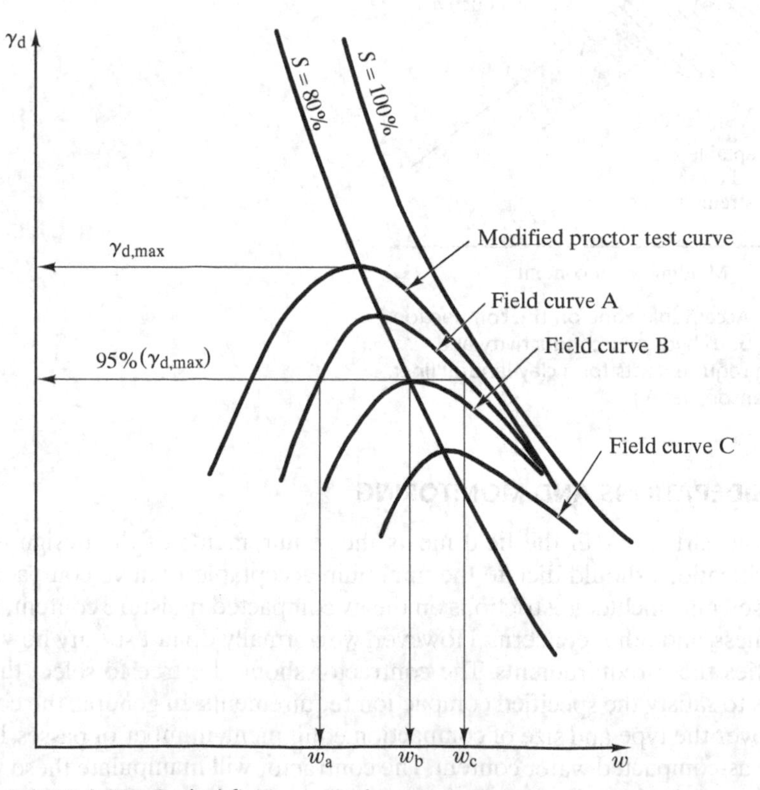

FIGURE 6.37 Typical field compaction curves.

is. Therefore, the contractor can optimize the compaction operation by selecting the optimal combination of compactive effort and range of required moisture content.

Example 6.2

Assume that the standard Proctor compaction curve obtained in Example 6.1 (Figure 6.30) is the same as the field compaction curve. Determine the range of as-compacted moisture content required to obtain a minimum relative compaction of 95% based on the standard Proctor test.

Solution:

For a C_R of 95%, $\gamma_d = (\gamma_{d,max})(95\%) = (19\ \text{kN/m}^3)(0.95) = 18.1\ \text{kN/m}^3$.

From Figure 6.30, the range of moisture content for $C_R > 95\%$ is:

7% to 16%

Lift Thickness

The lift thickness is the thickness of a lift (or layer) of soil to be compacted measured before compaction, called the *loose lift thickness*, or after compaction, called the *compacted lift thickness*. We should consider two requirements when specifying the lift thickness. First, the lift thickness should be small enough to ensure uniform compaction of the soil in each lift. Second, the lift thickness should not be too large because the influence of a roller decreases with depth and because of the associated costs. Typical lift thicknesses used in practice are between 200 and 450 mm (8 and 18 in.). Within this range, a larger lift thickness requires a heavier roller and/or more passes to achieve the same compactive effort.

Field Density Tests

The final link in assuring a quality compacted fill is to measure γ_d in the field after compaction, and compare the measured values to those required in the specifications using Equation 6.2. Several tests have been developed to make such measurements, and they are known as *field density tests*. All of these tests can be performed in the field, thus making it possible to immediately present the test results to the contractor. Such rapid feedback is important, because the contractor must rectify any inadequately compacted zones before they are buried by additional fill. A typical grading project requires dozens or even hundreds of field density tests.

Sand Cone Test One of the most common field density test methods is the *sand cone test* (ASTM D1556). The test procedure is essentially as follows (see ASTM for details):

1. Prepare a level surface in the fill and dig a cylindrical hole about 125 mm (5 in.) in diameter and about 125 mm (5 in.) deep. Save all of the soil that comes out of the hole and determine its weight, W.

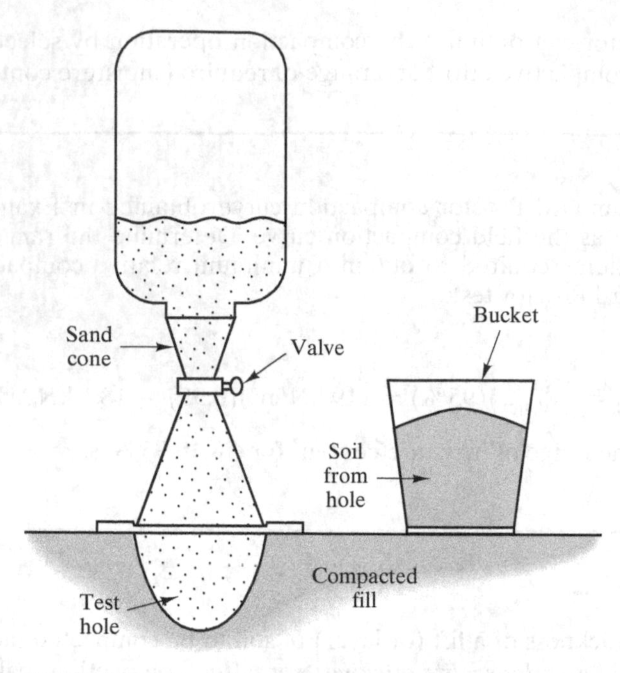

FIGURE 6.38 Use of a sand cone to measure the unit weight of a fill.

2. Fill the sand cone apparatus, shown in Figure 6.38, with a special free-flowing SP sand similar to that found in an hourglass. Then determine the weight of the cone and the sand, W_1.
3. Place the sand cone over the hole, as shown in Figure 6.39. Then open the valve and allow the sand to fill the hole and the cone.
4. Close the valve, remove the sand cone from the hole, and determine its new weight, W_2.

The volume of the hole, V, and unit weight of the fill, γ, are

$$V = \frac{W_1 - W_2}{\gamma_{sand}} - V_{cone} \tag{6.3}$$

$$\gamma = \frac{W}{V} \tag{6.4}$$

where:

V = volume of test hole

W_1 = initial weight of sand cone apparatus

W_2 = final weight of sand cone apparatus

FIGURE 6.39 A sand cone test being performed in the field.

γ_{sand} = unit weight of sand used in sand cone

V_{cone} = volume of sand cone below the valve

γ = unit weight of fill

W = weight of soil removed from test hole

Although a field engineer or soil technician can quickly measure γ in the field using the sand cone test, he or she still needs to measure the moisture content to obtain γ_d and C_R (using Equations 4.27 and 6.2, respectively). Normally this requires placing a soil sample in an oven overnight, as described in Chapter 4, but such a procedure would be unacceptably slow. Therefore, special rapid methods have been developed to measure the moisture content in the field. These include the following:

- "Stir-frying" the soil sample in a frying pan over a portable propane stove. This method is fast, and gives reasonably accurate results in clean sands and gravels. However, this heats the soil to a much higher temperature than in a laboratory oven, and thus can produce erroneous results in clays and organic soils. The computed moisture content in such soils will usually be too high.
- Mixing the soil with calcium carbide in a special sealed container known as a *speedy moisture tester*. The water in the soil reacts with the calcium carbide (CaC_2) to produce acetylene gas (C_2H_2), and this reaction continues until the water is depleted. A pressure gage measures the resulting pressure generated by

the production of this gas, and it is experimentally calibrated to compute the moisture content. The results are not as accurate as a standard laboratory moisture content test, but are usually sufficient for field density tests.

When carefully performed, the sand cone test is very accurate, and the equipment is inexpensive and durable.

Drive Cylinder Test Another field density test method is the *drive cylinder test* (ASTM D2937). It consists of driving a thin-wall steel tube into the soil using a special drive head and a mallet as shown in Figure 6.40. The cylinder is then dug out of the fill using a shovel, the soil is trimmed smooth, and it is weighed. The unit weight of the fill is then computed based on this weight and the volume of the cylinder, and the moisture content is determined as discussed earlier.

The drive cylinder test is much faster than the sand cone method and only slightly less precise. However, it is only suitable for fills with sufficient silt and clay to provide enough dry strength to keep the sample inside the cylinder. It is not satisfactory in clean sands, because they fall out too casily, or in gravelly soils.

Nuclear Density Test A third type of field density test is the *nuclear density test* (ASTM D2922). It is performed using a special device called the *nuclear moisture-density gauge*, as shown in Figure 6.41, that emits gamma rays and detects how they travel through the soil. The amount of gamma rays received back into the device correlates with the unit weight of the soil. The nuclear density test also measures the moisture content of the soil in a similar way using alpha particles.

Both the unit weight and moisture content measurements depend on empirical correlations, which ultimately must be programmed into the device. This allows it to directly display both parameters on digital electronic readouts.

Problems can be encountered during the nuclear tests of fills with unusual chemistries. Furthermore, regular calibrations of the device are required to maintain its accuracy. Because the equipment contains a source of radiation, it cannot be shipped

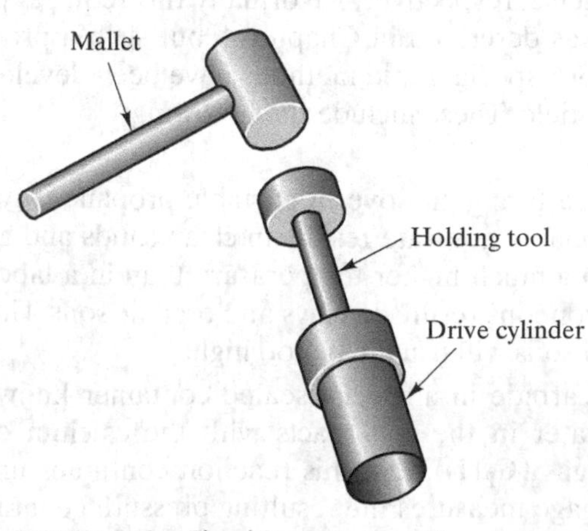

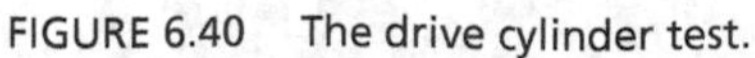
FIGURE 6.40 The drive cylinder test.

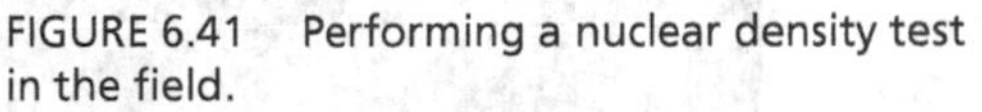

FIGURE 6.41 Performing a nuclear density test in the field.

through normal channels, such as in commercial aircraft or via mail. In addition, personnel must have special training before using this equipment.

In spite of its use of "hi-tech" equipment, the nuclear method gives results that are slightly less accurate than those from the sand cone test. This is because it is based on empirical correlations with the transmission of radiation, while the sand cone uses direct measurements of weight and volume. However, the nuclear test has sufficient accuracy for compaction assessments of normal fills and is faster to perform than the sand cone test. This saves time in the field, so its chief attraction is economy. Therefore, it has generally become the preferred method for many geotechnical firms and agencies.

Water Ring Test Sometimes engineers wish to place fills that contain cobbles and small boulders. For example, some earth dams contain zones made of such soils. Unfortunately, conventional field density test methods are not applicable because of the large particle sizes. One way to evaluate such fills is to use the *water ring test* shown in Figure 6.42. This test uses a much larger sample (the ring is typically about 1800 mm in diameter) and uses water instead of sand to measure the volume (ASTM D5030).

Example 6.3

A sand cone test has been performed in a compacted fill made with the soil described in Example 6.1. The test results were as follows:

Initial mass of sand cone apparatus = 5.912 kg
Final mass of sand cone apparatus = 2.378 kg
Mass of soil recovered from hole = 2.883 kg
Moisture content of soil from hole = 7.0%
Density of sand = 1300 kg/m^3
Volume of cone below valve = 1.114×10^{-3} m^3

(a)

(b)

FIGURE 6.42 Use of a water ring test to assess compaction in an earth dam with cobbles and boulders: (a) Hole has been dug and excavated soils have been weighed. (b) Hole has been lined with plastic and filled with water to measure its volume.

The project specifications require a relative compaction of at least 90% based on the standard Proctor test.

Compute γ_d and C_R and determine whether the project specifications have been met. If not, suggest a course of action.

Solution:

$$(M_{sand})_{cone+hole} = 5.912 \text{ kg} - 2.378 \text{ kg} = 3.534 \text{ kg}$$

$$V_{cone+hole} = \frac{(M_{sand})_{cone+hole}}{\rho_{sand}} = \frac{3.534 \text{ kg}}{1300 \text{ kg/m}^3} = 2.718 \times 10^{-3} \text{ m}^3$$

$$V_{hole} = V_{cone+hole} - V_{cone} = 2.718 \times 10^{-3} \text{ m}^3 - 1.114 \times 10^{-3} \text{ m}^3$$

$$= 1.604 \times 10^{-3} \text{ m}^3$$

$$W_{soil} = M_{soil}g = (2.883 \text{ kg})(9.81 \text{ m/s}^2)\left(\frac{1 \text{ kN}}{1000 \text{ N}}\right) = 2.828 \times 10^{-2} \text{ kN}$$

$$\gamma = \frac{W_{soil}}{V_{hole}} = \frac{2.828 \times 10^{-2} \text{ kN}}{1.604 \times 10^{-3} \text{ m}^3} = 17.63 \text{ kN/m}^3$$

$$\gamma_d = \frac{\gamma}{1 + w} = \frac{17.63 \text{ kN/m}^3}{1 + 0.070} = \mathbf{16.5 \text{ kN/m}^3}$$

$$C_R = \frac{\gamma_d}{(\gamma_d)_{max}} \times 100\% = \frac{16.5 \text{ kN/m}^3}{19.0 \text{ kN/m}^3} \times 100\% = \mathbf{86.8\%}$$

Conclusion and Recommendation

The relative compaction is less than the required 90%, so the specifications have not been met. This may be at least partially due to the low moisture content, which is well below optimum. Suggest ripping the soil, adding water, mixing, and recompacting.

6.6 SUITABILITY OF SOILS FOR USE AS COMPACTED FILL

When imported soils are required, we often have the opportunity to choose between several borrow sites, each with a different soil. Our selection is based on their engineering properties, the cost of importing them, and other factors. Often we only require that the imported soils be at least as good as those onsite, but sometimes other criteria might apply.

The suitability of various soil types that could be used to assist in selecting an import source are summarized below.

Gravels (GW, GP, GM, GC)

Gravels make good fills that have high strength and low compressibility. These fills also retain their strength when wetted. GW and GP soils have a high hydraulic conductivity that allows water to drain quickly, which is especially important in highways. Base material below pavements is always a GW or GP. GM and GC soils have substantially lower hydraulic conductivity, and do not drain nearly as well. Gravels easily compact to a high unit weight over a wide range of moisture contents, but are prone to caving if they must later be excavated, such as for utility trenches.

Sands (SW, SP, SM, SC)

Sands also make good fills that have high strength and low compressibility. SW and SP soils retain virtually all of their strength when wetted; SM and SC lose some strength, but still remain good. If sands are too wet or too dry, they can easily be brought to the optimum moisture content, and they can be compacted over a wide range of moisture contents. However, SP and SW soils are problematic when exposed at the ground surface, because vehicles can become stuck and grades are difficult to maintain. SP and SW soils also are especially prone to caving into small excavations such as utility trenches. All sandy soils are susceptible to surface erosion unless protected, such as with vegetation. SM and SC soils are nearly ideal for most applications, so long as their moderate hydraulic conductivity is acceptable.

Low-Plasticity Silts and Clays (ML and CL)

ML and CL soils are less desirable than SM or SC because they lose more strength when wetted and require more careful moisture control. They also are more difficult to dry if initial moisture content is far from the optimum, and are prone to frost heave problems in freezing climates. Nevertheless, they usually make acceptable fills.

High-Plasticity Silts and Clays (MH and CH)

MH and CH soils are the best choice when very low hydraulic conductivity is required, such as in landfill caps and liners. Otherwise, they are difficult to handle and compact, especially when the initial moisture content is above optimum. These soils expand when wetted, which can cause problems in pavements, lightly loaded foundations, flatwork concrete, and similar projects. They are a poor choice for retaining wall backfills because of their expansiveness, low wet strength, and low hydraulic conductivity.

Organic Soils (OL and OH)

Organic soils are very poor choices for compacted fills and are unsuitable for most applications. They are weak and compressible and difficult to compact. Never bring them in as an imported soil (except for landscaping purposes). If OL or OH soils are being generated from onsite cuts, try to place them away from critical areas and consider hauling them away.

Peat (Pt)

This soil is extremely poor. It makes fills that are weak and compressible, and are unsuitable for support of buildings or pavements. However, peat is ideal for use in landscape areas because of its fertility.

6.7 EARTHWORK QUANTITY COMPUTATIONS

Grading Plans

When planning earthwork construction, civil engineers begin with a topographic map of the existing ground surface, and then develop a new topographic map showing the proposed ground surface. These two maps are usually superimposed to produce a *grading plan*, such as the one in Figure 6.43. When the proposed elevations are higher than the existing ones, a fill will be required, whereas when they are lower, a cut becomes necessary.

The design engineer also uses these two sets of elevations to compute the anticipated quantities of cut and fill. Such computations can be performed by hand, but they are more likely done by computer, and the results are always expressed in terms of volume, such as cubic yards or cubic meters.

If extra soil needs to be hauled to the site to produce the required grades, the project has a *net import*. Conversely, when extra soil is left over and needs to be hauled away, it has a *net export*. Finally, when neither import nor export is required, the earthwork is said to be *balanced*. We usually strive to design the proposed grades such that the earthwork is close to being balanced because importing and exporting soils can become very expensive.

Bulking and Shrinkage

Let us consider a certain grading project that consists of excavating soil from one area, transporting it, and placing it as compacted fill in another area, as illustrated in Figure 6.44. The excavation will occur in an area called the *bank*. At this particular site, suppose each cubic yard of soil has a weight of 2700 lb (other sites would have different weights). The soil is loosened as it is removed and loaded into a dump truck, so the 2700 lb now occupies a volume of 1.25 yd^3. This is called the *loose* condition. Finally, it is placed and compacted and the volume shrinks to 0.80 yd^3; the *compacted* condition. These changes in volume are important to the contractor because they affect the equipment productivity, and to the design engineer because they affect the amount of fill produced by a given amount of cut.

The change in volume from the bank to the loose condition is called *bulking* or *swell* and depends on the unit weight in the bank, the soil type and moisture content, and other factors. Sands and gravels typically have about 10% bulking, while silts and clays usually have 30–40%. This means the contractor will need to consider bulking in arranging transport of a given volume of soil to be excavated.

The key to earthwork quantity computations is that, given a certain amount of soil, no matter how it may have been changed through such earthwork operations as excavation, moisture conditioning and compaction, the amount of solids in the soil remains the same. Given that the volume of solids remains unchanged, any change in soil volume during earthwork comes from a change in the voids volume of the soil. Using this observation, the net change in volume from the bank to compacted conditions, or the *shrinkage factor* $\Delta V / V_f$, can be shown to be given by

$$\frac{\Delta V}{V_f} = \left(\frac{\gamma_{d,f}}{\gamma_{d,c}} - 1 \right) \times 100\% \tag{6.5}$$

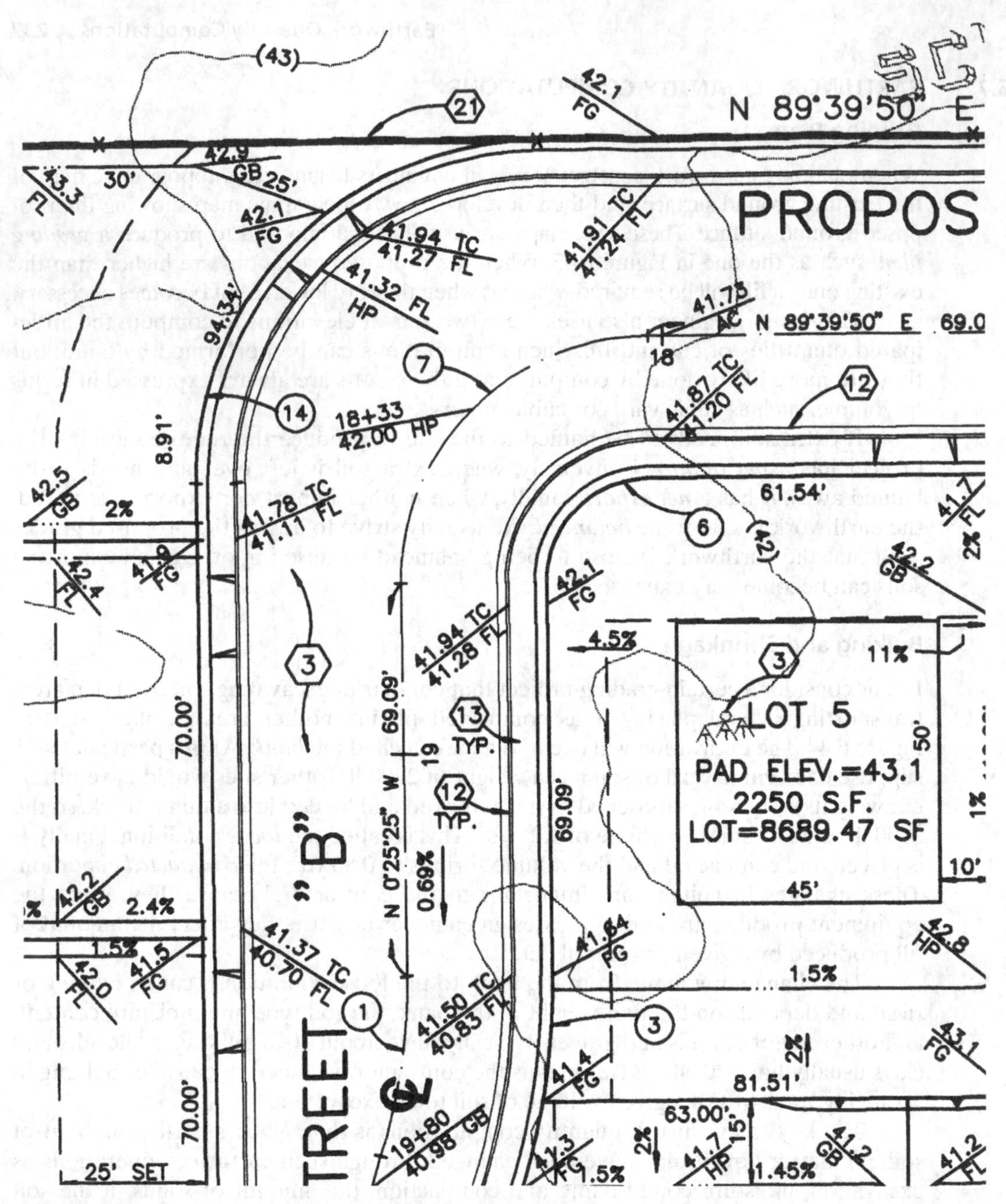

FIGURE 6.43 A small portion of grading plan for a proposed residential tract. The contour lines represent the existing conditions, while the spot elevations represent the proposed conditions. Proposed lots are located along the left and bottom right portion of the drawing. Key to notation: FG = Finish Grade; FL = Flow Line; GB = Grade Break; HP = High Point; TC = Top of Curb. Scale: 1 in. = 30 ft. All elevations and dimensions are in feet. (Algis Marciuska, Civil Engineering Department, California State Polytechnic University, Pomona.)

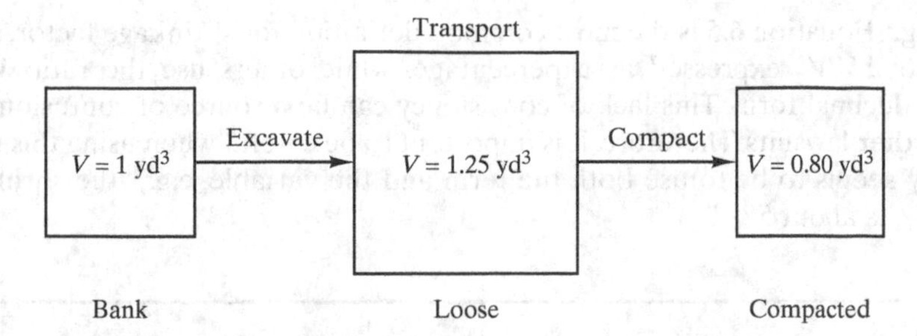

FIGURE 6.44 Changes in volume as soil is excavated, transported, and compacted. The numerical values are examples and would be different for each soil.

where:

ΔV = change in volume during grading
V_f = volume of fill
$\gamma_{d,f}$ = average dry unit weight of fill
$\gamma_{d,c}$ = average dry unit weight of cut (i.e., the bank soil)

A positive shrinkage factor, $\Delta V/V_f$, indicates a net decrease in volume, while a negative value indicates a net increase in volume, sometimes called *swelling*. These are not the same as shrinkage or swelling of expansive soils, which is due to an entirely different process. Most soils have a positive shrinkage factor because modern construction equipment usually compacts the soil to a dry unit weight greater than that in the bank.

For example, if a civil engineer plans a project for 100,000 m^3 of cut and 100,000 m^3 of fill without accounting for shrinkage, he or she will probably have an unpleasant surprise during construction. If the soils at this site have 10% shrinkage, the 100,000 m^3 removed from the cut area will produce only 90,910 m^3 of fill, which means 9090 m^3 of import will suddenly be required to reach the proposed grades. This will come as a surprise during construction and will be even more expensive than usual because of the urgency of the situation, idle workers and equipment, and the lack of money in the project budget. Such situations have been the source of many lawsuits. However, if the design engineer had properly accounted for shrinkage, the proposed grades could have been set accordingly (perhaps with 104,000 m^3 of cut and 94,550 m^3 of fill), or at least the need for imported soils would have been known in advance.

The value of $\gamma_{d,c}$ may be obtained from unit weight measurements on undisturbed samples from borings made in the cut area, while $\gamma_{d,f}$ is obtained from Proctor compaction test results and project specifications for C_R. However, the project specifications quote a *minimum* acceptable value for C_R, such as 90%, while Equation 6.5 requires the *average* value of $\gamma_{d,f}$. Grading contractors typically exceed the minimum requirements, so the shrinkage computations must be adjusted accordingly. Usually it is sufficient to compute $\gamma_{d,f}$ using a C_R value 2% higher than the minimum required in the project specifications. For example, if the specifications require $C_R \geq 90\%$, then use $\gamma_{d,f} = 0.92(\gamma_{d,max})$.

Although Equation 6.5 is the most common definition for shrinkage factor, some engineers use $\Delta V/V_c$ expressed as a percentage, while others use the ratio V_f/V_c expressed in decimal form. This lack of consistency can be a source of confusion, and has led to further lawsuits. Therefore, it is important to be careful when using this term. The best way seems to be to use both the term and the variable, e.g., "the shrinkage factor, $\Delta V/V_f$, is about 5%."

Example 6.4

Twelve undisturbed soil samples were obtained from borings in a proposed cut area. These samples had an average γ_d of 108 lb/ft³ and an average w of 9.1%. A Proctor compaction test performed on a representative bulk sample produced $\gamma_{d,max} = 124$ lb/ft³ and $w_o = 12.8\%$. A proposed grading plan calls for 12,000 yd³ of cut and 11,500 yd³ of fill, and the specifications call for a relative compaction of at least 90%.

a. Compute the shrinkage factor.

b. Compute the required quantity of import or export soils based on the unit weight of the cut.

c. Compute the weight of import or export in tons using the moisture content of the cut.

d. Compute the required quantity of water in gallons to bring the fill soils to the optimum moisture content.

Solution:

a. Using an average relative compaction of 92%:

$$\gamma_{d,f} = \gamma_{d,max} C_R = (124 \text{ lb/ft}^3)(0.92) = 114 \text{ lb/ft}^3$$

$$\frac{\Delta V}{V_f} = \left(\frac{\gamma_{d,f}}{\gamma_{d,c}} - 1\right) \times 100\% = \left(\frac{114}{108} - 1\right) \times 100\% = \mathbf{6\% \text{ (Shrinkage)}}$$

b.

$$\Delta V = \frac{\Delta V}{V_f} V_f = (6\%)(11{,}500 \text{ yd}^3) = 690 \text{ yd}^3$$

Cut required to produce 11,500 yd³ of fill = 11,500 yd³ + 690 yd³ = 12,190 yd³

Required import = 12,190 yd³ − 12,000 yd³ = **190 yd³**

c.

$$W_s = V\gamma_d = (190 \text{ yd}^3)\left(\frac{3 \text{ ft}}{1 \text{ yd}}\right)^3 (108 \text{ lb/ft}^3)\left(\frac{1\,T}{2000 \text{ lb}}\right) = 277\,T$$

$$W = W_s(1 + w) = (277\,T)(1 + 0.091) = \mathbf{302\,T}$$

d.

$$(W_s)_{fill} = (\gamma_d)_{fill}(V_{fill}) = (114 \text{ lb/ft}^3)(11{,}500 \text{ yd}^3)\left(\frac{3 \text{ ft}}{1 \text{ yd}}\right)^3 = 35.4 \times 10^6 \text{ lb}$$

Based on Equation 4.1, the weight of water to be added is the required increase in the moisture content, Δw, times the dry weight:

$$(W_w)_{added} = \Delta w\, W_s = (0.128 - 0.091)(35.4 \times 10^6 \text{ lb}) = 1.31 \times 10^6 \text{ lb}$$

Water has a unit weight of 8.34 lb/gal, so:

$$V_w = \frac{W_w}{\gamma_w} = \frac{1.31 \times 10^6 \text{ lb}}{8.34 \text{ lb/gal}} = \mathbf{157{,}000\ gal}$$

Comments: If our estimate of the shrinkage factor was in error, perhaps because the undisturbed samples were not truly representative, and the true shrinkage factor was 10%, the required import volume would become 650 yd^3 instead of the computed 190 yd^3. Such a situation would not be unusual in practice. Although this difference would represent a 340% increase in the required import quantity, it would only be 4% of the total fill volume. Therefore, it is important to keep in mind the potential effects of small errors in the computed shrinkage factor.

Sometimes the weights of fills cause significant settlement in the underlying natural soils. Some or all of this settlement may occur during construction, but much of it often occurs after construction. In either case, the earthwork quantity computations will be affected. We will discuss methods of predicting these settlements in Chapters 10 and 11.

6.8 LIGHTWEIGHT FILLS

Normal fills are very heavy. For example, a 2-m thick fill, which many people would hardly notice, has about the same weight as a six-story building. Usually this weight is not a major concern, but in some situations it can be problematic. For example,

- If the natural soils below the fill are soft, its weight will cause them to consolidate, thus producing large settlements at the ground surface. This can be especially problematic in fills leading to bridge abutments because the bridge (which is probably supported on pile foundations) does not settle with the abutment fill. See the example in Figure 10.2.
- The weight of a fill imposes new stresses on buried structures, such as pipelines or subway tunnels, and may overload them.
- The weight of a fill placed behind a retaining wall imposes stresses on the wall.
- The weight of a fill placed on a slope decreases the stability of that slope, possibly leading to a landslide.

When these issues are a concern, geotechnical engineers sometimes use special lightweight materials to build fills. These materials allow the fill to achieve the same elevation, without the adverse effects of excessive weight.

Various materials have been used as lightweight fills. When selecting the most appropriate material, the engineer needs to find the best compromise between cost and unit weight. For example, various cementitious materials with unit weights of 3.8–12.6 kN/m^3(24–80 lb/ft^3) are available, and can be pumped into place. Automobile tires chopped into small pieces also have been used as lightweight fill material (Whetten, et al., 1997). Another lighter, but more expensive, option is to use expanded polystyrene (EPS) and extruded polystyrene (XPS), otherwise known as *geofoam* (Horvath, 1995; Negussey, 1997). These are essentially the same as the materials commonly called *styrofoam*, except they are supplied in large blocks that are stacked on the ground as shown in Figure 6.45. The unit weight of geofoam used for geotechnical engineering purposes is only 1.25 lb/ft^3(0.20 kN/m^3), yet it has sufficient strength and stiffness to support heavy external loads, such as those from vehicles.

For example, a 20-ft deep fill placed against a bridge abutment would induce an increase in total vertical compressive stress, $\Delta\sigma_z$, of about 2400 lb/ft^2 in the underlying soils. If these soils are soft, the resulting settlement could be substantial. However, the same fill made from geofoam would induce a $\Delta\sigma_z$ of only 25 lb/ft^2 plus the weight of the soil cover and pavement. The resulting settlement would be far less.

Geofoam also has other uses, such as providing thermal insulation, compressible inclusions (i.e., soft spots to protect underground structures from excessive stresses), vibration damping, and fluid transmission through intentional voids in the foam (Horvath, 1995).

FIGURE 6.45 Geoform blocks being placed to produce a lightweight fill for a highway in Colorado. (BASF Corporation.)

6.9 DEEP FILLS

Deep compacted fills have been used for many years in the construction of earth dams. However, during the last third of the twentieth century, deep fills also became increasingly common in other kinds of projects, such as highways and residential developments. Fill thicknesses of 30 m or more became quite common, and these fills were constructed using the same techniques that had been successful in other more conventional fills. Unfortunately, some of these deep fills have experienced excessive settlements, which in some cases have damaged buildings and other structures (Rogers, 1992c; Noorany and Stanley, 1994).

The settlement problems with these deep fills stem from the following characteristics and processes:

- The fills are very heavy, and thus induce large compressive stresses in the lower parts of the fill and in the underlying soils.
- Loose natural soils were usually removed before placement of the fills, but these removals often were not as extensive as they should have been.
- Although the fills were typically placed at a moisture content near optimum, they slowly became wetter due to the infiltration of water from rain, irrigation, broken utility pipelines, and other sources.
- The wetting of the lower portions of the fill and the underlying natural soils often caused them to compress, causing settlements that varied with the fill thickness. When such fills were placed in old canyons, the thickness often changed quickly over short distances, thus causing large differential settlements (i.e., differences in settlements) over short distances.
- If the fill was made of expansive soils, the wetting caused heaving in the upper zones, which sometimes offset the compression of the underlying fill, and other times aggravated the differential settlements.

Unfortunately, these processes typically required several years to develop, which was long after structures and other improvements had been constructed on the fill. As a result, a large number of lawsuits ensued. These problems have resulted in new methods of designing and constructing such projects, including the following:

- The customary requirement for 90% relative compaction may not be sufficient for the lower portions of deep fills. In some cases, higher standards may be required.
- Removals of loose natural soils beneath proposed fills must be more aggressive than previously thought.
- The use of highly expansive soils needs to be more carefully controlled.
- Designs must be based on the assumption that some wetting will occur during the lifetime of the fill.

These effects may be evaluated by conducting an appropriate laboratory testing program using the proposed fill soils, then performing special analyses to predict the long-term settlements (Noorany et al., 1992). Based on the results of such a testing and analysis program, we can develop a design to keep fill settlements within tolerable limits.

SUMMARY

Major Points

1. Many civil engineering projects include making cuts and fills, and this work requires the active participation of geotechnical engineers.
2. Improperly designed or constructed fills often are problematic, but with proper design and construction, they provide reliable support for structures, highways, and other civil engineering works.
3. The advent of modern earthmoving equipment has made it possible to perform very large earthwork projects with high levels of reliability and efficiency at a very low cost.
4. Design engineers need to be familiar with earthwork construction methods and equipment.
5. Proper compaction is very important. The degree of compaction depends on the soil type, compaction method, compactive effort and the as-compacted moisture content.
6. For a given compactive effort, the compaction curve is a plot of the as-compacted dry unit weight against the as-compacted moisture content. Typically, a compaction curve is a bell-shaped curve with a maximum point that corresponds to the maximum dry unit weight and optimum moisture content.
7. The maximum dry unit weight, optimum moisture content, and the relative compaction are associated with a certain compactive effort in the laboratory or field. Typically, a higher compactive effort gives a higher maximum dry unit weight and a lower optimum moisture content.
8. Compaction is usually assessed using the modified Proctor test to determine the corresponding maximum dry unit weight and optimum moisture content, along with a series of field density tests to determine the as-compacted dry unit weight in the field, and hence, the relative compaction based on the modified Proctor test.
9. When soils are excavated, transported, and placed as compacted fill, their volume usually changes. Contractors need to consider these changes when estimating their equipment needs and productivity, and design engineers need to do so when developing grading plans.
10. The key to earthwork quantity computations is that, given a certain amount of soil, no matter how it may have been changed through such earthwork operations as excavation, moisture conditioning and compaction, the amount of solids in the soil does not change.
11. Sometimes it is useful to use lightweight materials, such as geofoam, to build fills. These materials produce less stress in the underlying soils, and thus can help control settlement problems.

Vocabulary

as-compacted dry unit weight
as-compacted moisture content
backhoe
blasting
borrow pit
bulking
bulldozer
clearing and grubbing
compacted fill
compaction
compaction curve
compactive effort
conveyor belt
cut
dispersed fabric
drive cylinder test
dump truck
earthwork
embankment
excavatability
excavation
excavator
field density test
fill
fine grading
flocculated fabric
geofoam
grading
grading plan
hoe
hydraulic fill
lift
loader
maximum dry unit weight
modified Proctor test
moisture conditioning
motor grader
nuclear density test
optimum moisture content
oriented fabric
pneumatic roller
Proctor compaction test
relative compaction
rippability
ripper
rough grading
sand cone test
scraper
sheepsfoot roller
shrinkage
shrinkage factor
smooth steel-wheel roller
standard Proctor test
tamping foot roller
tractor
vibratory roller
wagon
water ring test
water truck
zero air voids curve

QUESTIONS AND PRACTICE PROBLEMS

Section 6.1 Earthwork Construction Objectives

6.1 Describe the common objectives of earthwork in a civil engineering project.

6.2 Define cut and fill in the context of civil engineering earthwork and provide two examples of each.

Section 6.2 Construction Methods and Equipment

6.3 List, in order, the steps in conventional earthwork and describe the purpose of each step.

6.4 Use phase diagrams to illustrate the changes that occur in the three phases in a soil during earthwork.

6.5 You are preparing the earthmoving specifications for a project that will include a wide range of materials, from loose soil to hard rock. Because of this variation, you plan to use a classified excavation system so the contractor can give two unit prices, one for common excavation and another for rock. However, to avoid disagreements on classification, you also need to develop clear definitions for each type of excavation. What criteria might you use to differentiate between common and rock excavation?

6.6 What are the four methods used by mechanical compaction equipment to densify soils? Which of these methods are most effective in sands? In clays?

6.7 List the construction equipment typically required to perform earthwork.

6.8 A contractor needs to excavate several thousand cubic yards of silty clay at one end of a site, transport it to the other end, and place it as a compacted fill. The distance between the cut and fill areas is about 3000 ft. The soils in the cut area can be excavated with little or no ripping. However, they are very dry. Make a list of the construction equipment needed to complete this project.

6.9 Describe how the weight of compaction equipment, number of passes and lift thickness affect compaction of a given soil.

6.10 A developer plans to build a condominium complex on a site underlain by an old hydraulic fill. What special geotechnical problems need to be considered at this site? Why?

Section 6.3 Soil Compaction Concepts

6.11 Why do we use the dry unit weight as a measure of how compacted a soil is?

6.12 For a given compactive effort, describe how the dry unit weight of the compacted soil is affected by the as-compacted moisture content.

Section 6.4 Soil Compaction Standards and Specifications

6.13 How does the dry unit weight, γ_d, differ from the unit weight γ? Why is the Proctor method of assessing soil compaction based on the dry unit weight and not the unit weight?

6.14 A certain soil has a dry unit weight of 18.0 kN/m^3 and a specific gravity of solids of 2.67. Compute the moisture contents that correspond to $S = 80\%$ and $S = 100\%$.

6.15 Describe the differences between the standard and modified Proctor tests. Describe the $S = 100\%$ curve (zero air voids curve) and why it is important to the evaluation of compaction data.

6.16 Plot the $S = 100\%$ curves assuming G_s values of 2.5, 2.6, 2.7 and 2.8.

6.17 Plot the $S = 80\%$ curves assuming G_s values of 2.5, 2.6, 2.7 and 2.8.

6.18 Plot curves corresponding to S values of 60%, 70%, 80%, 90% and 100%, assuming a G_s of 2.65.

6.19 A compacted fill, currently under construction, will support a proposed supermarket. One of the field density tests in this fill gave a unit weight of 121 lb/ft^3 and a moisture content of 12.5%. A series of modified Proctor tests have also been performed on the fill soils per ASTM D1557, and they gave a maximum dry unit weight of 117 lb/ft^3 and an optimum moisture content of 13.0%. Compute the relative compaction based on the modified Proctor test and determine whether or not it satisfies the normal compaction specification for such a project.

6.20 A field density test has been conducted in a compacted fill that is currently under construction. The test results indicate a relative compaction based on the modified Proctor test of 101–103%, which made the construction manager believe the test must be incorrect. He believes it is impossible for the dry unit weight in the field to be greater than the maximum dry unit weight from the modified Proctor test. Prepare a 200–400 word memo explaining why the relative compaction in the field could indeed be greater than 100%.

6.21 You are a geotechnical engineer and have just received the results from some modified Proctor tests. In plotting the compaction curve, you notice that some of the data points are plotted to the right of the zero air voids curve. Is this cause for concern? Why do you think these points are plotted like this?

6.22 A series of modified Proctor tests have been performed on a soil that has a G_s of 2.68. The test results obtained are as follows:

		Moisture Content Test Results		
Point No.	**Weight of Compacted Soil + Mold (lb)**	**Mass of Can (g)**	**Mass of Can + Moist Soil (g)**	**Mass of Can + Dry Soil (g)**
1	8.73	22.13	207.51	202.30
2	9.07	25.26	239.69	225.27
3	9.40	19.74	253.90	230.64
4	9.46	23.36	250.93	219.74
5	9.22	20.28	301.47	250.95

The weight of the empty mold was 5.06 lb.

Plot the laboratory test results and the S = 80% and S = 100% curves, and then draw the modified Proctor compaction curve. Determine the maximum dry unit weight (lb/ft^3) and the optimum moisture content, based on the modified Proctor test.

Assuming that the field compaction curve is the same as the compaction curve obtained using the modified Proctor test, determine the range of as-compacted moisture content required in the field to obtain a relative compaction of at least (a) 90%, and (b) 95%.

6.23 A series of modified Proctor tests have been performed on a soil that has a G_s of 2.70. The test results obtained are as follows:

		Moisture Content Test Results		
Point No.	**Mass of Compacted Soil + Mold (kg)**	**Mass of Can (g)**	**Mass of Can + Moist Soil (g)**	**Mass of Can + Dry Soil (g)**
1	3.673	22.11	205.74	196.33
2	3.798	23.85	194.20	180.54
3	3.927	19.74	196.24	177.92
4	3.983	20.03	187.43	165.71
5	3.932	21.99	199.59	171.11

The mass of the empty mold was 1.970 kg.

Plot the laboratory test results and the S = 80% and S = 100% curves, and then draw the modified Proctor compaction curve. Determine the maximum dry unit weight (kN/m^3) and the optimum moisture content, based on the modified Proctor test.

Assuming that the S = 80% curve is the line of optima (a line that passes through the maximum points of compaction curves from different compactive efforts) and if the smallest possible compactive effort is to be used, determine the moisture content required to obtain a relative compaction based on the modified Proctor test of at least 95%.

Section 6.5 Field Considerations and Monitoring

6.24 A sand cone test has been performed in a recently compacted fill. The test results obtained are as follows:

Initial weight of sand cone + sand = 13.51 lb
Final weight of sand cone + sand = 4.26 lb

Weight of sand to fill cone = 2.12 lb

Weight of soil from hole + bucket = 12.42 lb

Weight of bucket = 1.21 lb

Moisture content test:

Mass of empty moisture content can = 23.11 g

Mass of moist soil + can = 273.93 g

Mass of oven-dried soil + can = 250.10 g

The sand used in the sand cone had a unit weight of 81.0 lb/ft^3, and the fill had a maximum dry unit weight of 121 lb/ft^3 and an optimum moisture content of 11.7%, based on the modified Proctor test. Compute the relative compaction based on the modified Proctor test.

Section 6.7 Earthwork Quantity Computations

6.25 Derive Equation 6.5.

6.26 A proposed building site requires 1200 yd^3 of imported fill. A suitable borrow site has been located, and the soils there have a shrinkage factor of 13%. How many cubic yards of soil must be excavated from the borrow site?

6.27 A contractor needs to excavate 50,000 yd^3 of silty clay and haul it with Caterpillar 69C dump trucks. Each truck can carry 30.9 yd^3 of soil per load, and operates on a 15-minute cycle. The job must be completed in five working days with the trucks working two 8-hour shifts per day. Using a bulking factor of 30%, how many trucks will be required?

6.28 A proposed grading plan requires 223,120 m^3 of cut and 206,670 m^3 of fill. Laboratory tests on a series of undisturbed samples from the cut area produced the following results:

Sample No.	Dry Unit Weight (kN/m^3)	Moisture Content (%)
3-1	17.3	9.1
3-2	17.7	9.5
5-1	16.8	8.9
5-2	17.1	7.2
8-1	16.0	12.0

A series of modified Proctor tests on representative bulk samples produced a maximum dry unit weight of 19.2 kN/m^3 and an optimum moisture content of 10.2%. The project specifications require a relative compaction based on the modified Proctor test of at least 90%.

(a) Determine the shrinkage factor and compute the required volume of import or export, if any. Use the dry unit weight of the cut when computing any import or export quantities.

(b) Determine the weight of any import or export soil, assuming it has a moisture content equal to the average moisture content in the cut. Express your answer in metric tons (1 metric ton = 1000 kg = 1 Mg).

6.29 A 3.0 ft deep cut is to be made across an entire 2.5-acre site. The average unit weight of this soil is 118 lb/ft^3, and the average moisture content is 9.6%. It also has a maximum dry unit weight of 122 lb/ft^3 and an optimum moisture content of 11.1%, based on the modified Proctor test. The excavated soil will be placed on a nearby site and compacted to an average relative compaction of 93%. Compute the volume of fill that will be produced, and express your answer in cubic yards.

6.30 A proposed highway is to pass through a hilly area and will require both cuts and fills. The horizontal alignment is fixed, but the vertical alignment can be adjusted within certain limits to make the earthwork balance. The design engineer has developed four trial vertical alignments, with A being the lowest one and D being the highest. The resulting earthwork requirements are as follows:

Trial Vertical Alignment	Cut Volume (m^3)	Fill Volume (m^3)
A	40,350	35,120
B	39,990	35,490
C	39,180	36,010
D	38,400	36,950

Using a shrinkage factor of 12%, determine which alignment would balance the earthwork. If none of the trial alignments work, then express your answer in terms of two of them (e.g., 20% of the way from C to D).

6.31 Make a copy of the grading plan in Figure 6.43, and then compare the existing and proposed grades. Using colored pencils, apply red shading to the fill areas and blue shading to the cut areas. Consider only the area south of the tract boundary (line 21). You may interpolate between and extrapolate beyond the two contour lines (the full drawing, which is much larger, includes many more contour lines). Finally, locate the area that will receive the greatest depth of fill, and determine this depth.

Comprehensive

6.32 A fill soil with a natural moisture content of 10% and an optimum moisture content based on the modified Proctor test of 14% is being used to construct a compacted fill. The contractor is placing this soil in 400-mm lifts, spraying the top with a water truck, and compacting it using a towed sheepsfoot roller. A soils technician has performed a series of field density tests in this fill and has found relative compaction values between 80 and 92%, based on the modified Proctor test. The measured moisture contents ranged from 10 to 23%. The specifications require a relative compaction based on the modified Proctor test of at least 90%, so the fill is not acceptable. What is wrong with the contractor's methods, and what needs to be done to remedy the problem?

6.33 A contractor needs to import 100,000 yd^3 (compacted volume) of soil to build a small earth dam. Two methods of hauling this soil are being considered, as follows:

Method A

Use model XL37 scrapers, each having a capacity of 20 yd^3. These scrapers will need to be pushed by a model BD12 bulldozer while they are loading soil at the borrow site, then can travel unassisted to the dam site and deposit the soil there. One BD12 will be required for every six scrapers, and the scrapers can work on a cycle time of 30 min. The labor and equipment costs are as follows:

XL37 scraper	\$250 per hour
BD12 bulldozer	\$200 per hour

Method B

Use model 98F wheel loaders at the borrow site to load the soil into model 356 dump trucks. Each dump truck has a capacity of 11 yd^3, and one loader will be required to service every five dump trucks. The dump trucks will then haul the soil to the dam site and deposit it there, which will require a cycle time of 20 min. The labor and equipment costs are as follows:

98F wheel loader	\$180 per hour
356 Dump truck	\$175 per hour

Either method is acceptable, so the choice between them will be based solely on cost. The soil has a bulking factor of 30% and a shrinkage factor of 12%. The hauling needs to be completed within 20 working days, using one 8-hour shift per day.

Determine how many scrapers, bulldozers, loaders, and dump trucks will be needed to complete the hauling in the required time, then compute the cost of each method. Based on the computed costs, select the better method for this project.

Note: These hourly rates are not necessarily representative of the actual labor and equipment costs and are for illustrative purposes only.

6.34 The proposed grading at a project site will consist of 25,100 m^3 of cut and 23,300 m^3 of fill and will be a balanced earthwork job. The cut area has an average moisture content of 8.3%. The fill will be compacted to an average relative compaction of 93% based on a maximum dry unit weight of 18.3 kN/m^3 and an optimum moisture content of 12.9% obtained from the modified Proctor test. Compute the volume of water in kiloliters that will be required to bring these soils to the optimum moisture content.

6.35 A well-graded silty sand with a maximum dry unit weight of 19.7 kN/m^3 and an optimum moisture content of 11.0%, obtained from the modified Proctor test, is being used to build a compacted fill. Two field density tests have been taken in the recently completed fill, but one of these tests has produced results that are definitely incorrect. Test A indicated a relative compaction of 85% and a moisture content of 8.9%, while Test B indicated a relative compaction of 98% and a moisture content of 14.9%. Which test is definitely incorrect? Why?

CHAPTER 7

Groundwater—Fundamentals and One-Dimensional Flow

All streams flow into the sea, yet the sea is never full. To the place the streams come from, there they return again.

Ecclesiastes 1:7 (NIV)

By the ancients, man has been called the world in miniature; and certainly this name is well bestowed, because inasmuch as man is composed of earth, water, air, and fire, his body resembles that of the earth; and as man has in him bones, the supports and framework of his flesh, the world has its rocks, the supports of the earth; as man has in him a pool of blood in which the lungs rise and fall in breathing, so the body of the earth has its ocean tide which likewise rises and falls every 6 hours, as if the world breathed; as in that pool of blood veins have their origin, which ramify all over the human body, so likewise the ocean sea fills the body of the earth with infinite springs of water. The body of the earth lacks sinews, and this is because the sinews are made expressly for movements and the world being perpetually stable, no movement takes place, and no movement taking place, muscles are not necessary. But in all other points they are much alike ... if the body of the earth were not like that of a man, it would be impossible that the waters of the sea—being so much lower than the mountains—could by their nature rise up to the summits of these mountains. Hence, it is to be believed that the same cause which keeps the blood at the top of the head in man keeps the water at the summits of the mountains.

Leonardo da Vinci (1452–1519) as quoted by Biswas (1970)

Karl Terzaghi once wrote "... in engineering practice, difficulties with soils are almost exclusively due not to the soils themselves, but to the water contained in their voids. On a planet without any water there would be no need for soil mechanics" (Terzaghi, 1939). The presence of water, or at least the potential for its presence, is a key aspect of most geotechnical analyses. Therefore, this is a topic worthy of careful study.

Specific water-related geotechnical issues include the following:

- The effect of water on the behavior and engineering properties of soil and rock
- The potential for water flowing into excavations
- The potential for pumping water through wells or other facilities
- The effect of water on the stability of excavations and embankments
- The resulting uplift forces on buried structures
- The potential for seepage-related failures, such as piping
- The potential for transport of hazardous chemicals along with the water.

In this chapter, we explore the fundamental principles of subsurface water. Chapter 8 continues this discussion and applies these principles to practical engineering problems.

7.1 HYDROLOGY

Hydrology is the study of water movements across the earth. It includes assessments of rainfall intensities, stream flows, and lake water levels, known as *surface water hydrology*, as well as studies of underground water, known as *groundwater hydrology*. These various movements are part of the grand process called the *hydrologic cycle*.

The Hydrologic Cycle

The movement of water across the earth is ultimately driven by energy received from the sun. Thus, the hydrologic cycle begins with water rising into the sky from open bodies of water through the process of *evaporation*, as shown in Figure 7.1. This process also draws water out of the near-surface soil, which dries the soil. A related process, called *transpiration*, acts through plants and draws water out of the ground through their roots. The two processes are sometimes conceptually combined and called *evapotranspiration*.

Water in the sky, which may be in the form of invisible water vapor or visible clouds, eventually falls to the earth as *precipitation* (rain, sleet, hail, and snow), much of which goes directly into the oceans. The precipitation that falls onto land and onto inland bodies of water becomes the source of virtually all surface water, much of which flows *overland* until it reaches streams or rivers, where it becomes *streamflow*.

However, a significant portion of the surface water soaks into the ground, either while it is flowing overland or after it has reached rivers or lakes. This *infiltration* recharges the groundwater. Water applied to the ground by irrigation also soaks in and

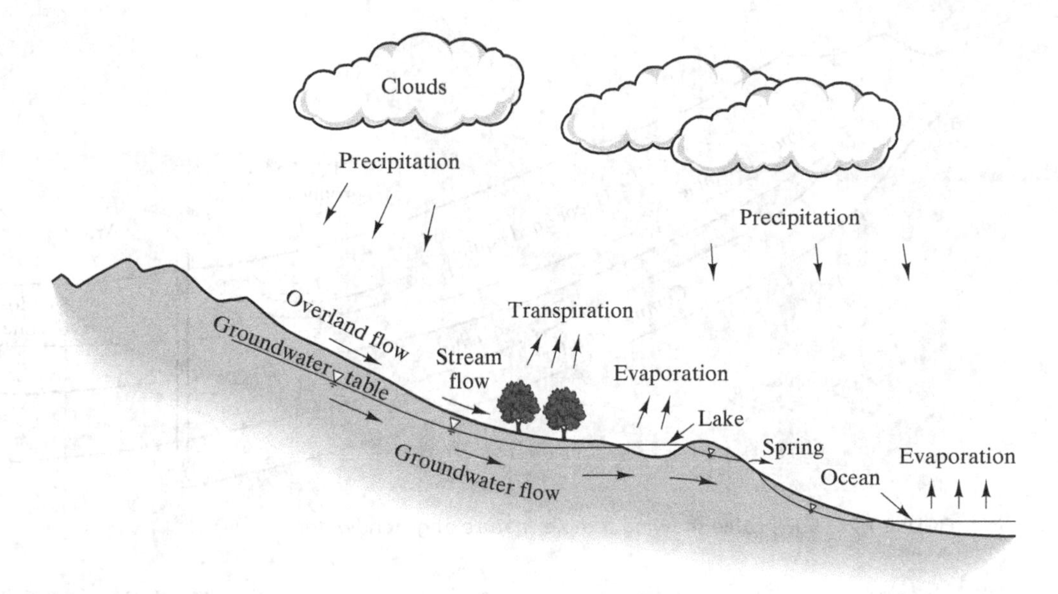

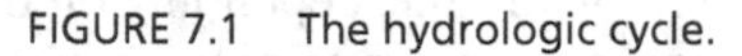
FIGURE 7.1 The hydrologic cycle.

can contribute to the groundwater. Although some of the infiltrated water remains near the ground surface, much of it penetrates down until it reaches the groundwater table, which is the fully saturated zone.

These processes occur both in mountains and lowlands, so the groundwater eventually builds up and gains enough potential energy to begin flowing through the ground. Eventually some of this water reappears at the ground surface as *springs*, or it seeps directly into rivers or lakes. There it joins overland flows that eventually lead to *sinks* (low spots in the land) or the ocean, where the hydrologic cycle begins anew.

Groundwater Hydrology

Geotechnical engineers are mostly interested in the portions of the hydrologic cycle that occur underground. The term *subsurface water* encompasses all underground water, virtually all of which is located within the soil voids or rock fissures. A very small percentage of subsurface water is located in underground caverns, but this special case is not of much interest to geotechnical engineers.

We use various kinds of information to describe and understand subsurface water. One of the most important is the *groundwater table* (also called the *phreatic surface*), which can be located by installing observation wells, as shown in Figure 3.22, and allowing the groundwater to seep into them until it reaches equilibrium. The water level inside these wells is, by definition, the groundwater table, where the pore water pressure is equal to zero. Soil profiles represent the groundwater table as a line marked with a triangle, as shown in Figure 7.2. It also can be presented in planview as a series of contour lines on a map. The groundwater table location is important, and finding it is one of the primary objectives of a site characterization program.

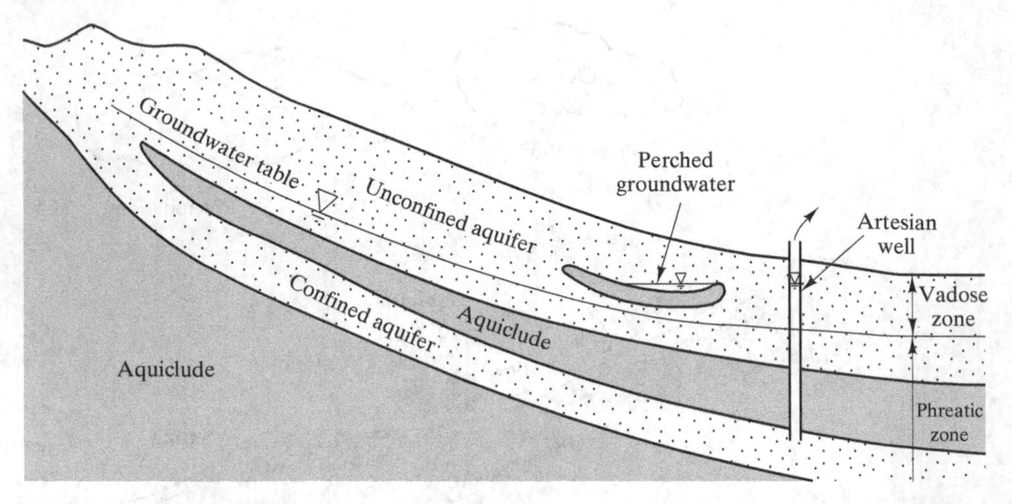

FIGURE 7.2 Soil profile showing complex nature of groundwater.

The groundwater table elevation often changes with time, depending on the season of the year, recent patterns of rainfall, irrigation practices, pumping activities, and other factors. At some locations, these fluctuations are relatively small (perhaps less than 1 m or 3 ft), while in other places the groundwater table elevation has changed by 20 m (60 ft) or more in only a year or two. Thus, the groundwater conditions encountered in an exploratory boring are not necessarily those we use for design. Often we need to use the observed conditions as a basis for estimating the worst-case conditions that are likely to occur during the project life.

Water also is usually present in the soil above the groundwater table, but the physical processes that control its movement are quite different. Thus, it is useful to divide subsurface water into two zones:

- The portion below the groundwater table is called the *phreatic zone*. This water is subjected to a positive pore water pressure as a result of the weight of the overlying water (and possibly due to other causes as well). Most subsurface water is in the phreatic zone.
- The portion above the groundwater table is called the *vadose zone*. This water has a negative pore water pressure, and is held in place by capillary action and other forces present in the soil.

Technically, only the water in the phreatic zone is true *groundwater*. However, we often use the term *groundwater* to describe all subsurface water.

Some soils, such as sands and gravels, can transmit large quantities of groundwater. These are known as *aquifers*, and are good candidates for wells. Other soils, such as clays, transmit water very slowly. They are known as *aquicludes*. Intermediate soils, such as silty sand, pass water at a slow-to-moderate rate and are called *aquitards*. All three categories of soil might be present in a single soil profile, so the distribution and flow of groundwater can be quite complex. For example, a *perched groundwater* condition can occur when an aquiclude separates two aquifers, as shown in Figure 7.2. In this case, there may be two or more groundwater tables.

An *unconfined aquifer*, such as the upper aquifer in Figure 7.2, is one in which the bottom flow boundary is defined by an aquiclude, but the upper flow boundary (the groundwater table) is free to reach its own natural level. The groundwater occupies the lower portion of the aquifer, just as water in a kitchen pot occupies the lower part of the pot. The zone of soil through which the water flows is called the *flow regime*. If more groundwater arrived at the site, the groundwater table in an unconfined aquifer would rise accordingly.

Conversely, a *confined aquifer*, such as the lower aquifer in Figure 7.2, is one in which both the upper and lower flow boundaries are defined by aquicludes. This type of aquifer is similar to a pipe that is flowing full. Most confined aquifers also are *artesian*, which means the water at the top of the aquifer is under pressure. People often drill wells into such aquifers, because the water will rise up through the aquiclude without pumping. If the water pressure is high enough, *artesian wells* deliver water all the way to the ground surface without pumping.

Figure 7.2 shows an example of an artesian condition. Groundwater enters the confined aquifer from the left side of the cross-section, and then travels down and to the right. By the time the water reaches the right side of the cross-section, it has developed an artesian condition. Artesian conditions are situations when the water inside the well casing rise above the groundwater table as shown in the well in Figure 7.2. Artesian conditions can occur only in confined aquifers.

7.2 PRINCIPLES OF FLUID MECHANICS

Our groundwater analyses will use the three-dimensional Cartesian coordinate system shown in Figure 7.3, where the x and y axes are in a horizontal plane and the z-axis is vertical. For some analyses, geotechnical engineers express vertical dimensions in terms of elevation (z positive upward). However, it is usually more convenient to work in terms of depth (z positive downward). For example, we often speak of depth below the ground surface, depth below the groundwater table, or depth below the bottom of

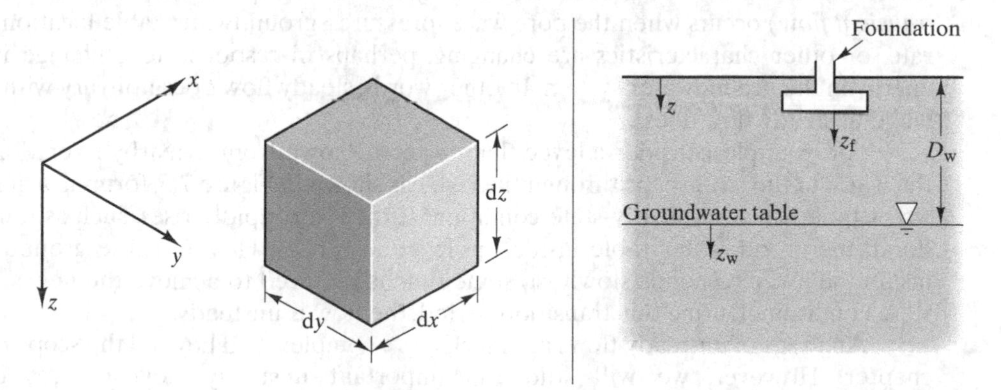

FIGURE 7.3 Coordinate system used in groundwater analysis, along with typical soil element.

a foundation. In addition, boring logs are always presented in terms of depth below the ground surface. Therefore, in this book, all z-values are expressed as depths with the positive direction downward, as shown in Figure 7.3. A z-value with no subscript indicates depth below the ground surface; z_w indicates depth below the groundwater table; and z_f indicates depth below a foundation or other applied load. The depth from the ground surface to the groundwater table is D_w.

Groundwater Flow Conditions

One-, Two-, and Three-Dimensional Flow For analysis, we need to distinguish between one-, two-, and three-dimensional flow conditions. A one-dimensional flow condition is one in which the velocity vectors are all parallel and of equal magnitude, as shown in Figure 7.4(a). In other words, the water always moves parallel to some axis and through a constant cross-sectional area.

Two-dimensional flow conditions are present when all of the velocity vectors are confined to a single plane, but vary in direction and magnitude within that plane. For example, the flow into a long excavation, as shown in Figure 7.4(b), might be very close to a two-dimensional condition described along a vertical plane through the excavation.

Three-dimensional flow is the most general condition. It exists when the velocity vectors vary in the x, y, and z directions. Flow into multiple wells, as shown in Figure 7.4(c), is an example. Most groundwater flow conditions are truly three-dimensional but for analysis purposes can often be simplified to one- or two-dimensions with satisfactory results.

Steady and Unsteady Flow

The term *steady-state condition* means a system has reached equilibrium. In the context of groundwater analyses, it means the flow pattern has been established and is not in the process of changing. We call this *steady flow* or *steady-state flow*. Under steady-state condition, the direction and velocity of fluid flow is constant with time and, therefore, the flow rate, Q, also remains constant with time.

In contrast, the *unsteady condition* (also known as the *transient condition*) exists when something is in the process of changing. For seepage problems, *unsteady flow* (or *transient flow*) occurs when the pore water pressures, groundwater table location, flow rate, or other characteristics are changing, perhaps in response to a change in the energy in the groundwater system. In other words, steady flow does not vary with time, while unsteady flow does.

For example, consider a levee that protects a town from a nearby river. Some of the water in the river seeps through the levee as shown in Figure 7.5, forming a groundwater table. This is a steady-state condition. If the river rapidly rises, such as during a flood, the groundwater table inside the levee also rises. However, the groundwater inside the levee responds slowly, so some time is required to achieve the new steady-state condition. During this transition period, the flow is unsteady.

Analyses of unsteady flows are much more complex, and beyond the scope of this chapter. However, we will study an important unsteady flow process called *consolidation* in Chapters 10 and 11.

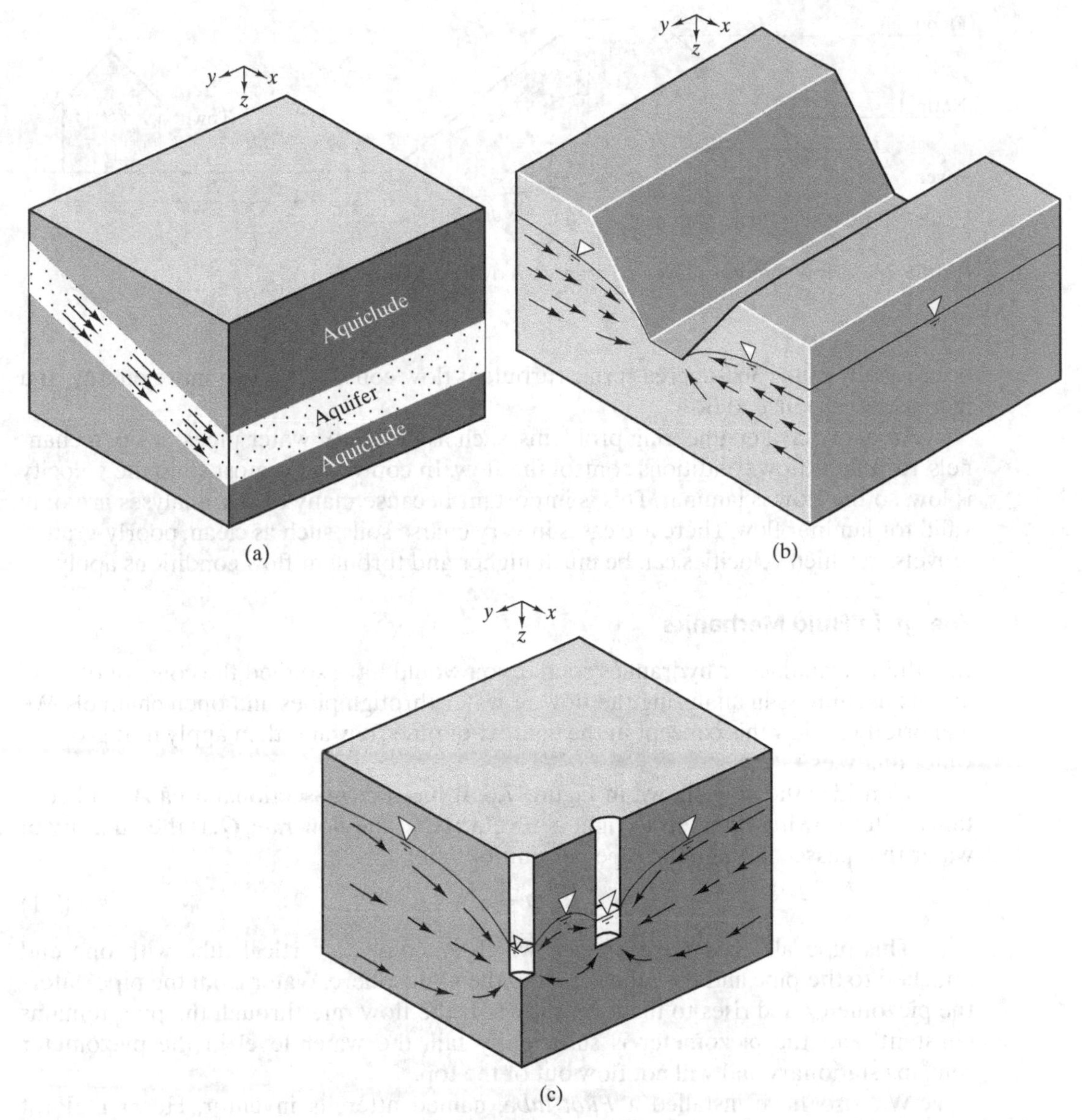

FIGURE 7.4 One-, two-, and three-dimensional flow conditions: (a) one-dimensional flow in a confined aquifer; (b) two-dimensional flow into a long excavation; (c) three-dimensional flow into a pair of wells.

Laminar and Turbulent Flow

Sometimes water flows in a smooth orderly fashion, known as *laminar flow*. This flow pattern occurs when the velocity is low, and is similar to cars moving smoothly along an interstate highway. The other possibility is called *turbulent flow*, which means the water swirls as it moves. This happens when the velocity is high, and might be compared to an interstate highway filled with vehicles who move too fast, weave back and forth, and

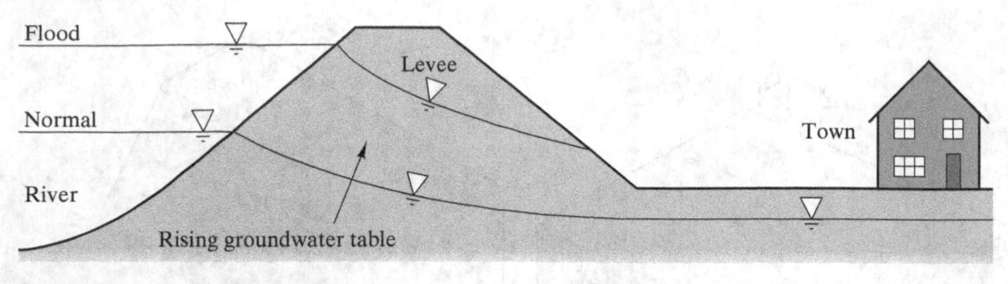

FIGURE 7.5 Flow through a levee adjacent to a river.

occasionally make 360-degree turns. Turbulent flow consumes much more energy, and increases resistance to flow.

In many civil engineering problems, such as design of water pipes or open channels, turbulent flow conditions control the flow. In contrast, for most soils, the velocity is low, so the flow is laminar. This is important because many of our analyses are only valid for laminar flow. There are cases in very coarse soils, such as clean, poorly-graded gravels, in which velocities can be much higher and turbulent flow conditions apply.

Energy in Fluid Mechanics

In a fluid mechanics or hydraulics course, you would have studied the concept of *head* and its usefulness in analyzing the flow of water through pipes and open channels. We will briefly review this concept in the context of pipe flow and then apply it to groundwater analyses.

Consider the pipe shown in Figure 7.6. It has a cross-sectional area A, and contains water flowing from left to right at a velocity v. The *flow rate*, Q, is the quantity of water that passes through the pipe per unit of time:

$$Q = vA \tag{7.1}$$

This pipe also has a *piezometer*, which is simply a vertical tube with one end attached to the pipe and the other open to the atmosphere. Water from the pipe enters the piezometer and rises to the level shown. If the flow rate through the pipe remains constant, and the piezometer is sufficiently tall, the water level in the piezometer remains stationary and will not flow out of the top.

We also have installed a *Pitot tube*, named after its inventor, Henri DePitot (1695–1771), which is similar to a piezometer except the tip is pointed upstream. The opening of the Pitot tube is normal to the direction of flow and receives a dynamic ramming effect from the flowing water. The water level in the Pitot tube is affected by the velocity of flow and thus is slightly higher than the water level in the piezometer because the opening of the piezometer is parallel to the direction of flow and the height of the water in the piezometer is not affected by the velocity of flow.

Finally, Figure 7.6 shows a horizontal datum elevation, which is the level from which elevations may be measured. This datum is arbitrary and might be set at sea level, the laboratory floor, or some other suitable location.

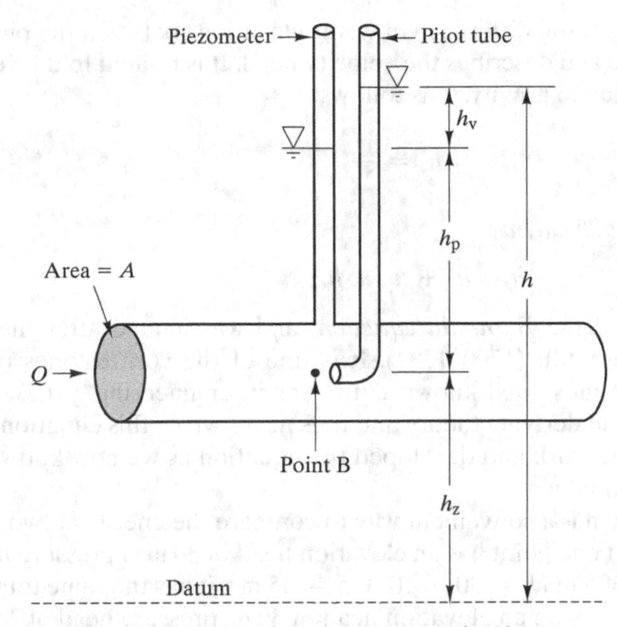

FIGURE 7.6 A pipe with a piezometer and a Pitot tube. These instruments measure the heads at Point B in the pipe.

Head

An element of groundwater, such as the one at Point B in Figure 7.6, contains energy in various forms, including the following:

- *Potential energy*, which is due to its elevation above the datum
- *Strain energy*, which is due to the pressure in the water (similar to the energy contained in a spring that has been compressed under an external load)
- *Kinetic energy*, which is due to its velocity.

We could express these energies using Joules, BTUs, or some other suitable unit. However, it is more convenient to do so using the concept of *head*, which is energy divided by the acceleration of gravity, g. This method converts each form of energy to the equivalent potential energy and expresses it as the corresponding height. Thus, we express these three forms of energy as follows:

- The *elevation head*, h_z, is the difference in elevation between the datum and the point, as shown in Figure 7.6. It describes the potential energy at that point.
- The *pressure head*, h_p, is the difference in elevation between the point and the water level in a piezometer attached to the pipe. It describes the strain energy.

- The *velocity head*, h_v, is the difference in water elevations between the piezometer and the Pitot tube and describes the velocity head. It is related to the velocity, v, and acceleration due to gravity, g, as follows:

$$h_v = \frac{v^2}{2g} \tag{7.2}$$

The sum of these is the *total head*, h:

$$h = h_z + h_p + h_v \tag{7.3}$$

Equation 7.3 is called the *Bernoulli equation*, and was named after the Swiss mathematician Daniel Bernoulli (1700–1782). It is one of the cornerstones of fluid mechanics and one of the most well-known equations in engineering, yet Bernoulli developed only part of the underlying theory and thus never wrote this equation. Later investigators completed the work and developed the equation as we now know it, but the credit has gone to Bernoulli.

The Bernoulli equation is a convenient way to compare the energy at two points. For example, if the water at one point has an elevation head of 30 m, a pressure head of 10 m, and a velocity head of 5 m ($h = 30 + 10 + 5 = 45$ m), it has the same total head as the water at another point with an elevation head of 39 m, pressure head of 2 m, and velocity head of 4 m ($h = 39 + 2 + 4 = 45$ m). Flow occurs only when there is a total head differential, so there would not be any flow of water between these two points.

Head Loss and Hydraulic Gradient

Figure 7.7 shows a pipe with piezometers and Pitot tubes at two points, A and B. Water always flows from a point of high total head to a point of low total head. Thus, the water

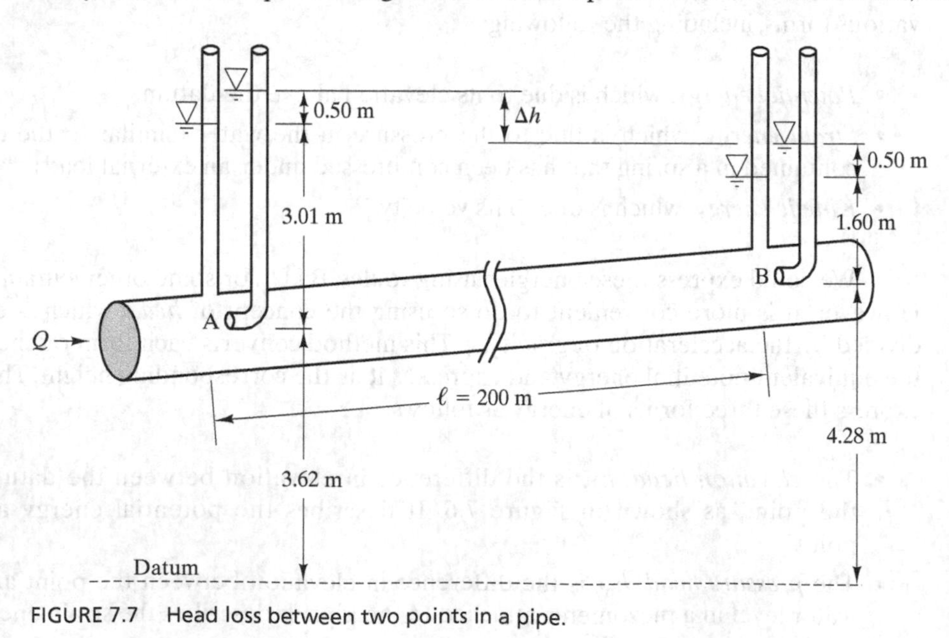

FIGURE 7.7 Head loss between two points in a pipe.

in this pipe must be flowing from Point A to Point B because the total head at A is greater than the total head at B. As the water flows from A to B, some of its energy is lost due to friction, a quantity known as the *head loss*, Δh, which in this case is equal to $h_A - h_B$.

The *hydraulic gradient*, i, is the change in total head per unit length in the direction of flow:

$$i = -\frac{dh}{dl} \tag{7.4}$$

where:

i = hydraulic gradient
h = total head
l = distance the water travels

The total head decreases as water moves downstream, or down gradient (i.e., $dh < 0$, $dl > 0$), so i is always a positive number. In addition, both h and l are lengths, so i is dimensionless. A large hydraulic gradient reflects extensive friction, and thus means either water is flowing very fast or the soil has a large resistance to flow (or a combination of the two).

Example 7.1

Piezometers and Pitot tubes have been installed at Points A and B in the pipe shown in Figure 7.7. The water levels under steady-state flow are as shown. Determine the following:

The elevation, pressure, velocity, and total heads at Points A and B
The head loss between Points A and B
The hydraulic gradient between Points A and B

Solution:

At Point A:

$h_z = \mathbf{3.62\ m}$
$h_p = \mathbf{3.01\ m}$
$h_v = \mathbf{0.50\ m}$
$h = h_z + h_p + h_v = 3.62\ \text{m} + 3.01\ \text{m} + 0.50\ \text{m} = \mathbf{7.13\ m}$

At Point B:

$h_z = \mathbf{4.28\ m}$
$h_p = \mathbf{1.6\ m}$
$h_v = \mathbf{0.50\ m}$
$h = h_z + h_p + h_v = 4.28\ \text{m} + 1.61\ \text{m} + 0.50\ \text{m} = \mathbf{6.38\ m}$

For pipe segment between A and B:

$$\Delta h = h_B - h_A = 6.39 \text{ m} - 7.13 \text{ m} = \mathbf{-0.75 \text{ m}}$$

$$i = -\frac{\Delta h}{\Delta l} = -\left(\frac{-0.74 \text{ m}}{200 \text{ m}}\right) = \mathbf{0.0038}$$

Note that the flow is from the point of high total head, A, to the point of lower total head, B, even though point A is lower than point B. In other words, the water is flowing uphill, but down gradient.

Application to Soil and Rock

Although velocity head is important in pipe and open channel flow, the velocity of water flow in soil is much lower, so the velocity head is very small (less than 5 mm). Thus, we can neglect it for practical soil seepage problems. Equation 7.3 then reduces to

$$h = h_z + h_p \tag{7.5}$$

Thus, the total head, h, in soil also may be defined as the difference in elevation between the datum and the water surface in the piezometer.

Field Instrumentation Geotechnical engineers sometimes install piezometers in the ground to measure the pressure head at different locations in the subsurface. Figure 7.8 shows a simple *open standpipe piezometer*, which is similar to the observation well shown in Figure 3.22 except the inside of the pipe is hydraulically connected only to a small response zone in the soil (as shown in Figure 7.8). It measures the average pressure head at the response zone. In comparison, an observation well is hydraulically open along nearly its entire length and simply gives the depth of the groundwater table.

If the groundwater conditions consist of an unconfined aquifer where the water is either stationary or flowing in a near-horizontal direction, and steady-state conditions apply, then the water level readings in a piezometer and an observation well at the same location should be virtually the same. However, if the groundwater table is not horizontal, artesian, or perched conditions are present, or if transient flow conditions exist, these two readings could be quite different.

Both observation wells and open standpipe piezometers are read by simply determining the water elevation in the standpipe. This may be done by simply lowering a cloth tape measure with a weight on the end, or with an electronic water level indicator as shown in Figures 3.22 and 7.9.

Other types of piezometers also are available, such as the pneumatic piezometer shown in Figure 7.10. This device is read by applying nitrogen gas under pressure through a tube and matching the pressure to the pore water pressure. Pneumatic piezometers are more complex than open standpipe piezometers, and therefore less reliable, but they can be placed in difficult locations where open standpipes cannot be installed. Similar units with electrical sensors also are available.

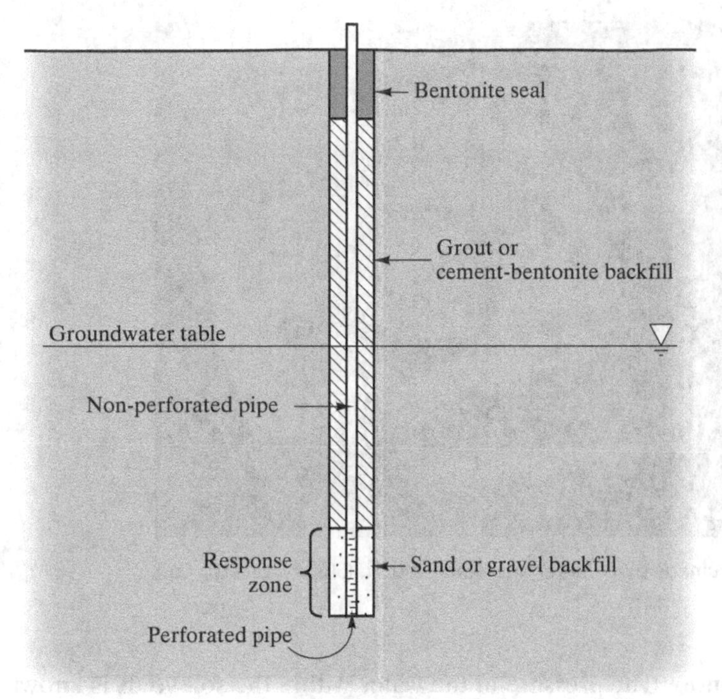

FIGURE 7.8 An open standpipe piezometer consists of a perforated pipe installed in a boring.

FIGURE 7.9 Filter tips for use on standpipe piezometers. The reel in the background is a probe to measure the water depth, as shown in Figure 3.22. (Slope Indicator Co.)

FIGURE 7.10 A pneumatic piezometer and readout unit. (Slope Indicator Co.)

Pore Water Pressure The pressure in the water within the soil voids is known as the *pore water pressure*, u. This is what some engineers call gage pressure (i.e., it is the difference between the absolute water pressure and atmospheric pressure). For points below the groundwater table, the pore water pressure is

$$u = \gamma_w h_p \tag{7.6}$$

where:

u = pore water pressure

h_p = pressure head

γ_w = unit weight of water = $62.4 \text{ lb/ft}^3 = 9.81 \text{ kN/m}^3$

It is simple to determine the pore water pressure at points where a piezometer is present. We simply subtract the average elevation of the response zone from the elevation of the water level in the piezometer to obtain the pressure head at the response zone, and then use Equation 7.6 to compute the pore water pressure. However, we usually do not have this luxury, and frequently need to compute u without the benefit of a piezometer.

To determine the pore water pressure at a point without a piezometer, we first need to determine if the pore water pressure is due solely to the force of gravity acting on the pore water. This is the case so long as the soil is not in the process of settling or shearing. We call this the *hydrostatic condition*, and the associated pore water pressure is the *hydrostatic pore water pressure*, u_h.

The next step is to determine if both of the following conditions also have been met:

- The aquifer is unconfined (i.e., the position of the groundwater table is not controlled by an overlying aquiclude).
- The groundwater is stationary or flowing in a direction within about 30° of the horizontal.

If all these conditions have been met (which they often are), then the pressure head is simply the difference in elevation between the groundwater table and the point where the pressure head is to be computed. Therefore, the pore water pressure is:

$$u = u_h = \gamma_w z_w \tag{7.7}$$

where:

u = pore water pressure

u_h = hydrostatic pore water pressure

γ_w = unit weight of water = 62.4 lb/ft^3 = 9.81 kN/m^3

z_w = depth from the groundwater table to the point

Example 7.2

Compute the pore water pressure at Points A and B shown in the Figure 7.11. Point A is located in the upper unconfined aquifer and Point B is located in the lower confined aquifer.

Solution:

Since Point A is in an unconfined aquifer, the pressure head at that location is simply equal to the distance below the phreatic surface, z_w, and the pore water pressure, u, is:

$$u = \gamma_w z_w$$

$$u = (62.4\ \text{lb/ft}^3)(260\ \text{ft} - 250\ \text{ft})$$

$$u = \mathbf{624\ lb/ft^2}$$

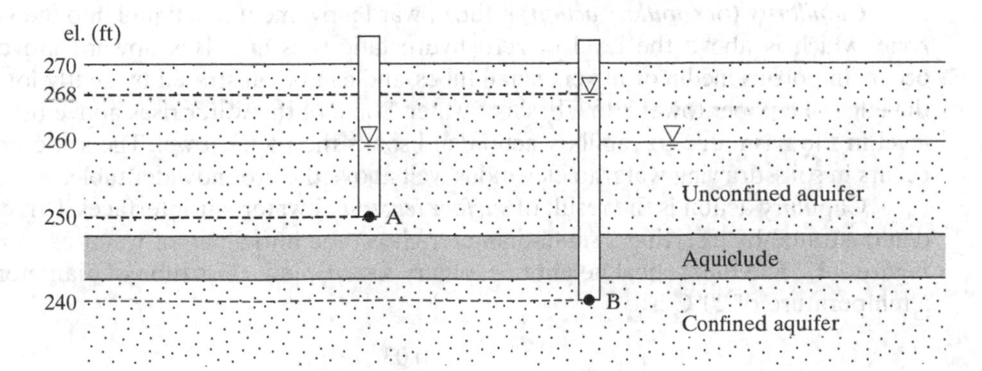

FIGURE 7.11 Soil profile for Example 7.2. el. = elevation.

Since Point B is located in a confined aquifer, it is necessary to install a piezometer to determine the pore water pressure. The pressure head, h_p, is equal to difference in elevation between Point B and the elevation to which the water rises in the piezometer and the pore water pressure, u, is:

$$u = \gamma_w h_p$$

$$u = (62.4 \text{ lb/ft}^3)(268 \text{ ft} - 240 \text{ ft})$$

$$u = \mathbf{1747.2\ lb/ft^2}$$

Note that the water in the piezometer located at Point B rises above the water table in the unconfined aquifer. This indicates that the lower aquifer is in an artesian condition.

If the soil is in the process of settling or shearing, then *excess pore water pressures* will be present as discussed in Chapters 10–12, and the hydrostatic condition does not exist. In this case, computing the pore water pressure becomes more complex. If the water is flowing at a significant angle from the horizontal, or if it is confined, then it is necessary to perform a two- or three-dimensional analysis as described later in this chapter.

Above the groundwater table, we normally consider the pore water pressure to be zero. In reality, surface tension effects between the water and the solid particles produce a negative pore water pressure above the groundwater table, and this negative pressure is sometimes called *soil suction*. Some advanced analyses sometimes consider soil suction, but they are beyond the scope of this book (see Fredlund and Rahardjo, 1993).

Capillarity Earlier in this chapter, we defined the groundwater table as the elevation to which water would rise in an observation well. Soils below the groundwater table are saturated, and have a pore water pressure defined by Equation 7.7. For many engineering problems, this simple model is sufficient. However, the real behavior of soils is rarely so simple. One important aspect not addressed by this model is capillarity.

Capillarity (or *capillary action*) is the upward movement of a liquid into the vadose zone, which is above the level of zero hydrostatic pressure. This upward movement occurs in porous media or in very small tubes, and can be illustrated by gently lowering the edge of a paper towel into a basin of water. Some of the water rises above the water level in the basin due to capillary action and soaks the paper towel. The same process occurs in soils, drawing water to elevations well above the groundwater table.

Capillary action is the result of *surface tension* at water–air interfaces. This can be demonstrated by inserting a small-diameter glass tube into a pan of water as shown in Figure 7.12. The theoretical height of capillary rise, h_c, in a glass tube of diameter d, at a temperature of 20°C is:

$$h_c = \frac{0.03}{d} \tag{7.8}$$

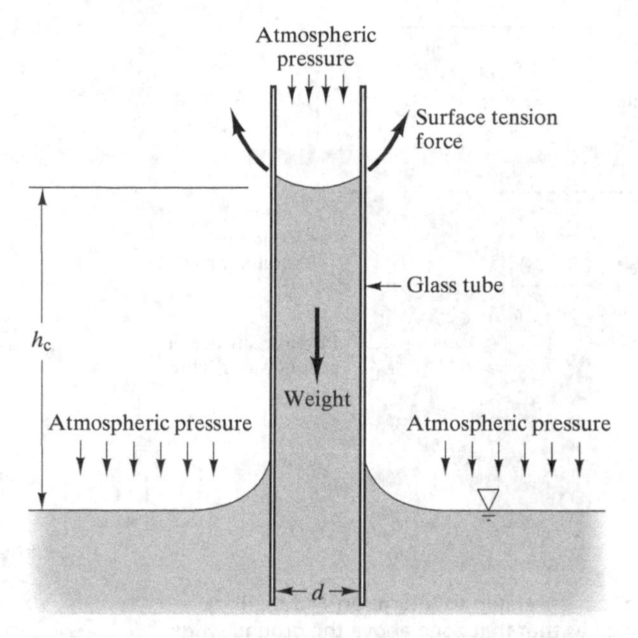

FIGURE 7.12 Capillary action demonstrated by a thin glass tube immersed in water.

where:

h_c = height of capillary rise (m)
d = diameter of glass tube (mm)

Capillary rise in soils is more complex because soils contain an interconnected network of pores of different sizes. However, using $0.2D_{10}$ as the equivalent d generally produces satisfactory results for sands and silts (Holtz and Kovacs, 1981). Thus, the height of capillary rise in these soils is approximately:

$$h_c = \frac{0.15}{D_{10}} \tag{7.9}$$

where:

h_c = height of capillary rise (m)
D_{10} = grain size corresponding to 10% finer (mm)

The theoretical capillary rise in clays can be in excess of 100 m, but in reality it is much less. Figure 7.13 shows the relationship among the vadose, phreatic, and capillary zones.

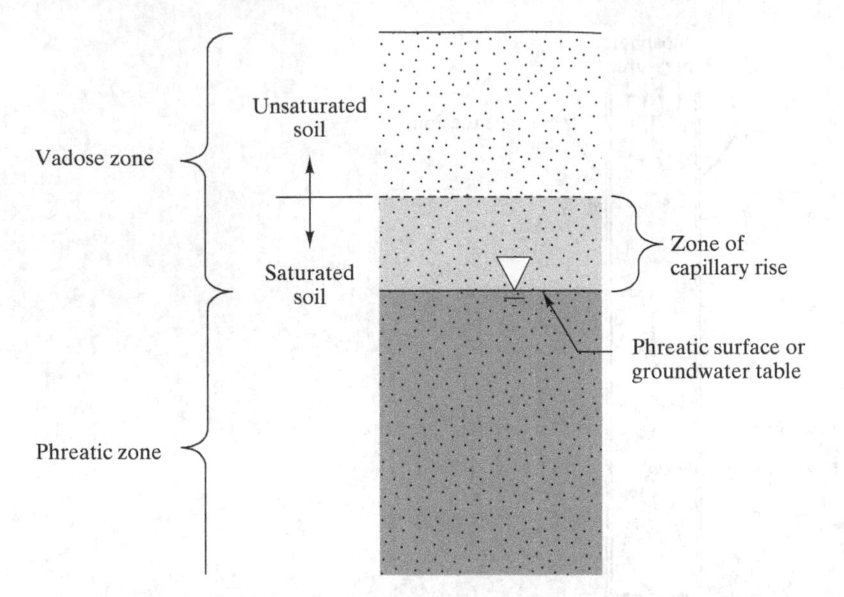

FIGURE 7.13 Definitions of soil zones related to saturation and capillary action. The vadose zone is defined as that that zone above the ground water table or phreatic surface. The phreatic zone is that zone below the phreatic surface. Notice that the soil can be saturated above the phreatic surface in the zone of capillary rise. The zone of capillary rise may be only a fraction of an inch in sands, but can be several feet in clay or silt soils.

7.3 ONE-DIMENSIONAL FLOW THROUGH SOIL

One-dimensional flow is the easiest condition to understand. It is used in laboratory test equipment, and some field conditions may be idealized as one-dimensional flow problems.

Geotechnical engineers often need to predict the flow rate, Q, through a soil. We could use Newton's law of friction, combined with the Navier–Stokes equations of hydrodynamics to describe one-dimensional flow through soils. However, the resulting formulas are very complex and thus impractical for normal geotechnical engineering analyses. Therefore, engineers use the simpler empirical method developed by the French engineer Darcy (1856).

Darcy's Law

Henry Darcy performed some of the first quantitative studies of the flow of water through sand (see image and text on page 270). He used the device shown in Figure 7.14, which is essentially the same as a modern constant head permeameter discussed later in this chapter. Using the piezometers on this device, Darcy measured the total head at the top and bottom of a sand column, and was thus able to compute the hydraulic gradient in the sand. He determined the flow rate by measuring the volume of water exiting the soil over a fixed period of time. After performing a number of experiments using different hydraulic gradients, Darcy observed that the flow rate was directly proportional to the

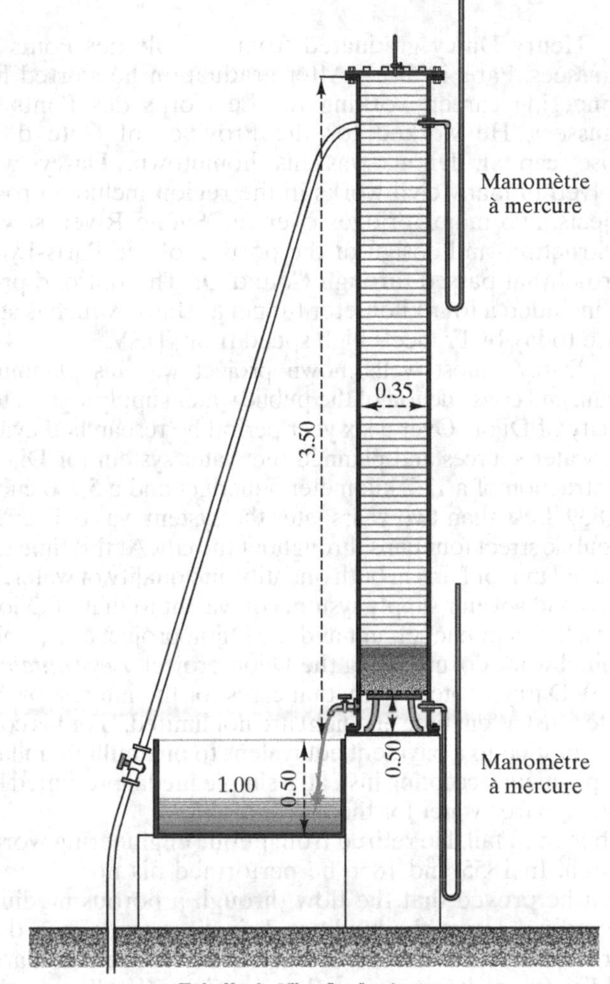

FIGURE 7.14 Laboratory device used by Henry Darcy in the 1850s to study water flow through sands. (Darcy, 1856.)

hydraulic gradient for a given sand. This relationship between hydraulic gradient and flow is known as *Darcy's law* and expressed as:

$$Q = kiA \tag{7.10}$$

where:

Q = flow rate
k = hydraulic conductivity (also known as coefficient of permeability)

Henry Darcy

Henry Darcy graduated from L'Ecole des Ponts et Chaussées, Paris in 1826. After graduation he started his engineering career working for Le Corps des Ponts et Chaussées. He worked for the Province of Côte d'Or whose capital, Dijon, was his hometown. Darcy was involved in many civil works in the region including road projects, two major bridges over the Saône River, sewer construction, and design of the portion of the Paris-Lyon railroad that passed through Côte d'Or. The railroad project included a four kilometer tunnel at Blaisy which is still in use today by France's high speed train, TGV.

Darcy's most well known project was his planning, design, and construction of the public water supply system for the city of Dijon. Over a six year period he researched available water sources and planned the water system for Dijon. Construction of a 12.7 kilometer aqueduct and a 5,700 cubic meter reservoir started in March 1839. Less than two years later the system was delivering clean water to major buildings and public street fountains throughout the city. At this time the Dijon water supply system far surpassed that of Paris in both quantity and quality of water. In fact, it would be 20 years before Paris had a water supply system equivalent to that of Dijon.

Darcy clearly understood himself as a public servant and the Dijon project as a public work in the truest sense. In his seminal work documenting the Dijon project, *Les Fontaines Publiques de la Ville de Dijon* (1856), Darcy wrote, "A city that cares for the interest of the poor class should not limit their water, just as daytime and light are not limited." For his contributions to this project, Darcy was entitled to a payment equivalent to one million dollars in today's currency. He refused this payment, accepting instead a simple medal presented by the City of Dijon along with the rights to free water for the rest of his life.

In the 1850s, Darcy's health began to fail. He retired from public engineering works and turned his attention to research. In 1855 and 1856 he performed his now famous sand column experiments in which he proved that the flow through a porous medium was directly proportional to the gradient across the medium. This discovery ushered in modern quantitative groundwater hydraulics (Simmons, 2006). During this period Darcy also conducted research and published works on water flow in pipes (leading to the Darcy-Weisback equation), open channel flow, and improvements to the Pitot tube. Darcy died of pneumonia on January 2, 1858 and was buried in Dijon.

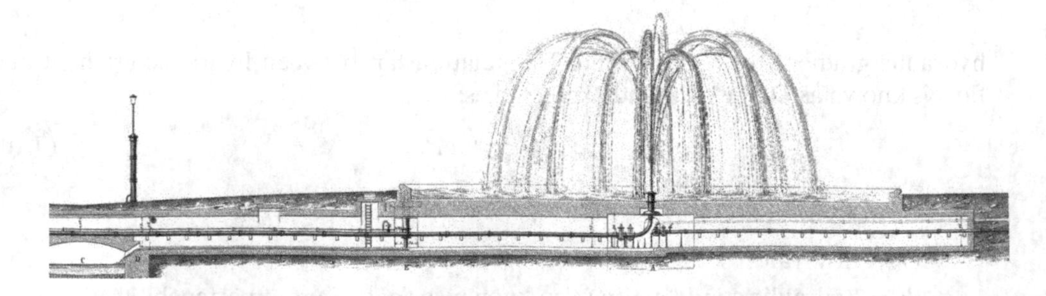

i = hydraulic gradient
A = area perpendicular to the flow direction

The cross-sectional area, A, in Equation 7.10 includes both the voids and the solids, even though the groundwater flows only through the voids. For example, to evaluate groundwater flowing through a $2 \text{ m} \times 4 \text{ m}$ zone of soil, we use $A = 8 \text{ m}^2$, even though only a fraction of this area is voids. This definition simplifies the computations and does not introduce any error in computations of Q so long as we use it consistently.

Calling Darcy's empirical relationship a "law" is rather pretentious. Nevertheless, it has been found to be valid for a wide range of soil types, from clays through coarse sands. The primary exceptions are clean gravels, where its accuracy is diminished because of the turbulent flow, and possibly in clays with low hydraulic gradients, where the absorption of water to the clay particles (discussed in Section 4.5) significantly influences flow.

Hydraulic Conductivity (Coefficient of Permeability) The *hydraulic conductivity*, k, depends on properties of both the soil and the liquid flowing through it, including the following:

Soil properties

- Void size (depends on particle size, gradation, void ratio, and other factors)
- Soil structure
- Void continuity
- Particle shape and surface roughness.

Liquid properties

- Density
- Viscosity.

Most practical problems deal with clean water or water contaminated with small quantities of other substances. Any variations in density and viscosity are small and can usually be ignored. Thus, we normally think of the hydraulic conductivity as being dependent only on the soil.

The most common unit of measurement for k is cm/s. However, many other units also are used, including ft/min, ft/yr, and even gallons/day/ft^2. Although k has units of length/time, the same as those used to describe velocity, it is not a measure of velocity. Table 7.1 presents typical values for different soil types.

Notice the extremely wide range of k values in Table 7.1. For example, clays typically have a k that is 1,000,000 times smaller than that of sands. Thus, according to Equation 7.10, Q also will be 1,000,000 times smaller. The low k in clays is due to the small particle size (and therefore small void size). It is not due to water being absorbed by the clay.

TABLE 7.1 Typical Values of Hydraulic Conductivity, k, for Saturated Soils

Soil Description	Hydraulic Conductivity, k (cm/s)	(ft/s)
Clean gravel	1 to 100	3×10^{-2} to 3
Sand-gravel mixtures	10^{-2} to 10	3×10^{-4} to 0.3
Clean coarse sand	10^{-2} to 1	3×10^{-4} to 3×10^{-2}
Fine sand	10^{-3} to 10^{-1}	3×10^{-5} to 3×10^{-3}
Silty sand	10^{-3} to 10^{-2}	3×10^{-5} to 3×10^{-4}
Clayey sand	10^{-4} to 10^{-2}	3×10^{-6} to 3×10^{-4}
Silt	10^{-8} to 10^{-3}	3×10^{-10} to 3×10^{-5}
Clay	10^{-10} to 10^{-6}	3×10^{-12} to 3×10^{-8}

Geotechnical engineers also use the term *coefficient of permeability* to describe k. Unfortunately, this term can generate some confusion because there are at least two similar parameters used to describe permeability:

- The *coefficient of permeability*, k, in Darcy's law describes the ease with which a given liquid flows through a certain soil. It depends on both the soil and the liquid, has a dimension of L/T, and is the same as *hydraulic conductivity*.
- The *intrinsic permeability* (also called the *specific permeability*) depends only on the soil. It has a dimension of L^2. This definition is used primarily by hydrogeologists and petroleum geologists.

Both parameters are frequently called *permeability* and the variables K and k have been used for both. In addition, the intrinsic permeability is sometimes measured with a special unit called *Darcys*, even though it is not the proper parameter for use in Equation 7.10. These inconsistencies in terminology can be a source of confusion. Whenever in doubt, check the units to determine which "permeability" is being used.

To avoid confusion, geotechnical engineers are gradually dropping the term *coefficient of permeability* and using *hydraulic conductivity* instead.

Confined Aquifer Flow In a confined aquifer of constant thickness, the flow is one dimensional so long as there are no rivers, lakes, or wells that affect the flow of water in the aquifer. In this case, we can compute the flow per unit width of the aquifer using Equation 7.10. Consider the following examples.

Example 7.3

Figure 7.15 shows a confined aquifer with two piezometers measuring pressure head at two different locations 1500 ft apart. The aquifer has a uniform thickness of 10.5 ft and a hydraulic conductivity of 2.4×10^{-4} ft/s. Assume the width of the aquifer into the page is 6 miles and the flow is one dimensional from left to right. Compute the total flow in the aquifer in ft^3/s.

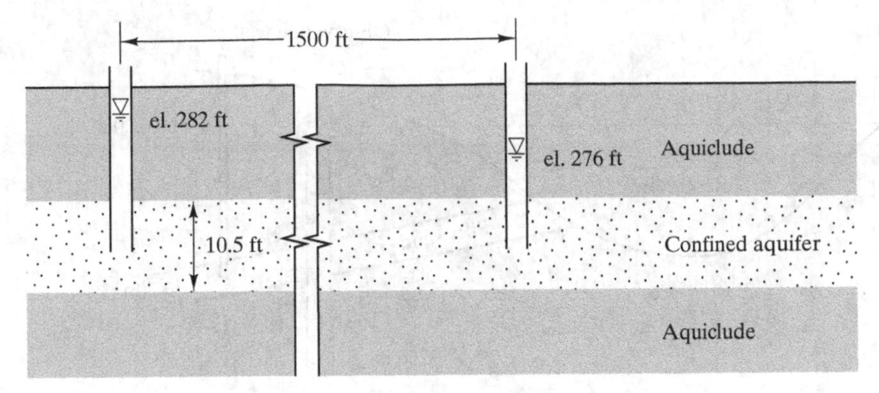

FIGURE 7.15 Soil profile for Example 7.3. el. = elevation.

Solution:

The flow in a confined aquifer is similar to the flow in a soil filled pipe, except in this case instead of a round pipe we have a rectangular-shaped aquifer with a cross-sectional area of 10.5 ft by 6 miles. Darcy's law applies.

Using Darcy's equation,

$$Q = kiA$$

$$i = \frac{\Delta h}{\Delta l} = \frac{6\text{ ft}}{1500\text{ ft}} = 0.004$$

$$Q = kiA$$

$$Q = (2.4 \times 10^{-4}\text{ ft/s})(0.004)(10.5\text{ ft})(6\text{ mile})\left(\frac{5280\text{ ft}}{1\text{ mile}}\right) = \mathbf{0.319\ ft^3/s}$$

Example 7.4

A 3.2 m thick silty sand stratum intersects one side of a reservoir as shown in Figure 7.16. This stratum has a hydraulic conductivity of 4×10^{-2} cm/s and extends along the entire 1000 m length of the reservoir. An observation well has been installed in this stratum as shown. Compute the seepage loss from the reservoir through this stratum.

Solution:

The observation well indicates a water level above the top of the silty sand strata. Therefore, it is a confined aquifer.

$$i = \frac{\Delta h}{\Delta l} = \frac{167.3\text{ m} - 165.0\text{ m}}{256\text{ m}} = 0.0090$$

$$A = (3.2\text{ m})(1000\text{ m}) = 3200\text{ m}^2$$

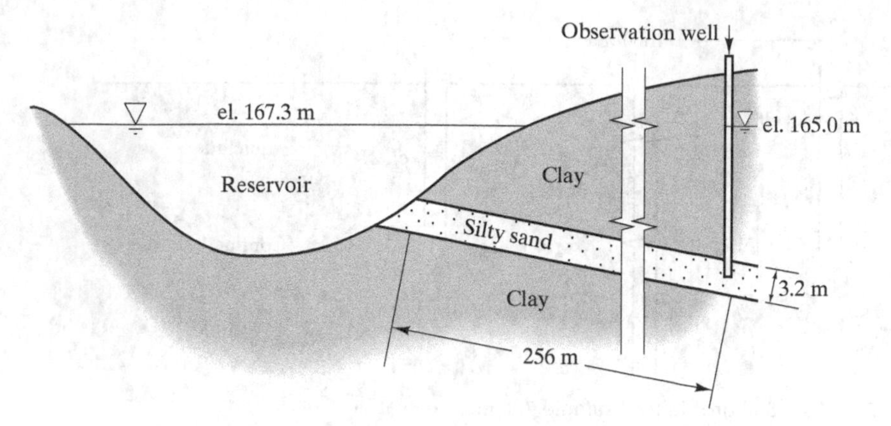

FIGURE 7.16 Cross-section through reservoir showing silty sand stratum. This cross-section is oriented parallel to the direction of flow. el. = elevation.

$$k = (4 \times 10^{-2} \text{ cm/s})\left(\frac{\text{m}}{100 \text{ cm}}\right)\left(\frac{3600 \text{ s}}{\text{hr}}\right)\left(\frac{24 \text{ hr}}{\text{day}}\right)\left(\frac{30 \text{ day}}{\text{mo}}\right) = 1000 \text{ m/mo}$$

$$Q = kiA = (1000 \text{ m/mo})(0.0090)(3200 \text{ m}^2) = \mathbf{30{,}000\ m^3/mo}$$

Apparent, Seepage, and True Velocities If we rearrange Darcy's law by dividing both sides of Equation 7.10 by the cross-sectional area of the soil, A, we get the following equation:

$$\frac{Q}{A} = ki \tag{7.11}$$

Noting that the left-hand side of Equation 7.11 has a dimension of L/T, similar to a velocity, we can rewrite Equation 7.11 as:

$$v_a = ki \tag{7.12}$$

where:

v_a = apparent velocity

The apparent velocity, sometimes called the *Darcian velocity*, is not the velocity at which the water moves through the soil. It is an artificial velocity computed when the flow rate is divided by the cross-sectional area of the soil. The area used in Darcy's equation is the total cross-sectional area of the soil, which includes both soil solids and the voids. However, the water flows only in the voids. We must account for the difference between the total cross-sectional area and the cross-sectional area of voids to determine how rapidly water actually flows through the soil. We use the term *seepage velocity*, v_s, to describe this velocity:

$$v_s = \frac{ki}{n_e} \tag{7.13}$$

where:

v_s = seepage velocity
k = hydraulic conductivity
i = hydraulic gradient
n_e = effective porosity

The *effective porosity*, n_e, represents the percentage of the cross-sectional area, A, that actually contributes to the flow.

The seepage velocity is especially important in geoenvironmental problems because it helps us determine how quickly contaminants travel through the ground. Note that the seepage velocity will always be greater than the apparent velocity since the effective porosity will always be less than 1, so a failure to account for this difference would result in underestimating the rate of contaminant transport.

In sandy soils, n_e is equal to the porosity n as defined in Equation 4.16. However, clayey soils contain a static layer of water around the particles, so the actual flow area is less than the void area. Design values of n_e in clays are best determined using special laboratory tests (Kim, Edil, and Park, 1997). If no test data is available, the Environmental Protection Agency uses $n_e = 0.10$ in clays (Brumund, 1995).

When viewed on a microscopic scale, the true velocity at which an element of water is moving within the soil will be greater than the seepage velocity because the actual flow path is a circuitous route around the soil particles, as shown in Figure 7.17(a). The

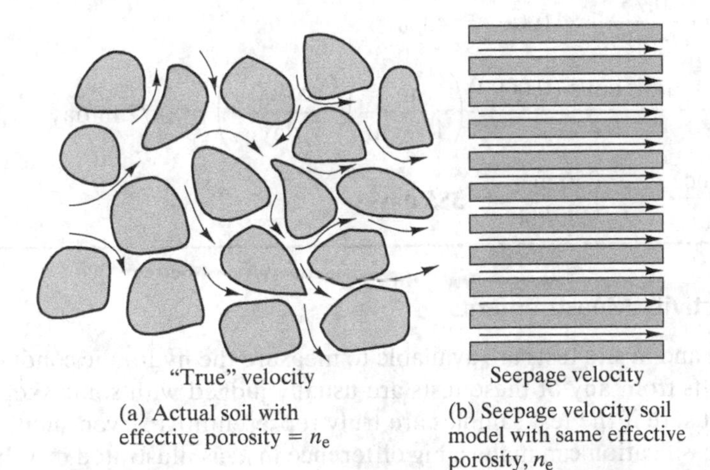

FIGURE 7.17 Comparison of true velocity and seepage velocity: (a) shows true velocity of pore water flowing around soil particles; the general direction of flow is from left to right; (b) shows the equivalent straight-line model used to compute the seepage velocity. The porosity of (a) and (b) must be the same.

seepage velocity is an "average" velocity viewed from a macroscale and is based on the equivalent straight-line movement as shown in Figure 7.17(b). However, the true velocity is primarily of theoretical interest, and the seepage velocity is more useful for solving practical contaminant transport problems.

Contaminant Transport Groundwater sometimes contains contaminants from fuel spills or industrial sources. In these cases, it is important to estimate how quickly the contaminants move through the ground. This is a relatively complex problem because the contaminants do not simply move along with the groundwater flow (a process called *advection*). In addition to advection, the contaminants can disperse and diffuse in the soils. In some cases, the contaminants can actually absorb onto or into soil particles. However, as a first-order estimate of the rate of transport, we can assume a contaminant undergoes only advection and moves at the same velocity as the groundwater.

Example 7.5

Assume a chemical solvent, which is denser than water is spilled into the reservoir in Example 7.4. How long will it take for the contaminant to reach the observation well assuming only advection occurs in the aquifer? The silty sand in the aquifer has a void ratio of 0.75.

Solution:

If only advection is occurring, then the contaminant will travel at the same velocity as the groundwater. The seepage velocity is the appropriate velocity to use in this case. Assume the effective porosity, n_e, is equal to the porosity for this silty sand. Using Equation 4.17,

$$n = \frac{e}{1+e} = \frac{0.75}{1+0.75} \times 100\% = 43\%$$

$$v_s = \frac{ki}{n_e} = \frac{4 \times 10^{-2}\ \text{cm/s}\ (0.0090)}{0.43}\left(\frac{\text{m}}{100\ \text{cm}}\right)\left(\frac{86{,}400\ \text{s}}{\text{day}}\right) = 0.72\ \text{m/day}$$

$$\text{time} = \frac{\text{distance}}{v_s} = \frac{256\ \text{m}}{0.72\ \text{m/day}} = \mathbf{355\ days}$$

Hydraulic Conductivity Measurements

Various laboratory and in situ tests are available to measure the hydraulic conductivity. However, the results from any of these tests are usually judged with some skepticism because we are not sure if the test samples are truly representative. Even small differences in the soil classification can make a big difference in k, as illustrated in Table 7.1. Thus, it is good to perform many tests and review the scatter in the results.

Another problem with laboratory tests is that the samples probably do not adequately represent small fissures, joints, sandy seams, and other characteristics in the field. In situ tests are better in this regard.

Even carefully conducted tests on good samples typically have a precision on the order of ±50% or more, and more routine tests have even less precision. Therefore, test results are normally reported to only one significant figure (e.g., 5×10^{-4} cm/s). Even then, we need to recognize the true hydraulic conductivity in the field may be substantially different from the test values.

Constant-Head Test The *constant-head test* is a laboratory hydraulic conductivity test that applies a constant head of water to each end of a soil sample in a *permeameter* as shown in Figure 7.18. The head on one end is greater than that on the other, so a flow is induced. We determine Q, i, and A from the test results, then compute k using Equation 7.10.

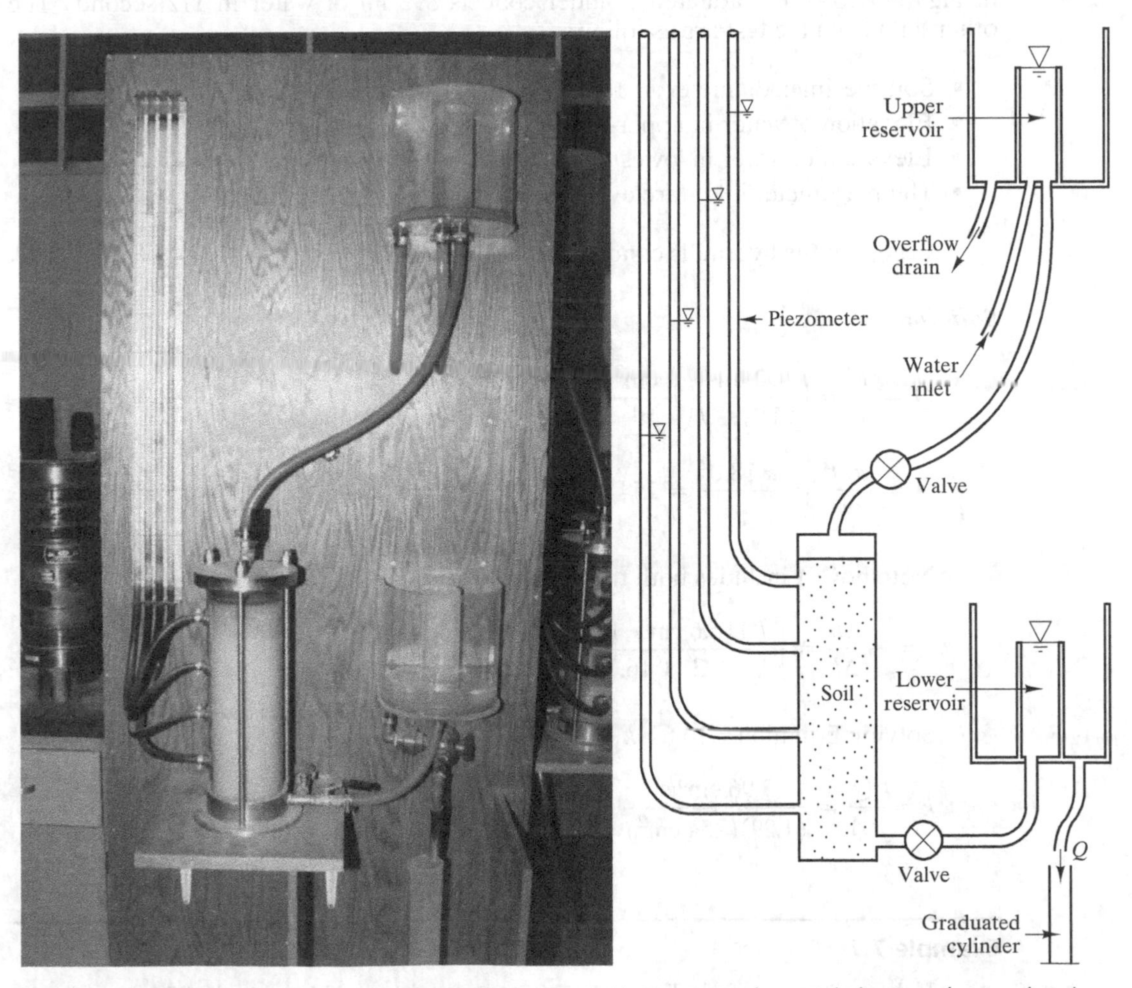

FIGURE 7.18 A constant-head permeameter. The upper and lower reservoirs contain inner and outer chambers that maintain constant heads. The piezometers measure the head at four locations in the soil sample.

Some constant-head permeameters do not have piezometers, so we must compute the hydraulic gradient, i, by dividing the head loss between the two reservoirs (i.e., the difference in their water surface elevations) by the height of the sample. This method implicitly assumes the head losses in the tubes, valves, etc., are very small compared to that in the soil. This is generally a poor assumption. It is better to use permeameters with piezometers and compute i by dividing the difference in the total heads by the distance between the piezometer inlets.

Example 7.6

A constant head test is performed using a permeameter similar to the one shown in Figure 7.18. The graduated cylinder collects 892 ml of water in 112 seconds. The other data from the test are as follows:

- Soil specimen diameter = 18.0 cm
- Elevation of water in upper-most piezometer = 181.0 cm
- Elevation of water in lowest piezometer = 116.6 cm
- The piezometer inlets are evenly spaced at 16.7 cm on center

Compute the hydraulic conductivity, k.

Solution:

$$Q = \frac{V}{t} = \left(\frac{892 \text{ ml}}{112 \text{ s}}\right)\left(\frac{1 \text{ cm}^3}{\text{ml}}\right) = 7.96\frac{\text{cm}^3}{\text{s}}$$

$$A = \frac{\pi D^2}{4} = \frac{\pi\, 18.0^2}{4} = 254 \text{ cm}^2$$

Note how A includes both the voids and the solids, as defined earlier.

$$i = -\frac{\Delta h}{\Delta l} = -\left(\frac{116.6 \text{ cm} - 181.0 \text{ cm}}{3 \times 16.7 \text{ cm}}\right) = 1.29$$

Solving Equation 7.10 for k,

$$k = \frac{Q}{iA} = \frac{7.96 \text{ cm}^3/\text{s}}{(1.29)(254 \text{ cm}^2)} = \mathbf{2 \times 10^{-2} \text{ cm/s}}$$

Example 7.7

If the soil specimen in Example 7.6 is sandy with a void ratio of 0.85, compute the seepage velocity through the specimen.

Solution:

The soil is sandy, so the effective porosity equals the porosity, n.

Using Equation 4.17,

$$n = \frac{e}{1+e} = \frac{0.85}{1+0.85} = 46\%$$

$$v_s = \frac{ki}{n_e} = \frac{(2 \times 10^{-2}\ \text{cm/s})(1.29)}{0.46} = \mathbf{5.6 \times 10^{-2}\ cm/s}$$

Falling-Head Test The *falling-head test* also is another laboratory test. It uses a standpipe on the upstream side as shown in Figure 7.19. In this test, the water in the standpipe is not replenished as it is in the constant-head reservoir. Thus, as the test progresses, the water level in the standpipe falls. This method is more suitable for soils with very low hydraulic conductivities, such as clays, where the flow rate is small and needs to be precisely measured.

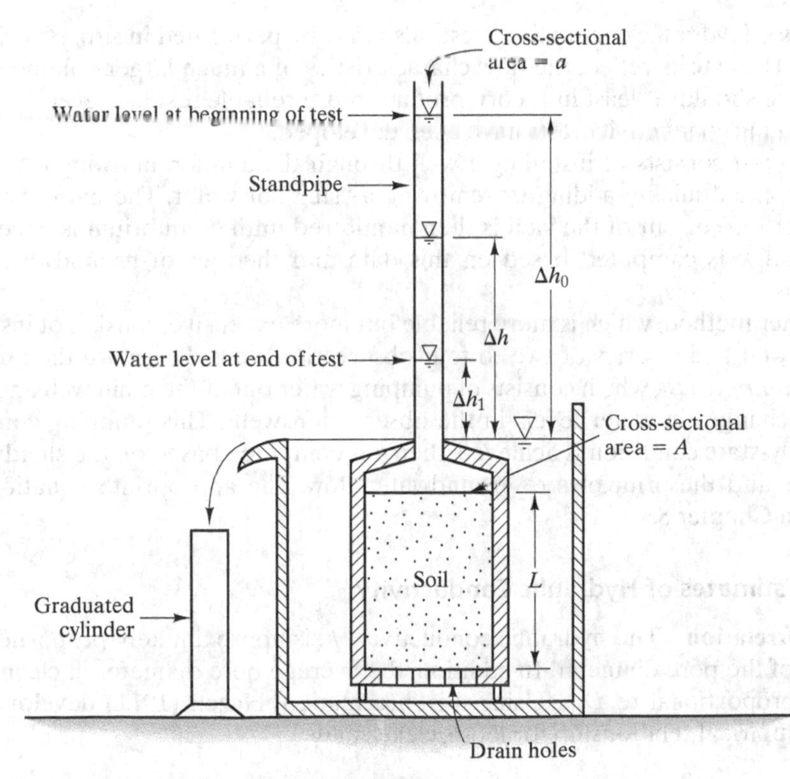

FIGURE 7.19 A falling-head permeameter.

The analysis of falling-head test results is more complex because the hydraulic gradient is not constant. This means the flow rate also is not constant (per Equation 7.10), so we must derive a new formula for k as follows:

Considering a head loss Δh_0 at the beginning of the test and Δh_1 after time t,

$$Q = kiA = k\frac{\Delta h}{L}A = -a\frac{\mathrm{d}(\Delta h)}{\mathrm{d}t}$$

$$\frac{kA}{L}\int_0^t \mathrm{d}t = -a\int_{\Delta h_0}^{\Delta h_1}\frac{\mathrm{d}(\Delta h)}{\Delta h}$$

$$\frac{kAt}{L} = -a\ln\left(\frac{\Delta h_1}{\Delta h_0}\right)$$

$$k = \frac{aL}{At}\ln\left(\frac{\Delta h_0}{\Delta h_1}\right) \tag{7.14}$$

Notice how the volume of water is determined by the change in water level in the standpipe. The graduated cylinder is used only as a check.

In Situ Tests Hydraulic conductivity tests also may be performed in situ, especially in sandy soils. These tests reflect the flow characteristics of a much larger volume of soil, and therefore should, at least in theory, produce more reliable results. Several kinds of in situ hydraulic conductivity tests have been developed.

A *slug test* consists of installing a well through the aquifer, allowing it to reach equilibrium, then quickly adding or removing a "slug" of water. The natural flow of groundwater into or out of the well is then monitored until equilibrium is once again achieved and k is computed based on this data and theories of groundwater flow around wells.

Another method, which is more reliable but more expensive, consists of installing a pumping well and a series of two to four observation wells. These are then used to conduct a *pumping test*, which consists of pumping water out of the main well and monitoring the changes in water levels in the observation wells. This pumping continues until a steady-state condition is achieved, then k is computed based on the steady-state water levels and the principles of groundwater flow. The appropriate equations are presented in Chapter 8.

Empirical Estimates of Hydraulic Conductivity

Hazen's Correlation The hydraulic conductivity, k, is approximately proportional to the square of the pore diameter. In addition, the average pore diameter in clean sands is roughly proportional to D_{10}. Using this information, Hazen (1911) developed the following empirical relationship for loose, clean sands:

$$k = CD_{10}^2 \tag{7.15}$$

where:

k = hydraulic conductivity (cm/s)
C = Hazen's coefficient = 0.8 to 1.2 (a value of 1.0 is commonly used)
D_{10} = diameter at which 10% of the soil is finer (mm) (also known as the *effective size*)

Note: Be sure to use the stated units for k and D_{10}.

The accuracy of Hazen's correlation is limited because it considers only a single measure of the grain size of a soil (D_{10}) and does not account for the distribution of grain sizes found in natural soils. Hazen's work was intended to be used in the design of sand filters for water purification, but can be used to estimate k in the ground. However, its applicability is limited to soils with 0.1 mm $< D_{10} <$ 3 mm and a coefficient of uniformity, $C_u < 5$ (Kashef, 1986).

Kozeny–Carman The Kozeny–Carman equation (Kozeny, 1927; Carman, 1956) is a semiempirical, semitheoretical equation for computing the hydraulic conductivity based on Darcy's law and the size and shape of pores in the soil. The equation is generally presented as:

$$k = \frac{\gamma}{\mu}\left(\frac{1}{T^2 S_0^2}\right)\left(\frac{e^3}{1+e}\right) \tag{7.16}$$

where:

γ = unit weight of pore fluid
μ = viscosity of pore fluid
T = dimensionless factor accounting for the shape of the pores
S_0 = specific surface of the soil particles, area per unit volume
e = void ratio of the soil

Laboratory testing and research (Carrier, 2003) have shown that T^2 can be taken to be approximately 5 for most soils. Using this value of T and the unit weight and viscosity of water, Equation 7.16 can be simplified to

$$k = 1.99 \times 10^{-4}\left(\frac{1}{S_0^2}\right)\left(\frac{e^3}{1+e}\right) \tag{7.17}$$

where:

k = hydraulic conductivity in cm/s
S_0 = specific surface of the soil particles, area per unit volume in 1/cm
e = void ratio of the soil

Using Equation 7.17, we need to know only the void ratio and specific surface to compute the hydraulic conductivity of a soil. If the soil was made up of uniform spheres then the specific surface would be

$$S_0 = \text{area/volume} = (\pi D^2)/(\pi D^3/6) = 6/D \tag{7.18}$$

TABLE 7.2 Shape Factor, S_F, for Use in Equation 7.20[a]

Particle Shape	Shape Factor (S_F)
Very angular	8.4
Angular	7.7
Subangular	7.4
Subrounded	6.6
Rounded	6.2
Well rounded	6.0

[a]Based on Fair and Hatch [1933] & Loudon [1952].

However, real soils are not made of uniform spheres, so we must replace the 6 in the numerator of Equation 7.18 with a shape factor, S_F, and Equation 7.18 becomes

$$S_0 = S_F/D \tag{7.19}$$

where the shape factor, S_F, comes from Table 7.2.

Now all that is necessary is to account for the different sizes of soil particles that exist in a natural soil. Carrier (2003) has shown that we can combine Equations 7.17, 7.19, and the results of a sieve analysis to rewrite the Kozeny–Carman equation as

$$k = 1.99 \times 10^4 \left(\frac{100\%}{\sum [f_i/(D_{li}^{0.404} \times D_{si}^{0.595})]} \right)^2 \left(\frac{1}{S_F} \right)^2 \left(\frac{e^3}{1+e} \right) \tag{7.20}$$

where:

k = hydraulic conductivity in cm/s
f_i = fraction of soil between two sieve sizes
S_F = shape factor (Table 7.2)
D_{li} = the size of the openings in the larger of the two sieves in cm
D_{si} = the size of the openings in the smaller of the two sieves in cm

Equation 7.20 can easily be setup in a spreadsheet. The Kozeny–Carman equation will provide much better estimates of the coefficient of hydraulic conductivity than Hazen's correlation because it accounts for the particle size distribution of the soil instead of using a single diameter to characterize the soil. This semiempirical formula is still limited to sands. It is not accurate for clays or for gravels.

Example 7.8

A poorly-graded, rounded, medium dense sand has a void ratio of 0.49 with the particle size distribution shown in the table below. Estimate the hydraulic conductivity using both Hazen's correlation and the Kozeny–Carman equation.

Sieve Opening (cm)	Sieve Size	Percent Passing
0.475	#4	100
0.2	#10	94
0.085	#20	70
0.0425	#40	55
0.015	#100	10
0.0075	#200	4

Solution:

Hazen's Correlation

$k = CD_{10}^2$

Assume C = 1

$k = (1)(0.15\text{ mm})^2$

$\mathbf{k = 2.3 \times 10^{-2}\ cm/s}$

Kozeny–Carman

Percent between #4 and #10

$$\frac{f_{\#4-\#10}}{D_{\#4}^{0.404} D_{\#10}^{0.595}} = \frac{100\% - 94\%}{0.475^{0.404} \times 0.2^{0.595}} = 21.12\%$$

Percent between #10 and #20

$$\frac{f_{\#10-\#20}}{D_{\#10}^{0.404} D_{\#20}^{0.595}} = \frac{94\% - 70\%}{0.2^{0.404} \times 0.085^{0.595}} = 199.34\%$$

Percent between #20 and #40

$$\frac{f_{\#20-\#40}}{D_{\#20}^{0.404} D_{\#40}^{0.595}} = \frac{70\% - 55\%}{0.085^{0.404} \times 0.0425^{0.595}} = 265.90\%$$

Percent between #40 and #100

$$\frac{f_{\#40-\#100}}{D_{\#40}^{0.404} D_{\#100}^{0.595}} = \frac{55\% - 10\%}{0.0425^{0.404} \times 0.015^{0.595}} = 1961.41\%$$

Percent between #100 and #200

$$\frac{f_{\#100-\#200}}{D_{\#100}^{0.404} D_{\#200}^{0.595}} = \frac{10\% - 4\%}{0.015^{0.404} \times 0.0075^{0.595}} = 601.66\%$$

$$\sum \frac{f_i}{D_{li}^{0.404} D_{si}^{0.595}} = 3049.42\%$$

$$k = 1.99 \times 10^4 \left(\frac{100\%}{\sum [f_i / (D_{li}^{0.404} \times D_{si}^{0.595})]} \right)^2 \left(\frac{1}{S_F} \right)^2 \left(\frac{e^3}{1 + e} \right)$$

$$k = 1.99 \times 10^4 \left(\frac{100\%}{3049.42\%}\right)^2 \left(\frac{1}{6.2}\right)^2 \left(\frac{0.49^3}{1 + 0.49}\right)$$

$$\mathbf{k = 4.4 \times 10^{-2}\ cm/s}$$

7.4 FLOW THROUGH ANISOTROPIC SOILS

Many natural soils, especially alluvial and lacustrine soils, contain thin horizontal stratifications that reflect their history of deposition. For example, there may be alternating layers of silt and clay, each only a few millimeters thick, as shown in Figure 7.20. The hydraulic conductivity in some layers is often much greater than in others, so groundwater flows horizontally much more easily than vertically. Such soils are said to be *anisotropic with respect to hydraulic conductivity*, so we need to determine two values of k: the horizontal hydraulic conductivity, k_x, and the vertical hydraulic conductivity, k_z, of the layered soil.

We can determine the equivalent horizontal and vertical hydraulic conductivity for layered soils by considering two constant-head permeameter tests of a layered soil as shown in Figure 7.21. For flow parallel to the soil layering, Figure 7.21(a), the flow must be horizontal. Since the flow is horizontal, no water may cross from one soil layer to another and we can treat each soil layer separately. The head loss in each layer is the same, Δh, as is the length of the flow path, L. Under these boundary conditions the gradient in each layer is equal:

$$i = i_1 = i_2 = i_3 = \frac{\Delta h}{L} \tag{7.21}$$

The flow in each layer will be different but the sum of the flow in each layer must be equal to the total flow:

$$Q = Q_1 + Q_2 + Q_3 = k_x i A \tag{7.22}$$

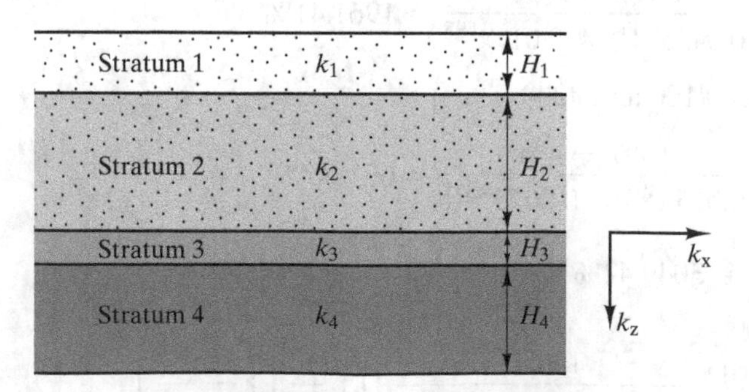

FIGURE 7.20 Flow of water through anisotropic soils.

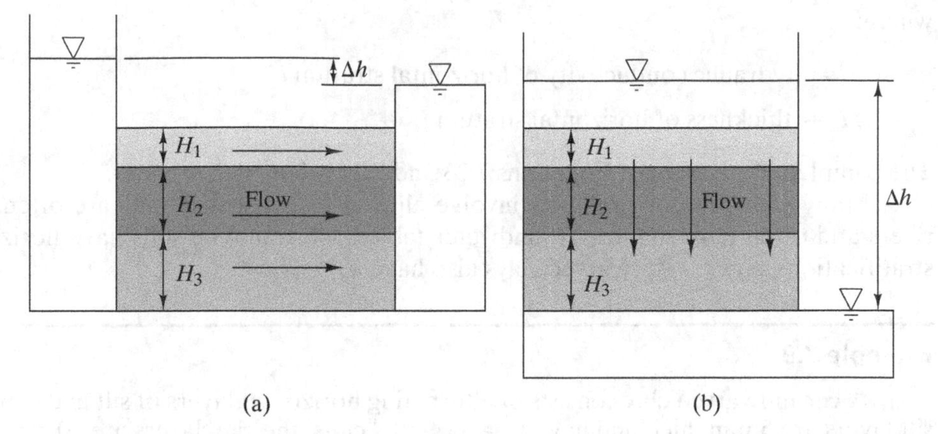

FIGURE 7.21 Constant-head permeameters illustrating (a) horizontal flow parallel to layering and (b) vertical flow normal to layering.

Using the above two observations, we can show that the equivalent horizontal hydraulic conductivity, k_x, for a system of i layers is

$$k_x = \frac{\sum k_i H_i}{\sum H_i} \tag{7.23}$$

where:

k_i = hydraulic conductivity of horizontal stratum i
H_i = thickness of horizontal stratum i

For the case of flow normal to the soil layering, we consider the permeameter shown in Figure 7.21(b). In this case the total head loss in the system must be equal to the sum of the head loss in each layer.

$$\Delta H = \Delta h_1 + \Delta h_2 + \Delta h_3 \tag{7.24}$$

And the flow in each layer must be the same and equal to the total flow, Q

$$Q_1 = Q_2 = Q_3 = Q = k_z i A \tag{7.25}$$

where:

k_z = equivalent vertical hydraulic conductivity

Using the above two observations, we can show that the equivalent vertical hydraulic conductivity, k_z, for a system of i layers is

$$k_z = \frac{\sum H_i}{\sum \left(\frac{H_i}{k_i} \right)} \tag{7.26}$$

where:

k_i = hydraulic conductivity of horizontal stratum i

H_i = thickness of horizontal stratum i

The complete derivation of Equations 7.23 and 7.26 is left to the reader.

Many groundwater problems involve alluvial soils because they are often near rivers and often have shallow groundwater tables. Most alluvial soils have horizontal stratifications, so $k_x > k_z$. Varved clays also have $k_x > k_z$.

Example 7.9

A certain varved clay consists of alternating horizontal layers of silt and clay. The silt layers are 5 mm thick and have $k = 3 \times 10^{-4}$ cm/s; the clay layers are 20 mm thick and have $k = 6 \times 10^{-7}$ cm/s. Compute k_x and k_z.

Solution:

From Equation 7.23,

$$k_x = \frac{\sum k_i H_i}{\sum H_i}$$

$$k_x = \frac{(3 \times 10^{-4}\text{ cm/s})(0.5\text{ cm}) + (6 \times 10^{-7}\text{ cm/s})(2\text{ cm})}{0.5\text{ cm} + 2\text{ cm}}$$

$$k_x = \mathbf{6 \times 10^{-5}\ cm/s}$$

From Equation 7.26,

$$k_z = \frac{\sum H_i}{\sum \left(\frac{H_i}{k_i}\right)}$$

$$k_z = \frac{0.5\text{ cm} + 2\text{ cm}}{\dfrac{0.5\text{ cm}}{3 \times 10^{-4}\text{ cm/s}} + \dfrac{2\text{ cm}}{6 \times 10^{-7}\text{ cm/s}}}$$

$$k_z = \mathbf{7 \times 10^{-7}\ cm/s}$$

Commentary: Water flowing horizontally moves through both types of soil in parallel. The ratio of thicknesses for the two strata is 20/5 = 4, but the ratio of k values is $3 \times 10^{-4}/6 \times 10^{-7} = 500$. Thus, the silt layers dominate the horizontal flow, even though they represent only 20% of the total cross-sectional area of flow. This is why the value of k_x is nearly equal to k of the silt. However, water flowing vertically must pass through all of the soil layers, and thus is controlled by the ones with the lowest hydraulic conductivity. Therefore, k_z is nearly equal to k of the clay.

SUMMARY

Major Points

1. The pores in a soil are interconnected and water is able to travel through these pores.
2. Hydrologically, the soil can be divided into the vadose zone above the groundwater table, and the phreatic zone below the groundwater table. The groundwater table is that location where the pore water pressure is equal to atmospheric pressure. In the phreatic zone below the groundwater table, the pore water is free to flow.
3. The energy in groundwater may be defined in terms of its total head, h, which is the sum of the elevation head, pressure head, and velocity head. In groundwater problems the velocity head is very small and can be ignored. Groundwater flow is driven by changes in total head. Water flows from locations of high head to locations of low head.
4. As water flows through soil it loses energy. The hydraulic gradient, i, describes the head loss per unit distance of groundwater travel.
5. The pore water pressure is the gage pressure of the pore water at a given location and can be computed as the pressure head times the unit weight of water.
6. In the vadose zone above the groundwater table, flow of water is controlled by capillary action which draws water into the smaller pore spaces and holds the water there. Water can rise well above the groundwater table through capillary action. In the zone of capillary rise, the pore water pressure will be negative (less than atmospheric pressure).
7. The flow of water through soil is usually described using Darcy's law, $Q = kiA$.
8. The hydraulic conductivity, k, in Darcy's law is a soil property that describes the ease with which water flows through a given soil. Hydraulic conductivity varies greatly from as low as 10^{-10} cm/s (3×10^{-12} ft/s) in clay to as high as 100 cm/s (3 ft/s) in gravel.
9. Hydraulic conductivity can be measured with laboratory tests, estimated using empirical correlations to particle size distribution of a soil, or measured in situ with field tests.
10. The seepage velocity describes the rate at which water flows through the ground. It is often used in contaminant transport analyses.
11. Because many soil profiles consist of horizontally bedded layers of different soils, the equivalent hydraulic conductivity of an aquifer is generally anisotropic with the equivalent horizontal hydraulic conductivity being several orders of magnitude greater than the equivalent vertical hydraulic conductivity.

Vocabulary

advection
anisotropic
apparent velocity
aquiclude
aquifer
aquitard
artesian condition
artesian well
Bernoulli equation
capillarity
capillary rise
confined aquifer
constant-head test
Darcy's law
effective porosity
elevation head
falling-head test
flow rate
flow regime
groundwater
groundwater hydrology
groundwater table
Hazen's correlation
hydraulic conductivity
hydraulic gradient
hydrologic cycle
hydrology
hydrostatic condition
hydrostatic pore water pressure
laminar flow
leachate
perched groundwater
permeameter
phreatic zone
piezometer
pore water pressure
pressure head
seepage velocity
steady-state condition
total head
unconfined aquifer
vadose zone

QUESTIONS AND PRACTICE PROBLEMS

Section 7.1 Hydrology

7.1 How does groundwater get into the ground?

7.2 Explain the difference between an aquifer, an aquiclude, and an aquitard.

7.3 Explain the difference between confined and unconfined groundwater flow.

Section 7.2 Principles of Fluid Mechanics

7.4 Explain the difference between steady-state and unsteady-state (transient) flow.

7.5 In most fluid flow applications the total head is the sum of the elevation head, pressure head and velocity head. In groundwater flow we generally assume the velocity head is zero. Why is this a safe assumption?

7.6 The water in a soil flows from Point K to Point L, a distance of 250 ft. Point K is at elevation 543 ft and Point L is at elevation 461 ft. Piezometers have been installed at both points, and their water levels are 23 ft and 74 ft, respectively, above the points. Compute the average hydraulic gradient between these two points.

7.7 Compute the pore water pressures at Points K and L in Problem 7.6.

7.8 The groundwater table in an unconfined aquifer is at a depth of 9.3 m below the ground surface. Assuming hydrostatic conditions are present, and the groundwater is virtually stationary, compute the pore water pressure at depths of 15.0 and 20.0 m below the ground surface.

7.9 An exploratory boring is being drilled. The soil encountered between the ground surface and a depth of 10 m has been dry sand, with no visible signs of groundwater. Then, at a depth of 10 m the soil changes to moist clay which becomes very wet at a depth of 12 m. At 12 m the soil changes back to silty sand which is very wet. The boring continues to a depth of 15 m. A piezometer is then installed in the lower silty sand layer. Within 2 days, the water in the piezometer had risen to a depth of only 8 m below the ground surface. Explain the groundwater conditions that have been encountered.

7.10 Compute the pore water pressures at the bottom of the piezometer in Problem 7.9.

7.11 Two small commercial buildings have been constructed at a site underlain by a sandy silt (ML) that has D_{10} = 0.03 mm. The groundwater table is at a depth of 6 ft. Both buildings have concrete slab-on-grade floors. In Building A, the slab was placed directly onto the natural soils, while Building B has a 4-in. layer of poorly-graded coarse gravel between the slab and the natural soils. Both buildings have vinyl floor coverings similar to those typically used in residential kitchens. Both buildings are now 3 years old.

Unfortunately, the tenant in Building A is having continual problems with the vinyl floors peeling up from the concrete slab. When the peeled sections are examined, moisture is always evident between the vinyl and the concrete. Curiously, the tenant in Building B has had no such problems, even though both buildings have the same floor covering. Could the problem in Building A be due to capillary action in the underlying soil? Explain why or why not. Also explain why Building B is not having any such problems.

7.12 A certain clayey zone has a zone of capillary rise of 4.5 m above the groundwater table. What is the pressure head and the pore pressure at a point 2 m above the groundwater table?

Section 7.3 One-Dimensional Flow Through Soil

7.13 A constant-head hydraulic conductivity test has been conducted on a 110-mm diameter, 270-mm tall fine sand specimen in a permeameter similar to the one shown in Figure 7.18. The upper and lower reservoir elevations were 2010 mm and 1671 mm above the lab floor. The piezometers, whose tips are spaced 200 mm apart, had readings of 1809 and 1578 mm, and the graduated cylinder collected 910 ml of water in 25 min 15 s. Using the best available data, compute the hydraulic conductivity. Does the result seem reasonable? Why or why not?

7.14 A falling-head hydraulic conductivity test has been conducted on a clay specimen in a permeameter similar to the one in Figure 7.19. The soil specimen was 97 mm in diameter and 20 mm tall. The standpipe had an inside diameter of 6.0 mm. The water level in the bath surrounding the specimen was 120 mm above the laboratory counter top and the water level in the standpipe fell from a height of 510 mm to 261 mm above the counter top in 46 hours 35 minutes. Compute the hydraulic conductivity. Does the result seem reasonable? Why or why not?

7.15 A falling-head hydraulic conductivity test has been conducted on a clay specimen in a permeameter similar to the one in Figure 7.19. The soil specimen was 4 in. in diameter and 1 in. tall. The standpipe had an inside diameter of 0.25 in. The water level in the bath surrounding the specimen was 5 in. above the laboratory counter top and the water level in the standpipe fill from a height of 20 in. to 10 in. above the counter top in 38 hours 12 minutes. Compute the hydraulic conductivity in ft/s. Does the result seem reasonable? Why or why not?

7.16 A certain 20-m thick sandy confined aquifer has a hydraulic conductivity of 2.4×10^{-2} cm/s and a void ratio of 0.91. Groundwater is flowing through this aquifer with a hydraulic gradient of 0.0065. How much time would be required for water to travel 1 km through this aquifer?

7.17 A tracer dye is injected into a 55-ft thick sandy gravel confined aquifer which has a hydraulic conductivity of 1.2×10^{-3} ft/s. The dye appears 14 days later in an observation well 75 ft away from the injection point. Compute the seepage velocity and estimate the hydraulic gradient in the aquifer if the porosity of the soil is 42%.

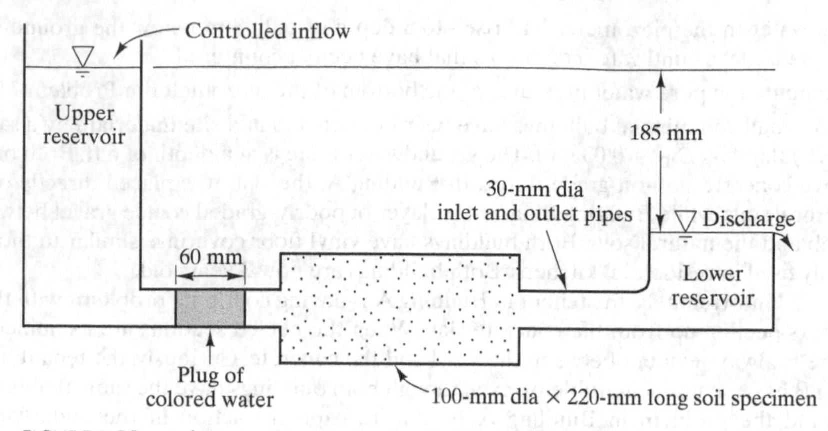

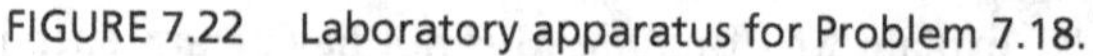
FIGURE 7.22 Laboratory apparatus for Problem 7.18.

7.18 The laboratory apparatus shown in Figure 7.22 maintains a constant head in both the upper and lower reservoirs. The soil sample is a silty sand (SM) with $k = 5 \times 10^{-3}$ cm/s and $w = 18.5\%$. Assume a reasonable value for G_s, then determine the time required for the plug of colored water to pass through the soil (i.e., from when the leading edge first enters the soil to when it begins to exit). Assume the colored water travels only through advection and it has the same unit weight and viscosity as plain water.

7.19 What are the pros and cons of using Hazen's equation versus the Kozeny-Carman equation for estimating hydraulic conductivity of a coarse-grained soil?

7.20 Which of the following methods would be the better way to determine k for a clean sand? Why?

(a) Place a soil sample in a constant-head permeameter, conduct a hydraulic conductivity test, and compute k using Equation 7.10.

(b) Conduct a sieve analysis and compute k using Equation 7.20.

7.21 May we use the Hazen correlation to estimate the hydraulic conductivity for soil C in Figure 4.13? Why or why not? If so, compute k.

7.22 Compute the hydraulic conductivity for soil C in Figure 4.13 using the Kozeny-Carmen equation (Equation 7.20).

7.23 Compute the hydraulic conductivity for soil B in Figure 4.13 using both Hazen's correlation the Kozeny-Carman equation (Equation 7.20).

Section 7.4 Flow Through Anisotropic Soils

7.24 Derive Equations 7.23 and 7.26.

7.25 A sandy soil with $k = 3 \times 10^{-2}$ cm/s contains a series of 5-mm thick horizontal silt layers spaced 300 mm on center. The silt layers have $k = 5 \times 10^{-6}$ cm/s. Compute k_x and k_z and the ratio k_x/k_z.

7.26 When drilling an exploratory boring through the soil described in Problem 7.25, how easy would it be to miss the silt layers? If we did miss them, how much effect would our ignorance have on computations of Q for water flowing vertically? Explain.

Comprehensive

7.27 Apparatus A, shown in Figure 7.23, consists of a single 20-mm diameter pipe and is subjected to a head difference of 60 mm. Apparatus B consists of four 10-mm diameter pipes connected in parallel. How will the flow rate through A compare with that through B? Explain.

7.28 On the basis of your observations in Problem 7.27, explain why saturated clays have a significantly lower hydraulic conductivity than saturated sands, even though the total void areas per square foot of soil are about the same for both.

7.29 An engineer is searching for a suitable soil to cap a sanitary landfill. This soil must have a hydraulic conductivity no greater than 1×10^{-8} cm/s. A soil specimen from a potential borrow site has been tested in a falling-head permeameter similar to the one in Figure 7.19. This specimen was 120 mm in diameter and 32 mm tall. The standpipe had an inside diameter of 8.0 mm. Initially, the water in the standpipe was 503 mm above the water in the water bath surrounding the specimen. Then, 8 hours 12 minutes later the water was 322 mm above the water in the water bath. Compute k and determine if this soil meets the specification.

7.30 An unlined irrigation canal is aligned parallel to a river, as shown in Figure 7.24. This cross-section continues for 4.25 miles. The soils are generally clays, but a 6-in. thick sand

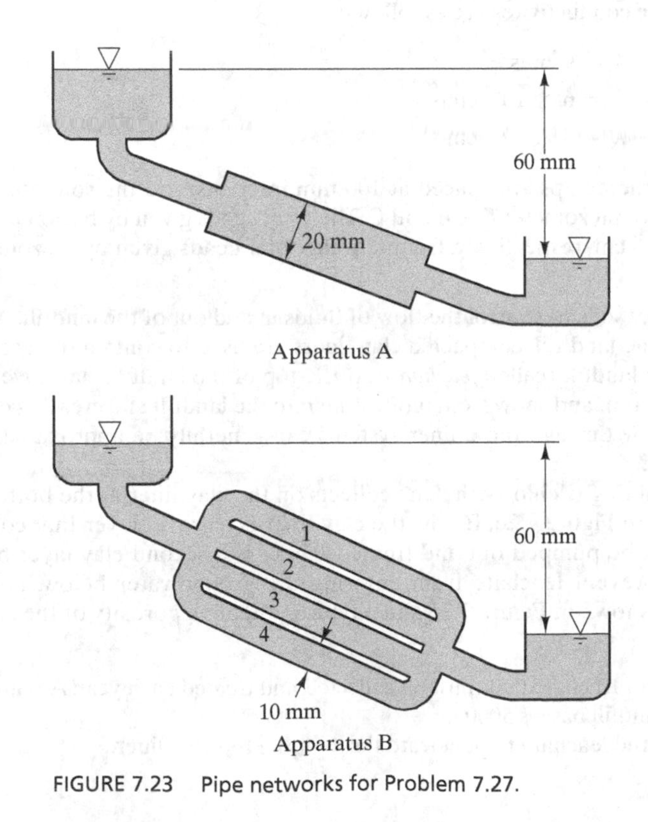

FIGURE 7.23 Pipe networks for Problem 7.27.

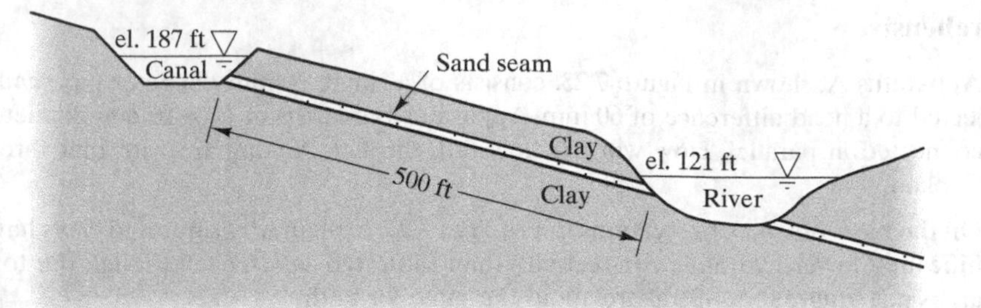

FIGURE 7.24 Cross-section for Problem 7.30. el. = elevation.

seam is present as shown. This sand has $k = 9 \times 10^{-2}$ cm/s. Compute the water loss from the canal to the river due to seepage through this sand layer and express your answer in acre-ft per month.

Note: One acre-foot is the amount of water that would cover one acre of ground to a depth of one foot, and thus equals 43,560 ft^3.

7.31 The constant-head permeameter shown in Figure 7.25 contains three different soils as shown. Their hydraulic conductivites are as follows:

Soil 1 – $k = 9$ cm/s
Soil 2 – $k = 6 \times 10^{-2}$ cm/s
Soil 3 – $k = 8 \times 10^{-3}$ cm/s

The four piezometer tips are spaced at 100 mm intervals, and the soil interfaces are exactly aligned with piezometer tips B and C. The total heads given by piezometers A and D are 98.9 and 3.6 cm, respectively. Compute the total heads given by piezometers B and C.

7.32 Landfills often use clay soils to control the flow of fluids in and out of the landfill. At the sides and bottom of the land fill, compacted clay liners are used to contain the accumulating fluid within the landfill (called *leachate*). At the top of the landfill, clay cover systems keep water from rain and snow from infiltrating into the landfill and creating excess leachate. The fluid flow through these liner systems can generally be approximated as one-dimensional flow.

Leachate containing trichloroethylene collects on the clay liner at the bottom of the landfill as shown in Figure 7.26. Below the clay liner is a gravel layer that collects the leachate so it can be pumped out and treated. There is a second clay layer below the gravel layer to prevent leachate from entering the groundwater below. For the hydraulic conditions shown in Figure 7.26 and a typical effective porosity of the clay of 0.10, compute:

(a) The total amount of leachate that must be collected and treated each year. Assume the total area of the landfill base is 5000 m^2.
(b) The time it takes the leachate to penetrate through the top clay liner.

FIGURE 7.25 Constant-head permeameter for Problem 7.31.

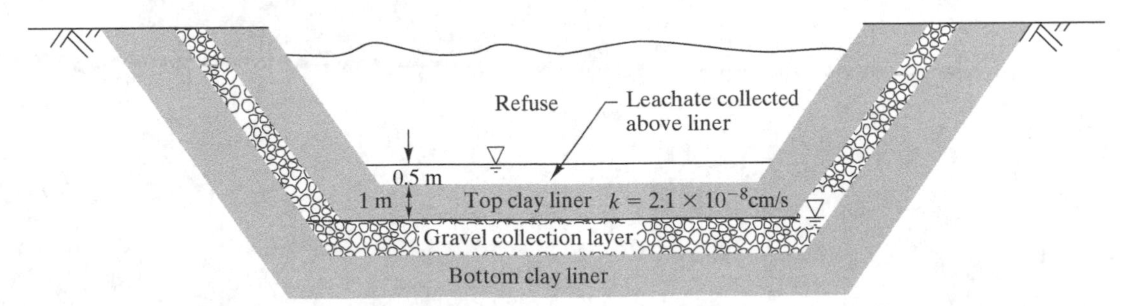

FIGURE 7.26 Cross-section of landfill liner for Problem 7.32.

7.33 The cross-section shown in Figure 7.27 consists of alternating layers of sand and silty sand. Assume the soils have subangular grain shapes and the particle size distributions shown below. Use the Kozeny-Carman method to find k for each soil. The void ratio for the silty sand is 0.63 and sand is 0.59. Then compute k_x and k_z for the layered system.

Sieve Opening (cm)	Sieve Size	Percent Passing	
		Sand	**Silty Sand**
0.475	#4	100	100
0.2	#10	86	92
0.085	#20	72	76
0.0425	#40	48	63
0.015	#100	8	23
0.0075	#200	4	14

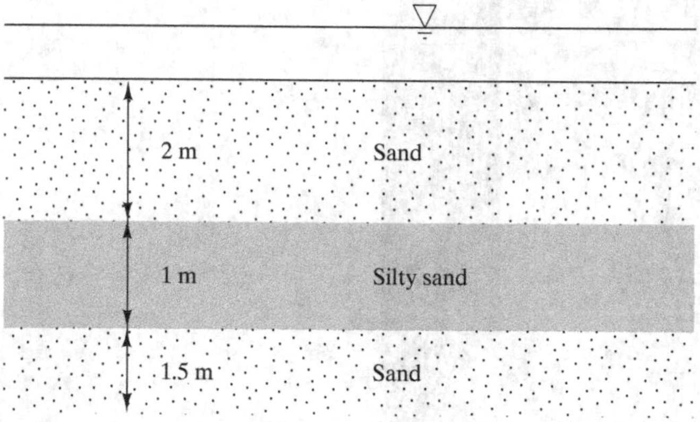

FIGURE 7.27 Cross-section for Problem 7.33.

CHAPTER 8

Groundwater—Multidimensional Flow and Applications

Boring—see Civil Engineers

Listing in the London telephone book

Civil Engineers are No Longer Boring

Headline in a London newspaper after the telephone company agreed to revise its method of referring readers to drilling and sampling companies

This chapter continues our discussion of groundwater, with more focus on applying the principles developed in Chapter 7 to practical engineering problems. Many of these problems require the analysis of two- and three-dimensional flow, so we will begin by expanding the analysis methods to accommodate these conditions.

8.1 MULTIDIMENSIONAL FLOW

In Chapter 7, we applied Darcy's Law to the problem of one-dimensional flow. However, most groundwater flow problems are two- or three-dimensional flow as shown in Figure 7.4 for the case of flow into a long excavation or flow into multiple wells. Inhomogeneous soil conditions also require the consideration of two- or three-dimensional flow.

We will start our analysis of multidimensional flow by considering two-dimensional flow in a vertical plane as illustrated in Figure 7.4(b) with the x axis oriented parallel to the direction of flow and the vertical z axis increasing downward. In some situations, two-dimensional analyses also can be performed in horizontal planes, but these instances are beyond the scope of this book. Once we have derived the solution for two-dimensional flow, we will expand the solution for the more general case of three-dimensional flow.

The LaPlace Equation

In general, Darcy's Law cannot be solved directly for two-dimensional flow because both i and A vary throughout the flow regime. Therefore, the analyses are more complex and need to incorporate a mathematical function called the *LaPlace Equation.*

We will begin our examination of the LaPlace Equation by considering a small element of soil in a vertical cross-section as shown in Figure 8.1, along with the following assumptions:

- Darcy's Law is valid.
- The soil is completely saturated ($S = 100\%$).
- The size of the element remains constant (i.e., no expansion or contraction).
- The soil is homogeneous (i.e., k is constant everywhere in the aquifer).
- The soil is isotropic (i.e., k is the same in all directions).

Using Equation 7.1, we will divide the flow of water into horizontal and vertical components, x and z. The total flows into and out of the element are then:

$$Q_{\text{in}} = (vAn_e)_{\text{in}} = Ln_e(v_x dz + v_z dx) \tag{8.1}$$

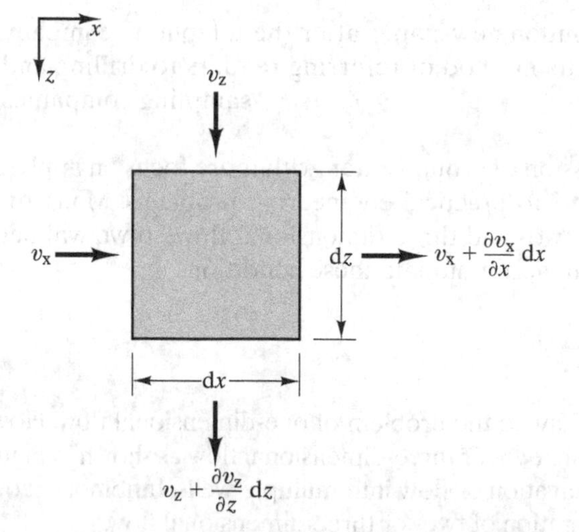

FIGURE 8.1 Element of soil for derivation of the LaPlace equation. The arrows indicate water flowing into or out of the element.

$$Q_{out} = (vAn_e)_{out} = Ln_e\left(\left(v_x + \frac{\partial v_x}{\partial x}dx\right)dz + \left(v_z + \frac{\partial v_z}{\partial z}dz\right)dx\right) \quad (8.2)$$

where v_x and v_z are the velocities in the x and z directions, respectively, L is the length of the element in the y direction, and n_e is the effective porosity.

The soil is saturated and its volume remains constant, so Q_{in} must equal Q_{out}:

$$\frac{\partial v_x}{\partial x}dxdz + \frac{\partial v_z}{\partial z}dzdx = 0 \quad (8.3)$$

which can be reduced to:

$$\frac{\partial v_x}{\partial x} + \frac{\partial v_z}{\partial z} = 0 \quad (8.4)$$

This means any change in velocity in the x direction must be offset by an equal and opposite change in the z direction.

Using Equation 7.13:

$$v = \frac{ki}{n_e} \quad (8.5)$$

$$v_x = -\frac{k}{n_e}\frac{\partial h}{\partial x} \quad (8.6)$$

$$v_z = -\frac{k}{n_e}\frac{\partial h}{\partial z} \quad (8.7)$$

Substituting Equations 8.6 and 8.7 into Equation 8.4 gives

$$\frac{\partial^2 h}{\partial x^2} + \frac{\partial^2 h}{\partial z^2} = 0 \quad (8.8)$$

This is the *LaPlace Equation* for two-dimensional flow. It describes the energy loss associated with flow through a medium, and is used to solve many kinds of flow problems, including those involving heat, electricity, and seepage. For three-dimensional flow, the LaPlace Equation becomes

$$\frac{\partial^2 h}{\partial x^2} + \frac{\partial^2 h}{\partial y^2} + \frac{\partial^2 h}{\partial z^2} = 0 \quad (8.9)$$

For problems with simple boundary conditions, it is possible to derive analytical solutions to these equations. Unfortunately, the boundary conditions associated with most practical seepage problems are much too complex, so geotechnical engineers must rely on alternative solution methods. The most common solutions are given below:

- Flow net solutions.
- Numerical solutions.

We will discuss each of these methods.

8.2 FLOW NET SOLUTION FOR TWO-DIMENSIONAL FLOW

The flow net solution is a graphical method of solving the two-dimensional LaPlace Equation. This solution has been attributed to Forchheimer (1917) and others. We will first develop it for soils that are homogeneous and isotropic with respect to permeability. Then, we will consider the anisotropic case where $k_z \neq k_x$.

Theory

Flow nets are based on two mathematical functions: the *potential function*, Φ, and the *flow function*, Ψ, (also known as the *stream function*). The potential function is defined as:

$$\Phi = -kh + C \tag{8.10}$$

where:

Φ = potential

k = hydraulic conductivity

h = total head

C = a constant

Combining with Equations 8.6 and 8.7 gives

$$\frac{\partial \Phi}{\partial x} = -k\frac{\partial h}{\partial x} = v_x n_e \tag{8.11}$$

$$\frac{\partial \Phi}{\partial z} = -k\frac{\partial h}{\partial z} = v_z n_e \tag{8.12}$$

Combining Equations 8.4, 8.11, and 8.12 would bring us back to Equation 8.8, which means the potential function satisfies the LaPlace Equation.

We can draw a curve in the cross-section such that Φ is constant everywhere along the curve. This is known as an *equipotential line* (even though it is a curve, not a line). Figure 8.2 shows a family of equipotential lines with $\Phi = \Phi_1, \Phi_2, \Phi_3$, etc. According to Equation 8.10, the total head also is constant along each equipotential line. Piezometers *D*, *E*, and *F* in Figure 8.2 have their lower ends located on the same equipotential line; note that the water in all three piezometers rises to the same level indicating the total head is the same along this equipotential line. Thus, this family of curves is similar to the contour lines on a topographic map, except that we are using a vertical section (not a plan view) and the lines represent equal total heads (not equal elevations).

Combining Equations 8.11 and 8.12 gives

$$d\Phi = \frac{\partial \Phi}{\partial x}dx + \frac{\partial \Phi}{\partial z}dz = n_e[v_x dx + v_z dz] \tag{8.13}$$

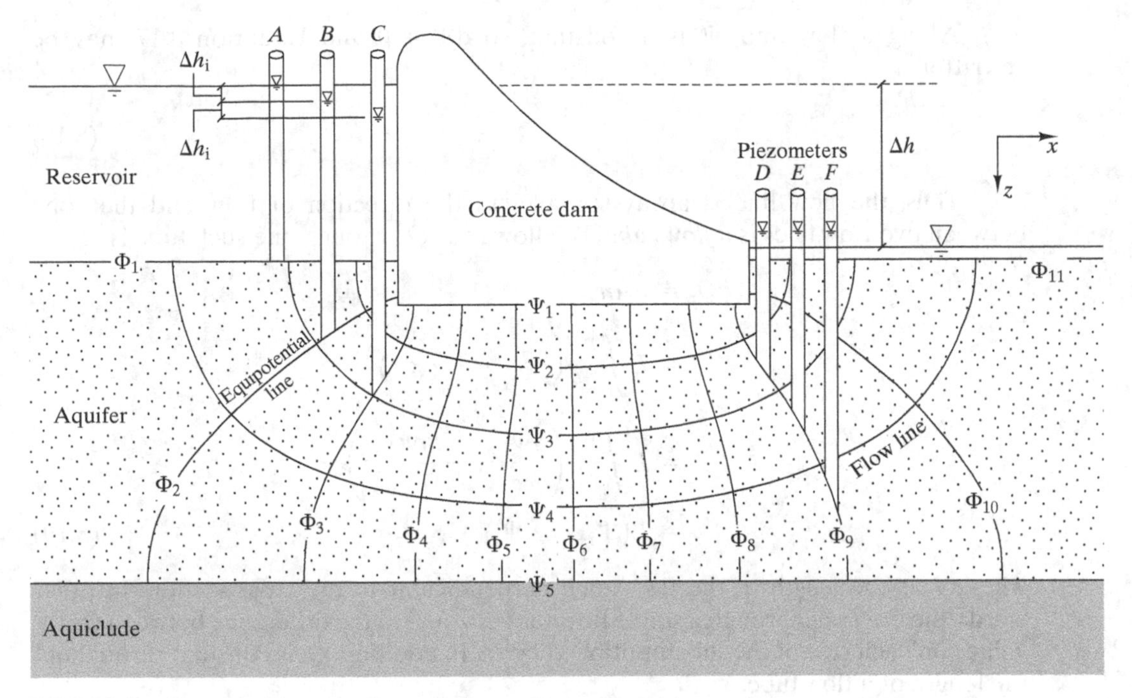

FIGURE 8.2 A sample flow net of seepage beneath a concrete dam.

Along an equipotential line, Φ is a constant, so $d\Phi = 0$ and Equation 8.13 may be rewritten as:

$$\frac{dz}{dx} = -\frac{v_x}{v_z} \tag{8.14}$$

The slope of the equipotential line is dz/dx, and the slope of the flow direction is v_z/v_x, so equipotential lines are always perpendicular to the direction of flow.

The flow function is the complement of the potential function:

$$-\frac{\partial \Psi}{\partial x} = -k\frac{\partial h}{\partial z} = v_z n_e \tag{8.15}$$

$$\frac{\partial \Psi}{\partial z} = -k\frac{\partial h}{\partial x} = v_x n_e \tag{8.16}$$

We also can draw a family of curves in the cross-section such that Ψ is constant everywhere along each curve. They are known as *flow lines* and follow lines of fluid flow. Figure 8.2 also shows a family of flow lines with $\Psi = \Psi_1, \Psi_2, \Psi_3$, etc. When presented together, these two families of curves (one family for Φ and one for Ψ) form a network known as a *flow net*.

Combining Equations 8.15 and 8.16 gives

$$d\Psi = \frac{\partial \Psi}{\partial x}dx + \frac{\partial \Psi}{\partial z}dz = n_e[-v_z dx + v_x dz] \tag{8.17}$$

Along a flow line, Ψ is a constant, so $d\Psi = 0$ and Equation 8.17 may be rewritten as

$$\frac{dz}{dx} = \frac{v_z}{v_x} \tag{8.18}$$

Thus, the flow line is always parallel to the direction of flow and the zone between two flow lines is a *flow tube*. The flow rate, Q, through one such tube is

$$\begin{aligned} Q_i &= vAn_e \\ &= L\int_{\psi_i}^{\psi_{i+1}} (-v_z dx + v_x dz) \\ &= L\int_{\psi_i}^{\psi_{i+1}} \left(\frac{\partial \Psi}{\partial x}dx + \frac{\partial \Psi}{\partial z}dz\right) \\ &= L(\Psi_{i+1} - \Psi_i) \end{aligned} \tag{8.19}$$

where L is the length of the flow zone perpendicular to the cross-section. In other words, the flow rate through a single flow tube is equal to the difference between the Ψ values on each side of the tube multiplied by L. In addition, Q_i is constant throughout the length of a flow tube.

If we draw N_F flow tubes ($N_F + 1$ flow lines) in such a way that each tube has the same Q_i, then the difference in Ψ across each tube is $\Delta\Psi/N_F$ and the total flow rate, Q, through the entire system is

$$Q = N_F L Q_i = N_F L \frac{\Delta\Psi}{N_F} = L\Delta\Psi \tag{8.20}$$

We also will call the difference in total head between successive equipotential lines an *equipotential drop*, and draw the equipotential lines such that the difference in Φ for each equipotential drop is $\Delta\Phi/N_D$, where N_D is the number of equipotential drops.

On the basis of Equation 8.10, $\Delta\Phi$ across the entire system is

$$\Delta\Phi = -k\Delta h \tag{8.21}$$

where Δh is the total head loss from one side of the cross-section to the other (between the first and last equipotential lines).

Figure 8.3 shows the intersections of two flow lines and two equipotential lines, forming a cell of the flow net. According to Equations 8.14 and 8.18, flow lines must intersect equipotential lines at right angles. The distances between these lines are a and b, as shown.

The velocity components, v_x and v_z, are then

$$v_x = v \cos \alpha \tag{8.22}$$

$$v_z = -v \sin \alpha \tag{8.23}$$

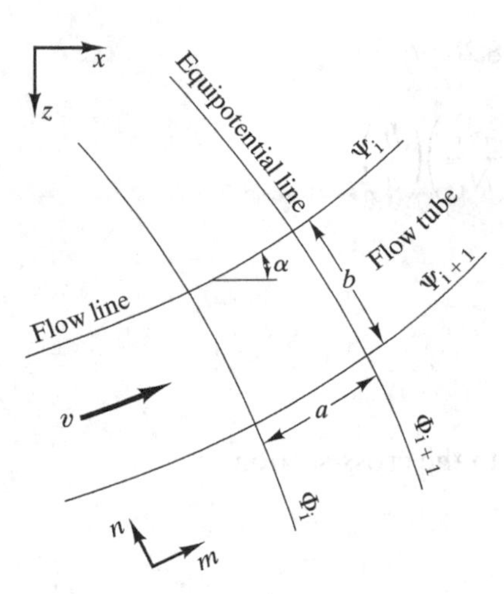

FIGURE 8.3 Intersection of flow lines and equipotential lines.

We also may write:

$$\begin{aligned}\frac{\partial \Phi}{\partial m} &= \frac{\partial \Phi}{\partial x}\frac{\mathrm{d}x}{\mathrm{d}m} + \frac{\partial \Phi}{\partial z}\frac{\mathrm{d}z}{\mathrm{d}m} \\ &= v_x n_e \frac{\mathrm{d}x}{\mathrm{d}m} + v_z n_e \frac{\mathrm{d}z}{\mathrm{d}m} \\ &= v n_e \cos\alpha \cos\alpha - v n_e \sin\alpha\,(-\sin\alpha) \\ &= v n_e (\cos^2\alpha + \sin^2\alpha) \\ &= v n_e \end{aligned} \tag{8.24}$$

$$\begin{aligned}\frac{\partial \Psi}{\partial n} &= \frac{\partial \Psi}{\partial x}\frac{\mathrm{d}x}{\mathrm{d}n} + \frac{\partial \Psi}{\partial z}\frac{\mathrm{d}z}{\mathrm{d}n} \\ &= -v_z n_e \frac{\mathrm{d}x}{\mathrm{d}n} + v_x n_e \frac{\mathrm{d}z}{\mathrm{d}n} \\ &= v n_e \sin\alpha(-\sin\alpha) + v n_e \cos\alpha\,(-\cos\alpha) \\ &= -v n_e (\sin^2\alpha + \cos^2\alpha) \\ &= -v n_e \end{aligned} \tag{8.25}$$

Thus, Equations 8.24 and 8.25 give

$$\frac{\partial \Psi}{\partial n} = \frac{-\partial \Phi}{\partial m} \rightarrow \frac{\Delta\Psi/N_F}{\Delta n} = \frac{-\Delta\Phi/N_D}{\Delta m} \rightarrow \frac{\Delta\Psi}{\Delta\Phi} = \frac{-N_F}{N_D}\frac{b}{a} \tag{8.26}$$

Note that $b/a = \Delta n/\Delta m$.

Finally, combining Equations 8.20, 8.21, and 8.26 produces

$$Q = kL\Delta h\left(\frac{N_F}{N_D}\right)\left(\frac{b}{a}\right) \tag{8.27}$$

where:

Q = flow rate

k = hydraulic conductivity

L = length of aquifer perpendicular to the cross-section

Δh = head loss through the flow net

N_F = number of flow tubes

N_D = number of equipotential drops

b/a = width-to-length ratio of cells (the length is measured parallel to the flow direction; see Figure 8.3)

In summary, we can graphically solve the LaPlace Equation by drawing a flow net consisting of a family of flow lines and equipotential lines. For this graphical solution to be correct, it must satisfy the following conditions.

- The flow net must be drawn to scale.
- Equipotential lines and flow lines must intersect at 90° angles.
- The cells in the flow net must have a constant aspect ratio (i.e. b/a = constant).
- Boundary conditions must be met.

If these conditions are met, the total flow through the flow net can be computed using Equation 8.27. In addition, the flow net will have the following two properties. First, the flow in each tube, Q_i, will be the same and

$$Q_i = \frac{Q}{N_F} \tag{8.28}$$

Second, the equipotential drop across each cell, Δh_i, will be the same and

$$\Delta h_i = \frac{\Delta h}{N_D} \tag{8.29}$$

Note that piezometers A, B, and C in Figure 8.2 are located on three successive equipotential lines and the total head drop between each successive pair of piezometers is the same and equal to Δh_i.

Example 8.1

Compute the total flow, Q, under the dam shown in Figure 8.2 in units of m^3/day. Assume the hydraulic conductivity of the aquifer soil, $k = 10^{-2}$ cm/s, the length of the dam perpendicular to the section, $L = 150$ m, and the total head change from the reservoir to the tail water, $\Delta h = 8$ m.

Solution:

Upon examining Figure 8.2, we note that each cell in the flow net is approximately square that is $b/a = 1$. We also note the number of flow channels, $N_F = 4$, and the number of equipotential drops, $N_D = 10$. Note that the number of flow channels is 1 less than the number of flow lines and, likewise, the number of drops is one less than the number of equipotential lines.

Now we must convert the hydraulic conductivity to the units of m/day.

$$10^{-2}\frac{\text{cm}}{\text{s}}\frac{\text{m}}{100\text{ cm}}\frac{86{,}400\text{ s}}{\text{day}} = 8.64\text{ m/day}$$

We can now use Equation 8.27 to compute the flow under the dam

$$Q = \left(8.64\frac{\text{m}}{\text{day}}\right)(150\text{ m})(8\text{ m})\left(\frac{4}{10}\right)(1) = \mathbf{4150\ m^3/day}$$

Flow Net Construction

Equation 8.27 allows us to compute Q from a correctly drawn flow net. The next problem is producing such a flow net. There are two classes of flow problems. The first are *confined flow* problems where all boundary conditions are known and the flow is confined between an upper and lower impervious boundary. Figure 8.2 is an example of a confined flow problem. The aquiclude forms a lower flow boundary and the concrete dam forms an upper flow boundary. The flow in the aquifer is confined between these two boundaries. The second case of *unconfined flow* problems occurs when the upper flow boundary is not known. Figure 8.6 show three flow nets for the unconfined case. In each of these flow nets, the location of the uppermost flow line is dependent on the flow itself. In Figure 8.6(c), for example, if the pumping from the wells is increased, the groundwater table will be drawn down further and the upper flow line will lower accordingly. This is similar to the case of a well in an unconfined aquifer.

Flow nets are created using a converging trial and error graphical technique. We will first describe how to create flow nets for the case of confined flow in homogeneous isotropic soils.

Confined Flow in Homogeneous Isotropic Soils All flow nets must comply with the following criteria:

a. No two flow lines can intersect.

b. No two equipotential lines can intersect.

c. Flow lines and equipotential lines must intersect at right angles.
d. The *b/a* (width-to-length) ratio must be the same for all cells. (There can be an exception near the ends of the flow area; see below.)
e. The flow and equipotential lines are smooth.

Using these criteria we will construct a flow net for the case where the confined flow conditions exist and the soil is homogeneous and isotropic using Figure 8.4 as an example:

1. **Draw cross-section to scale.** Because the flow net is a graphical solution it is critical that the two-dimensional cross-section be properly drawn to scale. This drawing should include the ground surface, the limits of the pervious zone, the groundwater table, and any free water surfaces (i.e., water levels in lakes or other bodies of water), and any other pertinent data. Be sure the vertical and horizontal scales are equal, and the cross-section is oriented parallel to the direction of flow. For example, when evaluating seepage under a dam, draw the cross-section perpendicular to its longitudinal axis. Figure 8.4(a) shows the cross-section of a sheet pile cofferdam driven into a layer of sandy soil. Note the scale shown on the figure. It is best to draw the cross-section in ink on velum rather than bond paper.
2. **Identify boundary conditions.** Next we must clearly identify all of the boundary conditions. In the case of confined flow, every physical boundary must be either a flow line or an equipotential line. Referring to Figure 8.4(a), line A–B is an equipotential line since the total head along this line is fixed and equal to the upstream head of water. Similarly, line E–F is an equipotential line. The other two boundaries in the problem, lines B–C–D–E and G–H are each a flow line.
3. **Select an integer value for N_F, the number of flow tubes.** Larger values of N_F produce more precise flow nets, but also require more effort to finalize. It is usually best to start with N_F values of 2 or 3. If more precision is needed, the flow tubes can be subdivided to create N_F of 4 or 6. This usually represents a good compromise between precision and effort. For our example, we have chosen to use $N_F = 2$ for our initial trial. We will refine the flow net later.
4. **Sketch the initial flow lines.** Use a soft pencil when sketching your flow net. Since this is a trial and error process you will be doing a lot of erasing. Using velum rather than regular bond paper will also make it easier to sketch and erase trial flow nets. First sketch in the flow lines based on the number of flow channels you have selected. In our case, we need one additional flow line to create two flow channels as shown in Figure 8.4(b).
5. **Add equipotential lines.** Sketch in the approximate locations of the equipotential lines following two key rules: (a) equipotential lines and flow lines must intersect at right angles and (b) the *b/a* ratio must be the same for each of the cells formed by the equipotential and flow lines. It is easiest to set $b/a = 1$ (so the cells are somewhat square in shape). The proper number of equipotential lines will come out automatically as you draw them while maintaining right angles and the requirement that $b/a = 1$. This method will not necessarily produce an integer value for N_D. You may end up with a partial cell at one or both ends of your flow

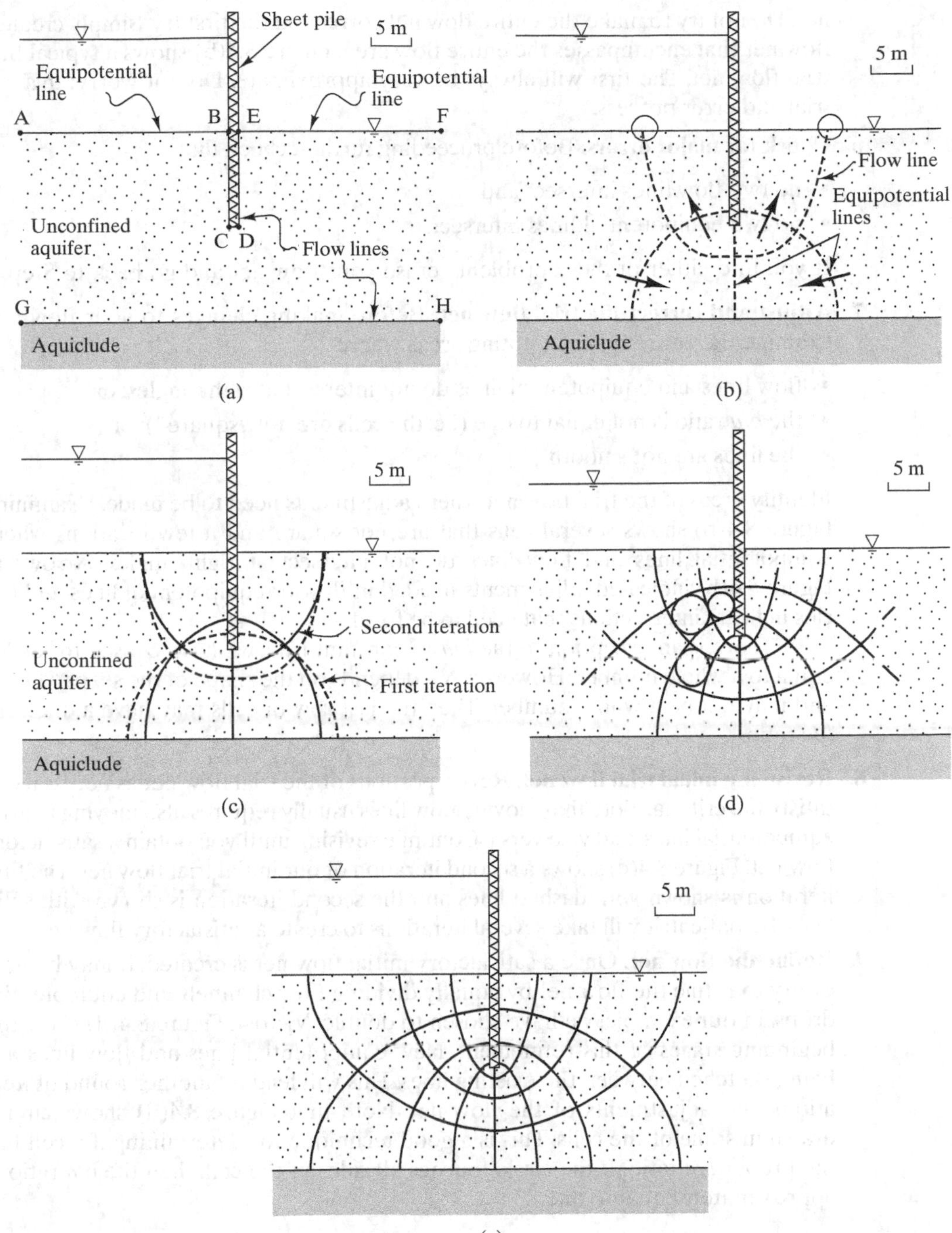

FIGURE 8.4 Example of flow net construction: (a) geometry and boundary conditions; (b) initial iteration of flow net; (c) adjustment of flow net in second iteration; (d) refining flow net by subdivision; and (e) final flow net.

net. Do not try to make the entire flow net correct on the first try. Simply create a flow net that encompasses the entire flow area. Figure 8.4(b) shows a typical first trial flow net. The first will always be very approximate. Do not worry; this is a trial and error process.

6. **Check for major errors.** Before proceeding further, check that

   - no two flow lines intersect and
   - no two equipotential lines intersect.

   If you find either of these problems, erase your flow net and go back to Step 4.

7. **Adjust and correct the trial flow net.** Before making changes to your flow net, examine the entire flow net noting areas where

   - flow lines and equipotential lines do not intersect at right angles; or
   - the b/a ratio is not equal to one (i.e. the cells are not "square"); or
   - the lines are not smooth.

   Identify areas of the trial flow net where adjustments need to be made. Examining Figure 8.4(b) shows several cells that are not square and a few locations where equipotential lines and flow lines do not intersect at right angles. Arrows in Figure 8.4(b) indicated adjustments needed in flow or equipotential lines and circles indicate intersections that need to be fixed.

   There is an exception to the b/a = constant rule: we have chosen to set N_F equal to a whole number. However, N_D depends on the shape of the seepage zone and may not be a whole number. Thus, the last row of cells may have a different b/a ratio.

8. **Revise the initial trial flow net.** Revise portions of the trial flow net as necessary to satisfy the criteria. Note that moving flow lines usually requires also moving nearby equipotential lines, and vice versa. Continue revising until you obtain a satisfactory flow net. Figure 8.4(c) shows a second iteration of our initial trial flow net. The first iteration is shown with dashed lines and the second iteration is shown with solid lines. Be patient; it will take several iterations to create a satisfactory flow net.

9. **Refine the flow net.** Once a satisfactory initial flow net is created, it may be necessary to refine the flow net by equally dividing flow channels and equipotential drops. In our example, we have chosen to double N_F to 4. Figure 8.4(d) show the beginning stages of this refinement. New equipotential lines and flow lines are being sketched between the existing lines. This will lead to another round of iterations and adjustments of the flow net. Note that Figure 8.4(d) shows circles drawn in some of the cells. This is a good technique for determining if a cell has the proper b/a ratio. If the circle touches all sides of the cell, then the b/a ratio is approximately equal to one.

This process of refining could go on indefinitely, but you will quickly reach a point of diminishing returns where large amounts of additional effort result in only nominal improvements to the flow net. Keep in mind that the uncertainty in k is probably much greater than that in N_F and N_D, so there is no need to be overly meticulous with the flow net. Figure 8.4(e) illustrates the final flow net with four flow tubes and eight equipotential drops.

It should be noted that when the cross-section is symmetrical about some vertical axis, it is easier to draw only half of the flow net. This could have been done in our example case, but for clarity we choose to demonstrate with the full flow net. If you choose to draw only half of a flow net, be sure to properly adjust N_F and/or N_D as needed to properly account for the full flow net. For example, if we drew only half of the flow net in our example, we would need to double N_D before inputting it into Equation 8.27.

Harr (1962) presents a modification of the flow net method called the *method of fragments*. For some problems, this method is quicker and easier than the conventional method. However, discussion of this method is beyond the scope of this book.

Confined Flow in Homogeneous Anisotropic Soils As discussed in Section 7.4, in natural soil deposits that are horizontally bedded, it is often the case that the soil can be effectively modeled as homogeneous anisotropic soil with a horizontal hydraulic conductivity greater than the vertical hydraulic conductivity. In this case, it is possible to create a flow net solution by transforming the cross-section into a space where the equivalent horizontal and vertical hydraulic conductivity are equal. We can then perform a normal flow solution in the transformed space and then inverse the transform to create the flow net in the original space. Using techniques similar to those presented in Section 7.4, we can show that the correct transform is to multiply all horizontal dimensions by the ratio $\sqrt{k_z/k_x}$.

This process is illustrated in Figure 8.5. Figure 8.5(a) shows a scaled cross section of a ship lock that might be located in a river or a canal. The natural soils under the lock are anisotropic with $k_x = 4k_z$. In this case, we must create a transformed section where the horizontal scale is reduced by the ratio of $\sqrt{k_z/k_x}$ or $\sqrt{k_z/(4k_z)} = 1/2$. (Alternatively, we could have chosen to expand the vertical scale by a factor of 2.) Figure 8.5(b) shows the transformed section and the flow net created using the procedure described earlier. Finally, the transformed section with the flow net is transformed back to the original space as shown in Figure 8.5(c). We can now compute the flow under the lock, Q, using Equation 8.27 and the transformed section in Figure 8.5(b), but we must use an equivalent hydraulic conductivity, k_{eq}, in the computation that accounts for the difference between the horizontal and vertical hydraulic conductivities. The equivalent hydraulic conductivity is computed as

$$k_{eq} = \sqrt{k_x k_z} \tag{8.30}$$

Unconfined Flow Unconfined flow problems differ from confined flow problems in that the upper flow line is unknown and must be determined during the analysis process. Figure 8.6 shows three different flow nets under unconfined flow conditions. In each case, the upper flow line also represents the phreatic surface (groundwater table). The location of the phreatic surface will depend on the flow conditions in the

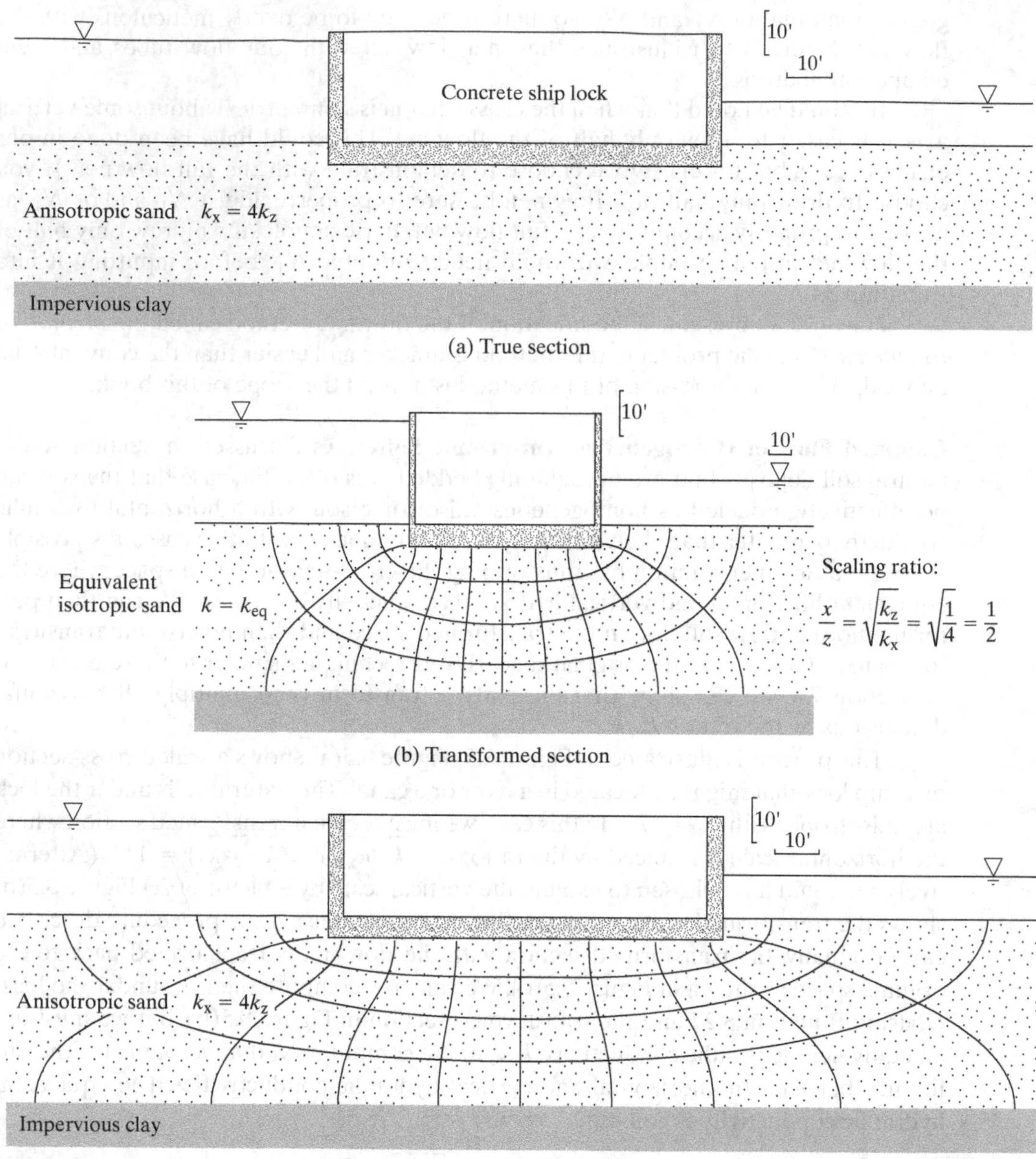

FIGURE 8.5 Illustration of flow net construction in uniform anisotropic soil: (a) cross-section geometry and boundary conditions at true scale; (b) transformed cross-section with completed flow net; in this case, $k_x = 4k_z$, so the transformed section has been produced using a vertical exaggeration factor of 2; and (c) cross-section at true scale with flow net.

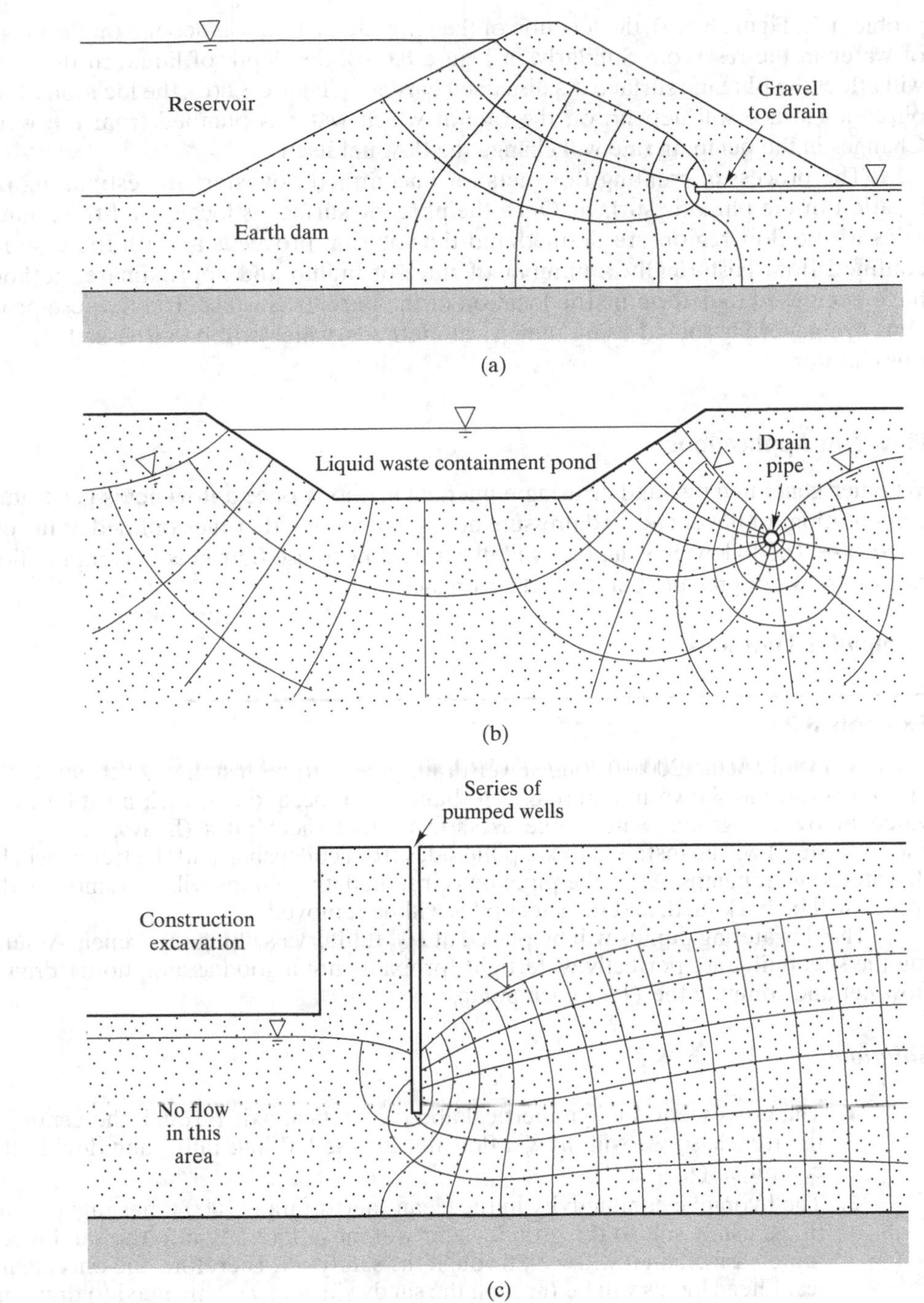

FIGURE 8.6 Sample flow nets for the case of unconfined flow.

problem. In Figure 8.6(a), the location of the phreatic surface will depend on the height of water in the reservoir. Similarly, in Figure 8.6(b), the depth of liquid in the pond will affect the phreatic surface. As discussed earlier, in Figure 8.6(c), the location of the phreatic surface will depend on the rate at which water is pumped from the wells. Changes in the pumping rate will change the flow net itself.

The process of creating flow nets for unconfined flow starts by estimating the location of the phreatic surface. Once the phreatic surface is located, all the boundaries of the flow region are defined and the solution proceeds as with the case for confined flow. Historically, a number of rules-of-thumb and approximate methods have been used to determine the location of the phreatic surface. Today, these problems are generally solved using numerical methods, which will be discussed later in this chapter.

Flow Net Applications

Whether generated by hand or using numerical methods, once a flow net is generated, computations of flow, head, and pressure are the same for either the confined or unconfined case. The following examples will illustrate how to use flow nets to compute flow rates, pore water pressures, and effective stresses.

Computing Flow Rates

Example 8.2

A 6-ft diameter, 2000-ft long sewer drain pipe is to be installed 24 ft below the ground surface as shown in Figure 8.7. To build this pipe, a long trench must be excavated below the groundwater table as shown. Steel sheet piles (heavy, corrugated sheets of steel) will be installed to keep the sides from collapsing, and the trench will be dewatered with pumps. After the pipeline is installed, the pumps will be removed, the trench will be backfilled, and the sheet piles will be removed.

The dewatering pumps will be placed at 100 ft intervals along the trench. Assuming the sheet piles are perfectly watertight (probably not a good assumption!), draw a flow net and compute the Q for each pump.

Solution:

1. The cross-section is symmetrical about a vertical axis through the center of the trench, so we will draw a flow net for the left half only, then double the computed Q.
2. The hydraulic conductivity in the clean sand strata is 1000 times higher than in the sandy silt, so the groundwater will move laterally into the sand much more easily than it will seep through the sandy silt. Therefore, the only significant head losses will be through the sandy silt, and we only need to draw the flow net through this stratum.
3. The resulting flow net, developed by the converging trial and error method described earlier, is shown on the right side of Figure 8.7. We chose to use

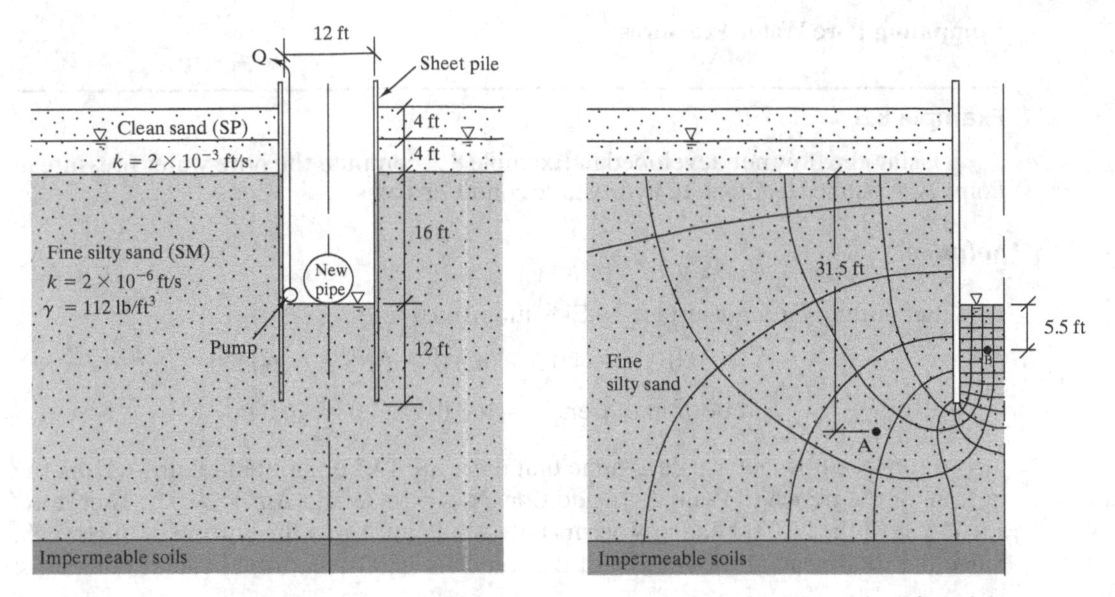

FIGURE 8.7 Cross-section of trench for Examples 8.2 and 8.3. The given data is on the left and the completed flow net is on the right.

$N_F = 4$ and $b/a = 1$, and counted $N_D = 16.33$ from the finished flow net. The uppermost equipotential drop is not from a full cell, so it only counts as one-third of a drop.

4. The flow rate per pump is then:

$$Q = 2kL\Delta h \frac{N_F}{N_D} \frac{b}{a}$$

$$= 2\,(2 \times 10^{-6}\ \text{ft/s})(100\ \text{ft})(20\ \text{ft})\left(\frac{4}{16.33}\right)(1)$$

$$= 2.0 \times 10^{-3}\ \text{ft}^3\text{/s} \times 449\ \text{gal/min per ft}^3\text{/s}$$

$$= \mathbf{0.88\ gal/min\ per\ pump}$$

Note the factor of two at the beginning of the equation; it accounts for the fact that the flow net was drawn for only half the trench.

According to Equation 7.1, the velocity is highest when the area is small, so the groundwater velocity will be highest when the flow tube is narrow. In this cross-section, the narrowest spot is near the tip of the sheet pile. Because the velocity here is high, there will be much more friction between the flowing water and the soil, so the hydraulic gradient also will be high. Thus, the equipotential lines also are close together in this area.

Computing Pore Water Pressures

Example 8.3

Using the flow net developed in Example 8.2, compute the pore water pressure at Point A. Assume steady-state hydrostatic conditions exist.

Solution:

Set datum: Ground surface = Elevation 100.0 ft

$$(h_z)_A = 100.0 \text{ ft} - 4.0 \text{ ft} - 4.0 \text{ ft} - 31.5 \text{ ft} = 60.5 \text{ ft}$$

$$h \text{ at bottom of trench} = 100.0 - 24.0 = 76.0 \text{ ft}$$

From the flow net, we determine that there are 13.2 equipotential drops from the bottom of the trench to Point A. In addition, $N_D = 16.33$ and $\Delta h = 20.0$ ft. Therefore, 13.2/16.33 of the 20.0 ft head loss occurs between Point A and the bottom of the trench. Since the total head at the bottom of the trench is 76.0 ft, the total head at Point A is:

$$h_A = 76.0 \text{ ft} + 20.0 \text{ ft}\left(\frac{13.2}{16.33}\right) = 92.2 \text{ ft}$$

$$(h_p)_A = h_A - (h_z)_A = 92.2 \text{ ft} - 60.5 \text{ ft} = 31.7 \text{ ft}$$

Thus, if a piezometer were present at Point A, the water would rise to a level 31.7 ft above the point.

$$u = \gamma_w (h_p)_A = (62.4 \text{ lb/ft}^3)(31.7 \text{ ft}) = \mathbf{1980 \text{ lb/ft}^2}$$

Computing Uplift Pressures on Structures When buried structures extend below the groundwater table, they are subjected to uplift pressures from the pore water. These pressures are comparable to those acting on the hull of a ship. Uplift pressures are important, especially if the structure does not have much weight. For example, buried tanks or vaults might experience uplift pressures that exceed their weight, which would cause them to lift out of their intended position.

Engineers once thought that full hydrostatic uplift pressures could occur only in sands, and only one-third to one-half of the full pressure would occur in clays (Terzaghi, 1936). Even Terzaghi once stated "we are not sure as to the extent to which hydrostatic uplift acts within a mass of plastic clay" (Terzaghi, 1929). Fortunately, he and others eventually realized full hydrostatic pore water pressures occur in all soils, so designs should be based on full hydrostatic uplift forces.

If the groundwater table is horizontal, or nearly so, then the uplift pressure at a point may simply be computed using Equation 7.7. This is true even though when viewed on a microscopic scale, only part of the submerged structure is in contact with the pore water (the remainder is in contact with the solid particles). Thus, the uplift force is the same as it would be if the structure was submerged to a comparable depth in a lake.

If the groundwater table is far from being horizontal, then significant head losses are occurring as the water flows from one side of the structure to the other. The ship lock in Figure 8.5 is an example. In this case, it is necessary to use the following procedure:

1. Draw a flow net in the soil beneath the structure.
2. Using the equipotential lines, compute the total head at several points along the base of the structure.
3. Determine the elevation head at the points used in Step 2.
4. Using Equation 7.5 and the data from Steps 2 and 3, determine the pressure head at each point.
5. Using Equation 7.6, determine the pore water pressure at each point. This is the uplift pressure.
6. Develop a plot of uplift pressure across the structure.

Example 8.4

A sewage pump is to be located in the proposed underground vault shown in Figure 8.8. The site will be temporarily dewatered during construction, but afterward, the groundwater table will be allowed to return to its natural level. The highest probable level during the life of the project is as shown.

The vault will be made of reinforced concrete and waterproofed so very little groundwater will enter. The vault and overlying soil has a mass of 30,000 kg, exclusive of the floor. A sump pump will eject any water that might accumulate inside.

The weight of the vault and the overlying soil must be sufficient to provide a factor of safety of at least 1.5 against buoyant uplift (i.e., these weights must be at least

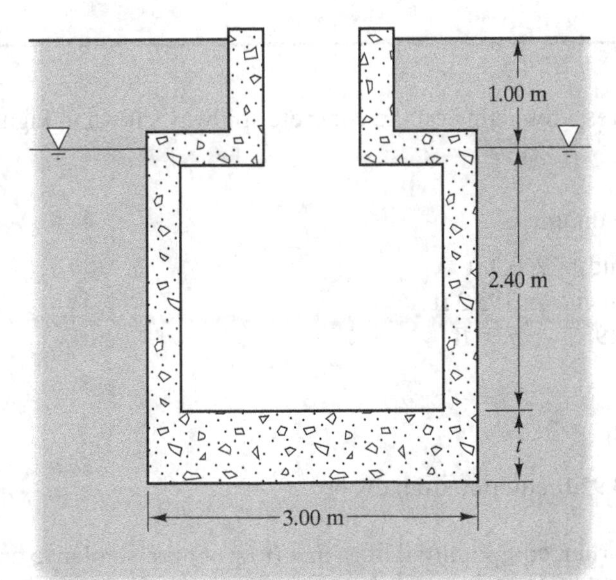

FIGURE 8.8 Cross-section for Example 8.4.

1.5 times the total uplift force). Write an equation for the buoyant uplift force, then compute the required base thickness, t. Neglect the weight of the pump and neglect any sliding friction between the sides of the vault and the soil.

Solution:

Uplift pressure

$$u = \gamma_w z_w = (9.81 \text{ kN/m}^3)(2.40 + t \text{ m}) = 23.5 + 9.81\, t \text{ kPa}$$

Uplift force

$$P = uA = (23.5 + 9.81t)(3^2) = 212 + 88.3t \text{ kN}$$

Weight

Use ρ concrete $= 2400 \text{ kg/m}^3$.

$$W = Mg$$

$$= [30{,}000 \text{ kg} + (2400 \text{ kg/m}^3)(3^2 \text{ m})t](9.81 \text{ m/s}^2)\left(\frac{1 \text{ kN}}{1000 \text{ N}}\right)$$

$$= 294 + 212t \text{ kN}$$

For factor of safety $= 1.5$,

$$1.5(212 + 88.3t) = 294 + 212t$$

$$t = \mathbf{300 \text{ mm}}$$

Example 8.5

Compute the uplift pressures acting on the concrete spillway shown in Figure 8.9.

Solution:

Using sea level as the datum:

h at upstream end $= 260.3$ ft
h at downstream end $= 193.3$ ft
$\Delta h = 260.3 - 193.3 = 67.0$ ft

From the flow net:

$$N_D = 17.0$$

$$\frac{\Delta h}{N_D} = \frac{67.0}{17.0} = 3.92 \text{ ft/equipotential drop}$$

Note: As discussed earlier, equipotential lines in a flow net are similar to contour lines in a topographic map. Therefore, $\Delta h/N_D$ is similar to the contour interval on a topographic map.

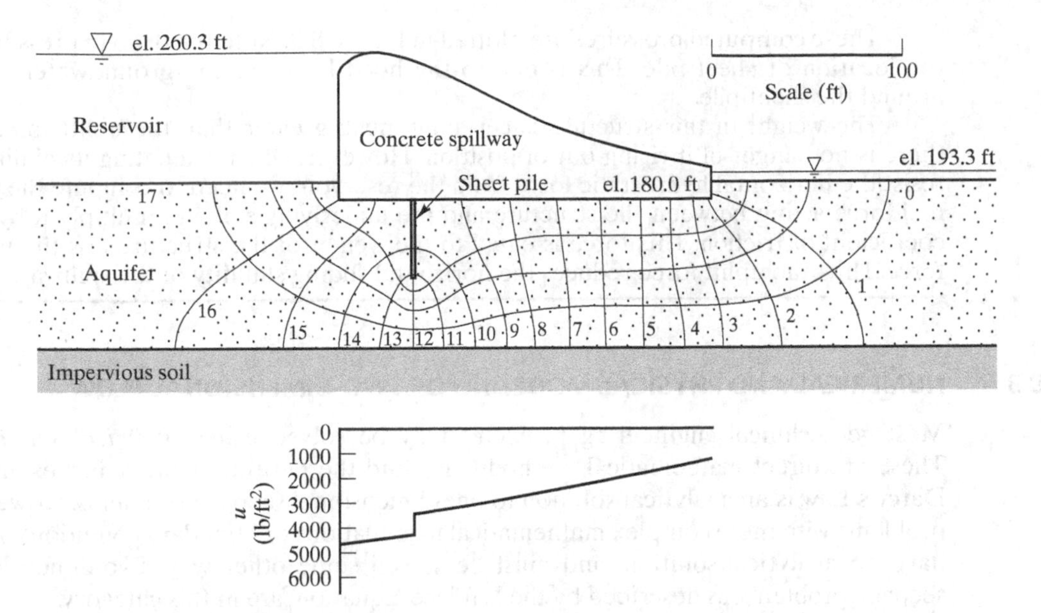

FIGURE 8.9 Cross-section for Example 8.5. el. = elevation.

For each equipotential line, compute h based on h at the downstream end, the number of equipotential drops, and $\Delta h/N_D$. Then compute h_p and u along the bottom of the structure and plot the results.

Equipotential line	**h (ft)**	**Along bottom of structure**		
		h_z (ft)	**h_p (ft)**	**u (lb/ft^2)**
0 (downstream)	193.3	—	—	—
1	197.2	—	—	—
2	201.2	180.0	21.2	1323
3	205.1	180.0	25.1	1566
4	209.1	180.0	29.1	1816
5	213.0	180.0	33.0	2059
6	216.9	180.0	36.9	2303
7	220.9	180.0	40.9	2552
8	224.8	180.0	44.8	2796
9	228.8	180.0	48.8	3045
10	232.7	180.0	—	—
11	236.7	180.0	—	—
12	240.6	180.0	—	—
13	244.5	180.0	—	—
14	248.5	180.0	—	—
15	252.4	180.0	72.4	4518
16	256.4	180.0	76.4	4767
17	260.3	—	—	—

These computed pressures are plotted in Figure 8.9. Note the drop in pressure at the location of sheet pile. This is due to the head losses as the groundwater flows around the sheet pile.

The weight of this structure is certainly much greater than the uplift force, so there is no danger of it rising out of position. However, when evaluating its ability to resist the horizontal hydrostatic force from the reservoir, we need to compute the normal force acting between the structure and the underlying soil and multiply it by the coefficient of friction. This force is equal to the weight of the structure less the uplift force. Thus, the uplift force reduces the horizontal sliding stability of the spillway.

8.3 NUMERICAL AND PHYSICAL MODELING OF TWO-DIMENSIONAL FLOW

Most geotechnical engineering problems may be solved using *analytical solutions*. These are direct mathematical methods that find the desired answers. For example, Darcy's Law is an analytical solution to one-dimensional seepage problems. However, problems with more complex mathematical formulations or boundary conditions often have no analytical solution, and must be solved some other way. Two-dimensional seepage problems, as described by the LaPlace Equation, are in this category.

Numerical Solutions

Although hand-drawn flow nets may be used to solve the LaPlace Equation, the widespread availability of digital computers has made numerical solutions the preferred method for many seepage problems.

Numerical solutions (also known as numerical methods) are mathematical techniques that solve complex problems by dividing them into small physical pieces and writing simpler equations that describe the functions within each piece and the relationships between the pieces. Several such techniques are available, and they are routinely used to solve a wide variety of engineering problems. The two most common techniques used for solving seepage problems are the *finite difference method* (*FDM*) and the *finite element method* (*FEM*).

Both the finite difference and finite element methods divide the flow regime into a large number of discrete elements, as shown in Figure 8.10. A typical solution will have hundreds or thousands of such elements. We then assume that the hydraulic gradients i_x and i_z are constant within each element, which allows us to directly apply Darcy's Law to describe the flow within an element. The next element has its own values of i_x and i_z, and the flow within that element also may be described using Darcy's Law. Finally, the flow rate, Q, exiting one side of an element is equal to the flow rate entering the side of the adjacent element. The elements in Figure 8.11 show these relationships.

This process generates thousands of simultaneous equations, and thus requires matrix algebra and a computer to solve. The final results represent an approximate solution whose precision depends on the number of elements and other factors. The output may be presented as a flow net, as shown in Figure 8.12, along with digital values of pore pressure in the center of each element, flow rate through the system, and other useful information.

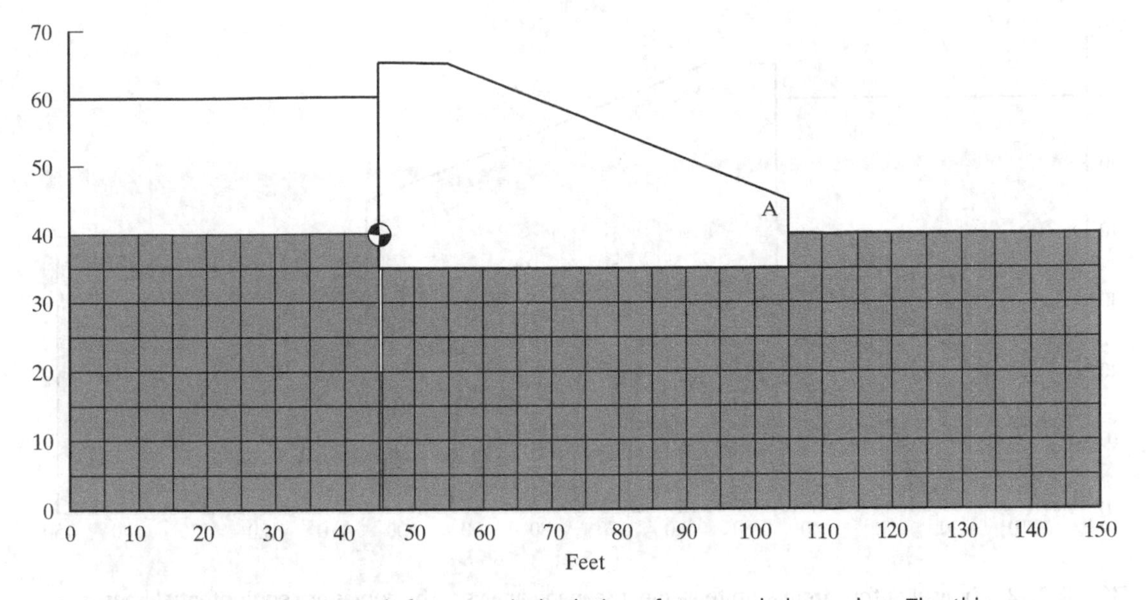

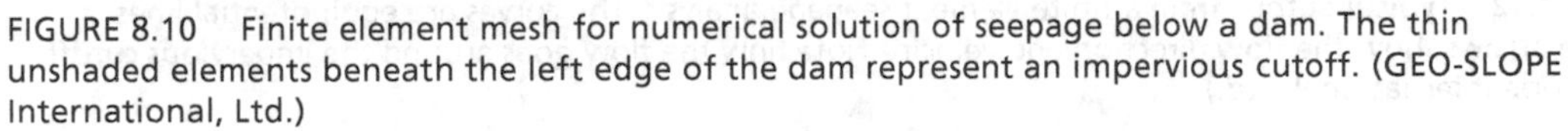

FIGURE 8.10 Finite element mesh for numerical solution of seepage below a dam. The thin unshaded elements beneath the left edge of the dam represent an impervious cutoff. (GEO-SLOPE International, Ltd.)

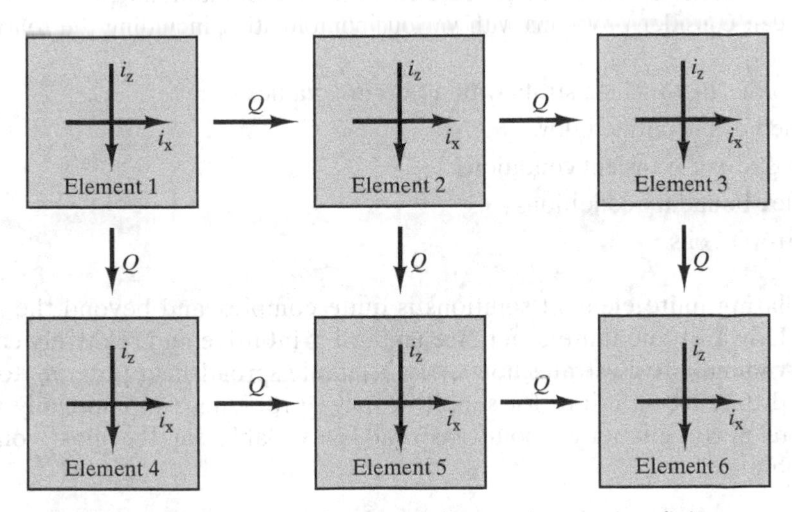

FIGURE 8.11 Relationships between adjacent elements in a finite element analysis. Each element has its own values of i_x and i_z, which are constant within that element. In addition, the Q exiting one side of an element is equal to the Q entering the side of the adjacent element.

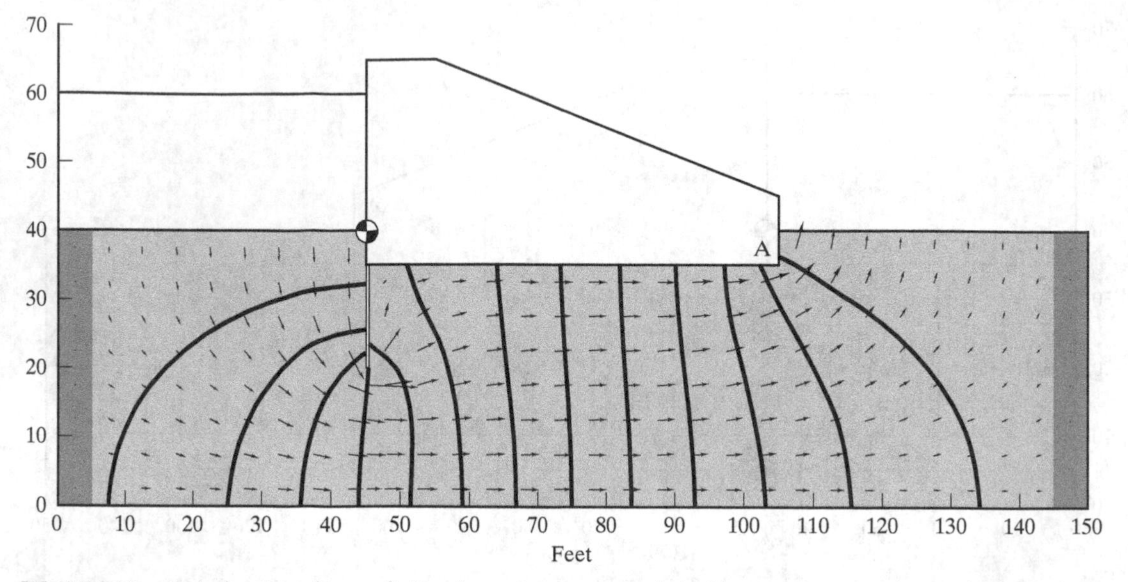

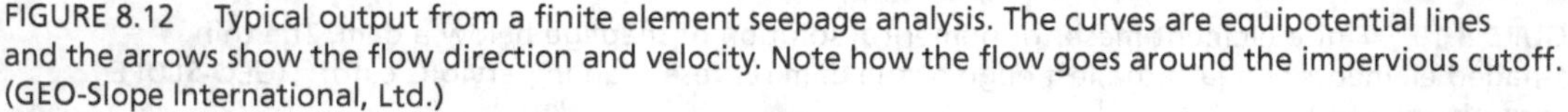
FIGURE 8.12 Typical output from a finite element seepage analysis. The curves are equipotential lines and the arrows show the flow direction and velocity. Note how the flow goes around the impervious cutoff. (GEO-Slope International, Ltd.)

Numerical methods are attractive because of their computational power and flexibility. They can consider problems with various complexities, including the following:

- Soil profiles that include strata with different k values
- Confined or unconfined flow
- Steady-state or transient conditions
- Complex boundary conditions
- Anisotropic soils.

Formulating finite element solutions is quite complex and beyond the scope of this text. However, the finite difference method is intuitive and relatively easy to formulate. Appendix A illustrates how to use a simple spreadsheet program to generate finite difference solutions for simple seepage problems. Commercially available software makes numerical solutions readily available for the most complex seepage problems.

Physical Models

It is possible to create a physical model that illustrates the flow lines in a flow net by injecting dye into the model, as shown in Figure 8.13. This method is not used for routine analyses, but can be useful for instructional purposes.

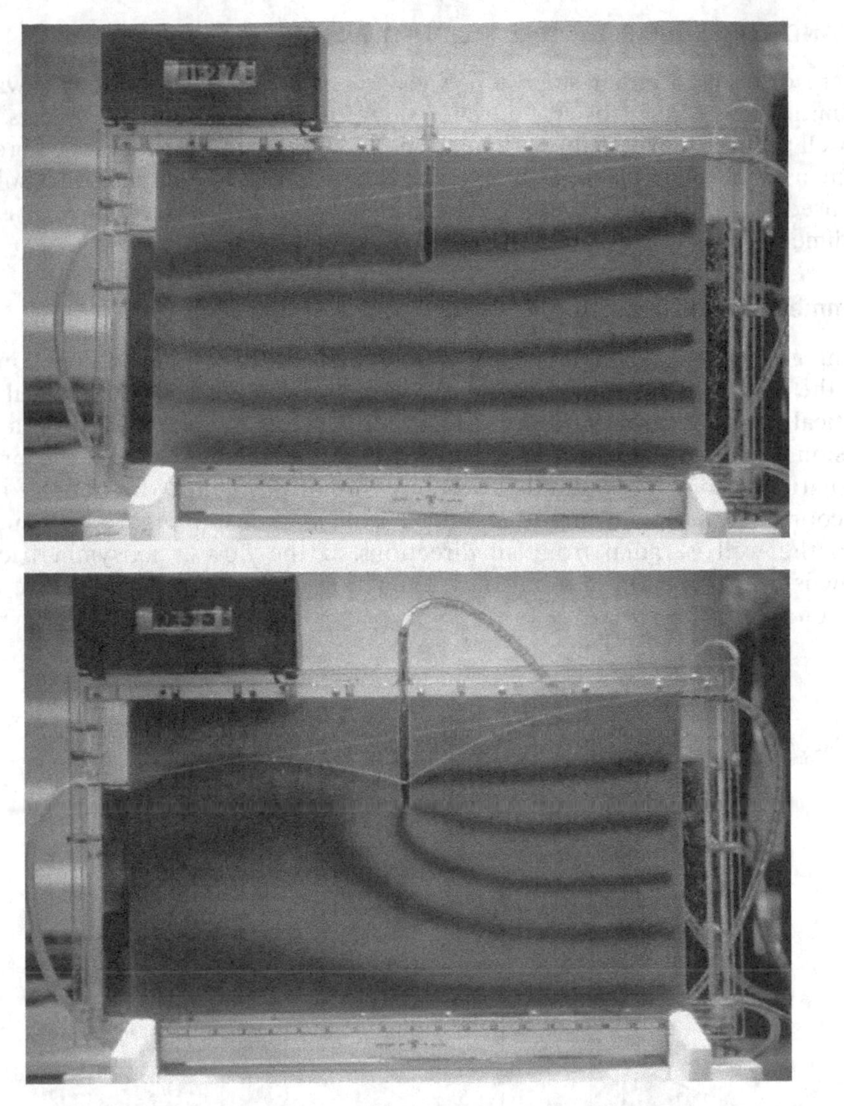

FIGURE 8.13 This laboratory model was constructed to illustrate groundwater flow. The top photograph shows steady-state conditions with the groundwater flowing from right to left. Dye has been injected along the right side of the model, and is being carried along by the flowing water, thus illustrating flow lines. In the bottom photograph, water is being pumped from the well. This alters the flow patterns, as reflected by the dye, even causing some of the water downstream of the well to reverse direction. The groundwater table in the vicinity of the well also has been drawn down. (National Ground Water Association, Westerville, OH.)

8.4 TWO- AND THREE-DIMENSIONAL FLOW TO WELLS

Wells are often used either singly or in groups or arrays, to both supply water to communities and to control groundwater flow for civil engineering projects. While many well problems involve three-dimensional flow, there are some cases where exact two-dimensional solutions are available and others where two-dimensional solutions can be used to approximate the three-dimensional flow. In the more complex cases, full three-dimensional solutions are necessary.

Axisymmetric Two-Dimensional Solutions for a Single Well

Pumping groundwater from a vertical well creates an axisymmetric flow condition where the flow is in a radial direction toward the well and is symmetrical about its vertical axis as shown in Figure 8.14. Although this may appear to be a three-dimensional flow problem, the presence of this symmetry simplifies the problem mathematically to a fairly simple two-dimensional problem. We choose to use a radial coordinate system centered about the well, as shown in Figure 8.14, since the flow to the well is radial from all directions. If the flow is axisymmetric, then the flow is not a function of θ but only a function of r and z and therefore a two-dimensional problem.

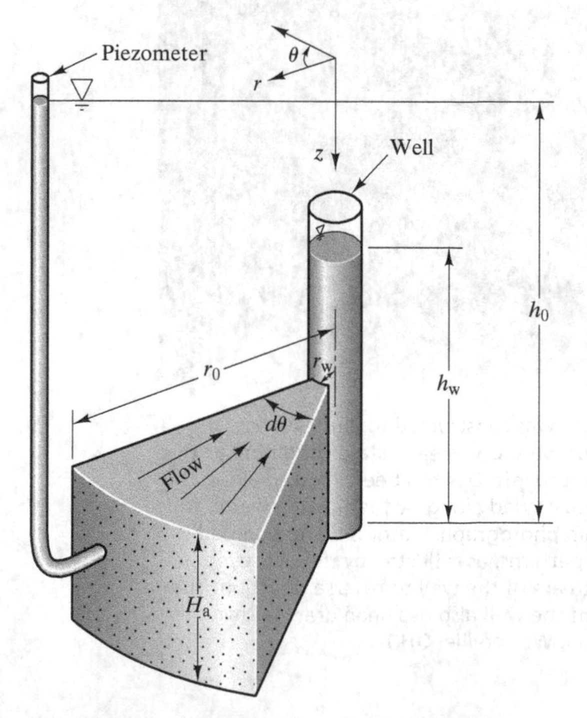

FIGURE 8.14 Axisymmetric radial flow geometry used for well problems.

Confined Aquifers A confined aquifer, as shown in Figure 8.15, is one that is sandwiched between two aquicludes. Thus, the upper and lower flow boundaries are fixed and the water flows through the entire depth of the aquifer.

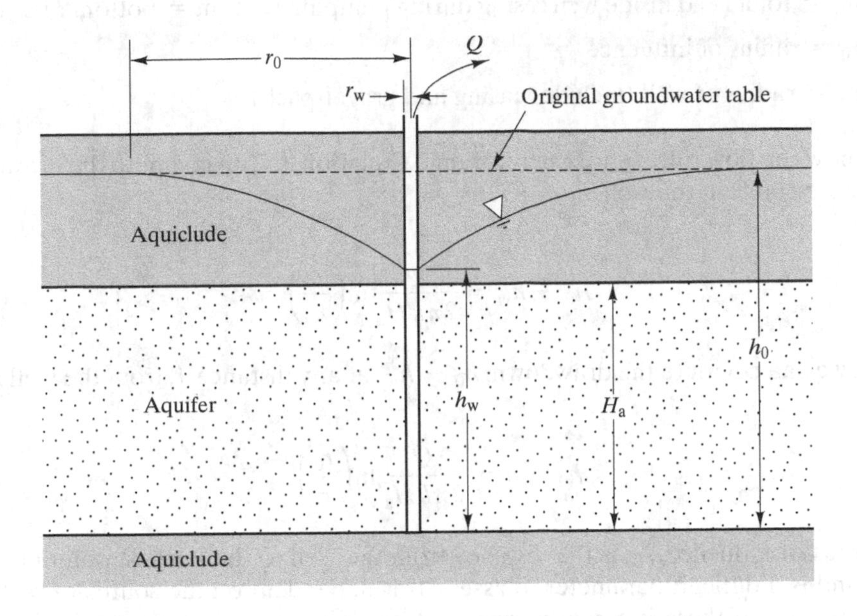

FIGURE 8.15 Well in a confined aquifer.

The flow rate, Q, produced by the well may be derived from Darcy's Law by considering a pie-shaped section of the aquifer as shown in Figure 8.14:

$$Q = \int_0^{2\pi} kiA$$

$$= \int_0^{2\pi} k \frac{dh}{dr} H_a \, r \, d\theta$$

$$Q = \frac{2\pi k H_a (h_0 - h_w)}{\ln\left(\dfrac{r_0}{r_w}\right)} \tag{8.31}$$

where:

Q = flow rate to well

k = hydraulic conductivity of aquifer

H_a = saturated thickness of aquifer

h_0 = total head in aquifer before pumping (datum = bottom of aquifer)

h_w = total head inside well casing during pumping (datum = bottom of aquifer)

r_0 = radius of influence

r_w = radius of well (includes casing and gravel-pack)

If we know the flow rate, Q, we can rearrange Equation 8.31 to compute the drawdown of the water table at the well, $h_0 - h_w$, as

$$h_0 - h_w = \frac{Q}{2\pi k H_a} \ln\left(\frac{r_0}{r_w}\right) \tag{8.32}$$

In fact, we can compute the drawdown, $h_0 - h_w$, at any distance, r, from the well as

$$h_0 - h = \frac{Q}{2\pi k H_a} \ln\left(\frac{r_0}{r}\right) \tag{8.33}$$

The radius of influence, r_0, is the distance from the well to the farthest point of drawdown, and is a difficult parameter to assess. It is dependent on the source of recharge for the aquifer. If there is a nearby river or large lake recharging the aquifer, then r_0 can be taken to be twice the distance from the well to the water boundary. If there is no such surface recharge body nearby, we must either perform a pump test to determine r_0 or use an empirical method. Sichart and Kyrieleis (1930) presented the following empirical formula that gives an approximate value of r_0:

$$r_0 = 300(h_0 - h_w)\sqrt{k} \tag{8.34}$$

r_0, h_0, and h_w must be expressed in the same units, and k must be in cm/s. This method is very approximate. Fortunately, even approximate estimates are often sufficient because ln (r_0/r_w) is always very large, and it is not overly sensitive to errors in r_0. For example, changing r_0/r_w from 1000 to 5000 increases ln (r_0/r_w) by a factor of only 1.2.

Unconfined Aquifers In an unconfined aquifer, as shown in Figure 8.16, the lower flow boundary is fixed, but the upper flow boundary is the groundwater table, which is free to seek its own level. The groundwater table is drawn down in the vicinity of the well, so the height of the flowing zone decreases as the water approaches the well.

Using a derivation similar to that for confined aquifers, we can develop the following formula:

$$Q = \frac{\pi k (h_0^2 - h_w^2)}{\ln(r_0/r_w)} \tag{8.35}$$

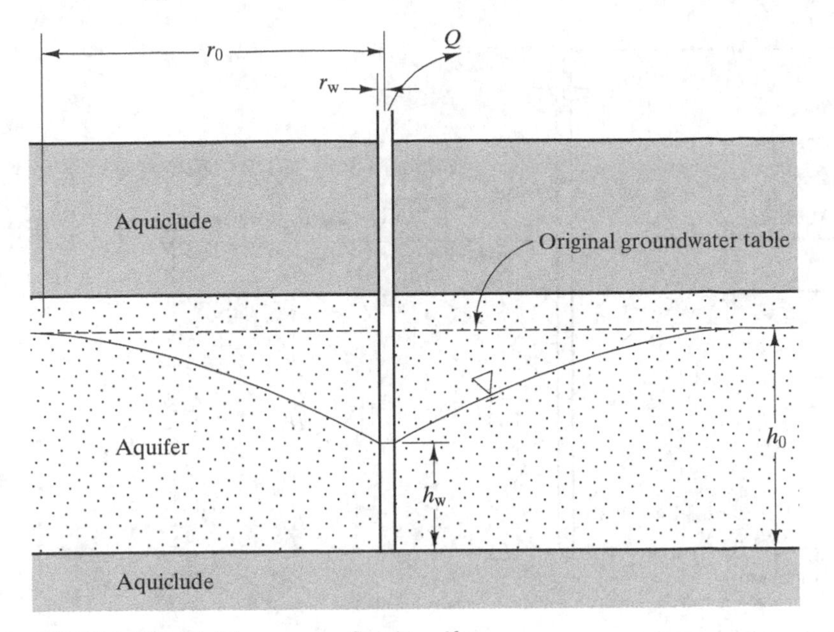

FIGURE 8.16 Well in an unconfined aquifer.

The height of the groundwater surface, h, at any distance from the well, r, is

$$h = \sqrt{h_0^2 - \frac{Q}{\pi k} \ln\left(\frac{r_0}{r}\right)} \tag{8.36}$$

Mixed Aquifers A mixed aquifer is one that was originally confined, but the portion near the well becomes unconfined because of the drawdown to the well as shown in Figure 8.17. This is the case when the water level inside the well casing is pumped to an elevation below the top of the confined aquifer.

Again, we can derive a formula for Q:

$$Q = \frac{\pi k(2H_a h_0 - H_a^2 - h_w^2)}{\ln(r_0/r_w)} \tag{8.37}$$

And the height of the groundwater table at any distance from the well, r, is

$$h = \sqrt{2H_a\, h_0 - H_a^2 - \frac{Q}{\pi k} \ln\left(\frac{r_0}{r}\right)} \tag{8.38}$$

Equations 8.31–8.38 are valid only when the well penetrates completely through the aquifer and is hydraulically open (i.e., has a well screen) throughout the aquifer. The flow rate to partially penetrating or partially open wells would be less (see Driscoll, 1986; Powers, 1992).

The formulas for all three aquifers may be used in anisotropic soils by using a transformed section as described earlier.

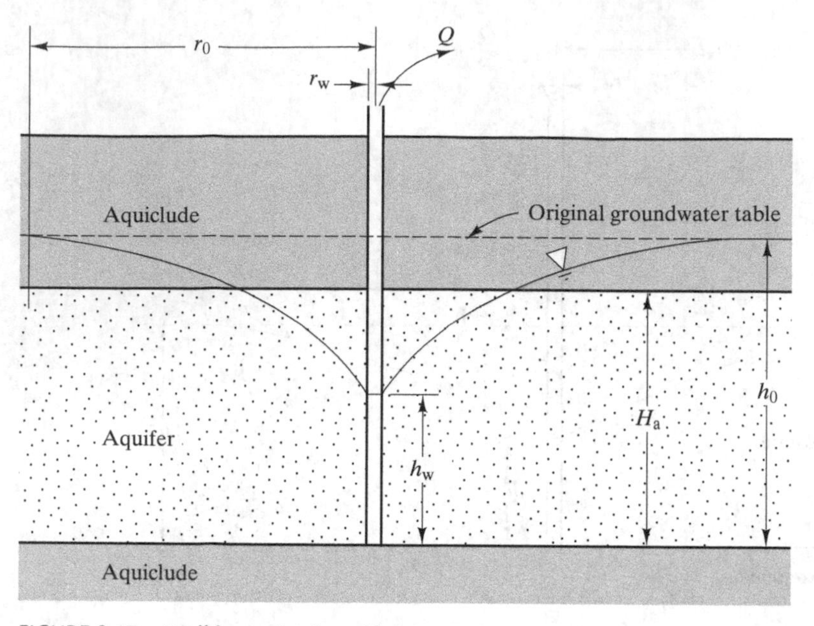

FIGURE 8.17 Well in a mixed aquifer.

Example 8.6

A municipal water supply well is to be installed in the aquifer shown in Figure 8.18. This well will have an 8-in. diameter casing in a 24-in. diameter boring with a submersible pump that draws the water level down to a depth of 62 ft below the ground surface. Compute the maximum flow rate that could be produced by this well.

Solution:

This is an unconfined aquifer, so we will use Equation 8.35.

$$h_0 = 55 \text{ ft} + 22 \text{ ft} - 56 \text{ ft} = 21 \text{ ft}$$

$$h_w = 55 \text{ ft} + 22 \text{ ft} - 62 \text{ ft} = 15 \text{ ft}$$

$$k = (6 \times 10^{-3} \text{ ft/s})\left(\frac{30.5 \text{ cm}}{1 \text{ ft}}\right) = 0.18 \text{ cm/s}$$

Since we do not have any information about recharge sources for the well, we will use Equation 8.34 to estimate r_0.

$$r_0 = 300\,(h_0 - h_w)\sqrt{k} = 300(21 \text{ ft} - 15 \text{ ft})\sqrt{0.18 \text{ cm/s}} = 760 \text{ ft}$$

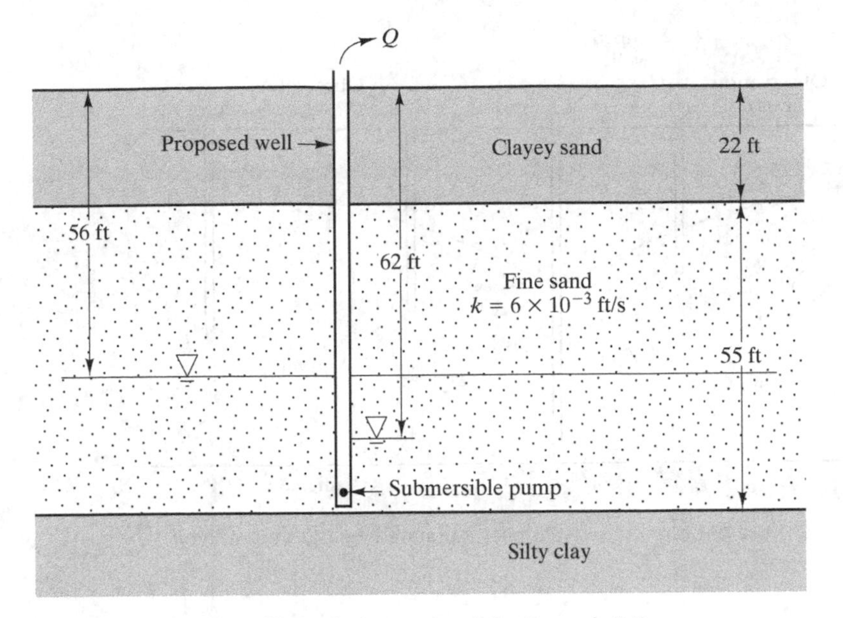

FIGURE 8.18 Soil profile and proposed well for Example 8.6.

Using a unit conversion factor of 449 gal/min per ft^3/s,

$$Q = \frac{\pi k\,(h_0^2 - h_w^2)}{\ln\left(\dfrac{r_0}{r_w}\right)}$$

$$= \frac{\pi(6 \times 10^{-3}\text{ ft/s})[(21\text{ ft})^2 - (15\text{ ft})^2]}{\ln\left(\dfrac{760\text{ ft}}{1\text{ ft}}\right)}$$

$$= 0.61\text{ ft}^3/\text{s}$$

$$= \mathbf{276\ gal/min}$$

Use of Pumped Wells to Conduct In Situ Permeability Tests Equations 8.31 through 8.38 are written in a form that computes the flow rate, Q, from the well, and thus could be used to design the well. These equations can also be rewritten such that k is the unknown, and used as a basis for conducting in situ permeability tests. This method should produce much more accurate values of k than those obtained from laboratory tests because it mobilizes flow through a much larger volume of soil, and it is testing soil in its in situ condition.

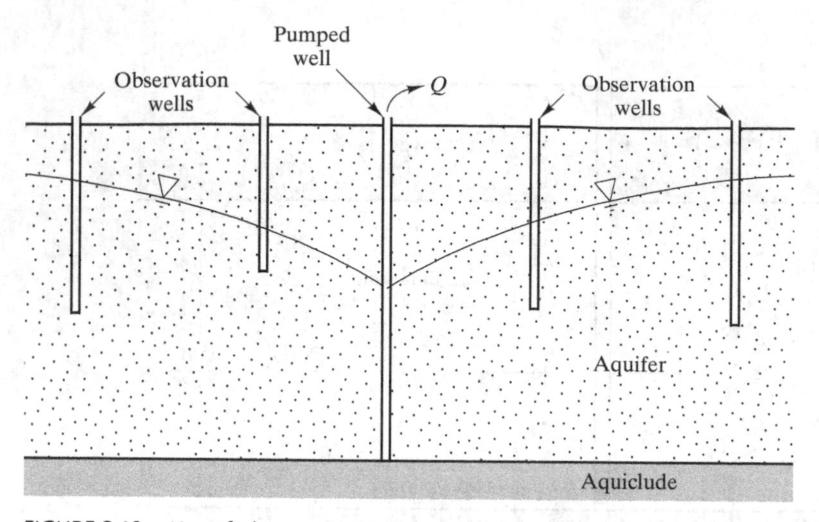

FIGURE 8.19 Use of observation wells near a pumped well to determine k in situ.

This methodology can be further developed by using two or more observation wells as shown in Figure 8.19 (Driscoll, 1986). These observation wells must be located at a distance less than r_0 from the pumped well. We can then rewrite Equations 8.31, 8.35, and 8.37 as follows:
For confined aquifers,

$$k = \frac{Q \ln(r_1/r_2)}{2\pi H_a(h_1 - h_2)} \tag{8.39}$$

For unconfined aquifers,

$$k = \frac{Q \ln(r_1/r_2)}{\pi(h_1^2 - h_2^2)} \tag{8.40}$$

For mixed aquifers,

$$k = \frac{Q \ln(r_1/r_2)}{\pi(2H_a h_1 - H_a^2 - h_2^2)} \tag{8.41}$$

where:

k = hydraulic conductivity of aquifer
Q = flow rate from pumped well

r_1 = radius from pumped well to farthest observation well (must be $< r_0$)

r_2 = radius from pumped well to nearest observation well

h_1 = total head in farthest observation well (datum = bottom of aquifer)

h_2 = total head in nearest observation well (datum = bottom of aquifer)

H_a = saturated thickness of aquifer

Again, these equations are valid only when the pumped well extends through the entire aquifer and is open to the groundwater through the entire aquifer.

This method of determining k is valid only after steady-state conditions have been achieved. This may be accomplished by continuously pumping from the pumped well and monitoring the water levels in the observation wells. When these water levels and the flow rate in the pumped well have both reached equilibrium, steady-state conditions have been achieved.

Example 8.7

A pumped well and two observation wells have been installed through a fine-to-medium sand as shown in Figure 8.20. A pump has been discharging water from the pumped well for a sufficient time to achieve steady-state conditions. The water levels in observation wells A and B were then observed to be 20 and 35 ft below the ground surface, respectively. Compute the hydraulic conductivity of the aquifer.

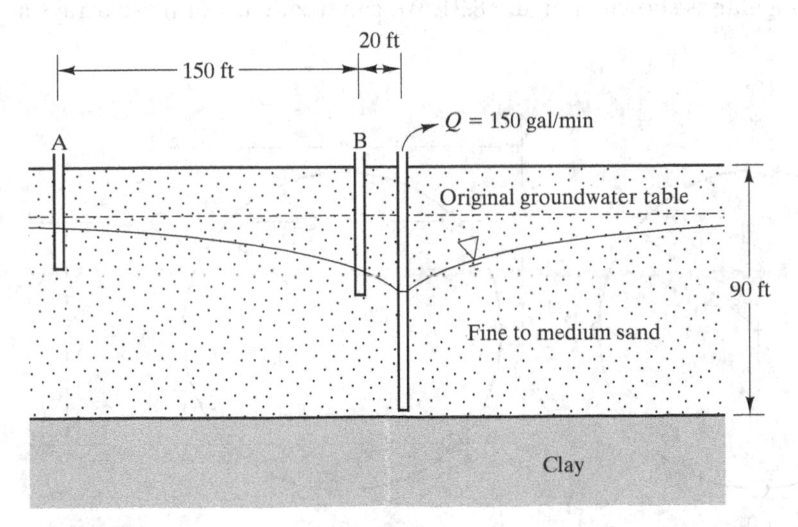

FIGURE 8.20 Cross-section for Example 8.7.

Solution:

This is an unconfined aquifer, so use Equation (8.40).

$$Q = (150 \text{ gal/min})\left(\frac{1 \text{ ft}^3}{7.48 \text{ gal}}\right)\left(\frac{1 \text{ min}}{60 \text{ s}}\right) = 0.33 \text{ ft}^3/\text{s}$$

$$h_1 = 90 - 20 = 70 \text{ ft}$$

$$h_2 = 90 - 35 = 55 \text{ ft}$$

$$k = \frac{Q \ln(r_1/r_2)}{\pi(h_1^2 - h_2^2)}$$

$$= \frac{(0.33 \text{ ft}^3/\text{s})\ln(170 \text{ ft}/20 \text{ ft})}{\pi[(70 \text{ ft})^2 - (55 \text{ ft})^2]}$$

$$= \mathbf{1 \times 10^{-4} \text{ ft/s}}$$

This answer seems reasonable, per Table 7.1.

Approximate Three-Dimensional Solutions for Well Arrays

It is common to us an array of wells when dewatering a construction site as discussed in Section 8.5 below. The groundwater flow in the case of multiple wells is a three-dimensional problem as illustrated in Figure 7.4(c). However, a number of approximate methods are available to solve such problems. These approximate methods are based on the two-dimensional solutions already presented in this chapter.

Equivalent Well Radius The two most common well array patterns are the circular or the rectangular as shown in Figure 8.21. We can treat each of these arrays as single

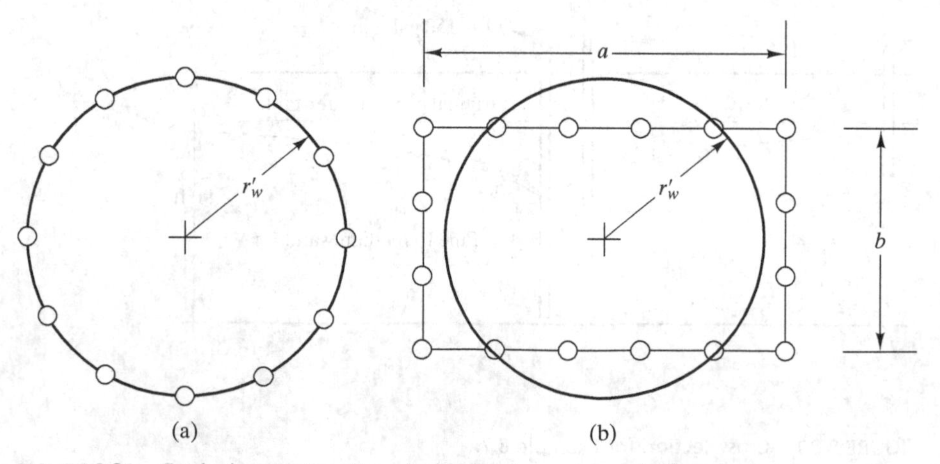

FIGURE 8.21 Equivalent well radius, r'_w: (a) circular arrays and (b) rectangular arrays.

equivalent well with an equivalent radius, r'_w, as shown in Figure 8.21. In the case of a circular array, r'_w is simply equal to the radius of the array. In the case of a rectangular array, r'_w is the radius of a circle that contains the same area as the rectangle and

$$r'_w = \sqrt{\frac{ab}{\pi}} \tag{8.42}$$

Flow, Q, can then be computed using Equations (8.31), (8.35), or (8.37), as appropriate to the type of aquifer, substituting r'_w for r_w. If the well spacing becomes large, the actual flow will be greater than that computed using this approximation. These approximations are accurate when the wells have a close spacing and when r_0 is much greater than r'_w. According to Powers (1992), the rectangular array approximation is appropriate when the aspect ratio a/b is less than 1.5.

Superposition for Drawdown Computation If the spacing of wells is not close or the array is in an irregular pattern, we can use superposition to compute the drawdown at any location. We do this by computing the drawdown of each well, assuming each act independently, and then add the drawdown of the individual wells to determine the total drawdown. In this case, the flow rate, Q, for each well is computed separately using Equation (8.31), (8.35), or (8.37) as appropriate.

Theoretically, superposition applies only to the case of confined aquifers. However, according to Powers (1992), the method provides satisfactory results for unconfined aquifers so long as the total drawdown is less than approximately 20% of the saturated thickness of the aquifer, H_a. If greater drawdown is needed, numerical modeling is required.

True Three-Dimensional Solutions

The above solutions work only when aquifers surrounding wells are homogeneous and isotropic and when the well arrays are simple like those described earlier. When the soil conditions become inhomogeneous or the well arrays are complicated, the two-dimensional solutions presented earlier are inadequate and a true three-dimensional solution is needed. In these circumstances, analytical solutions are almost never possible and it is necessary to use numerical solutions as described earlier in Section 8.3. Three-dimensional numerical models using either the finite element or finite difference methods are possible and a number of commercial computer programs support such solutions.

8.5 GROUNDWATER CONTROL

Engineers and contractors often need to control groundwater flows, either temporarily or permanently. Temporary controls are necessary when performing underground construction below the groundwater table; permanent controls may be needed to keep groundwater from reaching sensitive areas. These controls often consist of removing groundwater from certain zones using various *dewatering* technologies. Alternatively, groundwater control can sometimes be achieved by constructing groundwater barriers that block the flow.

Temporary Construction Dewatering

The most common type of temporary groundwater control is *construction dewatering*, which is performed to keep groundwater out of construction sites. Construction dewatering is in place only during the construction period, and is for the "convenience" of the contractor, so it is not shown on the design drawings or specifications. The contractor is responsible for designing the dewatering system, and normally does so by subcontracting this work to a dewatering specialist.

There are four basic categories of construction dewatering methods: open pumping, predrainage, cutoffs, and exclusion (Powers, 1992). The proper selection depends on many factors including:

- The soil conditions, especially the hydraulic conductivity
- The size and depth of the construction excavation, especially the depth below the groundwater table
- The planned method of excavation and ground support
- The type and proximity of nearby structures
- The type of structure being built
- The planned schedule
- The presence and characteristics of groundwater contaminants (if any).

Figure 8.22 shows an excavation that has been temporarily dewatered to permit construction of two pipelines.

Open Pumping Open pumping methods allow groundwater to enter the excavation, then direct it to low points known as sumps where it is pumped out. Figure 8.23 shows an open pumping system. This method is most appropriate in soils with a low to moderate hydraulic conductivity because these soils discharge controllable quantities of water into the excavation. When used in favorable conditions, open pumping is generally inexpensive and effective.

However, open pumping is inappropriate in soils with high hydraulic conductivity values, especially when the excavation extends to significant depths below the groundwater table. Such soils can produce very large flow rates that cannot be effectively collected or pumped. This method also is inappropriate when the excavation is potentially unstable as a result of boils, quicksand, or heave, or when the excavation is to be performed by scrapers or other equipment that has traction problems in wet soil.

Predrainage *Predrainage* encompasses several methods of intercepting groundwater before it reaches the excavation. Unlike open pumping, these methods allow the contractor to lower the groundwater table before excavation begins, which often solves equipment mobility problems. They also are suitable for a wider range of soil conditions.

FIGURE 8.22 This 11-m (36-ft) deep excavation was required to construct these two water supply pipelines. The excavation extends 10 m (33 ft) below the groundwater table, and a river is located just beyond the excavation on the left side of the photograph. Therefore, a temporary dewatering system has been installed to draw down the groundwater table. This system includes a series of pumped wells, including the one in the right foreground. (Photo courtesy of Foothill Engineering.)

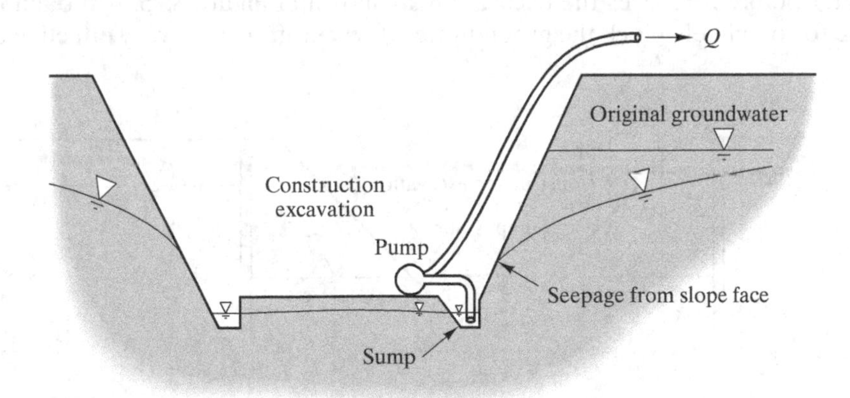

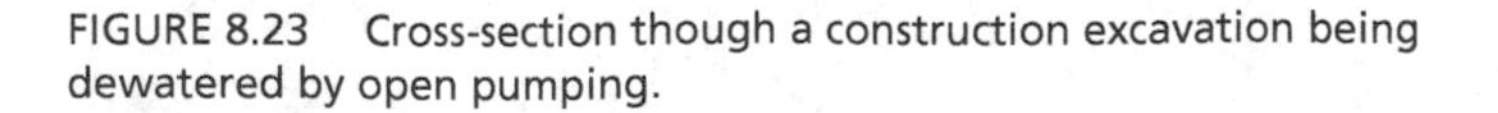

FIGURE 8.23 Cross-section though a construction excavation being dewatered by open pumping.

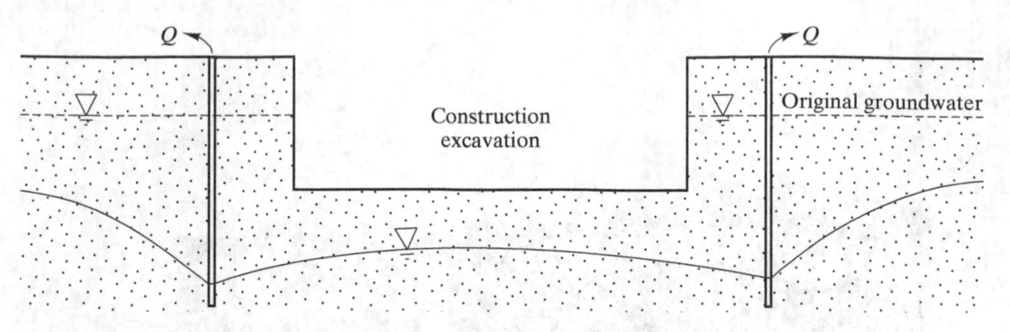

FIGURE 8.24 Cross-section of construction excavation being dewatered by predrainage.

Most predrainage systems use wells located a short distance outside the perimeter of the excavation, as shown in Figure 8.24. Several methods are available to extract water from the wells, including:

- *Pumped well systems* with a submersible pump in each well. This method can accommodate large flow rates, but requires a large investment in equipment. The excavation in Figure 8.22 used pumped wells.
- *Wellpoint systems* extracting the water by applying a vacuum to each well. This is much less expensive than installing individual pumps, but is limited to depths of not more than 5–6 m.
- *Ejector systems* using a nozzle and a venturi in each well to lift the water. These systems do not have the depth limitations of wellpoints, and often are less expensive than pumped well systems. However, they are inherently inefficient and thus quickly lose their cost advantage when the required flow rates are high.

Cutoffs Cutoffs (also known as seepage barriers) are intended to physically block the groundwater before it reaches the excavation, as shown in Figure 8.25. Although it is impossible to completely block the groundwater flow, cutoffs can be very effective and

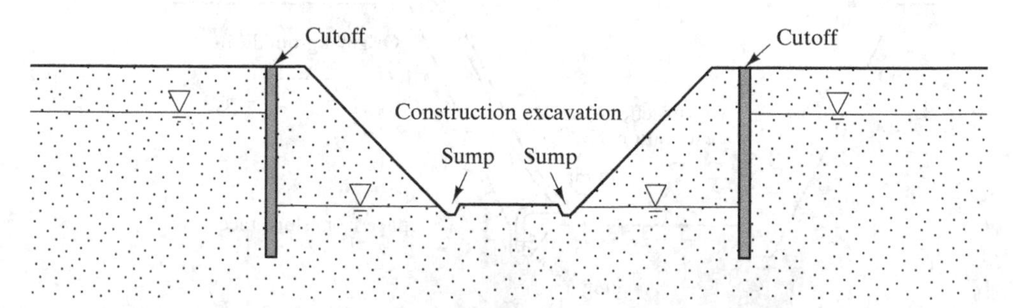

FIGURE 8.25 Use of a cutoff to block groundwater flow before it reaches a construction excavation.

typically reduce the water inflow to a small fraction of what it otherwise would have been. The small amount of groundwater that does eventually reach the excavation can usually be removed by open pumping.

Many methods and techniques have been used to cutoff groundwater. These include the following (Powers, 1992):

- *Steel sheet piles* are heavy corrugated steel sheets driven vertically into the ground. They are installed around the perimeter of a proposed excavation, then the soil inside is removed. The primary purpose of sheet piles is to provide structural support to keep the excavation from caving in. However, they also provide a partially effective groundwater cutoff.
- *Diaphragm walls* are constructed by excavating a trench (typically about 1 m wide) around the perimeter of the proposed excavation and filling it with concrete. To avoid caving, the trench is normally filled with bentonite slurry as it is being excavated, and the concrete is placed using a tremie extending through the slurry. These walls provide both structural support and groundwater control.
- *Slurry trenches* are constructed similar to diaphragm walls, except they are filled with low-permeability soils instead of concrete. They provide groundwater control, but do not provide structural support. Figure 8.26 shows a slurry trench under construction.
- *Secant drilled shaft walls* are drilled shaft foundations (large diameter borings filled with concrete) placed on close centers, thus creating a continuous underground wall.
- *Tremie seals* are concrete barriers placed at the bottom of excavations to block water and resist hydrostatic pressures. They are typically used in combination with sheet piles or diaphragm walls, and are often placed underwater using a tremie.
- *Permeation grouting* consists of injecting cement or special chemicals into the ground. This fills the voids and significantly reduces the hydraulic conductivity of the soil.
- *Ground freezing* consists of inserting refrigeration devices into the ground and freezing the groundwater. This method has been used successfully, and is very effective. It also is very expensive.

Exclusion Exclusion methods use compressed air to prevent groundwater from entering a construction excavation, as shown in Figure 8.27. This method has been used in tunnel and foundation construction because these excavations can be sealed and pressurized. The air pressure is maintained at a level approximately equal to the pore water pressure, thus preventing or at least greatly reducing seepage into the excavation. Workers, equipment, and excavated soil must pass through a decompression chamber located between the excavation and the atmosphere.

(a)

(b)

FIGURE 8.26 A slurry trench barrier under construction: (a) an excavator is digging the 20-m (65-ft) deep trench through a slurry and (b) then a bulldozer pushes in a clayey backfill.

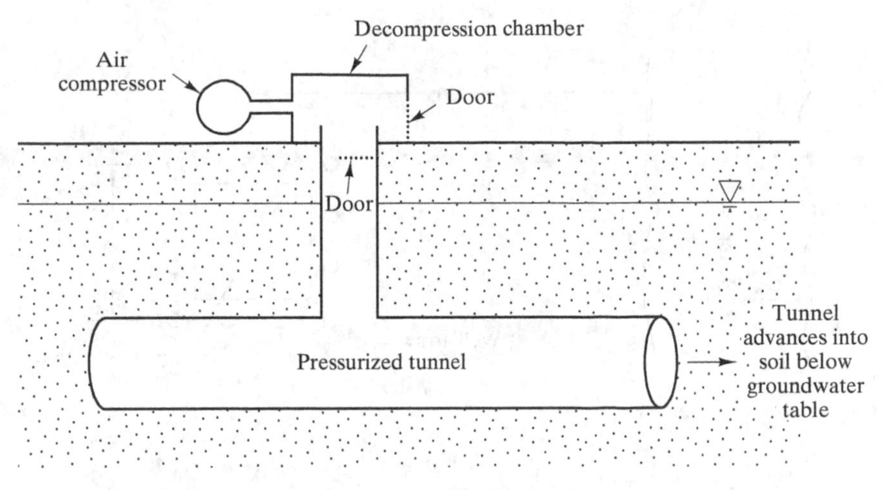

FIGURE 8.27 Tunnel construction under compressed air.

Example 8.8

An excavation for a power plant foundation is to be made near a river. A well array around the excavation is to be used to predrain the excavation site. The site plan and elevation are shown in Figure 8.28. The aquifer height must be drawn down from 90 ft. to 75 ft. in order to complete the excavation. Compute the total flow, Q, that will be required to draw the aquifer down to the required level.

Solution:

This problem can be modeled using the equivalent well radius method—the wells are spaced relatively close together and the radius of influence, r_0, is relatively large compared with the equivalent well radius, r'_w.

Since there is a river recharging the aquifer, r_0 is twice the distance to the river.

$$r_0 = 2(750) = 1500 \text{ ft}$$

The well array is square so we can use Equation 8.42 to compute r'_w.

$$r'_w = \sqrt{\frac{300^2}{\pi}} = 169 \text{ ft}$$

This is an unconfined aquifer so we use Equation 8.35 to compute the flow:

$$Q_{\text{total}} = \frac{\pi(2 \times 10^{-3} \text{ ft/s})(90^2 - 75^2)\text{ft}^2}{\ln(1500 \text{ ft}/169 \text{ ft})} = 7.12 \text{ ft}^3/\text{s}$$

$$= \mathbf{3{,}200 \text{ gal/min}}$$

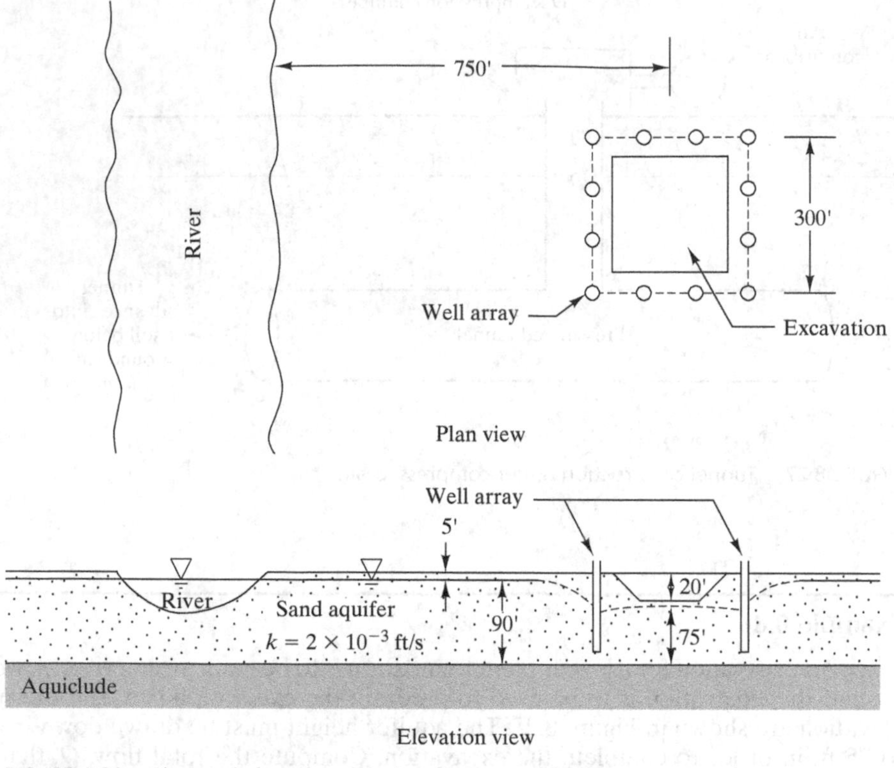

FIGURE 8.28 Plan and elevation drawing for Example 8.8.

This is the total flow for the well array. The flow for each pump will be

$$Q = \frac{3{,}200 \text{ gal/min}}{12} = 270 \text{ gal/min for each pump.}$$

Permanent Dewatering

Permanent dewatering measures are necessary when groundwater must be controlled throughout the useful life of a civil engineering project. Some of the construction dewatering methods also may be used for permanent dewatering. However, economics and reliability concerns generally prevent the use of extensive long-term pumping. The specific methods of permanently controlling groundwater depend on the application (Cedergren, 1989). The following sections describe some common techniques and concerns for specific applications.

Underground Structures Buildings with basements, tunnels, utility vaults, sewage pumping stations, and other structures need to be protected from flooding, especially when they extend below the groundwater table. Usually these structures have thick concrete walls that are reasonably watertight. In addition, the outside areas of these walls are usually coated with a waterproof material, and drain pipes are sometimes installed around the perimeter. However, in spite of these precautions, some water may enter the structure. Therefore, it is generally necessary to direct any such water to a sump and pump it out. Normally, these sump pumps do not need to operate continuously.

Retaining Walls If groundwater is allowed to build up behind retaining walls, the resulting hydrostatic pressures may be much greater than the lateral pressures due to the soil. Sometimes it becomes necessary to design for these pressures, but this can significantly increase the cost of the wall. The other option is to install drain pipes or weep holes (drainage holes near the bottom of the wall) to drain any groundwater, thus maintaining the groundwater table at a suitably low elevation.

Pavements Pavements for highways, parking lots, airports, and other facilities are very sensitive to the accumulation of water in the underlying soils, and many pavement failures may be attributed to poor drainage of the subgrade soils. Water inevitably passes through joints, cracks, and other openings in pavements, and in the case of certain open-graded asphalt, it may pass through the pavement itself. It is virtually impossible to stop water from reaching this zone. Therefore, engineers need to drain this water to a safe location.

Earth Slopes Permanent drainage measures are often installed in earth slopes to control the groundwater level and thus enhance stability. This method is discussed in more detail in Chapter 13.

Earth Dams Earth dams include extensive drainage facilities to control the groundwater flow through the dam. This enhances their stability.

8.6 CONTAMINANT CONTROL AND REMEDIATION

An important application of groundwater analysis is in the control of contaminants that have been inadvertently introduced into an aquifer. As discussed in Chapter 7, advection is a principal means of transporting contaminants in an aquifer, so controlling the flow of groundwater in a contaminated part of an aquifer is often an effective way of controlling the movement of contaminants. This section briefly presents a number of techniques for doing so and then treating the groundwater contaminants. The design of each of these control and remediation methods is beyond the scope of this book; however, such designs make use of the groundwater analysis techniques described in Chapters 7 and 8.

Containment

One option is to surround the contaminated soils with an impervious barrier to prevent the contaminants from traveling outside the containment zone. Containment is especially attractive when the cost or risk of removal is not acceptable. Containment is similar to the use of cutoffs discussed in Section 8.5 and include constructing a *slurry trench wall*, a *grout curtain*, or *sheet piles*, as shown in Figure 8.29. Other containment methods also have been used.

Slurry trench walls are built by excavating a trench while keeping it filled with a bentonite slurry (a combination of bentonite and water), as shown in Figure 8.26. The purpose of the slurry is to keep the sides of the trench from caving. Normally the excavation extends down to an impervious strata, which may be 20 m or more below the ground surface. Then, a clayey soil is pushed into the trench to fill it and displace most of the slurry. This mixture of clayey soil and the remaining bentonite forms the groundwater barrier. The process continues until the slurry trench wall has the desired length, which sometimes means placing it around the entire site.

Grout curtains are made by injecting cement grout into the soil to form an impervious barrier. Sheet piles are heavy, corrugated steel sheets that are driven into the ground to form a barrier.

The contaminated zone also may be covered with a compacted clay cap, which often includes one or more geosynthetic membranes. These caps are intended to reduce the infiltration of surface water and minimize the potential for human exposure to the waste.

Containment systems can only be built around the sides and top of the contaminated zone. In some cases, the bottom might be naturally contained by impervious strata, but even then containment systems are not completely effective. As with any underground construction, uncertainties will always be present, and leaks can occur in unexpected places. Therefore, containment systems are often accompanied by other remediation measures, and almost always include monitoring systems.

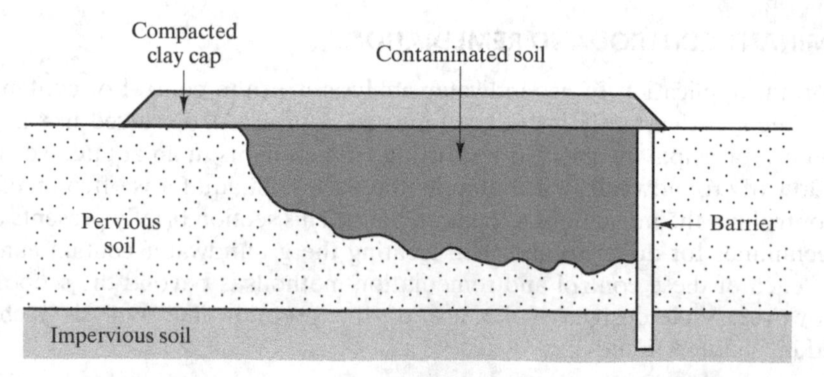

FIGURE 8.29 Use of containment barriers to block the flow of contaminants.

Pump-and-Treat

Pump-and-treat remediation consists of extracting the contaminated groundwater, passing it through above-ground treatment facilities, then discharging it back into the ground through injection wells. This is one of the most commonly used remediation methods.

A wide variety of treatment methods is available, depending on the type and concentration of contaminants in the extracted water. Contaminants removed from the water are then hauled to a suitable disposal site.

Usually wells are used to extract the contaminated groundwater, as shown in Figure 8.30. Their placement and pumping rates are usually chosen so all of the contaminated groundwater flows toward the wells, thus increasing the recovery and reducing the potential for further expansion of the plume. These systems may be designed with the aid of numerical models that are available as computer software.

When the hazardous materials float on top of the groundwater, a trench might be used to capture the contaminant as shown in Figure 8.31.

Bioremediation

Bioremediation is the engineered enhancement of natural biodegradation processes (Norris et al., 1994). It is most frequently performed in situ by providing favorable conditions for the biochemical processes. It also may be done ex situ as a treatment method on excavated soils.

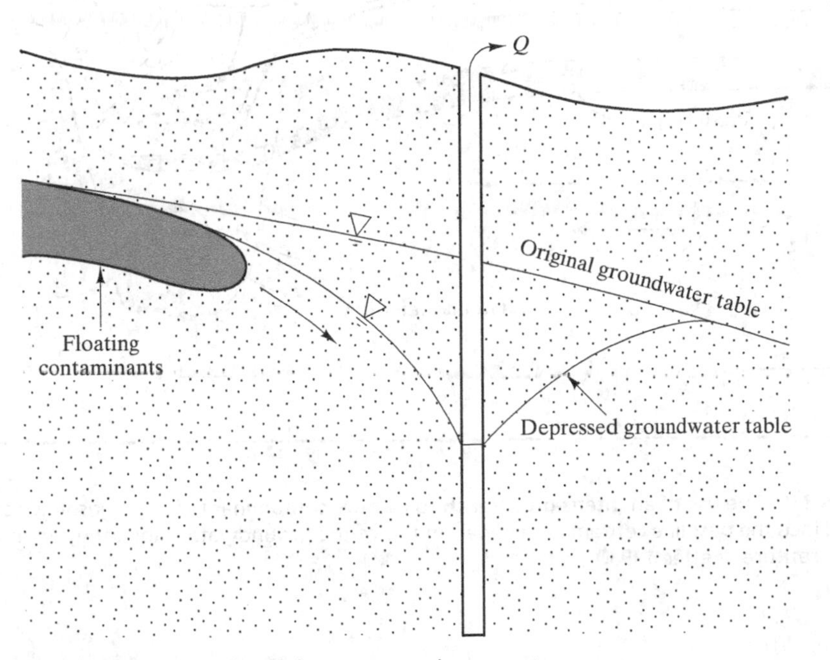

FIGURE 8.30 Use of wells in a pump-and-treat system.

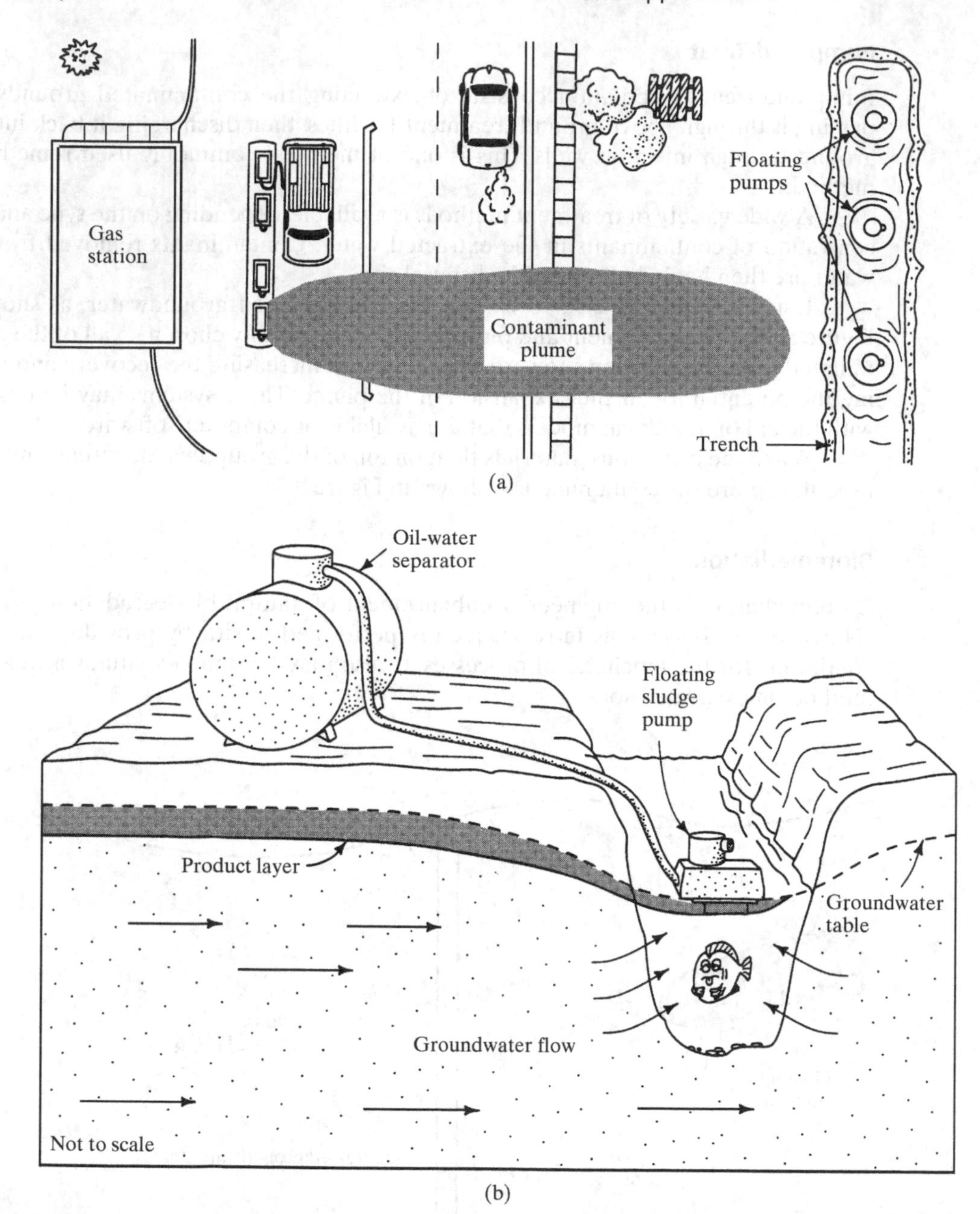

FIGURE 8.31 The use of an interceptor trench to capture contaminants from a gasoline service station. Since the gasoline contaminants float on top of the groundwater table, they are much easier to capture. (Fetter, 1993.)

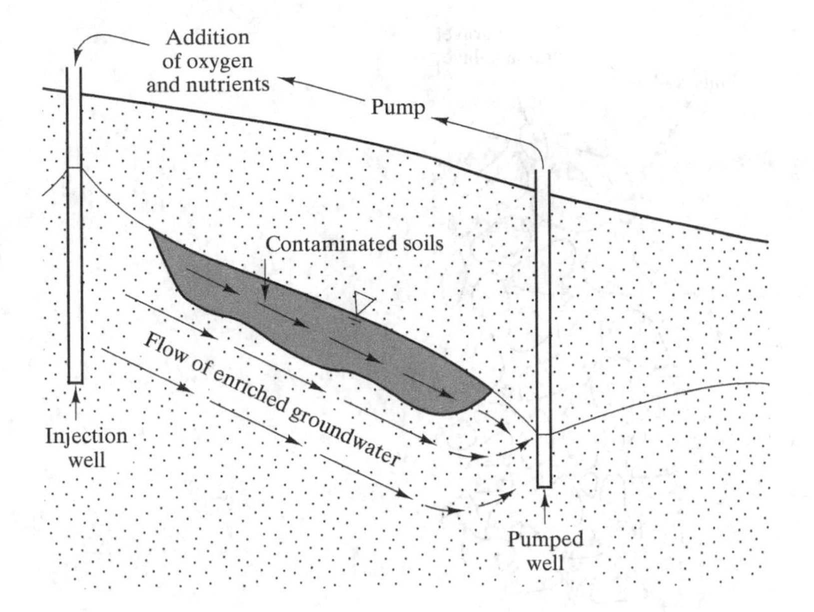

FIGURE 8.32 Typical in situ bioremediation system.

In situ bioremediation systems usually consist of wells that inject nutrients, an electron acceptor, and possibly other substances to promote biodegradation. Pumps arc installed in another set of wells to develop a hydraulic gradient across the site, thus distributing the injected materials, as shown in Figure 8.32.

8.7 SOIL MIGRATION AND FILTRATION

Geotechnical engineers often intentionally place highly pervious soils in key locations to capture and drain groundwater. Poorly-graded coarse gravels are especially useful in this regard because they have very high hydraulic conductivities. However, a problem occurs when we place such soils adjacent to finer-grained soils (and nearly everything is finer than coarse gravel!) because the seepage forces push the finer soils into the gravel drainage layer, as shown in Figure 8.33. We call this process *soil migration*.

This migration has at least two detrimental results. First, the drainage layer becomes clogged and no longer functions properly. This clogging can occur whenever a coarse soil is downstream of a significantly finer soil and the seepage forces (i.e., hydraulic gradients) are large enough to cause soil migration. Once the subsurface drainage system ceases to work, failure often follows. Second, the migrating soils leave voids in the upstream strata. In some cases, these voids can propagate for long distances, creating underground channels that lead to *piping failure* as shown in Figure 8.34. This is an important cause of failure in dams and levees (see the Teton Dam sidebar).

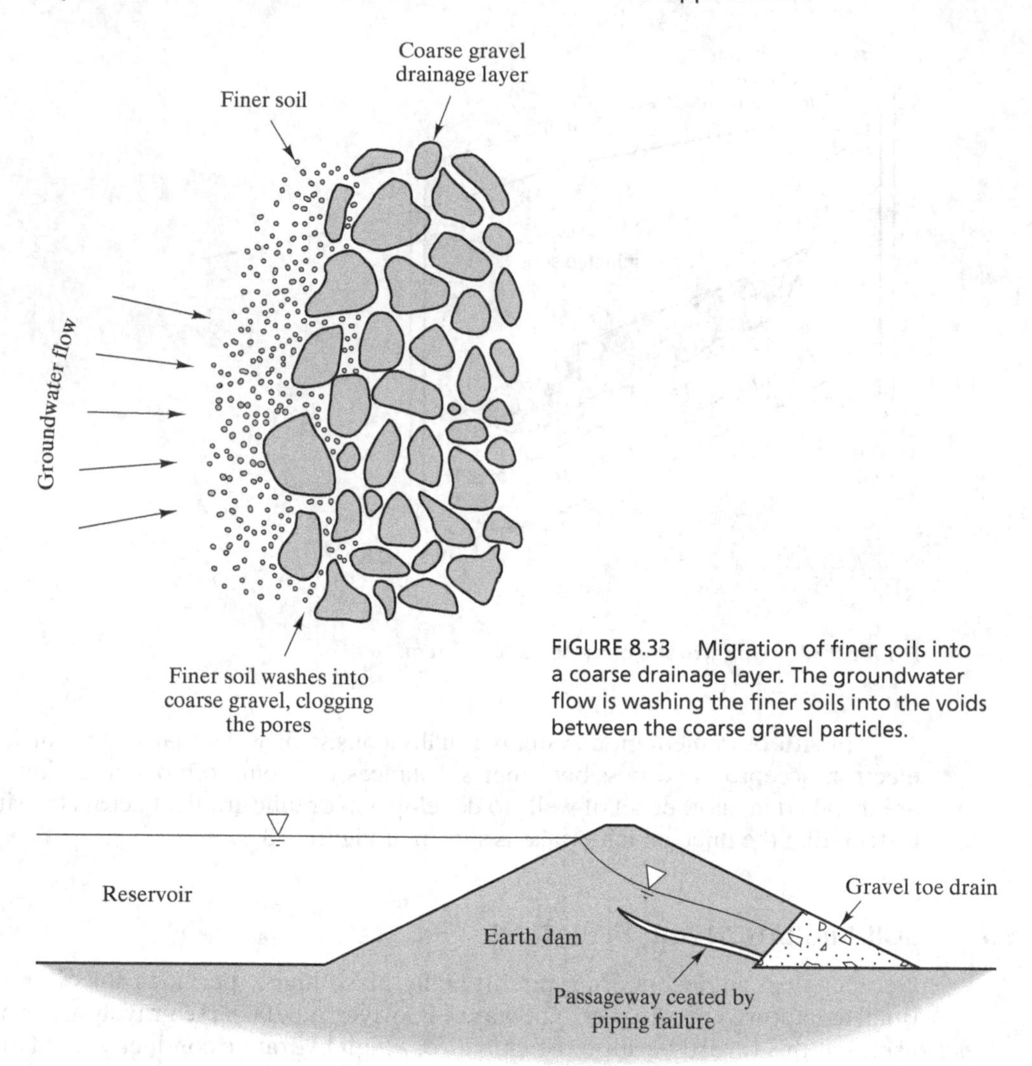

FIGURE 8.33 Migration of finer soils into a coarse drainage layer. The groundwater flow is washing the finer soils into the voids between the coarse gravel particles.

FIGURE 8.34 Piping failure in an earth dam.

Migration problems can be avoided by providing *filters* that are intended to pass water but retain potentially migrating soil. There are two principal types of filters: *graded soil filters* and *geosynthetic filters*.

Teton Dam Failure

Teton Dam was a 305-ft (93-m) high earth dam constructed on the Teton River in southeastern Idaho to provide flood control, power generation, and irrigation in the Teton Valley and the eastern Snake River plain. Construction of the dam started in February 1972. By the beginning of June 1975 the reservoir had nearly reached its capacity of 288,000 acre ft (356×10^6 cubic meters) of water (Solava and Delatte, 2003).

About 7:00 a.m. on June 5, inspectors noticed a damp channel approximately halfway down the embankment where it met the right abutment and muddy water was flowing out of the right abutment. The flows increased throughout the morning. At about 10:30 a.m. a large leak appeared in the face of the downstream embankment near the right abutment. Flow from this leak grew rapidly and started to erode a gully in the embankment. Bulldozers were dispatched to try and fill the growing hole in the dam (Independent Review Panel, 1976).

PHOTO 1 ~10:30 a.m. Leak in downstream face.

About 11:00 a.m. workers noticed a whirlpool forming in the reservoir just upstream of the embankment. At 11:30 a.m. a sinkhole formed on the downstream side of the dam swallowing two bulldozers—coworkers pulled the operators to safety with ropes they had tied around their waists. At 11:55 a.m. the crest of the dam collapsed and the reservoir started emptying, flooding the Teton Valley below the dam and dumping onto the Snake River plain downstream (Independent Review Panel, 1976).

PHOTO 2 ~11:30 a.m. Sinkhole forms in downstream face.

The ensuing flood killed 11 people and 13,000 head of cattle. Thousands of homes and businesses were lost in towns downstream. Cleanup efforts lasted the entire summer. The damages from the flood totaled hundreds of millions of dollars (Solava and Delatte, 2003).

Investigations of the failure concluded that piping, or internal erosion, of the dam core caused the failure. The geology of the region consists of a highly fractured ryolite tuff bedrock. The joints and fractures in the tuff are highly permeable to groundwater. The soils in the region are almost entirely wind-blown silts (loess). Because there was not a suitable source of clay for the dam's impervious core, designers used the local silt instead. However, this silt was difficult to compact, highly erodible, and lacked the ductility and plasticity of clay used in the core of most earth dams. The combination of a highly erodible core and fractured bedrock open to flow proved deadly (Smalley, 1992).

PHOTO 3 ~11:55 a.m. Dam breached and reservoir empties.

Sadly, the investigation panel concluded the failure was not due to unforeseeable conditions but because designers did not recognize the inherent problems within the geology and the materials, and did not account for these problems in their design (Independent Review Panel, 1976).

PHOTO 4 ~6:30 p.m. Flooding in City of Rexburg and surrounding farmland.

Photos 1-3 by Eunice Olson, courtesy of Note Dame University.

Photo 4 by Glade Walker, Bureau of Reclamation.

TABLE 8.1 Graded Soil Filter Design Criteria[a]

Soil Group No.	Soil Description	D_{15} Design Criteria
1	Fine silts and clays with >85% passing the #200 sieve	$D_{15} \leq 9d_{85}$, but not smaller than 0.2 mm
2	Silty and clayey sands and sandy silts and clays with 40–85% passing the #200 sieve	$D_{15} \leq 0.7$ mm
3	Silty and clayey sands and gravels with 15–39% passing the #200 sieve	$D_{15} \leq \left(\frac{40 - A}{25}\right)(4d_{85} - 0.7 \text{ mm}) + 0.7 \text{ mm}$
4	Silty and clayey sands and gravelly sands with ≤15% passing the #200 sieve	$D_{15} \leq 4d_{85}$

[a]Sherard, et al., 1984a, 1984b, 1985, 1989; SCS, 1986.

Graded Soil Filters

Graded soil filters consist of one or more layers of carefully graded soil placed between the potentially migrating soil and the drain. Each filter zone is fine enough to prevent significant migration of the upstream soil, yet coarse enough not to migrate into the downstream soil. The filter also must have a sufficiently high hydraulic conductivity to effectively transmit water to the drain.

Sherard et al. (1984a, 1984b, 1985, 1989) developed design criteria to ensure graded soil filters prevent soil migration. These criteria are based on the D_{15} size, as shown in Table 8.1, where:

D_{15} = the particle size at which 15% of the filter material is finer
d_{85} = the particle size at which 85% of the soil to be filtered is finer
A = the percentage of the soil to be filtered that passes the #200 sieve

For soil groups 1, 2, and 4, the values of d_{85} and A should be based only on the portion of the soil that passes the #4 sieve. In other words, if these soils contain plus #4 material, the particle size distribution curve needs to be reconstructed to what it would be if that material was not present. This adjustment is not necessary for soil group 3. To ensure the filter has a sufficiently high hydraulic conductivity, D_{15} of the filter must be greater than 4 times d_{15} of the soil being filtered (Cedergren, 1989).

The shape of the filter particle size distribution curve should be similar to that of the soil being filtered (i.e., their slopes on the particle size distribution chart should be about equal).

Other more detailed filter design criteria also are available (i.e., Reddi and Bonala, 1997).

Example 8.9

The design of a proposed earth dam is to include a gravel drain to control the groundwater inside the dam. This drain is to be protected with a graded soil filter zone. The particle size distribution curves of the soils used to build the core of the dam and the gravel used to build the drain are shown in Figure 8.35. Determine the range of acceptable filter material.

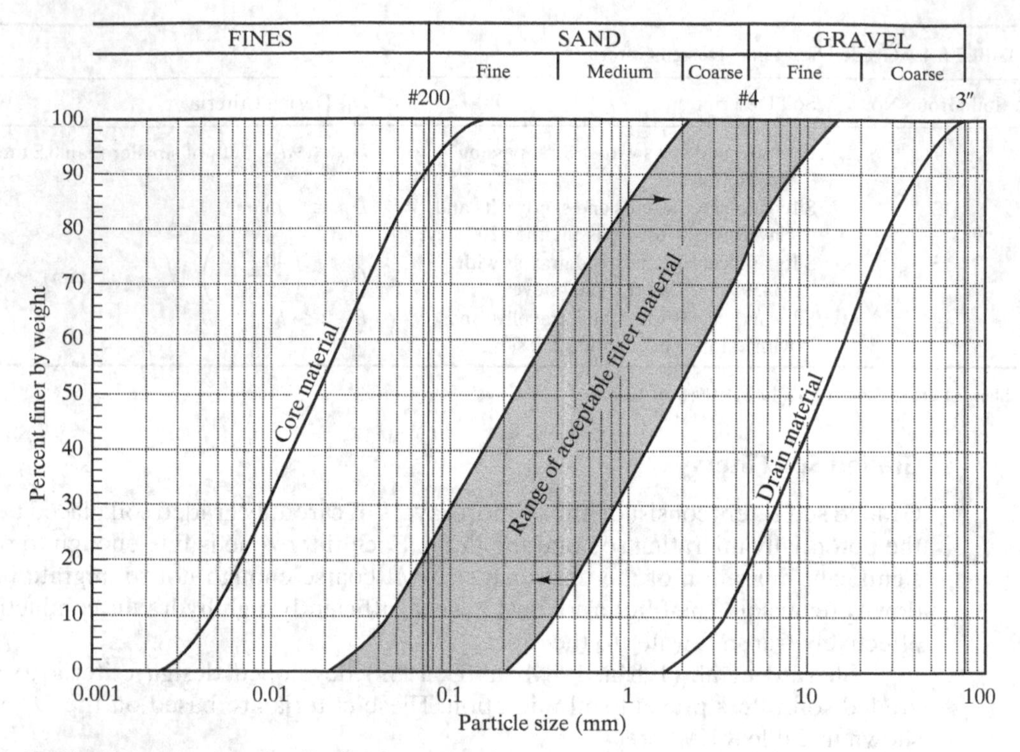

FIGURE 8.35 Particle-size distribution curves for Example 8.9.

Solution:

Migration of core soils into filter:

91% passing #200 sieve, so this is soil type 1 per Table 8.1

$d_{85} = 0.056$ mm

$D_{15} \leq 9d_{85} \rightarrow D_{15} \leq (9)(0.056 \text{ mm}) \rightarrow D_{15} \leq 0.50$ mm

Therefore, the filter soils must have $D_{15} \leq 0.50$ mm, as shown by the mark on Figure 8.35. We then draw a particle distribution curve parallel to the core material and check to see that D_{15} of the filter material is greater than four times D_{15} of the core material.

Migration of filter soils into drain:

We use the same analysis, but now D represents the drain and d represents the filter.

On the basis of the earlier results, the filter will be a type 4 soil

$D_{15} = 4.0$ mm

$D_{15} \leq 4d_{85} \rightarrow 4.0 \text{ mm} \leq (4)(d_{85}) \rightarrow 1.0 \leq d_{85}$

Therefore, the filter soils must have $d_{85} \geq 1.0$ mm, as shown by the mark on Figure 8.35. We then draw a particle distribution curve parallel to the drain material and check to see that D_{15} of the drain material is greater than four times D_{15} of the filter material.

On the basis of these criteria, the range of acceptable filter material is as shown in Figure 8.35.

Geosynthetic Filters

A wide variety of geosynthetic filter materials (often called *filter fabrics*) are now used as alternatives to graded soil filters. These are known as *geotextiles* (Koerner, 1998) and are supplied as rolls of fabric as shown in Figure 8.36.

Geotextile manufacturers provide various specifications for their materials to help the engineer select proper fabric for each application. One of these specifications is the *equivalent opening size* or *EOS* (also called *apparent opening size* or *AOS*), which is expressed either as O_{95} or as the equivalent sieve size. For example, a geotextile with EOS = AOS = #30 (O_{95} = 0.60 mm) has openings similar to those in a #30 sieve. Such a fabric would retain 95% of soil particles that have a diameter of 0.60 mm.

Carroll (1983) recommends selecting geotextiles for filtration based the following criterion:

$$O_{95} < (2 \text{ or } 3)(d_{85})_{\text{soil}} \tag{8.43}$$

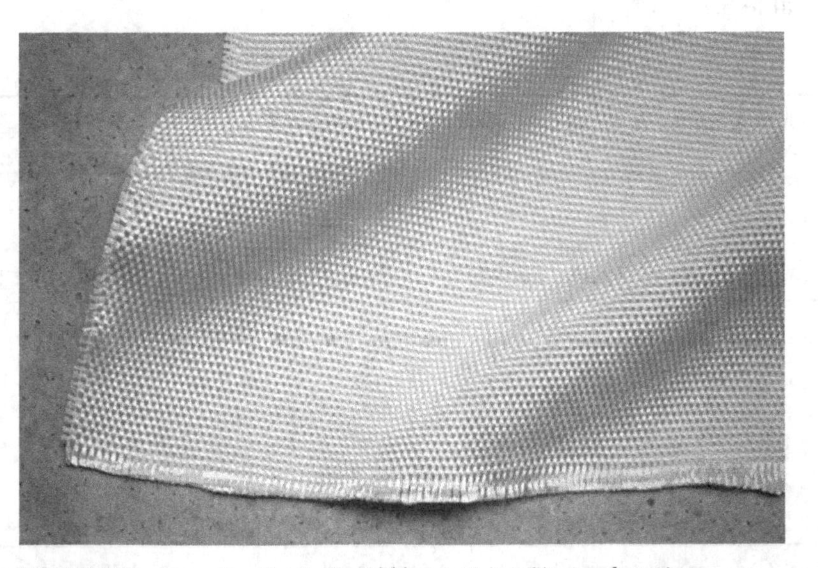

FIGURE 8.36 A geotextile that could be used as a filter or for other purposes. This one is a Tensar Vectra fabric. (Courtesy Tensar Earth Technologies, Inc., Atlanta, Georgia.)

The geotextile filter fabric also must have sufficient *permittivity* to pass the required groundwater flow rate. This parameter is a measure of the fabric's ability to pass water, and is defined as:

$$\psi = \frac{k_n}{t} \tag{8.44}$$

Combining with Darcy's Law Equation 7.8 gives

$$\psi = \frac{(Q/A)}{\Delta h} \tag{8.45}$$

where:

ψ = permittivity of the geotextile
k_n = hydraulic conductivity for flows normal to the fabric face
t = thickness of the geotextile
Q = flow rate normal to fabric
A = area of fabric
Δh = head loss through fabric

The minimum required permittivity may be determined by assigning an allowable head loss and using Equation 8.45. Measured permittivity values and other physical properties for various geotextiles from one manufacturer are presented in Table 8.2.

TABLE 8.2 Physical Properties of Geosynthetic Fabrics Manufactured by Evergreen Technologies, Inc.[a]

Product Designation	Thickness (mils)	Puncture Resistance (lb)	Apparent Opening Size (AOS) O_{95} (mm)	Apparent Opening Size (AOS) Sieve Size	Permittivity (s^{-1})
TG 1000	155	170	0.150	100	0.6
TG 800	120	145	0.177	80	1.0
TG 750	100	115	0.177	80	1.1
TG 700	85	100	0.210	70	1.3
TG 650	75	95	0.210	70	1.6
TG 600	65	80	0.250	60	1.8
TG 550	55	70	0.300	50	1.9
TG 500	45	60	0.300	50	2.3

Evergreen is one of the many manufacturers. Information on other products may be found in IFAI (1997) or by contacting the manufacturers.

[a]Courtesy of Evergreen Technologies, Inc., Atlanta, GA.

Example 8.10

A sandy silt with $D_{85} = 0.10$ mm is to be drained by a perforated drainage pipe surrounded by a 3/4-in. gravel and a filter fabric as shown in Figure 8.37. The estimated flow rate to this drain will be 5 gal/min per lineal foot (half of which is from each side), and the head loss through the fabric must not exceed 0.1 ft. Select an appropriate filter fabric from Table 8.2.

Solution:

Check maximum allowable AOS.

$$O_{95} < (2 \text{ or } 3)(D_{85})_{\text{soil}}$$
$$< (2 \text{ or } 3)(0.10 \text{ mm})$$
$$< 0.2 - 0.3 \text{ mm}$$

Check minimum permittivity requirement.

$$Q = (5 \text{ gal/min})\left(\frac{1 \text{ ft}^3/\text{s}}{449 \text{ gal/min}}\right) = 0.01 \text{ ft}^3/\text{s}$$

$$A = (0.5 \text{ ft} + 1.0 \text{ ft} + 0.5 \text{ ft})(1.0 \text{ ft}) = 2.0 \text{ ft}^2$$

$$\frac{Q}{A} = \frac{0.01 \text{ ft}^3/\text{s}}{2.0 \text{ ft}^2} = 0.005 \text{ ft/s}$$

$$\text{minimum required } \psi = \frac{Q/A}{\Delta h} = \frac{0.005 \text{ ft/s}}{0.1 \text{ ft}} = \mathbf{0.05 \ s^{-1}}$$

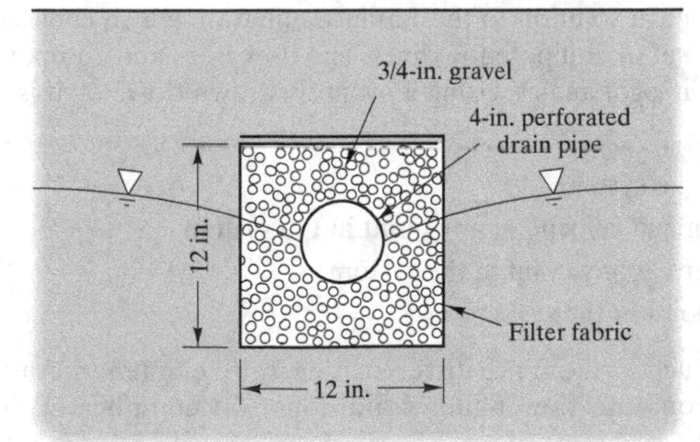

FIGURE 8.37 Proposed drain for Example 8.10.

Conclusion

Fabrics TG 750–TG 1000 definitely satisfy the AOS requirement, and the other fabrics marginally satisfy it. All of the fabrics satisfy the permittivity requirement by a wide margin. The costs of each are not given, so select the TG 1000 fabric, because it has the highest puncture strength, while still meeting the hydraulic requirements. However, if this fabric is too expensive or unavailable, the TG 800 or TG 750 also would be acceptable.

Occasionally, geotextile drains may be subject to soil clogging, which blocks the pores and prevents or severely restricts the flow of water. Although this is usually not a problem, the following conditions have been found to cause clogging (Koerner, 1998):

- Cohesionless sands and silts
- Gap-graded soils and
- High hydraulic gradients.

When all three of these conditions are present, it is better to use graded soil filters.

Clogging also can occur in soils where chemical precipitates tend to build up on the fabric, or when the groundwater contains a large concentration of biological matter (i.e., municipal landfill leachate).

SUMMARY

Major Points

1. Two- and three-dimensional flow is governed by Darcy's Law, but its solution is more complex and requires use of the LaPlace Equation.
2. The flow net is a graphical solution to the LaPlace equation which consists of a set of flow lines and a set of equipotential lines. The flow lines and equipotential lines always intersect at right angles. Using a properly drawn flow net, it is possible to compute:
   (a) Total flow rate in the system
   (b) Apparent and seepage velocity at any point in the system
   (c) Pore water pressure at any point in the system
   (d) Gradient at any point in the system.
3. Numerical methods such as the finite difference or finite element method are commonly used to compute flow under conditions too complicated to use flow nets.

4. Structures that extend below the groundwater table are subject to hydrostatic uplift pressures. These pressures may be computed using either flow nets or numerical solutions. It is important to compute uplift pressures on such structures in order to evaluate their stability.
5. Groundwater flow to a single well is a two-dimensional radially axisymmetric flow problem with simple closed form solutions.
6. Data collected from a pumped well along with two or more observation wells may be used to back-calculate the in situ hydraulic conductivity. Such measurements are generally more precise than laboratory permeability tests because they encompass a much larger volume of soil and they test the soil in situ avoiding the problem of sample disturbance.
7. Groundwater flow to an array of several wells can often be approximated as a two-dimensional axisymmetric problem.
8. True three-dimensional flow is much more difficult to analyze than simplified two-dimensional flow, and generally requires numerical solutions.
9. Groundwater flow patterns often need to be controlled or modified. This is most often done to either (a) facilitate underground construction or (b) control the flow of contaminants in the groundwater. Flow nets, well equations, and numerical methods can all be used to design and evaluate groundwater control systems.
10. When groundwater flows from a fine-grained soil to a coarser soil, it can cause the finer particles to migrate into the coarser zone. This migration can cause problems, such as piping, and thus may need to be controlled, especially in dams, levees, and other critical facilities. Soil migration also can clog drains, and thus needs to be controlled for that reason as well.
11. There are two ways to control soil migration: through the use of graded soil filters, or by installing filter fabrics. Both methods must satisfy certain design criteria.

Vocabulary

axisymmetric flow
bioremediation
confined flow
cutoffs
dewatering
drawdown
equipotential drop
equipotential line
equivalent opening size
equivalent well radius
exclusion
filtration
flow line
flow net
flow tube
geosynthetic filter
graded soil filter
LaPlace equation
numerical solution
open pumping
piping failure
predrainage
pump-and-treat
soil migration
three-dimensional flow
two-dimensional flow
unconfined flow
uplift pressure

QUESTIONS AND PRACTICE PROBLEMS

Section 8.2 Flow Net Solutions for Two-Dimensional Flow

8.1 One of the factors in Equation 8.27 is N_F, yet when drawing a flow net we *assume* a value for this parameter. How can this formula produce correct results when one of the factors is assumed (i.e., would assuming a higher N_F produce a higher computed value of Q?)?

8.2 The flow net in Figure 8.38 is incorrect. Explain why.

8.3 Compute the total flow rate under the dam shown in Figure 8.2 under the following conditions: $\Delta h = 15$ m, the soil beneath the dam has a hydraulic conductivity of 3×10^{-3} cm/s, and the length of the dam perpendicular to the section is 80 m.

8.4 Redraw the cross-section in Figure 8.39 to a scale of 1:500 (1 cm = 5 m), then draw a flow net that describes the seepage below this 150-m long concrete dam. Finally, compute the flow rate through this soil, expressed in liters per second, and the pore water pressure at Point A, which is at elevation 122.0 m.

8.5 Using the flow net from Problem 8.4, develop a plot of seepage flow rate versus the water elevation in the reservoir. Consider reservoir elevations between 126.1 m and 135.0 m.

8.6 The earth dam shown in Figure 8.40 is to be built on a gravelly sand with silt and cobbles. This dam will extend a distance of 850 ft perpendicular to the cross-section. To reduce the flow rate through these soils, a concrete cutoff wall will be built as shown. Redraw this cross-section to a scale of 1 in. = 100 ft, draw a flow net, and compute Q. Then, identify the area in the flow net that has the greatest hydraulic gradient.

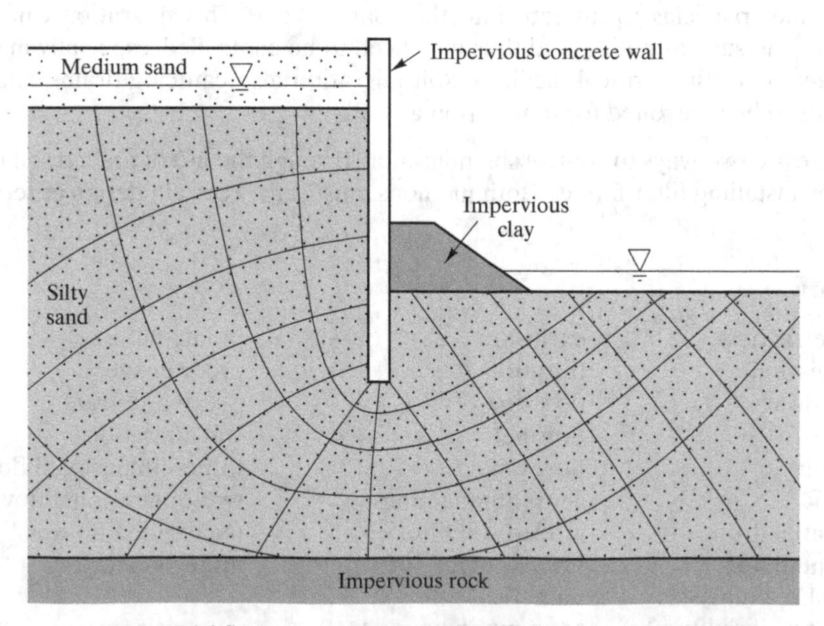

FIGURE 8.38 Trial flow net for Problem 8.2. Note: This flow net is not drawn correctly and should not be used as an example!

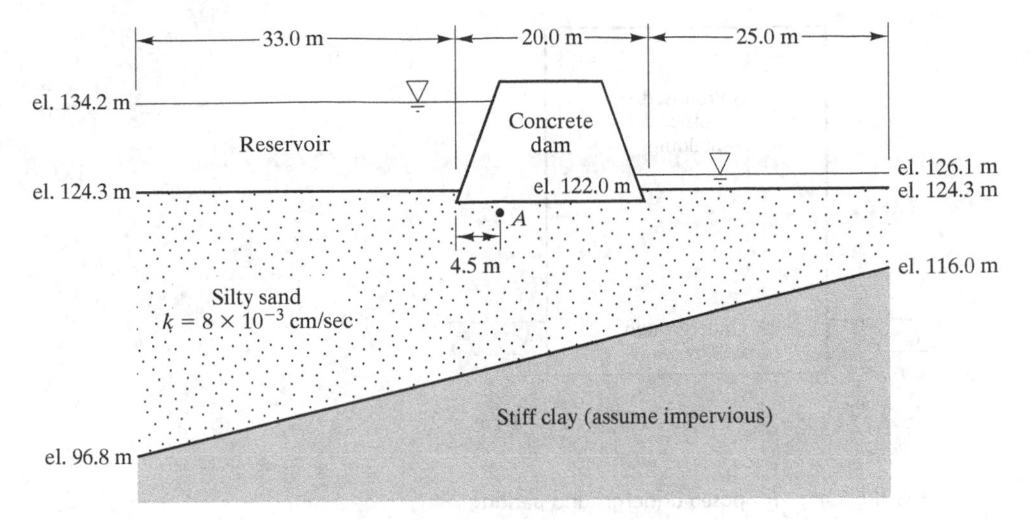

FIGURE 8.39 Cross-section of dam for Problems 8.4 and 8.5. el. = elevation.

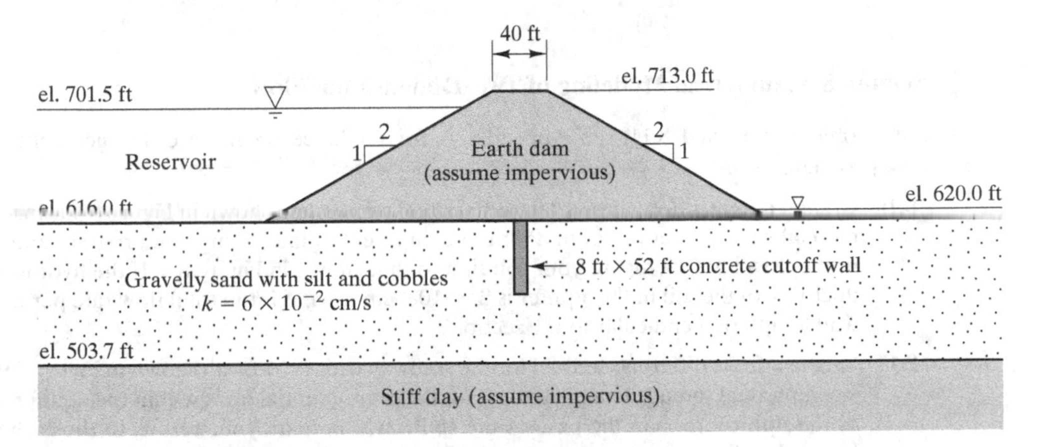

FIGURE 8.40 Cross-section for Problem 8.6. el. = elevation.

8.7 A proposed 20-story office building with three levels of underground parking will be supported on a concrete mat foundation, as shown in Figure 8.41. The bottom of this mat will be 40 ft below the street, and its plan dimensions will be 200 ft × 150 ft. The groundwater table is currently at a depth of 25 ft below the ground surface, but could rise to only 13 ft below the ground surface during the life of the building. Compute the total hydrostatic uplift force to be used in the design.

8.8 Compute the uplift forces acting on the dam in Figure 8.2 using the data in Problem 8.3. Draw a diagram of the uplift pressure acting on the dam and compute the total uplift force.

8.9 The sheet pile in Example 8.5 was located near the upstream end of the spillway. Would the total hydrostatic uplift force acting on the structure change if the sheet pile was located near the downstream end? Explain. Which position would be best? Why?

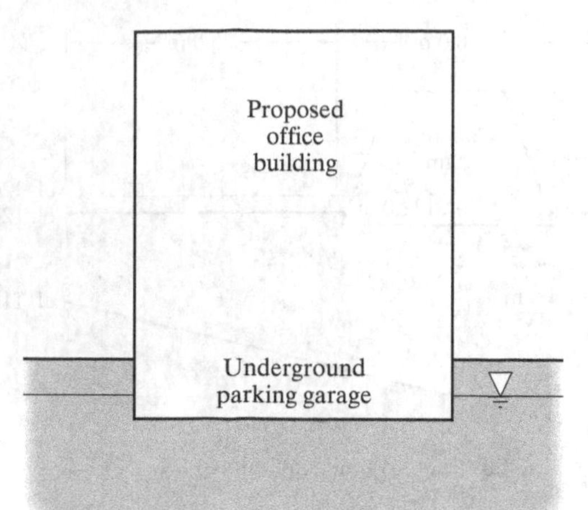

FIGURE 8.41 Proposed underground parking area for Problem 8.7.

Section 8.3 Numerical Modeling of Two-Dimensional Flow

For Problems 8.10 and 8.11, see Appendix A for guidance on finite difference solutions to flow problems.

8.10 Create a finite difference model for the sheet pile system shown in Figure 8.4 using a commercial spreadsheet program. Plot the equipotential lines using you program and sketch in the flow lines. Compare your solution to that shown in Figure 8.4. If the hydraulic conductivity of the soil in the aquifer is 2×10^{-4} cm/s, what is the total flow rate per meter of wall length perpendicular to the section?

8.11 Create a finite difference model for the spillway with cutoff wall shown in Figure 8.9 using a commercial spreadsheet program. Plot the equipotential lines within the aquifer as well as the uplift forces on the base of the spillway. Compare your results to those shown in Example 8.5.

Section 8.4 Two- and Three-Dimensional Flow to Wells

8.12 The proposed well shown in Figure 8.42 will be used to supply a municipal water system. Compute its pumping capacity with the groundwater level as shown, and express your answer in gallons per minute.

8.13 Using the well shown in Figure 8.42, if the drawdown is limited such that the level of water in the well is maintained at the top of the gravelly sand layer, what pumping rate is possible? Express your answer in gallons per minute.

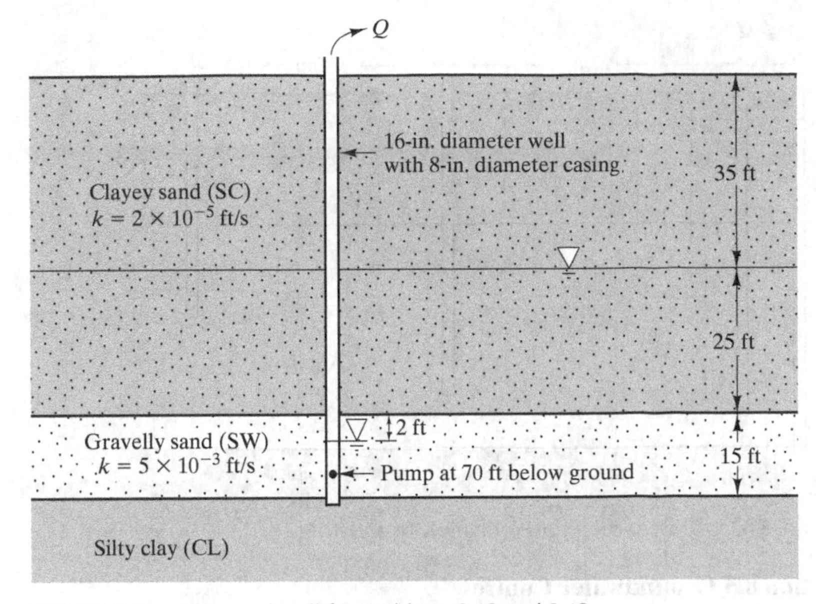

FIGURE 8.42 Proposed well for Problems 8.12 and 8.13.

8.14 After reaching steady-state conditions, the test well shown in Figure 8.43 is producing a flow rate of 17 l/s. The aquifer is an alluvial soil with interbedded medium-to-coarse sand and silty sand.

The water depths in the observation wells are as follows:

Well	Water Depth From Ground Surface (m)	
	Before Pumping	**During Pumping**
Pumping	16.9	26.0
Observation A	16.9	23.5
Observation B	16.9	18.1
Observation C	16.9	16.9

Using the best available data, compute the hydraulic conductivity of the soil in the aquifer. Is the computed k value reasonable? Explain why or why not.

8.15 For the well shown in Figure 8.43 assume the original depth from the ground surface to the phreatic surface is 16.9 m and the hydraulic conductivity of the soil in the aquifer is 10^{-2} cm/s What is the maximum pumping rate, Q, such that the distance from the ground surface to the phreatic surface at observation well B is no greater than 18 m? Assume $r_w = 0.06$ m.

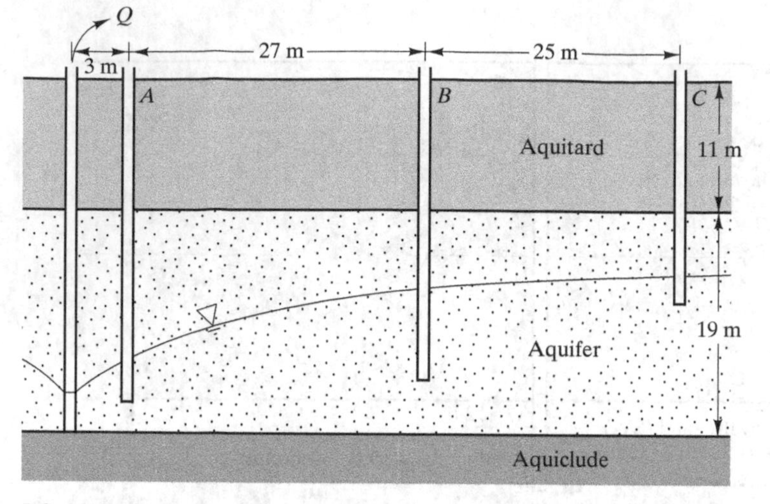

FIGURE 8.43 Cross-section for Problems 8.14 and 8.15.

Section 8.5 Groundwater Control

8.16 A construction site needs to be predrained in order to allow for the excavation shown in Figure 8.44. This will require dropping the groundwater table 5 m below its current location. From observations of the drawdown from other wells in the area, the radius of influence, r_0, is estimated to be 825 m. Compute the pumping rate required to lower the groundwater table using the well array shown.

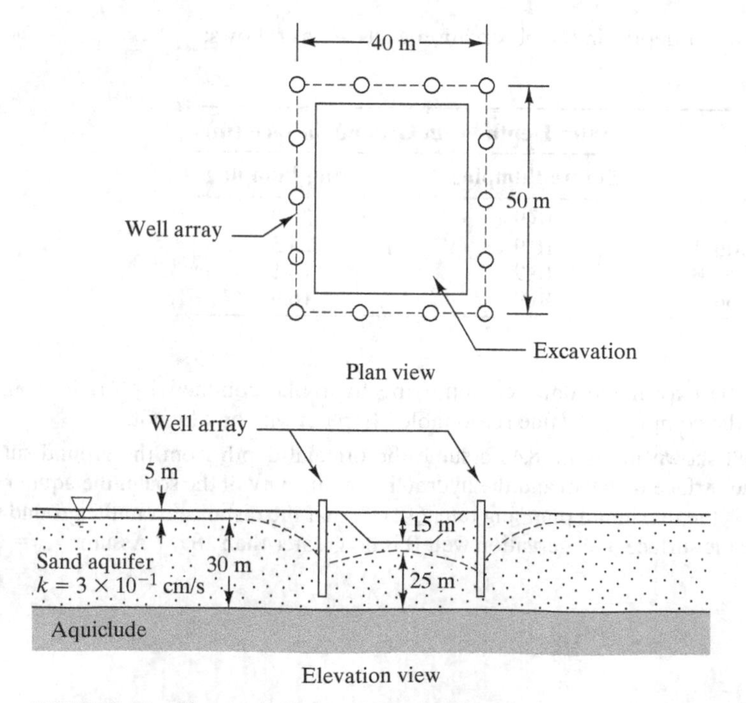

FIGURE 8.44 Plan and elevation drawing for Problem 8.16.

Section 8.6 Contaminant Control and Remediation

8.17 A contaminated soil site is to be capped and provided with a slurry wall barrier similar to the situation shown in Figure 8.29. A bioremediation process will be used to clean up the contaminated soil. This bioremediation process is expected to take 15 years. The slurry wall must be designed to prevent migration of the contaminants off of the site during the 15-year bioremediation process. If the regional hydraulic gradient is 0.0043 from left to right in Figure 8.29 and the clay in the slurry wall has a hydraulic conductivity (k) of 2×10^{-5} cm/s with an effective porosity (n_e) of 65%, how thick does the slurry wall have to be to contain the contaminants for 15 years?

Section 8.7 Soil migration and Filtration

8.18 A proposed levee is to be built using the soil described in Figure 8.45. This levee will include a toe drain similar to the one in Figure 8.34 to control the groundwater flow, and thus maintain adequate stability. This toe drain must be coarse enough to adequately collect and transmit the water, yet fine enough to provide sufficient filtration to prevent migration of the main levee soils. There will be no separate filter layer; the drain must act as the filter.

To maintain sufficient hydraulic conductivity, the drain must have no more than 3% passing the #200 sieve. In addition, to provide adequate filtration, it must meet the criteria described in this chapter. Determine the acceptable range of particle size distribution for this material and plot it on a particle size distribution curve.

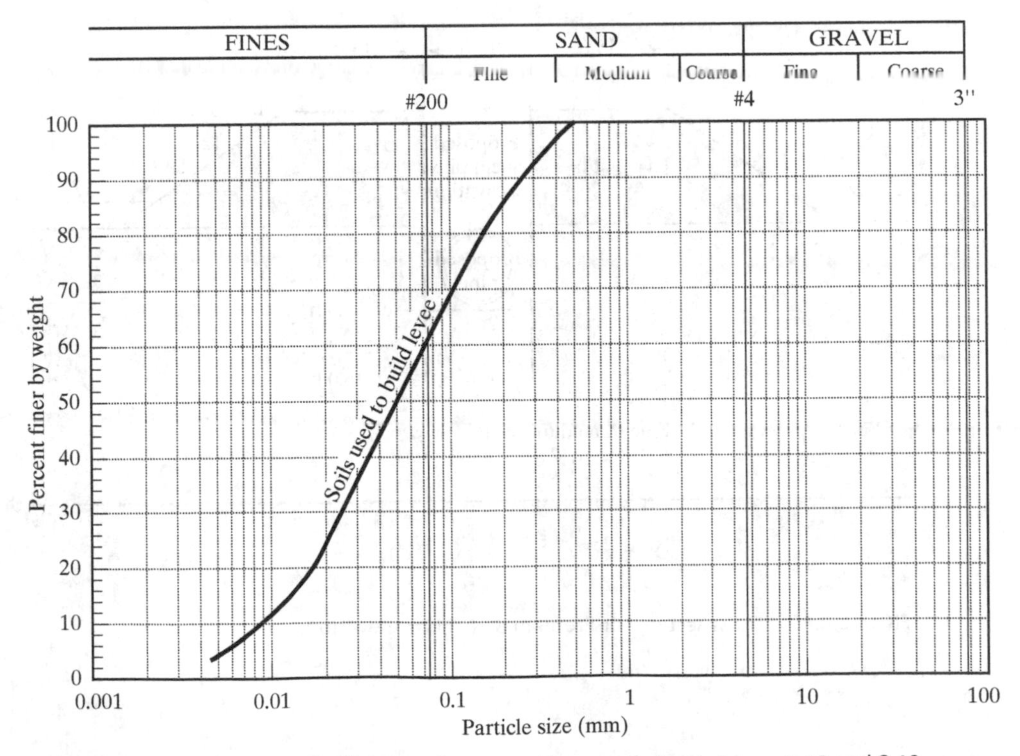

FIGURE 8.45 Particle size distribution of proposed levee soils for Problems 8.18 and 8.19.

8.19 An alternative design for the levee in Problem 8.18 uses a perforated pipe drain instead of the toe drain. This perforated pipe drain would be surrounded with gravel and wrapped with a filter fabric, similar to the one shown in Figure 8.37. The design flow rate into the drain is 80 gal/min per square foot, and the maximum acceptable head loss is 0.10 ft. Select an appropriate fabric from Table 8.2.

Comprehensive

8.20 Several years ago a state highway department built a highway across a shallow lake by placing the clayey fill shown in Figure 8.46. The top of this fill is above the high water level, thus protecting the highway from flooding.

It is now necessary to install a buried pipeline beneath the roadway. To install this pipe, the contractor plans to make a temporary excavation using steel sheet piles as shown. The contractor plans to use sump pumps at 50 ft intervals to maintain the water level at the bottom of the excavation. Once the pipe has been installed, the excavation will be backfilled and the pumps and sheet piles removed. You are to perform the following tasks in connection with this project:

(a) Recognizing that the proposed cross-section is symmetrical, redraw half of it to a scale of 1 in. = 10 ft and construct a flow net.

(b) Determine where the largest hydraulic gradient occurs and mark this spot on the cross-section.

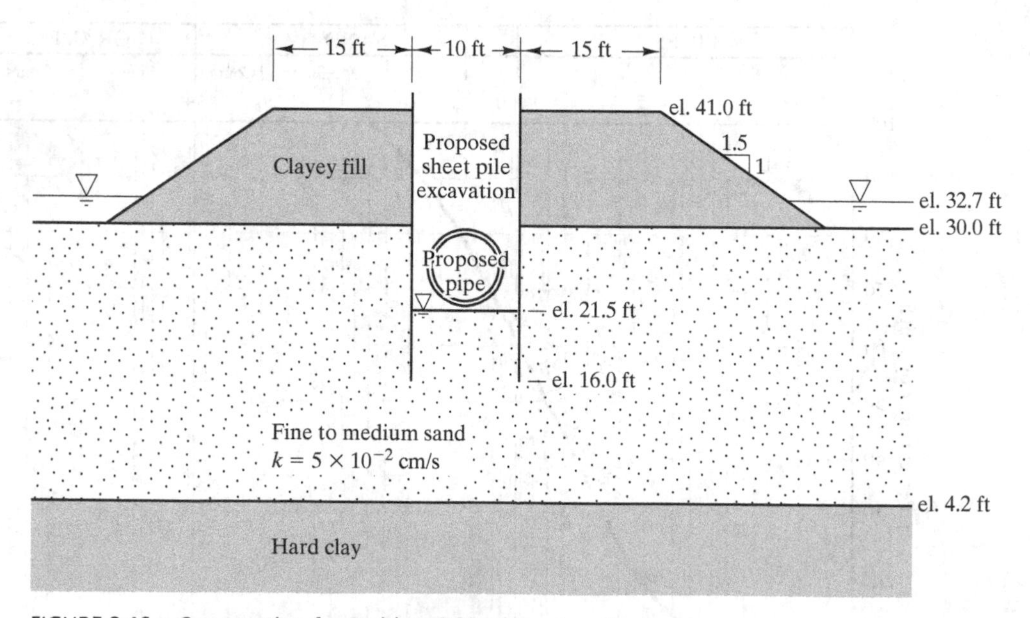

FIGURE 8.46 Cross-section for Problem 8.20. el. = elevation.

(c) Using the flow net, compute the minimum required capacity for each pump, expressed in gallons per minute.

(d) Describe two methods of reducing the flow rate into the excavation (and thus the required pump size). Explain how each of them would reduce Q.

8.21 Using the flow net from Problem 8.4, compute the uplift pressures acting on the bottom of the dam in Figure 8.39 and develop a plot similar to the one shown in Figure 8.9.

8.22 A 30-m wide, 40-m long, 8-m deep construction excavation needs to be made in a silty clay (CL). The groundwater table is at a depth of 2 m. The sides of the excavation will be sloped at an angle of about 1 horizontal to 1 vertical, and no sensitive structures or other improvements are nearby. Suggest an appropriate method of construction dewatering for this site, and explain the reason for your choice. Include statements of any assumptions, if any.

8.23 After the analysis described in Example 8.6 was completed, the well was installed to the depth indicated in Figure 8.18. However, when the pump was installed, it produced a flow rate of only 102 gal/min. Is the difference between this value and the computed flow rate within the normal range of uncertainty for these kinds of analyses? Explain. What portion of the analysis usually introduces the greatest error?

8.24 Compute Q for the cross-section in Figure 8.39 using the following hydraulic conductivities for the silty sand: $k_x = 5 \times 10^{-2}$ cm/s, $k_z = 4 \times 10^{-3}$ cm/s.

8.25 By observing the groundwater drawdown in the vicinity of a proposed well, an engineer had determined the radius of influence, r_0. This engineer then used Equation 8.34 to compute k for the aquifer. Write a 200–300 word memo to this engineer, explaining why this is not a good method of computing k, then suggest a better method.

8.26 A below-ground swimming pool was built 20 years ago and, until recently, has been performing satisfactorily. About 10 years ago it became necessary to temporarily drain the pool to clean out some algae. It was then refilled without any problems. However, the pool recently developed some cracks in its concrete shell, so it became necessary to drain it once again. Soon after it was drained, the pool rose out of the ground a distance of about 1 m. Provide a possible explanation for this behavior, and a possible explanation for why the pool did not rise out of the ground when it was drained the first time.

8.27 An excavation for a foundation is to be made in the upper clay soil as shown in Figure 8.47. In order to prevent the base of the excavation from being blown out by the uplift pressure of the confined aquifer below the clay layer, the construction contractor is planning to predrain the excavation using four wells as shown. The excavation is half a mile from a large lake that recharges the aquifer below the clay.

(a) Determine the elevation to which the groundwater table must be lowered to provide a factor of safety of 3 against uplift of the clay layer assuming the weight of the clay is the only resistance to uplift (i.e., ignore any strength in the clay).

(b) Assuming only three of the four wells are in operation, compute the pumping rate required to achieve the needed drawdown.

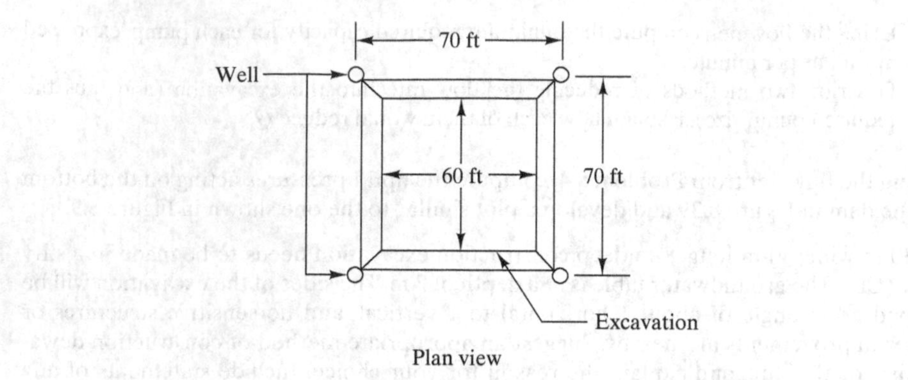

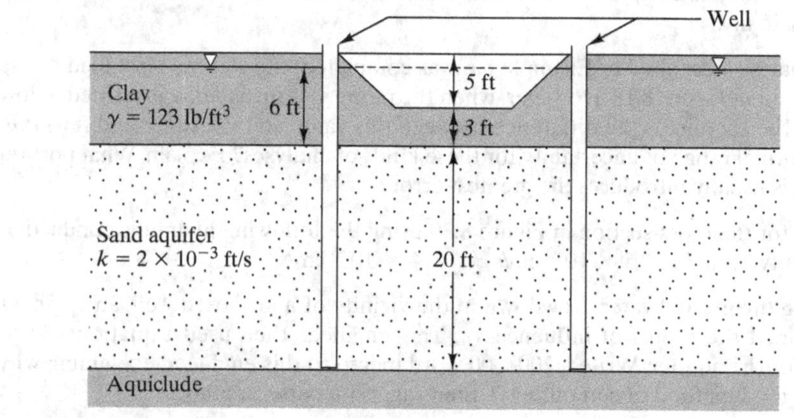

Elevation view

FIGURE 8.47 Plan and elevation drawing for Problem 8.27.

CHAPTER 9

Stress

In soil mechanics the accuracy of computed results never exceeds that of a crude estimate, and the principle function of theory consists of teaching us what and how to observe in the field.

Karl Terzaghi, 1936

Virtually all civil engineering projects impart loads onto the ground that supports them, and these loads produce compressive, shear, and possibly even tensile stresses. For example, when we construct a building, its weight is transmitted to the ground through the foundations, thus inducing stresses in the underlying strata. These stresses might cause problems, such as shear failure or excessive settlement, and thus are important to geotechnical engineers.

Additional stresses exist in the ground due to the weight of the overlying soil and rock. These stresses are also important and need to be considered in a wide range of geotechnical engineering problems. For example, the potential for slope stability problems, such as landslides, depends on the difference between these stresses and the strength of the soil or rock.

Many geotechnical problems depend on assessments of stresses in the ground, so this subject is worthy of careful study. Some of the methods we use to evaluate and describe these stresses are similar to those used with more conventional engineering materials, such as steel. However, because of the particulate nature of soils and the presence of water and/or air in the voids, we also need to introduce new concepts and methods.

This chapter discusses methods of analyzing stresses in the ground. Subsequent chapters, Chapters 10–17, will apply these methods to specific geotechnical problems.

9.1 SIMPLIFYING ASSUMPTIONS

Soil and rock are very complex materials that cannot be modeled mathematically without introducing certain simplifying assumptions. We must understand these assumptions before proceeding. For the purpose of computing stresses, we will assume that the soil or rock has the following characteristics:

1. It is a *continuous material*, not a particulate or discontinuous material, so stresses flow through soil and rock in much the same way as other more familiar materials such as steel. We assume there are no cracks or joints to disrupt the flow of stress, and that the stresses are uniform, not a microscopic patchwork of stressed particles with very small contact areas. This clearly is a simplification because the particulate nature of soil means the actual transfer of stresses through the solid particles in a soil is very complex. In reality, the stresses at the particle contact points are very high, while elsewhere they are much smaller. The stresses in rock also are complex due to the presence of joints and fissures. This assumption greatly simplifies the computations and is reasonable so long as the dimensions in our problem are large compared with the particle size and the joint spacing. Later in this chapter, our discussion of effective stress will modify this assumption to allow consideration of the distribution of stresses between the solid particles and the pore water.
2. It is *homogeneous*, that is, the relevant engineering properties are the same at all locations. In the context of stress analyses, we require only the modulus of elasticity, shear modulus, and Poisson's ratio (as defined later) to be constant. In other words, there are no "hard spots" and "soft spots," which can significantly alter the stress distribution. However, we will allow the unit weight to vary from place to place, and we will account for parts of the ground being above the groundwater table and parts below. This assumption virtually always introduces some error since few soils are so homogeneous. Later in this chapter, we will explore solutions that include two layers with different modulus values.
3. It is *isotropic*, that is, the engineering properties (in this context, the two moduli and Poisson's ratio) are the same in all directions. Many soils nearly meet this criterion, but some materials such as bedded sedimentary rock do not.
4. It has *linear elastic stress–strain properties*, that is, the stress is proportional to the strain (i.e., the stress–strain curve is a straight line and the material has not reached a yield point). This means a load applied at one point will induce some increment of stress, even a small one, everywhere in the ground, and that there will be a smooth pattern of stress distribution. This is quite different from a *plastic* material, which is one that has exceeded its yield point, and thus has reached its maximum stress-carrying capacity. The distribution of stresses in plastic materials is therefore quite different. This assumption is satisfactory for deformation analyses, where the strains are small, but may not be acceptable for certain strength analyses where the stresses are nearly equal to the strength.

These assumptions need to remain in effect only until we have computed the appropriate stresses in the ground. The follow-up analyses do not necessarily need to adhere to the same assumptions. For example, the settlement computations in Chapter 10 will be based on a nonlinear stress–strain relationship, and the strength analyses in Chapter 12 will consider the presence of cracks and joints.

9.2 MECHANICS OF MATERIALS REVIEW

Our simplifying assumptions have transformed soil and rock into something similar to standard engineering materials. Therefore, we will begin by reviewing principles of stress and strain that you learned in a mechanics of materials course and discussing how we will apply them to geotechnical problems.

When conducting stress analyses, we will continue to use the coordinate system introduced in Chapter 7, where the x and y axes are horizontal, and the z axis is vertical, as shown in Figure 9.1. Usually we evaluate stresses acting on a small element of soil or rock such as the one shown. This particular element is aligned with the axes, but other orientations also are possible.

Stress

Each face of the element is subjected to a *normal stress*, σ, which could be either *tension* or *compression*. Most engineers use the following sign convention to differentiate between the two:

$+$ = tension
$-$ = compression

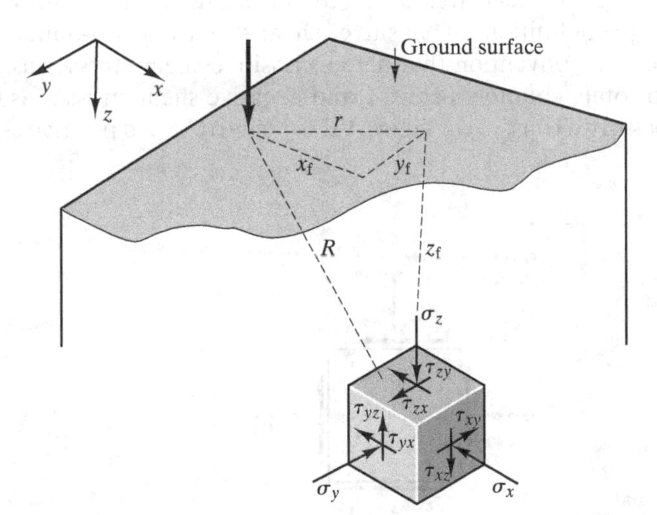

FIGURE 9.1 Element of soil in three-dimensional *x-y-z* space.

However, geotechnical engineers like to be different. We use the opposite sign convention:

+ = compression
− = tension

Although it can be confusing at first, there is a good reason for using this unorthodox sign convention: We almost always deal with compression. This is because of the nature of soil stress problems and because soil has a very low tensile strength. Thus, our sign convention avoids the continual use of negative stresses and eliminates the resulting mistakes that might occur.

When the element is aligned with the axes, as shown, the horizontal normal stresses (those acting on vertical planes) are σ_x and σ_y, and the vertical normal stress (which acts on a horizontal plane) is σ_z. Each face of the element is also subjected to a *shear stress*, τ, which we will divide into two perpendicular components as identified by two subscripts. For example, τ_{xz} is the component of shear stress in the x plane (i.e., the plane perpendicular to the x axis) acting in the z direction.

Although the normal stresses in the x, y, and z directions are independent of each other, the various shear stresses are not. To maintain static equilibrium, τ_{xz} and τ_{zx}, as shown in Figure 9.2, must be equal in magnitude and opposite in the sense of rotation they cause. The same relationship applies to the other pairs of shear stresses:

$$\tau_{xz} = -\tau_{zx} \tag{9.1}$$

$$\tau_{yz} = -\tau_{zy} \tag{9.2}$$

$$\tau_{xy} = -\tau_{yx} \tag{9.3}$$

Assigning the proper sign to shear stresses can be confusing, especially since engineers have proposed multiple definitions of "positive" shear stress. For two-dimensional problems, we will use the sign convention that defines positive shear stresses as those that cause the element to rotate counterclockwise, and negative shear stresses as those that cause it to rotate clockwise. Thus, τ_{zx} in Figure 9.2 is negative, τ_{xz} is positive.

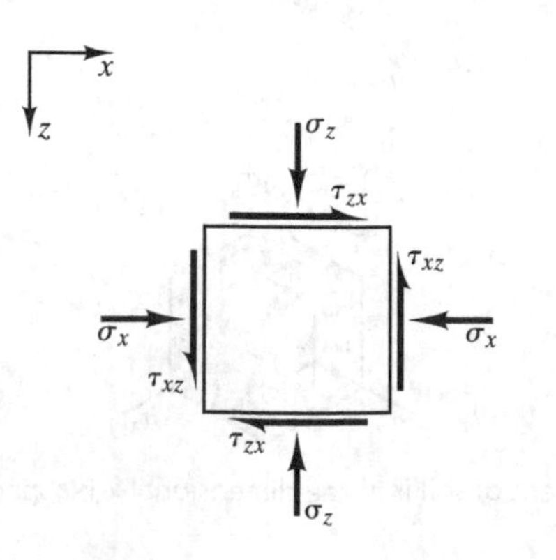

FIGURE 9.2 A two-dimensional soil element aligned with the x- and z-axes.

Because we are treating the ground as if it were a continuous material, not a particulate material, the area used to compute stresses is the total area of solids plus voids, not just the area of solids. For example, if a water tank has a total weight of 30,000 kN, and its base area is 300 m^2, then we say the vertical compressive stress in the soil immediately below the tank is 30,000 kN/300 m^2 = 100 kPa, even though the actual area of the soil solids is much less than 300 m^2. This means the real compressive stress within the soil particles is greater than 100 kPa, and the stresses at the particle contact points may be substantially greater. However, so long as we also define stress–strain properties and strengths in the same way, and the physical dimensions used in our analysis are large compared to the size of the individual soil particles, our computations will be essentially correct.

All stresses are expressed in units of force per area. When working with English measurements, geotechnical engineers normally use lb/ft^2 (or psf), except in some laboratory tests where lb/in.^{2} (or psi) is used (primarily because these are the units in pressure gages). With SI units, all soil stresses are expressed in kPa. Finally, geotechnical engineers in non-SI metric countries typically use kg$_f$/m^2, kg$_f$/cm^2, or bars (1 bar = 100 kPa).

Strain

When materials are subjected to a stress, they respond by deforming. Engineers quantify this deformation using the parameter *strain*. Normal stresses produce normal strains and shear stresses produce shear strains.

Normal strain, ε, is the change in length divided by the initial length, as shown in Figure 9.3:

$$\varepsilon = -\frac{dL}{L} \tag{9.4}$$

To be consistent with our sign convention for normal stresses, compressive strains are positive (i.e., the length becomes shorter, so $dL < 0$, which produces $\varepsilon > 0$).

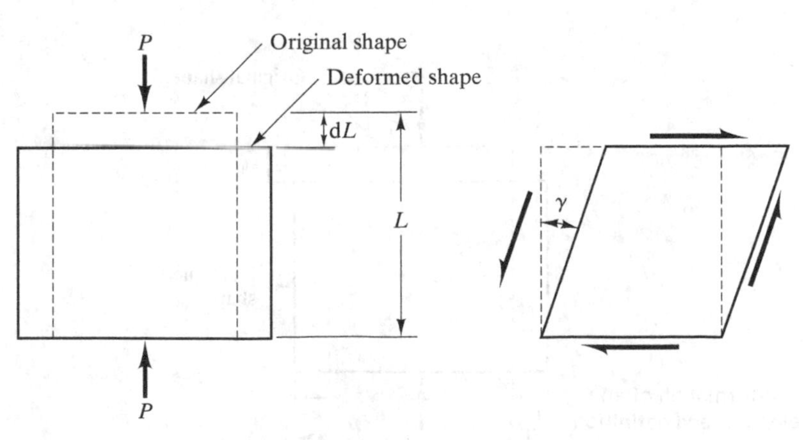

FIGURE 9.3 Definitions of normal and shear strain.

Shear strain, γ, is the angle of deformation (expressed in radians), shown in Figure 9.3.

Modulus of Elasticity, Shear Modulus, and Poisson's Ratio

To perform deformation analyses, we need to define the relationships between stress and strain. In linear elastic materials, these relationships are expressed using three parameters, as follows: the *modulus of elasticity*, E (also known as *Young's modulus*), the *shear modulus*, G (also known as the *modulus of rigidity*), and *Poisson's ratio*, ν. The modulus of elasticity is the ratio of normal stress to normal strain:

$$E = \frac{\sigma}{\varepsilon} \tag{9.5}$$

Thus, large values of E indicate a material that is very stiff and does not experience much deformation under an applied load, whereas low values indicate a soft material. The shear modulus has a similar definition:

$$G = \frac{\tau}{\gamma} \tag{9.6}$$

When a compressive load is applied to the element as shown in Figure 9.4, a compressive strain parallel to the stress, $\varepsilon_{\parallel}$, occurs. In addition, if the element is unconfined, a tensile strain perpendicular to the load, $\varepsilon_{\perp}$, also occurs. The ratio of these two strains is defined as *Poisson's ratio*, ν:

$$\nu = -\frac{\varepsilon_{\perp}}{\varepsilon_{\parallel}} \tag{9.7}$$

The magnitude of ν in elastic materials varies from 0 to 0.5. Those with $\nu = 0.5$ are said to be *incompressible* because the compression in the direction of the load is exactly matched by the expansion in the two perpendicular directions, resulting in no net volume change.

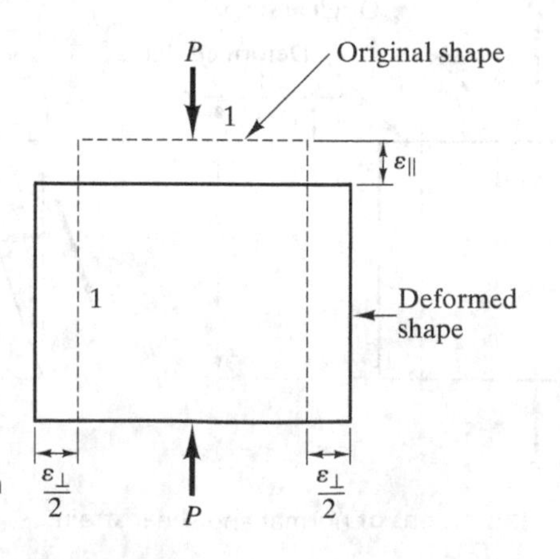

FIGURE 9.4 Deformation of an unconfined element and definition of Poisson's ratio.

TABLE 9.1 Typical Values of Poisson's Ratio for Soils and Rocks[a]

Soil or Rock Type	Poisson's Ratio, ν
Saturated soil, undrained condition	0.50
Partially saturated clay	0.30–0.40
Dense sand, drained condition	0.30–0.40
Loose sand, drained condition	0.10–0.30
Sandstone	0.25–0.30
Granite	0.23–0.27

[a]Adapted from Kulhawy, et al., 1983.

Although it is possible to measure ν of soil or rock in the laboratory, geotechnical engineers usually rely on tabulated values such as those in Table 9.1. These values are sufficiently precise for nearly all geotechnical analyses.

One-, Two-, and Three-Dimensional Analyses

Stresses propagate through soils in all three dimensions, so an analysis that keeps track of all the stresses identified in Figure 9.1 would be considered a *three-dimensional analysis*. Although such analyses are sometimes necessary, it is often appropriate to use more simplified methods. For example, a vertical cross-section through a long earth dam on a uniform soil deposit will be virtually constant along the entire length of the dam, as shown in Figure 9.5. Thus, we could reasonably evaluate such a problem using a *two-dimensional analysis* in a vertical plane oriented perpendicular to the dam axis.

Other scenarios can even be reduced to a *one-dimensional analysis*. For example, many problems require only the vertical stress, σ_z, due to the weight of the overlying ground. This requires only vertical dimensions and certain soil properties, making these one-dimensional analyses.

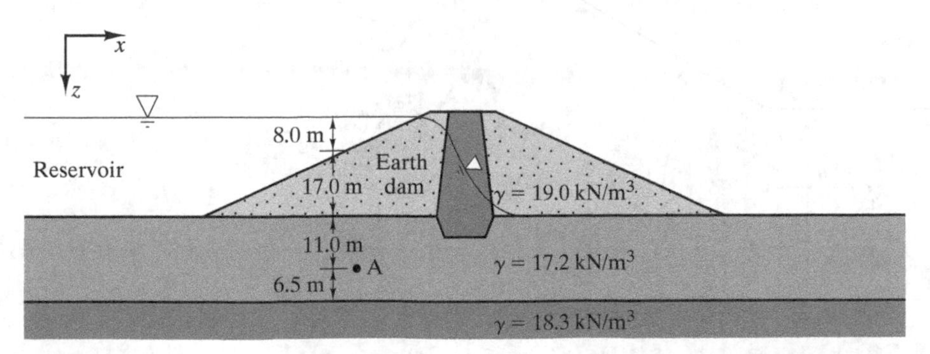

FIGURE 9.5 Use of a two-dimensional analysis to evaluate stresses in an earth dam.

9.3 MOHR CIRCLE ANALYSES

So far, we have considered only the stresses acting on vertical and horizontal planes. However, some problems may require computation of the stresses acting on other planes. Consider, for example, the Point A under slope shown in Figure 9.6. We might know the stresses acting in the x- and z-directions as shown in Figure 9.6(a), but, as we will learn in Chapter 13, we may need to know the stresses acting in the a- and b-directions as shown in Figure 9.6(b). Since the choice of coordinate system is arbitrary, then the stresses shown in Figure 9.6(a) must be equivalent to the stresses shown in Figure 9.6(b). The soil at Point A does not know what coordinate system we choose to use. It "feels" the same *state of stress* regardless of which coordinate system is used to define the normal and shear stresses. Therefore, there can only be one unique state of stress at Point A and it must be possible to represent that state of stress in any arbitrary

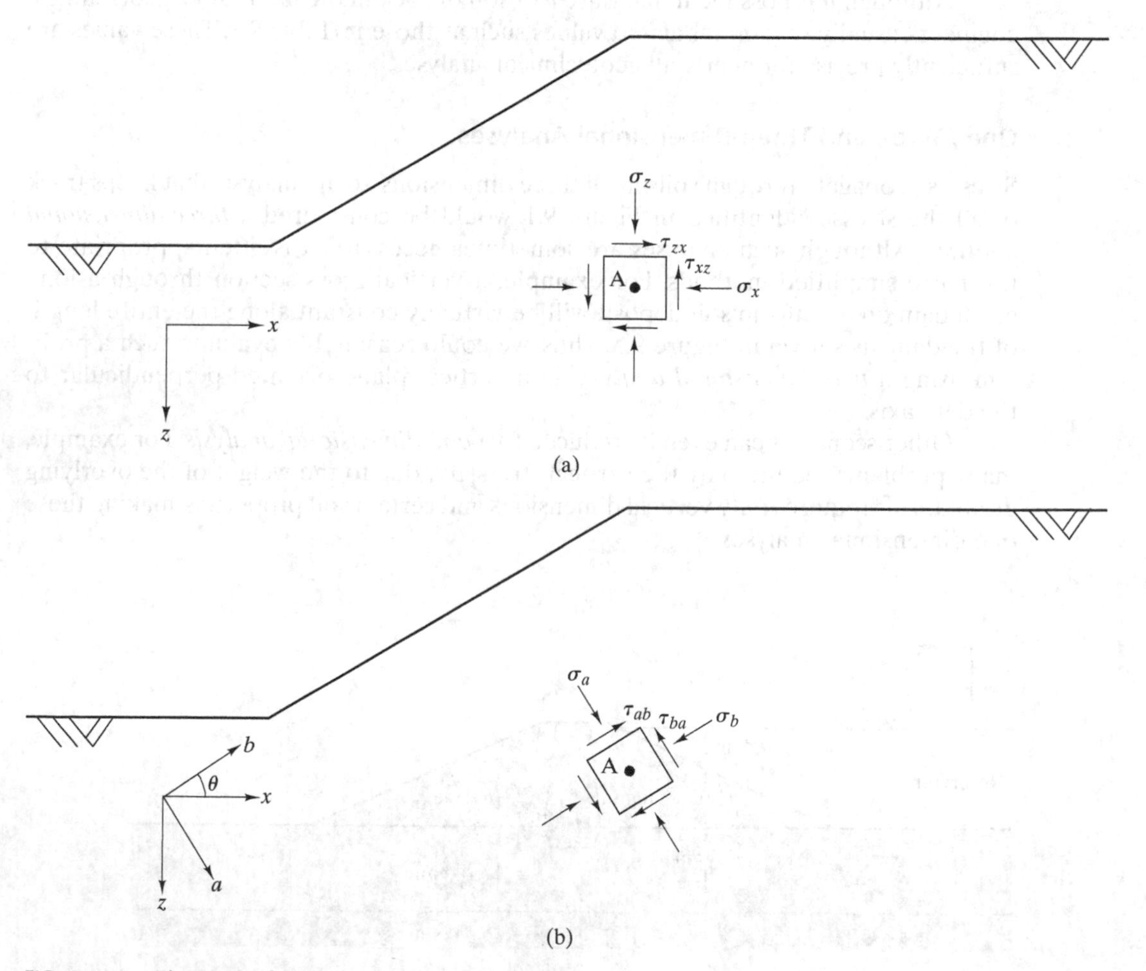

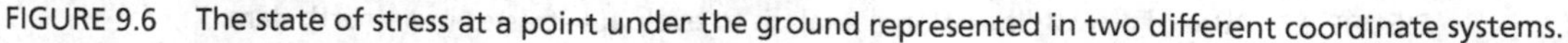

FIGURE 9.6 The state of stress at a point under the ground represented in two different coordinate systems.

coordinate system. It also must be possible to transform the shear and normal stresses from one coordinate system to another since they represent the same state of stress in either system. It is important to understand that the soil at any point under the ground is subjected to a certain state of stress; changing the coordinate system we use to compute shear and normal stresses does not change the state of stress at that point, it simply changes how we describe that state of stress. Put another way, at the Point A, the shear and normal stresses acting on any two orthogonal planes define the state of stress at this point. However, there are an infinite number of pairs of orthogonal planes that can pass through Point A, each with their own set of shear and normal stresses. Since the stresses acting on any two pairs of orthogonal planes must represent the same state of stress at Point A, there must be a way to transform stresses from one pair of orthogonal planes to any other pair of orthogonal planes.

We can use geometry and the rules of statics to derive formulas for transforming stresses from one coordinate system to another, or we can use a graphical representation called a *Mohr circle*, which was developed by the German engineer Otto Mohr (1835–1918). The graphical nature of the Mohr circle provides an intuitive feel for how stresses change from one coordinate system to another. We will use the Mohr circle with a particular method called the *pole method.*

A Mohr circle describes the two-dimensional state of stress at a point as a circle plotted on a σ versus τ diagram (sometimes called *a Mohr diagram*), as shown in Figure 9.7(a). In Figure 9.7(a), the location of the center and the radius of the circle represent the state of stress, and points on the circle represent the shear and normal stresses for axes systems rotated at different angles.

Mohr Circle Construction and the Pole Method

The equations defining a Mohr circle can be derived analytically. However, the concept can be understood using an illustration. Before creating a Mohr circle, we must first know the applied stresses in one coordinate system, that is, the shear and normal stresses on two orthogonal planes. For our illustration, we will use the soil element shown in Figure 9.2 and the x–y coordinate system in the same figure. We will assume we know the values of the stresses shown in Figure 9.2 and that σ_z is greater than σ_x. Starting from these known stresses, we will use the following steps to create our Mohr circle as shown in Figure 9.7(a).

- **Locate the center of the circle on the $\sigma-\tau$ plot.** The circle is always centered on the σ-axis at the point $(\sigma_m, 0)$, where σ_m is equal to the average of the normal stresses acting on the element, or

$$\sigma_m = \frac{\sigma_x + \sigma_z}{2} \tag{9.8}$$

- **Locate the points representing the known applied stresses and complete the circle.** The stresses acting on the z- and x-planes will be represented by two points lying on the circle. We can show theoretically that the two points will be directly opposite to one another. Referring to Figure 9.2, we note that τ_{zx} is negative

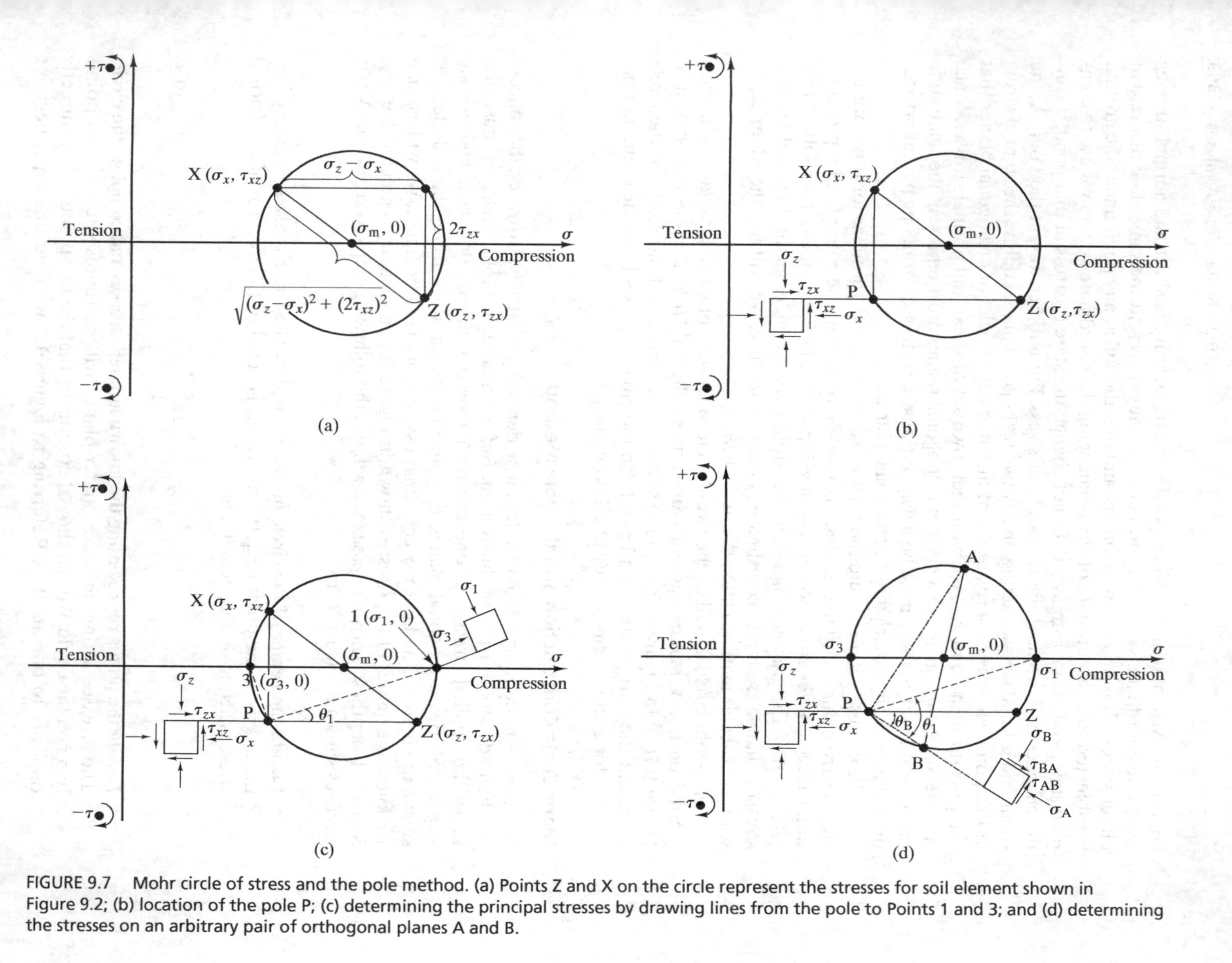

FIGURE 9.7 Mohr circle of stress and the pole method. (a) Points Z and X on the circle represent the stresses for soil element shown in Figure 9.2; (b) location of the pole P; (c) determining the principal stresses by drawing lines from the pole to Points 1 and 3; and (d) determining the stresses on an arbitrary pair of orthogonal planes A and B.

(rotates element clockwise) and τ_{xz} is positive (rotates element counterclockwise). Assuming that σ_z is greater than σ_x, we plot Point Z representing the stresses on the z-plane (the plane perpendicular to the the z-axis) and Point X representing stresses on the x-plane. Knowing the coordinates of Points Z and X, we can compute the radius of the circle, r, as

$$r = \frac{\sqrt{(\sigma_z - \sigma_x)^2 + (2\tau_{zx})^2}}{2} = \sqrt{\left(\frac{\sigma_z - \sigma_x}{2}\right)^2 + \tau_{zx}^2} \tag{9.9}$$

We have now created our Mohr circle and can locate the *pole*, or *origin of planes*.

- **Locate the pole.** The pole, also known as the origin of planes, is a special point on the Mohr circle that has the following property: A straight line drawn through the pole parallel to a given plane will intersect the Mohr circle at the point which represents the stresses acting on that plane. Since we know Point Z represents the stresses on the z-plane, we can locate the pole by drawing a horizontal line from Point Z to where it intersects the circle, Point P as shown in Figure 9.7(b). Similarly, if we draw a vertical line from Point X (parallel to the plane on which σ_x and τ_{xz} act), it also intersects the circle at Point P. Therefore, Point P is the pole. The location of the pole on a given Mohr circle is unique. Note also that the pole method only works if the sign conventions for compressive (positive for compression) and shear stresses (positive for counterclockwise rotation) given in Section 9.2 is followed.

We have now created our Mohr circle and located the pole. One advantage of the pole method is that we can draw the element on the figure in its proper orientation as shown in Figure 9.7(b). This makes it easy to ensure the stresses are shown in the proper directions, using the correct sign conventions.

Principal Stresses

If the soil element is rotated to a certain angle, the shear stresses will be zero on all four sides. The planes of this element are represented by Points 1 and 3 in Figure 9.7(c), and are known as *principal planes*. The stresses acting on them are known as *principal stresses*. The *major principal stress*, σ_1, also is the greatest normal stress that acts on any plane, whereas the *minor principal stress*, σ_3, is the smallest normal stress that acts on any plane. These two stresses act at right angles to each other. If we were conducting a three-dimensional analysis, there also would be an *intermediate principal stress*, σ_2, which acts at right angles to both σ_1 and σ_3. However, the vast majority of geotechnical analyses do not explicitly consider σ_2. From Figure 9.7(c), we can determine the magnitude of the principal stresses as

$$\sigma_1 = \frac{\sigma_x + \sigma_z}{2} + \sqrt{\left(\frac{\sigma_x - \sigma_z}{2}\right)^2 + \tau_{zx}^2} \tag{9.10}$$

$$\sigma_3 = \frac{\sigma_x + \sigma_z}{2} - \sqrt{\left(\frac{\sigma_x - \sigma_z}{2}\right)^2 + \tau_{zx}^2} \tag{9.11}$$

where:

σ_1 = major principal stress

σ_3 = minor principal stress

σ_x = horizontal stress

σ_z = vertical stress

τ_{zx} = shear stress acting on a horizontal plane

Using the pole we can determine the planes on which the principal stresses act by drawing lines from Point P through Points 1 and 3. This is illustrated by the dashed lines in Figure 9.7(c). Notice that we have drawn the element showing the correct orientation of the principal stresses by using the pole. The angle, θ_1, can be determined from Figure 9.7(c) as

$$\theta_1 = \frac{1}{2}\cos^{-1}\sqrt{\frac{1}{1 + [2\tau_{zx}/(\sigma_z - \sigma_x)]^2}} \tag{9.12}$$

Stresses on Other Planes

Once we have constructed our Mohr circle and found the pole, we can compute the stresses acting on any plane by simply drawing a line from the pole parallel to the plane on which we wish to know the stresses. The point where this line intersects the circle represents the stresses acting on that plane. Figure 9.7(d) illustrates this process. The Points A and B represent the stresses on two orthogonal planes with plane B rotated an angle of θ_B from the horizontal. Notice that we can find the angle between plane B and the plane of the major principal stress, θ_1, from the same figure. We can use this technique to find the stresses on any arbitrary plane. From Figure 9.7(d), we can show that the stresses acting on any pair of orthogonal planes are as given below:

$$\sigma = \frac{\sigma_1 + \sigma_3}{2} \pm \frac{\sigma_1 - \sigma_3}{2}\cos 2\theta \tag{9.13}$$

$$\tau = \frac{\sigma_1 - \sigma_3}{2}\sin 2\theta \tag{9.14}$$

where:

σ = normal stress acting on the plane of interest

σ_1 = major principal stress

σ_3 = minor principal stress

θ = the angle between the plane of interest and the major principal stress plane

τ = shear stress acting on the plane of interest

The "±" symbol will generate the two normal stress, σ_A and σ_B. The magnitude of the shear stress, τ, will be the same on both planes. We can use the Mohr circle to determine the direction of the shear stress. In Figure 9.7(d), for example, Point B is below the σ-axis, therefore τ_{BA} is negative (will rotate the element clockwise). Conversely, τ_{AB} is positive (will rotate the element counterclockwise).

The greatest shear stress, τ_{max}, occurs on the planes represented by points at the top and bottom of the Mohr circle. These planes are oriented at 45° angles from the principal planes, and the shear stress acting on them is equal to the radius of the Mohr circle:

$$\tau_{max} = \frac{\sigma_1 - \sigma_3}{2} \tag{9.15}$$

Since the Mohr circle is a graphical method, we could compute the stresses on any plane by simply drawing the Mohr circle to scale, locating the point on the Mohr circle corresponding to the plane using the pole method, and then scale the magnitudes of the stresses from this point on the circle. There is really no need to use the equations presented in this section. Before the advent of calculators and modern computers, this was the easiest way to determine stresses at a point. However, with the power and ease of computing equipment today, the best application of the Mohr circle is to use it to determine the directions of all the desired stresses (particularly the shear stresses) and then use the equations presented above to compute the magnitudes of the stresses. Thus, a simple sketch of the Mohr circle is generally sufficient and there is no longer any need to carefully draft it to an exact scale.

Example 9.1

The vertical and horizontal stresses at a certain point in a soil are as follows:

$$\sigma_x = 2100 \text{ lb/ft}^2$$
$$\sigma_z = 3000 \text{ lb/ft}^2$$
$$\tau_{zx} = 300 \text{ lb/ft}^2$$

a. Determine the magnitudes and directions of the major and minor principal stresses.

b. Determine the magnitude and directions of the maximum shear stress.

c. Determine the normal and shear stresses acting on a soil element inclined at 50° clockwise from the horizontal plane.

Solution:

a. The construction of the Mohr circle is shown in Figure 9.8(a). The center of the circle is located at

$$\sigma_m = \frac{\sigma_x + \sigma_z}{2} = \frac{2100 + 3000}{2} = 2550 \text{ lb/ft}^2$$

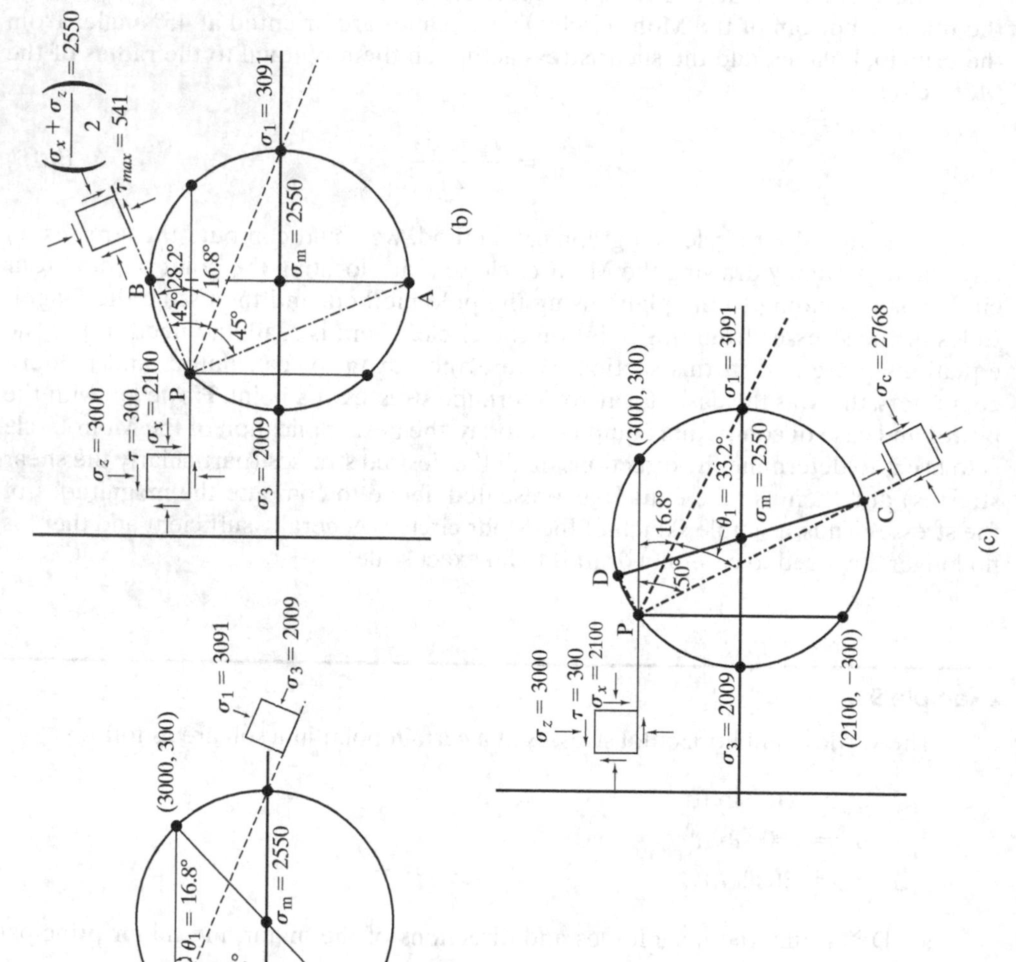

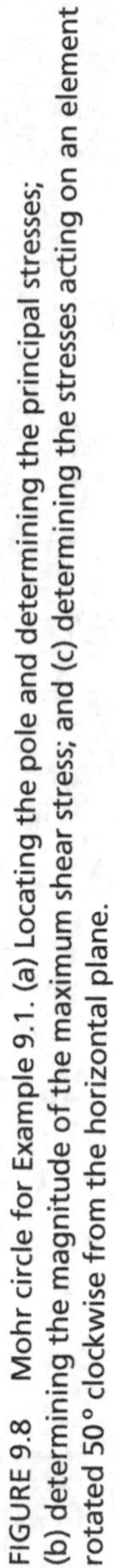
FIGURE 9.8 Mohr circle for Example 9.1. (a) Locating the pole and determining the principal stresses; (b) determining the magnitude of the maximum shear stress; and (c) determining the stresses acting on an element rotated 50° clockwise from the horizontal plane.

And the radius of the circle is given by

$$r = \sqrt{\left(\frac{\sigma_z - \sigma_x}{2}\right)^2 + \tau_{zx}^2}$$

$$= \sqrt{\left(\frac{3000 - 2100}{2}\right)^2 + 300^2} = 541 \text{ lb/ft}^2$$

We can then compute the principal stresses as

$$\sigma_1 = 2550 + 541 = \mathbf{3091\ lb/ft^2}$$

$$\sigma_3 = 2550 - 541 = \mathbf{2009\ lb/ft^2}$$

The directions of the principal stresses are shown in Figure 9.8(a) and the angle, θ_1, is given as

$$\theta_1 = \frac{1}{2}\cos^{-1}\sqrt{\frac{1}{1 + [2\tau_{zx}/(\sigma_z - \sigma_x)]^2}}$$

$$= \frac{1}{2}\cos^{-1}\sqrt{\frac{1}{1 + [2(300)/(3000 - 2100)]^2}} = \mathbf{16.8^\circ}$$

This equation does not give the sign for θ_1. However, based on Figure 9.8(a), we can see that major principal plane is oriented at an angle of 16.8° clockwise from the horizontal and the minor principal plane is oriented at an angle of (90 − 16.8) = 73.2° counterclockwise from the horizontal.

b. The maximum shear stress, τ_{max} occurs on the planes represented by the top and bottom of the Mohr circle shown as Points A and B in Figure 9.8(b):

$$\tau_{max} = \frac{\sigma_1 - \sigma_3}{2}$$

$$\tau_{max} = \frac{3091 - 2009}{2}$$

$$\tau_{max} = \mathbf{541\ lb/ft^2}$$

This stress acts on planes oriented at ±45° from the principal planes. These planes are 45 − 16.8 = **28.2° counterclockwise** and 45 + 16.8 = **61.8° clockwise** from the horizontal as shown in Figure 9.8(b).

c. The plane on which we wish to compute σ and τ is oriented at an angle of 50° clockwise from the horizontal. To locate this point, we draw a line from the pole at an angle of 50°. This line intersects the circle at Point C as shown in Figure 9.8(c). The plane normal to the plane represented by Point C is represented by Point D, which is directly across from Point C. The angle between the C plane and the major principal plane is

$$\theta_1 = 50 - 16.8 = 33.2^\circ$$

We now compute the normal and shear stresses at Points C and D:

$$\sigma_C = \frac{\sigma_1 + \sigma_3}{2} \pm \frac{\sigma_1 - \sigma_3}{2}\cos 2\theta_1$$

$$\sigma_C = \frac{3091 + 2009}{2} + \frac{3091 - 2009}{2}\cos(2 \cdot 33.2°) = \mathbf{2766\ lb/ft^2}$$

$$\sigma_D = \frac{3091 + 2009}{2} - \frac{3091 - 2009}{2}\cos(2 \cdot 33.2°) = \mathbf{2333\ lb/ft^2}$$

$$\tau = \frac{\sigma_1 - \sigma_3}{2}\sin 2\theta_1$$

$$\tau = \frac{3091 - 2009}{2}\sin(2 \cdot 33.2°) = \mathbf{496\ lb/ft^2}$$

The directions for the normal and shear stresses are shown in Figure 9.8(c). Note that the shear stress on the C plane is negative and the shear stress on the D plane is positive.

Stress Invariants

As discussed earlier, the Mohr circle represents the state of stress of a given soil element at a given point in the soil. Using the Mohr circle, we can rotate that element at any number of angles and compute the stresses acting on the element from a single Mohr circle. It follows then that there must be some quantities in common between the stresses computed on different orientations of a given soil element. These quantities that do not change while rotating the soil element at the same point are called *stress invariants* because they do not vary as we change the orientation of the soil element.

In two dimensions, there are two stress invariants. Using the Mohr circle, they are easy to identify. Mathematically, the stress invariants are (a) the point of the center of the circle and (b) the diameter of the circle. No matter which points on the circle we use to compute the stresses, the circle neither changes in location nor in diameter.

Conceptually these invariants have important meanings. The first invariant is called the *mean normal stress*, σ_m, or sometimes the *hydrostatic pressure*. The mean normal stress is a measure of the magnitude of the compressive stresses acting on a soil element. The greater the mean normal stress, the more the soil is being compressed or squeezed. In terms of principal stresses, it is computed as

$$\sigma_m = \frac{\sigma_1 + \sigma_3}{2} \tag{9.16}$$

which is the same as the normal stress at the center of the Mohr circle.

The second invariant is called the *deviator stress*, σ_d, or *deviatoric stress*. The deviator stress is a measure of the maximum shear stress acting on a soil element. The greater the deviator stress, the more shear the soil is subjected to. In terms of principal stresses, it is computed as

$$\sigma_d = \pm(\sigma_1 - \sigma_3) \tag{9.17}$$

which is equal to the diameter of the Mohr circle.

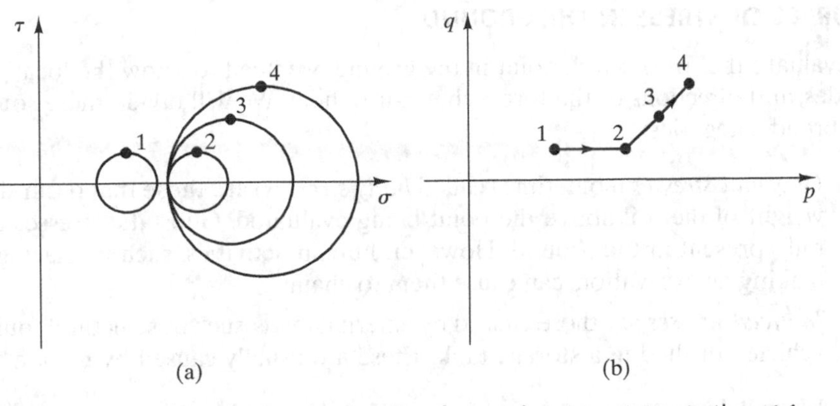

FIGURE 9.9 Stress paths showing a series of successive stress states starting at 1 and continuing through 4. Points 1 through 4 represent the same stress states in each diagram: (a) using Mohr circles and (b) using a *p*–*q* diagram.

By convention, the deviator stress is taken to be positive when the vertical stress exceeds the normal stress (when $\sigma_z > \sigma_x$) and negative when the horizontal stress exceeds the vertical stress (when $\sigma_z < \sigma_x$).

Stress Paths

The behavior of a soil is often dependent on the history of the stresses to which it has been subjected. For example, a soil that has been compressed and then unloaded will behave differently if it is recompressed than it did under the initial loading. We can evaluate the effects of loading history of a soil by keeping track of the changes in stress state with time through the use of *stress paths*. Figure 9.9 shows a stress path for a certain soil. As shown in Figure 9.9(a), the soil starts at stress state represented by the Mohr circle containing Point 1. The mean normal stress then increases from Point 1 to Point 2 without any increase in deviator stress. From Points 2 to 4, the major principal stress, σ_1, increases, whereas the minor principal stress, σ_3, remains constant.

As one can see from Figure 9.9(a), tracking the changes in stress states using Mohr circles can become confusing, particularly as the number of circles increases. Therefore, we often plot stress paths as *p*–*q* diagrams using the *p* and *q* variables shown in Figure 9.9(b); *p* and *q* are defined as follows:

$$p = \frac{\sigma_1 + \sigma_3}{2} \tag{9.18}$$

$$q = \frac{\sigma_1 - \sigma_3}{2} \tag{9.19}$$

Note that *p* is equal to the mean normal stress and *q* is equal to one-half of the deviator stress. Therefore, a stress path is the plot of the highest points of the Mohr circles representing successive states of stress of a soil element. A complete study of stress paths is beyond the scope of this book, but forms an important part of advanced studies in soil mechanics (see Holtz and Kovacs, 1981).

9.4 SOURCES OF STRESS IN THE GROUND

To evaluate the stresses at a point in the ground, we need to know the locations, magnitudes, and directions of the forces that cause them. We will divide these sources into two broad categories:

- *Geostatic stresses* (sometimes called *body stresses*) are those that occur due to the weight of the soil above the point being evaluated. Geostatic stresses are naturally present in the ground. However, human activities, such as placing a fill or making an excavation, can cause them to change.
- *Induced stresses* are those caused by external loads, such as structural foundations, vehicles, or fluid in a storage tank. These are usually caused by human activities.

We will discuss each category separately, then combine them using superposition. Our discussions will be limited to static stresses. Dynamic stresses, such as those produced by earthquakes, explosions, or machine vibrations, are beyond the scope of this text (see Dowding, 1996 and Kramer, 1996).

9.5 GEOSTATIC STRESSES

Geostatic stresses are caused by gravity acting on the soil or rock, so the direct result is a vertical normal stress, σ_z. This stress has a significant impact on the engineering behavior of soil, and is one we frequently need to compute. This vertical normal stress indirectly produces horizontal normal stresses and shear stresses, which also are important to geotechnical engineers.

Vertical Stresses

To compute the geostatic σ_z at a point below the ground surface, consider a column of soil that extends from the ground surface down to that point, as shown in Figure 9.10. This column intercepts soil strata with unit weights γ_1, γ_2, and γ_3, so its weight is:

$$W = \mathrm{d}x\mathrm{d}y \sum \gamma H \tag{9.20}$$

The geostatic vertical stress, σ_z at the bottom of the column is then

$$\sigma_z = \frac{W}{A} = \frac{\mathrm{d}x\mathrm{d}y \sum \gamma H}{A} = \sum \gamma H \tag{9.21}$$

where:

W = weight of the column

γ = unit weight of a soil stratum

H = thickness of a soil stratum

A = horizontal cross-sectional area of the column

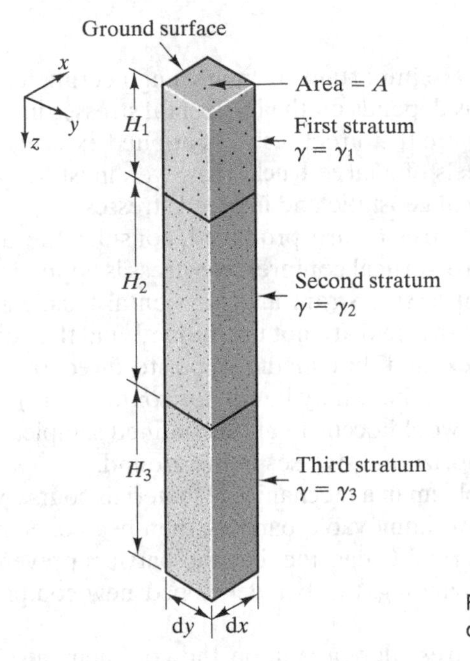

FIGURE 9.10 Imaginary column of soil to compute the geostatic σ_z.

Example 9.2

Compute σ_z at Point A in Figure 9.11.

Solution:

$$
\begin{aligned}
\sigma_z &= \sum \gamma H \\
&= (15.0\ \text{kN/m}^3)(2.0\ \text{m}) + (16.8\ \text{kN/m}^3)(2.5\ \text{m}) + (17.2\ \text{kN/m}^3)(3.6\ \text{m}) \\
&= \mathbf{134\ kPa}
\end{aligned}
$$

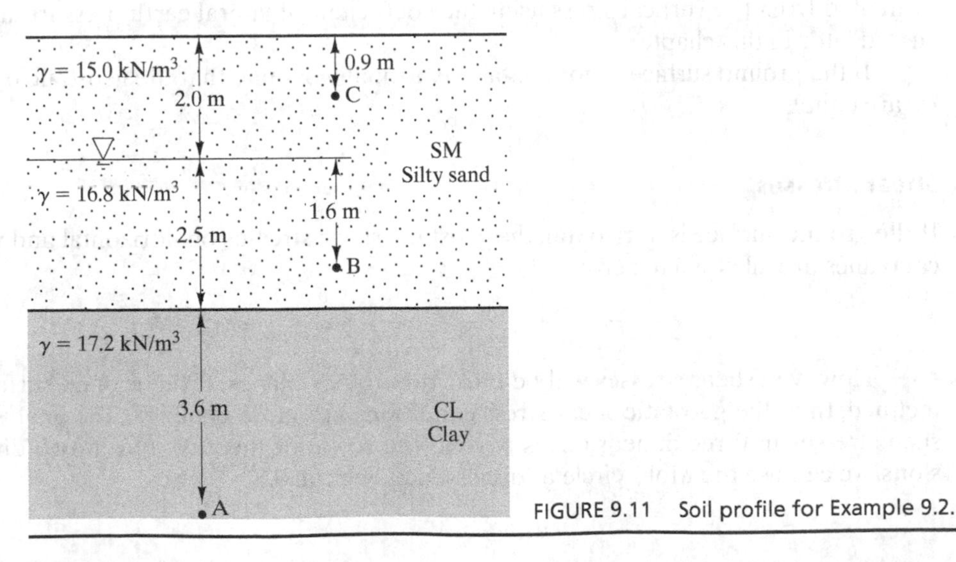

FIGURE 9.11 Soil profile for Example 9.2.

Horizontal Stresses

The horizontal stresses, σ_x and σ_y, also are important for many engineering analyses. For example, the design of retaining walls depends on the horizontal stresses in the soil being retained. Some horizontal stresses are the direct result of applied external loads, such as the braking forces from the wheels of a large truck. However, most horizontal stresses are indirectly produced by vertical geostatic and induced stresses.

To understand how these indirect stresses are produced, consider the unconfined element of soil in Figure 9.4. When a vertical compressive stress is applied to this element, it induces both a vertical compressive strain and horizontal tensile strains (per Equation 9.7). However, real soils in the field are not unconfined, and the adjacent elements of soil or rock also wish to expand, but in the opposite direction. These opposing forces may cancel each other (i.e., there may be no horizontal strain), or the horizontal strain may be much less than would occur in an unconfined sample. Either way, the result will be the formation of horizontal stresses in the ground.

You may recall solving a similar problem in a mechanics of materials course, where a metal bar was tightly fitted between two immovable barriers, then heated. Normally the bar would become longer due to thermal expansion, but the barriers prevented it from doing so. Therefore, instead of expanding, the bar developed new compressive stresses.

Because the horizontal geostatic stress depends upon the confinement of surrounding soil and rock, it is dependent upon many factors including soil stiffness, loading history, and regional geologic forces. For these reasons it is much more difficult to determine horizontal than vertical stress. We normally characterize the horizontal soil stress in relation to the vertical stress using the concept of the *lateral earth pressure coefficient.* This concept is presented in later in Section 9.8. Chapter 17 covers the theory of lateral earth pressure in more detail.

If analyses require accurate determinations of horizontal stress, field measurements are necessary. Horizontal stresses can be measured in situ using the pressuremeter test (PMT), the dilatometer test (DMT), or other methods. Alternatively, it can be estimated from the vertical stress using the coefficient of lateral earth pressure, as discussed later in this chapter.

If the ground surface is horizontal, we normally assume that the geostatic σ_x and σ_y are equal.

Shear Stresses

If the ground surface is horizontal, the geostatic shear stresses on horizontal and vertical planes are all equal to zero:

$$\tau_{xz} = \tau_{zx} = \tau_{yz} = \tau_{zy} = \tau_{xy} = \tau_{yx} = 0 \tag{9.22}$$

However, shear stresses will be present on other planes. If the ground surface is inclined, then the geostatic shear stress conditions are more complex. The analysis of such stresses in three dimensions is beyond the scope of this text, but in two dimensions, we can use the Mohr circle as discussed in Section 9.3.

9.6 INDUCED STRESSES

Civil engineering projects often introduce external loads onto the ground, thus producing induced stresses. These loads include structural foundations, vehicles, tanks, stockpiles, and many others. The resulting induced stresses are often significant, and can be the source of excessive settlement, shear failure, or other problems. To distinguish induced stresses from geostatic stresses and to emphasize that they are a change in the stress resulting from the external load, we will use the prefix Δ to identify induced stresses, for example, $\Delta\sigma_z$ represents an induced vertical normal stress.

Boussinesq's Method

The French mathematician Joseph Boussinesq (1842–1929) developed a method of computing induced stresses in an infinite elastic half-space due to an applied external load (Boussinesq, 1885). The term *infinite elastic half-space* means that the linear elastic material extends infinitely in all directions beneath a plane (which in our case is the ground surface). Boussinesq solved the problem where the *point load*, *P*, is perpendicular to this plane (in our case, this means the load is vertical), as shown in Figure 9.1. According to his solution, such a load will induce the following stresses at a point in the ground:

$$\Delta\sigma_x = \frac{P}{2\pi}\left[\frac{3x_f^2 z_f}{R^5} - (1 - 2\nu)\left(\frac{x_f^2 - y_f^2}{Rr^2(R + z_f)} + \frac{y_f^2 z_f}{R^3 r^2}\right)\right] \tag{9.23}$$

$$\Delta\sigma_y = \frac{P}{2\pi}\left[\frac{3y_f^2 z_f}{R^5} - (1 - 2\nu)\left(\frac{y_f^2 - x_f^2}{Rr^2(R + z_f)} + \frac{x_f^2 z_f}{R^3 r^2}\right)\right] \tag{9.24}$$

$$\Delta\sigma_z = \frac{3Pz_f^3}{2\pi R^5} \tag{9.25}$$

$$\Delta\tau_{zx} = -\Delta\tau_{xz} = \frac{3Pz_f^2 x_f}{2\pi R^5} \tag{9.26}$$

$$\Delta\tau_{yx} = -\Delta\tau_{xy} = \frac{P}{2\pi}\left[\frac{3x_f y_f z_f}{R^5} - (1 - 2\nu)\left(\frac{(2R + 2)x_f y_f)}{(R + z_f)^2 R^3}\right)\right] \tag{9.27}$$

$$\Delta\tau_{yz} = -\Delta\tau_{zy} = \frac{3Pz_f^2 y_f}{2\pi R^5} \tag{9.28}$$

$$R = \sqrt{x_f^2 + y_f^2 + z_f^2} \tag{9.29}$$

$$r = \sqrt{x_f^2 + y_f^2} \tag{9.30}$$

The parameters x_f, y_f, and z_f in Equations 9.23 through 9.30 are the distances from the load to the point, as shown in Figure 9.1. The values of x_f and y_f may be either positive or negative, but z_f is always positive because the point is always below the load. When performing two-dimensional analyses in the x–z plane, we only need Equations 9.23, 9.25, and 9.26.

Example 9.3

The dimensions in Figure 9.1 are $x_f = 10.0$ ft, $y_f = 0.0$ ft, and $z_f = 15.0$ ft. The load P is 132 k, and the soil is a partially saturated clay. Compute the induced $\Delta\sigma_x$, $\Delta\sigma_z$, and $\Delta\tau_{zx}$ in the soil element.

Solution:

Per Table 9.1, use $\nu = 0.35$.

$$R = \sqrt{x_f^2 + y_f^2 + z_f^2}$$

$$= \sqrt{10.0^2 + 0.0^2 + 15.0^2}$$

$$= 18 \text{ ft}$$

$$r = \sqrt{x_f^2 + y_f^2}$$

$$= \sqrt{10.0^2 + 0.0^2}$$

$$= 10 \text{ ft}$$

$$\Delta\sigma_x = \frac{P}{2\pi}\left[\frac{3x_f^2 z_f}{R^5} - (1 - 2\nu)\left(\frac{x_f^2 - y_f^2}{Rr^2(R + z_f)} + \frac{y_f^2 z_f}{R^3 r^2}\right)\right]$$

$$= \frac{132{,}000}{2\pi}\left[\frac{3(10.0)^2(15.0)}{18.0^5} - (1 - 2(0.35))\left(\frac{10.0^2 - 0.0^2}{18.0(10.0)^2(18.0 + 15.0)} + \frac{0.0^2(15.0)}{(18.0)^3(10.0)^2}\right)\right]$$

$$= \mathbf{39\ lb/ft^2}$$

$$\Delta\sigma_z = \frac{3Pz_f^3}{2\pi R^5}$$

$$= \frac{3(132{,}000)(15.0)^3}{2\pi(18.0)^5}$$

$$= \mathbf{113\ lb/ft^2}$$

$$\Delta\tau_{zx} = \frac{3Pz_f^2 x_f}{2\pi R^5}$$

$$= \frac{3(132{,}000)(15.0)^2(10.0)}{2\pi(18.0)^5}$$

$$= \mathbf{75\ lb/ft^2}$$

Application to Line Loads Although Boussinesq developed formulas only for point loads, others have extended them to other loading conditions. The most simple extension is to a *line load*, which is a vertical load distributed evenly along a horizontal line. We express such loads using the parameter P/b, where P is the vertical load and b is a unit length along the line (i.e., $P/b = 100$ kN/m).

If we consider a line load of infinite length oriented parallel to the y axis, and integrate Equation 9.25 over its length, we obtain

$$\Delta\sigma_z = \frac{2z_f^3 P/b}{\pi(x_f^2 + z_f^2)^2} \tag{9.31}$$

where x_f and z_f are the horizontal and vertical distances from line to the point at which $\Delta\sigma_z$ is to be computed.

Application to Area Loads The most common loading condition for geotechnical analyses is the *area load*, which is one distributed evenly across a horizontal area. Examples include spread footing foundations, tanks, wheel loads, stacked inventory in a warehouse, and small fills. We define the contact pressure between this load and the ground as the *bearing pressure*, q:

$$q = \frac{P}{A} \tag{9.32}$$

where:

q = bearing pressure
P = applied vertical load
A = area upon which the load acts

In the case of spread footing foundations, P must include both the column load and the weight of the foundation.

For area loads, we normally compute the induced stress in relation to the bearing pressure using the equation

$$\Delta\sigma_z = I_\sigma q \tag{9.33}$$

where:

$\Delta\sigma_z$ = the induced vertical stress
I_σ = the influence factor for vertical stress
q = bearing pressure of the surface load

The influence factor represents a decimal fraction of the applied bearing stress present at a point under the ground. The influence factor will take on different forms depending on the nature of the applied surface stress and the model used to compute the induced stress. The induced stresses beneath the loaded area can be computed using an extension of the Boussinesq equations.

Analytical Solutions Sometimes it is possible to integrate the Boussinesq equations over the area to produce new equations. Newmark (1935) used this method with Equation 9.25 to develop the following analytical solution for the vertical induced stress at a depth z_f beneath the corner of a loaded rectangle of width B and length L, as shown in Figure 9.12:

If $B^2 + L^2 + z_f^2 < B^2L^2/z_f^2$,

$$I_\sigma = \frac{1}{4\pi}\left[\left(\frac{2BLz_f\sqrt{B^2 + L^2 + z_f^2}}{z_f^2(B^2 + L^2 + z_f^2) + B^2L^2}\right)\left(\frac{B^2 + L^2 + 2z_f^2}{B^2 + L^2 + z_f^2}\right) + \pi - \sin^{-1}\frac{2BLz_f\sqrt{B^2 + L^2 + z_f^2}}{z_f^2(B^2 + L^2 + z_f^2) + B^2L^2}\right] \tag{9.34}$$

Otherwise,

$$I_\sigma = \frac{1}{4\pi}\left[\left(\frac{2BLz_f\sqrt{B^2 + L^2 + z_f^2}}{z_f^2(B^2 + L^2 + z_f^2) + B^2L^2}\right)\left(\frac{B^2 + L^2 + 2z_f^2}{B^2 + L^2 + z_f^2}\right) + \sin^{-1}\frac{2BLz_f\sqrt{B^2 + L^2 + z_f^2}}{z_f^2(B^2 + L^2 + z_f^2) + B^2L^2}\right] \tag{9.35}$$

where:

I_σ = influence factor for vertical stress at a point beneath the corner of the loaded rectangle
B = width of loaded rectangle
L = length of loaded rectangle
z_f = vertical distance from loaded rectangle to the point (always > 0)

Notes:

1. The $\sin^{-1}$ term must be expressed in radians.
2. Newmark's solution is often presented as a single equation with a $\tan^{-1}$ term, but that equation is incorrect when $B^2 + L^2 + z_f^2 < B^2L^2/z_f^2$.
3. It is customary to use B as the shorter dimension and L as the longer dimension.

Using the principle of superposition, as described later, and Equations 9.33–9.35, we can compute $\Delta\sigma_z$ at any point beneath a rectangular loaded area.

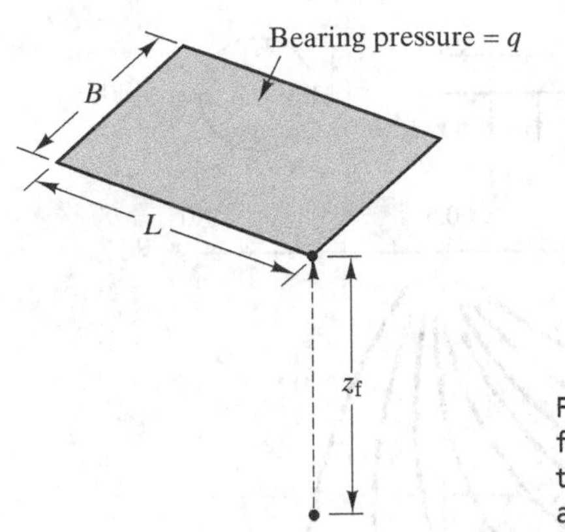

FIGURE 9.12 Newmark's solution for induced vertical stresses beneath the corner of a rectangular loaded area.

Numerical Solutions If the shape of the loaded area is too complex, it becomes necessary to use a numerical solution to compute the induced stresses. The term *numerical solution* (or *numerical method*) refers to a class of problem-solving methods that use a series of simplified equations assembled in a way that approximately models the actual system. For example, we could divide a loaded area into a large number of subareas, then replace the total load by point loads acting at the centers of the subareas and compute the induced stresses using the Boussinesq formulas and superposition (discussed later in this section). Many engineering disciplines use these methods to develop solutions to problems that otherwise would be very difficult or impossible. The *finite element method* and the *finite difference method* are examples of numerical solutions. We discussed the finite difference method and its application to seepage analyses in Chapter 8.

Chart Solutions Another option is to perform a series of computations using either analytical or numerical methods and express the results in nondimensional charts. We then can use these charts to compute stresses in the soil. Figures 9.13 through 9.16 are some of many such charts that have been developed. The curves in Figures 9.13 through 9.15, which connect points of equal induced stress, are sometimes called *pressure bulbs* or *stress bulbs*.

These charts are easy to use, but do not have the flexibility or computational accuracy of a properly implemented numerical solution. Other charts also have been developed for more complex loading conditions (see U.S. Navy, 1982; Poulos and Davis, 1974). However, the ready availability of computer-based numerical solutions has replaced the more complex chart solutions. In spite of the availability of numerical solutions, the charts provide a visual sense of how the stresses are distributed. For example, Figure 9.13 contains the solution for a circular loaded area. Similarly, Figures 9.14 and 9.15 contains the solutions for a square loaded area and and a semi-infinite strip load (such as a long embankment). Figures 9.14 and 9.15 clearly show that the stresses induced by a strip load exceed those induced by a square loaded area of the same width. Notice also that stresses induced by a strip load extend to a much greater depth than those induced by a square loaded area. Finally, Figure 9.16 presents influence factors for loads under the corner of a rectangular loaded area.

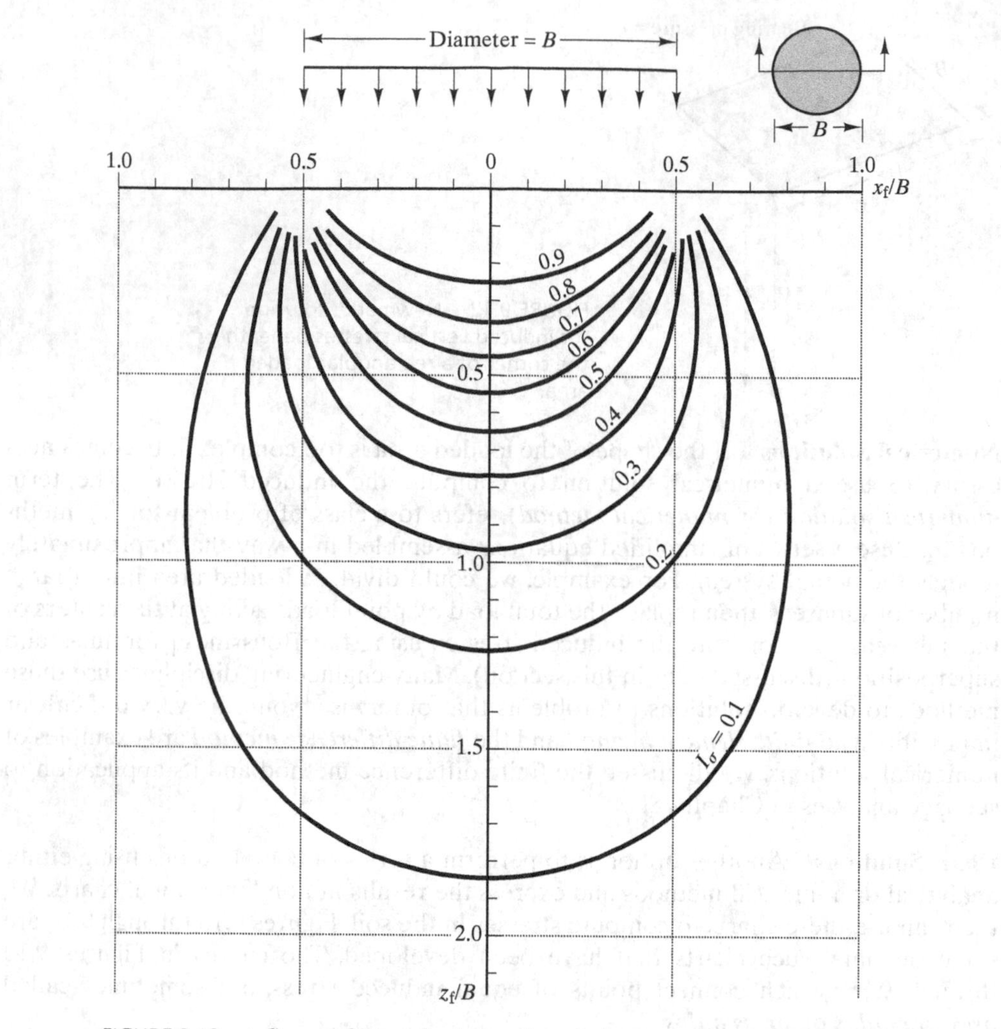

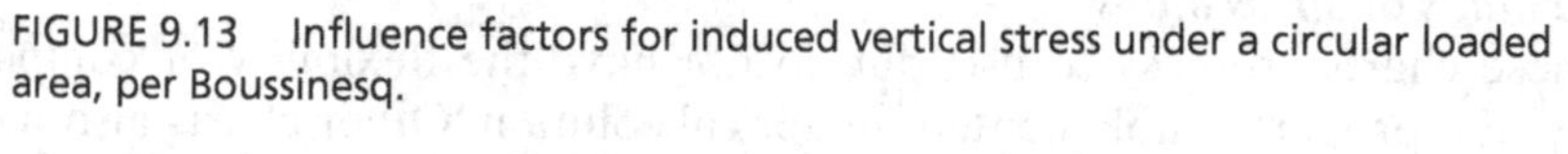

FIGURE 9.13 Influence factors for induced vertical stress under a circular loaded area, per Boussinesq.

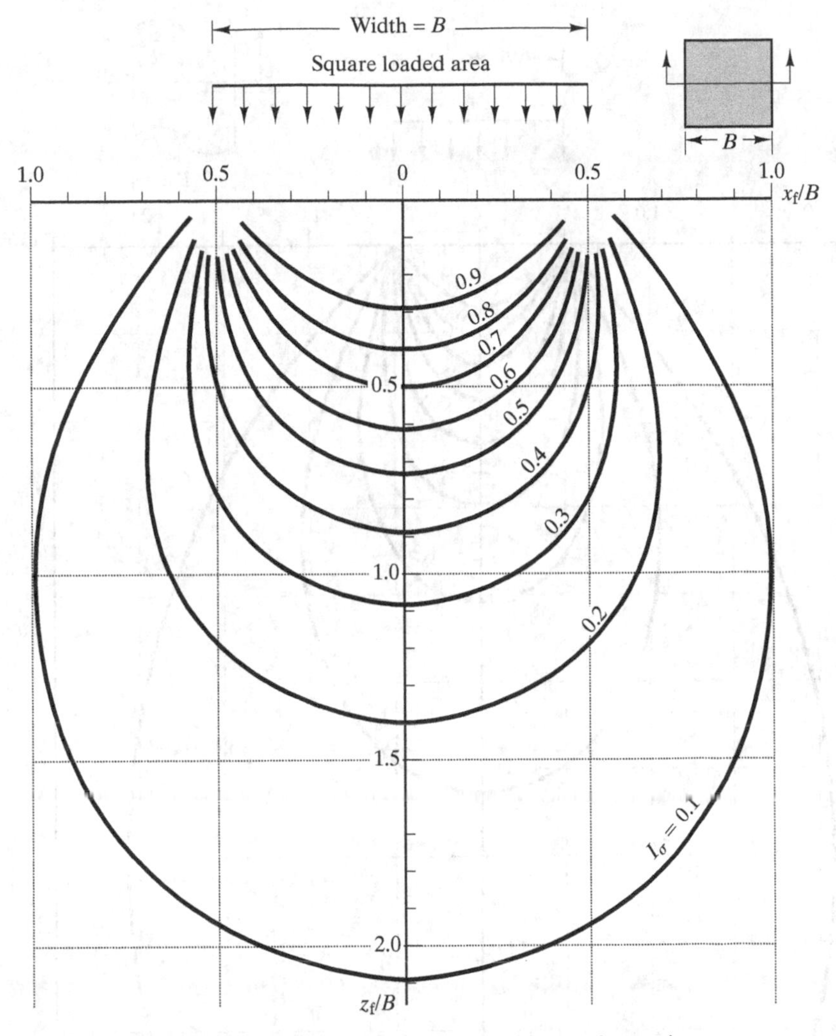

FIGURE 9.14 Influence factors for induced vertical stress beneath a square loaded area, per Boussinesq.

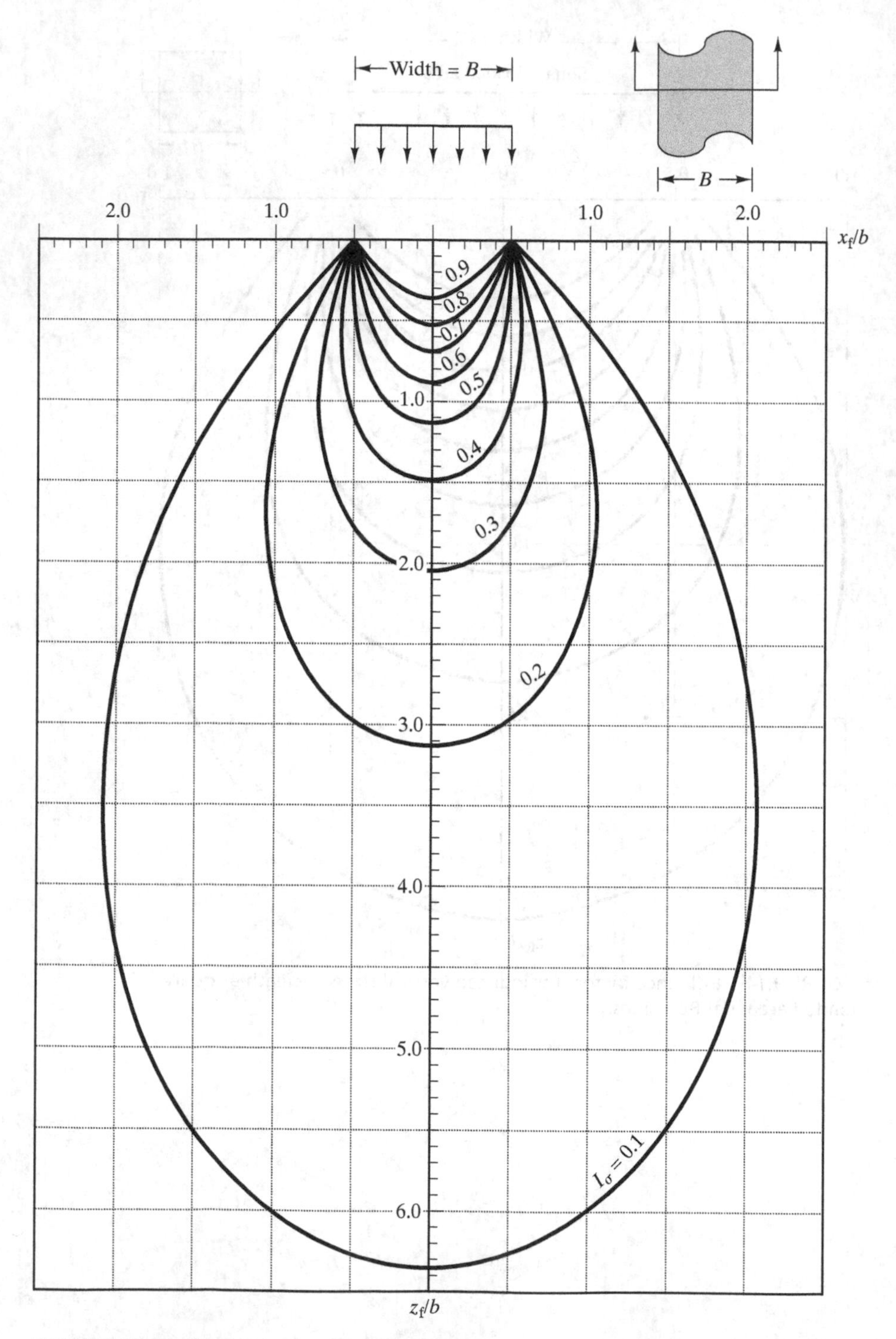

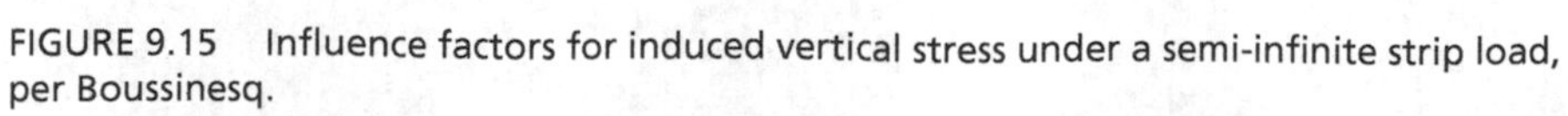
FIGURE 9.15 Influence factors for induced vertical stress under a semi-infinite strip load, per Boussinesq.

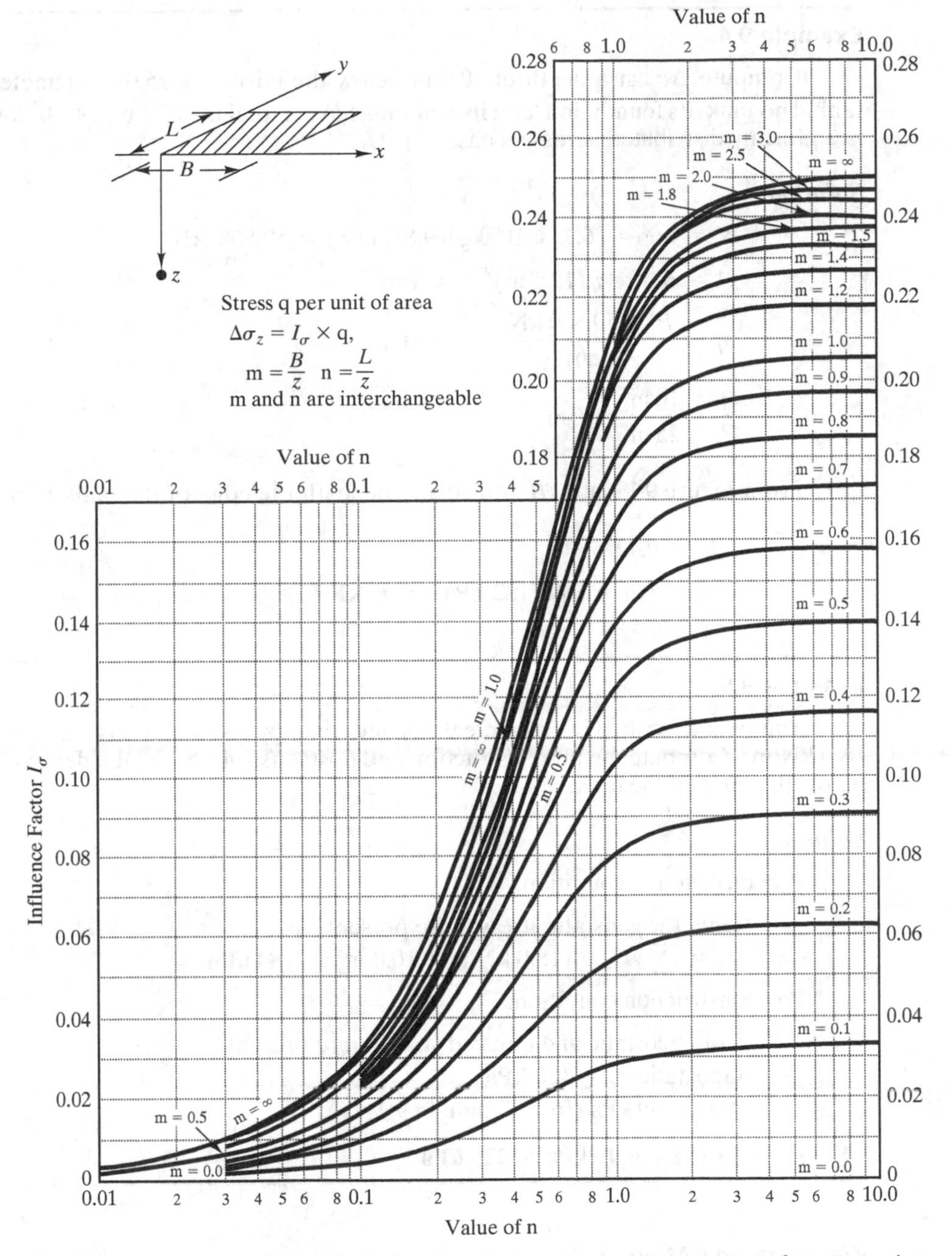

FIGURE 9.16 Influence factors for induced vertical stress beneath the corner of a rectangular loaded area. (U.S. Navy (1982).)

Example 9.4

Compute $\Delta\sigma_z$ at a depth of 10.0 m below the edge of a 25.0-m diameter water tank. The tank, its foundation, and its contents have a total mass of 6.1×10^6 kg, which is uniformly distributed across its base.

Solution:

$$W = mg = (6.1 \times 10^6 \text{ kg})(9.81 \text{ m/s}^2) = 59{,}800 \text{ kN}$$
$$A = \pi r^2 = \pi(12.5 \text{ m})^2 = 491 \text{ m}^2$$
$$q = \frac{P}{A} = \frac{59{,}800 \text{ kN}}{491 \text{ m}^2} = 122 \text{ kPa}$$
$$\frac{z_f}{B} = \frac{10 \text{ m}}{25 \text{ m}} = 0.40$$

From Figure 9.13, at $z_f/B = 0.40$ and beneath the edge of the tank, $x_f/B = 0.50$,

$$I_\sigma = 0.40$$
$$\Delta\sigma_z = I_\sigma q = 0.40(122 \text{ kPa}) = \mathbf{49\ kPa}$$

Example 9.5

The tank described in Example 9.4 is underlain by a soil that has a unit weight of 18.0 kN/m^3. Compute the preconstruction and postconstruction σ_z at a depth of 10.0 m below the edge of the tank.

Solution:

Preconstruction condition:

Only the geostatic stresses are present.

$$\sigma_z = \sum \gamma H = (18.0 \text{ kN/m}^3)(10.0 \text{ m}) = \mathbf{180\ kPa}$$

Postconstruction condition:

Both geostatic and induced stresses are present.

Geostatic $\sigma_z = 180$ kPa

$\Delta\sigma_z = 49$ kPa (from Example 9.4)

$$\sigma_z = 180 \text{ kPa} + 49 \text{ kPa} = \mathbf{229\ kPa}$$

Westergaard's Method

Westergaard (1938) solved the same problem Boussinesq addressed, but with slightly different assumptions. Instead of using a perfectly elastic material, he assumed one that contained closely spaced horizontal reinforcement members of infinitesimal thickness, such that the horizontal strain is zero at all points. This model may be a more precise representation of layered soils where some layers are much stiffer than others.

Terzaghi (1943) presented the following formula for σ_z due to a vertical point load, P, based on Westergaard's method:

$$\Delta\sigma_z = \frac{PC}{2\pi z_f^2}\left[\frac{1}{C^2 + (r/z_f)^2}\right]^{1.5} \tag{9.36}$$

$$C = \sqrt{\frac{1-2\nu}{2(1-\nu)}} \tag{9.37}$$

The Westergaard solution produces σ_z values equal to or less than the Boussinesq values. As ν increases, the computed stress becomes smaller, eventually reaching zero at $\nu = 0.5$. Although some geotechnical engineers prefer Westergaard, at least for certain soil profiles, Boussinesq is more commonly used and appropriate for most problems.

Approximate Methods

Sometimes it is useful to have simple approximate methods of computing stresses in soil. The widespread availability of computers has diminished the need for these methods, but they still are useful when a quick answer is needed, or when a computer is not available.

The following approximate formulas compute the influence factor for induced vertical stress, I_σ, beneath the center of an area load. They produce answers that are within 5% of the Boussinesq values, which is more than sufficient for virtually all practical problems.

For circular loaded areas (Poulos and Davis, 1974),

$$I_\sigma = 1 - \left(\frac{1}{1 + (B/2z_f)^2}\right)^{1.5} \tag{9.38}$$

For square loaded areas,

$$I_\sigma = 1 - \left(\frac{1}{1 + (B/2z_f)^2}\right)^{1.76} \tag{9.39}$$

For continuous loaded areas (also known as *strip loads*) of width B and infinite length,

$$I_\sigma = 1 - \left(\frac{1}{1 + (B/2z_f)^{1.38}}\right)^{2.60} \tag{9.40}$$

For rectangular loaded areas of width B and length L,

$$I_\sigma = 1 - \left(\frac{1}{1 + (B/2z_f)^{1.38+0.62B/L}}\right)^{2.60-0.84B/L} \tag{9.41}$$

where:

I_σ = influence factor for vertical stress beneath the center of a loaded area

z_f = depth from bottom of loaded area to point

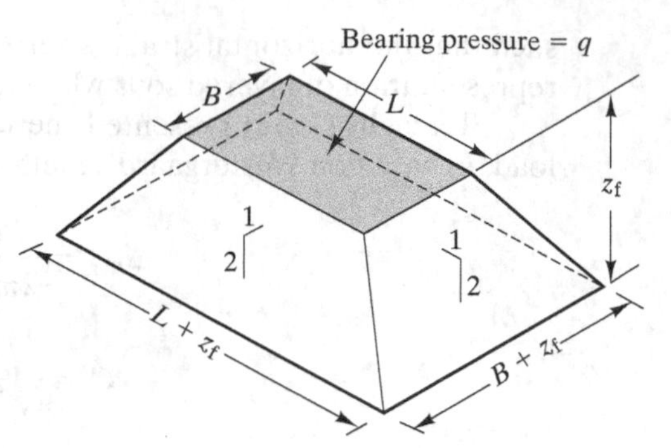

FIGURE 9.17 Use of 1:2 method to compute the average $\Delta\sigma_z$ at a specified depth below a loaded area.

B = width or diameter of loaded area

L = length of loaded area

A commonly used approximate method is to draw surfaces inclined downward at a slope of 1 horizontal to 2 vertical from the edge of the loaded area, as shown in Figure 9.17. To compute the $\Delta\sigma_z$ at a depth z_f below the loaded area, simply draw a horizontal plane at this depth, compute the area of this plane inside the inclined surfaces, and divide the total applied load by this area. The $\Delta\sigma_z$ computed by this method is an estimate of the average $\Delta\sigma_z$ across this area, and is most often used for approximate settlement computations.

When applied to a rectangular loaded area of $B \times L$, the 1:2 method produces the following formula for the influence factor for vertical stress, I_σ, at a depth z_f:

$$I_\sigma = \frac{BL}{(B + z_f)(L + z_f)} \tag{9.42}$$

The primary advantage of this method is that Equation. 9.42 can easily be derived from memory by simply applying the principles of geometry.

9.7 SUPERPOSITION

Since we have assumed the soil or rock is a linear elastic material, we can take advantage of the principle of superposition when computing σ and τ. This means problems that have multiple sources of stress may be evaluated by assessing each source separately, then adding the results. For example, if a certain point in the ground is subjected to geostatic stresses plus induced stresses from three different sources, we could perform four separate stress analyses (one for the geostatic and one for each of the induced stresses), then sum the results. This procedure greatly simplifies the analysis.

Example 9.6

Two square footings carry vertical loads as shown in Figure 9.18. Compute the magnitude of σ_z at Point A under the edge of the smaller footing, considering both geostatic and induced stresses.

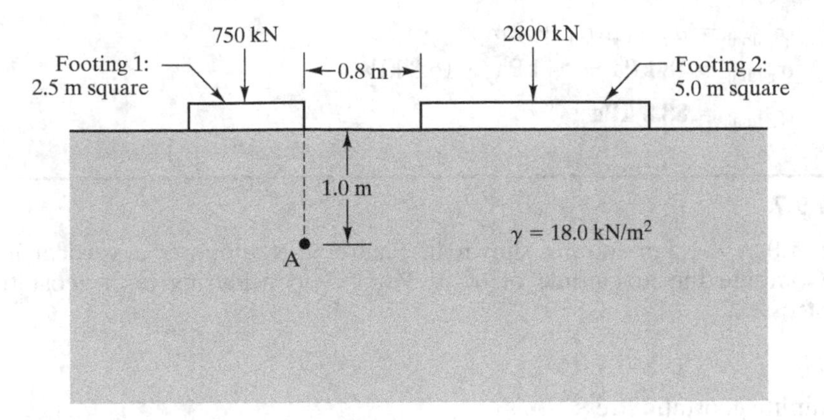

FIGURE 9.18 Profile for Example 9.6.

Solution:

Compute geostatic stress

$$\sigma_z = \Sigma \gamma H = (18.0\ \text{kN/m}^3)(1.0\ \text{m}) = 18\ \text{kPa}$$

Compute bearing pressure

$$q_1 = \frac{P}{A} = \frac{750\ \text{kN}}{(2.5\ \text{m})^2} = 120\ \text{kPa}$$

$$q_2 = \frac{P}{A} = \frac{2800\ \text{kN}}{(5.0\ \text{m})^2} = 112\ \text{kPa}$$

Compute relative z and x coordinates for point of interest.

Footing 1:

$$\frac{z_f}{B} = \frac{1.0\ \text{m}}{2.5\ \text{m}} = 0.40$$

Beneath the edge of the footing $x_f/B = 0.50$

From Figure 9.14, at $z_f/B = 0.40$ and $x_f/B = 0.50$

$I_\sigma = 0.45$

$\Delta\sigma_{z1} = I_\sigma q_1 = 0.45\ (120\ \text{kPa}) = 54\ \text{kPa}$

Footing 2:

$$\frac{z_f}{B} = \frac{1.0\ \text{m}}{5.0\ \text{m}} = 0.20$$

$$\frac{x_f}{B} = \frac{3.3\ \text{m}}{5.0\ \text{m}} = 0.66$$

From Figure 9.14, at $z_f/B = 0.20$ and $x_f/B = 0.66$

$I_\sigma = 0.15$

$\Delta\sigma_{z2} = I_\sigma q_2 = 0.15\ (112\ \text{kPa}) = 16.8\ \text{kPa}$

$$\sigma_{z\,\text{final}} = \sigma_z + \Delta\sigma_{z1} + \Delta\sigma_{z2}$$
$$\sigma_{z\,\text{final}} = 18\text{ kPa} + 54\text{ kPa} + 16.8\text{ kPa}$$
$$\sigma_{z\,\text{final}} = \mathbf{88.8\ kPa}$$

Example 9.7

The 1.0 m × 1.5 m footing shown in Figure 9.19 supports a vertical load of 475 kN. Compute the magnitude of σ_z at Point A, considering both geostatic and induced stresses.

Solution:

Compute geostatic stress

$$\sigma_z = \Sigma\gamma H = (17.0\text{ kN/m}^3)(1.2\text{ m}) = 20.4\text{ kPa}$$

Compute induced stress

Equations 9.34 and 9.35 compute σ_z below the corner of a loaded rectangle, but Point A is not beneath the corner. Therefore, it is necessary to compute the stress beneath a fictitious footing I + II, which is 1.0 m wide and 2.0 m

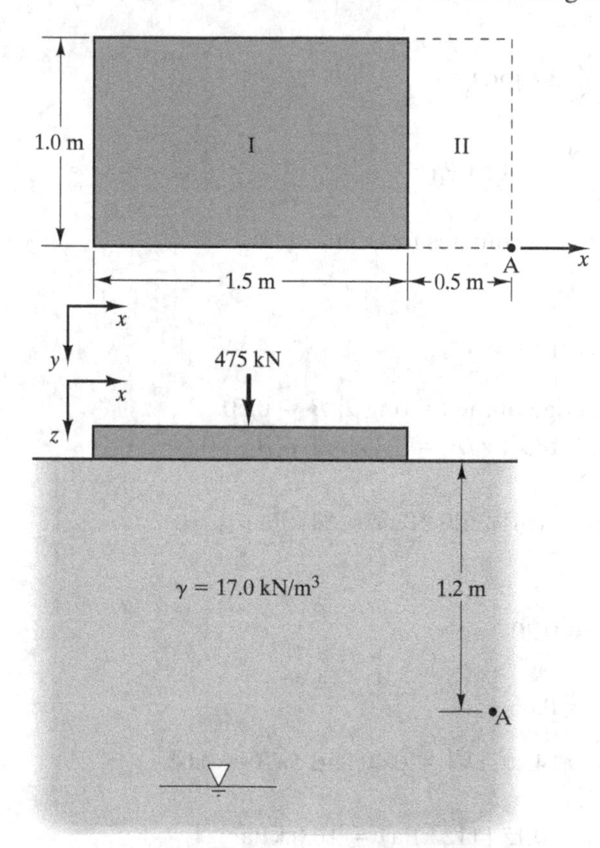

FIGURE 9.19 Plan and profile for Example 9.7.

long, and beneath a second fictitious footing II, which is 0.5 m wide and 1.0 m long. Both of these fictitious footings have a corner over Point A, and both have the same bearing pressure as the real footing. By superposition, the true σ_z at Point A is the difference between the σ_z values from these two footings.

$$q = \frac{P}{A} = \frac{475 \text{ kN}}{(1 \text{ m})(1.5 \text{ m})} = 317 \text{ kPa}$$

Solving for footing I + II

$B = 1.0, L = 2.0, z_f = 1.2$
$B^2 + L^2 + z_f^2 = 6.44, B^2L^2 = 4$; therefore, use Equation 9.35
$I_\sigma = 0.182$
$\Delta\sigma_z = (0.182)(317) = 57.7 \text{ kPa}$

Solving for footing II

$B = 0.5, L = 1.0, z_f = 1.2$
$B^2 + L^2 + z_f^2 = 2.69, B^2L^2 = 0.25$; therefore, use Equation 9.35
$I_\sigma = 0.098$
$\Delta\sigma_z = (0.098)(317) = 31.1 \text{ kPa}$

By superposition, $\Delta\sigma_z = 57.7 \text{ kPa} - 31.1 \text{ kPa} = 26.6 \text{ kPa}$
Combined results,

$\sigma_z = 20.4 \text{ kPa} + 26.6 \text{ kPa} = \mathbf{47.0 \text{ kPa}}$

9.8 EFFECTIVE STRESSES

The compressive stress, σ, computed using the techniques described thus far, is carried partially by the solid particles and partially by the pore water. Geotechnical engineers call it the *total stress* because it is the sum of the stresses carried by these two phases in the soil. Although the total stress can be very useful, we gain even more insight by dividing it into two parts:

- The *effective stress*, σ', which is the portion carried by the solid particles, and
- The *pore water pressure*, u, which is the portion carried by the pore water. This is the same u we discussed in Chapters 7 and 8.

Karl Terzaghi was the first to recognize the importance of effective stress, and it has since become one of the most important concepts in geotechnical engineering.

Submerged Sphere Analogy

To understand the physics of soil particles under the groundwater table and the differences between total and effective stresses, let us consider the sphere resting on a scale as shown in Figure 9.20. It has a volume of 0.100 m^3 as determined by measuring its diameter and a weight of 2.60 kN as determined by the scale.

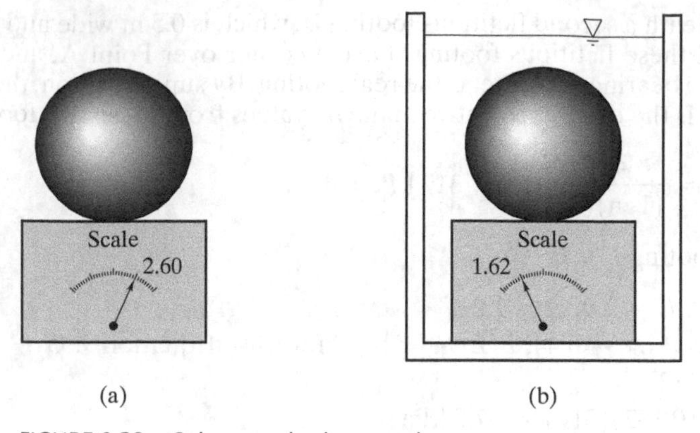

FIGURE 9.20 Submerged sphere analogy.

Then, we take the scale and the sphere and place them into a tank of water, as shown. In this new environment, the sphere is subjected to a buoyancy force, F_B, equal to the weight of the displaced water:

$$F_B = V\gamma_w$$
$$F_B = (0.100 \text{ m}^3)(9.8 \text{ kN/m}^3)$$
$$F_B = 0.98 \text{ kN} \quad (9.43)$$

The contact force between the sphere and the scale is thus reduced to

$$F = 2.60 \text{ kN} - 0.98 \text{ kN}$$
$$F = 1.62 \text{ kN} \quad (9.44)$$

The weight of the sphere has not changed, but it is now being supported partially by the scale and partially by the water.

The contact forces between soil particles above the groundwater table are similar to that between the dry sphere and the scale, whereas soils below the groundwater table are similar to the submerged sphere and scale. Buoyancy forces act on the soil solids the same way they act on the sphere. Therefore, the particle contact forces in an element of soil that is initially above the groundwater table will decrease if the groundwater rises above that element.

Interparticle Forces and Vertical Effective Stress

From the submerged sphere analogy, we see how σ in an element of soil below the groundwater table is distributed between the solid particles and the pore water. As discussed earlier, the portion carried by the solid particles is known as the *effective stress*, σ', while the portion carried by the pore water is equal to the *pore water pressure*, u. Figure 9.21 illustrates the division of total stress into effective stress and pore water pressure. Figure 9.21(a) shows a saturated soil element carrying a total vertical stress, σ_z, which is carried by both the pore water pressure, u, and the effective stress, σ'_z. Figure 9.21(b) is a close-up of an interparticle contact. The effective stress is carried by the solid soil particles through interparticle contact forces as shown. Notice how the

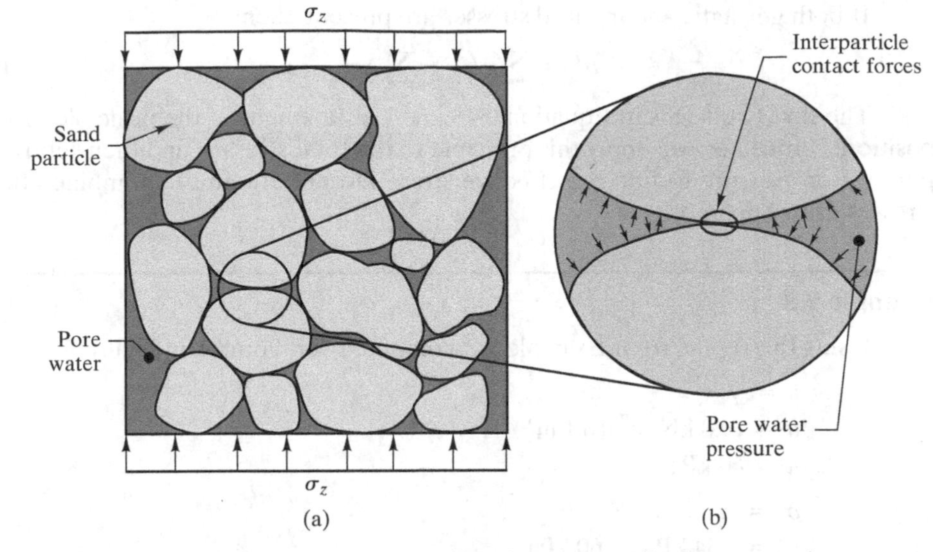

FIGURE 9.21 Illustration of the division of total stress in to effective stress and pore water pressure. Panel (a) shows a saturated soil element carrying a total vertical stress of σ_z. This total stress is carried by both the pore water pressure, u, and the effective stress, σ'_z. Panel (b) is a close-up of an interparticle contact. The effective stress is the portion carried by the interparticle contact forces.

pore water pressure pushes against the soil particles, thereby reducing the interparticle forces. The total stress then must be the sum of the pore water pressure and the effective stress carried by interparticle contact forces.

Under most unconfined hydrostatic conditions, we can compute u below the groundwater table using Equation 7.7. Then, we can compute the effective stress using

$$\sigma' = \sigma - u \tag{9.45}$$

or

$$\sigma'_z = \sigma_z - u \tag{9.46}$$

where:

σ'_z = vertical effective stress

σ_z = vertical total stress

u = pore water pressure

If only geostatic stresses are present, we can combine Equations 9.21 and 9.46 to produce:

$$\sigma'_z = \sum \gamma H - u \tag{9.47}$$

Note how the distribution of force between the solids and water is *not* proportional to their respective cross-sectional areas.

If both geostatic and induced stresses are present, then:

$$\sigma'_z = \sum \gamma H + \sum \Delta\sigma_z - u \tag{9.48}$$

The first two terms in Equation 9.48 are a restatement of the principle of superposition. Notice how we apply this principle to the total stresses, and then subtract the pore water pressure to find the effective stress. Do not attempt to combine effective stresses using superposition.

Example 9.8

Using the results from Example 9.2, compute σ'_z at Point A in Figure 9.11.

$$u = \gamma_w z_w$$
$$u = (9.8 \text{ kN/m}^3)(6.1 \text{ m})$$
$$u = 60 \text{ kPa}$$

$$\sigma'_z = \sigma_z - u$$
$$\sigma'_z = 134 \text{ kPa} - 60 \text{ kPa}$$
$$\sigma'_z = \mathbf{74\ kPa}$$

Commentary: A vertical compressive stress of 134 kPa is present at Point A, 74 kPa of this stress is being carried by the solid particles, and 60 kPa by the pore water.

The principle of effective stress is the key to understanding many aspects of soil behavior. In the following chapters, we will see how settlement and strength analyses are normally based on effective stresses, not on total stresses.

Horizontal Effective Stress

The horizontal effective stresses, σ'_x and σ'_y, are related to the horizontal total stresses as follows:

$$\sigma'_x = \sigma_x - u \tag{9.49}$$

$$\sigma'_y = \sigma_y - u \tag{9.50}$$

If multiple sources of stress need to be combined, do so by using superposition with the total stresses, then subtract the pore water pressure.

The ratio of the horizontal to vertical effective stresses is defined as the *coefficient of lateral earth pressure*, K. For geostatic stresses beneath a level ground surface, we normally assume K in the x direction is equal to that in the y direction:

$$K = \frac{\sigma'_x}{\sigma'_z} = \frac{\sigma'_y}{\sigma'_z} \tag{9.51}$$

However, if the ground surface is inclined, or if induced stresses are present, K may be different in the x and y directions. The value of K varies from about 0.3 to 3. We will discuss methods of evaluating it in Chapter 17.

Values of K may reflect the existing or preconstruction condition or they may reflect conditions after some load has been applied. When the values reflect the existing natural conditions we use the term *lateral earth pressure at rest* and the symbol K_0. If a new load, such as that from a foundation, is to be applied, the lateral earth pressure will usually change because of the induced horizontal and vertical effective stresses. Depending on the type of load, K in the x direction also may be different than in the y direction. This can cause some confusion when solving stress problems, because the given K may only be used to evaluate the preconstruction stresses. Example 9.9 illustrates the proper way to solve such problems.

Example 9.9

A proposed vertical point load of 90.0 k is to be applied to the ground surface 3.0 ft south and 4.0 ft east of Point A in Figure 9.22. Compute all of the total and effective stresses acting on the vertical and horizontal planes at Point A. Consider both the geostatic and induced stresses, and use a coordinate system with the x and y axes oriented in the east and south directions, respectively.

Note: The x and y axes do not always need to be aligned with cardinal compass directions. Often it is more convenient to orient them parallel and perpendicular to a proposed structure or slope.

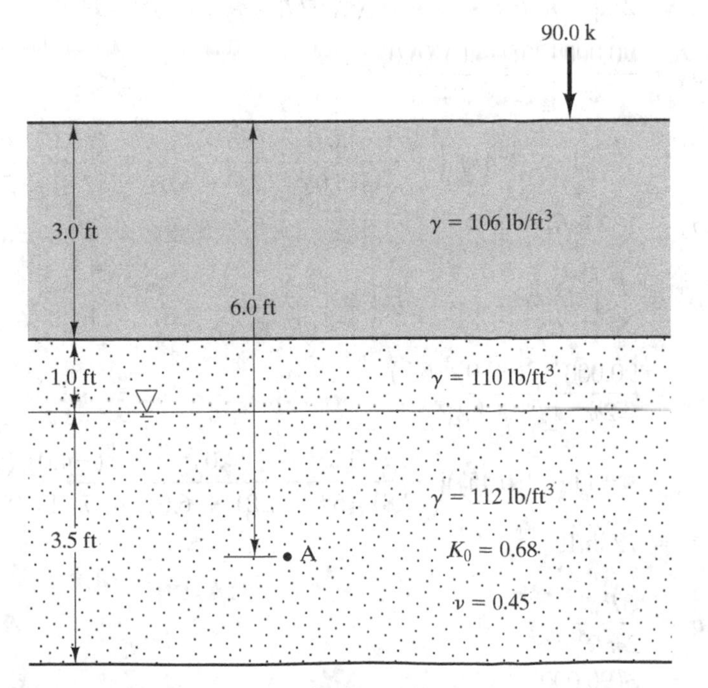

FIGURE 9.22 Cross-section for Example 9.9.

Solution:

Geostatic stresses (initial condition):

$$\sigma_z = \Sigma\gamma H$$
$$= (106\ \mathrm{lb/ft^3})(3\ \mathrm{ft}) + (110\ \mathrm{lb/ft^3})(1\ \mathrm{ft}) + (112\ \mathrm{lb/ft^3})(2\ \mathrm{ft}) = 652\ \mathrm{lb/ft^2}$$
$$u = \gamma_w z_w = (62.4\ \mathrm{lb/ft^3})(2\ \mathrm{ft}) = 125\ \mathrm{lb/ft^2}$$
$$\sigma'_z = \sigma_z - u = 652\ \mathrm{lb/ft^2} - 125\ \mathrm{lb/ft^2} = 527\ \mathrm{lb/ft^2}$$
$$\sigma'_x = \sigma'_y = K_0\sigma'_z = (0.68)(527\ \mathrm{lb/ft^2}) = 358\ \mathrm{lb/ft^2}$$
$$\sigma_x = \sigma_y = \sigma'_x + u = 358\ \mathrm{lb/ft^2} + 125\ \mathrm{lb/ft^2} = 483\ \mathrm{lb/ft^2}$$

Since the ground surface is horizontal:

$$\tau_{xz} = \tau_{zx} = \tau_{yz} = \tau_{zy} = \tau_{xy} = \tau_{yx} = 0$$

Induced stresses:

Using the Boussinesq equations:

$$x_f = -4\ \mathrm{ft};\ y_f = -3\ \mathrm{ft};\ z_f = 6\ \mathrm{ft}$$
$$R = \sqrt{x_f^2 + y_f^2 + z_f^2} = \sqrt{(-4.0)^2 + (-3.0)^2 + 6.0^2} = 7.81\ \mathrm{ft}$$
$$r = \sqrt{x_f^2 + y_f^2} = \sqrt{(-4.0)^2 + (-3.0)^2} = 5.0\ \mathrm{ft}$$

$$\Delta\sigma_x = \frac{P}{2\pi}\left[\frac{3x_f^2 z_f}{R^5} - (1 - 2\nu)\left(\frac{x_f^2 - y_f^2}{Rr^2(R + z_f)} + \frac{y_f^2 z_f}{R^3 r^2}\right)\right]$$

$$\Delta\sigma_x = \frac{90{,}000}{2\pi}\left[\frac{3(-4.0)^2(6.0)}{7.81^5} - (1-2(0.45))\left(\frac{(-4.0)^2 - (-3.0)^2}{(7.81)(5.0)^2(7.81 + 6.0)} + \frac{(-3.0)^2(6.0)}{(7.81)^3(5.0)^2}\right)\right]$$

$$\Delta\sigma_x = 132\ \mathrm{lb/ft^2}$$

$$\Delta\sigma_y = \frac{P}{2\pi}\left[\frac{3y_f^2 z_f}{R^5} - (1 - 2\nu)\left(\frac{y_f^2 - x_f^2}{Rr^2(R + z_f)} + \frac{x_f^2 z_f}{R^3 r^2}\right)\right]$$

$$\Delta\sigma_y = \frac{90{,}000}{2\pi}\left[\frac{3(-3.0)^2(6.0)}{7.81^5} - (1-2(0.45))\left(\frac{(-3.0)^2 - (-4.0)^2}{(7.81)(5.0)^2(7.81 + 6.0)} + \frac{(-4.0)^2(6.0)}{(7.81)^3(5.0)^2}\right)\right]$$

$$\Delta\sigma_y = 72\ \mathrm{lb/ft^2}$$

$$\Delta\sigma_z = \frac{3Pz_f^3}{2\pi R^5}$$

$$\Delta\sigma_z = \frac{3(90{,}000)(6.0)^3}{2\pi(7.81)^5} = 319\ \mathrm{lb/ft^2}$$

$$\Delta\tau_{zx} = -\Delta\tau_{xz} = \frac{3Pz_f^2x_f}{2\pi R^5} = \frac{3(90{,}000)(6.0)^2(-4.0)}{2\pi(7.81)^5} = -213 \text{ lb/ft}^2$$

$$\Delta\tau_{yx} = \frac{P}{2\pi}\left[\frac{3x_fy_fz_f}{R^5} - (1-2\nu)\left(\frac{(2R+2)x_fy_f)}{(R+z_f^2)R^3}\right)\right]$$

$$\Delta\tau_{yx} = \frac{90{,}000}{2\pi}\left[\frac{3(-4.0)(-3.0)(6.0)}{7.81^5}\right.$$

$$\left. - (1-2(0.45))\left(\frac{(2(7.81)+2)(-4.0)(-3.0))}{(7.81+6.0^2)(7.81)^3}\right)\right]$$

$$\Delta\tau_{yx} = 92 \text{ lb/ft}^2$$

$$\Delta\tau_{yz} = -\Delta\tau_{zy} = \frac{3Pz_f^2y_f}{2\pi R^5} = \frac{3(90{,}000)(6.0)^2(-3.0)}{2\pi(7.81)^5} = -160 \text{ lb/ft}^2$$

Overall stresses (by superposition):

$$\sigma_x = 483 \text{ lb/ft}^2 + 132 \text{ lb/ft}^2 = \mathbf{615\ lb/ft^2}$$
$$\sigma_y = 483 \text{ lb/ft}^2 + 72 \text{ lb/ft}^2 = \mathbf{555\ lb/ft^2}$$
$$\sigma_z = 652 \text{ lb/ft}^2 + 319 \text{ lb/ft}^2 = \mathbf{971\ lb/ft^2}$$
$$\sigma'_x = \sigma_x - u = 615 \text{ lb/ft}^2 - 125 \text{ lb/ft}^2 = \mathbf{490\ lb/ft^2}$$
$$\sigma'_y = \sigma_y - u = 555 \text{ lb/ft}^2 - 125 \text{ lb/ft}^2 = \mathbf{430\ lb/ft^2}$$
$$\sigma'_z = \sigma_z - u = 971 \text{ lb/ft}^2 - 125 \text{ lb/ft}^2 = \mathbf{846\ lb/ft^2}$$
$$\tau_{zx} = -\tau_{xz} = \mathbf{-213\ lb/ft^2}$$
$$\tau_{yx} = -\tau_{xy} = \mathbf{92\ lb/ft^2}$$
$$\tau_{yz} = -\tau_{zy} = \mathbf{-160\ lb/ft^2}$$

Notes:

1. The K_0 value given in the problem statement was applied only to the initial conditions, which in this case consisted of the geostatic stresses only. It would not be correct to apply K_0 to the $\sigma'_z = 846 \text{ lb/ft}^2$ value, because after the applied load, the at-rest conditions no longer apply.
2. We compute the proposed stresses by first combining the geostatic and induced total stresses by superposition, then subtracting the pore water pressure.

Mohr Circles Based on Effective Stresses

We can use the Mohr circle to represent effective stresses as well as total stresses. As an illustration, Figure 9.23 shows the effective and total stress Mohr circles for the stresses in the x–z plane for Example 9.9 both before and after application of the surface load. Note that the Mohr circle for effective stress has the same diameter as that for total stress, but it is offset horizontally by a distance equal to the pore water pressure (in this

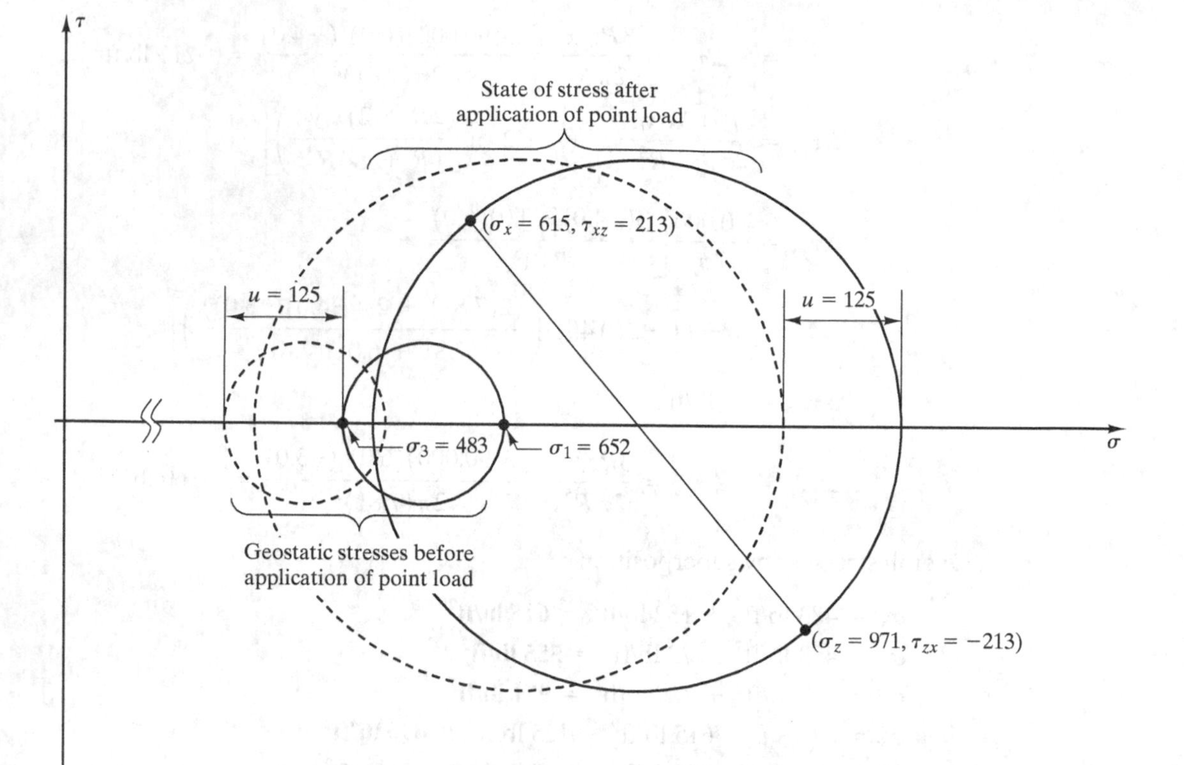

FIGURE 9.23 Total and effective stress Mohr circles representing stresses on the z- and x-planes from Example 9.9. The small circles represent the geostatic stresses before the application of the point load. The large circles represent stresses after application of the point load. The solid circles represent total stresses and the dashed circles represent effective stresses.

case, 125 lb/ft^2). The diameter of the circle has not changed because the shear stress is not affected by the pore water pressure. This is because the pore water cannot carry a static shear stress, so the solid particles must carry all of the shear stress. In other words, the principle of effective stress applies only to normal stresses, not to shear stresses.

Stresses Beneath Bodies of Water

Sometimes geotechnical engineers need to compute stresses in soils beneath bodies of water, such as lakes, rivers, or oceans. For example, this might be necessary while designing the foundation for a bridge or when evaluating the stability of a slope that extends underwater. Although this case appears confusing at first, we simply need to apply Equation 9.46. This is best illustrated using Figure 9.24(a) and (b). These two soil profiles are the same except in Figure 9.24(a), the groundwater table is at the ground surface and in Figure 9.24(b), the water level is at a height, h, above the ground surface. We will use Equation 9.46 in each case to compute the vertical effective stress, σ'_z.

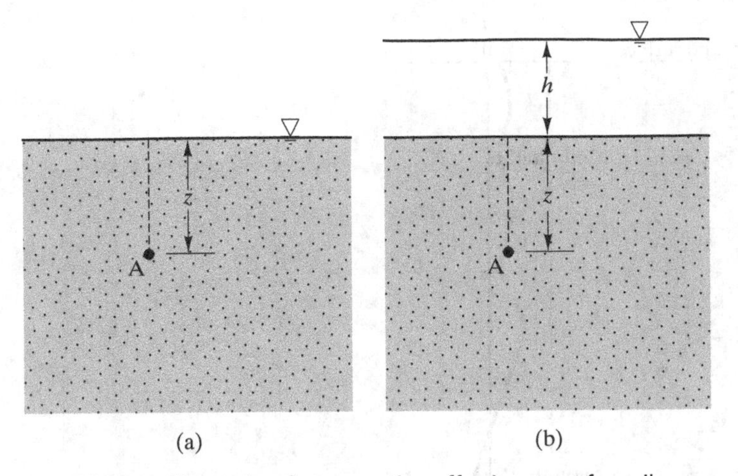

FIGURE 9.24 Illustration for computing effective stress for soil beneath bodies of water.

For case (a), with the groundwater table at the ground surface, at Point A,

$$\sigma_z = \gamma z$$
$$u = \gamma_w z$$
$$\sigma'_z = \gamma z - \gamma_w z = z(\gamma - \gamma_w)$$

For case (b), with the water level at a height of h above the ground surface

$$\sigma_z = \gamma z + \gamma_w h$$
$$u = \gamma_w (z + h)$$
$$\sigma'_z = (\gamma z + \gamma_w h) - \gamma_w (z + h) = z(\gamma - \gamma_w)$$

Notice that the water above the ground surface does not change the effective stress because it contributes equally to both the total stress and the pore water pressure.

Finally, if we recall that the buoyant unit weight, γ_b, of a soil is equal to the total unit weight less the unit weight of water, then when the groundwater level is at or above the ground surface, we can determine the effective vertical stress using the equation.

$$\sigma'_z = z\gamma_b \tag{9.52}$$

Stress Conditions with Negative Pore Water Pressures

Most geotechnical analyses assume that the pore water pressure above the groundwater table is zero. Thus, according to Equation 9.45, the effective stress equals the total stress. However, this is a simplification of the truth. In reality, these soils generally have *negative pore water pressures* (also known as *soil suction*), which means the effective stress is greater than the total stress.

In Chapter 7, we distinguished between the phreatic zone, below the water table, and the vadose zone, above the water table. Within the vadose zone, the soil remains saturated in the zone of capillary rise as shown in Figure 9.25. In the zone of capillary rise, the pore water pressure may be computed using Equation 7.7, except z_w is now negative.

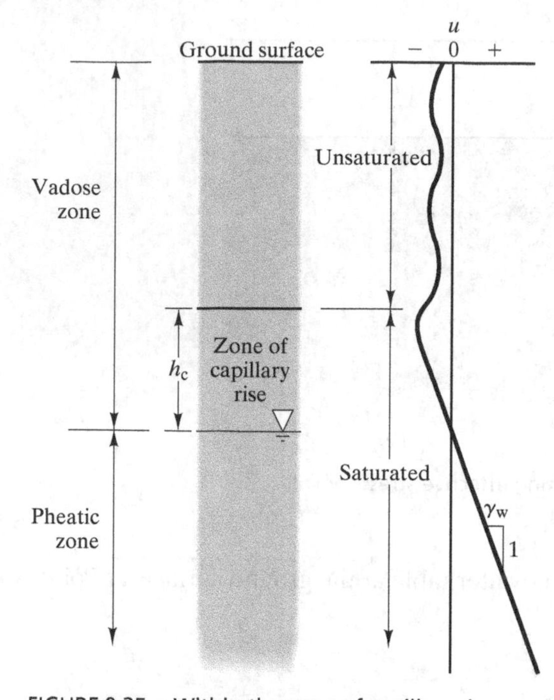

FIGURE 9.25 Within the zone of capillary rise, a plot of pore water pressure versus elevation is simply an extension of the plot below the groundwater table. Above the zone of capillary rise, the pore water accumulates near the particle contact points and surface tension forces develop between these pockets of pore water and the adjacent solid particles. These forces produce tensile stresses (negative pore water pressures) in the water. The magnitude of the negative pore water pressures in this zone depends on the soil type and other factors, and can be much greater (i.e., more negative) than those in the capillary zone.

This produces a negative u, as shown in Figure 9.25. In other words, the pore water in this zone may be visualized as a column of water held in tension by the capillary forces.

In the vadose zone above the zone of capillary rise, the degree of saturation falls well below 100% and the situation is more complicated. In the unsaturated zone, the pore water collects into small capillary zones adjacent to the particle contact points as shown in Figure 9.26. Surface tension forces develop between this water and the particles, creating tensile forces (negative pore water pressures) in the water. However, in the air pockets, the pressure is atmospheric. The pore water pressure can no longer be computed using Equation 7.7 and the traditional definition of effective stress using Equation 9.45 is no longer valid. The transition between saturated and unsaturated zones is often poorly defined. Negative pore water pressures can be measured in situ or in the laboratory using a variety of techniques (Fredlund and Rahardjo, 1993), and these measurements have been used to develop more rational explanations of soil behavior above the groundwater table.

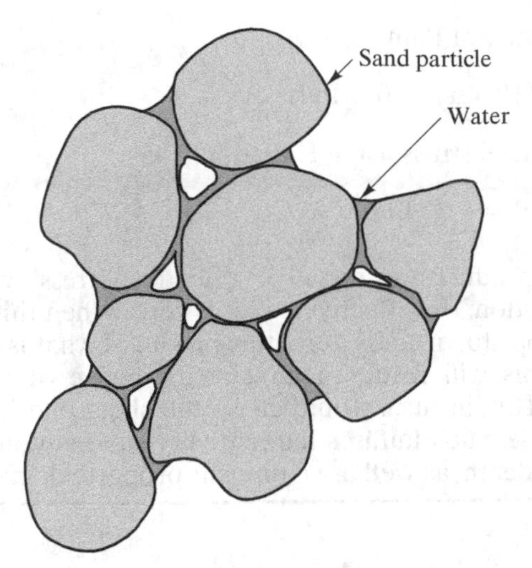

FIGURE 9.26 Above the capillary zone, the pore water retreats to the particle contact points. Surface tension forces develop between this water and the soil particles, producing a negative pore water pressure in these capillary fringes formed around contacts. In the air pockets, the pressure is atmospheric.

9.9 EFFECTIVE STRESS UNDER STEADY-STATE FLOW

Thus far, the analyses in this chapter have considered only the case where the groundwater is static and the pore water pressure can be computed using Equation 7.7. When groundwater is flowing under steady-state conditions, the concept of effective stress applies and Equation 9.45 is still applicable. However, we must now compute the pore water pressure using techniques presented in Chapter 8. Example 9.10 illustrates this process.

Example 9.10

Compute the total and effective vertical stress at Point B in Figure 8.7(b).

Solution:

The total vertical stress at Point B is

$$\sigma_z = \gamma z = 112 \text{ lb/ft}^2 \ (5.5 \text{ ft}) = 616 \text{ lb/ft}^2$$

Compute the pore water pressure at Point B using the flow net in Figure 8.7(b). The total head loss from the outside of the trench to the inside, Δh, is 20 ft. There are 16.33 equipotential drops in the flow net, so the head loss in each equipotential drop is computed as

$$h_i = \left(\frac{\Delta h}{N_D}\right) = \left(\frac{20}{16.33}\right) = 1.225 \text{ ft/drop}$$

Point B is 4 equipotential drops from the free water surface at the base of the trench. So a piezomenter placed at Point B will rise $4h_i$ or $4(1.225) = 4.90$ ft above the base of the trench. Therefore, the pressure head at Point B is given by

$$h_{pB} = 5.5 \text{ ft} + 4.90 \text{ ft} = 10.4 \text{ ft}$$

Therefore, the pore water pressure at Point B is

$$u_B = \gamma_w h_{pB} = 62.4 \text{ lb/ft}^3(10.4 \text{ ft}) = 649 \text{ lb/ft}^2$$

Now compute the effective vertical stress using Equation 9.45

$$\sigma'_z = \sigma_z - u = 616 - 649 = -33 \text{ lb/ft}^2$$

Commentary: This computation produces a negative effective stress, which is physically impossible! In such a situation, the effective stress is zero. When this occurs in a sand, the shear strength also drops to virtually zero, this produces what is called a *quick condition* or a *quicksand.* This will result in an upward heave of the soils immediately below the excavation. This, in turn, would cause the sheet piles to move inward and the excavation to collapse. Such failures happen very suddenly, and have been the cause of serious injury and death, as well as significant property damage.

Problems with heave, such as the one described in Example 9.10, can be avoided by keeping the vertical effective stress well above zero. We accomplish this by maintaining the hydraulic gradient at acceptably low values (perhaps by extending the sheet piles or installing dewatering wells) or by covering the excavation with a highly pervious surcharge fill, such as gravel, which adds to the total stress, but does not contribute significantly to the seepage force.

Seepage Force

Another method for computing effective stresses under steady-state flow is to use the concept of *seepage force*. In this concept, as the groundwater is moving through the soil, it imparts a drag force, called *a seepage force* on the solid particles:

$$j = i\gamma_w \tag{9.53}$$

where:

j = seepage force per unit volume of soil
i = hydraulic gradient
γ_w = unit weight of water

For example, if water is flowing through a certain soil with a hydraulic gradient of 0.15, the seepage force will be equal to $(0.15)(9.8 \text{ kN/m}^3) = 1.5 \text{ kN/m}^3$ and will act in the same direction the water is flowing.

The hydraulic gradient in soils is usually small enough that seepage forces may be ignored. However, seepage forces can be important if the water is flowing upward with a large i, as shown in Figure 9.27. This is because the seepage force now acts in the opposite direction of gravity, and thus reduces the effective stress. In this case, Equation 9.47 may be rewritten as:

$$\sigma'_z = \sum[(\gamma - j)H] - u_{static} \tag{9.54}$$

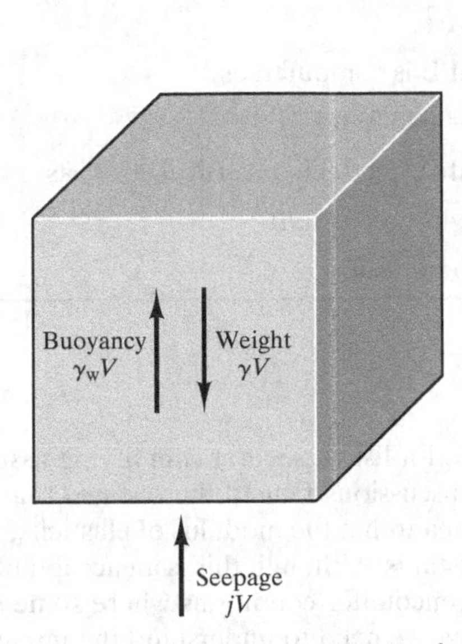

FIGURE 9.27 Forces acting on an element of soil when the seepage forces are acting vertical upward.

where:

H = the thickness of the stratum
γ = the total unit weight of the soil
u_{static} = the static pore pressure computed as if water was not flowing

Seepage forces also may be important in some slope stability problems, especially if the groundwater is flowing toward the slope face and the hydraulic gradient is high.

Example 9.11

Compute the total and effective vertical stress at Point B in Figure 8.7(b) using seepage forces.

Solution:

From Point B to the base of the trench, the cells in the flow net are essentially all of the same size. This indicates the hydraulic gradient in this area is constant. From Example 9.10, we computed the head loss from Point B to the base of the trench to be 4.90 ft. From this information, we can compute the hydraulic gradient in this area:

$$i = \frac{\Delta h}{l} = \frac{4.90}{5.5} = 0.891$$

The seepage force is computed as

$$j = 0.891(62.4) = 55.6 \text{ lb/ft}^2$$

The static pore pressure at Point B is computed as

$$u_{\text{static}} = \gamma_w z = 62.4(5.5) = 343 \text{ lb/ft}^2$$

Using Equation 9.54, we compute the effective vertical stress as

$$\sigma'_z = (112 - 55.6)5.5 - 343 = -33 \text{ lb/ft}^2$$

So both methods produce the same result.

9.10 STRESSES IN LAYERED STRATA

The beginning of this chapter included a list of several simplifying assumptions that have governed the remainder of our discussions. One of these stated that the ground is homogeneous, which in this context meant that the modulus of elasticity, E, shear modulus, G, and Poisson's ratio, ν, are constants. Although this is an acceptable assumption for many soil profiles, sometimes we encounter conditions where some strata are significantly stiffer than others. Therefore, we need to understand the impact of such differences, and in some cases be able to quantify them.

One common condition consists of a soil layer underlain by a much stiffer bedrock ($E_1 < E_2$) as shown in Figure 9.28. In this case, there is less spreading of the load, so the induced stresses in the soil are greater than those computed by Boussinesq. Conversely, if we have a stiff layer underlain by a softer soil ($E_1 > E_2$), the load spreading is enhanced and the induced stresses are less than the Boussinesq values.

Sometimes we can use this behavior to our advantage, such as with highway pavements. The pavement and the underlying aggregate base course are much stiffer than the soils that support them, so they spread the wheel loads over a larger area of soil. This decreases the induced stresses, and thus enhances the soil's load-carrying capacity.

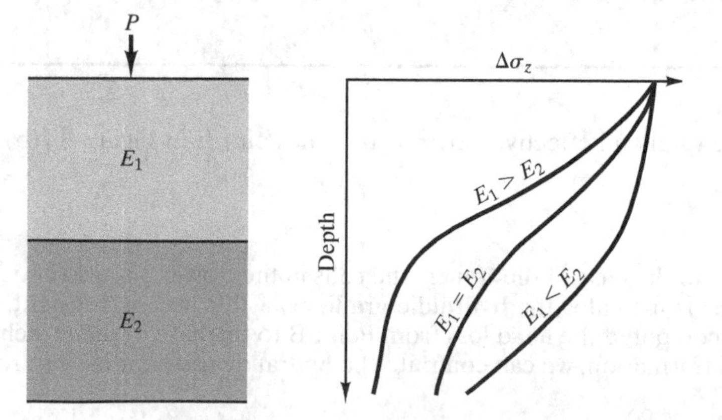

FIGURE 9.28 Distribution of $\Delta\sigma_z$ with depth in layered profiles.

SUMMARY

Major Points

1. Soil and rock are much more complex than traditional engineering materials; so, a true mathematical model to describe the distribution of stresses would be far too difficult to use in practice. Therefore, we treat soil and rock as if they were continuous, homogeneous, isotropic, and linear elastic materials. The error introduced by these simplifying assumptions is acceptably small for most practical analyses.
2. Our analyses consider both normal stress, σ, and shear stress, τ, but we use a sign convention opposite to that of most engineers: Compression is positive and tension is negative.
3. Although we can keep track of stresses in all three dimensions, thus performing a three-dimensional analysis, many problems can be simplified by using two-dimensional or even one-dimensional analyses.
4. For a given point under the ground, there is a unique state of stress. However, the magnitude of shear and normal stresses will depend upon the orientation of the planes used to compute the stresses.
5. A Mohr circle may be used to represent the state of stress in two dimensions. Using the pole method, it is possible to compute the shear and normal stress acting on any plane passing through a point.
6. In a two-dimensional analysis, there are two stress invariants: the mean normal stress, σ_m, respresented by the center of the Mohr circle and the deviator stress, σ_d, respesented by the diameter of the Mohr circle. Regardless of what orientation is chosen to compute the shear and normal stresses, these invariants do not change.
7. In a two-dimensional analysis, the major principal stress at a point in the ground is the maximum normal stress acting on any plane through that point and the minor principal stress is the smallest normal stress acting on any plane through that point. The principal planes are orthogonal. The state of stress at a point is uniquely defined by the major principle stress, the minor principle stress and the orientation of the principal planes.
8. Stresses in the ground can be divided into two components based on their source: geostatic stresses due to the weight of the ground itself, and induced stresses due to applied external loads.
9. We generally compute induced stresses using elastic theory using either equations or charts.
10. We use the principle of superposition to determine the combined effect of geostatic stress and induced stress from one or more applied loads.
11. The concept of effective stress is one of the most important concepts in geotechnical engineering. The total stress in a soil is carried partly by the solid particles through interparticle contacts and partly by the pore water through pore water pressure. That portion of the stress carried through interparticle contacts is called the *effective stress*. Understanding the concept of effective stress is critical because much of soil behavior is governed by effective stress rather than total stress.

12. Horizontal effective stress in soil is usually computed in relation to the vertical effective stress using the coefficient of lateral earth pressure, K, which is the ratio of the effective horizontal stress to the effective vertical stress.
13. When groundwater is flowing through a soil, we can compute the effective stress either by computing pore water pressures using techniques developed in Chapter 8 and then subtracting this pore water pressure from the total stress or we can use the seepage force concept. Both methods give the same results.
14. Pore water pressures generated during seepage can significantly reduce the effective stress especially when the direction of flow is vertically upward. If the effective stress reaches zero, quick conditions occur which can lead to heave and a loss of soil strength.

Vocabulary

analytical solution
area load
bearing pressure
body stress
Boussinesq's method
coefficient of lateral earth pressure
compression
continuous material
deviator stress
effective stress
geostatic stress
heave
homogeneous
incompressible
induced stress
infinite elastic half-space
interparticle force
intermediate principal stress
isotropic
line load
major principal stress
mean normal stress
minor principal stress
modulus of elasticity
modulus of rigidity
Mohr circle
normal strain
normal stress
numerical solution
point load
Poisson's ratio
pole
pore water pressure
pressure bulb
principal planes
principal stresses
quick condition
quick sand
seepage force
shear modulus
shear strain
shear stress
soil suction
strain
stress
stress bulb
stress invariant
stress path
superposition
tension
total stress
Westergaard's method
Young's modulus

QUESTIONS AND PRACTICE PROBLEMS

Section 9.1 Mechanics of Materials

9.1 A 0.500 ft × 0.500 ft × 0.500 ft cube of soil is subjected to a vertical compressive force of 500 lb. This force is being applied to the top of the cube. As a result of this force, the cube compresses to a height of 0.450 ft. Compute the vertical normal stress, the vertical normal strain, and the Young's modulus of the soil.

9.2 A soil has Young's modulus of 27,000 kPa and Poisson's ratio of 0.3. A cylindrical sample of the soil 0.10 m in diameter and 0.2 m tall is subject to a vertical stress of 320 kPa. Compute the vertical normal strain, vertical deformation, horizontal normal strain, and horizontal deformation.

Section 9.2 Mohr Circle Analyses

9.3 The major and minor principal stresses at a certain point in the ground are 450 and 200 kPa, respectively. Draw the Mohr circle for this point, compute the maximum shear stress, τ_{max}, and indicate the points on the Mohr circle that represent the planes on which τ_{max} acts.

9.4 The stresses at a certain point in the ground are $\sigma_x = 210$ kPa, $\sigma_z = 375$ kPa, $\tau_{zx} = 75$ kPa, and $\tau_{xz} = -75$ kPa. Draw the Mohr circle for this point and determine the following:

(a) The location of the pole
(b) The mean normal and deviator stress
(c) The magnitudes and directions of the principal stresses
(d) The magnitude and directions of the maximum shear stress
(e) The normal and shear stresses acting on a plane inclined 55° clockwise from the horizontal.

9.5 The major principal stress at a certain point is 4800 lb/ft^2 and acts vertically. The minor principal stress is 3100 lb/ft^2. Draw the Mohr circle for this point, locate the pole, then compute the normal and shear stresses acting on a plane inclined 26° counterclockwise from the horizontal.

9.6 A certain element of soil is subject to a mean normal stress of 420 kPa and a deviator stress of 280 kPa. The major principal plane is rotated 30° counterclockwise from the horizontal. Draw the Mohr circle for this soils element, locate the pole, and compute the normal and shear forces acting on the horizontal and vertical planes.

9.7 A laboratory soil specimen is initially subject to principal stresses of 3700 lb/ft^2 and 2300 lb/ft^2. During testing, the major principal stress is increased while the minor principal stress is kept the same. What is the major principal stress when the deviator stress has reached 1800 lb/ft^2? Draw the Mohr circles for the initial and final stress states on a single figure.

9.8 A cylindrical specimen of soil is placed in a special testing device which applies vertical and horizontal normal stresses to the specimen. No shear stresses are applied on the vertical and horizontal planes so they are always principal planes. The following loading sequence is applied:

Load Step	Vertical Normal Stress σ_z (lb/in.^{2})	Horizontal Normal Stress σ_x (lb/in.^{2})
1	12	12
2	24	12
3	36	24
4	48	24
5	48	36

Plot the stress path followed during this load sequence. Create two separate plots, one showing the stress path using Mohr circles and a separate plot showing the the stress path in p–q space.

Sections 9.5 and 9.6 Geostatic and Induced Stresses

9.9 A certain sandy soil has a total unit weight of 118 lb/ft^3. What is the vertical normal stress, σ_z, at a point 15 ft below the ground surface?

9.10 Compute the vertical normal stress, σ_z, at points A, B, and C in Figure 9.11.

9.11 Using the soil profile in Figure 9.11, develop a plot of σ_z versus depth. Consider depths between 0 and 10 m.

9.12 A vertical point load of 50.0 k acts upon the ground surface at coordinates $x = 100$ ft, $y = 150$ ft. Using a Poisson's ratio of 0.40, compute the induced stresses $\Delta\sigma_x$, $\Delta\sigma_z$, and $\Delta\tau_{xz}$ at a point 3 ft below the ground surface at $x = 104$ ft, $y = 150$ ft.

9.13 A vertical line load of 75 kN/m acts upon the ground surface. Assuming this load extends for a very long distance in both directions, compute the induced vertical stress, $\Delta\sigma_z$, at a point 1.5 m horizontally off the line (measured perpendicularly to the line) and 2.0 m below the line.

9.14 A grain silo is supported on a 20.0 by 50.0 m mat foundation. The total weight of the silo and the mat is 180,000 kN. Using Boussinesq's method, compute the induced vertical stress, $\Delta\sigma_z$, in the soil at a point 15.0 m below the center of the mat. First use the full analytical solutions given by Equations 9.34 and 9.35. Repeat the computation using the approximate method given by Equation 9.41. Finally, repeat the computation using Figure 9.16. Compare the results from these three methods and comment on whether the differences are significant.

9.15 For the grain silo described in Problem 9.14, compute the induced vertical stress, $\Delta\sigma_z$, in the soil at a point at the midpoint of the long edge of the mat and 10 m below the ground surface. Use both analytical solutions given by Equations 9.34 and 9.35 and the chart method using Figure 9.16 and compare the results.

Section 9.7 Superposition

9.16 A dilatometer test (an in situ test described in Chapter 3) has been conducted at a depth of 3.20 m in a soil that has a level ground surface and a unit weight of 19.2 kN/m^3. According to this test, the horizontal geostatic σ_x at this point is 48 kPa. A proposed vertical point load of 1100 kN is to be applied to the ground surface at a point 1.10 m west of the test location. Using a Poisson's ratio of 0.37, compute the total σ_x, σ_z, and τ_{zx} at the test point after the load is applied.

9.17 The circular tank in Figure 9.29 imparts a bearing pressure of 3000 lb/ft^2 onto the soil below.

(a) Compute the geostatic vertical stress, σ_z, at Point A. This is the stress that existed before the tank was built.

(b) Using Figure 9.13, compute the induced vertical stress, $\Delta\sigma_z$ at Point A due to the weight of the tank.

(c) Combine the results from *a* and *b* to find the total σ_z at Point A after the tank is built.

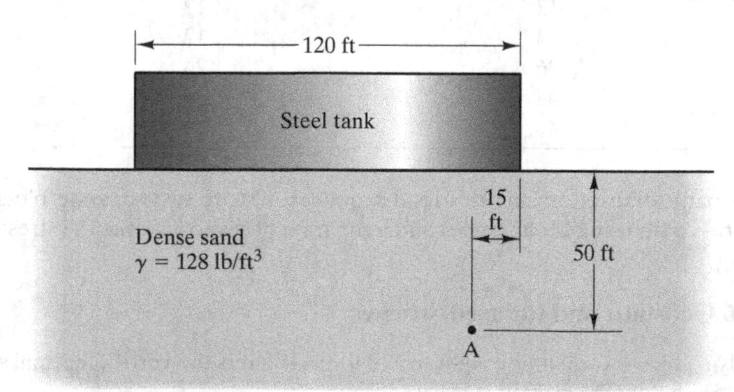

FIGURE 9.29 Storage tank and soil profile for Problem 9.17.

9.18 A second identical circular tank is constructed to the right of tank in Figure 9.29. The center to center spacing of the two tanks is 190 ft. Both tanks impart a bearing pressure of 3000 lb/ft^2 onto the soil below. Compute the induced vertical stress, $\Delta\sigma_z$, due to both tanks at a point midway between the two tanks at a depth of 120 ft.

Section 9.8 Effective Stresses

9.19 At a certain site the soil profile consists of a sandy soil with a total unit weight of 108 lb/ft^3 above the water table and 127 lb/ft^3 below the water table. The groundwater table is at a depth of 8 ft. Compute the total vertical stress, σ_z, and the effective vertical stress, σ_z', at a point 17 ft below the ground surface.

9.20 A lake with a water depth of 12 m is underlain by a soil with a total unit weight of 18.2 kN/m^3. Compute the total vertical stress, σ_z, and the effective vertical stress, σ_z' , at a point 8 m below the bottom of the lake.

9.21 Develop a plot of σ_z, u and σ_z' versus depth for the soil profile in Figure 9.11. Consider depths from 0 to 10 m, assume hydrostatic conditions are present, and assume $u = 0$ above the groundwater table. Plot depth, z, on the vertical axis, with zero at the top and increasing downward. This method of plotting the data is easier to visualize, because depth on the plot is comparable to depth in a cross-section.

9.22 Compute the values of σ_x, σ_x', σ_z, σ_z', and τ_{zx} at Point B in Figure 9.11. The coefficient of lateral earth pressure in the silty sand is 0.60. Draw both the total and effect stress Mohr circles for Point B on the same figure.

9.23 According to an in situ soil suction measurement, the pore water pressure at Point C in Figure 9.11 is −5.0 kPa. Compute the vertical effective stress at this point.

9.24 Use the "overall stresses" data in the x–z plane from Example 9.9 to perform the following computations:

(a) Draw the Mohr circles for total and effective stresses for this point and identify the pole and the locations on the circle that represent the vertical and horizontal stresses.

(b) Compute σ_1, σ_3, σ_1', σ_3', and τ_{max}.

(c) Determine the angle between the major principal stress and the vertical, then prepare as a sketch showing the orientation of the major and minor principal stresses with respect to the vertical.

(d) Compute σ, σ', and τ that act on a plane inclined at an angle of 45° clockwise from the horizontal.

Section 9.9 Effective Stress Under Steady-State Flow

9.25 Water is flowing vertically upward in sand with a saturated unit weight of 20 kN/m^3. At what gradient will quick conditions occur?

9.26 Compute the total vertical stress, σ_z, and the effective vertical stress, σ_z', at Point A in Figure 8.7.

Comprehensive

9.27 Using $K = 0.61$ and assuming the major principal stress acts vertically, compute the following at Point A in Figure 9.30.

(a) $\sigma_x, \sigma_z, \sigma_x', \sigma_z', \sigma_1, \sigma_3, \sigma_1', \sigma_3', u$

(b) Mohr circles for total and effective stresses

(c) σ, σ', and τ on the plane shown in the figure.

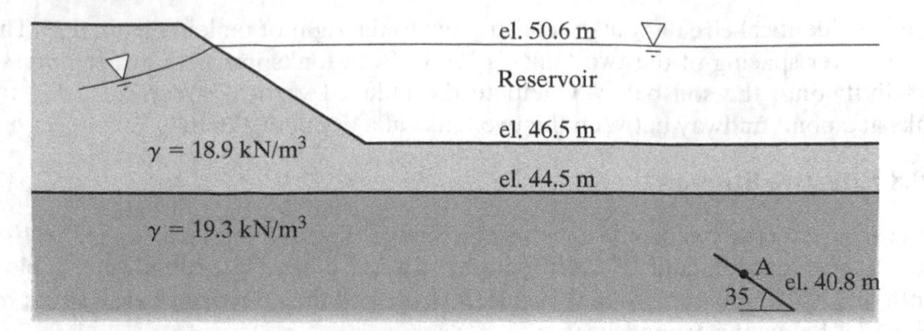

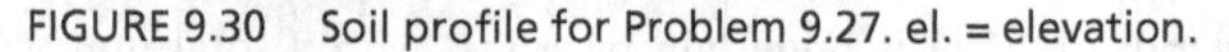
FIGURE 9.30 Soil profile for Problem 9.27. el. = elevation.

9.28 Using the approximate methods represented by Equations 9.39 and 9.40, create a spreadsheet to compute the induced verticle stress, $\Delta\sigma_z$, under the center of a square loaded area and an infinitely long strip load. Using your spreadsheet develop a plot $\Delta\sigma_z$ as a function of depth under the center of a square and a strip loads of the same width, B, with the same applied stress at the surface. Explain the differences between the stress induced by the square load and the strip load.

9.29 Create a spreadsheet to compute the induced normal stresses $\Delta\sigma_x$, $\Delta\sigma_y$, and $\Delta\sigma_z$, at any arbitrary point due to a point load at the ground surface using Boussinesq's method.

9.30 A vertical point load P is to be applied to a level ground surface. The underlying soil has the following preconstruction characteristics:

Groundwater table: 5.5 ft below the ground surface
Unit weight above the groundwater table = 121 lb/ft³
Unit weight below the groundwater table = 124 lb/ft³
$K = 0.87$
$\nu = 0.33$

The horizontal total stress, σ_x, at a point 8 ft below the ground surface and 3 ft east of the point of load application must not exceed 1000 lb/ft². Compute the maximum allowable value of P using the spreadsheet developed in Problem 9.29.

9.31 A 21.0-m diameter oil tank is to be built on a soil that has γ = 18.4 kN/m³, K = 0.60, and ν = 0.40. The tank and its contents have a total mass of 3.70 × 10⁶ kg, and the bottom of the tank is flush with the ground surface. The groundwater table is at a great depth. Compute the geostatic vertical total and effective stress, the induce vertical stress and final total and effective vertical stresses for the following two points:

(a) 8.0 m below the center of the tank
(b) 8.0 m below the east edge of the tank.

9.32 A proposed 5 ft × 5 ft spread footing foundation will support an office building. The column load plus the weight of the foundation will be 80 k, and the bottom of the foundation will be 2 ft below the ground surface. The unit weight of the soil is 121 lb/ft³ and the groundwater table is at a depth of 5 ft. Develop a plot of the vertical effective stress below the center of this foundation versus depth (after the footing has been placed and loaded). Consider depths from the bottom of the footing to 15 ft below the bottom of the footing.

9.33 The data in the following table were obtained from three borings at a certain site. The ground surface is level, and the groundwater table is at a depth of 3.7 m below the ground surface.

Develop a representative one-dimensional design soil profile for this site, similar to the one in Figure 3.38. Then develop plots of total vertical stress, pore water pressure, and effective vertical stress versus depth. All three plots should be superimposed on the same diagram, with the vertical axis (depth) increasing in the downward direction.

To develop the one-dimensional design soil profile, convert the information from the table into three boring logs. Then compare these logs, looking for similar soil types, and combine them into a single representative profile. Then use the γ_d and w values to compute the average unit weight for each stratum. Keep in mind that computations of the total stress are based on the unit weight, γ, not the dry unit weight, γ_d.

Boring	Depth (m)	Soil Classification	Dry Unit Weight (kN/m³)	Moisture Content (%)
1	0.6	Medium sand (SP)	18.1	8.2
1	1.2	Fine to medium sand (SW)	17.9	8.0
1	2.1	Medium sand (SP)	18.7	8.9
1	2.7	Silty sand (SM)	18.4	10.3
1	3.3	Silty sand (SM)	18.5	11.0
1	4.3	Sandy gravel (GW)	19.6	12.0
1	5.2	Gravel (GP)	19.9	11.4
1	6.1	Sandy silt (ML)	17.1	19.5
1	6.7	Silty clay (CL)	16.5	21.7
1	7.6	Silty clay (CL)	16.3	22.0
2	0.9	Fine sand (SP)	17.6	7.5
2	1.5	Fine sand (SP)	17.4	9.1
2	2.1	Fine to medium sand (SW)	18.7	9.5
2	2.7	Fine sand (SP)	18.2	9.9
2	3.7	Sandy silt (ML)	17.6	11.9
2	4.9	Gravel (GP)	19.9	11.0
2	6.1	Silt (ML)	16.2	22.8
3	1.2	Fine to medium sand (SW)	18.2	8.0
3	2.7	Silty sand (SM)	17.8	8.1
3	4.3	Gravelly sand (SW)	19.2	13.4
3	5.8	Silty clay (CL)	16.0	23.2
3	7.3	Clay (CL)	15.4	25.9

9.34 A point load is to be applied near an existing retaining wall as shown in Figure 9.31. Develop plots of the geostatic and induced horizontal contact pressure, $\Delta\sigma_x$, acting on the wall at the following immediately adjacent to the point load.

9.35 When combining stresses from multiple sources, Equation 9.35 instructs us to combine the total stresses using superposition, then subtract the pore water pressure. Why would it be incorrect to compute the various effective stresses, then combine them by superposition?

9.36 An excavation similar to the one in Figure 8.7 has recently been constructed and dewatered. Unfortunately, this excavation is beginning to show signs of incipient heave and/or quicksand problems. An analysis similar to the one in Example 9.8 confirms that this is a potential problem.

As an emergency measure, the contractor is proposing to remove the dewatering pumps, and fill the excavation with water. Evaluate this proposal and prepare a 200–300 word essay describing why this method would or would not provide temporary relief from the heave and quicksand problem.

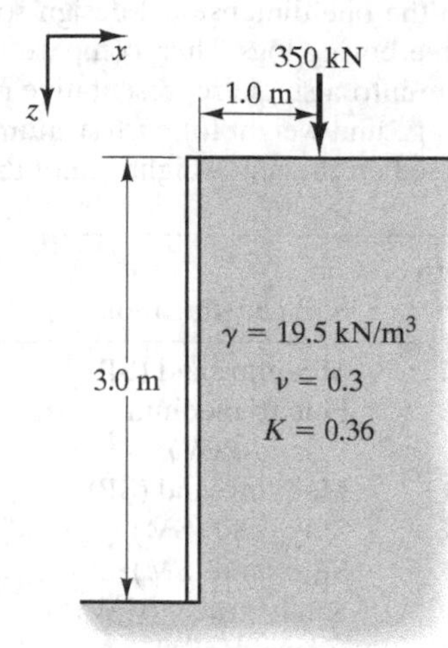

FIGURE 9.31 Cross section of retaining wall for Problem 9.34.

9.37 A point load and a square area load are to be applied to the ground surface as shown in Figure 9.32. Develop a plot of σ'_z versus depth below Point A. This plot should contain two curves: one that represents the pre-construction condition (i.e., without the applied loads) and one that represents the post-construction condition. The plot should extend from the ground surface to the bottom of the fat clay stratum.

9.38 A truck stop is to be built on a parcel of land adjacent to a major highway. During the planning stage of this project, the engineers found an existing 6 ft by 6 ft concrete box culvert under the proposed truck parking area, as shown in Figure 9.33. The project engineer is concerned that the weight of the parked trucks may overstress it, and has asked you to compute the vertical pressures acting on the top of the culvert. The results of your analyses will be provided to a structural engineer, who will then develop shear and moment diagrams and determine if the culvert can safely support the weight of the trucks.

(a) Compute the vertical pressure acting on the top of the culvert due to the weight of the overlying soil without any trucks. This is the same as the geostatic vertical stress at this depth, and represents the current condition. Use a unit weight of 120 lb/ft^3 and assume the groundwater table is at a depth of 45 ft.

(b) Compute the vertical pressure acting on the top of the culvert due to the wheel loads from a parked truck. This is the same as the induced vertical stress in the soil. Base your computations on two axles 48 in. apart, with the truck aligned parallel to the culvert. Perform all computations in the x–z plane of the first axle (i.e., $y = 0$ for the first axle, and $y = 48$ in. for the second). Each axle carries a total vertical load of 18,000 lb, which is evenly divided among its four wheels. You may assume each wheel acts as a point load. Repeat this computation for several values of x along the top of the culvert, then present your results in the form of a pressure diagram.

(c) Using superposition, combine the results from parts a and b.

Note: The culvert is stiffer than the soil, so the Boussinesq solution gives an approximate solution to this problem. A more precise analysis would need to consider the ratio of

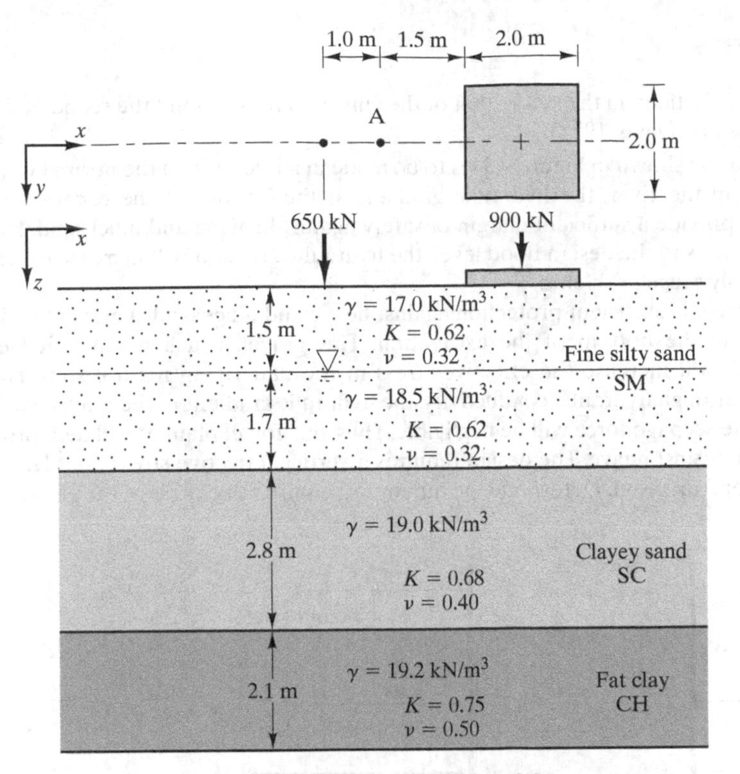

FIGURE 9.32 Plan and profile view for Problem 9.37.

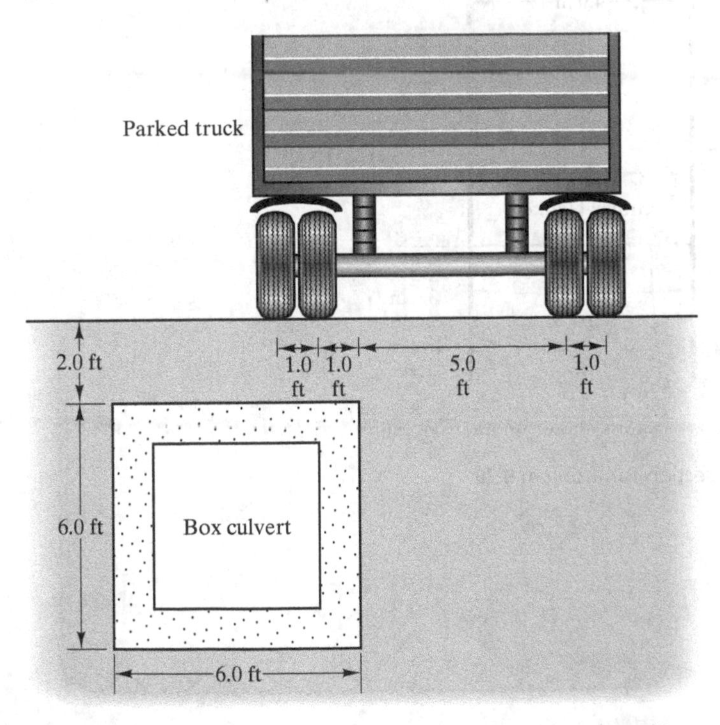

FIGURE 9.33 Existing box culvert below a proposed truck parking area for Problem 9.38.

modulus of elasticity in the soil to that of the culvert, and is beyond the scope of this book (see Poulos and Davis, 1974).

9.39 The excavation shown in Figure 9.34 is to be made in a river. When the normal water level is present in the river, the hydraulic gradient at the bottom of the excavation is low enough to provide a sufficient margin of safety against heaving and quicksand. However, if the river rises to the design flood level, the hydraulic gradient will increase to 1.1, which will probably cause problems.

To provide sufficient protection against heave and quicksand, a gravel blanket is to be placed in the bottom of the excavation. This gravel, which has a unit weight of 20.2 kN/m^3, will increase the effective stress in the underlying natural soils. However, because of its high hydraulic conductivity, the hydraulic gradient in the gravel will be very small, so the seepage force will be negligible. Thus, the gravel blanket will help protect the excavation against heave. The design requires a vertical effective stress of at least 25 kPa in the upper 3 m of soil. Determine the minimum required thickness of the gravel blanket.

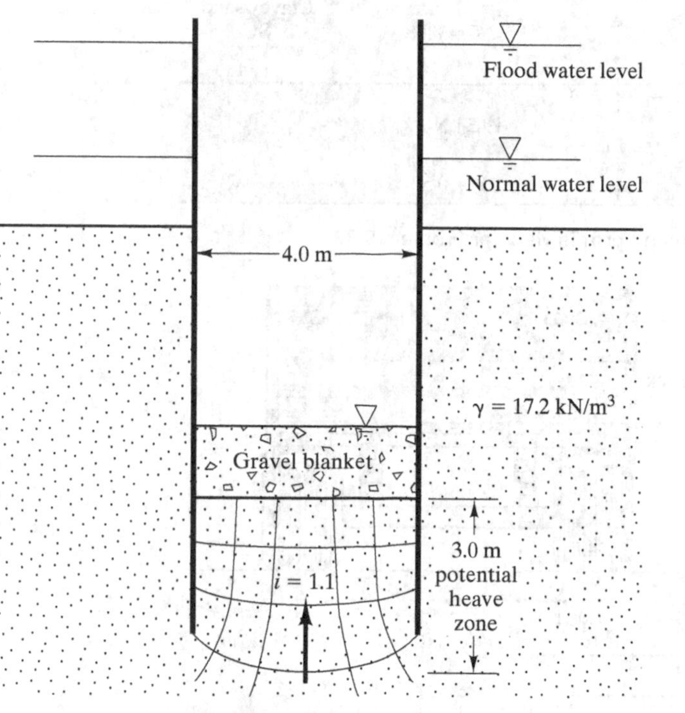

FIGURE 9.34 Cross-section for Problem 9.39.

CHAPTER 10

Compressibility and Settlement

Less than 10 years ago the Foundation Committee of a well-known engineering society decided, at one of its meetings, that the word "settlement" should be avoided in public discussions, because it might disturb the peace of mind of those who are to be served by the engineering profession.

Karl Terzaghi, 1939

In a mechanics of materials course, you learned that changes in the normal stresses in any object always produce corresponding normal strains, and the integration of these strains over the length of the object is its deformation or displacement. This principle also applies to soils, and is important because many civil engineering projects impart loads onto the ground, which produce corresponding increase in the vertical effective stress, σ'_z. Just as in any other material, these stress increases induce vertical strains, ε_z, in the soil, which cause the ground surface to move downward. We call this downward movement *settlement*. When settlement occurs over a large area, it is sometimes called *subsidence*. Whenever σ'_z increases, there always will be a corresponding settlement, δ. Thus, the issue facing a geotechnical engineer is not *if* settlements will occur, but rather the magnitudes of these settlements and how they compare with tolerable limits.

This chapter discusses the various factors that influence settlement and presents methods of predicting its magnitude. Chapter 11 continues these discussions and addresses the rate of settlement, then Chapters 14 and 15 apply these methods to the design of structural foundations. Engineers use the results of these settlement analyses to design structures and other civil engineering projects. For example, if an analysis indicates that the weight of a proposed building would cause excessive settlement in the soils below, with corresponding damage to the building, the geotechnical engineer may decide to use a different kind of foundation system, such as piles, that penetrate deeper into the ground to a harder and less compressible stratum.

Case Studies

Some of the most dramatic examples of soil settlements are found in Mexico City. Parts of the city are underlain by one of the most troublesome soils in any urban area of the world, a very soft lacustrine clay that was deposited in the former Lake Texcoco. Its engineering properties are as follows (Hiriart and Marsal, 1969):

Moisture content, w	Average 281%, maximum 500%
Liquid limit, w_L	Average 289, maximum 500
Plastic limit, w_P	Average 85, maximum 150
Void ratio, e	Average 6.90

A comparison of these values with the typical ranges given in Tables 4.5 and 4.8 demonstrates that this is an extraordinary soil. For example, the very high void ratio indicates that it contains, by volume, nearly seven times as much water as solids! Another of its important properties is an extremely high compressibility.

As the city grew, municipal water demands increased and many wells were installed through this clay and into deeper water-bearing sand layers. These activities resulted in a significant drop in the groundwater levels, which, as we will discuss later in this chapter, caused an increase in the effective stress. Because the clay is so compressible, and the stress increase was so large, the resulting settlements became a serious problem. Between 1898 and 1966, parts of the city settled by 6–7 m (Hiriart and Marsal, 1969)! At times, the rate of settlement has been as great as 1 mm/day. Fortunately, Mexican geotechnical engineers, such as Dr. Nabor Carrillo, recognized the connection between groundwater withdrawal and settlement, and convinced government authorities to prohibit pumping in the central city area.

In addition to widespread settlements in this area due to groundwater withdrawal, local settlements also have occurred beneath heavy structures and monuments. Their weight increased the stresses in the underlying soil, causing it to settle. One example is the Palacio de las Bellas Artes (Palace of Fine Arts), shown in Figure 10.1. It was built between 1904 and 1934, and experienced large settlements even before it was completed. By 1950, the palace and the immediately surrounding grounds were about 3 m lower than the adjacent streets (Thornley et al., 1955), which necessitated building new stairways from the street down to the building area.

As a result of these problems, geotechnical engineers in Mexico City have developed techniques for safely supporting large structures without the detrimental effects of excessive settlement. One of these, the 43-story Tower Latino Americana, is discussed in *Foundation Design: Principles and Practices* (Coduto, 2001), the companion volume to this book. This building is across the street from the Palace of Fine Arts, and has been performing successfully since its completion in the mid-1950s.

The Tower of Pisa in Italy is another example of excessive settlement. In this case, one side has settled more than the other, a behavior we call *differential settlement,* which gives the tower its famous tilt. *Foundation Design: Principles and Practices* (Coduto, 2001) also explores this case study.

Settlement problems are not limited to buildings. For example, the highway bridge shown in Figure 10.2 is underlain by a soft clay deposit. This soil is not able to

FIGURE 10.1 By 1950, the Palace of Fine Arts in Mexico City has settled about 3 m more than the surrounding streets.

FIGURE 10.2 The approach fills adjacent to this bridge in California have settled. However, the bridge, being supported on pile foundations, has not. Note the abrupt change in grade in the sidewalk, and the asphalt patch between the two signs. This photograph was taken about 12 years after the bridge was built.

support the weight of the bridge, so pile foundations were installed through the clay into harder soils below and the bridge was built on the piles. These foundations protect it from large settlements. It was also necessary to place fill adjacent to the bridge abutments so the roadway could reach the bridge deck. These fills are very heavy, so their weight increased σ'_z in the clay, causing it to settle. When this photograph was taken, about 12 years after the bridge was built, the fill had settled about 1 m, as shown by the sidewalk in the foreground.

10.1 PHYSICAL PROCESSES

Since settlement is a displacement caused by a change in stress, we can compute the magnitude of settlement if we know (a) the magnitude of the change in stress and (b) the stress–strain properties of soil. For many materials, such as steel and concrete, this is a relatively simple problem solved by measuring the stress–strain properties of the material (e.g., Young's modulus and Poisson's ratio), determining the applied stresses and then computing the strains and displacements. However, soils, because of their particulate and multiphase makeup, have a much more complicated stress–strain behavior that reflects multiple physical processes, some of which are time-dependent. Related to these various processes, we can define three different types of settlement:

- ***Distortion settlement***, δ_d, is the settlement that results from lateral movements of the soil in response to changes in σ'_z. Distortion settlement occurs without volume change and is similar to the Poisson's effect where an object loaded in the vertical direction expands laterally. Distortion settlements primarily occur when the load is confined to a small area, such as a structural foundation, or near the edges of large loaded areas, such as embankments.
- ***Consolidation settlement*** (also known as *primary consolidation settlement*), δ_c, occurs when a soil is subjected to an increase in σ'_z and the individual particles respond by rearranging into a tighter packing. The volume of the solid particles themselves remains virtually unchanged, so the change in the total volume (and the resulting strain) is due solely to a decrease in the volume of the voids, V_v. The pore water is, for practical purposes, incompressible, so if the soil is saturated ($S = 100\%$), this reduction in V_v can occur only if some of the pore water is squeezed out of the soil. All soils experience some consolidation when they are subjected to an increase in σ'_z, and this is usually the most important source of settlement. In some cases, this process occurs as quickly as the load is applied, but in other situations it occurs much more slowly. The rate of consolidation will depend on the degree of saturation of the soil, its hydraulic properties, and other factors.
- ***Secondary compression settlement***, δ_s, is primarily due to particle reorientation, creep, and decomposition of organic materials. This particle reorientation causes a reduction in the volume of the voids, like consolidation settlement. However, unlike consolidation settlement, secondary compression settlement is not due to changes in σ'_z; it occurs at a constant σ'_z. Secondary compression is

always time-dependent and can be significant in highly plastic clays, organic soils, and sanitary landfills, but it is negligible in sands and gravels.

The settlement at the ground surface, δ, is the sum of these three components:

$$\delta = \delta_d + \delta_c + \delta_s \tag{10.1}$$

Because soil settlement can have both time-dependent and nontime-dependent components, it is often categorized in terms of *short-term settlement* (or *immediate settlement*), which occurs as quickly as the load is applied, and *long-term settlement* (or *delayed settlement*), which occurs over some longer period. Many engineers associate consolidation settlement solely with the long-term settlement of clay soils. However, this is not strictly true as pointed out by Salgado (2008). Consolidation is related to volume change due to change in effective stress regardless of the soil type or time required for the volume change. Table 10.1 illustrates the relationships among soil type, sources of settlement, and their time dependence.

Other sources of settlement, such as that from underground mines, sinkholes, or tunnels, also can be important, but they are beyond the scope of our discussion.

10.2 CHANGES IN VERTICAL EFFECTIVE STRESS

Most settlement is due to changes in the vertical effective stress, so we will begin by examining these changes. The *initial vertical effective stress*, σ'_{z0}, at a point in the soil is the value of σ'_z before the event that causes settlement occurs. The *final vertical effective stress*, σ'_{zf}, is the value after the settlement process is complete. Notice how settlement analyses are based on changes in effective stress, not total stress.

The value of σ'_{z0} may be computed using the techniques described in Chapter 9. Usually the initial condition consists of geostatic stresses only and thus are evaluated using Equation 9.47.

The method of computing σ'_{zf} depends on the kind of event that is causing the stresses to increase. The most common events are placement of a fill, placement of an external load, and changes in the groundwater table elevation.

TABLE 10.1 Time-Rates and Magnitudes of Soil Settlement Processes[a]

	Soil Types			
	Clays and Silts		Sands and Gravels	
Time Frame	Process	Magnitude	Process	Magnitude
Short-term	Distortion	Negligible to small	Distortion Consolidation	Negligible to small Small to moderate
Long-term	Consolidation Secondary compression	Moderate to large Small to large	Secondary compression	Negligible to small

[a]Adapted from Salgado, 2008.

Stress Changes Due to Placement of a Fill

When a fill is placed on the ground, σ'_z in the underlying soil increases due to the weight of the fill. If the length and width of the fill are large compared to the depth of the point at which we wish to compute the stresses, the fill can be called an *areal fill*. When the point is beneath the central area of the fill, then we compute σ'_{zf} by simply adding another layer to the $\Sigma \gamma H$ of Equation 9.47. Therefore,

$$\sigma'_{zf} = \sigma'_{z0} + \gamma_{fill} H_{fill} \tag{10.2}$$

where:

σ'_{z0} = initial vertical effective stress

σ'_{zf} = final vertical effective stress

γ_{fill} = unit weight of the fill

H_{fill} = thickness of the fill

Unless stated otherwise, you may assume that all fills discussed in this book are areal fills satisfying these criteria and that Equation 10.2 is valid.

If the width or length of the fill are less than about twice the depth to the point at which the stresses are to be computed, or if this point is near the edge of the fill, then we need to evaluate the fill as an area load using the techniques described in Chapter 9, and compute the change in σ'_z using Equation 9.33.

Example 10.1

A 5.0-ft thick fill is to be placed on a site underlain by medium clay, as shown in Figure 10.3. Compute σ'_{z0} and σ'_{zf} at Point A.

Solution:

$$\sigma'_{z0} = \sum \gamma H - u$$

$$\sigma'_{z0} = (98\ \text{lb/ft}^3)(1.6\ \text{ft}) + (100\ \text{lb/ft}^3)(4.4\ \text{ft}) - (62.4\ \text{lb/ft}^3)(4.4\ \text{ft})$$

$$\sigma'_{z0} = \mathbf{322\ lb/ft^2}$$

$$\sigma'_{zf} = \sigma'_{z0} + \gamma_{fill} H_{fill}$$

$$\sigma'_{zf} = 322\ \text{lb/ft}^2 + (122\ \text{lb/ft}^3)(5.0\ \text{ft})$$

$$\sigma'_{zf} = \mathbf{932\ lb/ft^2}$$

Commentary: The placement of this fill will eventually cause σ'_z at Point A to increase from 322 to 932 lb/ft². The value of σ'_z at other depths in the natural soil will also increase, causing a vertical strain ε_z. As a result, the top of the natural ground will sink from elevation 10.6 ft to some lower elevation. Thus, the placement of a 5.0-ft thick fill will ultimately produce a ground surface that is less than 5.0 ft higher than the initial ground surface elevation.

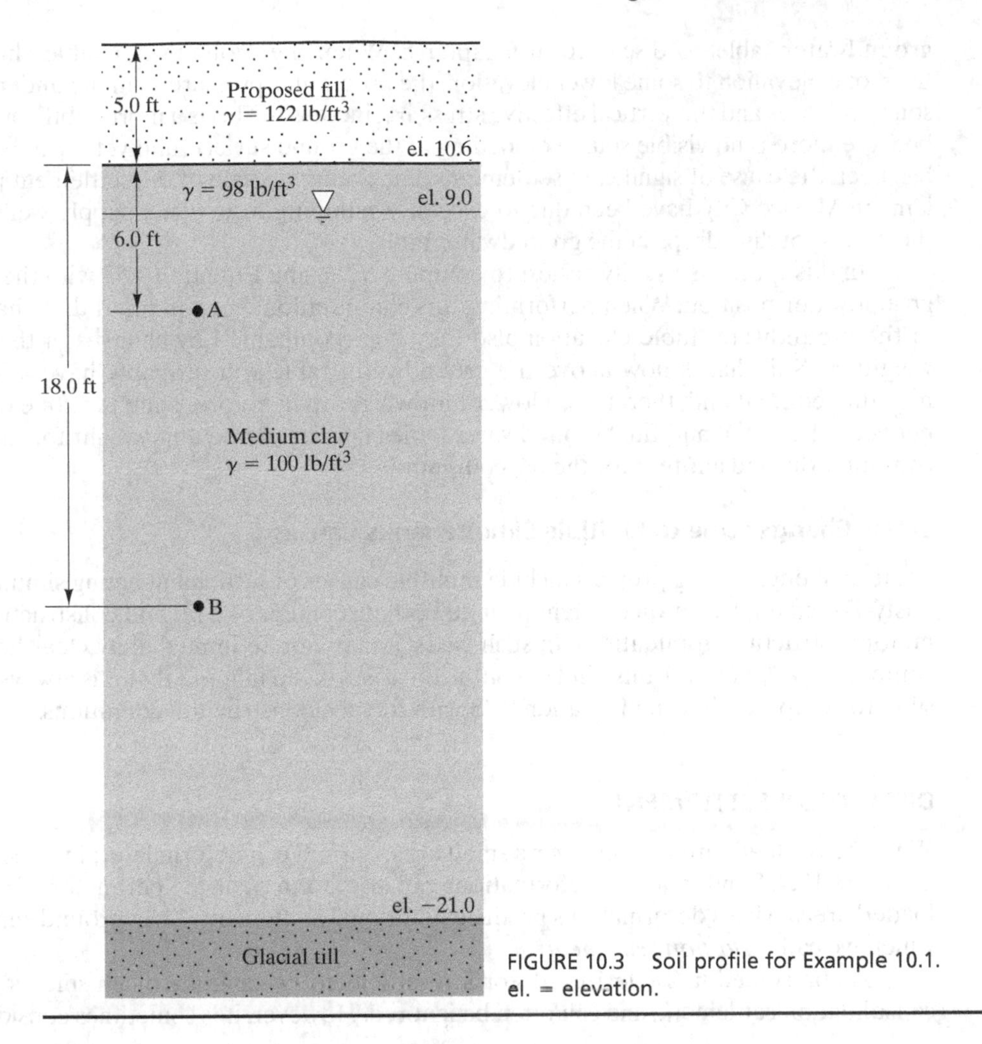

FIGURE 10.3 Soil profile for Example 10.1. el. = elevation.

Stress Changes Due to Placement of an External Load

External loads, such as structural foundations, also produce increases in σ'_z. In this case, σ'_{zf} is

$$\sigma'_{zf} = \sigma'_{z0} + \Delta\sigma_z \tag{10.3}$$

where $\Delta\sigma_z$ is the induced vertical stress computed using the techniques described in Chapter 9. This computation may be performed by hand using the equations in Section 9.6, or by programming the relevant equations into a spreadsheet.

Stress Changes Due to Changes in the Groundwater Table Elevation

Sometimes natural events or construction activities produce changes in the groundwater table elevation. For example, pumping from wells causes a drop in the nearby groundwater table, as discussed in Chapter 8. When the groundwater table changes from one elevation to some lower elevation, the pore water pressure, u, in the underlying soils decreases and the vertical effective stress, σ'_z, increases. This is a more subtle process because there is no visible source of loading at the ground surface, and yet it can be and has been the cause of significant settlements. For example, some of the settlement problems in Mexico City have been due to excessive pumping from water supply wells and the corresponding drops in the groundwater table.

In this case, it is usually easiest to compute σ'_{zf} using Equation 9.47 with the final groundwater position. When performing this computation, keep in mind that changes in the groundwater table elevation also may be accompanied by changes in the unit weight, γ. Soil that is now above the groundwater table will probably have a lower moisture content and, therefore, a lower unit weight than before. Thus, the zone of soil between the initial and final groundwater tables may have one unit weight for the σ'_{z0} computation, and another for the σ'_{zf} computation.

Stress Changes Due to Multiple Simultaneous Causes

Some civil engineering projects include multiple causes of settlement acting simultaneously. For example, a project might include both placement of a fill and construction of multiple structural foundations. In such cases, it may not be immediately clear how to compute σ'_{zf}. Whenever this kind of confusion arises, keep in mind that it is always possible to compute σ'_{zf} using Equation 9.48 with the postconstruction conditions.

10.3 DISTORTION SETTLEMENT

When heavy loads are applied over a small area, the soil can deform laterally, as shown in Figure 10.4. Similar lateral deformations can also occur near the perimeter of larger loaded areas. These deformations produce additional settlement at the ground surface, which we call *distortion settlement*.

As presented in Table 10.1, distortion settlement is generally much smaller than consolidation settlement, and can often be ignored. However, it is sometimes considered

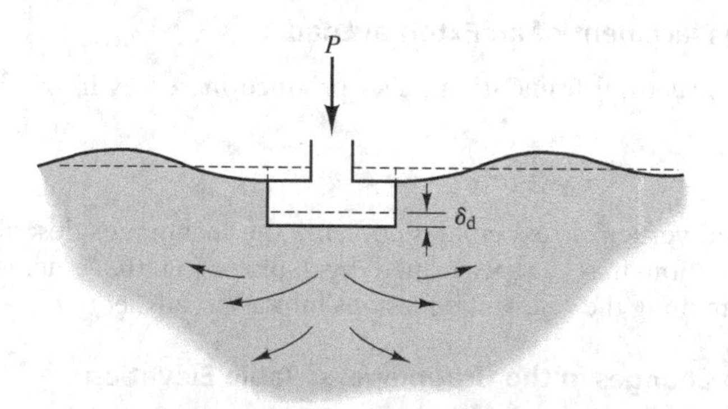

FIGURE 10.4 Distortion settlement beneath a small loaded area.

in the design of spread footing foundations. When distortion settlements need to be considered, they are normally computed using solutions based on elastic theory, as discussed in *Foundation Design: Principles and Practices* (Coduto, 2001). For the problems covered in this book, we will assume that distortion settlements are negligible.

10.4 CONSOLIDATION SETTLEMENT—PHYSICAL PROCESSES

We use the term *consolidation* to describe the pressing of soil particles into a tighter packing in response to an increase in effective stress, as shown in Figure 10.5. We assume that the volume of solids remains constant (i.e., the compression of individual particles is negligible); only the volume of the voids changes. The resulting settlement is known as *consolidation settlement*, δ_c. This is the most important source of settlement in soils, and its analysis is one of the cornerstones of geotechnical engineering.

Consolidation analyses usually focus on saturated soils ($S = 100\%$), which means the voids are completely filled with water. Both the water and the solids are virtually incompressible, so consolidation can occur only as some of the water is squeezed out of the voids. We can demonstrate this process by taking a saturated kitchen sponge and squeezing it; the sponge compresses, but only as the water is pushed out. This relationship between consolidation and pore water flow was qualitatively recognized as early as 1809 when the British engineer Thomas Telford placed a 17-m deep surcharge fill over a soft clay "for the purpose of squeezing out the water and consolidating the mud" (Telford, 1830; Skempton, 1960). The American engineer William Sooy Smith

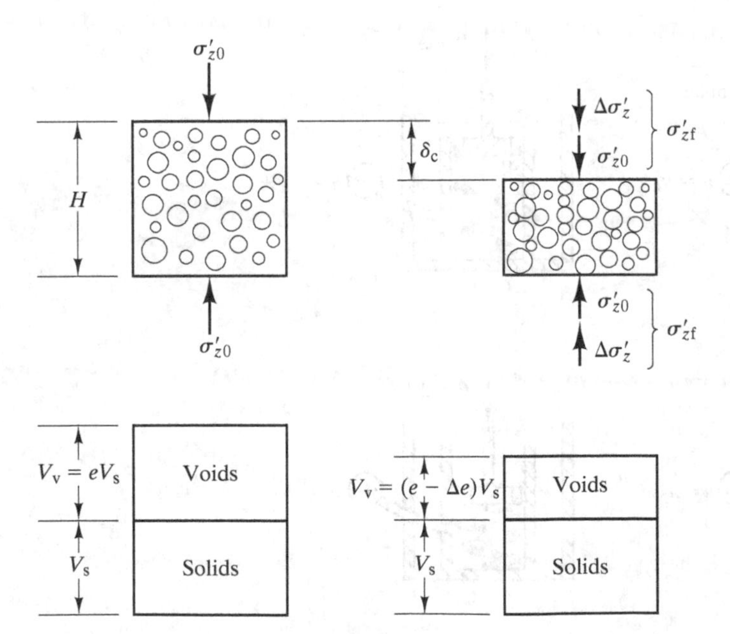

FIGURE 10.5 Consolidation of soil particles under the influence of an increasing vertical effective stress.

also recognized that "slow progressive settlements result from the squeezing out of the water from the earth" (Sooy Smith, 1892).

The first laboratory soil consolidation tests appear to have been performed around 1910 in France by J. Frontard. He placed samples of clay in a metal container, applied a series of loads with a piston, and monitored the resulting settlement (Frontard, 1914). Although these tests provided some insight, the underlying processes were not yet understood. About the same time, the German engineer Forchheimer developed a crude mathematical model of consolidation (Forchheimer, 1914).

Karl Terzaghi, who was one of Forchheimer's former students, made the major breakthrough. He was teaching at a college in Istanbul, and began studying the soil consolidation problem. This work, which he conducted between 1919 and 1923, produced the first clear recognition of the principle of effective stress, which led the way to understanding the consolidation process. Terzaghi's *Theory of consolidation* (Terzaghi, 1921, 1923a, 1923b, 1924, 1925a, and 1925b) is now recognized as one of the major milestones of geotechnical engineering. Although this theory includes several simplifications, it has been validated and is considered to be a good representation of the field processes. We will discuss it here and in Chapter 11.

Piston and Spring Analogy

To understand the physical process of consolidation and its relationship with the flow of pore water, let us consider the mechanical piston and spring analogy shown in Figure 10.6. This device consists of a piston and spring located inside a cylinder. The

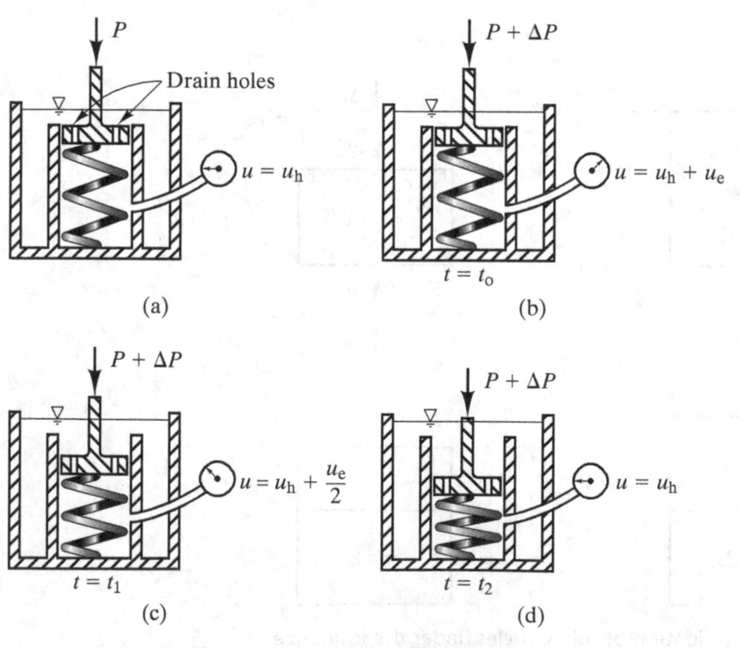

FIGURE 10.6 Piston and spring analogy.

cylinder is filled with water and small drain holes are present in the piston. These components represent an element of soil at some depth in the ground, with the spring representing the soil solids, the water representing the pore water, and the drain holes representing the soil voids through which the pore water must flow.

We will begin with the piston in static equilibrium under a certain vertical load, P, as shown in Figure 10.6(a). The assembly is submerged in a tank, so the water is subjected to a hydrostatic pressure that represents the hydrostatic pore water pressure, u_h, in the soil (see Equation 7.7). In addition, the water pressures on the top and bottom are equal, so the applied load on the piston is carried entirely by the spring. This load divided by the cross-sectional area of the cylinder represents the initial vertical effective stress, σ'_{z0}.

Then, at time $= t_0$, we apply an additional load ΔP to the piston, as shown in Figure 10.6(b). This represents the additional total vertical stress $\Delta\sigma_z$ in a soil, such as that induced by a new fill. It causes a very small downward movement of the piston, but this movement is resisted by both the spring and the water. The water is much stiffer than the spring, so it carries virtually all the additional load and the water pressure increases. This additional pressure is known as *excess pore water pressure*, u_e. Thus, the water pressure, u, inside the cylinder now equals $u_h + u_e$.

The water pressure (and the total head) inside the cylinder is now greater than that outside, so some of the water begins to flow through the holes. These holes are very small, so the flow rate through them is also small, but eventually a certain quantity of water passes through. The outward flow of water allows the piston to move farther down, thus compressing the spring and relieving some of the load from the water. This process represents the gradual transfer of stress from the pore water to the soil solids. Note the relationship between compression of the spring and dissipation of the excess pore water pressure. Understanding this relationship is key to solving the consolidation problem.

At time $= t_1$, as shown in Figure 10.6(c), half of $\Delta\sigma_z$ has been transferred to the soil solids and half is still being carried by the pore water. The process continues until the spring has compressed sufficiently to accommodate the original effective stress plus the additional stress as shown in Figure 10.6(d) (time $= t_2$). The excess pore water pressure is now zero, so flow through the holes ceases. We have returned to static equilibrium, but the piston is in a lower position than before. This change in position represents the vertical strain in that element of soil in the field.

Processes in the Field

The initial buildup of excess pore water pressures in soils is more complex than the piston and spring analogy because it depends on changes in both the mean normal stress, σ_m, and the deviator stress, σ_d, and on certain empirical coefficients known as *Skempton's pore pressure parameters A* and *B* (Skempton, 1954). However, we will simplify the problem by assuming the excess pore water pressure, u_e, immediately after loading is equal to $\Delta\sigma_z$.

This increase in pore water pressure produces a hydraulic gradient in the soil, causing some of the pore water to flow away. As each increment of water is discharged, the solid particles consolidate and begin to carry part of the new load, just as the spring compressed in our analogy. Thus, $\Delta\sigma_z$ is gradually transferred from the pore water to the soil solids, and the vertical effective stress $\Delta\sigma'_z$ rises. Eventually, all of the new load is carried by the solids, the pore water pressure returns to its hydrostatic value, and the flow of pore water ceases.

This transfer of load from water to solids is one of the most important processes in geotechnical engineering.

Example 10.2

The element of soil at point A in Figure 10.7 is initially subjected to the following stresses:

$$\sigma_{z0} = \sum \gamma H - u$$
$$\sigma_{z0} = (18.7\text{ kN/m}^3)(1.0\text{ m}) + (19.0\text{ kN/m}^3)(2.0\text{ m}) + (16.5\text{ kN/m}^3)(4.8\text{ m})$$
$$\sigma_{z0} = 136\text{ kPa}$$

$$u = \gamma_w z_w$$
$$u = (9.8\text{ kN/m}^3)(6.8\text{ m})$$
$$u = 67\text{ kPa}$$

$$\sigma'_{z0} = \sigma_{z0} - u$$
$$\sigma'_{z0} = 136 - 67\text{ kPa}$$
$$\sigma'_{z0} = 69\text{ kPa}$$

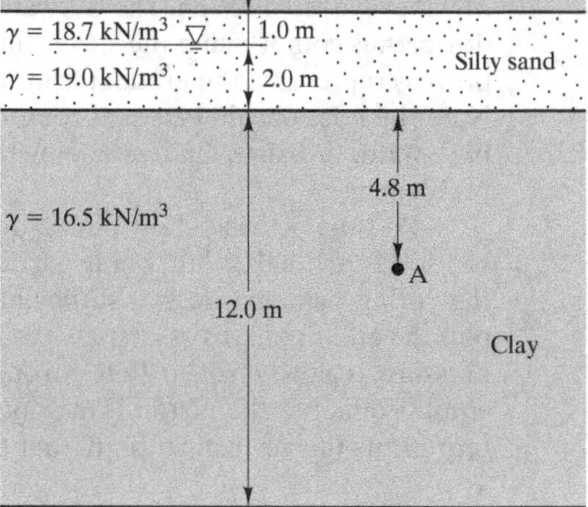

FIGURE 10.7 Soil profile for Example 10.2.

These conditions are illustrated on the left side of the plots in Figure 10.8. Then, we place a 5-m deep fill that has a unit weight of 19.5 kN/m³. This increases the vertical total stress to

$$\sigma_{zf} = \sigma_{z0} + \gamma_{fill} H_{fill} = 136\text{ kPa} + (19.5\text{ kN/m}^3)(5.0\text{ m}) = 234\text{ kPa}$$

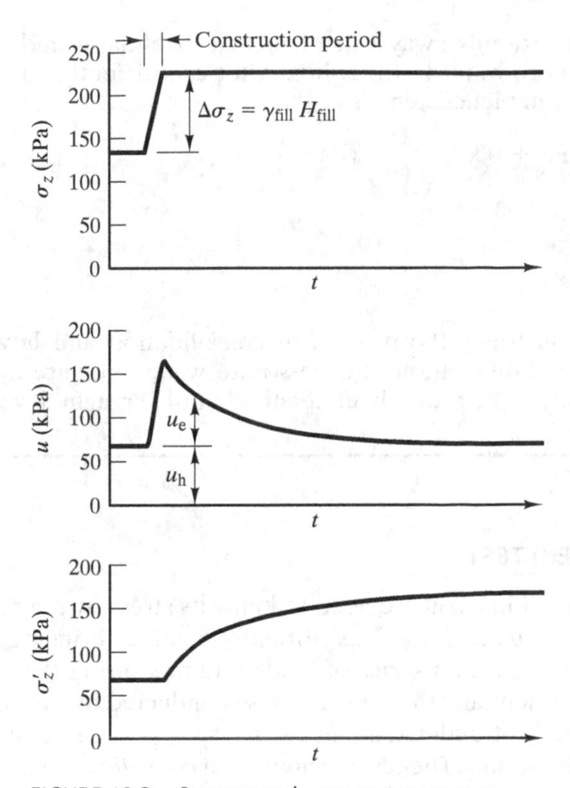

FIGURE 10.8 Stresses and pore water pressure at Point A in Example 10.2. σ_z steadily increases during the construction period, and then remains constant. Initially, it causes an equal increase in u, but as the excess pore water drains, the load is gradually transferred to the soil solids, causing σ'_z to increase.

Notice the jumps in the curves in Figure 10.8. Initially, the applied load is carried entirely by the pore water, so the pore water pressure becomes

$$u = u_h + u_e = u_h + \gamma_{fill} H_{fill} = 67 \text{ kPa} + (19.5 \text{ kN/m}^3)(5.0 \text{ m}) = 165 \text{ kPa}$$

but the vertical effective stress remains unchanged at

$$\sigma'_z = \sigma_z - u$$
$$\sigma'_z = 234 - 165 \text{ kPa}$$
$$\sigma'_z = 69 \text{ kPa}$$

As some of the pore water drains away, this element consolidates and $\Delta\sigma_z$ is gradually transferred from the pore water to the solids. After a sufficiently long time, $u_e = 0$ and the consolidation is complete. Then

$$u = u_h + u_e = 67 \text{ kPa} + 0 \text{ kPa} = 67 \text{ kPa}$$

$$\sigma'_z = \sigma_z - u$$
$$\sigma'_z = (234 - 67) \text{ kPa}$$
$$\sigma'_z = 167 \text{ kPa}$$

Commentary: This example illustrates the process of consolidation and how it is intimately tied to the buildup and dissipation of excess pore water pressures. It also illustrates why this process could not be properly understood until Terzaghi developed the principle of effective stress.

10.5 CONSOLIDATION (OEDOMETER) TEST

To predict consolidation settlement in a soil, we need to know its stress–strain properties (i.e, the relationship between σ'_z and ε_z). This normally involves bringing a soil sample to the laboratory, subjecting it to a series of loads, and measuring the corresponding settlements. This test is essentially the same as those conducted by Frontard in 1910, but now we have the benefit of understanding the physical processes, and thus can more effectively interpret the results. The test is known as a *consolidation test* (also known as an *oedometer test*), and is conducted in a *consolidometer* (or *oedometer*) as shown in Figure 10.9.

We are mostly interested in the engineering properties of natural soils as they exist in the field, so consolidation tests are usually performed on high-quality "undisturbed" samples. It is fairly simple to obtain these samples in soft to medium clays, and the test results are reliable. However, it is virtually impossible to obtain high-quality undisturbed samples in uncemented sands, so we use empirical correlations or in situ tests instead of consolidation tests to assess the stress–strain properties, as discussed in Section 10.7.

It is also important for samples that were saturated in the field to remain so during storage and testing. If the sample is allowed to dry, a process we call *desiccation*, negative pore water pressures will develop and may cause irreversible changes in the soil.

Sometimes engineers need to evaluate the consolidation characteristics of proposed compacted fills, and do so by performing consolidation tests on samples that have been remolded and compacted in the laboratory. These tests are usually less critical because a well-compacted fill generally has a low compressibility.

Test Procedure

The soil specimen, which has the shape of an upright cylinder, is placed in the consolidometer, and surrounded by a brass or stainless steel ring. The purpose of this ring is to maintain zero horizontal strain, thus producing *one-dimensional consolidation*.

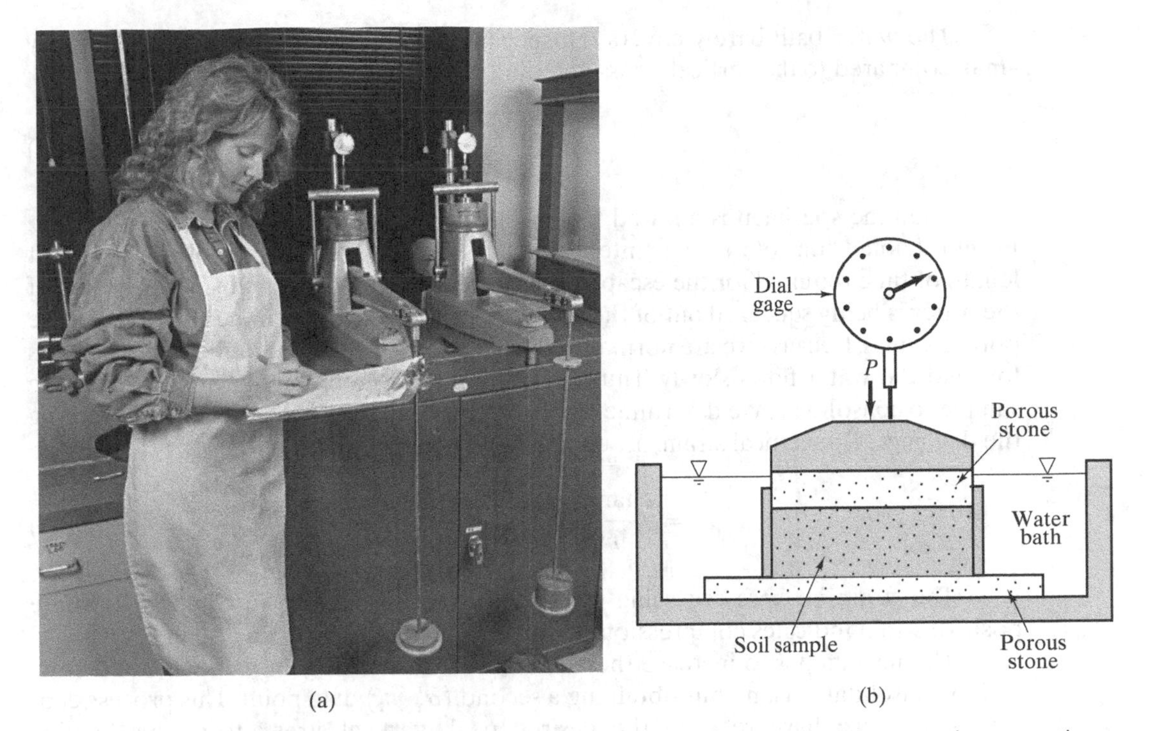

FIGURE 10.9 (a) Performing consolidation tests in the laboratory. The two consolidometers shown use the weights in the foreground to load the samples. (b) Cross-section of a consolidometer.

Porous stones are placed above and below the specimen. These stones are manufactured products that are strong enough to carry the applied loads, yet porous enough to allow water to pass through freely.

The specimen, rings, and porous stones are submerged in a water bath. This keeps the soil saturated, thus simulating the worst-case conditions in the field. A dial gage (or comparable electronic device) is placed above the specimen to measure its compression as the test progresses.

The test begins by applying a vertical normal load, P. It produces a vertical effective stress of

$$\sigma'_z = \frac{P}{A} - u \tag{10.4}$$

where:

σ'_z = vertical effective stress

P = applied load

A = cross-sectional area of soil specimen

u = pore water pressure inside soil specimen

The water bath barely covers the specimen, so the pore water pressure is very small compared to the vertical stress and thus may be ignored. Thus

$$\sigma'_z = \frac{P}{A} \tag{10.5}$$

Then the specimen is allowed to consolidate. While conducting these early tests, Frontard noted "one of the most interesting facts which have been revealed is the great length of time required for the escape of the excess water." During this period, some of the water is being squeezed out of the voids, and must pass through the soil to reach the porous stones. Because we are normally testing clayey soils, the hydraulic conductivity is low and the water flows slowly. Thus, several hours or more may be required for the sample to consolidate. We determine when the consolidation is complete by monitoring the dial gage. The vertical strain, ε_z, upon completion of consolidation is

$$\varepsilon_z = \frac{\text{change in dial gage reading}}{\text{initial height of sample}} \tag{10.6}$$

The strain is expressed using the sign convention defined in Chapter 9, where positive strain indicates compression. We now have one $(\sigma'_z, \varepsilon_z)$ data point.

The next step is to increase the normal load to some higher value and allow the soil to consolidate again, thus obtaining a second $(\sigma'_z, \varepsilon_z)$ data point. This process continues until we have reached the desired peak vertical stress; from this loading sequence we obtain the loading curve ABC in Figure 10.10. We then incrementally unload the sample and allow it to rebound, thus producing the unloading curve CD in Figure 10.10. The stress–strain curve for soil is decidedly nonlinear when plotted in an arithmetic space, as shown in Figure 10.10(a). However, when presented on a semilogarithmic plot, as shown in Figure 10.10(b), the same data are much easier to interpret. In semilogarithmic space, the stress-strain curve has three distinct segments and each is nearly linear, as shown in Figure 10.10(b). The segment AB is known as the *reloading curve* or *recompression curve*, BC is the *virgin curve*, and CD is the *unloading curve* or *rebound curve*. Because the semilogarithmic plot is easier to interpret, stress-strain curves for soil are almost always presented in semilogarithmic space.

Methods of Presenting Consolidation Test Results

Geotechnical testing laboratories use two different methods of presenting consolidation test results: a strain plot or a void ratio plot. The test results that arrive on a geotechnical engineer's desk could be presented in either form or both, so it is important to recognize the difference, and be able to use either method.

The first method uses a plot of ε_z versus log σ'_z, and thus is a direct representation of the data obtained in the laboratory. The horizontal and left axes of Figure 10.10(b) use this method. A strain plot is the most straightforward approach, because the purpose of a consolidation test is to measure the stress–strain properties of the soil.

The second method presents the data as a plot of void ratio, e, versus log σ'_z as shown in the horizontal and right axes in Figure 10.10(b). This was the method Terzaghi

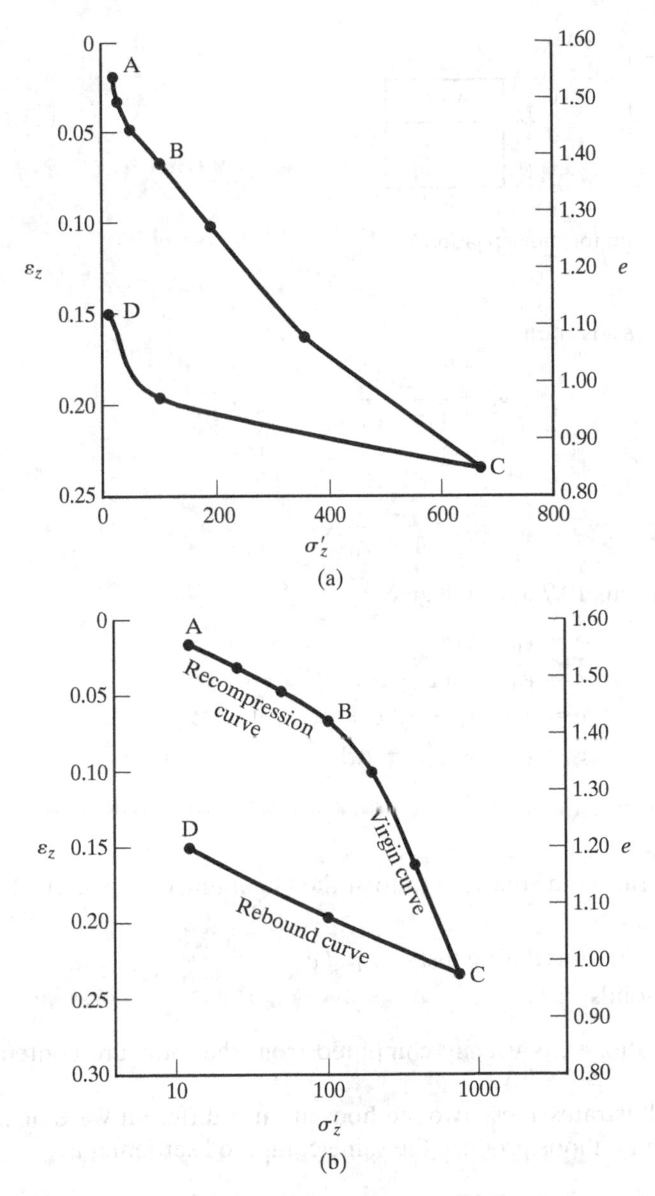

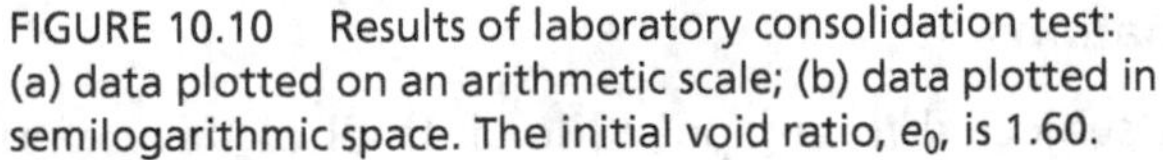
FIGURE 10.10 Results of laboratory consolidation test: (a) data plotted on an arithmetic scale; (b) data plotted in semilogarithmic space. The initial void ratio, e_0, is 1.60.

used, presumably because it emphasizes the reduction in void size that occurs during consolidation. To compute the void ratio at various stages of the test, we need to develop an equation that relates void ratio with strain. Using Figure 10.11,

$$eV_s = e_0V_s - \Delta eV_s \tag{10.7}$$

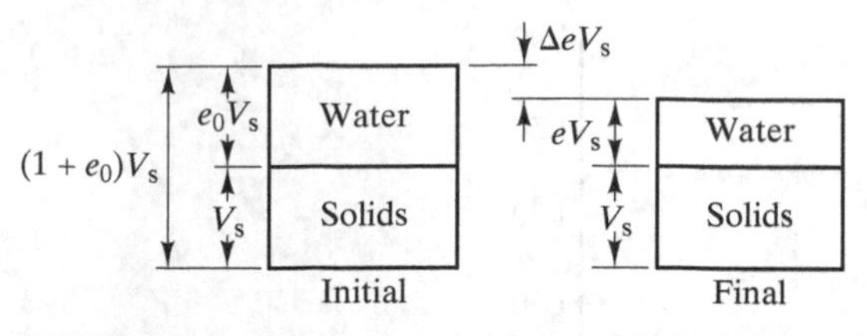

FIGURE 10.11 Phase diagram for the derivation of Equations 10.7–10.9.

the vertical strain, ε_z, is then

$$\varepsilon_z = \frac{\Delta e V_s}{(1 + e_0)V_s}$$

which simplifies to

$$\varepsilon_z = \frac{\Delta e}{1 + e_0} \tag{10.8}$$

Combining Equations 10.7 and 10.8 gives

$$e = e_0 - \Delta e$$
$$e = e_0 - \varepsilon_z(1 + e_0)$$
$$e = 1 + e_0 - \varepsilon_z(1 + e_0) - 1$$
$$e = (1 - \varepsilon_z)(1 + e_0) - 1 \tag{10.9}$$

where:

e = void ratio
e_0 = initial void ratio (i.e., the void ratio at the beginning of the test)
ε_z = vertical strain
Δe = change in void ratio during test = $e_0 - e$
V_s = volume of solids

The initial void ratio, e_0, is usually computed from the moisture content using Equation 4.26.

As Figure 10.10 illustrates, these two methods are just different ways of expressing the same data. Both methods produce the same computed settlements.

Plastic and Elastic Deformations

All materials deform when subjected to an applied load. If all of this deformation is retained when the load is released, it is said to have experienced *plastic deformation*. Conversely, if the material returns to its original size and shape when the load is released, it is said to have experienced *elastic deformation*. For example, we could illustrate plastic deformation by bending a copper wire and elastic deformation by bending a rubber hose. The copper will retain nearly all of its deformation, while the rubber will not.

Soil exhibits both plastic and elastic deformations, which is why we see two slopes in the loading and unloading curves in Figure 10.10(b). To understand this behavior, let

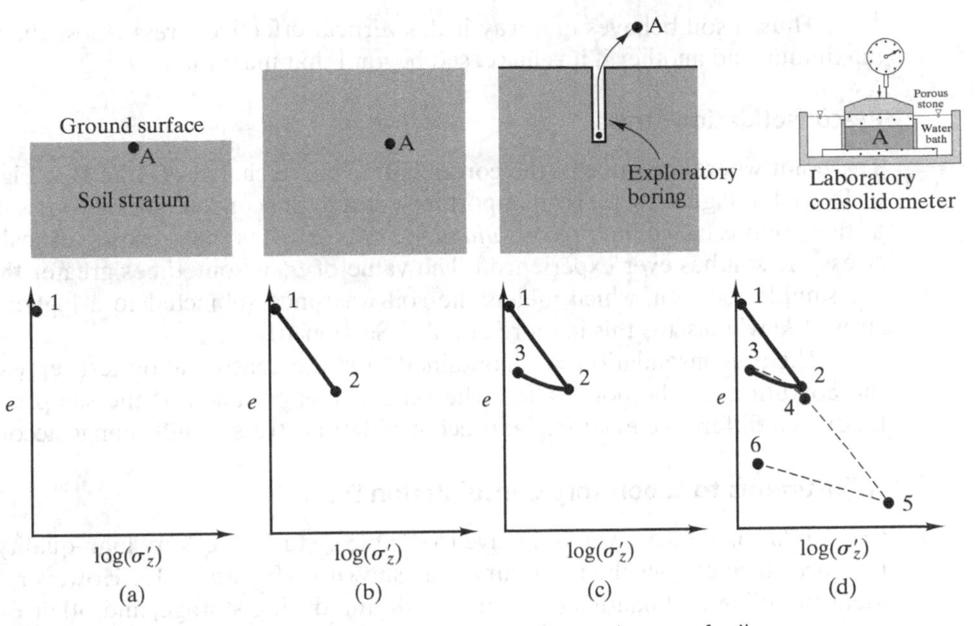

FIGURE 10.12 Soil profile and consolidation history for an element of soil.

us consider soil element A in the profile shown in Figure 10.12, and consolidation data for this element, also shown in Figure 10.12.

In Figure 10.12(a), the element of soil has recently been deposited and is described by Point 1 on the plot. The effective stress is low and the void ratio is high. Then, additional deposition occurs and the element becomes progressively buried by the newly deposited soil. The effective stress increases and the void ratio decreases (i.e., consolidation occurs). At this stage, both plastic and elastic deformations are occurring. Point 2 in Figure 10.12(b) describes the conditions that existed just prior to our drilling and sampling effort.

Then, we drill an exploratory boring, obtain an undisturbed sample from Element A, and take it to the laboratory. This process removes the overburden stress, so the effective stress drops and the sample expands slightly (i.e., the void ratio increases) as described by Point 3 in Figure 10.12(c). This expansion reflects the elastic portion of the compression that occurred naturally in the field. Although there has been some elastic rebound, most of the compression was plastic, so the unloading Curve 2–3 is much flatter than the loading Curve 1–2.

Once a test specimen from the sample has been installed in the laboratory consolidometer, we once again load it and produce Curve 3–4–5 as shown in Figure 10.12(d). The initial part, Curve 3–4, has already been defined as the reloading curve. It is nearly parallel to Curve 1–2 and reflects elastic compression only. The effective stress is less than the maximum past effective stress, so no new plastic deformation occurs. However, when the curve reaches Point 4, its slope suddenly changes, and Curve 4–5 reflects new plastic deformations, which occur only when the effective stress is higher than ever before. This is the virgin curve defined earlier. Finally, we unload the specimen in the lab and form Curve 5–6, the unloading curve. It is nearly parallel to Curves 2–3 and 3–4, and reflects the elastic component only.

Thus, a soil behaves one way if the vertical effective stress is less than the past maximum, and another if it is increased beyond that maximum.

Preconsolidation Stress

The point where the slope of the consolidation curve changes (Point B in Figure 10.10 or Point 4 in Figure 10.12) is an important event in the consolidation process. The stress at this point is called the *preconsolidation stress*, σ'_c. It is the greatest vertical effective stress the soil has ever experienced. The value of σ'_c is sometimes greater than σ'_{z0} at the sample location, which means the soil was once subjected to a higher effective stress. We will discuss this in more detail in Section 10.6.

The preconsolidation stress obtained from the consolidation test represents only the conditions at the point where the sample was obtained. If the sample had been taken at a different elevation, the preconsolidation stress would change accordingly.

Adjustments to Laboratory Consolidation Data

Consolidation tests are very sensitive to sample disturbance. Very high-quality samples produce distinct consolidation curves as shown in Figure 10.13. However, less than ideal sampling and handling techniques, drying during storage, and other effects can alter the sample and make the test results more obscure and difficult to interpret. It is especially difficult to obtain σ'_c from poor-quality samples because the curve near the transition between the recompression and virgin curves becomes much more rounded. Thus, it is best to be very careful with samples intended for consolidation tests.

Casagrande (1936) and Schmertmann (1955) developed methods of adjusting laboratory consolidation test results in an attempt to compensate for nominal sample

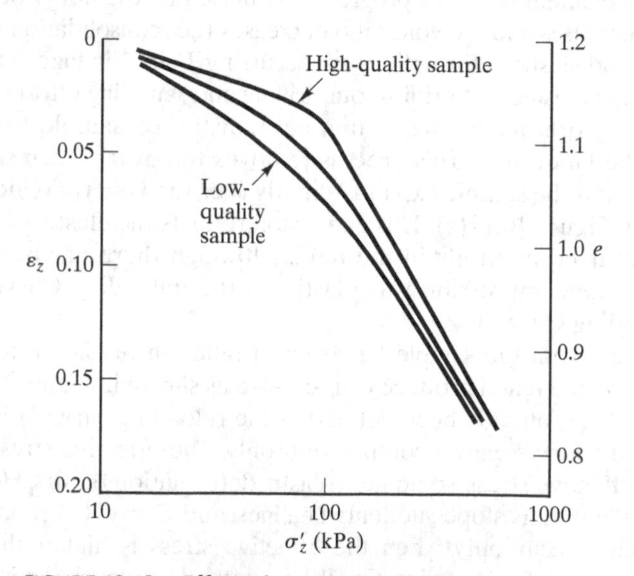

FIGURE 10.13 Effect of sample disturbance on consolidation test results.

disturbance effects. Both methods were developed primarily for soft clays, and are often more difficult to implement for stiffer soils.

The Casagrande procedure determines the preconsolidation stress, σ'_c, from laboratory data. Implement this method as follows, and as illustrated in Figure 10.14:

1. Locate the point of minimum radius on the consolidation curve (Point A).
2. Draw a horizontal line through Point A.
3. Draw a line tangent to the laboratory curve at Point A.
4. Draw the line that bisects the angle formed by the lines from Steps 2 and 3.
5. Extend the straight portion of the virgin curve upward until it intersects the line drawn in Step 4. This identifies Point B. The vertical effective stress at Point B is the preconsolidation stress, σ'_c. Note that e or ε_z corresponding to Point B is not used in any analysis.

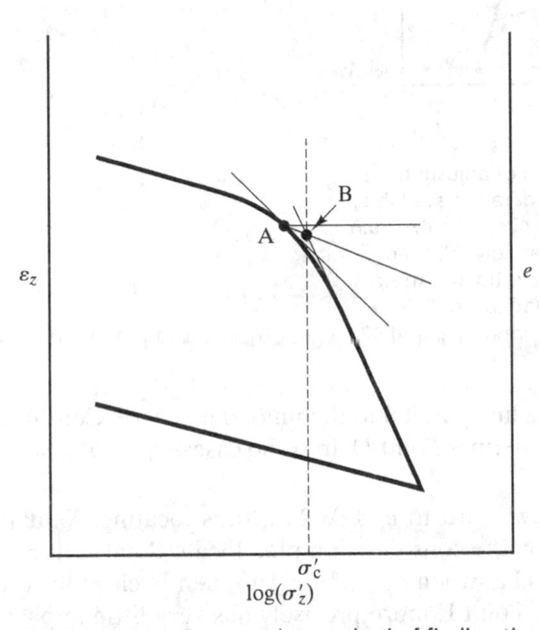

FIGURE 10.14 Casagrande's method of finding the preconsolidation stress.

Sample disturbance also affects the slope of the curves, so the Schmertmann procedure is an attempt to reconstruct the field consolidation curve (as illustrated in Figure 10.15). This procedure is performed as follows:

1. Determine σ'_c using the Casagrande procedure.
2. Compute the initial vertical effective stress, σ'_{z0}, at the sample depth. This is the vertical effective stress prior to placement of the proposed load.
3. Draw a horizontal line at $e = e_0$ (or $\varepsilon_z = 0$) from the vertical axis to σ'_{z0}. This locates Point C.

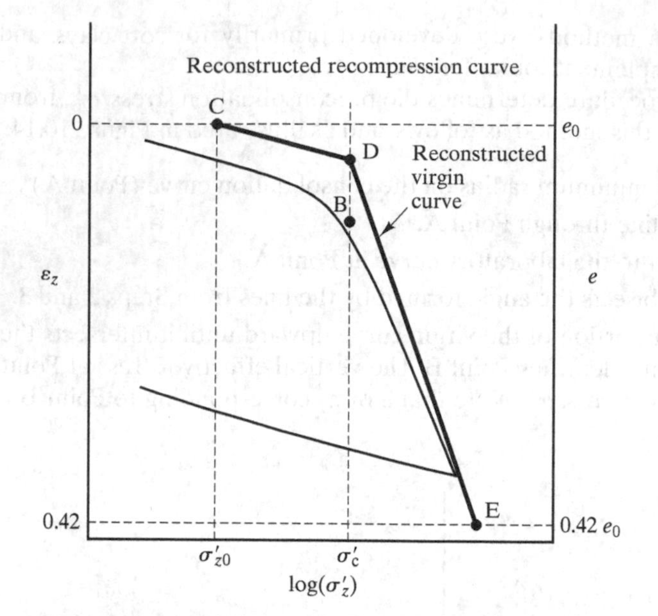

FIGURE 10.15 Schmertmann's method of adjusting consolidation test results. If void ratio data are available, then locate Point E at a void ratio of $0.42e_0$. If only strain data are available (i.e., no void ratios are given), then locate Point E at $\varepsilon_z = 0.42$. If both void ratio and strain data are given, use the void ratio data to locate Point E, even though this point may not correspond to $\varepsilon_z = 0.42$.

4. Beginning at Point C, draw a line parallel to the unloading curve. Continue to the right until reaching σ'_c. This defines Point D. In some cases, $\sigma'_c \approx \sigma'_{z0}$, so this step becomes unnecessary.
5. Extend the virgin curve downward to $e = 0.42\ e_0$, thus locating Point E. If no void ratio data is included on the consolidation plot, locate Point E at $\varepsilon_z = 0.42$, which is the same as $e = 0.42\ e_0$ when $e_0 = 2$ and sufficiently close for other initial void ratios (i.e., locating Point E more precisely has very little impact on the results of Schmertmann's construction).
6. Draw a line connecting points D and E. This is the reconstructed virgin curve.

The final result of the Casagrande and Schmertmann constructions is a bilinear function when plotted on a semilogarithmic diagram.

Soil Compressibility

The slopes on the consolidation plot reflect the *compressibility* of the soil. Steep slopes mean a given increase in σ'_z will cause a large strain (or a large change in void ratio), so such soils are said to be *highly compressible.* Conversely, shallow slopes indicate that the same increase in σ'_z will produce less strain, so the soil is *slightly compressible.*

Although we could use graphical constructions on these plots to determine the strain that corresponds to a certain increase in effective stress, it is much easier to do so mathematically as follows:

The slope of the virgin curve is defined as the *compression index*, C_c:

$$C_c = -\frac{de}{d \log \sigma'_z} \tag{10.10}$$

There is a potential point of confusion here, because geotechnical engineers also use the variable C_c to represent the coefficient of curvature, as defined in Equation 4.36. However, these are two entirely separate parameters. The compression index is a measure of the compressibility, while the coefficient of curvature describes the shape of the particle size distribution curve.

The reconstructed virgin curve is a straight line (on a semilogarithmic e versus. $\log \sigma'_z$ plot), so we can obtain a numerical value for C_c by selecting any two points, a and b, on this line and rewriting Equation 10.10 as

$$C_c = \frac{e_a - e_b}{(\log \sigma'_z)_b - (\log \sigma'_z)_a} \tag{10.11}$$

Alternatively, if the data are plotted only in $\varepsilon_z - \sigma'_z$ form (i.e., no void ratio data are given), then the slope of the virgin curve is:

$$\frac{C_c}{1 + e_0} = \frac{(\varepsilon_z)_b - (\varepsilon_z)_a}{(\log \sigma'_z)_b - (\log \sigma'_z)_a} \tag{10.12}$$

where the parameter $C_c/(1 + e_0)$ is called the *compression ratio*.

If the reconstructed virgin curve is sufficiently long, it is convenient to select Points a and b such that $(\sigma'_z)_b = 10\,(\sigma'_z)_a$. This makes the denominator of Equations 10.11 and 10.12 equal to 1, which simplifies the computation. This also demonstrates that C_c could be defined as the reduction in void ratio per tenfold increase (one log-cycle) in effective stress, as shown in Figure 10.16. Likewise, $C_c/(1 + e_0)$ is the strain per tenfold increase in effective stress.

In theory, the reloading and unloading curves have nearly equal slopes, but the rebound curve is more reliable because it is less sensitive to sample disturbance effects. This slope, which we call the *recompression index*, C_r, is defined in the same way as C_c and can be found using Equation 10.13 with Points c and d on the decompression curve:

$$C_r = \frac{e_c - e_d}{(\log \sigma'_z)_d - (\log \sigma'_z)_c} \tag{10.13}$$

If the data are plotted on a strain diagram, then the slope is $C_r/(1 + e_0)$, which is the *recompression ratio*:

$$\frac{C_r}{1 + e_0} = \frac{(\varepsilon_z)_d - (\varepsilon_z)_c}{(\log \sigma'_z)_d - (\log \sigma'_z)_c} \tag{10.14}$$

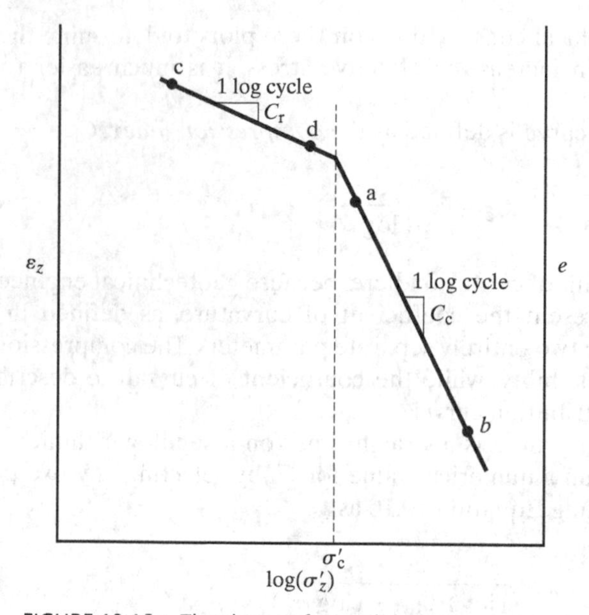

FIGURE 10.16 The slopes of consolidation curves on a semilogarithmic e versus σ'_z plot are C_c and C_r. The break in slope occurs at the preconsolidation stress, σ'_c.

Kulhawy and Mayne (1990) compared C_c and C_r values obtained from laboratory consolidation tests and from a theoretical soil called *modified cam clay* with the plasticity index, I_p, and found the following empirical correlations:

$$C_c = \frac{I_p}{74} \tag{10.15}$$

$$C_r = \frac{I_p}{370} \tag{10.16}$$

Most soils probably have C_c and C_r values within about ±50% of those predicted by Equations 10.15 and 10.16. These equations are useful for checking the reasonableness of laboratory test results and for performing preliminary analyses. However, final designs normally require actual laboratory tests on samples from the project site.

C_r values in saturated clays from conventional consolidation tests are typically about twice the true C_r in the field (Fox, 1995). This difference is due to the expansion of air bubbles in the pore water when the soil is unloaded during sampling and storage. This error is acceptable for most projects because the laboratory C_r is low enough that it does not produce large computed settlements (most consolidation settlement is due to C_c) and because it is conservative. However, when more precise measurements of C_r are needed, the consolidation tests can be performed in a special *backpressure consolidometer* that overcomes this problem.

Example 10.3

A consolidation test has been performed on a sample of soil obtained from Point A in Figure 10.7. The test results are shown in Figure 10.10. Compute σ'_c using Casagrande's method, then adjust the test results using Schmertmann's method. Finally, compute C_c and C_r.

Solution:

Stresses at sample depth:

From Example 10.2:

$$\sigma'_{z0} = 69 \text{ kPa}$$

From the Casagrande construction (Figure 10.17),

$$\sigma'_c = \mathbf{140 \ kPa}$$

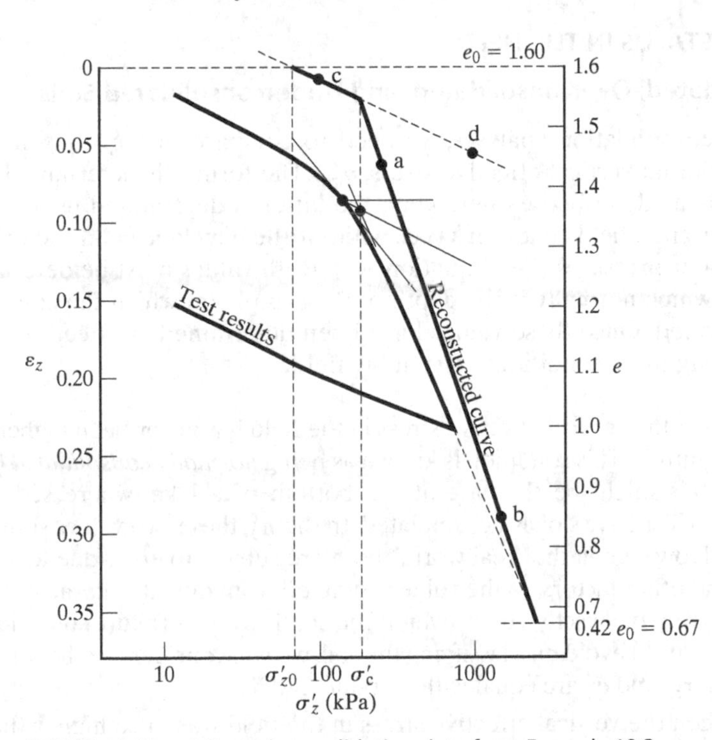

FIGURE 10.17 Adjusted consolidation data from Example 10.3.

Slopes of the reconstructed lines (Figure 10.17):

$$C_c = \frac{e_a - e_b}{(\log \sigma'_z)_b - (\log \sigma'_z)_a}$$

$$C_c = \frac{1.43 - 0.84}{\log 1580 \text{ lb/ft}^2 - \log 250 \text{ lb/ft}^2}$$

$$C_c = \mathbf{0.74}$$

$$C_r = \frac{e_c - e_d}{(\log \sigma'_z)_d - (\log \sigma'_z)_c}$$

$$C_r = \frac{1.58 - 1.46}{\log 1000 \text{ lb/ft}^2 - \log 100 \text{ lb/ft}^2}$$

$$C_r = \frac{0.12}{1}$$

$$C_r = \mathbf{0.12}$$

10.6 CONSOLIDATION STATUS IN THE FIELD

Normally Consolidated, Overconsolidated, and Underconsolidated Soils

When performing consolidation analyses, we need to compare the preconsolidation stress, σ'_c, with the initial vertical effective stress, σ'_{z0}. The former is determined from laboratory test data, as described earlier, while the latter is determined using Equation 9.47 with the original field conditions (i.e., without the new load) and the original hydrostatic pore water pressures (i.e., Equation 7.7). Both values must be determined at the same depth, which normally is the depth of the sample on which the consolidation test was performed. Once these values have been determined, we need to assess which of the following three conditions exist in the field:

- If $\sigma'_{z0} \approx \sigma'_c$, then the vertical effective stress in the field has never been higher than its current magnitude. This condition is known as being *normally consolidated* (NC). For example, this might be the case at the bottom of a lake, where sediments brought in by a river have slowly accumulated. In theory, these two values should be exactly equal. However, in the "real world" both are subject to error due to sample disturbance and other factors, so the values obtained from our site characterization program will rarely be exactly equal, even if the soil is truly normally consolidated. Therefore, in order to avoid misclassifying the soil, we will consider it to be normally consolidated if σ'_{z0} and σ'_c are equal within about $\pm 20\%$.
- If $\sigma'_{z0} < \sigma'_c$, then the vertical effective stress in the field was once higher than its current magnitude. This condition is known as being *overconsolidated* (OC) or *preconsolidated*. There are many processes that can cause a soil to become overconsolidated, including (Brumund et al., 1976):
  - Extensive erosion or excavation such that the ground surface elevation is now much lower than it once was.
  - Surcharge loading from a glacier, which has since melted.

- Surcharge loading from a structure, such as a storage tank, which has since been removed.
- Increase in the pore water pressure, such as from a rising groundwater table.
- Desiccation (drying) due to evaporation, plant roots, and other processes, which produces negative pore water pressures in the soil (Stark and Duncan, 1991).
- Chemical changes in the soil, such as the accumulation of cementing agents.
- Aging effects.

The term *overconsolidated* can be misleading because it implies there has been excessive consolidation. Although there are a few situations, such as cut slopes, where heavily overconsolidated soils can be less desirable, overconsolidation is almost always a good thing.

- If $\sigma'_{z0} > \sigma'_c$, the soil is said to be *underconsolidated*, which means the soil is still in the process of consolidating under a previously applied load. We will not be dealing with this case.

Table 10.2 gives a classification of soil compressibility based on $C_c/(1 + e_0)$ for normally consolidated soils or $C_r/(1 + e_0)$ for overconsolidated soils.

TABLE 10.2 Classification of Soil Compressibility[a]

$\frac{C_c}{1+e_0}$ or $\frac{C_r}{1+e_0}$	Classification
0–0.05	Very slightly compressible
0.05–0.10	Slightly compressible
0.10–0.20	Moderately compressible
0.20–0.35	Highly compressible
>0.35	Very highly compressible

[a]For soils that are normally consolidated, base the classification on $C_c/(1 + e_0)$. For soils that are overconsolidated, base it on $C_r/(1 + e_0)$.

Example 10.4

Using the consolidation test results developed in Example 10.3, determine whether the soil at point A in Figure 10.7 is normally consolidated or overconsolidated. The proposed fill has not yet been placed.

Solution:

At sample depth,

$\sigma'_{z0} = 69$ kPa   Per Example 10.2

$\sigma'_{z0} = 140$ kPa   Per Example 10.3

$\sigma'_{z0} < \sigma'_c$ by more than 20%, so **the soil is overconsolidated**

Overconsolidation Margin and Overconsolidation Ratio

The σ'_c values from the laboratory only represent the preconsolidation stress at the sample depth. However, we sometimes need to compute σ'_c at other depths (i.e., in a soil strata with the same geologic origin). To do so, compute the *overconsolidation margin*, σ'_m, using σ'_{z0} at the sample depth and the following equation:

$$\sigma'_m = \sigma'_c - \sigma'_{z0} \tag{10.17}$$

Table 10.3 presents typical values of σ'_m.

The overconsolidation margin should be approximately constant throughout in a stratum with common geologic origins. Therefore, we can compute the preconsolidation stress at other depths in that stratum by using Equation 10.17 with σ'_{z0} at the desired depth.

Another useful parameter is the *overconsolidation ratio* or OCR:

$$OCR = \frac{\sigma'_c}{\sigma'_{z0}} \tag{10.18}$$

Unlike the overconsolidation margin, OCR varies as a function of depth, and therefore cannot be used to compute σ'_c at other depths in a stratum. For normally consolidated soils, OCR = 1 and does not vary with depth.

TABLE 10.3 Typical Ranges of Overconsolidation Margins

Overconsolidation Margin, σ'_m		
(kPa)	(lb/ft^2)	Classification
0	0	Normally consolidated
0–100	0–2000	Slightly overconsolidated
100–400	2000–8000	Moderately overconsolidated
400	8000	Heavily overconsolidated

10.7 COMPRESSIBILITY OF SANDS AND GRAVELS

The principles of consolidation apply to all soils, but the consolidation test described in Section 10.5 and the methods of assessing consolidation status in the field, as described in Section 10.6, are primarily applicable to clays and silts. It is very difficult to perform reliable consolidation tests on most sands because they are more prone to sample disturbance, and this disturbance has a significant effect on the test results. Clean sands are especially troublesome. Gravels have similar sample disturbance problems, plus their large grain size would require very large samples and a very large consolidometer.

Fortunately, sands and gravels subjected to static loads are much less compressible than silts and clays, so it is often sufficient to use estimated values of C_c or C_r in lieu of laboratory tests. For sands, these estimates can be based on the data gathered by Burmister (1962) as interpreted in Table 10.4. He performed a series of consolidation

TABLE 10.4 Typical Consolidation Properties of Saturated Normally Consolidated Sandy Soils at Various Relative Densities[a]

Soil Type	$C_c/(1+e_0)$					
	$D_r = 0\%$	$D_r = 20\%$	$D_r = 40\%$	$D_r = 60\%$	$D_r = 80\%$	$D_r = 100\%$
Medium to coarse sand, some fine gravel (SW)	–	–	0.005	–	–	–
Medium to coarse sand (SW/SP)	0.010	0.008	0.006	0.005	0.003	0.002
Fine to coarse sand (SW)	0.011	0.009	0.007	0.005	0.003	0.002
Fine to medium sand (SW/SP)	0.013	0.010	0.008	0.006	0.004	0.003
Fine sand (SP)	0.015	0.013	0.010	0.008	0.005	0.003
Fine sand with trace fine to coarse silt (SP-SM)	–	–	0.011	–	–	–
Find sand with little fine to coarse silt (SM)	0.017	0.014	0.012	0.009	0.006	0.003
Fine sand with some fine to coarse silt (SM)	–	–	0.014	–	–	–

[a]Adapted from Burmister, 1962.

tests on samples reconstituted to various relative densities. Engineers can estimate the in situ relative density using the methods described in Chapter 4, then select an appropriate $C_c/(1 + e_0)$ from this table. Note that all of these values are "very slightly compressible" as defined in Table 10.2.

For saturated overconsolidated sands, $C_r/(1 + e_0)$ is typically about one-third of the values listed in Table 10.4, which makes such soils nearly incompressible. Compacted fills can be considered to be overconsolidated, as can soils that have clear geologic evidence of preloading, such as glacial tills. Therefore, many settlement analyses simply consider the compressibility of such soils to be zero. If it is unclear whether a soil is normally consolidated or overconsolidated, it is conservative to assume it is normally consolidated.

Very few consolidation tests have been performed on gravelly soils, but the compressibility of these soils is probably equal to or less than those for sand, as listed in Table 10.4.

Another characteristic of sands and gravels is their high hydraulic conductivity, which means any excess pore water drains very quickly. Thus, the rate of consolidation is very fast, and typically occurs nearly as fast as the load is applied. Thus, if the load is due to a newly placed fill, the consolidation of these soils may have little practical significance.

However, there are at least two cases where consolidation of coarse-grained soils can be very important and needs more careful consideration:

1. **Loose sandy soils subjected to dynamic loads, such as those from an earthquake.** They can experience very large and irregular settlements that can cause serious damage. Kramer (1996) discusses methods of evaluating this problem.

2. **Sandy or gravelly soils that support shallow foundations.** Structural foundations are often very sensitive to settlement, so we often conduct more precise assessments of compressibility. These are usually done using in situ tests, such as the SPT or CPT, and often expressed in terms of a modulus of elasticity, E, instead of C_c or C_r. Special analyses based on in situ test results are available to predict such settlements, as discussed in Chapter 15.

10.8 CONSOLIDATION SETTLEMENT PREDICTIONS

The purpose of performing consolidation tests is to define the stress–strain properties of the soil and thus they allow us to predict consolidation settlements in the field. We perform this computation by projecting the laboratory test results (as contained in the parameters C_c, C_r, e_0, and σ'_c) back to the field conditions. For simplicity, the discussions of consolidation settlement predictions in this chapter consider only the case of one-dimensional consolidation, and we will be computing only the ultimate consolidation settlement.

One-dimensional consolidation means only vertical strains occur in the soil (i.e., $\varepsilon_x = \varepsilon_y = 0$). We can reasonably assume this is the case when at least one of the following conditions exist (Fox, 1995):

1. The width of the loaded area is at least four times the thickness of the compressible strata.
2. The depth to the top of the compressible strata is at least twice the width of the loaded area.
3. The compressible strata lie between stiffer soil strata whose presence tends to reduce the magnitude of horizontal strains.

In this context, "compressible strata" refers to the strata that have a C_c or C_r large enough to contribute significantly to the settlement.

The most common one-dimensional consolidation problems are those that evaluate settlement due to the placement of a long and wide fill or due to the widespread lowering of the groundwater table. Many other problems, such as foundations, may also be idealized as being one-dimensional.

The *ultimate consolidation settlement*, $\delta_{c,ult}$, is the value of δ_c after all of the excess pore water pressures have dissipated, which may require many years or even decades. Chapter 11 explores this topic in more detail, and presents methods of developing time-settlement curves.

Normally Consolidated Soils ($\sigma'_{z0} \approx \sigma'_c$)

If $\sigma'_{z0} \approx \sigma'_c$, the soil is, by definition, normally consolidated. Thus, the initial and final conditions are as shown in Figure 10.18, and the compressibility is defined by C_c, the slope of the virgin curve.

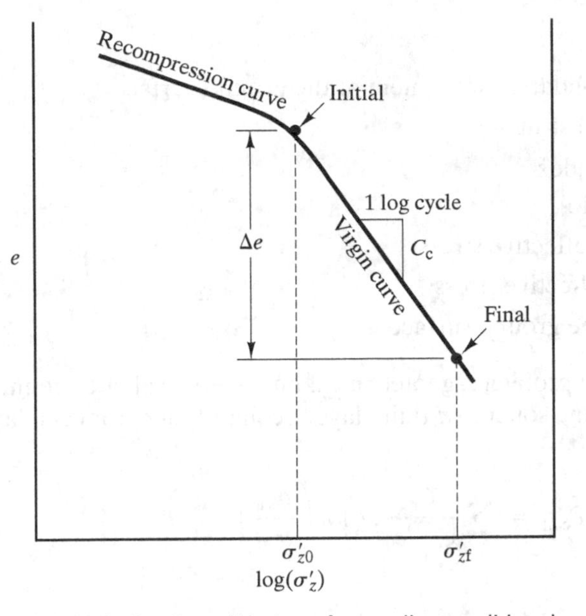

FIGURE 10.18 Consolidation of normally consolidated soils.

Rewriting Equation 10.10 gives

$$\mathrm{d}e = -C_c \mathrm{d} \log \sigma'_z \tag{10.19}$$

$$\Delta e = -C_c \log\left(\frac{\sigma'_{zf}}{\sigma'_{z0}}\right) \tag{10.20}$$

Combining with Equation 10.8 gives the vertical strain in the element of soil, ε_z:

$$\begin{aligned} \varepsilon_z &= -\frac{\Delta e}{1 + e_0} \\ \varepsilon_z &= \frac{C_c}{1 + e_0} \log\left(\frac{\sigma'_{zf}}{\sigma'_{z0}}\right) \end{aligned} \tag{10.21}$$

Integrating over the depth of the soil gives the consolidation settlement at the ground surface, δ_c:

$$\begin{aligned} \delta_{c,ult} &= \int \varepsilon_z dz \\ \delta_{c,ult} &= \int \frac{C_c}{1 + e_0} \log\left(\frac{\sigma'_{zf}}{\sigma'_{z0}}\right) dz \end{aligned} \tag{10.22}$$

where:

$\delta_{c,ult}$ = ultimate consolidation settlement at the ground surface
ε_z = vertical normal strain
C_c = compression index
e_0 = initial void ratio
σ'_{z0} = initial vertical effective stress
σ'_{zf} = final vertical effective stress
z = depth below the ground surface

For nearly all practical problems, geotechnical engineers evaluate the integral in Equation 10.22 by dividing the soil into n finite layers, computing δ_c for each layer, and summing

$$\delta_{c,ult} = \sum \frac{C_c}{1 + e_0} H \log\left(\frac{\sigma'_{zf}}{\sigma'_{z0}}\right) \tag{10.23}$$

where:

H = thickness of the soil layer

When using Equation 10.23, compute σ'_{z0} and σ'_{zf} at the midpoint of each layer.

Overconsolidated Soils—Case I ($\sigma'_{z0} < \sigma'_{zf} \leq \sigma'_c$)

If neither σ'_{z0} nor σ'_{zf} exceed σ'_c, the entire consolidation process occurs on the recompression curve as shown in Figure 10.19. The analysis is thus identical to that for normally consolidated soils except we use the recompression index, C_r, instead of the compression index, C_c:

$$\delta_{c,ult} = \sum \frac{C_r}{1 + e_0} H \log\left(\frac{\sigma'_{zf}}{\sigma'_{z0}}\right) \tag{10.24}$$

Overconsolidated Soils—Case II ($\sigma'_{z0} < \sigma'_c < \sigma'_{zf}$)

If the consolidation process begins on the recompression curve and ends on the virgin curve, as shown in Figure 10.19, then the analysis must consider both C_c and C_r:

$$\delta_{c,ult} = \sum \left[\frac{C_r}{1 + e_0} H \log\left(\frac{\sigma'_c}{\sigma'_{z0}}\right) + \frac{C_c}{1 + e_0} H \log\left(\frac{\sigma'_{zf}}{\sigma'_c}\right)\right] \tag{10.25}$$

This condition is quite common, because many soils that might appear to be normally consolidated from a geologic analysis actually have a small amount of overconsolidation (Mesri et al., 1994).

When using Equation 10.25, σ'_{z0}, σ'_c, and σ'_{zf} must be computed at the midpoint of each layer. This means σ'_c will need to be computed using Equation 10.17.

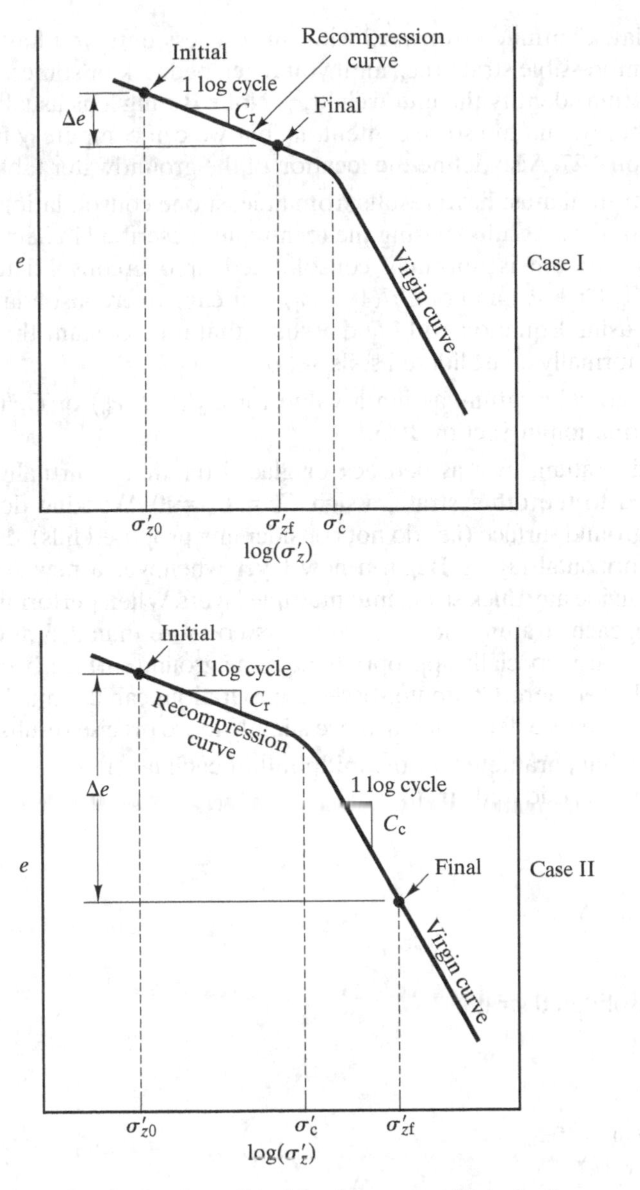

FIGURE 10.19 Consolidation of overconsolidated soil.

Ultimate Consolidation Settlement Analysis Procedure

Use the following procedure to compute $\delta_{c,ult}$:

1. Beginning at the original ground surface, divide the soil profile into strata, where each stratum consists of a single soil type with common geologic origins. For example, one stratum may consist of a dense sand, while another might be a

soft-to-medium clay. Continue downward with this process until you have passed through all the compressible strata (i.e., until you reach bedrock or some very hard soil). For each stratum, identify the unit weight, γ. Note: Boring logs usually report the dry unit weight, γ_d, and moisture content, w, but we can compute γ from this data using Equation 4.27. Also define the location of the groundwater table.

2. Each clay or silt stratum must have results from at least one consolidation test (or at least estimates of these results). Using the techniques described in Section 10.4, determine if each stratum is normally consolidated or overconsolidated, then assign values for $C_c/(1 + e_0)$ and/or $C_r/(1 + e_0)$. For each overconsolidated stratum, compute σ'_m using Equation 10.17 and assume that it is constant throughout that stratum. For normally consolidated soils, set $\sigma'_m = 0$.
3. For each sand or gravel stratum, assign a value for $C_c/(1 + e_0)$ or $C_r/(1 + e_0)$ based on the information in Section 10.5.
4. For any very hard stratum, such as bedrock or glacial till, that is virtually incompressible compared to the other strata, assign $C_c = C_r = 0$. Working downward from the original ground surface (i.e., do not consider any proposed fills), divide the soil profile into horizontal layers. Begin a new layer whenever a new stratum is encountered, and divide any thick strata into multiple layers. When performing computations by hand, each stratum should have layers no more than 2–5 m (5–15 ft) thick. Thinner layers are especially appropriate near the ground surface, because the strain is generally larger there. Computer-based computations can use much thinner layers throughout the entire depth, and achieve slightly more precise results.
5. Tabulate the following parameters at the midpoint of each layer:

   For normally consolidated strata,

   σ'_{z0}
   σ'_{zf}
   $C_c/(1 + e_0)$
   H

   For overconsolidated strata:

   σ'_{z0}
   σ'_{zf}
   $\sigma'_c = \sigma'_{z0} + \sigma'_m$
   $C_c/(1 + e_0)$
   $C_r/(1 + e_0)$
   H

   It is not necessary to record these parameters in incompressible strata.

   Normally we compute σ'_{z0} and σ'_{zf} using the hydrostatic pore water pressures (Equation 7.7) with no significant seepage force, and this is the only case we will consider in this book. However, if preexisting excess pore water pressures or significant seepage forces are present, they should be evaluated. Sometimes this may require the installation of piezometers to obtain accurate information on the in situ pore water pressures.

6. Using Equations 10.23–10.25, compute the consolidation settlement for each layer, then sum to find $\delta_{c,ult}$. Note that each layer will not necessarily use the same equation. If σ'_c is only slightly greater than σ'_{z0} (perhaps less than 20% greater), then it may not be clear if the soil is truly overconsolidated, or if the difference is only an apparent overconsolidation due to uncertainties in assessing these two values. In such cases, it is acceptable to use either Equation 10.23 (normally consolidated) or 10.25 (overconsolidated case II).

Example 10.5

A 3.0-m deep compacted fill is to be placed over the soil profile shown in Figure 10.20. A consolidation test on a sample from point A produced the following results:

$$C_c = 0.40$$
$$C_r = 0.08$$
$$e_0 = 1.10$$
$$\sigma'_c = 70.0 \text{ kPa}$$

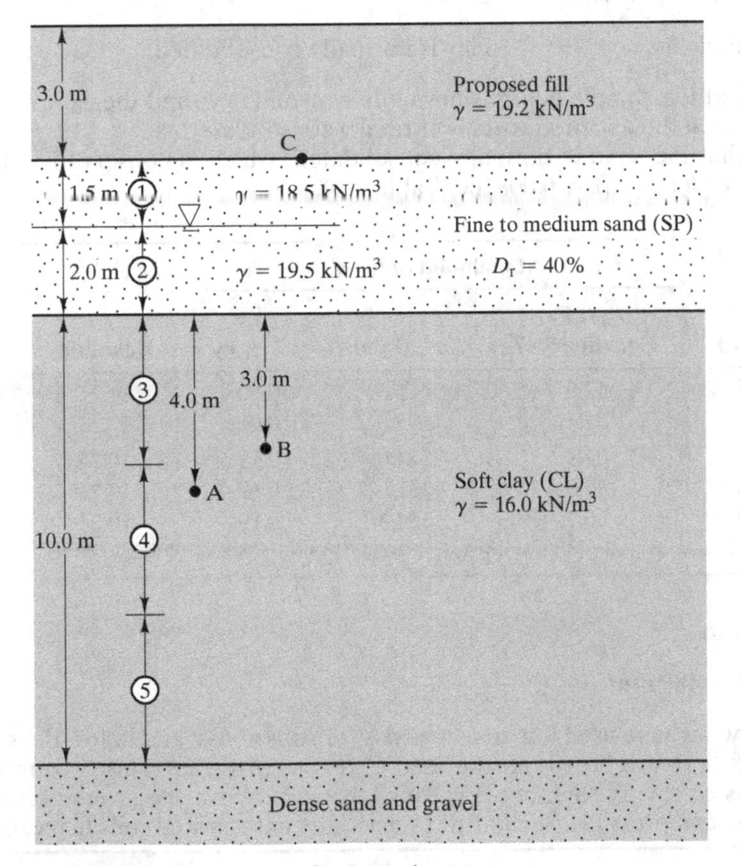

FIGURE 10.20 Soil profile for Example 10.5.

This sample is representative of the entire soft clay stratum. Compute the ultimate consolidation settlement due to the weight of this fill.

Solution:

Using Equation 10.2:

$$\sigma'_{zf} = \sigma'_{z0} + \gamma_{\text{fill}} H_{\text{fill}}$$
$$\sigma'_{zf} = \sigma'_{z0} + (19.2\ \text{kN/m}^3)(3.0\ \text{m})$$
$$\sigma'_{zf} = \sigma'_{z0} + 57.6\ \text{kPa}$$

Compute the initial vertical stress at sample location, using Equation 9.47:

$$\sigma'_{z0} = \sum \gamma H - u$$
$$\sigma'_{z0} = (18.5\ \text{kN/m}^3)(1.5\ \text{m}) + (19.5\ \text{kN/m}^3)(2.0\ \text{m}) + (16.0\ \text{kN/m}^3)(4.0\ \text{m}) - (9.8\ \text{kN/m}^3)(6.0\ \text{m})$$
$$\sigma'_{z0} = 72.0\ \text{kPa}$$

$$\frac{C_c}{1 + e_0} = \frac{0.40}{1 + 1.10} = 0.190$$

At the sample $\sigma'_c \approx \sigma'_{z0} \rightarrow \therefore$ clay is normally consolidated.

If the soil at the sample depth is normally consolidated, and the sample is truly representative, then the entire stratum is normally consolidated.

Assume that the sand is normally consolidated, which is conservative. For the sand strata, use $C_c/(1 + e_0) = 0.008$, per Table 10.4.

		At midpoint of layer				
Layer	***H* (m)**	**σ'_{z0} (kPa) Equation 9.47**	**σ'_{zf} (kPa)**	**$\frac{C_c}{1+e_0}$**	**Equation**	**$\delta_{c,ult}$ (mm)**
1	1.5	13.9	71.5	0.008	10.23	8
2	2.0	37.4	95.0	0.008	10.23	6
3	3.0	56.4	114.0	0.19	10.23	174
4	3.0	75.0	132.6	0.19	10.23	141
5	4.0	96.7	154.3	0.19	10.23	154
						$\delta_{c,ult} = 483$

Round off to

$$\delta_{c,ult} = \mathbf{480\ mm}$$

Notice how we have used the same analysis for soils above and below the groundwater table, and both are based on saturated $C_c/(1 + e_0)$ values. This is conservative (although in this case, very slightly so) because the soils above the groundwater table are probably less compressible. Section 10.11 discusses unsaturated soils in more detail.

Example 10.6

An 8.5-m deep compacted fill is to be placed over the soil profile shown in Figure 10.21. Consolidation tests on samples from points A and B produced the following results:

	Sample A	Sample B
C_c	0.25	0.20
C_r	0.08	0.06
e_0	0.66	0.45
σ'_c	101 kPa	510 kPa

Compute the ultimate consolidation settlement due to the weight of this fill.

Solution:

Using Equation 10.2,

$$\sigma'_{zf} = \sigma'_{z0} + \gamma_{fill} H_{fill}$$
$$\sigma'_{zf} = \sigma'_{z0} + (20.3\ \text{kN/m}^3)(8.5\ \text{m})$$
$$\sigma'_{zf} = \sigma'_{z0} + 172.6\ \text{kPa}$$

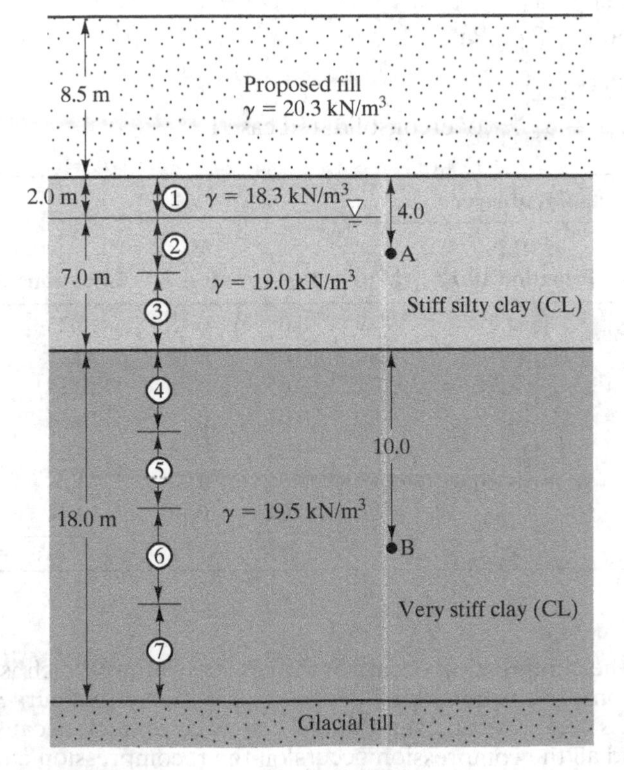

FIGURE 10.21 Soil profile for Example 10.6.

Applying Equation 9.47 to sample A,

$$\sigma'_{z0} = \sum \gamma H - u$$
$$\sigma'_{z0} = (18.3 \text{ kN/m}^3)(2.0 \text{ m}) + (19.0 \text{ kN/m}^3)(2.0 \text{ m}) - (9.8 \text{ kN/m}^3)(2.0 \text{ m})$$
$$\sigma'_{z0} = 55.0 \text{ kPa}$$

$$\sigma'_{zf} = \sigma'_{z0} + 172.6 \text{ kPa}$$
$$\sigma'_{zf} = 55.0 \text{ kPa} + 172.6 \text{ kPa}$$
$$\sigma'_{zf} = 227.6 \text{ kPa}$$

$$\sigma'_{z0} < \sigma'_c < \sigma'_{zf} \therefore \text{ overconsolidated case II}$$

$$\sigma'_m = \sigma'_c - \sigma'_z = 101 - 55 = 46 \text{ kPa}$$

Therefore, σ'_c at any depth in the stiff silty clay stratum is equal to $\sigma'_z + 46$ kPa.

For sample B,

$$\sigma'_{z0} = \sum \gamma H - u$$
$$\sigma'_{z0} = (18.3 \text{ kN/m}^3)(2.0 \text{ m}) + (19.0 \text{ kN/m}^3)(7.0 \text{ m}) + (19.5 \text{ kN/m}^3)(10.0 \text{ m}) - (9.8 \text{ kN/m}^3)(17.0 \text{ m})$$
$$\sigma'_{z0} = 198.0 \text{ kPa}$$

$$\sigma'_{zf} = \sigma'_{z0} + 172.6 \text{ kPa}$$
$$\sigma'_{zf} = 198.0 \text{ kPa} + 172.6 \text{ kPa}$$
$$\sigma'_{zf} = 370.6 \text{ kPa}$$

$$\sigma'_{z0} < \sigma'_c \text{ and } \sigma'_{zf} \leq \sigma'_c \therefore \text{ overconsolidated case I}$$

		At midpoint of layer						
Layer	H (m)	σ'_{z0} (kPa) Equation 9.47	σ'_c (kPa) Equation 10.17	σ'_{zf} (kPa)	$\frac{C_r}{1+e_0}$	$\frac{C_c}{1+e_0}$	Equation	$\delta_{c,ult}$ (mm)
1	2.0	18.3	64.3	190.9	0.05	0.15	10.25	196
2	3.0	50.4	96.4	223.0	0.05	0.15	10.25	206
3	4.0	82.6	128.6	255.2	0.05	0.15	10.25	217
4	4.0	120.4	–	293.0	0.04	0.14	10.24	62
5	4.0	159.2	–	331.8	0.04	0.14	10.24	51
6	5.0	202.8	–	375.4	0.04	0.14	10.24	53
7	5.0	251.4	–	424.0	0.04	0.14	10.24	45
								$\delta_{c,ult}$ = 830

$$\delta_{c,ult} = \mathbf{830 \text{ mm}}$$

Notice how most of the compression occurs in the upper stratum, which is overconsolidated case II (i.e., some of the compression occurs along the virgin curve). The lower stratum has much less compression even though it is twice as thick because it is overconsolidated case I and all the compression occurs on the recompression curve.

Example 10.7

The groundwater table in the soil profile shown in Figure 10.22 is currently at the elevation labeled "initial." A proposed dewatering project will cause it to drop to the elevation labeled "final." Compute the resulting ultimate consolidation settlement.

A consolidation test performed on a sample from point A produced the following results:

$$C_c/(1 + e_0) = 0.14$$
$$C_r/(1 + e_0) = 0.06$$
$$\sigma'_c = 3000 \text{ lb/ft}^2$$

Solution:

At sample A

$$\sigma'_{z0} = \sum \gamma H - u$$
$$\sigma'_{z0} = (121 \text{ lb/ft}^3)(7.0 \text{ ft}) + (125 \text{ lb/ft}^3)(6.0 \text{ ft}) + (127 \text{ lb/ft}^3)(18.0 \text{ ft}) + (110 \text{ lb/ft}^3)(5.0 \text{ ft}) - (62.4 \text{ lb/ft}^3)(29.0 \text{ ft})$$
$$\sigma'_{z0} = 2623 \text{ lb/ft}^2$$

$$\sigma'_{z0} \approx \sigma'_c \therefore \text{ normally consolidated}$$

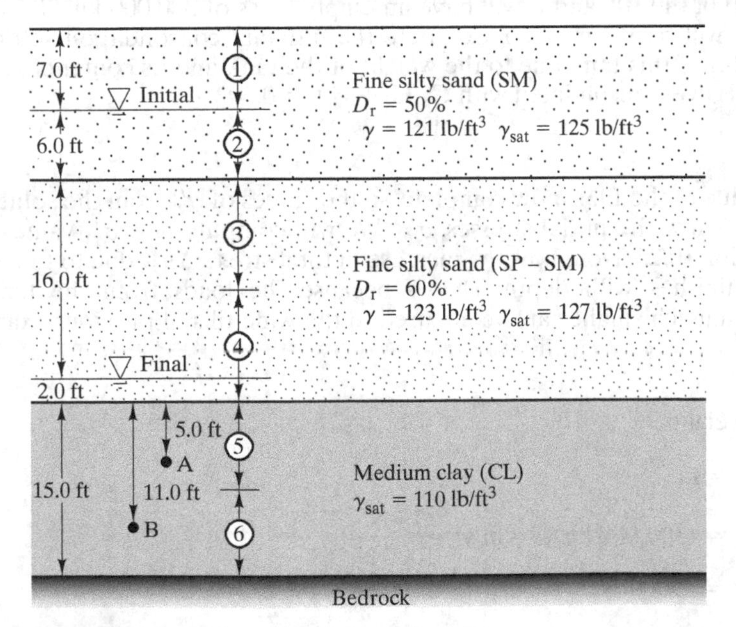

FIGURE 10.22 Soil profile for Example 10.7.

		At midpoint of layer				
Layer	H (ft)	σ'_{z0} (lb/ft^2)	σ'_{zf} (lb/ft^2)	$\frac{C_c}{1+e_0}$	Equation	$\delta_{c,ult}$ (in.)
1	7.0	423	423	0.011	10.23	0.0
2	6.0	1035	1210	0.011	10.23	0.1
3	9.0	1513	2126	0.009	10.23	0.1
4	9.0	2095	3233	0.009	10.23	0.2
5	7.0	2552	3837	0.14	10.23	2.1
6	8.0	2909	4194	0.14	10.23	2.1
						$\delta_{c,ult} = 4.6$

$\delta_{c,ult} = \mathbf{4.6\ in.}$

Notice how most of the compression occurs in the normally consolidated clay, even though it remains below the groundwater table. The cause of settlement is an increase in the effective stress, not drying, and the clay is most susceptible to this increase because it has the highest $C_c/(1 + e_0)$.

Example 10.8

After the settlement due to the fill described in Example 10.5 is completed, a 20-m diameter, 10-m tall cylindrical steel water tank is to be built. The bottom of the tank will be at the top of the fill, and it will have an empty mass of 300,000 kg. Ultimately, the water inside will be 9.5 m deep. Compute the ultimate consolidation settlement beneath the center of this tank due to the weight of the tank and its contents. Assume that the new fill is overconsolidated with $C_r/(1 + e_0) = 0.002$.

Strategy

The settlement due to the fill is now complete, so the values of σ'_{zf} from the solution of Example 10.5 are now the initial stresses, σ'_{z0}. We will compute new σ'_{zf} values using $\Delta\sigma_z$ from Equation 10.25. Note how the increase in total stress, $\Delta\sigma_z$, due to the weight of the tank diminishes with depth. This is because the loads from the tank are distributed over a much smaller area compared to the wide fills of previous examples.

Solution:

Compute weights:

$$W_{tank} = m$$

$$W_{tank} = (300{,}000\ \text{kg})(9.8\ \text{m/s}^2)\frac{1\ \text{kN}}{1000\ \text{N}}$$

$$W_{tank} = 2900\ \text{kN}$$

$$W_{water} = V_{tank}\gamma_w$$

$$W_{water} = \frac{\pi B^2 H}{4}\gamma_w$$

$$W_{water} = \frac{\pi(20.0 \text{ m})^2(9.5 \text{ m})}{4}(9.8 \text{ kN/m}^3)$$

$$W_{water} = 29{,}200 \text{ kN}$$

The weight of the water is much greater than that of the empty tank, so it is reasonable for us to assume the bearing pressure q is constant across the bottom of the tank.

$$q = \frac{W}{A}$$

$$q = \frac{2900 \text{ kN} + 29{,}200 \text{ kN}}{\pi(20.0 \text{ m})^2/4}$$

$$q = 102 \text{ kPa}$$

		At midpoint of layer								
Layer	H (m)	σ'_{z0} (kPa) Equation 9.47	z_f (m)	$\Delta\sigma_z$ (kPa) Equation 9.38	σ'_{zf} (kPa) Equation 10.3	$\frac{C_c}{1+e_0}$	$\frac{C_r}{1+e_0}$	Equation	$\delta_{c,ult}$ (mm)
1	3.0	28.8	1.5	101.7	130.5		0.002	10.24	4
2	1.5	71.5	3.7	97.7	169.2	0.008		10.23	4
3	2.0	95.0	5.5	90.6	185.6	0.008		10.23	5
4	3.0	114.0	8.0	77.1	191.1	0.19		10.23	128
5	3.0	132.6	11.0	60.7	193.3	0.19		10.23	93
6	4.0	154.3	14.5	45.1	199.4	0.19		10.23	85
									$\delta_{c,ult}$ = 319

Round off to

$$\delta_{c,ult} = \mathbf{320\ mm}$$

Commentary

1. If the tank were built immediately after the fill was placed, then σ'_{z0} would be the same as in Example 10.5, and everything else would remain unchanged. Such a solution would illustrate the use of superposition of stresses.
2. The values of $\Delta\sigma_z$ beneath the edge of the tank are less than those beneath the center (see Figure 9.13). Thus, the consolidation settlement will also be less and the bottom of the tank will settle into a dish shape. The difference between these two settlements is called *differential settlement*. We will discuss differential settlements in more detail in Chapter 15.

Example 10.9

Compute the ultimate consolidation settlement due to the weight of the fill shown in Figure 10.23.

Solution:

		At midpoint of layer							
Layer	H (ft)	σ_{z0} (lb/ft²) Equation 9.47	σ'_c (lb/ft²) Equation 10.17	$\Delta\sigma_z$ (lb/ft²)	σ'_f (lb/ft²)	$\frac{C_c}{1+e_0}$	$\frac{C_r}{1+e_0}$	Equation	$\delta_{c,ult}$ (ft)
1	5	287.5	–	1440	1727.5	0.01	0.003	10.23	0.038
2	3	659.9	–	1440	2099.9	0.01	0.003	10.23	0.015
3	5.5	856.5	1356.5	1440	2296.5	0.16	0.05	10.25	0.256
4	5.5	1079.8	1579.8	1440	2519.8	0.16	0.05	10.25	0.224
5	5.5	1303.1	1803.1	1440	2743.1	0.16	0.05	10.25	0.199
6	5.5	1526.4	2026.4	1440	2966.4	0.16	0.05	10.25	0.179
7	6.5	1789.5	6789.5	1440	3229.5	0.15	0.05	10.24	0.083
8	6.5	2080.7	7080.7	1440	3520.7	0.15	0.05	10.24	0.074

Final result: $\delta_{c,ult}$ = **1.07 ft (round off to 1.1 ft)**

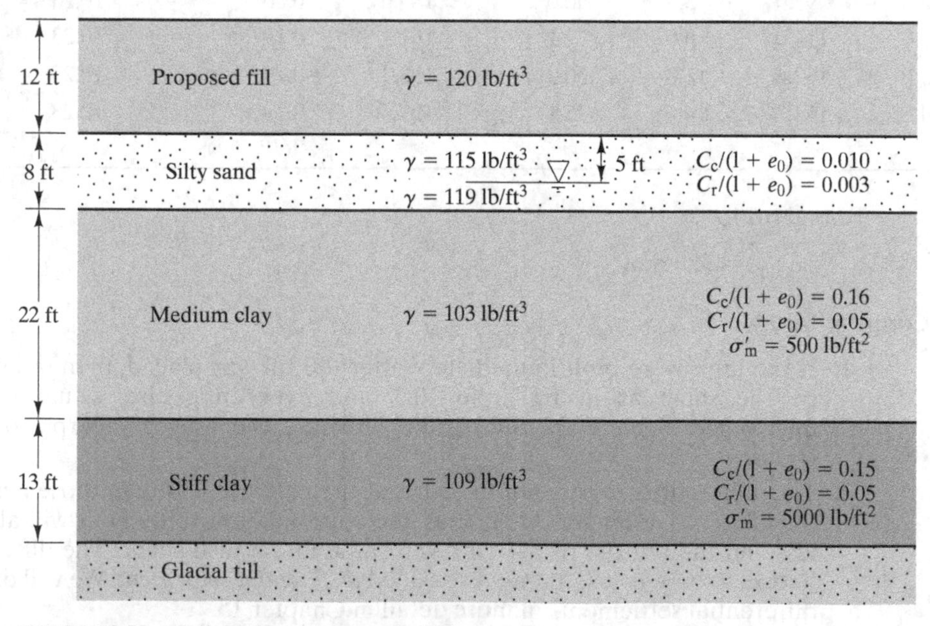

FIGURE 10.23 Soil profile for Example 10.9.

10.9 SECONDARY COMPRESSION SETTLEMENT

Once the excess pore water pressures have dissipated, consolidation settlement ceases. However, some soils continue to settle anyway. This additional settlement is due to *secondary compression* and occurs under a constant effective stress. We do not fully understand the physical basis for secondary compression, but it appears to be due to particle rearrangement, creep, and the decomposition of organics. Highly plastic clays, organic soils, and sanitary landfills are most likely to have significant secondary compression. Secondary compression is negligible in sands and gravels.

The *secondary compression index*, C_α, defines the rate of secondary compression. It can be defined either in terms of either void ratio or strain:

$$C_\alpha = -\frac{\mathrm{d}e}{\mathrm{d}\log t} \tag{10.26}$$

$$\frac{C_\alpha}{1 + e_\mathrm{p}} = -\frac{\mathrm{d}\varepsilon_z}{\mathrm{d}\log t} \tag{10.27}$$

where:

C_α = secondary compression index
e = void ratio
e_p = void ratio at end of consolidation settlement (can generally use $e_\mathrm{p} = e_0$ without introducing significant error)
ε_z = vertical strain
t = time

Design values for C_α are normally determined while conducting a laboratory consolidation test. The consolidation settlement occurs very rapidly in the lab (because of the short drainage distance), so it is not difficult to maintain one or more of the load increments beyond the completion of consolidation settlement. The change in void ratio after this point can be plotted against log time to determine C_α. See Chapter 11 for more details on the interpretation of laboratory test results.

Another way of developing design values of C_α is to rely on empirical data that relate it to the compression index, C_c. These data are summarized in Table 10.5.

TABLE 10.5 Empirical Correlation Between C_α and C_c[a]

Material	C_α/C_c
Granular soils, including rockfill	0.02 ± 0.01
Shale and mudstone	0.03 ± 0.01
Inorganic clays and silts	0.04 ± 0.01
Organic clays and silts	0.05 ± 0.01
Peat and muskeg	0.06 ± 0.01

[a]Terzaghi, Peck, and Mesri, 1996.

The settlement due to secondary compression is

$$\delta_s = \frac{C_\alpha}{1 + e_p} H \log\left(\frac{t}{t_p}\right) \tag{10.28}$$

where:

δ_s = secondary compression settlement
H = thickness of compressible strata
t = time after application of load
t_p = time required to complete consolidation settlement (in theory this is infinite, but for practical problems we can assume it occurs when 95% of the consolidation in the field is complete).

We assume the secondary compression settlement begins at time t_p.

Usually secondary compression settlement is much smaller than consolidation settlement, and thus is not a major consideration. However, in some situations, it can be very important. For example, the consolidation settlement in sanitary landfills is typically complete within a few years, while the secondary compression settlement continues for many decades. Secondary compression settlements on the order of 1% of the refuse thickness per year have been measured in a 10-year-old landfill (Coduto and Huitric, 1990).

Significant structures are rarely built on soils that have the potential for significant secondary compression. However, highways and other transportation facilities are sometimes built on such soils.

Example 10.10

The soft clay described in Example 10.5 has $C_\alpha/(1 + e_p) = 0.018$. Assuming that the consolidation settlement will be 95% complete 40 years after the fill is placed, compute the secondary compression settlement that will occur over the next 30 years.

Solution:

$$\delta_s = \frac{C_\alpha}{1 + e_p} H \log\left(\frac{t}{t_p}\right)$$

$$\delta_s = (0.018)(10{,}000 \text{ mm}) \log\left(\frac{40 \text{ yr} + 30 \text{ yr}}{40 \text{ yr}}\right)$$

$$\delta_s = 40 \text{ mm}$$

This is approximately one-tenth of the consolidation settlement of 480 mm, as computed in Example 10.5.

A complete example including both consolidation and secondary compression settlements is included in Chapter 11.

10.10 CRUSTS

Soft fine-grained soil deposits, such as those often found in wetlands, frequently have a thin crust near the ground surface, as shown in Figure 10.24. These crusts are typically less than 2 m (7 ft) thick, and are formed when the upper soils temporarily dry out. This drying process is called *dessication* and causes these soils to become overconsolidated. Thus, profiles that contain crusts have less settlement than identical profiles without crusts.

The presence of crusts has a significant impact on settlement computations, even if they are much thinner than the underlying compressible soils. In addition, variations in the crust thickness across a site can be a significant source of differential settlement. Thus, site characterization studies need to carefully evaluate the thickness and compressibility of crusts.

10.11 SETTLEMENT OF UNSATURATED SOILS

Thus far we have treated unsaturated soils using the techniques developed for saturated soils, except we have set $u = 0$ (e.g., Example 10.5). However, some unsaturated soils are prone to other kinds of settlement problems, especially if they become wetted sometime after construction.

One of these problems occurs in certain kinds of clay that are known as *expansive soils*. These clays expand when they become wetted, and contract when dried. Another kind of problematic soil is called a *collapsible soil*, which compresses when it is wetted.

Although the soft saturated soils generally have the worst problems with settlement, expansive and collapsible soils also can be problematic, especially in arid and semiarid climates.

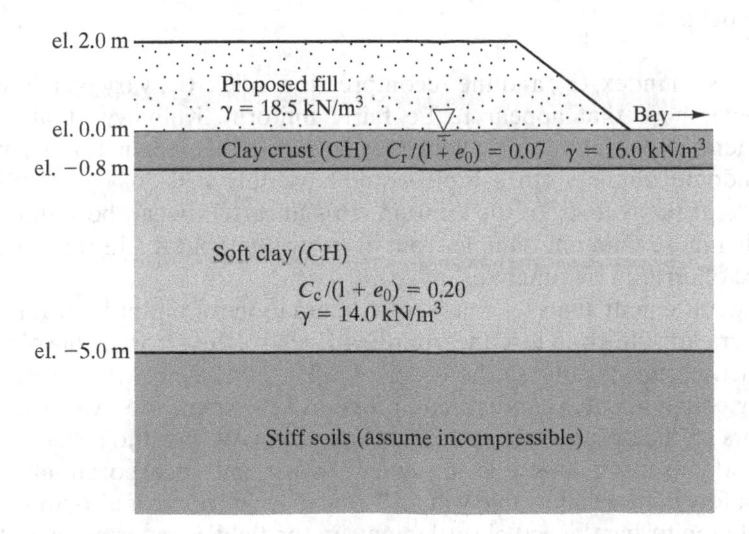

FIGURE 10.24 Typical crust near the ground surface in an otherwise normally consolidated clay. el. = elevation.

10.12 HEAVE DUE TO UNLOADING

Our discussions thus far have considered only settlement of soils in response to an increased load. Another possibility is heave (negative settlement) due to a decreased load, such as an excavation. In this case, $\sigma'_{zf} < \sigma'_{z0}$. The soil will heave according to the rebound curve in Figure 10.10, which has a slope of C_r, so we compute the heave using Equation 10.24. Because the soil is expanding, the excess pore water pressure is negative, causing pore water to be drawn into the voids. This is the opposite of the process described in Figure 10.8. There are other processes that can also cause heave in soils. The most notable one is the swelling of expansive clays.

10.13 ACCURACY OF SETTLEMENT PREDICTIONS

As with all other geotechnical analyses, settlement predictions are subject to many errors. These include

- Differences between the soil profile used in the analysis and the real soil profile, especially the proper identification of crusts
- Differences between the engineering properties of the soil samples and the average properties of the strata they represent (i.e., are they truly representative?)
- Sample disturbance
- Errors introduced due to testing techniques in the laboratory
- Errors in assessing σ'_c
- The assumption that consolidation in the field is one-dimensional (i.e., there is no horizontal strain)
- Differences between Terzaghi's theory of consolidation and the real behavior of soils in the field.

The compression index, C_c, and the recompression index, C_r, vary widely within soil deposits, even those that appear to be fairly uniform. Kulhawy et al. (1991) reported coefficients of variation in C_c values of 26–52%. This means that C_c values from a single randomly obtained soil sample would have only a 30–56% probability of being within 20% of the true C_c of the stratum. This uncertainty can be significantly reduced by testing more than one sample from each stratum, but it still represents an important source of error in our analyses.

Fortunately, settlement analyses consist of a summation for multiple strata, which introduces an averaging effect on test uncertainties. Even so, the error in consolidation settlement predictions is typically on the order of ±25–50%, even with careful sampling and testing. Analyses of secondary compression settlement are even less accurate, having errors on the order of about ±75% (Fox, 1995). We need to consider these potential errors when setting allowable settlement values, and incorporate an appropriate factor of safety in these allowable values. These margins of error also underscore the usefulness of monitoring the actual settlements in the field, comparing them to the predicted settlements, and, if necessary, modifying the designs accordingly.

SUMMARY

Major Points

1. Settlement can be caused by several different physical processes. We have considered three: distortion settlement, δ_d, consolidation settlement, δ_c, and secondary compression settlement, δ_s.
   - **(a)** Distortion settlement is due to the horizontal movements of soil, and occurs primarily when the loaded area is small and the bearing pressure is high, such as in the loading of a structural foundation.
   - **(b)** Consolidation is caused by an increase in the vertical effective stress from an initial value, σ'_{z0}, to a final value, σ'_{zf}. This change causes the solid particles to move into a tighter packing, which results in a vertical strain and a corresponding settlement.
   - **(c)** Secondary compression occurs without any change in the vertical effective stress. It is the result of particle rearrangement, creep, decomposition of organic materials, and other processes. This settlement is time-dependent. It is much slower than consolidation settlement and continues at an ever-decreasing rate with time.
2. Settlement can also be characterized based on the time frame in which the settlement occurs. Short-term, or immediate, settlement occurs as quickly as the load is applied to the soil. Long-term settlement occurs some time after the application of the load.
   - **(a)** In clay and silt soils, short-term settlement is composed of distortion settlement. Long-term settlement is composed of consolidation settlement and secondary compression settlement. The long-term settlement generally greatly exceeds the short-term settlement in these soils.
   - **(b)** In sand and gravel soils, short-term settlement is composed of distortion settlement and consolidation settlement. Long-term settlement is composed of secondary compression settlement and is generally much smaller than the short-term settlement.
3. If the soil is saturated, which is the case in most consolidation analyses, the applied load is first carried by the pore water. This causes a temporary increase in the pore water pressure. This increase is called an *excess pore water pressure*, u_e. The presence of this pressure induces a hydraulic gradient in the soil, forcing some pore water to flow out of the voids, thus reducing the excess pore water pressure. After some period of time, the excess pore pressure dissipates, $u_e \rightarrow 0$, and the applied load is transferred to the solid particles increasing the effective vertical stress, $\sigma'_{z0} \rightarrow \sigma'_{zf}$.
4. Sands and gravels have high hydraulic conductivities. In these soils consolidation occurs as quickly as the loads are applied, and therefore, leads to short-term settlement. In clay or silt soils, the hydraulic conductivity is many orders of magnitude lower which greatly slows the dissipation of excess pore water pressure and slows

the consolidation process. In these soils consolidation leads to long-term settlement that may continue for years or decades.

5. We measure the stress–strain properties of a soil by conducting consolidation tests in the laboratory on undisturbed samples. Stress–strain curves for soils are plotted as void ratio versus log of effective stress (e vs. $\log \sigma'_z$) or strain versus log of effective stress (ε vs. $\log \sigma'_z$). When plotted in semilogarithmic space, the stress–strain curve can usually be modeled as a bilinear curve with the following parameters:

    e_0 = initial void ratio
    σ'_c = preconsolidation stress
    C_c = compression index
    C_r = recompression index

    The preconsolidation stress, σ'_c, is the greatest vertical effective stress the soil has ever experienced at the point where the sample was obtained. The preconsolidation stress also represents the point where the recompression curve intersects the virgin curve. The parameters C_r and C_c define the slope of the recompression curve and virgin curve, respectively.

6. Normally consolidated soils are those that have never experienced a vertical effective stress significantly greater than the present vertical effective stress; that is $\sigma'_{z0} \approx \sigma'_c$. Conversely, overconsolidated soils are those that have experienced higher stresses than the present vertical effective stress; that is $\sigma'_{z0} < \sigma'_c$.
7. The amount of overconsolidation in a give soil can be measured by either the overconsolidation margin, σ'_m, which is the difference between the preconsolidation stress and the current vertical effective stress ($\sigma'_m = \sigma'_c - \sigma'_{z0}$), or the overconsolidation ratio, OCR, which is the ratio of the preconsolidation stress to the current vertical effective stress ($\text{OCR} = \sigma'_c/\sigma'_{z0}$).
8. Normally consolidated soils often have an overconsolidated crust near the ground surface. It is important to recognize the presence of these crusts in the site characterization program.
9. Using the parameters from the consolidation test and the changes in effective stress in the field we can compute the consolidation settlement.
10. The greatest consolidation settlements occur in soft clays. Sandy and gravelly soils are usually much less compressible. In addition, it is nearly impossible to obtain sufficiently undisturbed samples to conduct reliable consolidation tests on sands and gravels, so the compressibility of these soils is determined by empirical correlations or by in situ tests.
11. In a soil profile with both sandy or gravelly soils and clay soils, the consolidation of the sandy or gravelly soils can often be ignored because the compressibility of the clay soil will greatly exceed that of the sandy or gravelly soils.
12. If the soil is unloaded, the vertical effective stress decreases ($\sigma'_{zf} < \sigma'_{z0}$) and the soil will swell or heave instead of compressing.

13. Settlement predictions are subject to several sources of error. Even careful predictions of consolidation settlement typically have a precision on the order of ±25–50%. Predictions of secondary compression settlement typically are even less accurate, with errors on the order of ±75%.

Vocabulary

collapsible soil
compressibility
compression index
compression ratio
consolidation
consolidation settlement
consolidation test
consolidometer
crust
desiccation
differential settlement
distortion settlement
elastic deformation
excess pore water pressure
expansive soil
heave
immediate settlement
long-term settlement
normally consolidated
oedometer
one-dimensional consolidation
overconsolidated
overconsolidation margin
overconsolidation ratio
plastic deformation
porous stones
preconsolidated
preconsolidation stress
rebound curve
recompression curve
recompression index
recompression ratio
secondary compression index
secondary compression settlement
settlement
short-term settlement
subsidence
theory of consolidation
ultimate consolidation settlement
virgin curve

QUESTIONS AND PRACTICE PROBLEMS

Section 10.2 Changes in Vertical Effective Stress

10.1 The current σ'_z at a certain point in a saturated clay is 181 kPa. This soil is to be covered with a 2.5-m thick fill that will have a unit weight of 19.3 kN/m^3. What will be the value of σ'_z at this point immediately after the fill is placed, before any consolidation has occurred? What will it be after the consolidation settlement is completed?

10.2 A circular tank 50 ft in diameter is constructed on top of a saturated soil with a saturated unit weight of 112 lb/ft^3. The total weight of the tank when filled is 6125 k. The groundwater table is at the ground surface. Compute σ'_z a point 25 ft below the center of the tank immediately after the tank is placed and filled, before any consolidation has occurred. What will σ'_z be at the same point after consolidation settlement is completed?

10.3 A 4.0-m thick fill with a unit weight of 20.1 kN/m^3 is to be placed on the soil profile shown in Figure 9.11. Develop plots of σ'_{z0} and σ'_{zf} versus depth. The plot should extend from the original ground surface to a depth of 10.0 m.

10.4 The groundwater table at a certain site was at a depth of 10 ft below the ground surface, and the vertical effective stress at a point 30 ft below the ground surface was 2200 lb/ft^2. Then a series of wells was installed, which caused the groundwater table to drop to a depth of 25 ft below the ground surface. Assuming the unit weight of the soil above and below the groundwater table are equal, compute the new σ'_z at this point.

10.5 The fill shown in Figure 10.21 was placed many years ago and consolidation in both clay layers due to the fill placement is complete. The fill was placed near a river that has since been dammed. Filling of the reservoir behind the dam has raised the local groundwater table to a point 2 m below the top of the fill. The raising of the water table has been slow and has not induced any excess pore water pressures in the clay soils. Compute σ'_z at points A and B due to the rise of the groundwater table. Assume the saturated unit weight of the fill material is 21.1 kN/m^3.

10.6 A 1.00 m^3 element of soil is located below the groundwater table. When a new compressive load was applied, this element consolidated, producing a vertical strain, ε_z, of 8.5%. The horizontal strain was zero. Compute the volume of water squeezed out of this soil during consolidation and express your answer in liters.

Section 10.5 Consolidation Test

10.7 A consolidation test is being performed on a 3.50-in diameter saturated soil specimen that had an initial height of 0.750 in. and an initial moisture content of 38.8%.

(a) Using $G_s = 2.69$, compute the initial void ratio, e_0.

(b) At a certain stage of the test, the normal load P was 300 lb. After the consolidation at this load was completed, the specimen height was 0.690 in. Compute σ'_z (expressed in lb/ft^2), ε_z, and e.

10.8 A consolidation test of a soft marine silty clay produced the following data:

σ'_z(lb/ft^2)	50	250	500	1000	2000	4000	8000	16,000	4000	500
ε_z	0.007	0.028	0.059	0.097	0.145	0.189	0.235	0.278	0.271	0.253

The initial void ratio of the sample was 1.24 and σ'_{z0} at the sample depth was 270 lb/ft^2.

(a) Plot these data on an arithmetic diagram similar to that in Figure 10.10(a).

(b) Plot these data on a semilogarithmic diagram similar to that in Figure 10.10(b).

(c) Using Casagrande's method, find σ'_c.

(d) Using Schmertmann's method adjust the test results.

(e) Determine C_c and C_r.

(f) The soil has a plasticity index of 17. Based on Kulhawy and Mayne's correlations, do the consolidation test results seem reasonable?

10.9 A consolidation test on a sample of clay produced the following data:

σ'_z (kPa)	8	16	32	64	128	256	512	1024	16
ε_z	0.032	0.041	0.051	0.069	0.109	0.173	0.240	0.301	0.220

The initial void ratio of the sample was 1.21, and σ'_{z0} at the sample depth was 40 kPa.

(a) Plot these data on an arithmetic diagram similar to that in Figure 10.10(a)

(b) Plot these data on a semilogarithmic diagram similar to that in Figure 10.10(b).

(c) Using Casagrande's method, find σ'_c.

(d) Using Schmertmann's method, adjust the test results.

(e) Determine C_c and C_r.

(f) The soil has a plasticity index of 23. Based on Kulhawy and Mayne's correlations, do the consolidation test results seem reasonable?

Section 10.6 Consolidation Status in the Field

10.10 Before the placement of any fill, a consolidation test was performed on a soil sample obtained from Point B in Figure 10.20. The measured preconsolidation stress was 88 kPa. Determine whether the soil is normally consolidated or overconsolidated, then compute the overconsolidation margin and overconsolidation ratio at Point B.

Note: These computations are based on the initial conditions, and thus should not include the weight of the proposed fill.

10.11 A saturated, normally consolidated, 1000-year-old fine-to-medium sand has an SPT $N_{60} = 12$ at a depth where the vertical effective stress is about 1000 lb/ft^2 and $D_{50} =$ 0.5 mm. Using the techniques described in Chapters 3 and 4, determine the relative density of this soil, then estimate $C_c/(1 + e_0)$ based on Table 10.4.

10.12 A consolidation test has been performed on a sample obtained from Point A in Figure 10.3. The measured preconsolidation stress was 1500 lb/ft^2.

(a) Determine if the soil is normally consolidated or overconsolidated.
(b) Compute the overconsolidation margin and the overconsolidation ratio.
(c) Compute σ_c' at Point B.

10.13 A consolidation test has been performed on a sample obtained from a saturated clay at a point 6.5 m below the ground surface. The groundwater table is at the ground surface and the unit weight of the clay is 18.5 kN/m^3. The measured preconsolidation stress was 260 kPa.

(a) Determine if the soil is normally consolidated or overconsolidated.
(b) Compute the overconsolidation margin and the overconsolidation ratio.
(c) Compute σ_c' at depth of 12 m in the same soil.

10.14 Laboratory consolidation test for all soils have both a recompression curve and a virgin curve. This is true even for normally consolidated soils. Explain how the lab test for a normally consolidated soil can have a recompression curve even though the soil in the field has never been preloaded.

Section 10.8 Consolidation Settlement Predictions

10.15 A 2-m thick fill with an in place unit weight of 19.2 kN/m^3 is to be placed on top of the soil profile shown in Figure 10.25. Both the sand and the clay are normally consolidated.

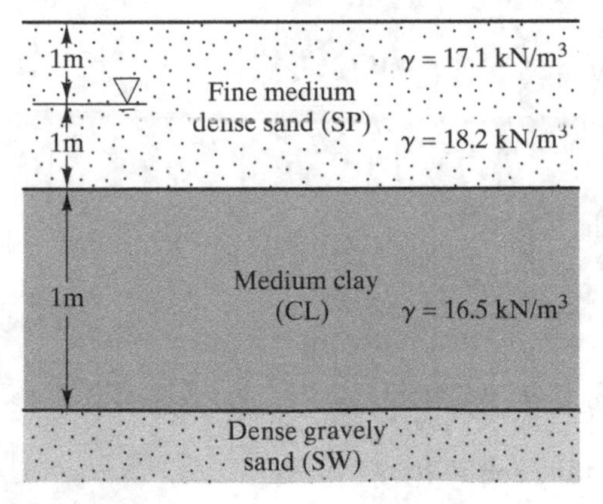

FIGURE 10.25 Soil Profile for Problem 10.15.

The sand is fine, poorly-graded and medium dense with $C_c/(1 + e_0)$ estimated to be 0.009. Lab consolidation tests on the clay produced $C_c/(1 + e_0) = 0.18$. Compute the total consolidation settlement due to the placement of the fill. Would it be acceptable to ignore the settlement generated by consolidation of the sand layer?

10.16 A 5.0-ft thick fill is to be placed on the soil profile as shown in Figure 10.3. A consolidation test performed on a sample obtained from Point B produced the following results: $C_c = 0.27$, $C_r = 0.10$, $e_0 = 1.09$, $\sigma'_c = 760$ lb/ft^2. Compute the ultimate consolidation settlement due to the weight of this fill and determine the ground surface elevation after the consolidation is complete.

Note: The first layer in your analysis should extend from the original ground surface to the groundwater table.

10.17 A 4.0-m thick fill is to be made of a soil with a Proctor maximum dry unit weight of 19.4 kN/m^3 and an optimum moisture content of 13.0%. This fill will be compacted at optimum moisture content to an average relative compaction of 92%. The underlying soils are as shown in Figure 10.26. Consolidation tests were performed at Points A and B, with the following results:

Sample	C_c	C_r	e_0	σ'_c (kPa)
A	0.59	0.19	1.90	75
B	0.37	0.14	1.21	100

The silty sand is normally consolidated. Determine the ultimate consolidation settlement due to the weight of this fill.

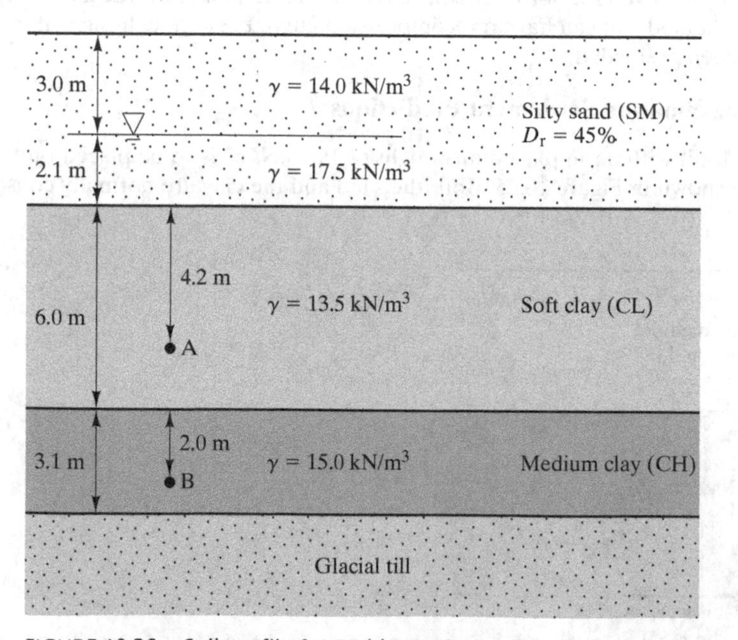

FIGURE 10.26 Soil profile for Problems 10.17 and 10.30.

10.18 The owner of the land shown in the profile in Example 10.5 has decided not to build the proposed fill. Instead, the land will be used for farming. To provide irrigation water, a series of shallow wells will be drilled into the sand, and these wells will cause the groundwater table to drop to the bottom of the sand layer (i.e., 2.0 m below its current position). Compute the ultimate consolidation settlement due to this drop in groundwater. Do you think such a settlement will adversely affect the farming?

10.19 A certain site is underlain by the soil profile shown in Figure 10.22 with the groundwater table at the initial location. The groundwater table will remain at this location, but a 20.0-ft deep fill with a unit weight of 119 lb/ft^3 is to be placed. The only consolidation data available is from a test conducted on a sample from Point B. The test results are as follows: $e_0 = 1.22$, $C_c = 0.31$, $C_r = 0.09$, $\sigma'_c = 3800$ lb/ft^2. Determine the ultimate consolidation settlement due to the weight of this proposed fill. Assume the sands are normally consolidated.

10.20 Using the data in Example 10.8, compute the consolidation settlement at the edge of the tank. Then compute the differential settlement, which is the difference between the settlement at the center and the edge.

Hint: Compute $\Delta\sigma_z$ using Figure 9.13.

Section 10.9 Secondary Compression Settlement

10.21 Using the data from Example 10.10, develop a plot of secondary compression settlement versus time for the period 40–100 years after completion of the fill. Is the rate of secondary compression settlement increasing or decreasing with time?

Section 10.12 Heaving Due to Unloading

10.22 Point C in Figure 10.20 was originally at elevation 12.00 m, but it dropped to elevation 11.52 m as a result of the consolidation settlement described in Example 10.5. Now that the consolidation is complete, the fill is to be removed. Compute the new elevation of Point C after the natural soils rebound in response to the fill removal. Ignore any secondary compression settlement.

10.23 In Example 10.7 the total settlement due to the lowering of the groundwater table was computed to be 4.6 in. It has been several years after the groundwater table was lowered and consolidation is complete. Now the upper 30 ft of the soil profile is to be excavated to create underground parking for a mid-rise building. How much will the base of the excavation heave due to removal of the soil? If the total mass of the structure being constructed is 65% of the mass of the soil removed, what will be the net settlement or heave of the structure?

Comprehensive

10.24 A road embankment is being placed across a shallow section of a bay. The existing profile consists of 1 m of water over a 5-m thick normally consolidated clay soil which overlies a very dense and stiff gravelly sand. A consolidation test on the clay generated the following results: $C_c = 0.21$, $e_0 = 1.21$. The embankment material is expected to be place at a unit weight of 18.1 kN/m^3. Determine the thickness of the embankment such that the final elevation of the embankment is 2 m above the water level. This will require an iterative solution.

10.25 Develop a spreadsheet that can compute one-dimensional consolidation settlement due to the weight of a fill. This spreadsheet should be able to accommodate a fill of any unit weight and thickness, underlain by multiple compressible soil strata. It also should be able to accommodate both normally consolidated and overconsolidated soils. As the computer does all of the computations, the spreadsheet should use at least 50 layers. Once the spreadsheet is completed, use it to solve Examples 10.5 and 10.6. Submit printouts of both analyses.

10.26 A cross-section through a tidal mud flat area is shown in Figure 10.24. This site is adjacent to a bay, is subject to varying water levels according to tides, and is occasionally submerged when heavy runoff from nearby rivers raises the elevation of the water in the bay. For analysis purposes, use a groundwater table at the ground surface, as shown. A crust has formed in the upper 0.8 m of soil due to dessication (drying) and is stiffer than the underlying soil. This crust is overconsolidated case I, and the soils below are normally consolidated. The proposed fill is required to protect the site from future flooding, and thus permit construction of a commercial development. Use the spreadsheet developed in Problem 10.25 to determine the ultimate consolidation settlement due to the weight of this proposed fill. Then, consider the possibility that the crust was not recognized in the site characterization program, and perform another analysis using $C_c/(1 + e_0) = 0.20$ and $\gamma = 14.0\ \text{kN/m}^3$ for the entire 5.0 m of clay. Compare the results of these two analyses and comment on the importance of recognizing the presence of crusts.

10.27 Explain the difference between normally consolidated soil and overconsolidated soils, and give examples of geologic conditions that would form each type.

10.28 What types of natural soils are best suited for testing in a consolidometer? Why? Which are not well suited? Why?

10.29 According to the results from a consolidation test, the preconsolidation stress for a certain soil sample is 850 lb/ft^2. The in situ vertical effective stress at the sample location is 797 lb/ft^2, and the proposed load will cause σ_z to increase by 500 lb/ft^2. Which equation should be used to compute the consolidation settlement, 10.23, 10.24, or 10.25? Why?

10.30 A 3.0-m thick fill with a unit weight of 18.1 kN/m^3 is to be placed on the soil profile shown in Figure 10.26. Consolidation test results at Points A and B are as stated in Problem 10.17, except that σ'_c at point B is now 200 kPa. Using Equation 10.21 with C_c or C_r as appropriate, develop a plot of vertical strain, ε_z versus depth from the original ground surface to the top of the glacial till. How does this curve vary within a given soil stratum? Why? Does it suddenly change at the strata interfaces? Why?

10.31 Considering the variation of strain with depth, as found in Problem 10.30, does a 1-m thick layer near the top of a stratum contribute more or less to the consolidation settlement than a 1-m thick stratum near the bottom? Explain. Does this finding support the statement in Section 10.10 that "The presence of crusts has a significant impact on settlement computations, even if they are much thinner than the underlying compressible soils"? Explain.

10.32 A shopping center is to be built on a site adjacent to a tidal mud flat. The ground surface elevation is +0.2 m, and the groundwater table is at the ground surface. The underlying soils consist of 7.3 m of medium clay with $C_c/(1 + e_0) = 0.18$, $C_r/(1 + e_0) = 0.06$, $\sigma'_m = 0$, and $\gamma = 15.1\ \text{kN/m}^3$. The clay stratum is underlain by relatively incompressible stiff soils.

In order to provide sufficient flood control protection, a fill must be placed on this site before the shopping center is built, thus maintaining the entire site above the highest flood level. This fill will have a unit weight of 19.0 kN/m^3. According to a hydrologic study, the fill must be thick enough so that the ground surface elevation is at least + 1.8 m after all of the consolidation settlement is complete. Use the spreadsheet developed in Problem 10.25 to determine the required ground surface elevation at the end of construction. Assume all of the settlement occurs after construction.

10.33 A consolidation test has been performed on a sample of lodgement till from a region that was once covered with a glacier. The current vertical effective stress at the sample location is 1800 lb/ft^2 and the measured preconsolidation stress is 32,500 lb/ft^2.

(a) Assuming the glacier was in place long enough for complete consolidation to occur, and assuming the ground surface and groundwater table elevations have remained unchanged, compute the maximum thickness of the glacier. The specific gravity of glacial ice is about 0.87.

(b) Although glacial ice was present for a very long time, it also extended over very large areas, so the required drainage distance for the excess pore water was very long. As a result, all of the excess pore water pressures may not have dissipated (Chung and Finno, 1992). Therefore, our assumption that complete consolidation occurred may not be accurate. If so, would our computed thickness be too large or too small? Explain.

10.34 A highway is to be built across a wetlands with the soil profile shown in Figure 10.27 below. These wetlands are subject to flooding, so a fill must be placed to keep the pavement above the highest flood level. According to a hydrologic analysis, the roadway must be at elevation 7.0 ft or higher to satisfy this requirement. Sandy fill material that has a compacted unit weight of 122 lb/ft^3 is available from a nearby borrow site.

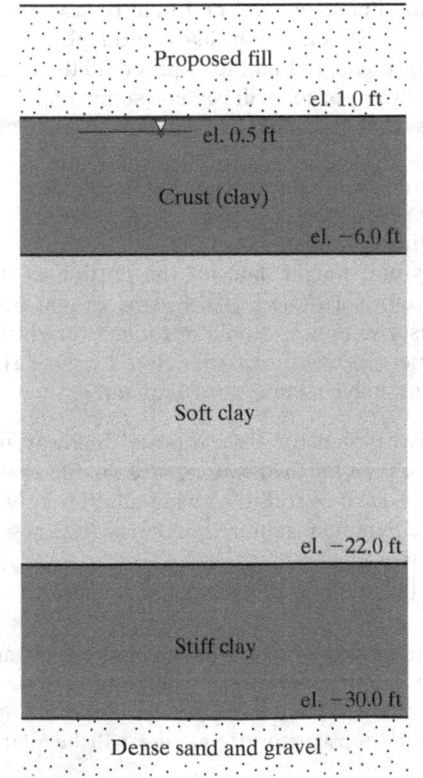

FIGURE 10.27 Soil profile for Problem 10.34. el. = elevation.

A subsurface exploration program has been completed at this site, and laboratory tests have been performed. The results of this program are tabulated below:

Depth (ft)	Dry unit weight (lb/ft^3)	Moisture content (%)	$C_c/(1 + e_0)$	$C_r/(1 + e_0)$	σ'_c(lb/ft^2)
2.0	95	28.6	0.13	0.06	3000
7.5	89	33.0	0.16	0.06	550
13.0	92	30.5	0.12	0.05	850
24.0	93	29.9	0.14	0.07	4800

All depths are measured from the original ground surface.

The natural soils will settle under the weight of the proposed fill. Approximately 25 years will pass before this settlement is complete. Therefore, the road must be built at an elevation higher than 7.0 ft so that after the settlement is complete it is at 7.0 ft. The pavement thickness is 0.5 ft, so the top of the fill must remain at or above elevation 6.5 ft.

(a) Use the spreadsheet developed in Problem 10.25 to determine the required elevation of the roadway immediately after construction. Assume that no settlement occurs during construction, and the pavement has the same unit weight as the fill.

Hint: As the fill thickness becomes greater, the settlement increases. Thus, this problem requires a trial-and-error solution. You will need to estimate the required fill thickness, then compute the settlement and final roadway elevation. Try to have one trial that produces a road elevation that is too high, and another that produces one too low. Then interpolate to find the required fill thickness.

Note 1: As the fill settles, the lower portion will become submerged below the groundwater table, so σ'_{zf} will be less than predicted by Equation 10.2. However, we will ignore this effect.

Note 2: Laboratory tests have been performed on two samples of the soft clay. Combine these two sets of test results, then assign γ, $C_c/(1 + e_0)$, $C_r/(1 + e_0)$, and σ'_m values that apply to the entire stratum.

Note 3: We do not have any unit weight data for the portion of the crust above the groundwater table. Therefore, assume it is the same as that below the groundwater table. In this case, this assumption should introduce very little error.

(b) Is our assumption regarding the submergence of the fill (per Note 1 in part a) conservative or unconservative? Is this a reasonable assumption? Explain.

10.35 An engineer has suggested an alternative design for the proposed highway in Problem 10.34. This design consists of using geofoam for the lower part of the fill, as shown in Figure 6.45. It will be covered with at least 1.0 ft of soil to provide a buffer between the pavement and the geofoam. Compute the minimum required geofoam thickness so that the roadway will always be at elevation 7.0 or higher.

Hint: The geofoam is an extra "hidden" layer between the fill and the natural ground surface with $\gamma = 0$.

10.36 A series of prefabricated dual-bore steel tubes similar to the one in Figure 10.28 are to be installed in an underwater trench to form a tunnel. The trench will be in seawater, which has a unit weight of 64.0 lb/ft^3. The tubes will be floated into position, and sunk into place by temporarily flooding the interior. Then, nonstructural concrete will be placed into

FIGURE 10.28 This prefabricated tunnel section is part of the Central Artery Project in Boston. It was floated into position, and then sunk to the bottom of the bay. (Photograph by Peter Vanderwarker, courtesy of the Central Artery/Tunnel Project.)

chambers along the tube to act as ballast, and the inside will be pumped dry. The completed tube will be 80 ft wide, 300 ft long, and 40 ft tall, and weigh 32,000 tons exclusive of buoyant forces. Finally, the tube will be covered with soil, producing the cross-section shown in Figure 10.29.

(a) The interior of the tube will be dewatered after the concrete is placed, but before the trench is backfilled. Once this is done, will the tube remain at the bottom of the trench, or will it float up to the water surface?

(b) After the trench is backfilled, what will be the net $\Delta\sigma_z$ in the soft clay? Assume $\Delta\sigma_z$ is constant with depth.

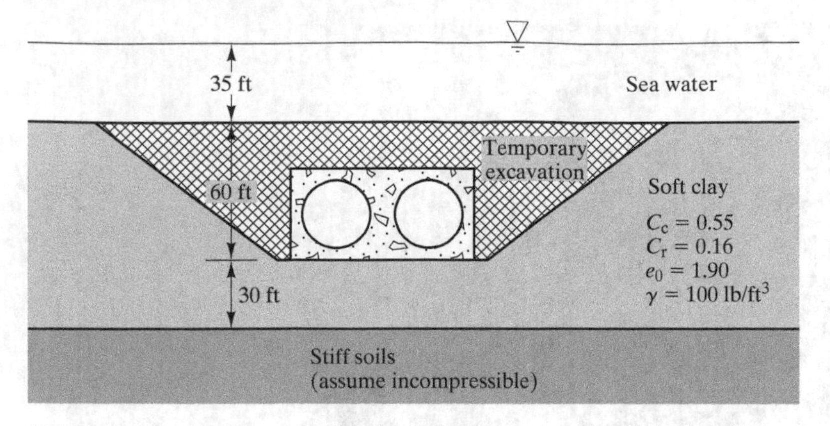

FIGURE 10.29 Final cross-section for underwater tunnel as described in Problem 10.36.

(c) Using the final cross-section, compute the ultimate consolidation settlement or heave of the tube due to $\Delta\sigma'_z$ in the soft clay. Assume no heave occurs during construction.

(d) The weakest parts of the completed tunnel will be the connections between the tube sections. In order to avoid excessive flexural stresses at these connections, the structural engineer has specified a maximum allowable differential settlement or differential heave of 5 in. along the length of the tube (the term *allowable* indicates this value already includes a factor of safety). An evaluation of the soil profile suggests the differential settlement or heave will be no more than 50% of the total. Has the structural engineer's criteria been met?

10.37 A proposed building is to have three levels of underground parking, as shown in Figure 10.30. To construct this building, it will be necessary to make a 10.0-m excavation, which will need to be temporarily dewatered. The natural and dewatered groundwater tables are as shown, and the medium clay is normally consolidated. The chief geotechnical engineer is concerned that this dewatering operation may cause excessive differential

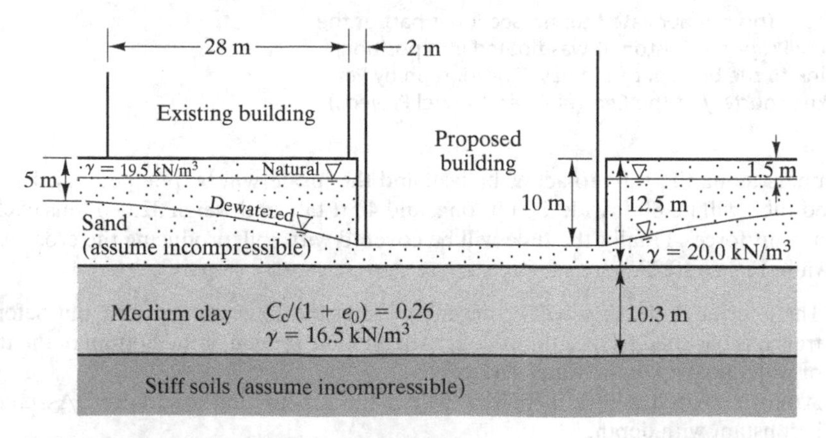

FIGURE 10.30 Soil profile for Problem 10.37.

settlements in the adjacent building and has asked you to compute the anticipated differential settlement across the width of this building. Assume the wall is perfectly rigid, and thus does not contribute to any settlement problems, and that the maximum allowable differential settlement from one side of the building to the opposite side is 50 mm. Neglect any loss in σ_z' below the existing building due to the removal of soil from the excavation. Discuss the implications of your answer.

10.38 The Palacio de las Bellas Artes in Mexico City, shown in Figure 10.1, is an interesting example of large consolidation settlement. It is supported on a 1.8- to 3.0-m thick mat foundation which is approximately 65 m wide and 115 m long. The average bearing pressure between the bottom of this mat and the supporting soil is 115 kPa (Ledesma, 1936).

The soil conditions beneath the palace are too complex to describe in detail here. However, we can conduct an approximate analysis using the following simplified profile:

Depth (m)	Description	γ (kN/m^3)	$\frac{C_c}{1+e_0}$
0–5	Sandy fill	17.5	0
5–45	Normally consolidated soft clays	11.5	0.53
>45	Stiff soils		0

For our simplified analysis, use a groundwater table at a depth of 5 m, and assume the bottom of the mat is at the original ground surface. In addition, assume the fill has been in pace for a very long time, so the consolidation settlement due to the weight of the fill is complete.

Divide the soft clay zone beneath the center of the building into five layers of equal thickness. Then, compute $\Delta\sigma_z$ at the midpoint of each layer using the methods described in Chapter 9. Finally, compute the ultimate consolidation settlement beneath the center of the palace due to its own weight.

10.39 A fill is to be placed at a proposed construction site, and you need to determine the ultimate consolidation settlement due to its weight. Write a 200–300 word essay describing the kinds of field exploration, soil sampling, and laboratory testing you will need to perform to generate the information needed for this analysis. Your essay should describe specific things that need to be done, and what information will be gained from each activity.

CHAPTER 11

Rate of Consolidation

Ut tensio sic vis (as to stretch, so the force).

Robert Hooke's 1678 description of the relationship between stress and strain, now known as *Hooke's Law*

When static loads are applied to structural members, such as beams or columns, the resulting deformations occur virtually as fast as the loads are applied. For example, when floor loads are applied to a steel beam, we assume that all of the resulting deflection occurs immediately. However, deformations in soils often occur much more slowly, especially in saturated clays. Many years, or even decades, may be required for the full settlement to occur in a soil, so geotechnical engineers often need to evaluate both the magnitude and the rate of consolidation settlement. Therefore, this chapter extends the analyses performed in Chapter 10 and develops methods to produce settlement versus time plots.

11.1 TERZAGHI'S THEORY OF CONSOLIDATION

Karl Terzaghi's most significant contribution to geotechnical engineering was his *theory of consolidation*, which he developed in Istanbul between 1919 and 1923 (Terzaghi, 1921, 1923a, 1923b, 1924, 1925a, and 1925b). Although others had studied the consolidation problem and made useful contributions, it was Terzaghi's work that properly identified and quantified the underlying physical processes. During this time, he identified the principle of effective stress, which became the key to understanding the consolidation process. Terzaghi's academic training as a mechanical engineer was very useful, because the physical processes that control consolidation are analogous to certain thermodynamic processes. In fact, he was teaching thermodynamics while conducting his consolidation experiments, which probably helped inspire the new theory.

Review of the Consolidation Process

The consolidation process, as discussed in Chapter 10, begins when the placement of a fill or some other load produces an increase in the vertical total stress, $\Delta\sigma_z$. Initially, this increase is carried entirely by the pore water, thus producing an excess pore water pressure, u_e. In one-dimensional consolidation analyses, the initial value of u_e equals $\Delta\sigma_z$. Thus, the vertical effective stress, σ'_z, immediately after loading is unchanged from its original value, σ'_{z0}.

The excess pore water pressure produces a localized increase in the total head, and thus induces a hydraulic gradient. Therefore, some of the pore water begins to flow away from the zone that is being loaded. This flow causes the excess pore water pressure to slowly dissipate, the vertical effective stress to increase, and the soil to consolidate. After sufficient time has elapsed, $u_e \rightarrow 0$, $\sigma'_z \rightarrow \sigma'_{zf}$, and the consolidation settlement, $\delta_c \rightarrow \delta_{c,ult}$. Terzaghi's theory of consolidation quantifies this process.

It is important to recognize that this theory is not simply an empirical description of settlement data obtained in the field; it is a rational method based on a physical model of the consolidation process. This is an important distinction, because it illustrates the difference between organized empiricism and the development of more fundamental understandings of soil behavior.

The various soil parameters needed to implement the theory of consolidation are normally obtained from a site characterization program, including laboratory consolidation tests, and thus are subject to many sources of error (i.e., are the samples truly representative?, what are the effects of soil disturbance?, and so on). Therefore, it does not give exact answers. However, the validity of this theory has been confirmed, and it is the basis for nearly all time–settlement computations.

Assumptions

To keep the computational process from becoming too complex, the theory of consolidation is based on certain simplifying assumptions regarding the compressible stratum:

1. The soil is homogeneous ($C_c/(1 + e_0)$, $C_r/(1 + e_0)$ and k are constant throughout).
2. The soil is saturated ($S = 100\%$).
3. The settlement is due entirely to changes in the void ratio, and these changes occur only as some of the pore water is squeezed out of the voids (i.e., the individual solid particles and the water are incompressible).
4. Darcy's Law (Equation 7.10) is valid.
5. The applied load causes an instantaneous increase in vertical total stress, $\Delta\sigma_z$. Afterward, the vertical total stress, σ_z, at all points remains constant with time.
6. Immediately after loading, the excess pore water pressure, u_e, is constant with depth, and equal to $\Delta\sigma_z$. This is generally true when the load is due to an extensive fill, but not when it is from a smaller loaded area, such as from a foundation.
7. The *coefficient of consolidation*, c_v, as defined below, is constant throughout the soil, and remains constant with time. In normally consolidated soils, c_v is

$$c_v = \left(\frac{2.30\sigma'_z k}{\gamma_w}\right)\left(\frac{1 + e}{C_c}\right) \tag{11.1}$$

where:

c_v = coefficient of consolidation
k = hydraulic conductivity
e = void ratio
C_c = compression index
γ_w = unit weight of water
σ'_z = vertical effective stress

For overconsolidated soils, substitute C_r for C_c in Equation 11.1.

8. The consolidation process is one-dimensional, as discussed below.

Some of these assumptions, such as number 4, are very realistic. Others, such as number 7, are only approximately correct and are intended to simplify the analysis. Some of these assumptions may be modified, with corresponding changes in the solution to Terzaghi's theory, but these enhancements are beyond the scope of this book. For most practical problems, the error introduced by these assumptions is acceptable. Section 11.5, later in this chapter, discusses the sources and probable magnitudes of errors in consolidation analyses.

One-Dimensional Consolidation

Terzaghi's theory assumes that the excess pore water flows only vertically, either up or down, and consolidation occurs only in the vertical direction. In other words, there is no horizontal drainage and no horizontal strain. This condition is called *one-dimensional consolidation*, and is shown in Figure 11.1. This assumption is most valid when the width

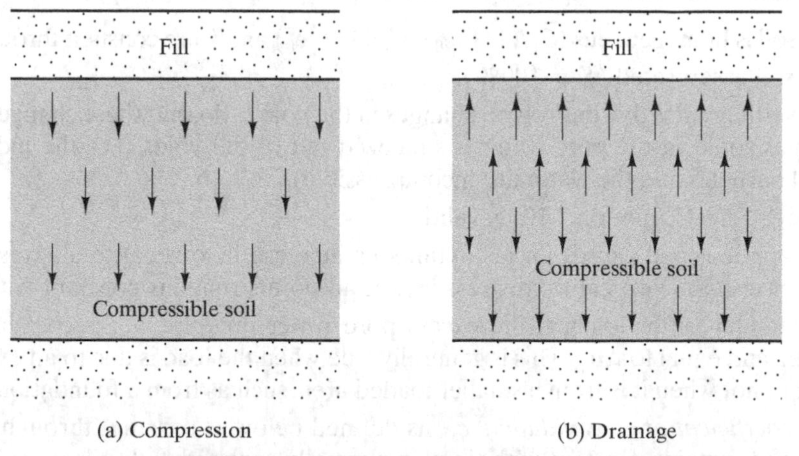

FIGURE 11.1 One-dimensional consolidation has the following characteristics: (a) the consolidation settlement is assumed to occur only in the vertical direction; and (b) the excess pore water is assumed to escape only by flowing vertically.

of the loaded area (such as a fill) is at least several times greater than the depth to the bottom of the consolidating strata.

One of the important parameters in one-dimensional consolidation analyses is the *maximum drainage distance*, H_{dr}. This is the longest distance any molecule of excess pore water must travel to move out of the consolidating soil. There are two possibilities, as shown in Figure 11.2.

- If the strata above and below are much more permeable than the consolidating soil (i.e., they have a much greater hydraulic conductivity, k), then the excess pore water will drain both upward and downward. This condition is known as *double drainage* and H_{dr} is equal to half the thickness of the consolidating strata.
- If the stratum below is less permeable, such as bedrock, then all of the excess pore water must travel upward, a condition known as *single drainage*. In this case, H_{dr} equals the thickness of the consolidating strata.

In both cases, H_{dr} is measured in a straight line, even though on a microscopic scale the actual flow path is a circuitous one that winds around the individual soil particles. We do it this way to be consistent with the definition used in Darcy's Law.

The value of H_{dr} has a significant effect on the time required to complete the consolidation process. All else being equal, this time is proportional to H_{dr}^2. Thus, if a 6-m thick stratum requires 10 years to fully consolidate, a 12-m stratum of the same soil (double the thickness) would require 40 years (four times as long).

Derivation of the One-Dimensional Consolidation Equation

In Section 10.4, we discussed how the application of the additional total stress $\Delta\sigma_z$ onto the ground induces an excess pore water pressure, u_e. Immediately after the application of this load, which is assumed to occur instantaneously, u_e is constant with depth (according to Assumption 6). The rate of consolidation depends on the dissipation of

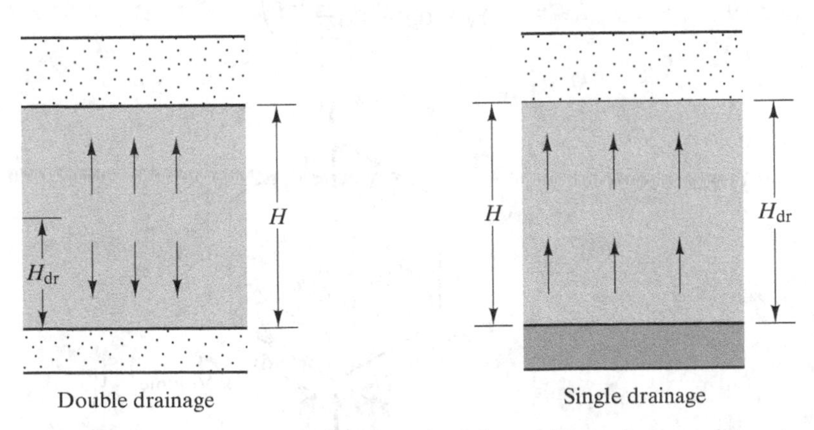

FIGURE 11.2 Computation of the length of the maximum drainage distance, H_{dr}, for one-dimensional consolidation problems.

these excess pore water pressures and the corresponding transfer of stress to the solid particles. Therefore, we can derive the governing equation by first examining the dissipation of excess pore water pressure in a typical soil element as shown in Figure 11.3. Then we will examine the consolidation that occurs in this element as the stresses are transferred. Finally, we will combine these two processes into one equation.

We will assume that the excess pore water flows upward through the element. At the top of the element, the hydraulic gradient is

$$i_z = \frac{dh}{dz} \tag{11.2}$$

where z is the depth to the top of the element. Unlike Equation 7.4, this equation does not have a negative sign because the water is flowing in the z-direction (upward).

The elevation head at the top of the element remains constant with time. Only the pressure head, h_p, changes, and this change is due only to changes in the excess pore water pressure, u_e. Therefore, using Equation 7.6 we obtain

$$i_z = \frac{dh_p}{dz} = \frac{1}{\gamma_w}\frac{du_e}{dz} \tag{11.3}$$

The hydraulic gradient varies with depth, as defined by

$$\frac{di}{dz} = \frac{1}{\gamma_w}\frac{d^2u_e}{dz^2} \tag{11.4}$$

Therefore, the hydraulic gradient at the bottom of the element is

$$i_{z+dz} = i_z + \frac{di}{dz}dz$$

$$i_{z+dz} = \frac{1}{\gamma_w}\left(\frac{du_e}{dz} + \frac{d^2u_e}{dz^2}dz\right) \tag{11.5}$$

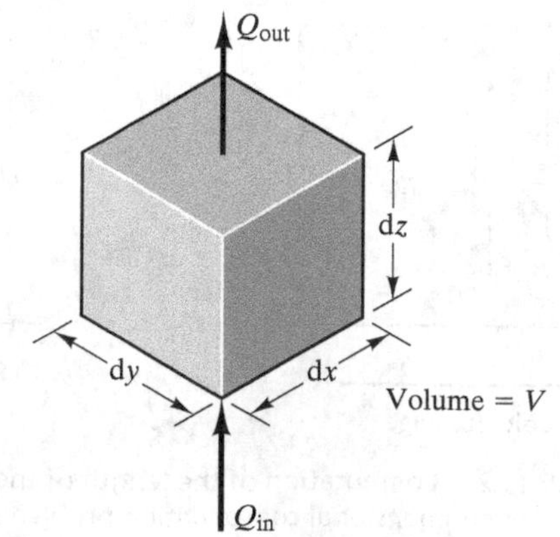

FIGURE 11.3 Soil element used to derive Terzaghi's theory of consolidation.

Using Darcy's Law (per Assumption 4),

$$Q = \frac{dV}{dt} = kiA \Rightarrow dV = kiA\, dt \tag{11.6}$$

$$dV_{in} = k\frac{1}{\gamma_w}\left(\frac{du_e}{dz} + \frac{d^2u_e}{dz^2}dz\right) dx\, dy\, dt \tag{11.7}$$

$$dV_{out} = k\frac{1}{\gamma_w}\frac{du_e}{dz}dx\, dy\, dt \tag{11.8}$$

$$dV = dV_{in} - dV_{out}$$

$$dV = \frac{k}{\gamma_w}\frac{d^2u_e}{dz^2}dx\, dy\, dz\, dt \tag{11.9}$$

Next, we will consider the relationship between volume changes and excess pore water pressures inside the sample. According to Equation 10.10,

$$C_c = -\frac{de}{d\log\sigma'_z} \tag{11.10}$$

may be rewritten as

$$de = -\frac{C_c}{2.30\sigma'_z}d\sigma'_z \tag{11.11}$$

Changes in effective stress are due solely to changes in excess pore water pressure, so

$$d\sigma'_z = -du_e \tag{11.12}$$

$$de = \frac{C_c}{2.30\sigma'_z}du_e \tag{11.13}$$

An extension of Equation 10.8 gives the following formula for the vertical strain, ε_z:

$$d\varepsilon_z = \frac{de}{1+e} \tag{11.14}$$

The change in volume is then

$$dV = -d\varepsilon_z A\, dz$$

$$dV = -\frac{de}{1+e}dx\, dy\, dz$$

$$dV = \frac{C_c}{(2.30\sigma'_z)(1+e)}dx\, dy\, dz\, du_e \tag{11.15}$$

Finally, we combine Equations 11.1, 11.9, and 11.15. The excess pore water pressure, u_e, now varies with both depth z and time t, so we have a partial differential equation:

$$\frac{\partial u_e}{\partial t} = c_v \frac{\partial^2 u_e}{\partial z^2} \tag{11.16}$$

where:

u_e = excess pore water pressure

t = time

c_v = coefficient of consolidation (a constant, per Assumption 7)

z = vertical distance below the ground surface

Equation 11.16 is called the *one-dimensional consolidation equation*. It is similar to Fick's law of thermal diffusion, which mechanical engineers use to analyze the flow of heat through a material and the resulting temperature changes.

Solution of the One-Dimensional Consolidation Equation

The solution of Equation 11.16 requires the establishment of two boundary conditions for z and one initial condition for u_e. For the single drainage condition with z_t = depth to the top of the compressible stratum, and H = the thickness of the compressible stratum, they are as follows:

1. At $z = z_t$, $u_e = 0$ at all times (the excess pore water pressure is always zero at the top of the compressible stratum). This appears to be in violation of Assumption 6, and more correctly describes the conditions immediately after $t = 0$. This initial dissipation of excess pore water pressures sets up a hydraulic gradient that permits the process to continue.
2. At $z = z_t + H$, $i = du_e/dz = 0$ (the hydraulic gradient is zero at the bottom of the compressible stratum). This is because we are considering the single drainage condition, and there is no flow at the lower boundary.
3. At $t = 0$, $u_e = \Delta\sigma_z$ (immediately after placement of the load, the applied vertical stress is carried entirely by the excess pore water pressure, is equal to the change in total stress, $\Delta\sigma_z$, and is constant with depth). This is a restatement of Assumption 6.

For the double drainage case, simply add a mirror image of the drainage model to the bottom half of the compressible stratum.

An analytical solution based on these boundary conditions produces the following infinite series formula for u_e at any point in the compressible strata at any time after the application of the load (Means and Parcher, 1963):

$$\frac{u_e}{\Delta\sigma_z} = \sum_{N=0}^{\infty}\left(\frac{4}{(2N+1)\pi}\sin\left[\frac{(2N+1)\pi}{2}\left(\frac{z_{dr}}{H_{dr}}\right)\right]e^{-\left[\frac{(2N+1)^2\pi^2}{4}T_v\right]}\right) \tag{11.17}$$

Note that this solution is presented in terms of three dimensionless parameters:

$\frac{u_e}{\Delta\sigma_z}$ is excess pore pressure at a given depth relative to the initial induced stress at that point

$\frac{z_{dr}}{H_{dr}}$ is the depth into the consolidating layer relative to the maximum drainage distance

T_v is a dimensionless measure of time called the time factor and defined as

$$T_v = \frac{c_v t}{(H_{dr})^2} \tag{11.18}$$

where:

u_e = excess pore water pressure
$\Delta\sigma_z$ = change in vertical total stress due to applied load
z_{dr} = vertical distance from point to nearest drainage boundary
H_{dr} = maximum drainage distance (for single drainage, H_{dr} = thickness of the compressible stratum; for double drainage, H_{dr} = half the thickness of the compressible stratum)
e = base of natural logarithms = 2.7183
c_v = coefficient of consolidation
t = time since application of load

and the sine term in Equation 11.17 must be in radians.

The series in Equation 11.17 generally converges for N less than 10, though sometimes more terms are needed.

Application of the One-Dimensional Consolidation Equation

Equation 11.17 describes the ratio of the excess pore water pressure, u_e, to the instantaneous increase in total stress, $\Delta\sigma_z$, at a point in a soil. This ratio is defined as a function of relative depth of the point, (z_{dr}/H_{dr}), and the time factor, T_v. Immediately after this load is applied, $(u_e/\Delta\sigma_z) = 1$ at all points, then it gradually diminishes with time, eventually becoming equal to zero. In other words, at the end of this process the excess pore water pressures will have dissipated and the groundwater will have returned to the hydrostatic condition. The time required for this process and the applicability of this equation to practical problems depend upon many factors, most importantly the soil properties, characterized by c_v, and the maximum drainage distance, H_{dr}.

Clays and Silts The excess pore water pressures dissipate only as some of the pore water flows away from the zone of soil that is being loaded. Clays and silts have a low hydraulic conductivity, k, so water flows very slowly through these soils, and a long time is required to return to the hydrostatic condition. The theory of consolidation reflects this through the use of a low coefficient of consolidation, c_v (per Equation 11.1). Using these low c_v values in Equations 11.17 and 11.18 demonstrates that years or even decades will be required to fully dissipate the excess pore water pressures and return to the hydrostatic condition, especially when the maximum drainage distance, H_{dr}, is large.

In this case, our assumption that the load is applied instantaneously is not too far from the truth, because the duration of construction is probably very short compared to the time required to dissipate the excess pore water pressures. Therefore, the analyses described in this chapter are applicable to these soils. We will use these methods to compute the dissipation of excess pore water pressures and thus develop plots of consolidation settlement versus time.

Sands and Gravels The hydraulic conductivity, k, of sands and gravels is much greater than that of clays and silts, so their time–settlement behavior is correspondingly different. According to Table 7.1, k in sands is typically about 1,000,000 times greater than that in clays, and according to Equation 11.1, c_v is proportional to k. If we place these high c_v values in Equations 11.17 and 11.18, it becomes clear that the excess pore water pressures dissipate very quickly, perhaps in a few minutes or less. This is much faster than the rate of construction, so the consolidation settlement occurs virtually as fast as the load is applied.

Therefore, it is not necessary to conduct rate of consolidation analyses in sandy and gravelly soils. We simply compute the ultimate consolidation settlement, $\delta_{c,ult}$, using the methods described in Chapter 10 and assume that it occurs as quickly as the load is applied.

Example 11.1

Consider the soft clay strata in Example 10.5. According to a laboratory consolidation test, $c_v = 0.0021\ \text{m}^2/\text{day}$. Compute the hydrostatic, excess, and total pore water pressures at Point B, 2000 days after the placement of the fill.

Solution:

$$\Delta\sigma_z = 57.6\ \text{kPa}$$
$$u_h = \gamma_w z_w = (9.8\ \text{kPa})(5.0\ \text{m}) = 49.0\ \text{kPa}$$

The soils above and below the clay are much more permeable than the clay. Therefore, use double drainage.

$$H_{dr} = \frac{10.0\ \text{m}}{2} = 5.0\ \text{m}$$
$$z_{dr} = 3.0\ \text{m}$$
$$\frac{z_{dr}}{H_{dr}} = \frac{3.0}{5.0} = 0.6$$

Using Equations 11.17 and 11.18,

N	$u_e/\Delta\sigma_z$
0	0.680
1	0.004
2	0.000
Sum	0.684

This time, only three increments of N are required for the series to converse.

$$u_e = (0.684)(57.6) = 39.4 \text{ kPa}$$
$$u = u_h + u_e = (49.0 + 39.4) \text{ kPa} = 88.4 \text{ kPa}$$
$$u_h = \mathbf{49.0\ kPa}$$
$$u_e = \mathbf{39.4\ kPa}$$
$$u = \mathbf{88.4\ kPa}$$

To develop a complete plot of pore pressure versus depth, the computations in Example 11.1 would need to be repeated at many different depths. This would be tedious to do by hand, but easy to do with a computer. Since Equation 11.17 is presented in dimensionless parameters we need to perform these computations only once. The dimensionless parameters will allow us to use this solution for different consolidation problems.

Just such a solution is shown in Figure 11.4, which presents plots of $u_e/\Delta\sigma_z$ for various values of T_v. It was developed from Equation 11.17. Notice how the consolidation process (i.e., the dissipation of excess pore water pressures) occurs very quickly at the top and bottom because the excess pore water drains most easily there. However, the process is much slower in the center because it is farther from the drainage boundaries.

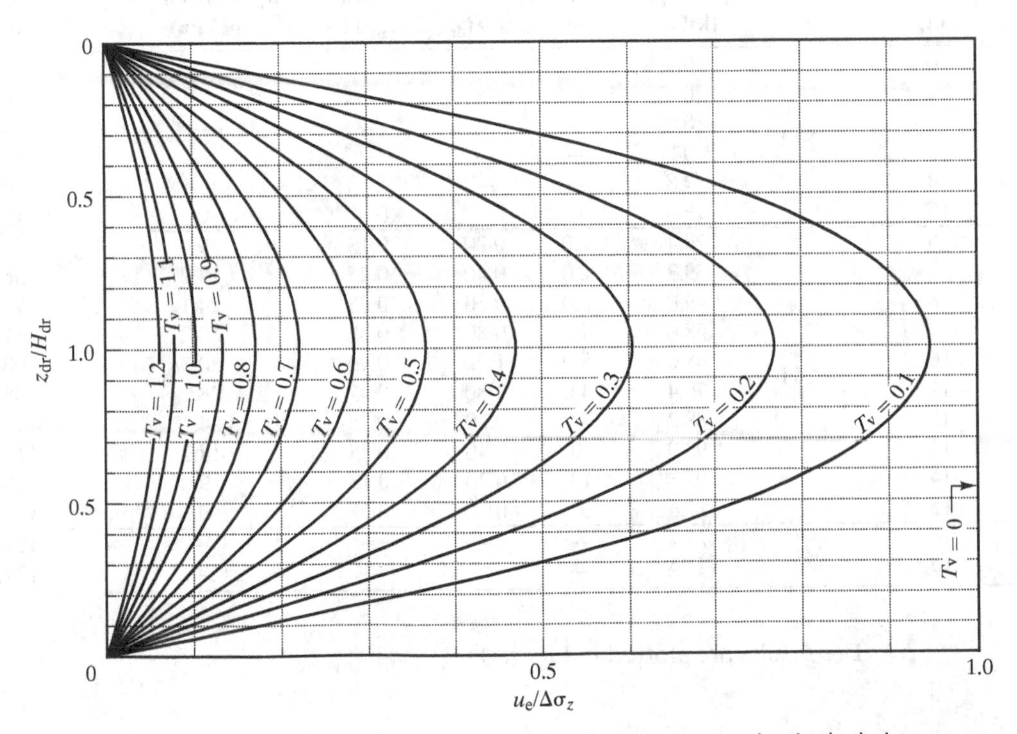

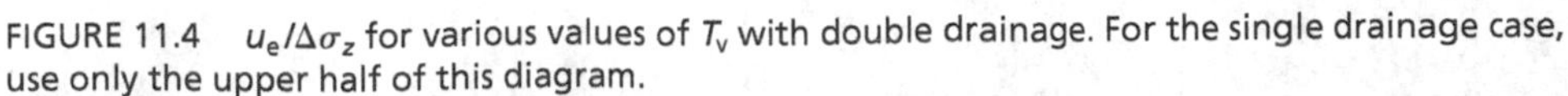
FIGURE 11.4 $u_e/\Delta\sigma_z$ for various values of T_v with double drainage. For the single drainage case, use only the upper half of this diagram.

Example 11.2

A fill is to be placed on the soil profile shown in Figure 11.5. Using the curves in Figure 11.4, develop a plot of u_h, u_e, and u versus depth at $t = 10$ year after the placement of the fill.

Solution:

The hydraulic conductivities of the SM and ML strata are much greater than that of the CH strata, so the double drainage condition exists.

$$H_{dr} = \frac{10.0\text{ m}}{2} = 5.0\text{ m}$$

$$T_v = \frac{c_v t}{H_{dr}^2} = \frac{(3 \times 10^{-3}\text{ m}^2/\text{d})(10\text{ yr})(365\text{ d/yr})}{(5.0\text{ m})^2} = 0.438$$

$$\Delta\sigma_z = \gamma_{fill} H_{fill} = (19.7\text{ kN/m}^3)(3.5\text{ m}) = 68.9\text{ kPa}$$

Depth from original ground surface (m)	Soil	$u_h = \gamma_w z_w$ (kPa)	z_{dr} (m)	z_{dr}/H_{dr}	$u_e/\Delta\sigma_z$ from Fig. 11.4	$u_e = 68.9\, u_e/\Delta\sigma_z$ (kPa)	$u = u_h + u_e$ (kPa)
0	SM	0	—	—	—	—	
1		0	—	—	—	—	0
2		0	—	—	—	—	0
3		0	—	—	—	—	0
4		9.8	—	—	—	—	9.8
5		19.6	0	0	0	0	19.6
6	CH	29.4	1.0	0.20	0.13	9.0	38.4
7		39.2	2.0	0.40	0.25	17.2	56.4
8		49.0	3.0	0.60	0.35	24.1	73.1
9		58.8	4.0	0.80	0.41	28.2	87.0
10		68.6	5.0	1.00	0.44	30.3	98.9
11		78.4	4.0	0.80	0.41	28.2	106.6
12		88.2	3.0	0.60	0.35	24.1	112.3
13		98.0	2.0	0.40	0.25	17.2	115.2
14		107.8	1.0	0.20	0.13	9.0	116.8
15		117.6	0	0	0	0	117.6
16	ML	127.4	—	—	—	—	127.4
17		137.2	—	—	—	—	137.2

The results are plotted in Figure 11.5.

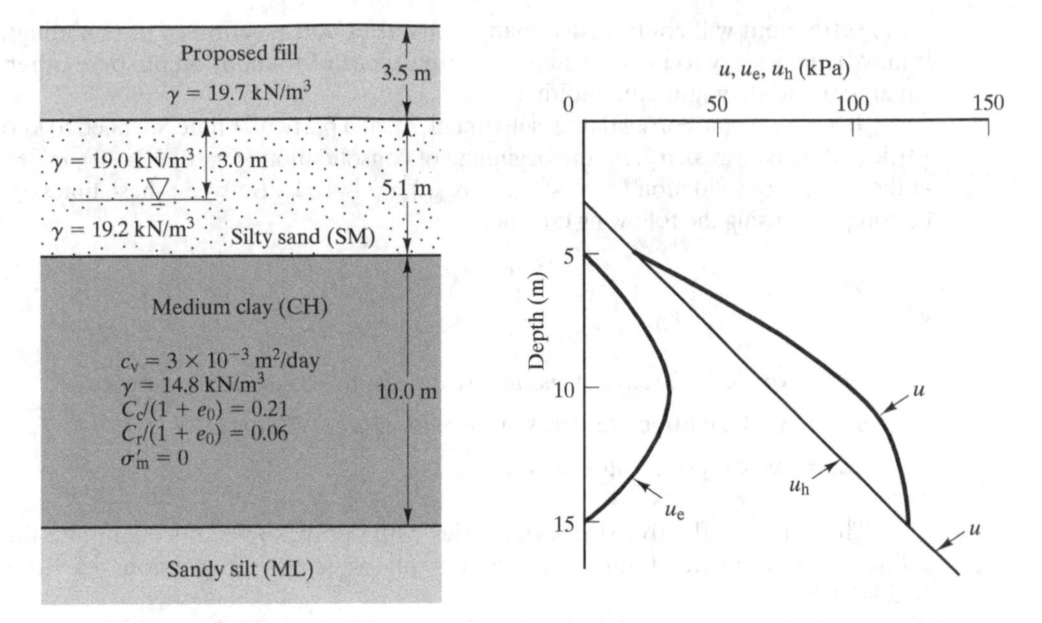

FIGURE 11.5 Soil profile and results for Example 11.2.

Commentary:

1. Due to the high hydraulic conductivity (or high c_v) in the SM stratum, the potential rate of drainage is probably high relative to the rate of loading, so there will not be any significant excess pore water pressures. Some excess pore water pressures might be present in the ML stratum during the early stages of consolidation, but they will have dissipated long before $t = 10$ years. Therefore, excess pore water pressures are present only in the CH stratum.
2. If single drainage conditions had been present in the CH strata, then the z_{dr}/H_{dr} values would range from 0 at the top of the stratum to 1 at the bottom, and u_e values would be correspondingly higher.

11.2 CONSOLIDATION SETTLEMENT VERSUS TIME COMPUTATIONS

Now that we have Equation 11.17 and are able to compute excess pore water pressures as a function of depth and time, we can also compute consolidation settlement, δ_c, as a function of the time since loading, t. Plots of anticipated settlement versus time are very valuable to geotechnical engineers because they help us plan appropriate mitigative measures. For example, if the weight of a proposed fill will produce a certain amount of settlement, but this settlement will be virtually complete before construction of any buildings begins, then its impact on the buildings will be minimal. However, if the settlement will continue for many years after construction of the buildings, then it may be necessary to provide some different types of foundation or some other measures to avoid damaging the buildings.

To compute the consolidation settlement, δ_c, at a particular time, we need to know the vertical effective stress, σ'_z. At the beginning of consolidation ($t = 0, \delta_c = 0$), $\sigma'_z = \sigma'_{z0}$; at the end of consolidation ($t = \infty, \delta_c = \delta_{c,ult}$), $\sigma'_z = \sigma'_{zf}$. Between these times, σ'_z may be computed using the following equation:

$$\sigma'_z = \sigma'_{zf} - u_e \tag{11.19}$$

where:

σ'_z = vertical effective stress at any time in the consolidation process

σ'_{zf} = vertical effective stress at the end of consolidation

u_e = excess pore water pressure

The vertical effective stress, σ'_z, varies with depth, so we must compute the consolidation settlement at time t using the following revised versions of Equations 10.23–10.25.

For normally consolidated soils ($\sigma'_{z0} \approx \sigma'_c$):

$$\delta_c = \sum \frac{C_c}{1 + e_0} H \log\left(\frac{\sigma'_z}{\sigma'_{z0}}\right) \tag{11.20}$$

For overconsolidated soils—Case I ($\sigma'_{z0} < \sigma'_z \leq \sigma'_c$):

$$\delta_c = \sum \frac{C_r}{1 + e_0} H \log\left(\frac{\sigma'_z}{\sigma'_{z0}}\right) \tag{11.21}$$

For overconsolidated soils—Case II ($\sigma'_{z0} < \sigma'_c < \sigma'_z$):

$$\delta_c = \sum \left[\frac{C_r}{1 + e_0} H \log\left(\frac{\sigma'_c}{\sigma'_{z0}}\right) + \frac{C_c}{1 + e_0} H \log\left(\frac{\sigma'_z}{\sigma'_c}\right)\right] \tag{11.22}$$

where:

δ_c = consolidation settlement at time t

C_c = compression index

C_r = recompression index

e_0 = initial void ratio

σ'_{z0} = initial vertical effective stress

σ'_z = vertical effective stress at time t

σ'_c = preconsolidation stress

H = thickness of soil strata

Equations 11.20–11.22 must be selected carefully, because the appropriate choice may vary with both depth and time. For example, a given point in the soil may be overconsolidated Case I during the early stages of consolidation, then change to overconsolidated Case II when σ'_z reaches σ'_c.

It is also helpful to define a new parameter, the *degree of consolidation*, *U*, which is the percentage of the ultimate consolidation settlement that has occurred at a certain time after loading:

$$U = \frac{\delta_c}{\delta_{c,ult}} \times 100\% \tag{11.23}$$

where:

U = degree of consolidation (percent)
δ_c = consolidation settlement
$\delta_{c,ult}$ = ultimate consolidation settlement

We will consider two ways to develop time–settlement curves: one that explicitly considers the dissipation of pore water pressures but requires a computer, and another simplified method that may be solved by hand.

Computer-Based Solution

The analysis described thus far may be implemented using a computer as follows:

1. Divide the compressible stratum into horizontal layers.
2. Using Equation 7.7, compute the hydrostatic pore water pressure, u_h, at the midpoint of each layer.
3. Using the pore water pressure from Step 2, compute the initial vertical effective stress, σ'_{z0}, at the midpoint of each layer.
4. Select an appropriate time, *t*, after placement of the load.
5. Using Equation 11.17, compute the excess pore water pressure, u_e, at the midpoint of each layer.
6. Add the values obtained from Steps 2 and 5 to find the pore water pressure, *u*, at the midpoint of each layer ($u = u_h + u_e$).
7. Using the pore water pressure from Step 6, compute the vertical effective stress, σ'_z, at the midpoint of each layer. This is the σ'_z at time *t*.
8. Using Equation 11.20, 11.21, or 11.22, as appropriate, compute the consolidation settlement for each layer and sum. This is δ_c at time *t*. If Equation 11.22 is to be used, it will be necessary to compute σ'_c at the midpoint of each layer using Equation 10.17.
9. Repeat Steps 4 through 8 using new values of *t* until an acceptable δ_c versus *t* plot has been obtained.

This is a type of numerical solution, and its precision depends on the number of layers used in the computations. At least 50 layers are typically necessary to develop reasonable plots.

Simplified Solution

The second method of computing the consolidation settlement introduces an additional simplifying assumption: for the purpose of computing the dissipation of u_e, we will assume that the vertical strain, ε_z, due to consolidation in the compressible strata is proportional to the vertical effective stress, σ'_z. In other words, the stress–strain curve is linear. This is clearly not true—it is a logarithmic relationship, as discussed in Chapter 10. However, this assumption simplifies the computations in a significant way because the vertical strain becomes proportional to the drop in excess pore water pressure ($\varepsilon_z \propto -u_e$). Thus, a certain drop in u_e at one depth in the compressible stratum produces the same strain as an equal drop in u_e at another depth.

This assumption can be confusing in that it applies only to the settlement rate computation. The value of $\delta_{c,ult}$ remains unchanged, and is still based on the nonlinear Equations 10.23–10.25. The only difference in the results obtained from this simplified method and the more precise computer-based solution is the shape of the time–settlement curve.

With this new simplifying assumption, U becomes equal to half the area to the right of each T_v curve in Figure 11.4. Therefore, we can develop a unique relationship between U and T_v, as shown in Figure 11.6. This relationship also may be represented by the following fitted equations (adapted from Terzaghi, 1943):

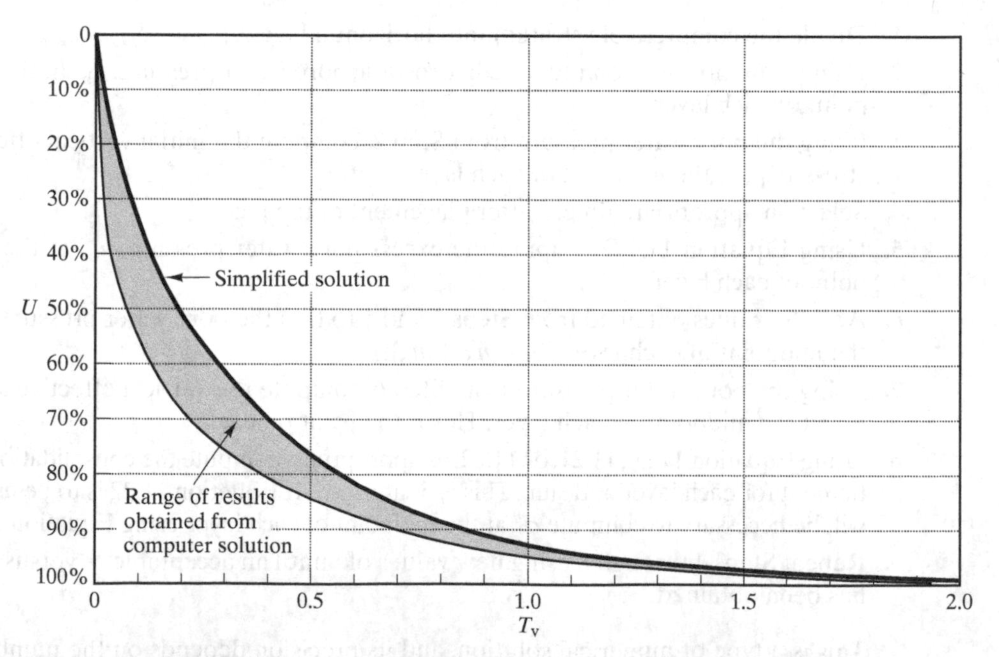

FIGURE 11.6 The solid line is the U versus T_v function for the simplified analysis of one-dimensional consolidation. The shaded area represents the range of values obtained from the more precise computer-based solution.

For $T_v \leq 0.217$ ($U \leq 52.6\%$),

$$U = \sqrt{\frac{4T_v}{\pi}} \times 100\% \qquad (11.24)$$

For $T_v > 0.217$ ($U > 52.6\%$),

$$U = \left[1 - 10^{-\left(\frac{0.085 + T_v}{0.933}\right)}\right] \times 100\% \qquad (11.25)$$

where:

U = degree of consolidation (percent)
T_v = time factor

Figure 11.6 also shows the range of U versus T_v values obtained from the more precise computer solution described earlier. The difference between these two methods is often small, but it can be quite significant, especially during the early stages of consolidation. The simplified solution is also presented in Table 11.1.

TABLE 11.1 Relationship between degree of consolidation, U, and the time factor, T_v using the simplified solution (Equations 11.24 and 11.25)

U (%)	T_v	U (%)	T_v	U (%)	T_v
1%	0.000079	35%	0.0962	75%	0.477
2%	0.000315	40%	0.126	80%	0.567
5%	0.00196	45%	0.159	85%	0.684
10%	0.00785	50%	0.196	90%	0.848
15%	0.0177	55%	0.239	95%	1.13
20%	0.0314	60%	0.286	97%	1.34
25%	0.0491	65%	0.340	98%	1.50
30%	0.0707	70%	0.403	99%	1.78

Example 11.3

A fill is to be placed over the soil profile shown in Figure 11.7. Both the SP and CL/ML strata are normally consolidated. Develop a plot of consolidation settlement versus time using the simplified solution.

Solution:

Sand stratum:

The consolidation settlement in the sand stratum will occur as quickly as the load is applied. Therefore, we only need to compute the ultimate consolidation settlement.

At midheight in the sand stratum,

$$\sigma'_{z0} = \sum \gamma H - u$$

$$\sigma'_{z0} = (17.8 \text{ kN/m}^3)(1.2 \text{ m}) + (18.1 \text{ kN/m}^3)(0.95 \text{ m}) - (9.8 \text{ kN/m}^3)(0.95 \text{ m})$$

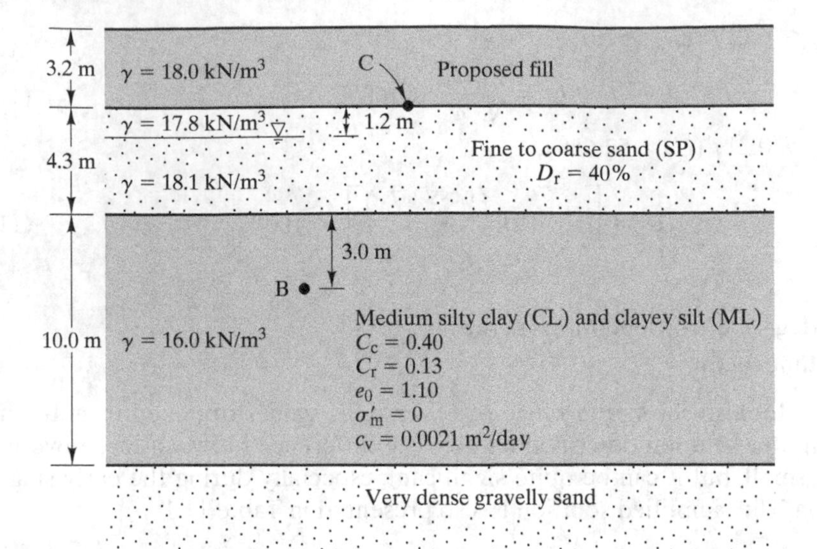

FIGURE 11.7 Soil profile for Example 11.3.

$$\sigma'_{z0} = 29.2 \text{ kPa}$$
$$\sigma'_{zf} = \sigma'_{z0} + \gamma_{\text{fill}} H_{\text{fill}}$$
$$\sigma'_{zf} = 29.2 \text{ kPa} + (18.0 \text{ kN/m}^3)(3.2 \text{ m})$$
$$\sigma'_{zf} = 86.8 \text{ kPa}$$

Per Table 10.4, $C_c/(1 + e_0) = 0.007$

$$\delta_{c,\text{ult}} = \sum \frac{C_c}{1 + e_0} H \log\left(\frac{\sigma'_{zf}}{\sigma'_{z0}}\right)$$

$$\delta_{c,\text{ult}} = (0.007)(4.3 \text{ m})\log\left(\frac{86.8 \text{ kPa}}{29.2 \text{ kPa}}\right)$$

$$\delta_{c,\text{ult}} = 14 \text{ mm}$$

Gravelly sand stratum:

This soil is classified as very dense, so we can assume that its compressibility is negligible compared to that of the other strata.

Silty clay/clayey silt stratum:

From a hand analysis similar to the ones in Chapter 10, we determine ultimate consolidation settlement, $\delta_{c,\text{ult}}$, of the clay stratum is 471 mm.

Using Equation 11.18 to compute T_v and Figure 11.6 (or Equations 11.24 and 11.25) to find U, we obtain the following results:

Time				Settlement (mm)		
(days)	(years)	T_v	U (%)	Sand Strata	Clay Strata	Total
2		0.00017	1	14	5	19
5		0.00042	2	14	9	23
10		0.00084	3	14	14	30
20		0.0017	5	14	24	38
50		0.0042	7	14	33	47
100		0.0084	10	14	47	61
200		0.017	15	14	71	85
500	1.4	0.042	23	14	108	122
1,000	2.7	0.084	33	14	155	169
1,400	3.8	0.12	39	14	184	198
2,000	5.5	0.17	47	14	221	235
3,200	8.8	0.27	58	14	273	287
5,000	13.7	0.42	71	14	334	348
7,100	19.5	0.60	82	14	386	400
10,000	27.4	0.84	90	14	424	438
14,000	38.4	1.2	96	14	452	466
20,000	54.8	1.7	99	14	466	480

Commentary: Although most of the settlement occurs in the first 25 years, the consolidation process continues for at least another 25 years.

If we use the computer-based solution as described earlier to solve this problem, we can compute the following settlement versus time data.

Time		Settlement (mm)		
(days)	(years)	Sand Strata	Clay Strata	Total
2		14	7	21
5		14	12	26
10		14	18	32
20		14	26	40
50		14	41	55
100		14	58	72
200		14	81	95
500	1.4	14	127	141
1,000	2.7	14	178	192
1,400	3.8	14	209	223
2,000	5.5	14	247	261
3,200	8.8	14	304	318
5,000	13.7	14	361	375
7,100	19.5	14	402	416
10,000	27.4	14	434	448
14,000	38.4	14	455	469
20,000	54.8	14	466	480

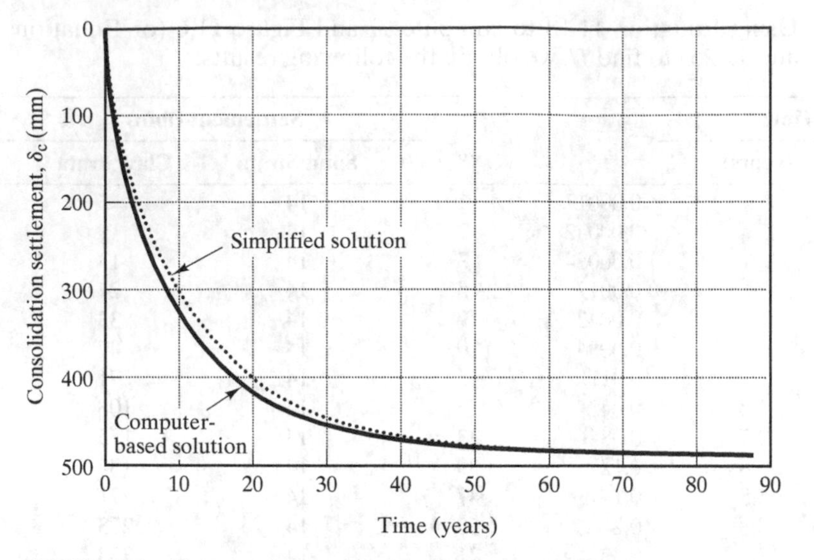

FIGURE 11.8 Time–settlement plots for Example 11.3.

The results of both the simplified and more accurate computer-based solutions are plotted in Figure 11.8. The greatest difference between these results is about 12%. Although this difference may seem excessive, in reality other uncertainties in the analysis, particularly the definition of the stratigraphy, H_{dr}, and c_v, introduce greater errors, so the simplified solution is sufficiently accurate for most practical problems.

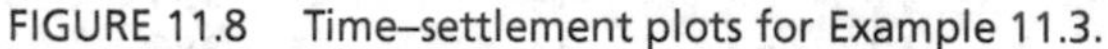

Correction for Construction Period

In reality, most loads applied to soils do not occur instantaneously. They are usually applied during some construction process that may last for weeks or months. For example, loads due to the weight of new fills are imparted only as fast as the fill is constructed.

A simple method of computing settlements during and after the construction period is to assume that the load is applied at a uniform rate, then adjust the time t in the settlement computations as follows:

For $t \leq t_c$:

$$t_{adj} = \frac{t}{2} \tag{11.26}$$

For $t > t_c$:

$$t_{adj} = t - \frac{t_c}{2} \tag{11.27}$$

where:

t = time since beginning of construction

t_c = duration of construction period (i.e., time at the end of construction)

t_{adj} = adjusted time

Then, perform the settlement rate computation using t_{adj} and the value of H_{fill} present at time t. For example, if a 6-m thick fill is to be placed at a uniform rate over a period of 30 days, we would compute the settlement at $t = 20$ days using $t_{adj} = 20/2 = 10$ days and $H_{fill} = 6$ m (20/30) $= 4$ m (i.e., the amount of fill present at $t = 20$ days).

Olson (1977) provided a more exact (and more complicated) analytical solution to the problem of time-dependant loading. If the construction period is a significant portion of the consolidation time, then the adjusted-time method given above will not give reasonable results; in this case, Olson's solution should be used.

Example 11.4

A proposed fill is to be placed on the soil profile shown in Figure 11.9. The fill will be placed at a uniform rate over a period of 6 months. Develop a time–settlement curve considering this construction period.

Solution:

H_{fill} increases linearly from 0 at $t = 0$ to 12 ft at $t = 180$ days.
SM/ML stratum:

Assume that the consolidation settlement occurs as quickly as the fill is placed (i.e., no significant excess pore water pressures).

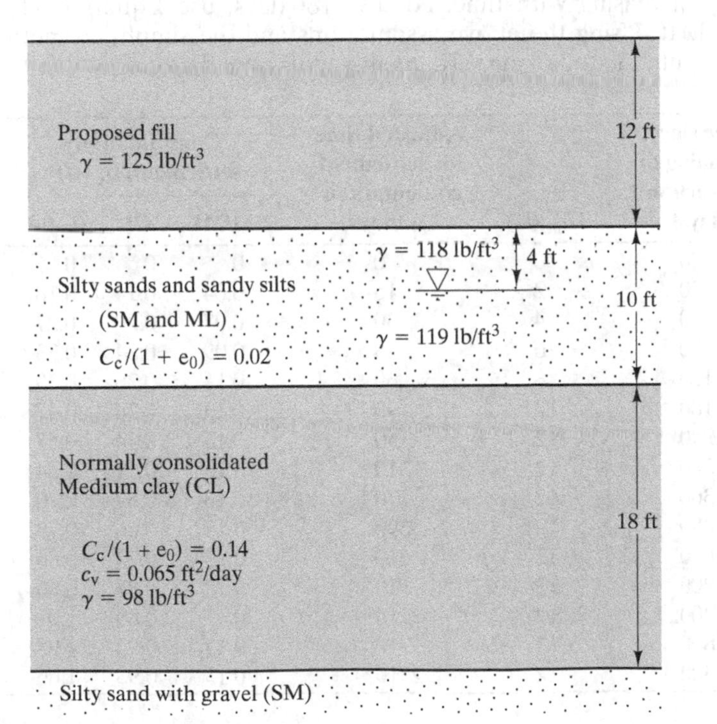

FIGURE 11.9 Soil profile for Example 11.4.

Divide into two layers: layer 1 is the upper 4 ft and layer 2 is the lower 6 ft. For layer 1,

$$\sigma'_{z0} = \sum \gamma H - u = (118 \text{ lb/ft}^3)(2 \text{ ft}) - 0 = 236 \text{ lb/ft}^2$$

$$\delta_{c,ult} = \sum \frac{C_c}{1+e_0} H \log\left(\frac{\sigma'_{zf}}{\sigma'_{z0}}\right) = (0.02)(4)\log\left(\frac{236 + \gamma_{fill}H_{fill}}{236}\right)$$

For layer 2,

$$\sigma'_{z0} = \sum \gamma H - u = (118 \text{ lb/ft}^3)(4 \text{ ft}) + (119 \text{ lb/ft}^3)(3 \text{ ft}) - (62.4 \text{ lb/ft}^3)(3 \text{ ft}) = 642 \text{ lb/ft}^2$$

$$\delta_{c,ult} = \sum \frac{C_c}{1+e_0} H \log\left(\frac{\sigma'_{zf}}{\sigma'_{z0}}\right) = (0.02)(4)\log\left(\frac{642 + \gamma_{fill}H_{fill}}{642}\right)$$

Compute δ_c for each time using the corresponding H_{fill}. The results of this computation are shown in the following table.

CL stratum:

The consolidation in this stratum will continue well beyond the construction period, and needs to be computed using one-dimensional consolidation theory. For $t = 0$ to 180 days, use Equation 11.26 to find t_{adj}, and use H_{fill} linearly increasing with time. For $t > 180$ days, use Equation 11.27 and $H_{fill} = 12$ ft. Using the above assumptions and the simplified method, we can compute the following settlement versus time data.

Time since beginning of construction t (days)	H_{fill} (ft)	Adjusted time for settlement computations t_{adj} (days)	Consolidation settlement, δ_c (ft)		
			SM/ML	CL	Total
0	0	0	0	0	0
30	2	15	0.04	0.14	0.18
60	4	30	0.07	0.20	0.27
90	6	45	0.09	0.24	0.33
120	8	60	0.11	0.28	0.39
180	12	90	0.13	0.34	0.47
240	12	150	0.13	0.44	0.57
300	12	210	0.13	0.51	0.64
360	12	270	0.13	0.57	0.70
480	12	390	0.13	0.66	0.79
600	12	510	0.13	0.73	0.86
900	12	810	0.13	0.83	0.96
1200	12	1110	0.13	0.88	1.01
1800	12	1710	0.13	0.92	1.05
2400	12	2310	0.13	0.93	1.06

These results are plotted in Figure 11.10.

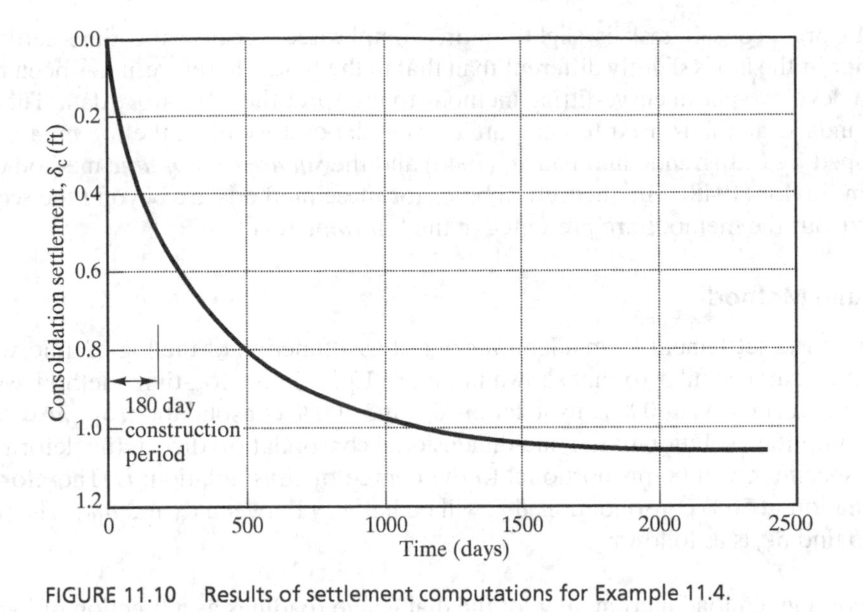

FIGURE 11.10 Results of settlement computations for Example 11.4.

11.3 THE COEFFICIENT OF CONSOLIDATION, c_v

Terzaghi's theory lumps all of the soil properties (other than the drainage distance) into one parameter, the coefficient of consolidation, c_v, as defined in Equation 11.1. Of the various parameters in this equation, the hydraulic conductivity, k, varies most widely, and thus is the most important factor. Therefore, c_v is very small in clays and very large in sands.

We must have some means of measuring c_v before we can perform time–settlement analyses. One method of doing so might be to assess each of the parameters in Equation 11.1 and calculate c_v, but this is rarely done. Instead, engineers usually measure the rate of consolidation in a laboratory consolidation test and back-calculate c_v by performing a time–settlement analysis in reverse. Because H_{dr} in the lab is very small, the rate of consolidation is much higher than that in the field. Nevertheless, c_v should, in theory, be equal to the field value.

In principle, it should be a simple matter to obtain c_v from laboratory time–settlement data. We can rewrite Equation 11.18 as

$$c_v = \frac{T_v H_{dr}^2}{t} \tag{11.28}$$

If we use our simplifying assumption that deformation is proportional to the degree of consolidation, then the U versus T_v relationship is exactly as shown by the solid line in Figure 11.6 and Equations 11.24 and 11.25. Thus, we might expect to simply select an appropriate point on the laboratory time–settlement plot, identify the corresponding values of U, t, and T_v, and use Equation 11.28 to compute c_v.

In practice, this task is slightly more complicated because the time–settlement behavior in the lab is slightly different than that in the field. Therefore, it has been necessary to develop special curve-fitting methods to interpret the laboratory data. There are two standard methods used to compute c_v using laboratory data: the *log time method* developed by Casagrande and Fadam (1940) and the *square root of time* method developed by Taylor (1948). The theoretical bases for these methods are beyond the scope of this text, but the methods are presented in the following text.

Log-Time Method

If we plot the settlement from a laboratory test as a function of the log of time, we will generate a curve similar to that shown in Figure 11.11. In the log-time method, we first find the deformation at 0% consolidation, δ_0, and 100% consolidation, δ_{100}. According to our simplified solution to the one-dimensional consolidation theory, the deformation of the specimen will be proportional to the degree of consolidation, U. Therefore, the deformation at 50% consolidation, δ_{50}, will be halfway between δ_0 and δ_{100}. The procedure to find δ_{50} is as follows:

1. For a given load increment, plot the dial gauge readings as a function of log-time and connect with a smooth curve as shown in Figure 11.12.
2. To estimate the point of 100% primary consolidation, extend the linear tail of the curve back toward the y-axis (line 1 in Figure 11.12). Then draw a line tangent to the point of inflection in the central portion of the curve (line 2 in Figure 11.12). The point where these two lines intersect is taken to be the point at which primary

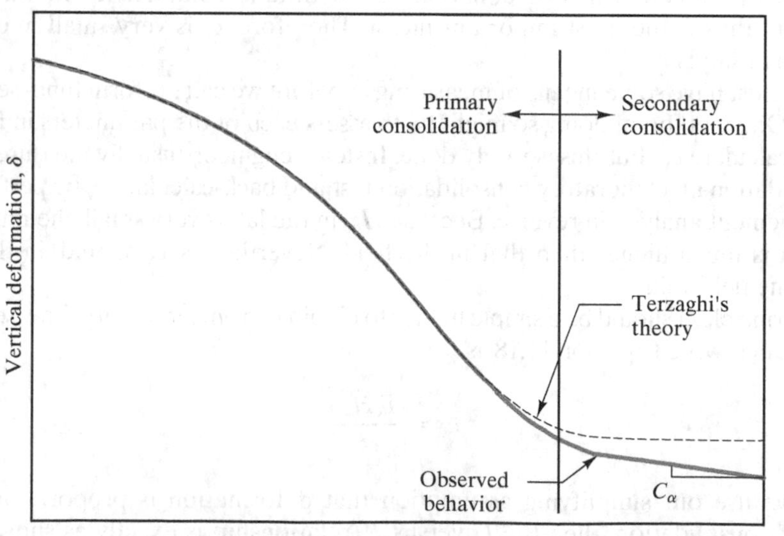

FIGURE 11.11 Typical plot of vertical deflection versus log-time showing difference between Terzaghi's theory and observed behavior and the effects of secondary consolidation.

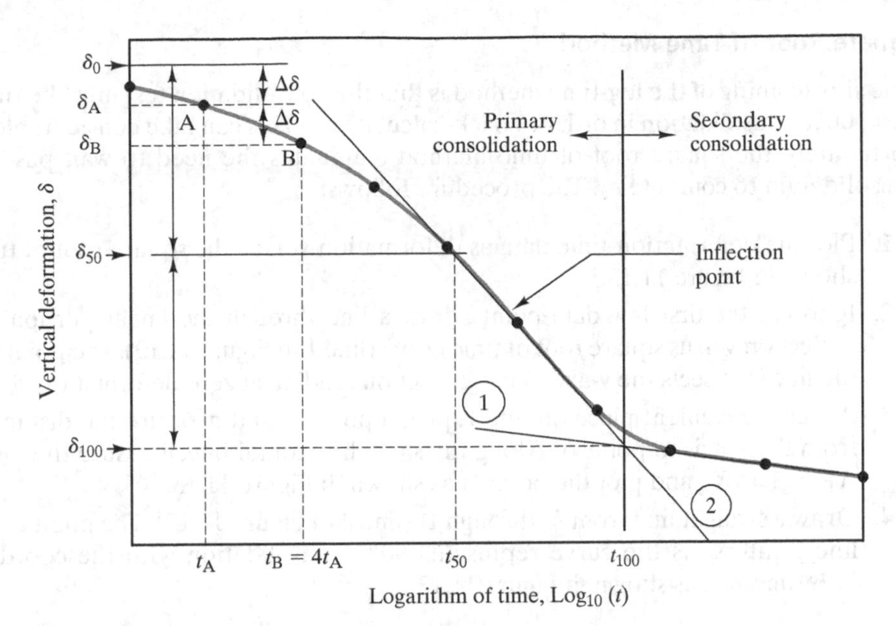

FIGURE 11.12 Typical plot of vertical deflection versus log-time for a consolidation test showing procedure for determining δ_{100}, δ_{50}, and t_{50} using Casagrande's method.

consolidation ends, or the point where the degree of consolidation is 100%. Determine δ_{100} from the point where these lines intersect.

We cannot extend the plot of deformation versus log-time back to a time of zero since zero time does not exist on the semilog plot. Instead we will take advantage of the fact that the early part of the curve has a parabolic shape. Using this assumption, the procedure continues.

3. Select a time t_A in the early portion of the curve as shown in Figure 11.12. Then compute the time t_B such that $t_B = 4t_A$.
4. Compute $\Delta\delta$, the difference between the dial gauge readings at t_A and t_B.
5. Compute the deformation at time zero, δ_0, as $\delta_0 = \delta_A - \Delta\delta$ as shown in Figure 11.12.
6. Compute δ_{50} as $\delta_{50} = (\delta_{100} - \delta_0)/2$ and determine t_{50} from δ_{50} as shown in Figure 11.12.
7. Compute the thickness of the sample at t_{50} using the original thickness and δ_{50}. The maximum drainage distance, H_{dr}, will be one-half of the thickness since the sample is doubly drained.
8. Insert t_{50} and H_{dr}, into Equation 11.28 and compute c_v at 50% consolidation:

$$c_v = \frac{T_{50} H_{dr}^2}{t_{50}} \tag{11.29}$$

where:

T_{50} = time factor at 50% consolidation,
= 0.196, from Figure 11.6 or Equation 11.24.

Square Root of Time Method

One shortcoming of the log-time method is that the consolidation test must be run well past 100% consolidation in order to back-calculate t_{50}. This can take considerable time. Fortunately, the square root of time method eliminates the need to wait past 100% consolidation to compute c_v. The procedure follows:

1. Plot the deformation-time data as deformation versus the square root of time, as shown in Figure 11.13.
2. Ignoring the first few data points, draw a line through the linear portion of the deflection versus square root of time curve (line 1 in Figure 11.13). The point where the line intersects the *y*-axis is the deflection reading at zero deformation, δ_0.
3. At any convenient place on line 1, pick a point A and measure the distance X_A from the *y*-axis to point A. Along the same horizontal line, measure the distance $X_B = 1.15X_A$ and plot the point B as shown in Figure 11.13.
4. Draw a straight line from δ_0 through B (line 2 in Figure 11.13). The point at which line 2 intersects the curve represents 90% consolidation with the coordinates $\sqrt{t_{90}}$ and δ_{90}, as shown in Figure 11.13.

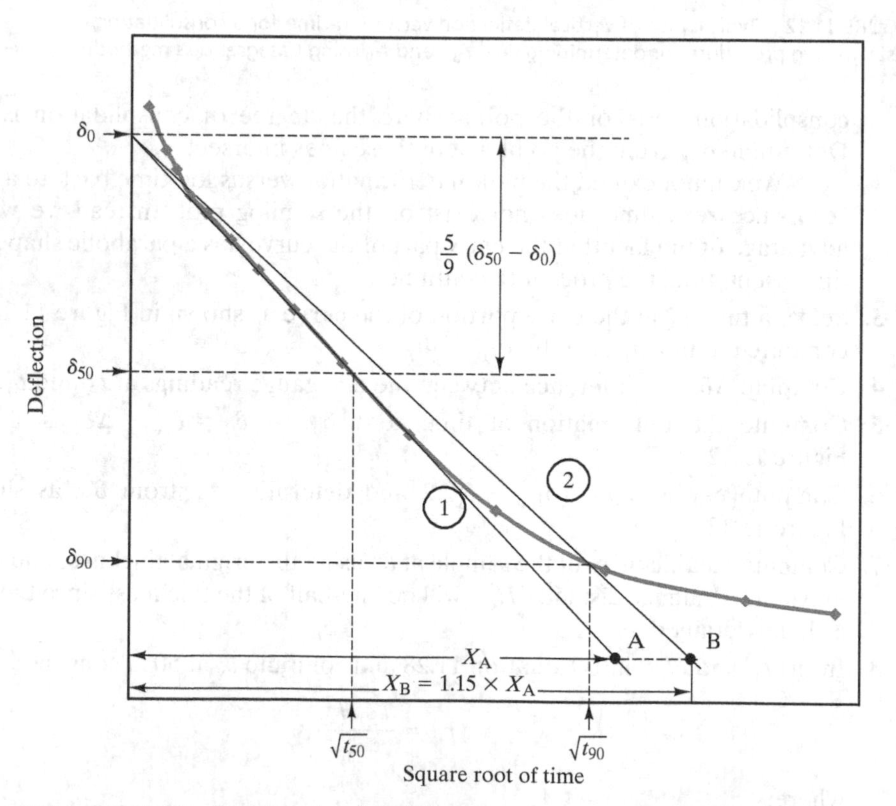

FIGURE 11.13 Typical plot of vertical deflection versus square root of time procedure for determining δ_{90}, δ_{50}, and t_{50} using Taylor's method.

5. Compute δ_{50} using as $\delta_{50} = \delta_0 + \frac{5}{9}(\delta_{90} - \delta_0)$ and graphically determine $\sqrt{t_{50}}$ as shown in Figure 11.13.
6. Using the value of t_{50} determined earlier, and the maximum drainage distance of the sample, H_{dr}, compute c_v using Equation 11.29.

Most geotechnical engineers prefer the square root of time method because it permits the next load to be placed as soon as t_{90} has been reached, whereas the logarithm of time method requires the load be left on long enough to identify t_{100}. Since consolidation tests are very slow anyway, typically requiring days to complete, this difference can have a significant impact on the cost of performing the test.

Figure 11.14 presents an approximate correlation between c_v and the liquid limit. Although this diagram is not a substitute for performing laboratory tests, it may be used to check test results for reasonableness and for preliminary estimates. Table 11.2 provides a list of some measured values. Notice that there is some disagreement between these two references.

Terzaghi's one-dimensional consolidation equation (Equation 11.16) is based on c_v being a constant. However, in reality it varies with effective stress as described in Equation 11.1. This effect can be seen by computing c_v for each of the several load increments in the consolidation test. Figure 11.15 shows measured values at different effective stresses for several soils. Note the sudden change in c_v at the preconsolidation

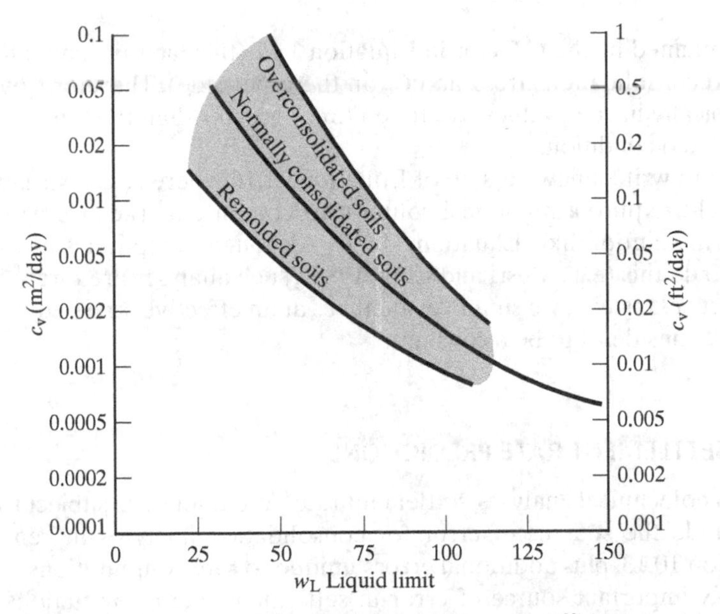

FIGURE 11.14 Approximate correlation between coefficient of consolidation, c_v, and liquid limit, w_L. Undisturbed normally consolidated soils typically plot near the center of the shaded zone, undisturbed overconsolidated soils in the upper portion, and remolded soils in the lower portion. (U.S. Navy, 1982.)

TABLE 11.2 Measured Values of c_v[a]

Soil	c_v	
	m²/day	ft²/day
Boston blue clay (CL) (Ladd and Luscher, 1965)	0.033 ± 0.016	0.33 ± 0.16
Organic silt (OH) (Lowe, Zaccheo, and Feldman, 1964)	0.00016–0.00082	0.0016–0.0082
Glacial lake clays (CL) (Wallace and Otto, 1964)	0.00055–0.00074	0.0055–0.0074
Chicago silty clay (CL) (Terzaghi and Peck, 1967)	0.00074	0.0074
Swedish medium sensitive clays (CL-CH) (Holtz and Broms, 1972)		
Laboratory	0.0003–0.0006	0.003–0.006
Field	0.0006–0.003	0.006–0.03
San Francisco bay mud (CL)	0.0016–0.0033	0.016–0.033
Mexico City clay (MH) (Leonards and Girault, 1961)	0.0008–0.0014	0.008–0.014

[a]Adapted from Holtz and Kovacs, 1981.

stress, which is explained by the C factor in Equation 11.1 (this factor is equal to C_r on one side of the preconsolidation stress, and C_c on the other side). Therefore, overconsolidated soils typically have c_v values five to ten times greater than the same soils in a normally consolidated condition.

It is possible to write a new version of Equation 11.16 where c_v is a variable, but this equation would require a numerical solution to compute u_e (i.e., we would not have a closed form solution like Equation 11.17). Although computer software has been developed to do this (e.g., Mesri and Choi, 1985), such analyses are rarely, if ever, performed in practice. Instead, we simply evaluate c_v at an effective stress equal to the σ'_z in the field, and consider it to be a constant.

11.4 ACCURACY OF SETTLEMENT RATE PREDICTIONS

As with all other geotechnical analyses, settlement rate predictions are subject to many errors. These include the sources of error for consolidation analyses in general, as described in Section 10.13, plus additional errors unique to rate computations.

An especially important source of error in settlement rate computations is our assessment of the maximum drainage distance, H_{dr}. The time required to reach full consolidation is proportional to $(H_{dr})^2$, as discussed earlier, so even small errors in this value can produce significant changes in the computed settlement rate. This problem can be quite insidious, because compressible clay strata that appear to be homogeneous

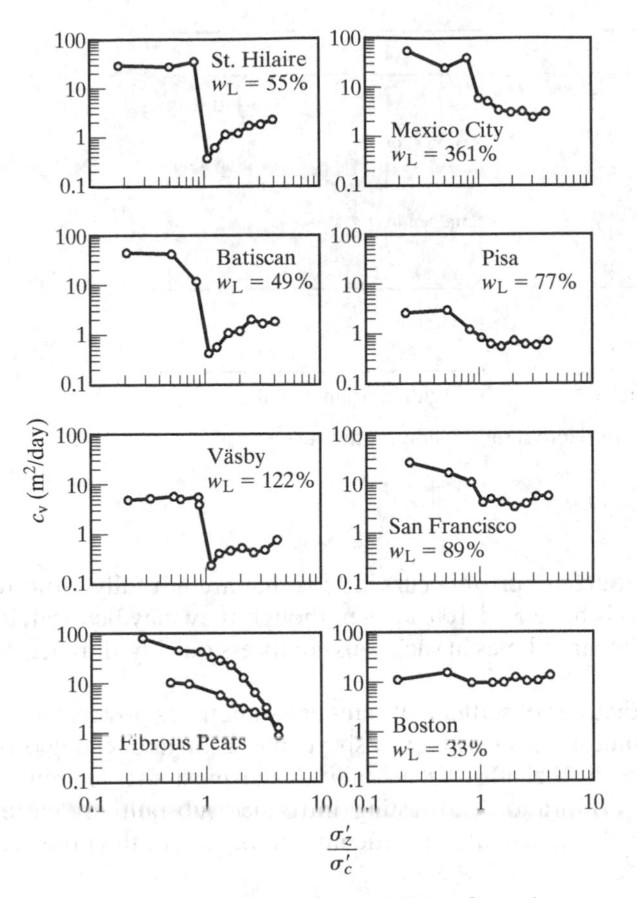

FIGURE 11.15 Coefficient of consolidation for various soils as a function of effective stress. (Terzaghi et al., 1996.)

contain thin horizontal sandy seams. If these layers are continuous, or nearly so, the horizontal hydraulic conductivity can be much greater than the vertical hydraulic conductivity as discussed in Chapter 7 ($k_x \gg k_z$). Therefore, the excess pore water in the clay will move up or down to the nearest sand seam, and then escape horizontally through the seam as shown in Figure 11.16.

For example, consider an 8-m thick clay stratum with double drainage. If it contains thin continuous sand seams every 2 m, then H_{dr} is really 1/4 of the apparent value, and the time required for consolidation will be $(1/4)^2 = 1/16$ of the computed value. This illustrates the importance of identifying small details when logging exploratory borings.

The analysis method presented in this chapter was also based on purely vertical drainage. This assumption loses its validity when the loaded area is small, such as a structural foundation. In such cases, much of the drainage is horizontal, even if no sandy seams exist, and the consolidation settlement is correspondingly faster.

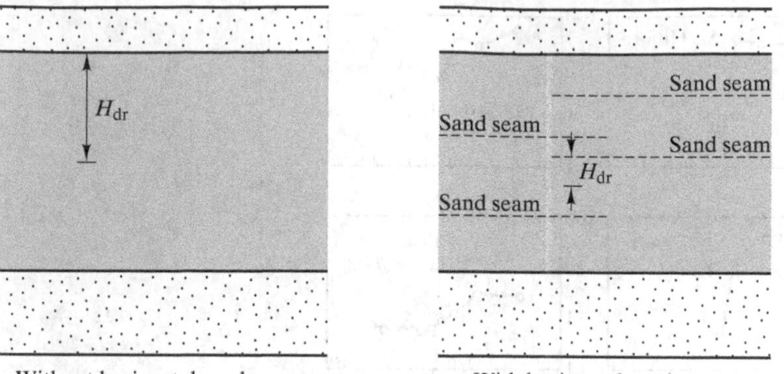

FIGURE 11.16 Effect of thin horizontal sand seams on the rate of consolidation.

Another potential source of error occurs in soils that are not fully saturated. For example, some organic soils have $S < 100\%$, even though they may be located below the groundwater table. The air bubbles in such soils compress quickly, thus accelerating the consolidation process.

Fairly accurate predictions of settlement rates are sometimes possible when wide loads are placed at very uniform sites with extensive exploration and testing programs. However, nonuniformities in the subsurface conditions at most sites, combined with economic limitations on exploration and testing, introduce substantially more error. Thus, it is not unusual for the actual rate of settlement to be half of the predicted rate, or twice the predicted rate.

11.5 CONSOLIDATION MONITORING

Predictions of both the rate and magnitude of settlement can be seriously in error, so engineers usually use conservative estimates when assessing the impact of these settlements on proposed construction. Often, such conservative estimates are acceptable because they do not have a serious impact on the final design. However, when the predicted settlements are large, such conservative designs can be very expensive. In such cases, geotechnical engineers sometimes install instruments in the ground to monitor the actual settlements, especially during the early stages of consolidation. We then use the data from these instruments to update our preconstruction settlement computations. These updated predictions are much more reliable because they have been calibrated based on the actual field performance, and thus may be used to justify less conservative designs.

There are two approaches to monitoring consolidation in the field. The first monitors only settlements, while the second monitors both settlements and pore water pressures.

Monitoring Settlement Only

A basic monitoring program consists of installing survey monuments at the ground surface and periodically measuring their elevation using a total station or other conventional land surveying equipment. These monuments need to be firmly fastened to the ground, and must not be subject to surficial soil movements, vandalism, or damage from construction equipment.

Another method of monitoring settlements is to install remote devices such as that shown in Figures 11.17 and 11.18. This one consists of a water reservoir installed at a fixed elevation with a tube leading to the sensing unit, which is buried in the ground. A pressure transducer at the sensing unit measures the pressure in the water, which permits computation of the difference in elevation between the reservoir and the sensor. As the sensor settles, the pressure changes and the new elevation can be computed. This device is installed before the fill is placed, thus allowing settlements to be monitored both during and after construction.

Settlement monitoring devices that can be installed in bore holes are also available. These permit monitoring settlements as a function of depth. The Sondex device, shown in Figures 11.19 and 11.20, is one such device.

Although settlement data are very useful, their interpretation is not always easy. For example, Figure 11.21 shows the preconstruction prediction of settlement versus time from Example 11.3, along with the first 1000 days of observed settlements obtained from a settlement plate installed at Point C in Figure 11.7. The observed

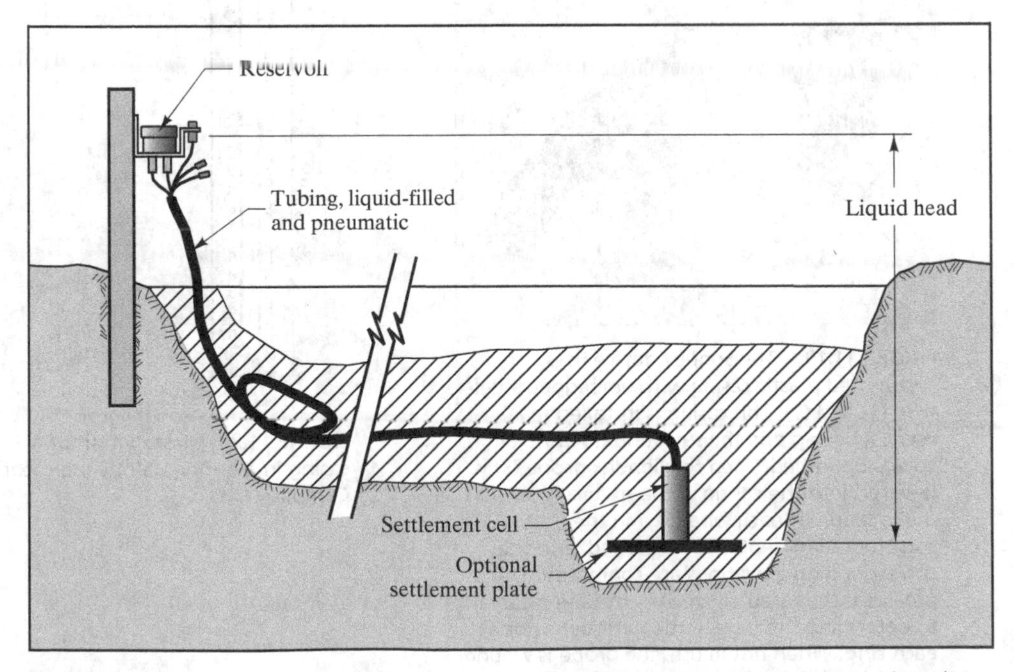

FIGURE 11.17 Installation of remote reading settlement plate. The reservoir is located on the ground that is not settling. (Slope Indicator Company.)

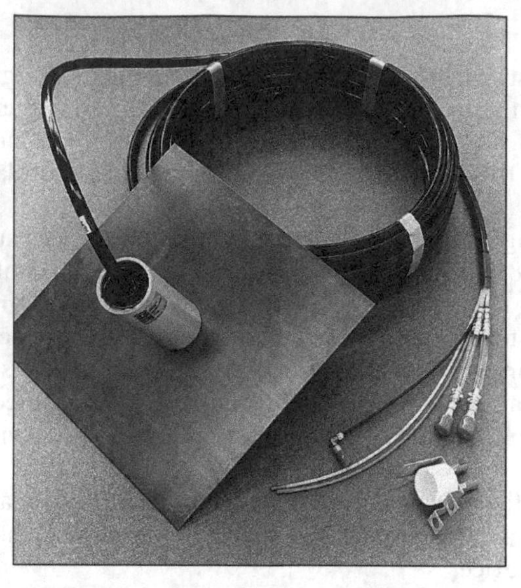

FIGURE 11.18 Remote reading settlement plate. (Slope Indicator Company.)

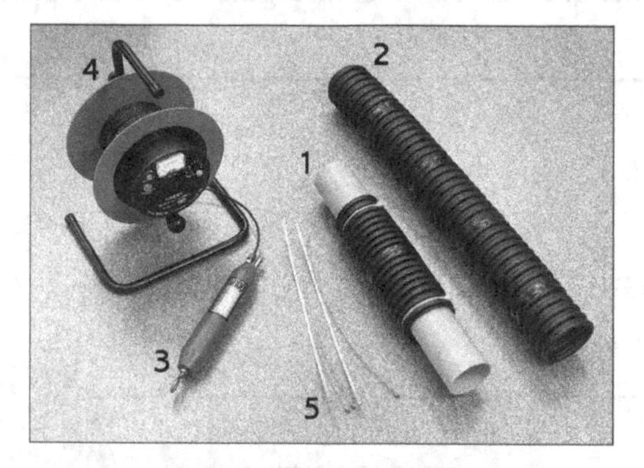

FIGURE 11.19 The Sondex device uses corrugated plastic casing with embedded steel wire (1) and (2). This casing is installed in a vertical boring and the surrounding annular void is backfilled. Then the Sondex probe (3) is lowered into the casing. It magnetically senses the location of each wire (5) and the corresponding depths are recorded, and the corresponding elevations are computed. This process is repeated at convenient time intervals to determine the time–settlement behavior at each ring. When not in use, the probe is wound into the reel (4) and stored in a safe location. (Slope Indicator Company.)

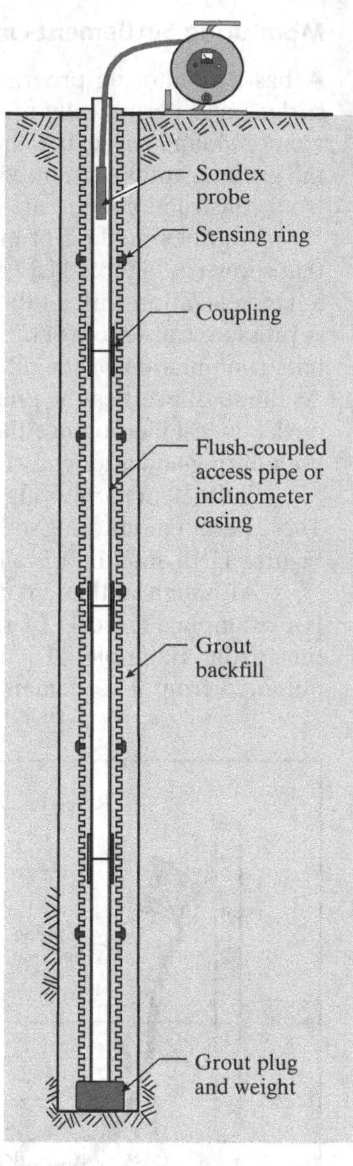

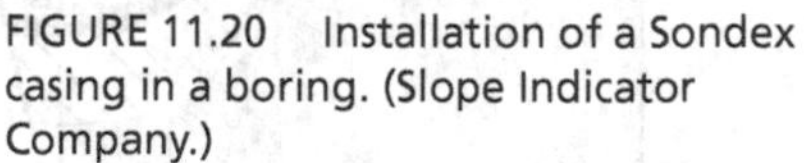

FIGURE 11.20 Installation of a Sondex casing in a boring. (Slope Indicator Company.)

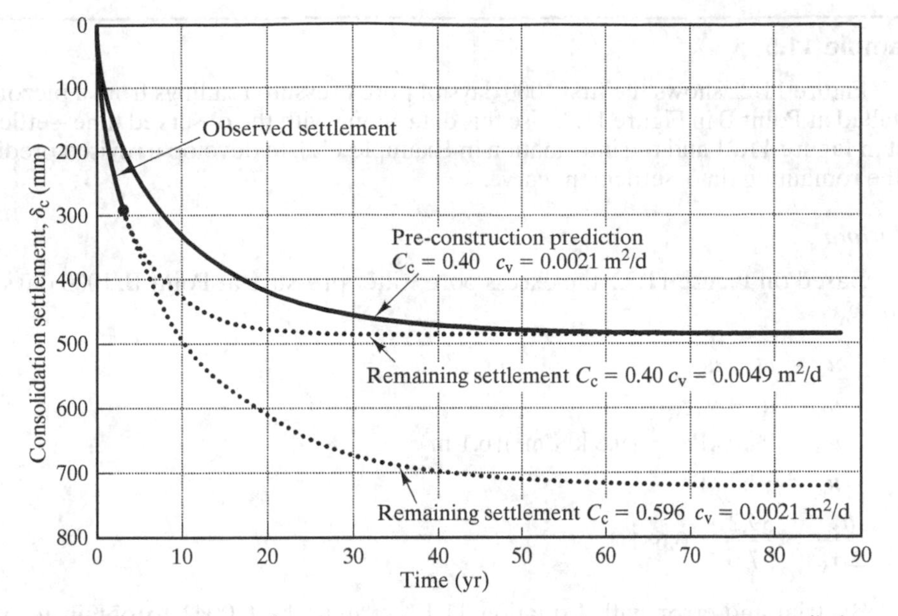

FIGURE 11.21 Observed settlement data compared to the preconstruction prediction. The dotted lines show many possible predictions of the remaining settlement.

settlement is greater than the original prediction, but it is not clear whether this is because the ultimate settlement will be greater than predicted, the rate of settlement is faster than predicted, or both. Thus, there are many reasonable interpretations of the remaining settlement. Two such interpretations are shown: one that assumes that all of the error is in the C_c value, and the other one that assumes that all the error is in c_v.

Monitoring Both Settlement and Pore Water Pressure

Fortunately, much of the ambiguity in interpreting settlement data can be overcome by also installing piezometers in the natural ground. It is best to install them before the fill is placed, perhaps by using remote reading units such as the one shown in Figure 7.10. Initially these piezometers record the hydrostatic pore water pressure. Then, as the fill is placed, the pore water pressure will rise, reflecting the excess pore pressures. These excess pressures will eventually dissipate, and after a sufficiently long time the pore pressure will return to its hydrostatic value.

Using this pore pressure data, we can compute the field value of c_v. Then, by combining this value with the observed time–settlement curve, we obtain an unambiguous plot of the remaining time–settlement behavior. Although this analysis still contains some error, it is more precise than those made without piezometer data, and far superior to analyses based only on laboratory tests. Example 11.5 illustrates this technique.

Example 11.5

Figure 11.22 shows the first 1000 days of pore pressure readings from a piezometer installed at Point B in Figure 11.7. Use this data, along with the observed time–settlement plot in Figure 11.21 and the information in Example 11.3, to develop a revised prediction of the remaining time–settlement curve.

Solution:

Based on Figure 11.22, the excess pore water pressure at Point B, 1000 days after loading is

$$u_e = u - u_h$$

$$u_e = u - \gamma_w z_w$$

$$u_e = 99.0 \text{ kPa} - (9.8 \text{ kN/m}^3)(6.1 \text{ m})$$

$$u_e = 39.2 \text{ kPa}$$

$$\frac{u_e}{\Delta\sigma_v} = \frac{39.2}{57.6} = 0.681$$

By trial-and-error with Equation 11.17, c_v must be 0.0043 to obtain $u_e/\Delta\sigma_v = 0.681$. This is about twice the laboratory value of 0.0021, and explains why the consolidation is occurring faster than anticipated.

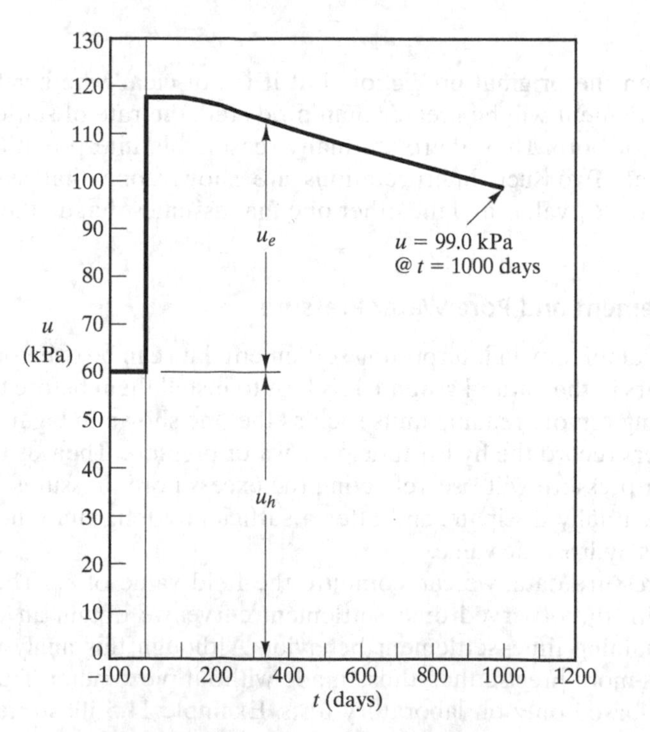

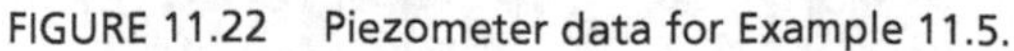

FIGURE 11.22 Piezometer data for Example 11.5.

Per Figure 11.21, the observed consolidation settlement at $t = 1000$ days (2.7 years) is 280 mm. Because 14 mm of this settlement occurred in the SP stratum (per Example 11.3), 280 − 14 = 266 mm must have occurred in the CL/ML stratum. Using $c_v = 0.0021$, we determine the value of $C_c/(1 + e_0)$ required to produce 266 mm of settlement in 1000 days. Using the simplified solution, by trial-and-error, we determine that the value of $C_c/(1 + e_0)$ must be approximately 0.20. Thus, $C_c = (0.20)(1 + 1.1) = 0.42$. This is slightly higher than the 0.40 obtained from the laboratory tests.

Using these revised values of c_v and C_c, and the simplified method, we can develop the revised time–settlement curve shown in Figure 11.23.

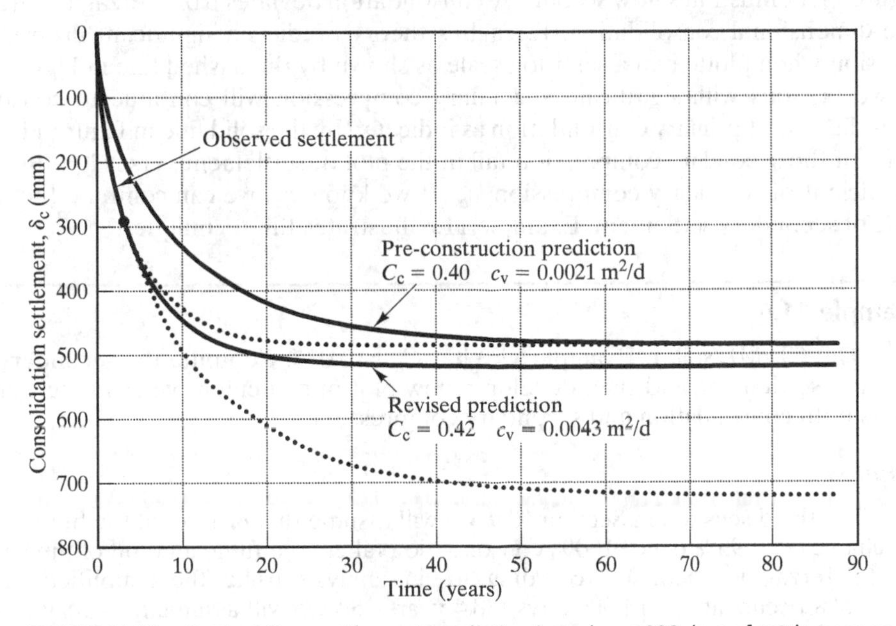

FIGURE 11.23 Revised time-settlement prediction based on 1000 days of settlement and pore water pressure data.

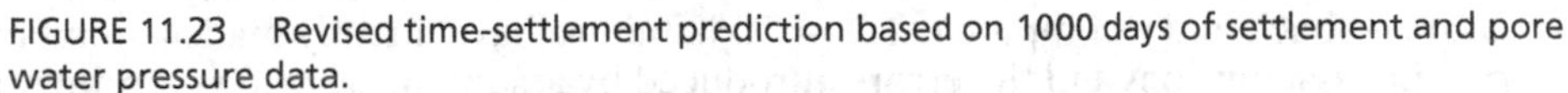

This methodology of making preconstruction predictions and then revising them based on observed performance is known as the *observational method* (Peck, 1969), a term coined by Terzaghi. This method is very useful in geotechnical engineering, and is one of the ways we can overcome many of the accuracy problems in our analyses without resorting to overly conservative designs. The observational method can be applied to any branch of civil engineering, but it is especially useful for geotechnical engineers.

11.6 OTHER SOURCES OF TIME DEPENDENT SETTLEMENT

The discussions in this chapter have been based on the assumption that dissipation of excess pore water pressures is the sole cause of time dependency in soil settlement. Although this mechanism is probably the most important one, others also exist.

Secondary Compression

Settlements due to secondary compression, as discussed in Section 10.9, are by definition independent of excess pore water pressures. Secondary compression is initiated by changes in a soil that occur during consolidation, but it continues long after excess pore pressures have dissipated. Secondary compression will be most significant in highly compressible soils which are loaded past the preconsolidation stress (see Table 10.5). When secondary compression is significant, it may be evaluated by developing a time–settlement plot that begins when the consolidation settlement is complete. Figure 11.11 illustrates how secondary consolidation deviates from Terzaghi's theory of one-dimensional consolidation. Terzaghi's theory predicts a significant drop in compression when plotted on a semi-log scale as shown by the dashed line in Figure 11.11. However, soils with significant secondary compression will continue to consolidate after the end of primary consolidation as indicated by the solid line in Figure 11.11. The slope of the secondary compression tail in the plot of settlement versus log-time is the coefficient of secondary compression, c_α. If we know c_α, we can compute the magnitude of secondary settlement. Example 11.6 illustrates this technique.

Example 11.6

The CL/ML soil in Example 11.3 has $C_\alpha = 0.015$. Compute the secondary compression settlement, and then develop a new plot of settlement versus time that considers both consolidation and secondary compression.

Solution:

Per the discussion in Section 10.9, we will assume that t_p is equal the time required to achieve $U = 95\%$ (since 100% consolidation takes an infinite amount of time according to Terzaghi's theory). According to an analysis using the simplified method, $U = 96\%$ occurs at $t = 14{,}000$ days (38.4 years). So we will assume $t_p = 38$ years. (We could go back and recompute t at $U = 95\%$ but we have already made a number of simplifying assumptions and the errors introduced by assuming $t_p = 38$ years are not large.)

Using Equation 10.28,

$$\delta_s = \frac{C_\alpha}{1 + e_p} H \log\left(\frac{t}{t_p}\right)$$

$$\delta_s = \frac{0.015}{1 + 1.10}(10{,}000 \text{ mm}) \log\left(\frac{t}{38}\right)$$

$$\delta_s = 71 \log\left(\frac{t}{38}\right)$$

For example, at $t = 54.8$ years, we computed the consolidation settlement as $466 + 14 = 480$ mm. Using the earlier equation, the secondary compression settlement is

$$\delta_s = 71 \log\left(\frac{54.8}{38}\right) = 11 \text{ mm}$$

and the total settlement is (480 + 11) = 491 mm. Using this method, we compute the total settlement, including secondary, any time after 38 years. This is shown in Figure 11.24. Comparing this plot with Figure 11.8 demonstrates that secondary compression is not a major issue at this site.

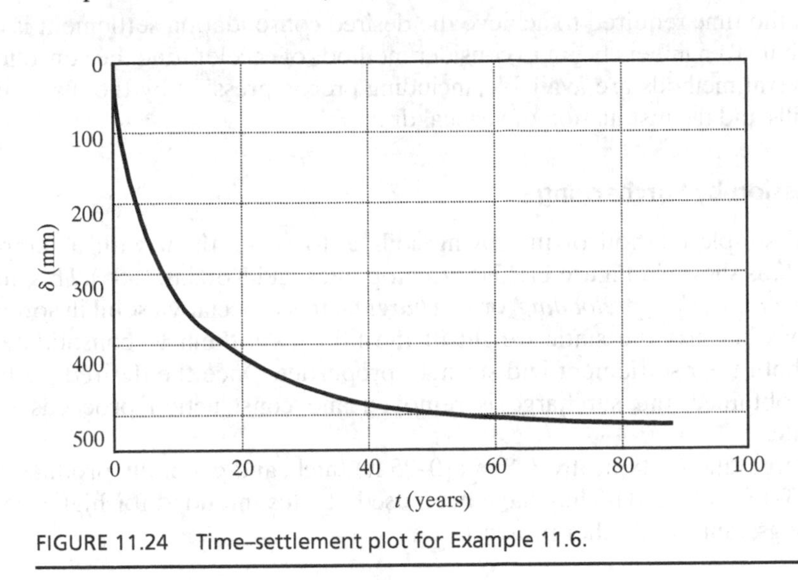

FIGURE 11.24 Time–settlement plot for Example 11.6.

Other Sources of Volume Change

The analyses discussed in this chapter also has assumed that the soil is saturated ($S = 100\%$) and remains so. Unsaturated soils are subject to volume changes if they become wetted, and initially saturated soils may change in volume if they are dried. In the case of expansive soils, wetting causes swelling, while drying causes shrinkage. Another type of soil, called a *collapsible soil*, shrinks when it is wetted. Any of these processes can produce significant settlement or heaving at the ground surface and the timing of these movements depends primarily on when the soil becomes wetted or dried.

This chapter also has considered only static loads. Vibratory loads, such as from heavy machinery or earthquakes, also can cause settlement, especially in loose sands. Static loads that cycle over longer periods, such as those due to the annual loading and unloading of grain silos, also can cause additional settlements (Coduto, 2001).

11.7 METHODS OF ACCELERATING SETTLEMENTS

In theory, an infinite time is required to achieve $U = 100\%$. Fortunately, we do not need to wait that long in practice! Normally, construction of structures and other facilities can begin once U reaches a value such that the remaining settlement is less than some maximum allowable settlement. It is not necessary to wait until all the consolidation settlement is completed. Nevertheless, even the time required to achieve this level of consolidation is sometimes excessive, perhaps requiring many years or even decades. This is especially likely when the compressible stratum is soft (high C_c) and thick (high H_{dr}).

When the time required to achieve the desired consolidation settlement is excessive, geotechnical engineers begin to consider methods of accelerating the consolidation process. Several methods are available, including precompression by the placement of surcharge fills and the installation of vertical drains.

Precompression by Surcharging

An old and simple method of improving soils is to cover them with a temporary *surcharge fill*, as shown in Figure 11.25 (Stamatopoulos and Kotzias, 1985). This method is called *precompression*, *preloading*, or *surcharging*. It is especially useful in soft clayey and silty soils because the static weight of the fill causes them to consolidate, thus improving both their settlement and strength properties. Once the desired properties have been obtained, the surcharge is removed and construction proceeds on the improved site.

Surcharge fills are typically 3–8 m (10–25 ft) thick, and generally produce settlements of 0.3–1.0 m (1–3 ft). They have been used at sites intended for highways, runways, buildings, tanks, and other projects.

FIGURE 11.25 This surcharge fill will remain in place until the underlying soils have settled. The smokestack and crane in the background are located behind the fill.

Precompression has many advantages, including:

- It requires only conventional earthmoving equipment, which is readily available. No special or proprietary construction equipment is needed.
- Any grading contractor can perform the work.
- The results can be effectively monitored by using appropriate instrumentation (especially piezometers) and ground level surveys.
- The method has a long track record of success.
- The cost is comparatively low, so long as soil for preloading is readily available and can readily be discarded at the end of the process.

However, there also are disadvantages, including:

- The surcharge fill generally must extend horizontally at least 10 m (33 ft) beyond the perimeter of the planned construction. This may not be possible at confined sites.
- The transport of large quantities of soil onto the site may not be practical, or may have unacceptable environmental impacts (i.e., dust, noise, traffic) on the adjacent areas.
- The surcharge must remain in place for months or years, thus delaying construction. However, the process can be accelerated as described subsequently.

Vertical Drains

We showed in Section 11.1 that the time required to achieve a certain level of consolidation is proportional to H_{dr}^2, where H_{dr} is the maximum drainage distance. Thus, if the stratum of compressible soil is very thick, the time required to achieve the desired consolidation may be excessive. In some cases, this time can easily be several years or even decades, even with a surcharge fill. Very few projects can accommodate such long delays. Therefore, when precompression is used on thick compressible soils, we generally need to employ some means of accelerating the consolidation process.

The most effective way of accelerating soil consolidation is to reduce H_{dr} by providing artificial paths for the excess pore water to escape. This can be done by installing vertical drains, as shown in Figure 11.26. The excess pore water within the compressible soil now drains horizontally to the nearest vertical drain, a much shorter distance than before. In addition, most soft clays contain thin horizontal sandy or silty seams, so the horizontal hydraulic conductivity, k_x, is typically much higher than the vertical value, k_z. This further increases the rate of consolidation. Thus, the time required to achieve the required degree of consolidation can typically be reduced from several years (or even decades!) to only a few months. Vertical drains also may be used with only the permanent fill, thus eliminating the expense of a surcharge fill.

The excess pore water pressures generated during consolidation provide the head to drive water through the vertical drains. Once consolidation is complete, the excess pore water pressures become zero and drainage ceases.

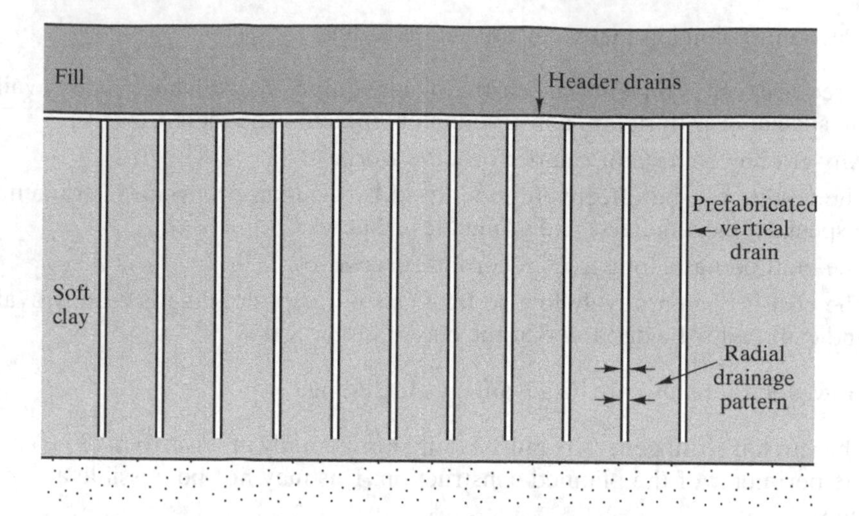

FIGURE 11.26 Use of vertical drains to accelerate consolidation.

The earliest vertical drains consisted of a series of borings filled with sand. These *sand drains* were expensive to construct, so engineers developed another method: *prefabricated vertical drains* (also known as *wick drains* or *band drains*). They consist of corrugated or textured plastic ribbons surrounded by a geosynthetic filter cloth as shown in Figure 11.27. Most are about 100 mm (4 in.) wide and about 5 mm (0.2 in.) thick. These drains are supplied on spools, and are inserted into the ground using special

(a) (b)

FIGURE 11.27 (a) A typical prefabricated vertical drain; (b) prefabricated vertical drains after installation. The drains are the small strips extending out of the ground. Wider header drains also have been installed on the ground surface to collect water from the vertical drains and carry it to a discharge location. The site is now ready to be covered with the fill. (American Wick Drain Corporation.)

FIGURE 11.28 Equipment used to install prefabricated vertical drains. (American Wick Drain Corporation.)

equipment that resembles a giant sewing machine, as shown in Figure 11.28. Prefabricated vertical drains are considerably less expensive than sand drains, and thus have become the preferred method on nearly all projects.

The required spacing of vertical drains is determined by a radial drainage analysis, and represents a compromise between construction cost and rate of consolidation. Typically they are spaced about 3 m (10 ft) on center, which means hundreds of drains are usually required.

Although these methods are expensive, their cost often can be justified because they permit construction to begin years or even decades earlier than would otherwise be possible.

SUMMARY

Major Points

1. Consolidation is initiated by the application of a load which increases total stress on a soil. In saturated soils this change in total stress initially creates excess pore pressure, u_e, which in turn causes pore water to flow out of the soil. The rate at which the pore water flows out of the soil will control the rate at which consolidation occurs.
2. Terzaghi's theory of consolidation describes the rate at which excess pore water pressure dissipates in saturated soils and can be used to predict the rate of consolidation and thereby the rate of settlement at the ground surface. The theory has a number of assumptions among which the most critical are: water flow and consolidation occur only in a vertical direction, the soil is homogeneous and its properties do not change over time, and the load is applied instantaneously.
3. The solution to Terzaghi's theory of consolidation provides the excess pore pressure as a function of both time and depth within the compressible layer. The key parameters in the solution are as follows:

   (a) c_v, the coefficient of consolidation; a soil property that measures the relative speed at which a soil will consolidate. The greater c_v is, the faster a soil will consolidate.

   (b) H_{dr}, the maximum drainage distance of the compressible layer.

   (c) T_v, the time factor, a dimensionless measure of time equal to $(c_v\, t)/(H_{dr})^2$.

4. Because T_v is inversely proportional to $(H_{dr})^2$, H_{dr} dramatically affects the rate of consolidation. If H_{dr} doubles, the time to reach a given degree of consolidation will increase by a factor of 4.
5. A dimensionless solution to Terzaghi's one-dimensional consolidation equation is possible and is presented in Equation 11.17 and Figure 11.4.
6. The rate of settlement is related to the average degree of excess pore water pressure dissipation across the height of the compressible layer. If we make the additional assumption that the stress-strain curve for the soil is linear, then there is a closed form solution for the settlement as a function of time. This simplified solution is presented in Equations 11.24 and 11.25 and Figure 11.6. More accurate computations of settlement versus time require the use of numerical solutions.
7. Theoretically it will take an infinite amount of time for a soil to reach 100% consolidation. For practical purposes geotechnical engineers consider consolidation to be complete when it has reached a theoretical level of 95–98%. This corresponds to at time factor, T_v, of 1.1 to 1.5.
8. The coefficient of consolidation, c_v, is normally determined from laboratory consolidation tests. The two methods available for computing c_v are Casagrande's log-time method and Taylor's square root of time method. Taylor's method is generally preferred because it allows for much shorter lab tests.

9. Because of limitations in Terzaghi's theory and the determination of in situ soil properties, settlement rate predictions are not extremely accurate. Therefore, geotechnical engineers often install settlement and pore water pressure monitoring equipment in the field. Data collected from this instrumentation may be used to update the laboratory values of C_c and C_v, which then may be used to produce revised time-settlement curves. This technique of updating calculations based on performance data is an example of the observational method developed by Ralph Peck.
10. Secondary compression is a time dependent source of settlement that is independent of excess pore water pressures. It is generally initiated by consolidation, but continues after all the excess pore pressures have dissipated.
11. Consolidation settlement can be accelerated by either preloading or installing vertical drains.
    - **(a)** Preloading accelerates consolidation by placing a surcharge load greater than the final design load on this soil. The settlement under the surcharge load is monitored and when the settlement reaches a predetermined value, the surcharge is removed and construction of the final structure can begin. While preloading will decrease both the magnitude and rate of settlement of the final structure, the preloading process itself can take months or years.
    - **(b)** Vertical drains can dramatically increase the rate of consolidation by reducing the maximum drainage distance. This method is often combined with surcharging.

Vocabulary

coefficient of consolidation
degree of consolidation
double drainage
log-time method (for computing c_v)
maximum drainage distance
observational method
one-dimensional consolidation equation
single drainage
square root of time method (for computing c_v)
precompression
surcharge fill
theory of consolidation
time factor
vertical drain
wick drain

QUESTIONS AND PRACTICE PROBLEMS

Section 11.1 Terzaghi's Theory of Consolidation

11.1 A 12.0-m thick clay stratum with double drainage is to be subjected to a $\Delta\sigma_z$ of 75 kPa. The coefficient of consolidation in this soil is 3.5×10^{-3} m²/d. Using Equation 11.17, compute the hydrostatic, excess, and total pore water pressure at a point 2.7 m above the bottom of this stratum 10 years after placement of the load.

11.2 Solve Question 11.1 using Figure 11.4.

11.3 For the soil profile and loading conditions described in Problem 11.1, how long will it take for the excess pore water pressure to reach one half the initial excess pore water pressure. Will the average degree of consolidation for the entire clay layer be less than, equal to, or greater than 50% at this time? Explain.

11.4 Repeat Problem 11.1 but assume the clay stratum is drained only at the top. Compare the pore water pressures computed for this case of single drainage with the pore water pressures computed for the case of double drainage computed in Problem 11.1.

11.5 A 20-ft thick fill with a unit weight of 120 lb/ft^3 is to be placed on the soil profile shown in Figure 11.29. Assuming the fill is placed instantaneously, use the curves in Figure 11.4 to develop a plot of u_e versus depth at $t = 1.5$ years. Plot depth on the vertical axis, increasing downward, and consider depths from the original ground surface to the bottom of the CL stratum.

11.6 Use Equation 11.17 to compute the hydrostatic, excess, and total pore water pressures at Point F in Figure 11.29 at $t = 1, 2, 4, 8,$ and 16 years after placement of the fill. Then use this data to develop a plot of $u_h, u_e,$ and u at this point versus time. All three curves should be on the same diagram, with time on the horizontal axis.

11.7 Using the soil profile in Figure 11.29, develop a spreadsheet that solves Equation 11.17 at 1.0 ft depth intervals through the entire soft clay stratum. Use summations for N. Then use this spreadsheet to develop a curve of excess pore water pressure versus depth at $t = 6$ years after construction. Submit a printout of the spreadsheet, and a plot of the excess pore water pressure curve.

Note for those who may wish to develop spreadsheets or other software for more general solutions: The natural exponent term in Equation 11.17 may cause difficulties for some programming languages when they attempt to take e to a large negative power.

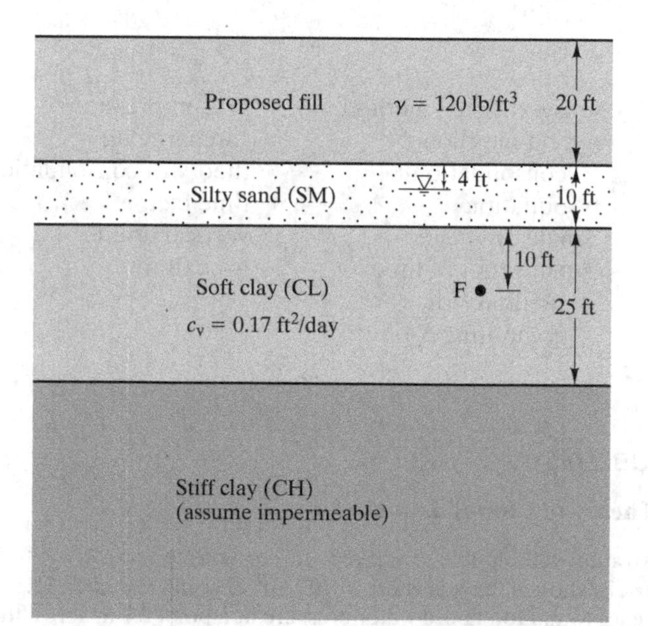

FIGURE 11.29 Soil profile for Problems 11.5–11.7 and 11.10.

However, these difficulties appear to occur only when N has risen to values beyond those necessary for the summation. Therefore, avoid such difficulties by terminating the summation whenever the exponent term generates an error, or when the increment of N produces a negligible change in the summation.

Section 11.2 Consolidation Settlement versus Time Computations

11.8 Consider the proposed fill and soil profile shown in Figure 11.5, except replace the sandy silt strata with an impervious bedrock. Using the simplified solution, compute the consolidation settlement at $t = 15$ years after placement of the fill. The ultimate consolidation settlement is 0.50 m. Do not apply any correction for the construction period.

11.9 For the situation described in Problem 11.8, how long will it take to reach 95%, 98%, and 99% of the ultimate consolidation settlement? Use the simplified method. The owner has asked you "How long will the settlement take?" How would you reply?

11.10 For the proposed fill shown in Figure 11.29, assume the ultimate consolidation settlement is 1.6 ft. The owner wants to build a structure on top of the fill. The structure can withstand a total settlement of 4 in. How long must the owner wait after placement of the fill before building the structure on top of the fill? Use the simplified method.

11.11 A fill is to be placed on the soil profile shown in Figure 11.30. The groundwater table is level with the original ground surface. Use the simplified method to develop a plot of consolidation settlement versus time. Continue the plot until $U > 99\%$. Do not apply any correction for the construction period.

Note: As consolidation settlement occurs, some of the fill will become submerged beneath the groundwater table. The resulting buoyant force will reduce σ'_{zf} and thus reduce the consolidation settlement. However, this effect is small for this problem and may be ignored.

11.12 A shopping center is to be built on the fill described in Problem 11.11. The proposed buildings and other facilities can tolerate a settlement due to the weight of the fill of no more than 2 in. Therefore, once the fill has been placed, it will be necessary to wait until

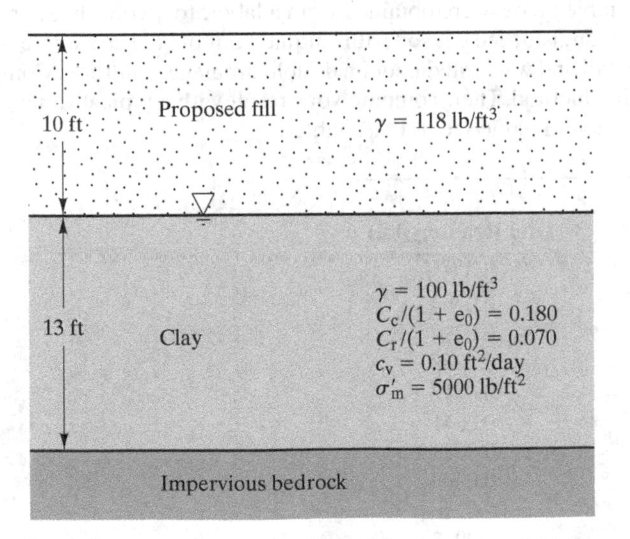

FIGURE 11.30 Soil profile for Problems 11.11 and 11.12.

enough settlement has occurred that the remaining settlement will be less than 2 in. Only then may the building construction begin.

Assuming the fill will be placed at a uniform rate from May 1 to June 1, determine the earliest start date for the building construction. Apply a correction for the construction period. For this problem, consider only settlement due to the weight of the fill. Do not consider settlement due to the weight of the buildings.

Section 11.3 The Coefficient of Consolidation

11.13 The data shown in the table below were obtained from a laboratory consolidation test on a normally consolidated undisturbed MH soil with a liquid limit of 65. The sample was 62 mm in diameter, 25 mm tall and was tested under a double drainage condition. Compute c_v using the log-time fitting method. Then, compare your result with a typical value of c_v for this soil and determine if your value seems reasonable.

Time Since Loading (HH:MM:SS)	Dial Reading (mm)
00:01:01	7.21
00:03:16	7.74
00:08:35	8.40
00:16:39	9.01
00:30:15	9.60
00:59:17	10.11
01:54:29	10.35
4:02:30	10.45
8:20:00	10.52

11.14 Repeat Problem 11.13 using the square root of time fitting method. Compare the results to those found using log-time method.

11.15 The data shown in the table below were obtained from a laboratory consolidation test on a normally consolidated undisturbed CL soil with a liquid limit of 38. The sample was 2.50 in. in diameter, 0.75 in. tall and was tested under a double drainage condition. Compute c_v using the log-time fitting method. Then, compare your result with a typical value of c_v for this soil and determine if your value seems reasonable.

Time Since Loading (HH:MM:SS)	Dial Reading (in.)
0:00:03	0.0755
0:00:08	0.0764
0:00:15	0.0774
0:00:30	0.0789
0:01:00	0.0812
0:02:00	0.0841
0:04:00	0.0872
0:09:00	0.0899
0:25:30	0.0918
0:49:00	0.0923
2:52:00	0.0930

11.16 Repeat Problem 11.15 using the square root of time fitting method. Compare the results to those found using log-time method.

Section 11.5 Consolidation Monitoring

11.17 A proposed fill is to be placed on the soil profile shown in Figure 11.31.

(a) Using the laboratory test results shown in this figure, develop a time–settlement plot. Do not apply any correction for the construction period. The medium clay is normally consolidated.

(b) A piezometer has been installed at Point A and a remote-reading settlement plate at Point B. Measurements from these instruments made 2580 days after placement of the fill indicated a pore water pressure of 1975 lb/ft^2 and a settlement of 1.20 ft. Using the technique described in Example 11.5, back-calculate the values of $C_c/(1 + e_0)$ and c_v, and compare them to the laboratory values.

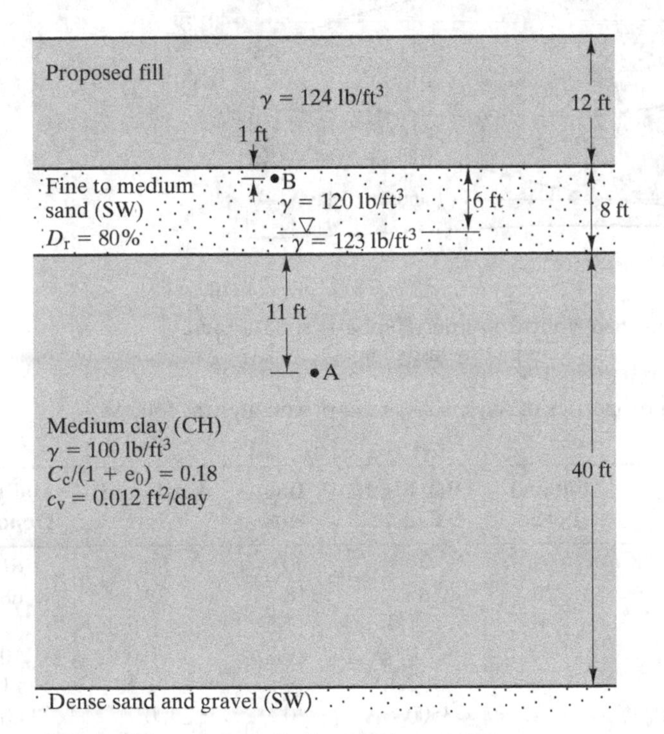

FIGURE 11.31 Soil profile for Problems 11.17, 11.18 and 11.31.

Comprehensive

11.18 The clay stratum in Figure 11.31 has $C_c/(1 + e_0) = 0.16$, $c_v = 0.022$, and $C_\alpha/(1 + e_p) = 0.017$. Develop a time–settlement plot for $t = 0$ to 75 years considering both consolidation and secondary compression. Do not apply any corrections for the construction period.

11.19 The CL/ML stratum in Figure 11.7 contains thin horizontal sand seams spaced about 1 m apart. Using this new information, reevaluate the computation in Example 11.3 and develop a revised time–settlement plot. Compare this plot with the ones in Figure 11.8 and explain why they are different.

11.20 Most of the international airport in San Francisco, California, is built on fill placed in San Francisco Bay. A cross-section through one portion of the airport is shown in Figure 11.32.

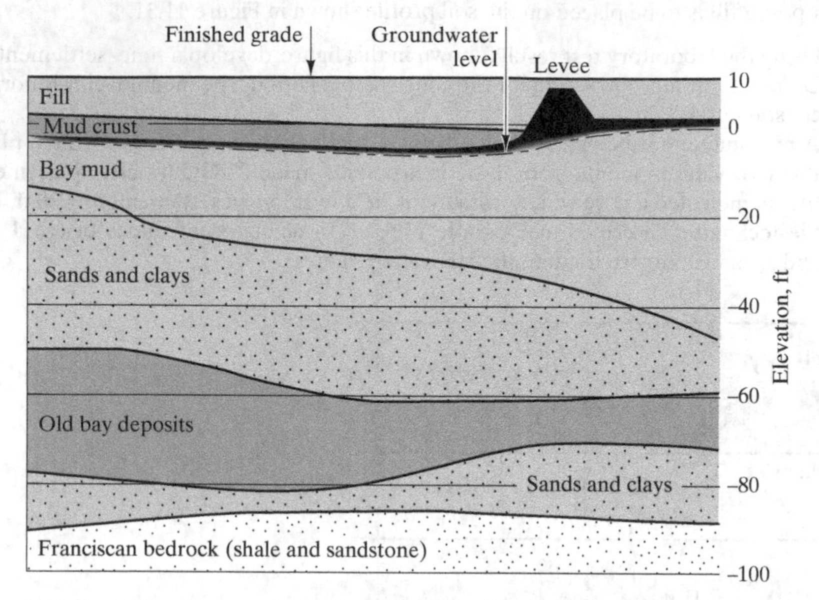

FIGURE 11.32 Cross-section at San Francisco Airport. (Roberts and Darragh, 1962.) The groundwater table is indicated by the dashed line.

The engineering properties of these soils are approximately as follows:

	Fill and Levee	Bay Mud Crust	Bay Mud	Sands and Clays	Old Bay Deposits
Dry unit weight, γ_d (lb/ft^3)	108	49	40	80	61
Moisture content, w (%)	20	82	118	29	68
Compression index, C_c	0	1.0	1.3	0.5	1.2
Recompression index, C_r	0	0.09	0.17	0.09	0.14
Initial void ratio, e_0		2.40	3.25	1.10	1.70
Overconsolidation margin, σ'_m (lb/ft^2)		3500	0	2800	2800
Coefficient of consolidation, c_v (ft^2/yr)		130	7	300	5

Note: Some of these values are from Roberts and Darragh; others have been estimated by the authors.

(a) Compute the ultimate consolidation settlement at various points along the cross-section. Then develop a plot of ultimate consolidation settlement versus horizontal position. When performing these computations, ignore the presence of the levee and any consolidation that may have already occurred due to its weight (in reality, these earlier settlements would increase the amount of differential settlement in this area, which could be worse than if the levee was never there).

(b) Develop plots of settlement versus time for the left and right ends of the cross-section. Assume all of the settlement in the crust and in the "sands and clays" strata will occur during construction, and assume both the bay mud and old bay deposits have double drainage.

Hint: Perform separate time-settlement computations for the bay mud and old bay deposits strata, then add the ultimate settlements from the other strata.

11.21 The information presented in Figure 11.31 has the following uncertainties:

Depth to bottom of proposed fill	±1 ft
Depth to bottom of SW stratum	±1 ft
Depth to bottom of CH stratum	±2 ft
Unit weights	±10%
Relative density	±15% (i.e., D_r = 68–92%)
$C_c/(1 + e_0)$	±20%
c_v	±35%

Considering these tolerances, compute the lower bound solution and upper bound solution for δ_c at t = 10 yr and $\delta_{c,ult}$. The lower bound solution is that which uses the best possible combination of factors, while the upper bound uses the worst possible combination.

11.22 The soil profile at a certain site includes an 8.5-m thick stratum of saturated normally consolidated medium silty clay. This soil has a unit weight of 16.4 kN/m^3. A remote reading settlement plate has been installed a short distance below the natural ground surface and a remote reading piezometer has been installed at the midpoint of the silty clay. The initial readings from these instruments indicate a ground surface elevation of 7.32 m and a pore water pressure of 52 kPa. The initial vertical effective stress at the top of the silty clay stratum was 50 kPa.

Then a 2.1-m deep fill with a unit weight of 18.7 kN/m^3 was placed on this site. A second set of readings made 220 days after placement of this fill indicate an elevation of 6.78 m and a pore water pressure of 77 kPa. Assuming all of the other soil strata are incompressible, and single drainage conditions exist in the silty clay, compute the values of $C_c/(1 + e_0)$ and c_v, then develop a plot of consolidation settlement versus time. This plot should extend from U = 0% to U = 95%. Finally, mark the point on this plot that represents the conditions present when the second set of readings was made.

11.23 The analysis in Example 11.5 did not explicitly consider the possibility that the drainage distance H_{dr} used in the original analysis was not correct. Does the adjustment of c_v based on piezometer data, as described in this example, implicitly consider H_{dr}? Explain.

11.24 An engineer in your office is planning a drilling and sampling program at a site that has a thick stratum of soft to medium clay. The information gathered from this program, along with the associated laboratory test results, will be used in various geotechnical analyses, including evaluations of consolidation rates. This engineer has submitted the plan to you for your review and approval.

The engineer expects the clay stratum will be very uniform, and therefore is planning to obtain only a few samples. These samples will then be used to conduct laboratory consolidation tests. Although this plan will probably be sufficient to characterize the consolidation properties of the clay, you are concerned that thin sandy layers might be present in the clay, and that they might not be detected unless more samples are obtained. Write a 200–300 word memo to this engineer explaining the need to search for possible sandy layers, and the importance of these layers in consolidation rate analyses.

11.25 A piezometer has been installed near the center of a 20-m thick stratum of saturated clay. A fill was then placed over the clay and the measured pore water pressure in the piezometer increased accordingly. However, 6 months after the fill was placed, the piezometer reading has not changed. Does this behavior make sense?

11.26 Explain how a time-settlement analysis could be used to estimate how long a surcharge fill must remain in place.

CHAPTER 12

Soil Strength

Most of the properties of clays, as well as the physical causes of those few properties that have been investigated, are unknown. We know nothing about the elasticity of clays, or the conditions that determine their water capacity, or the relations between their water content and their viscosity, or the earth pressure that they exert and not even about the physical causes of the swelling of wetted clays. As a consequence, the civil engineer, dealing with this important material, is at the mercy of some unreliable empirical rules, and laboratory work carried out with clays leads only to a mass of incoherent facts.

Karl Terzaghi, 1920

Terzaghi's statement was an accurate assessment of soil mechanics as it existed in 1920. Engineers had very little understanding of soil behavior, and soil strength was one of the most mysterious aspects, especially in clays. Although some researchers had performed soil strength tests, they did not fully understand what to do with the data once it had been obtained. Practicing engineers had to rely on empirical rules, intuition, and engineering judgment, which often were not adequate. As a result, unexplainable failures were far too common.

Fortunately, our knowledge and understanding of soil strength is now much better than it was in 1920. A large amount of research has been performed, and the results of this work have been successfully applied to practical engineering problems. Therefore, geotechnical designs that rely on soil strength assessments are now much more reliable.

12.1 STRENGTH ANALYSES IN GEOTECHNICAL ENGINEERING

The *strength* of a material is the greatest stress it can sustain. If the stress exceeds the strength, failure occurs. For example, structural engineers know the tensile yield strength of A36 structural steel is 36,000 lb/in.^{2} (248 MPa), so they must be sure that the tensile stresses in such members are less than this value. In practice, the working stresses must be substantially less to provide a sufficient factor of safety against failure. Such strength analyses can be performed for tensile, compressive, or shear stresses.

Although tensile strength analyses are an important part of structural engineering, geotechnical engineers rarely perform them because soil has very little tensile strength. Even rock masses cannot sustain tension over significant volumes because of the presence of cracks and fissures. There are a few occasions where tensile failures do occur such as tensile cracking near the top of incipient landslides, and heaving induced by upward seepage forces. However, the geometry of most geotechnical problems is such that nearly all of the ground is in compression.

Geotechnical engineering practice also differs from structural engineering practice in our assessment of compressive failures. Structural engineers define compressive strengths for various materials and design structural members accordingly, but soil and rock do not fail in compression per se, so we do not perform such analyses. Although the introduction of large compressive stresses may result in failure, empirical evidence tells us that the ground is actually failing in shear, not in compression. Therefore, nearly all geotechnical strength analyses evaluate shear only.

Many geotechnical engineering problems require an assessment of shear strength, with the shear failure modes shown in Figure 12.1. They include the following:

- **Earth slopes**—When the ground surface is inclined, gravity produces large geostatic shear stresses in the soil or rock. If these stresses exceed the shear strength, a landslide occurs.
- **Structural foundations**—Loads from a structure, such as a building, are transferred to the ground through structural foundations. This produces both compressive and shear stresses in the nearby soil. The latter could exceed the shear

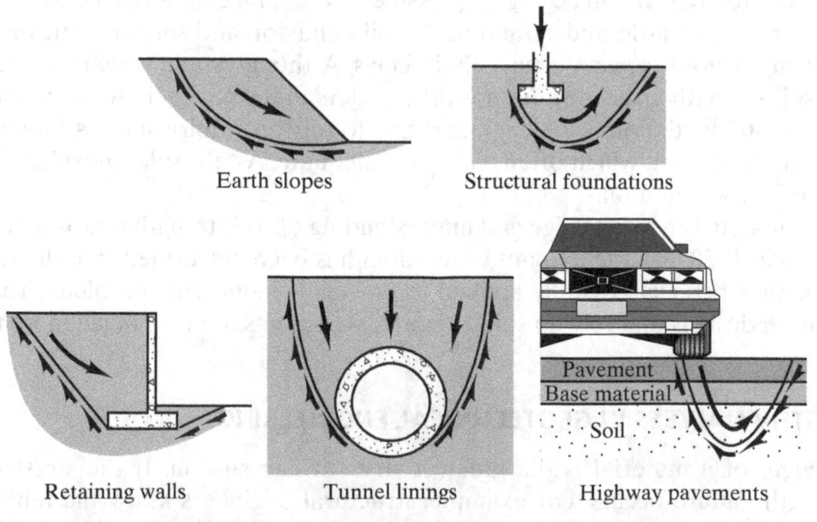

FIGURE 12.1 Typical applications of strength analyses in soils.

strength, thus producing a shear failure. This is known as a *bearing capacity failure* and could cause the structure to collapse.

- **Retaining walls**—The weight of soil behind a retaining wall produces shear stresses in that soil. Its shear strength resists some of this stress, and the wall resists the rest. Thus, the load carried by the wall depends on the shear strength of the retained soil.
- **Tunnel linings**—Tunnels in soil or weak rock normally require linings of steel or concrete for support. Such linings must resist pressures exerted by the surrounding ground, thus keeping the tunnel from collapsing. The magnitude of these pressures depends on the shear strength of the surrounding soil or rock.
- **Highway pavements**—Wheel loads from vehicles spread through the pavement and into the ground below. These loads produce shear stresses that could cause a shear failure. Engineers often place layers of well-graded gravel (known as *aggregate base material*), high-quality soils, or other materials between the pavement and the natural ground. These materials are stronger and stiffer, and help transfer the loads into the ground with much less potential for failure.

We will discuss some of these problems in more detail later in this book.

12.2 SHEAR FAILURE IN SOILS

The shear strength of common engineering materials, such as steel, is controlled by their molecular structure. Failure of these continuous materials generally requires breaking the molecular bonds that hold the material together, and thus depends on the strength of these bonds. For example, steel has very strong molecular bonds and thus a high shear strength, whereas plastic has much weaker bonds and a correspondingly lower shear strength.

However, the physical mechanisms that control the shear strength of a soil are much different. As discussed in Chapter 4, soil is a discontinuous particulate material made up of various sized particles. Shearing through soil is resisted not by the internal strengths of the particles in the soil, but by the interactions between the particles. These particle interactions include particle rearrangement (particles sliding and rolling past each other), as shown in Figure 12.2, and minor particle crushing. It is the work done to overcome the resistance to particle rearrangement and crushing that translates to the resistance to shearing. Because particle crushing is usually minor, the shear strength depends primarily on the rearrangement and not on the internal strengths of the particles.

Water and air, the other two phases in the soil besides the solid particles, provide no inherent resistance to shearing; however, as explained later, the water and air in the voids may affect the particle interactions and therefore indirectly affect the shear strength.

In the field, shear failure in a soil mass is usually confined to a narrow zone along a failure surface, for example, the sliding surface of a landslide shown in Figure 12.2. In this localized shear zone, shear failure occurs when the stresses between the particles are such that the particles slide or roll past each other as shown in Figure 12.2. We divide the shear strength from these interactions into two broad categories: *frictional strength* and *cohesive strength*.

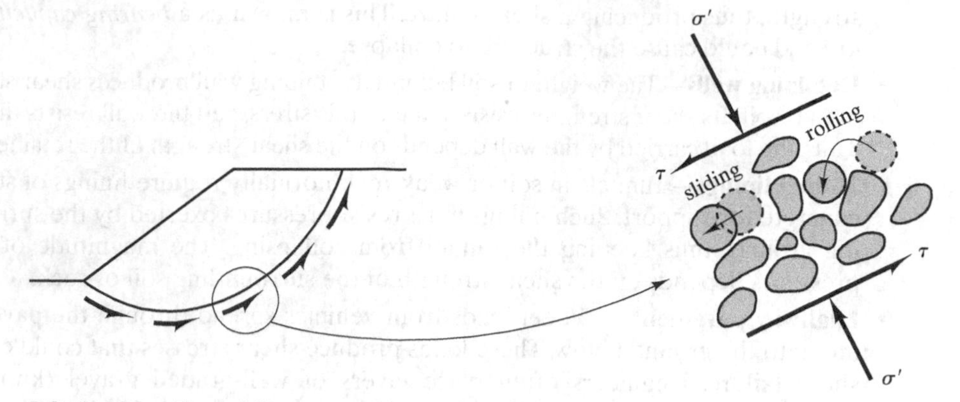

FIGURE 12.2 Shear failures occur in a soil when the shear stresses are large enough to make the particles roll and slide past each other.

Frictional Strength

Frictional strength in soils is similar to classic sliding friction from basic physics. The maximum friction that resists sliding is equal to the normal force multiplied by the coefficient of friction, μ, as shown in Figure 12.3. However, instead of using the coefficient of friction, μ, geotechnical engineers customarily describe frictional strength using the *effective friction angle* (or *effective angle of internal friction*), ϕ', where:

$$\phi' = \tan^{-1} \mu \tag{12.1}$$

In addition, we find it more convenient to work in terms of stresses instead of forces, so the shear strength, s, due to friction is:

$$s = \sigma' \tan \phi' = (\sigma - u) \tan \phi' \tag{12.2}$$

where:

s = shear strength
σ' = effective normal stress acting on the shear plane
ϕ' = effective friction angle
u = pore water pressure

Equation 12.2 is analogous to the friction law in physics. Notice how this equation uses the effective stress, not the total stress. This is an important distinction! We express strength in terms of the effective stress because fundamentally the effective stress, which describes the normal stresses between the solid particles and is analogous to the normal force between the contacting surfaces in the friction law, controls the shear strength.

In addition, it can be seen from Equation 12.2 that although the pore water has no static shear strength, it may indirectly affect the shear strength by affecting the magnitude of the effective stress because the effective stress is the difference between the total stress and the pore water pressure for a saturated soil.

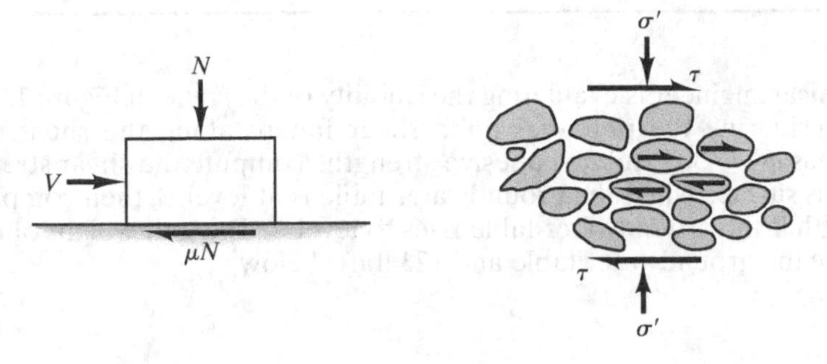

FIGURE 12.3 Comparison between friction on a sliding block and frictional strength in soil.

All soils have frictional strength. The value of ϕ' depends on both the frictional properties of the surfaces of the individual particles and the interlocking between particles. These are affected by many factors including:

- **Mineralogy**—Soil may contain particles that are made of many different minerals, and some minerals slide more easily than others. For example, the effective friction angle of *quartz sands* (sands containing primarily particles made of quartz) is typically 30–36°, whereas *micaceous sands* (sands containing significant quantitics of mica, which is much smoother than quartz) have a smaller ϕ'. The presence of clay minerals in a sand can greatly lower its ϕ' because clay minerals typically have low effective friction angles (ϕ' values as low as 4° have been measured in pure montmorillonite).
- **Particle shape**—The friction angle of sands with angular particles is much higher than that of sands with rounded ones.
- **Gradation**—Well-graded soils typically have more interlocking between the particles, and thus a higher friction angle, than soils that are poorly-graded. For example, GW soils typically have ϕ' values about 2° higher than comparable GP soils.
- **Void ratio**—Decreasing the void ratio, such as by compacting a soil with a sheepsfoot roller, also increases interlocking, which results in a higher ϕ'.
- **Organic material**—Organics introduce many problems, including a decrease in the friction angle.

The impact of water on frictional strength is especially important, and many shear failures are induced by changes in the groundwater conditions. Many people mistakenly believe that water-induced changes in shear strength are primarily due to lubrication effects. Although the process of wetting some dry soils can induce lubrication, the resulting decrease in ϕ' is actually very small. Sometimes the introduction of water has an antilubricating effect (Mitchell and Soga, 2005) and causes a small increase in ϕ'. However, focusing on these small effects tends to obscure another far more important process, which is illustrated in Example 12.1.

Example 12.1

A geotechnical engineer is evaluating the stability of the slope in Figure 12.4. This evaluation is considering the potential for a shear failure along the shear surface shown. The soil has $\phi' = 30°$ and no cohesive strength. Compute the shear strength at Point A along this surface when the groundwater table is at level B, then compute the new shear strength if the groundwater table rises to level C. The unit weight of the soil is 120 lb/ft^3 above the groundwater table and 123 lb/ft^3 below.

Solution:

Groundwater table at level B:

$$u = \gamma_w z_w = (62.4 \text{ lb/ft}^3)(20 \text{ ft}) = 1248 \text{ lb/ft}^2$$

$$\sigma_z' = \sum \gamma H - u = (120 \text{ lb/ft}^3)(26 \text{ ft}) + (123 \text{ lb/ft}^3)(20 \text{ ft}) - 1248 \text{ lb/ft}^2$$
$$= 4332 \text{ lb/ft}^2$$

The potential shear surface at Point A is horizontal, so $\sigma' = \sigma_z'$ and

$$s = \sigma' \tan \phi' = (4332 \text{ lb/ft}^2) \tan 30° = \mathbf{2501 \text{ lb/ft}^2}$$

Groundwater table at level C:

$$u = \gamma_w z_w = (62.4 \text{ lb/ft}^3)(32 \text{ ft}) = 1997 \text{ lb/ft}^2$$

$$\sigma_z' = \sum \gamma H - u = (120 \text{ lb/ft}^3)(14 \text{ ft}) + (123 \text{ lb/ft}^3)(32 \text{ ft}) - 1997 \text{ lb/ft}^2$$
$$= 3619 \text{ lb/ft}^2$$

$$s = \sigma' \tan \phi' = (3619 \text{ lb/ft}^2) \tan 30° = \mathbf{2089 \text{ lb/ft}^2}$$

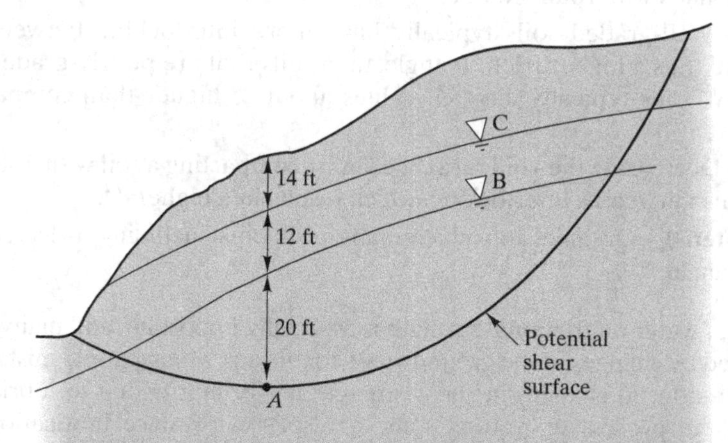

FIGURE 12.4 Cross-section of a potential landslide.

Commentary: The total stress at Point A increases slightly when the groundwater table rises from level B to level C due to the greater unit weight of the zone of soil that becomes saturated. However, this is more than offset by the increase in pore water pressure, so the effective stress is reduced, along with a corresponding decrease in the shear strength. It is quite possible that this loss in shear strength would be sufficient to induce a shear failure (landslide) in this slope. Additional analyses, which we will discuss in Chapter 13, are necessary to determine if such a failure will occur.

Example 12.1 illustrates the important impact of pore water pressure on shear strength. This is the primary way water impacts the frictional strength, and is one of the reasons a thorough understanding of the principle of effective stress is so important in understanding soil behavior.

Cohesive Strength

Some soils have shear strength even when the effective stress, σ', is zero, or at least *appears* to be zero. This strength is called the *cohesive strength*, and we describe it using the variable c', the *effective cohesion*. If a soil has both frictional and cohesive strength, then Equation 12.2 becomes:

$$s = c' + \sigma' \tan \phi' = c' + (\sigma - u) \tan \phi' \tag{12.3}$$

where:

s = shear strength
c' = effective cohesion
σ' = effective normal stress acting on the shear surface
ϕ' = effective friction angle
u = pore water pressure

There are two types of cohesive strength: true cohesion and apparent cohesion (Mitchell and Soga, 2005).

True cohesion is shear strength that is truly the result of bonding between the soil particles. These bonds include the following:

- *Cementation* is chemical bonding due to the presence of cementing agents, such as calcium carbonate ($CaCO_3$) or iron oxide (Fe_2O_3) (Clough et al., 1981). Even small quantities of these agents can provide significant cohesive strengths. *Caliche* is an example of a heavily cemented soil that has a large cohesive strength. Cementation also can be introduced artificially using Portland cement or special chemicals.
- *Electrostatic and electromagnetic attractions* hold particles together. However, these forces are very small and probably do not produce significant shear strength in soils.
- *Primary valence bonding* (*adhesion*) is a type of cold welding that occurs in clays when they become overconsolidated.

Apparent cohesion is shear strength that appears to be caused by bonding between the soil particles, but is really frictional strength in disguise. Sources of apparent cohesion include the following:

- *Negative pore water pressures* that have not been considered in the stress analysis. These negative pore water pressures are present in soils above the groundwater table due to capillary action or surface tension, as shown in Figures 9.25 and 9.26. However, many geotechnical engineering analyses ignore these pressures (i.e., we assume $u = 0$, even though it is really <0), so the effective stress is actually greater than we think it is (see Equation 9.45). The shear strength that corresponds to this additional effective stress thus appears to be cohesive, even though it is really frictional. This is the primary reason unsaturated clays appear to have "cohesive" strength and why moist unsaturated "cohesionless" sands can stand in vertical cuts or be used to build a sand castle on the beach.
- *Negative excess pore water pressures due to dilation.* Some soils tend to *dilate* or expand when they are sheared. In saturated soils, this dilation draws water into the voids. However, sometimes the rate of shearing is high relative to the rate at which water can flow (i.e., the voids are trying to expand more rapidly than they can draw in the extra water). This is especially likely in saturated clays because their hydraulic conductivities are so low. When this occurs, large negative excess pore water pressures can develop in the soil. We will explore this phenomenon in more detail later in Section 12.3.

  The term *excess pore water pressure* was defined in Chapter 10. It is an additional pore water pressure, either positive or negative, that is superimposed onto the hydrostatic pore water pressure. The excess pore water pressure is due to contraction or expansion of the voids of a saturated soil. In this case of dilation, the voids are expanding, so $u_e < 0$ and u becomes less than the hydrostatic value, u_h.

  If these negative excess pore water pressures are considered in our strength analysis, we can compute an accurate value of σ' and thus an accurate value of s using Equation 12.2. However, if we consider only the hydrostatic pore water pressure, our computed value of σ' will be too low. As a result, the soil will appear to have cohesive strength because its actual strength is larger than predicted by the frictional strength alone.
- *Apparent mechanical forces* are those due to particle interlocking, and can develop in soils where this interlocking is very difficult to overcome. The result is additional apparent cohesion.

Geotechnical engineers often use the term *cohesive soil* to describe clays. Although this term is convenient, it also is very misleading (Santamarina, 1997). Most of the so-called cohesive strength in clays is really apparent cohesion due to pore water pressures that are negative, or at least less than the hydrostatic pore water pressure. In such soils, it is better to think of "cohesive strength" as a mathematical idealization rather than a physical reality.

For design, where the cohesive strength really does reflect bonding between the soil particles, for example, in cemented soils, we may rely on the cohesive strength in some cases. However, in other cases, it is wise to ignore this source of strength if it may disappear during the life of the project. For example, if the cementing agent is water-soluble, it may disappear if the soil becomes wetted during the life of the project.

Definition of Failure

Another important difference between shear strength assessments of soils and those for more traditional engineering materials lies in our definition of failure. For example, with steel, we usually define failure as the point where the stress–strain curve becomes plastic and nonlinear (the yield strength) or when rupture occurs (the ultimate strength). However, in soils, the stress–strain curve is nonlinear and plastic from the very beginning, and there is no rupture point in the classic sense. Therefore, we must use other means of defining shear strength.

Soils typically have two kinds of shear stress–strain curves, as shown in Figure 12.5. A *ductile* soil has a curve that plateaus to a well-defined peak shear stress as shown, and we can use this value as the design shear strength. However, a *brittle* soil has a curve that shows two strengths, the *peak strength*, which corresponds to the highest point on the curve, and the *residual strength* (or *ultimate strength*), which is lower than the peak strength and occurs at a much larger shear strain. Either value could be used for design, depending on the kind of problem being evaluated.

Sands and gravels have shear stress–strain curves that are either ductile or only mildly brittle, so the difference between peak and residual strength is small and can be neglected. However, some clays have very brittle curves, so the distinction between peak and residual strength becomes very important, as discussed later in this chapter.

The problem of defining failure is further complicated by the lack of a unique stress–strain curve. For example, a soil tested under certain conditions (i.e., intermediate principal stress, strain constraint, rate of strain, etc.) produces a certain stress–strain curve and thus a certain shear strength, yet the same soil tested under different conditions can produce a different curve and a different strength. We attempt to overcome this problem by using test conditions that simulate the field conditions or by using standardized test conditions and calibrating the results with observed behavior in the field.

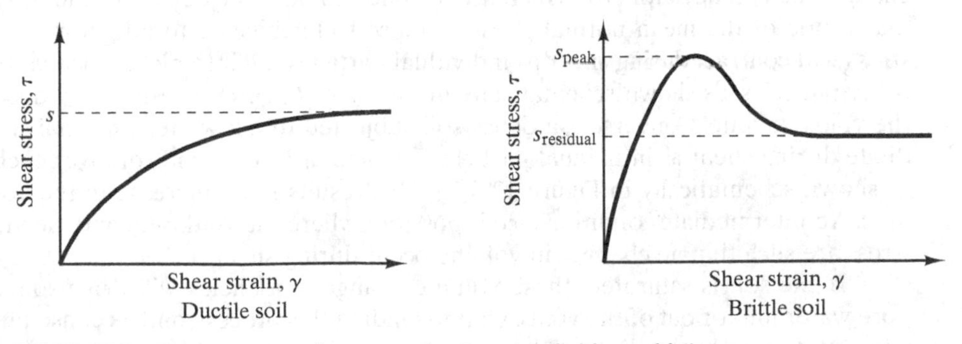

FIGURE 12.5 Shear stress-strain curves of soil, and definitions of failure.

12.3 THE DRAINED AND THE UNDRAINED CONDITIONS

Excess pore water pressures generated from loading a soil will impact the effective stress, and thus the shear strength of the soil, so the generation and dissipation of these excess pore water pressures are important considerations in shear strength evaluations. Geotechnical engineers often evaluate shear strengths by considering two limiting conditions: *the drained condition* and *the undrained condition*. Before describing these conditions, we will discuss the volume change tendencies in a soil during loading that lead directly to excess pore water pressure generation.

Volume Change during Loading

When soils are loaded (or unloaded) there is always volume change. In theory, these volume changes can be separated into two distinct parts: (a) contraction (compression) or dilation (expansion) due to changes in the hydrostatic (all-around) mean normal stress, σ_m (Equation 9.16), and (b) contraction or dilation due to changes in the deviator stress, σ_d (Equation 9.17).

Volume Change Due to Mean Normal Stress When a soil is subjected to an increase or decrease of the hydrostatic (all-around) mean normal stress, the soil skeleton formed by the soil particles will contract or dilate, respectively. In the Mohr-circle model, the mean normal stress corresponds to the distance between the center of the circle and the origin. Because the stress state in this case is hydrostatic, no shear stresses are induced in the soil, and the Mohr circle is represented by a point on the σ-axis. As the mean normal stress increases, this point moves farther to the right and the soil skeleton contracts. Conversely, as the mean normal stress decreases, this point moves to the left and the soil skeleton dilates.

Volume Change Due to Deviator Stress When a soil is subjected to an increase in the deviator stress, shear stresses are induced in the soil. These induced shear stresses will cause the soil particles to rearrange and can lead to volume change. In the Mohr-circle model, this corresponds to an increase in the diameter of the circle without changing the location of the center of the circle. The magnitude and sign of the volumetric change due to a deviator stress increase depends on both the density of the soil and the magnitude of the mean normal stress. A loose soil subjected to a high mean normal stress will contract during shear as individual particles fall into niches located between other particles, as shown schematically in Figure 12.6(a). This results in a decrease in the voids volume. Conversely, a dense soil subjected to a low mean normal stress will dilate during shear as individual particles become dislodged and roll over each other, as shown schematically in Figure 12.6(b). This results in an increase in the voids volume. An intermediate condition also is possible, where the void ratio and mean normal stress are such that no changes in volume occur during shear.

If the soil is saturated, these volume change tendencies will drive some of the pore water into or out of the voids. Understanding this process, and its consequences, is a key part of understanding soil behavior.

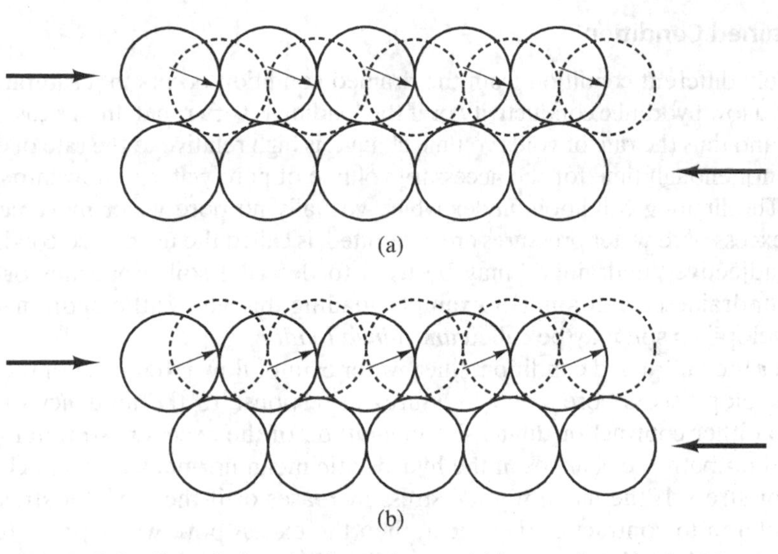

FIGURE 12.6 Shear-induced volume change: (a) loose soil tends to contract during shear; (b) dense soil tends to dilate during shear.

The Drained Condition

We define the drained condition as a limiting condition under which there is no excess pore water pressure in the soil. The important consequence of the drained condition is that the pore water pressure is equal to the hydrostatic pore water pressure and is thus easily computed. Therefore, the effective stress and strength also are easily computed.

The drained condition can develop in one of two ways:

1. If the soil is saturated, its hydraulic conductivity is sufficiently high, and the rate of loading (and therefore the rate of voids volume change) is sufficiently low, then the necessary volume of water can easily flow into or out of the voids. Any excess pore water pressures generated during loading will be quickly dissipated. Most loads are applied fairly slowly. For example, the dead load from a building is generated as the building is constructed, a process that typically occurs over weeks or months. If the soil is a sand or gravel, the hydraulic conductivity is sufficiently high that the pore water has enough time to move as needed. Thus, sands and gravels under static loading are usually under the drained condition, and the shear strength may be evaluated using the hydrostatic pore water pressure.
2. If a long time has elapsed after the end of loading, any excess pore water pressures generated during loading will have dissipated. In this case, the drained condition will be reached in the soil in the long term regardless of what type of soil it is.

The adjective "drained" may be used to describe soil properties or loading under the drained condition. For example, the shear strength of a soil under the drained condition may be called its *drained strength*, and loading that allows the drained condition to develop may be called *drained loading*.

The Undrained Condition

A completely different condition from the drained condition occurs in a saturated soil if the soil has a low hydraulic conductivity or if the loading is very rapid. In this case, the rate of loading, and thus the rate of voids volume change, is high relative to the rate of drainage, so there is not enough time for the necessary volume of pore water to flow into or out of the voids. The limiting condition, under which virtually no pore water movement takes place and excess pore water pressures are generated, is called the undrained condition.

The adjective "undrained" may be used to describe soil properties or loading under the undrained condition. For example, loading that allows the undrained condition to develop in a soil may be called *undrained loading*.

Under the undrained condition, since water cannot flow into or out of the soil, the soil will develop excess pore water pressures in response to the tendency of the soil skeleton to either contract or dilate. The magnitude of the excess pore water pressure will depend on both the changes in the hydrostatic mean normal stress and changes in the deviator stress. If the mean normal stress increases or if the deviator stress causes the soil skeleton to contract during shear, then the excess pore water pressure will be positive because the soil wants to push water out of the pores to contract. Conversely, if the mean normal stress decreases or if the deviator stress causes the soil skeleton to dilate during shear, then the excess pore water pressure will be negative because the soil wants to draw water into the pores to dilate. In either case, the excess pore water generation due to the combination of volume change tendencies and low hydraulic conductivity can have a significant impact on the effective stress and thus on the shear strength.

The hydraulic conductivity of clays is several orders of magnitude smaller than that of sands (see Table 7.1), so even typical rates of loading, such as those from placement of a fill or construction of a building, are very high relative to the rate of drainage. Thus, saturated clays are most often assumed to be under the undrained condition during the loading or construction period in the short term and their shear strength must be analyzed accordingly. Thus, the geotechnical engineer must directly or indirectly account for the excess pore water pressures generated during loading. However, a long time after the end of construction, all the excess pore water pressures will have dissipated, and the condition in clays reaches the drained condition. Therefore, for clays, the condition for shear strength evaluation changes from the undrained condition in the short term to the drained condition in the long term.

An especially interesting situation occurs when sands, for which the drained condition can normally be assumed, are subjected to very rapid loading, such as from machine vibrations or from an earthquake. In this case, the rate of loading may be very high relative to the rate of drainage, thus producing the undrained condition.

Similarly, if the rate of loading is extremely low, the condition in the short term even for clays, which normally is undrained, can be assumed to be drained.

Intermediate Drainage Conditions

The actual drainage conditions in the field are often such that some, but not all, of the necessary pore water movement occurs. Thus, the real drainage condition may be somewhere between the drained and undrained conditions. Although a geotechnical engineer could attempt to evaluate these intermediate drainage conditions, this is rarely done in practice. Instead, we nearly always conduct strength evaluations assuming either the drained or the undrained condition exists in the field.

12.4 MOHR–COULOMB FAILURE CRITERION

Some soils have both frictional and cohesive strengths, so we need to combine these two sources into a single all-purpose strength formula. Nearly all geotechnical analyses do this using the *Mohr–Coulomb failure criterion*, which is expressed as Equation 12.3. This formula allows us to project the test data back into our analyses of existing or proposed field conditions, which may be done using either effective stress analyses or total stress analyses.

Effective Stress Analysis

The shear strength of a soil is developed only by the solids of the soil; the water and air, representing the other two phases of the soil, have no shear strength. Therefore, the shear strength is fundamentally controlled by the effective stress, σ', because the effective stress is the portion of the total stress, σ, carried by the solid particles. This is why Equations 12.2 and 12.3 were written in terms of the effective stress.

If we perform a series of laboratory strength tests, each at a different value of σ', the typical results are shown in Figure 12.7. The vertical axis on this plot is shear stress, τ, and the curve represents the shear strength, s, which is the magnitude of τ at failure, plotted against σ' at failure.

For sands and gravels, these plots are nearly linear within the range of stresses normally encountered in the field. However, for clays they are typically slightly nonlinear, as shown. These nonlinear plots are inconvenient because they introduce more complexity into the analyses. Despite this nonlinear behavior, we nearly always use an idealized linear function by conducting tests at σ' values comparable to those expected in the field and fitting the data points with a straight line. We say the τ-intercept of this line is the effective cohesion, c', and the slope of the line gives the effective friction angle, ϕ'. In reality, the relative contributions of cohesive and frictional strengths are much more complex. For example, this c' value is really a combination of true cohesion and the mathematics of fitting a straight line to a nonlinear function. In some cases, it also may include some apparent cohesion. Nevertheless, this idealized representation

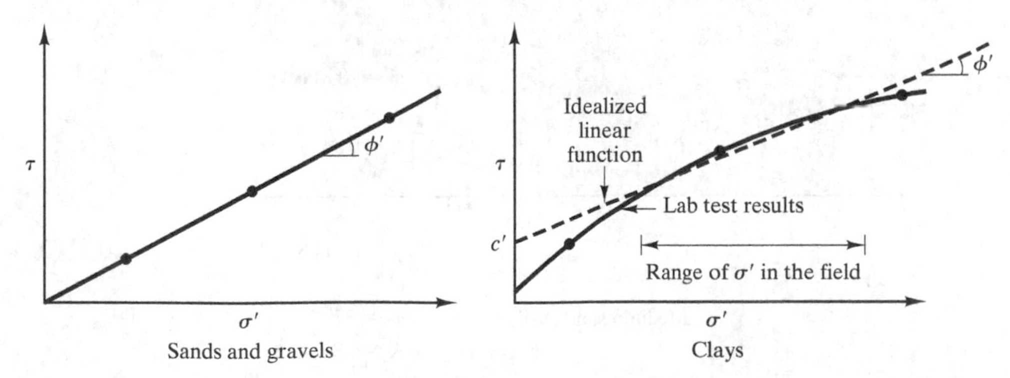

FIGURE 12.7 Shear strength as a function of effective stress. Each data point represents the results of a laboratory test.

is adequate for the vast majority of practical problems. The straight lines in Figure 12.7 are known as the *Mohr–Coulomb failure criterion*, and may be expressed mathematically using Equation 12.3. Because it is based on effective stresses, it can be more precisely called the *effective stress Mohr–Coulomb failure criterion.*

Shear strength is defined as the shear stress at failure, so planes through points in the soil that have (σ', τ) values that plot below the Mohr–Coulomb line theoretically will not fail in shear, while those that plot on or above the line will fail. We often call this line a *failure envelope* because it encloses the combinations of stresses that will not cause failure.

Once c' and ϕ', the effective stress strength parameters, have been determined, we can evaluate the shear strength in the field using Equation 12.3. We compute the effective stresses in the field, assuming the drained or undrained condition. For the drained case, we can calculate the effective stresses using the techniques described in Chapter 9, along with the hydrostatic or steady state groundwater conditions that represent the worst conditions anticipated during the life of the project.

Example 12.2

Samples have been obtained from both soil strata in Figure 12.8 and brought to a soil mechanics laboratory. A series of shear strength tests were then performed on both samples and plotted in diagrams similar to those in Figure 12.7. The c' and ϕ' values obtained from these diagrams are shown in Figure 12.8. Using this data and assuming the drained condition applies, compute the shear strength on horizontal and vertical planes at Points A, B, and C.

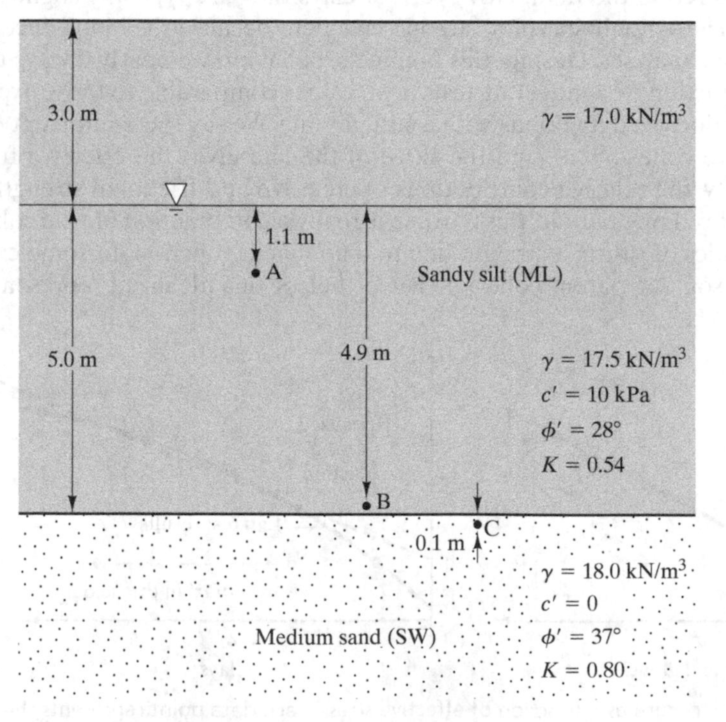

FIGURE 12.8 Soil profile for Example 12.2.

Solution:

Point A—horizontal plane

$$\sigma'_z = \sum \gamma H - u$$

$$\sigma'_z = (17.0\ \text{kN/m}^3)(3.0\ \text{m}) + (17.5\ \text{kN/m}^3)(1.1\ \text{m}) - (9.8\ \text{kN/m}^3)(1.1\ \text{m})$$

$$\sigma'_z = 59.5\ \text{kPa}$$

$$s = c' + \sigma' \tan \phi'$$

$$s = 10\ \text{kPa} + (59.5\ \text{kPa}) \tan 28^\circ$$

$$s = \mathbf{41.6\ kPa}$$

Point A—vertical plane

$$\sigma'_x = K\sigma'_z = (0.54)(59.5\ \text{kPa}) = 32.1\ \text{kPa}$$

$$s = c' + \sigma' \tan \phi'$$

$$s = 10\ \text{kPa} + (32.1\ \text{kPa}) \tan 28^\circ$$

$$s = \mathbf{27.1\ kPa}$$

Using similar computations:

Point B—vertical plane	$s =$ **54.4 kPa**
Point B—horizontal plane	$s =$ **68.1 kPa**
Point C—vertical plane	$s =$ **35.5 kPa**
Point C—horizontal plane	$s =$ **57.2 kPa**

Commentary: At each point, the shear strength on a vertical plane is less than that on a horizontal plane because $K < 1$. In addition, the shear strength at Point B is greater than that at Point A because the effective stress is greater. The strength at Point C is even higher than at Point B because it is in a new stratum with different c', ϕ', and K values. Thus, the strength would increase gradually with depth within each stratum, but change suddenly at the boundary between the two strata.

Example 12.3

Draw the shear strength envelope for the ML stratum in Figure 12.8, then plot the upper half of the Mohr circle for Point A on this diagram. Assume the principal stresses act vertically and horizontally.

Solution:

Using the results from Example 12.2, we develop Figure 12.9.

Commentary: In this case, the entire Mohr circle plots below the strength envelope. Therefore, the shear stress on all planes through Point A is less than the shear strength, and no shear failure will occur. However, if the Mohr circle touches the envelope, such as the circle shown in Figure 12.10, then a shear failure will occur on the plane represented by this point on the circle. This method of presenting stresses and strengths is named after the German engineer Otto Mohr (1835–1918) and the French scientist Charles Augustin de Coulomb (1736–1806). Neither of them drew diagrams like this to describe soil strength, but the underlying concepts are based on their work.

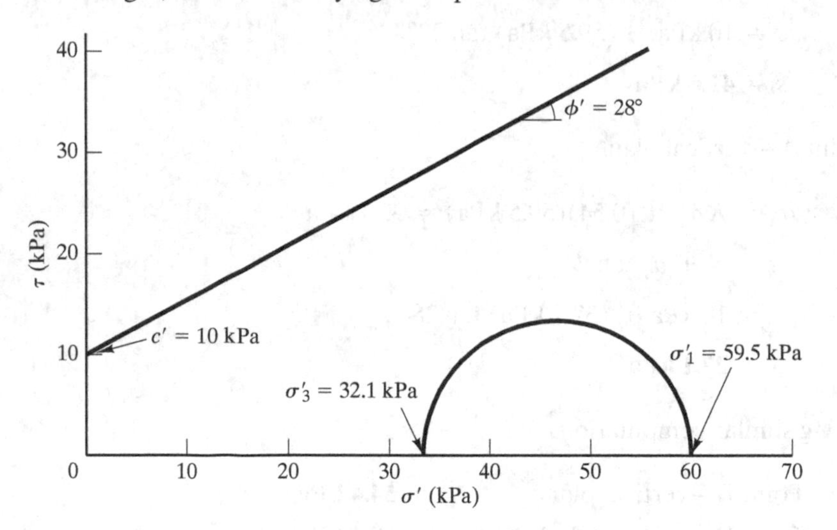

FIGURE 12.9 Failure envelope and Mohr circle for Example 12.3.

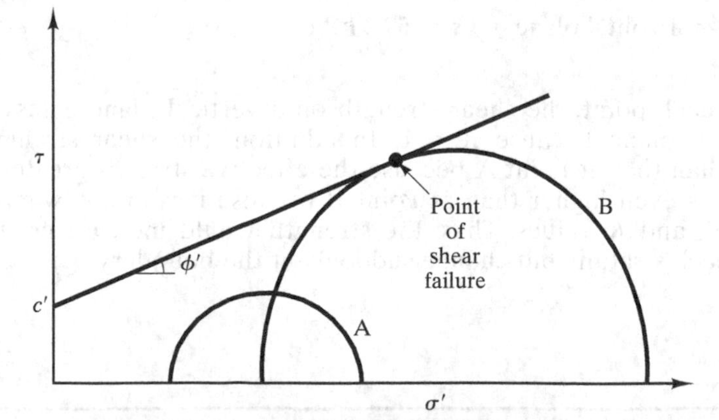

FIGURE 12.10 Shear failure occurs when the Mohr circle is large enough to touch the failure envelope. Thus, no failure will occur at the point represented by Circle A, but failure will occur at the point represented by Circle B.

The ratio of the shear strength, s, on a specific plane to the shear stress, τ, on that plane is defined as the *factor of safety*, F:

$$F = \frac{s}{\tau} \tag{12.4}$$

Normally, we define some minimum acceptable factor of safety, and then we check proposed designs to verify that this criterion has been met.

Example 12.4

An 18-in. diameter storm-drain pipe is to be installed under a highway by jacking as shown in Figure 12.11. A mound of soil will be placed as shown to provide a reaction for the jacks. Then, the first section of pipe will be pushed into the ground below the highway. An auger will clean out the soil collected inside, and then additional sections will be added, pushed in, and cleaned out one at a time until the pipe reaches the opposite side of the highway. This method allows the highway to remain in service while the pipe is being installed. The alternative would be to dig a trench, lay the pipe, and backfill, but this would require temporarily closing the highway.

At times the jack must apply a 300 k load to press the pipe into the ground. It will react against a steel plate placed on the soil mound. The load-carrying capacity of this plate is controlled by the shear strength of the adjacent soil. The soil has $c' = 400\ \text{lb/ft}^2$, $\phi' = 29°$, and $\gamma = 120\ \text{lb/ft}^3$. Will the soil beyond the plate be able to resist this load with a factor of safety of 1.5?

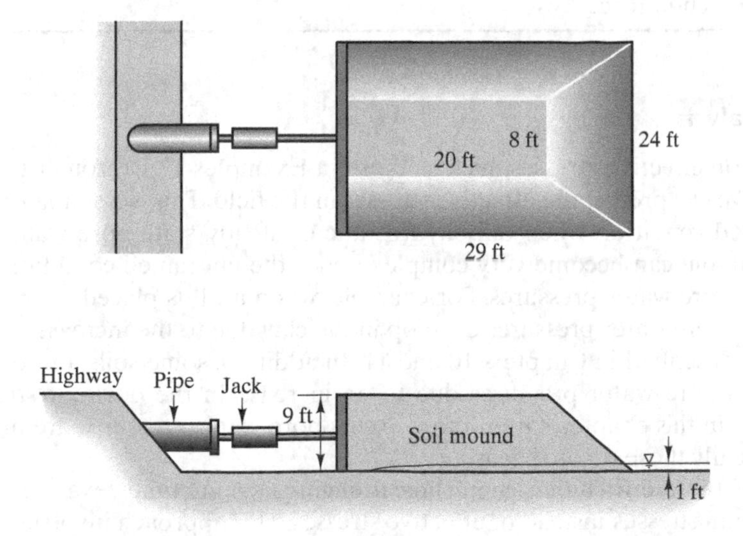

FIGURE 12.11 Proposed soil mound and jacking arrangement for Example 12.4.

Solution:

Soil mound:

$$A_{top} = (20\text{ ft})(8\text{ ft}) = 160\text{ ft}^2$$

$$A_{bottom} = (29\text{ ft})(24\text{ ft}) = 696\text{ ft}^2$$

$$A_{average} = \frac{160\text{ ft}^2 + 696\text{ ft}^2}{2} = 428\text{ ft}^2$$

$$W = (428\text{ ft}^2)(9\text{ ft})(120\text{ lb/ft}^3) = 462{,}000\text{ lb}$$

$$A_{shear} = A_{bottom} = 696\text{ ft}^2$$

$$\sigma_z = \frac{W}{A} = \frac{462{,}000\text{ lb}}{696\text{ ft}^2} = 664\text{ lb/ft}^2$$

$$u = \gamma_w z_w = (62.4\text{ lb/ft}^3)(1\text{ ft}) = 62\text{ lb/ft}^2$$

$$\sigma'_z = \sigma_z - u = 664\text{ lb/ft}^2 - 62\text{ lb/ft}^2 = 602\text{ lb/ft}^2$$

$$s = c' + \sigma' \tan\phi'$$

$$s = 400\text{ lb/ft}^2 + (602\text{ lb/ft}^2)\tan 29^\circ$$

$$s = 734\text{ lb/ft}^2$$

$$\tau = \frac{P}{A} = \frac{300{,}000\text{ lb}}{696\text{ ft}^2} = 431\text{ lb/ft}^2$$

$$F = \frac{s}{\tau} = \frac{734\text{ lb/ft}^2}{431\text{ lb/ft}^2} = \mathbf{1.7 > 1.5 \therefore OK}$$

According to this analysis, the soil mound is sufficiently large to provide the necessary reaction force.

Total Stress Analysis

Analyses based on effective stresses, such as those in Examples 12.1 through 12.4, are possible only if we can predict the effective stresses in the field. This is a simpler matter under the drained condition when only hydrostatic or steady state pore water pressures are present, but can become very complex under the undrained condition when there are excess pore water pressures. For example, when a fill is placed over a saturated clay, excess pore water pressures develop in the clay due to the increase in mean normal stress as described in Chapters 10 and 11. In addition, some soils also develop additional excess pore water pressures due to an increase in the deviator stress, as described earlier in this chapter. Often these excess pore water pressures are difficult to predict, especially those due to shearing.

Because of these difficulties, geotechnical engineers sometimes evaluate problems based on total stresses instead of effective stresses. This approach involves reducing the laboratory data in terms of total stresses and expressing the failure criterion in terms of the total stress parameters c_T and ϕ_T. Equation 12.3 then needs to be rewritten in terms of the total stress as

$$s = c_T + \sigma \tan \phi_T \tag{12.5}$$

where:

s = shear strength
c_T = total cohesion
σ = total stress acting on the shear surface
ϕ_T = total friction angle

This is called the *total stress Mohr–Coulomb failure criterion.*

The total stress analysis method assumes the excess pore water pressures developed in the laboratory test are the same as those in the field, and thus are implicitly incorporated into c_T and ϕ_T. This assumption introduces some error into the analysis, but it becomes an unfortunate necessity when we cannot predict the magnitudes of excess pore water pressures in the field. It also demands the laboratory tests be conducted in a way that simulates the field conditions as closely as possible.

The shear strength of a soil really depends on effective stresses, so total stress analyses are less desirable than effective stress analyses, and the results need to be viewed with more skepticism. However, there are many times when we must use total stress analyses because we have no other practical alternative.

Sections 12.5 and 12.6 discuss in more detail how the Mohr–Coulomb failure criterion can be used to obtain the shear strengths of different types of saturated soils, saturated sands and gravels in Section 12.5 and saturated clays in Section 12.6.

12.5 SHEAR STRENGTH OF SATURATED SANDS AND GRAVELS

Under typical static loading conditions, little or no excess pore water pressures are generated in clean sands and gravels because their hydraulic conductivities are so high. Changes in the state of stress of the soil (both mean normal stress and deviator stress) cause the voids to expand or contract and water easily flows in or out as necessary. Therefore, the pore water pressure, u, is equal to the hydrostatic pore water pressure (Equation 7.7), and shear strength analyses may be based on the drained condition and effective stresses.

Under dynamic loads, such as during an earthquake, a sand may not drain quickly enough to dissipate all excess pore water pressures. In such cases, we may evaluate the shear strength assuming the undrained condition applies. We can still use effective stress analyses if we can estimate the magnitudes of the excess pore water pressures generated during the dynamic loading.

Determining c' and ϕ'

If no cementing agents or clay are present, saturated sands and gravels should have $c' = 0$, as supported by empirical evidence. We determine the friction angle, ϕ', by conducting field or laboratory tests, as discussed later in this chapter. Figure 12.12 presents typical ϕ' values, which may be used for preliminary estimates or for checking test data.

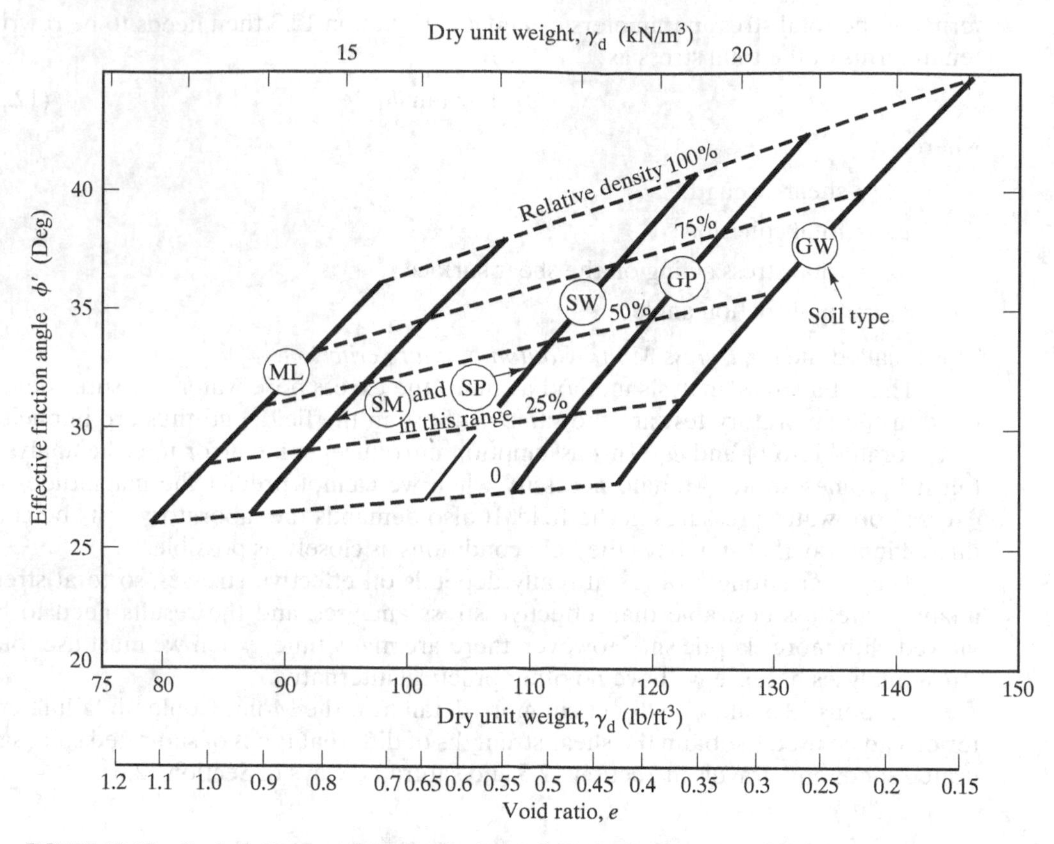

FIGURE 12.12 Typical ϕ' values for cohesionless soils without clay or cementing agents. (Adapted from U.S. Navy, 1982.)

Notice how the ϕ' values in Figure 12.12 increase as the unit weight increases. This is one of the reasons for compacting soils. This figure also illustrates that gravels are generally stronger than sands, and well-graded soils are generally stronger than poorly-graded ones. The presence of large amounts of silt decreases ϕ', primarily because these particles are smoother and have lower coefficients of friction.

If cementing agents or overconsolidated clays are present (i.e., SC soils), then c' will be greater than zero. However, engineers are generally reluctant to rely on this additional strength, especially if it is from clay, if the cementing agents are water-soluble, or if there is concern that it may be an apparent cohesion. For design purposes, most engineers either ignore the cohesive strength of such soils or use design values less than the measured c' value.

Selecting the Proper Value of σ'

We are performing an effective stress analysis, so the shear strength is defined by Equation 12.3. The value of ϕ' is determined by testing, and c' is usually zero, but what should we use for the effective stress, σ'? In most geotechnical design problems, as the state of stress of the soil changes, the shear and normal stresses change simultaneously, so σ' at the beginning of loading is different from that at the end.

To understand which σ' to use for design, study the plots in Figure 12.13. These plots describe the conditions at a point in the soil below a spread footing foundation.

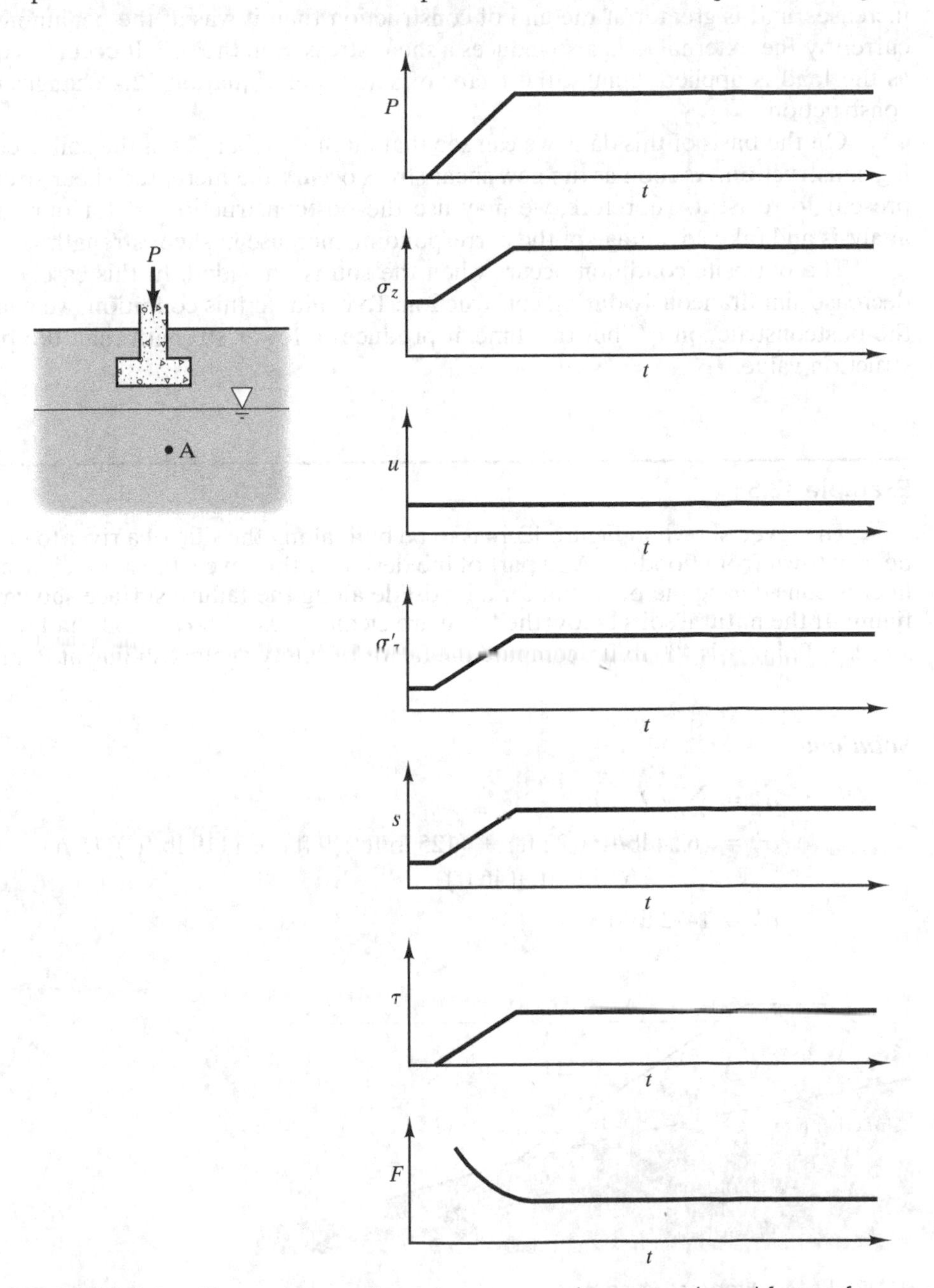

FIGURE 12.13 Changes in normal and shear stresses, shear strength, and factor of safety with time at Point A in a saturated sand below a structural foundation.

As the external load, P, is applied to the footing, perhaps over a construction period of a few weeks, the vertical total stress, σ_z, at point A increases accordingly. The pore water pressure, u, remains virtually constant because of the rapid drainage, so the vertical effective stress, σ_z' increases at the same rate as σ_z. Thus, the shear strength, s, also increases, and is greater at the end of construction than it was at the beginning. Concurrently, the external load also induces a shear stress, τ, in the soil. It occurs as quickly as the load is applied. Finally, the factor of safety per Equation 12.4 changes during construction.

On the basis of this data, we can see that all of the changes in the soil occur during construction. As soon as the new shear stress occurs, the increased shear strength is present to resist it. Therefore, we may use the postconstruction σ' for our strength analysis and take advantage of the corresponding increase in shear strength.

The opposite condition occurs when the soil is unloaded. In this case, σ' and s decrease simultaneously during construction. To evaluate this condition, we again use the postconstruction σ', but this time, it produces a lower strength than the preconstruction value.

Example 12.5

The levee shown in Figure 12.14 is to be built along the side of a river to protect a nearby town from flooding. As a part of the design of this levee, the geotechnical engineer is considering the potential for a landslide along the failure surface shown in the figure. If the natural soils below the levee are clean sands with $\phi' = 34°$ and the shear stress at Point A is 400 lb/ft^2, compute the factor of safety against sliding at Point A.

Solution:

$$\sigma_z' = \sum \gamma H - u$$

$$\sigma_z' = (62.4 \text{ lb/ft}^3)(22 \text{ ft}) + (125 \text{ lb/ft}^3)(9 \text{ ft}) + (119 \text{ lb/ft}^3)(15 \text{ ft}) - (62.4 \text{ lb/ft}^3)(46 \text{ ft})$$

$$\sigma_z' = 1412 \text{ lb/ft}^2$$

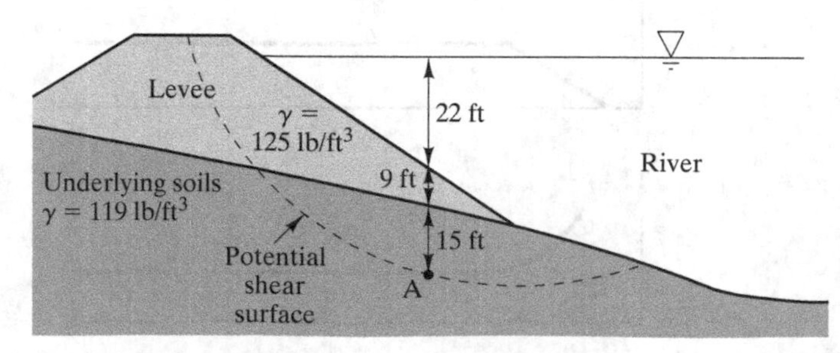

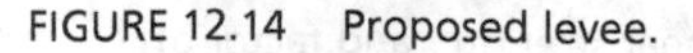
FIGURE 12.14 Proposed levee.

The shear surface at Point A is nearly horizontal, so assume $\sigma' = \sigma'_z$

$$s = c' + \sigma' \tan \phi'$$

$$s = 0 + (1412 \text{ lb/ft}^2) \tan 34°$$

$$s = 953 \text{ lb/ft}^2$$

$$F = \frac{s}{\tau} = \frac{953 \text{ lb/ft}^2}{400 \text{ lb/ft}^2} = \mathbf{2.4}$$

Commentary:

1. This analysis was based on effective stresses with the postconstruction σ'.
2. The computed factor of safety of 2.4 would be satisfactory.
3. Real slope stability analyses need to consider the factor of safety along the entire shear surface, not just at one point. We will discuss methods of performing such analyses in Chapter 13.

Soil Liquefaction

As discussed earlier, the loading during earthquakes is sometimes so rapid that even cohesionless soils cannot drain quickly enough and the undrained condition applies. This is especially problematic in loose, saturated sands because they tend to compress when loaded (Lee, 1965), which normally would force some water out of the voids. However, because the loading occurs so quickly, the water cannot easily drain away and positive excess pore water pressures develop instead. As these pressures build up, both the effective stress and the strength decrease (see Equation 12.3). Sometimes the effective stress drops to zero, which means the soil loses virtually all its shear strength and thus behaves as a dense liquid. We call this phenomenon *soil liquefaction*.

Soil liquefaction can cause extensive damage, so geotechnical engineers working in seismically active areas need to be aware of the soil conditions where this phenomenon is likely to occur.

Quicksand

Section 9.9 discussed seepage forces and the unfortunate consequences that can occur when water flows upward through a soil and the seepage forces oppose the gravitational forces. According to Equation 9.54, upward seepage forces can become large enough that the vertical effective stress drops to zero. This can cause heaving, as illustrated in Examples 9.10 and 9.11, especially if the seepage forces significantly exceed the gravitational forces. Another possibility is *quicksand*, which occurs in sandy soils when upward seepage produces a σ'_z close to zero. c' also is zero, so these soils have virtually no shear strength and behave as a heavy fluid.

Although true quicksand can occur in natural settings, most conditions identified as quicksand are actually just very loose saturated sand. However, quicksand is a very real danger in dewatered excavations and other constructed facilities where upward seepage forces have been artificially created. This strength loss can trigger the failure of shoring systems and other facilities, possibly resulting in property damage and loss of life.

12.6 SHEAR STRENGTH OF SATURATED CLAYS

The evaluation of the shear strength of a clay is more difficult than that of a sand or gravel because:

- Clay particles undergo more significant changes and therefore cause greater volume change during loading.
- The low hydraulic conductivity impedes the flow of water into and out of the voids, so the undrained condition applies in the short term and significant excess pore water pressures often develop in the soil.
- In the long term, water does flow into or out of the soil, allowing the excess pore water pressures generated during loading to dissipate virtually completely, and the drained condition applies.

In addition, saturated clays are generally weaker than sands and gravels, and thus are more often a cause of problems.

Volume Changes and Excess Pore Water Pressures in Clays

When a load such as from a fill or structural foundation is applied to a saturated clay, the three-dimensional state of stress of the soil changes, as discussed in Chapter 9. The total vertical stress, σ_z, and the shear stress, τ, increase, as shown on the plots in Figure 12.15. The plots in Figure 12.15(a) describe various changes in the clay during loading under the drained condition, which applies if the rate of loading is extremely low. These plots are similar to those in Figure 12.13 for loading of a sand, and the discussion previously of the plots in Figure 12.13 applies to these plots. However, if the same load is applied to the clay at a normal rate, the clay will be under the undrained condition during loading in the short term. The plots in Figure 12.15(b) describe this condition. Notice the spike in the u plot, which illustrates the immediate build-up of excess pore water pressures under the undrained condition in the short term and the gradual dissipation of the excess pore water pressures over time, reaching the drained condition in the long term. As a result, the increase in σ'_z is much slower than in sands.

Some construction projects cause a decrease in the mean normal stress and an increase in the deviator stress. This occurs, for example, when we make a sloped excavation. In this case, the postconstruction σ'_z and s are less than the preconstruction values, as shown in Figure 12.16. If the undrained condition prevails, negative excess pore water pressures are present.

Strictly speaking, the amount of excess pore water pressure depends on the change in the three-dimensional state of stress of the soil. As explained earlier in Section 12.3, the excess pore water pressure can be split into two parts, one induced by an all-around hydrostatic mean normal stress that does not produce any shear stresses in the soil and one induced by a deviator stress that produces some shear stresses. The induced shear stresses and the corresponding shear strains also can produce volume

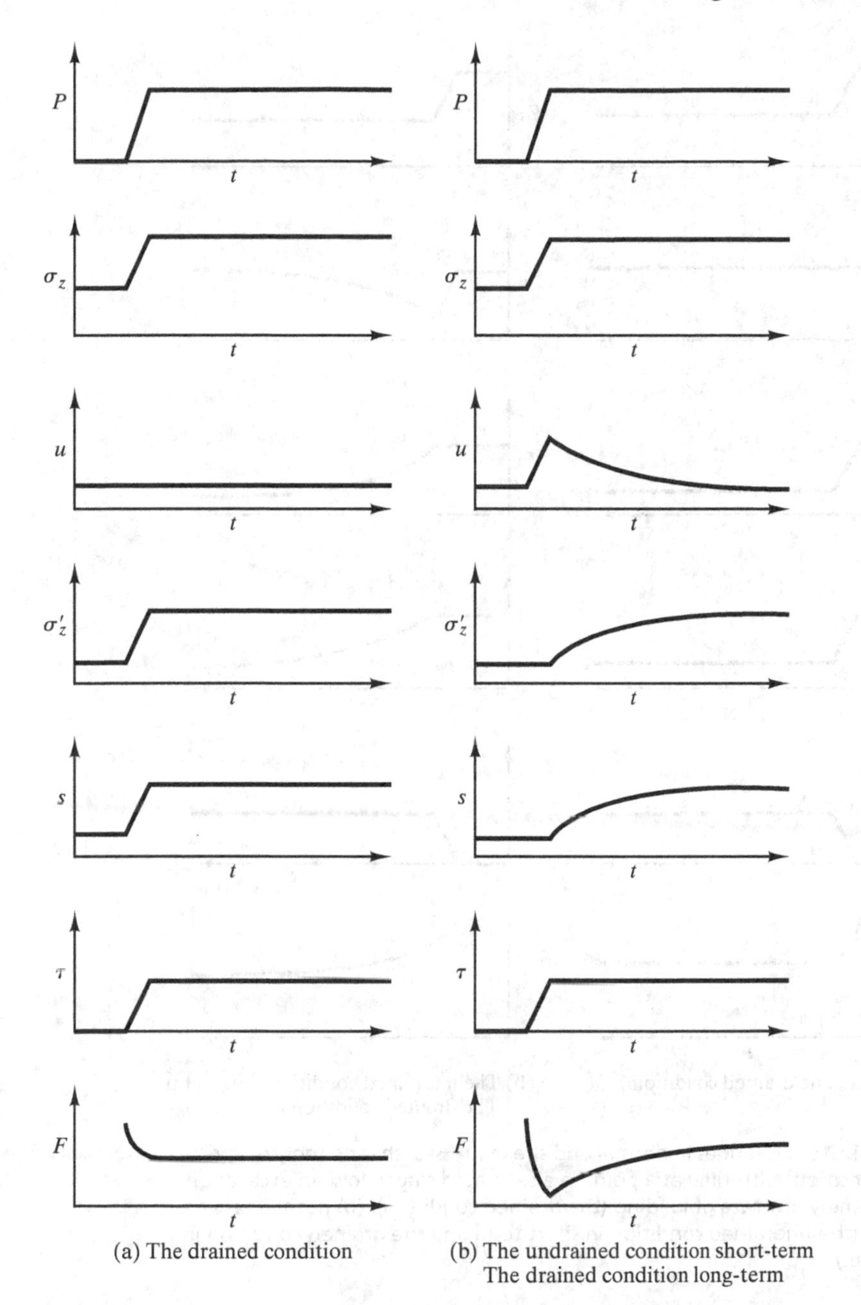

FIGURE 12.15 Changes in normal and shear stresses, shear strength, and factor of safety with time at a point in a saturated clay below a fill or structural foundation: (a) extremely low rate of loading (the drained condition); (b) normal rate of loading (the undrained condition in short term and the drained condition in long term).

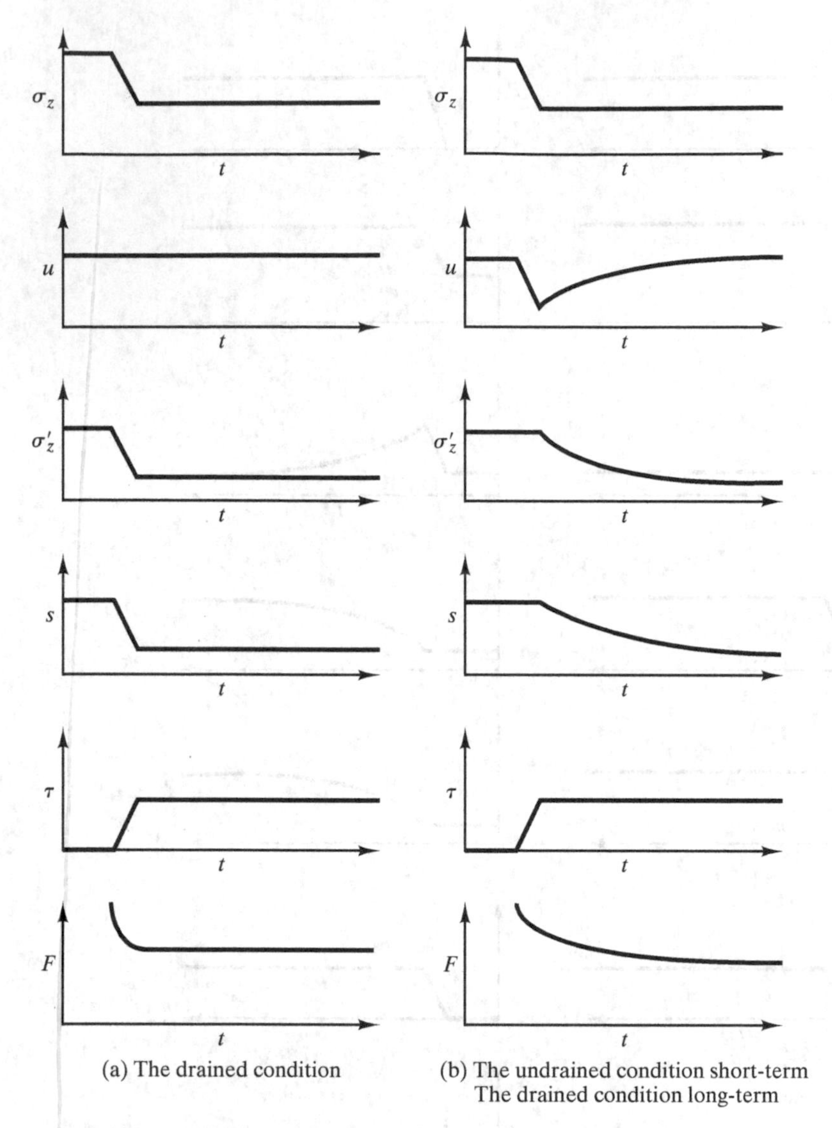

FIGURE 12.16 Changes in normal and shear stresses, shear strength, and factor of safety with time at a point in a saturated clay below an excavation: (a) extremely low rate of loading (the drained condition); (b) normal rate of loading (the undrained condition in short term and the drained condition in long term).

changes in soil. These changes are in addition to those due to changes in the mean normal stress. Normally consolidated clays usually contract as they are sheared, especially if the mean normal stress is high, thus forcing some water out of the voids. Conversely, overconsolidated clays dilate as they are sheared, especially when the mean normal stress is low, and draw additional water into the voids. Once again, if the rate of loading is low relative to the rate of drainage, there will be plenty of time for the water to flow in or out and the drained condition will prevail. Conversely, if the rate of loading is high relative to the rate of drainage, then the undrained condition will prevail in the short term. When the undrained condition prevails, excess pore water pressures develop in the soil.

Usually the induced mean normal stress and the induced deviator stress develop simultaneously due to a change in the overall state of stress of the soil, so the pore water pressure at a given time is as given below:

$$u = u_{\mathrm{h}} + u_{\mathrm{e,normal}} + u_{\mathrm{e,shear}} \tag{12.6}$$

where:

u = pore water pressure
u_{h} = hydrostatic pore water pressure (per Equation 7.7)
$u_{\mathrm{e,normal}}$ = excess pore water pressure due to induced mean normal stress
> 0 if subjected to increases in σ_{m}
< 0 if subjected to decreases in σ_{m}
$u_{\mathrm{e,shear}}$ = excess pore water pressure due to induced deviator stress
> 0 if soil tends to contract under increased σ_{d}
< 0 if soil tends to dilate under increased σ_{d}

Usually, $u_{\mathrm{e,normal}}$ dominates over $u_{\mathrm{e,shear}}$. Thus, most soils that are being loaded, such as by fills or foundations, have a net positive u_{e}, whereas those that are being unloaded, such as by an excavation, usually have a net negative u_{e}. These processes are most pronounced in soft clays because they experience large volume reductions due to consolidation.

When performing shear strength analyses, it is important to properly assess the drainage conditions that will occur in the field because this assessment determines how we will define the shear strength. There are three possibilities:

Case 1—Under the drained condition.
Case 2—Under the undrained condition with positive excess pore water pressures.
Case 3—Under the undrained condition with negative excess pore water pressures.

Each of these possibilities is discussed below.

Case 1: Shear Strength under the Drained Condition Shear strength under the drained condition is the easiest to evaluate because there are no excess pore water pressures. We simply evaluate c' and ϕ' using an appropriate test, compute σ' based on the hydrostatic pore water pressures and postconstruction conditions, and determine s using the effective stress Mohr–Coulomb failure criterion (Equation 12.3). We use this method to evaluate shear strength in two situations: (a) when the loading is very slow with respect to drainage and no excess pore water pressures are generated; and (b) a long time after loading when all excess pore water pressures have dissipated. The second case is the most common. It is almost always appropriate to evaluate the long-term scenario assuming the drained condition applies.

Figure 12.17 shows typical effective stress Mohr–Coulomb failure envelopes for both normally consolidated and overconsolidated clays under the drained condition. As shown by empirical evidence, clays that are normally consolidated and uncemented normally have $c' = 0$. However, overconsolidation produces a true cohesion, as discussed earlier. These two plots join together when the effective stress equals the preconsolidation stress, σ'_c. Figure 12.18 presents typical ϕ' values for clays, which may be used for preliminary analyses or for checking test results.

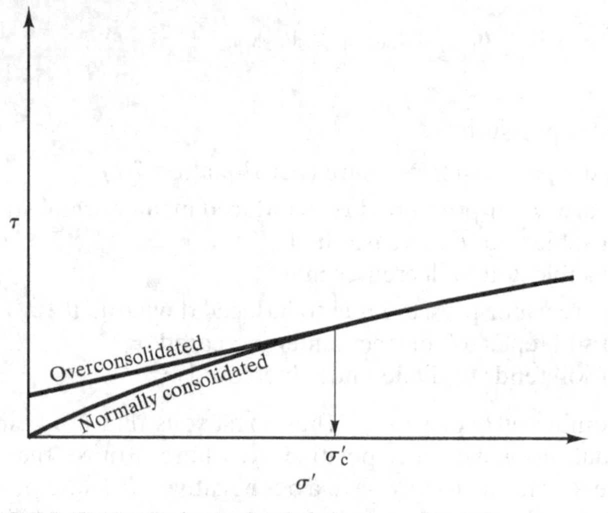

FIGURE 12.17 Mohr-Coulomb failure envelopes for saturated, uncemented clays and silts under the drained condition.

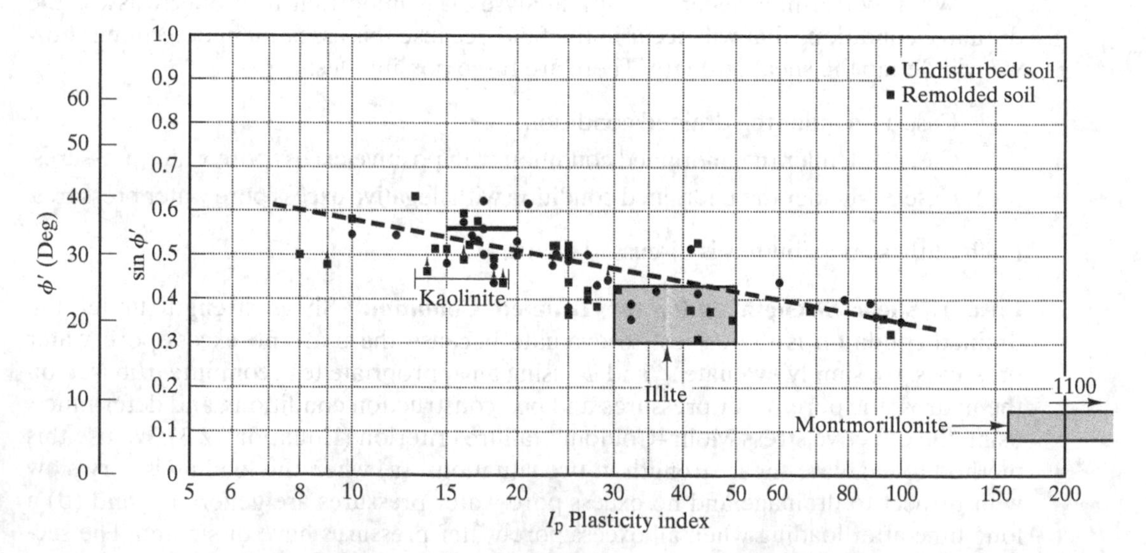

FIGURE 12.18 Typical effective friction angles for normally consolidated, saturated clays and silts. (From Fundamentals of Soil Behavior, 2nd ed. by J.K. Mitchell, Copyright © 1993; used by permission of John Wiley & Sons.)

Example 12.6

The natural earth slope shown in Figure 12.19 has been in its present configuration for a very long time. A slope stability analysis is to be performed on the potential failure surface shown in the figure. What shear strength should be used in this analysis?

Solution:

This slope has been in place for a long time, and no loads are being added or removed, so the drained condition should apply and we can use an effective stress analysis based on c' and ϕ' with the hydrostatic pore water pressures. We could obtain c' and ϕ' by performing appropriate laboratory tests on undisturbed samples (as described later in this chapter), and compute σ' based on the cross section, and then use the effective stress Mohr–Coulomb failure criterion (Equation 12.3) to calculate s.

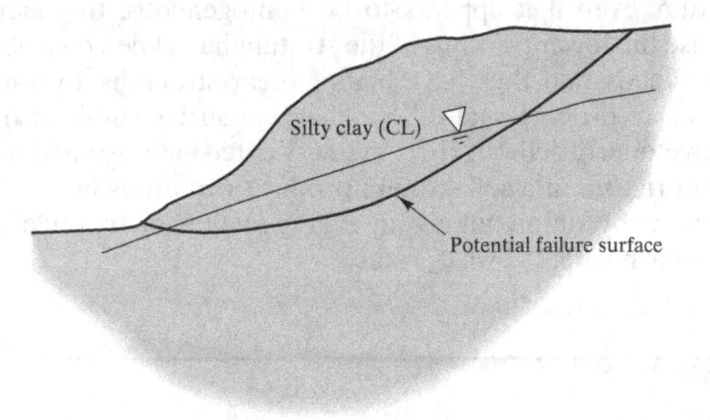

FIGURE 12.19 Cross-section of natural slope for Example 12.6.

Case 2: Shear Strength under the Undrained Condition with Positive Excess Pore Water Pressures The rate of construction for most projects is high relative to the rate of drainage in saturated clays, so the undrained condition prevails during and immediately after construction. If the new construction causes an increase in the mean normal stress, then the excess pore water pressures will be positive. This is the most common of the three cases. Examples include the construction and loading of structural foundations and the construction of embankments.

The plots in Figure 12.15(b) show changes in various soil parameters with time below a structural foundation being built on a saturated clay. The excess pore water pressures build up during construction under the undrained condition in the short term, and then slowly dissipate over time, reaching the drained condition in the long term. Since the excess pore pressures are positive, the lowest factor of safety occurs immediately after construction, when the shear strength is still low but the full shear stresses are already present. We want to be sure that this factor of safety is adequate, so we need to evaluate the end-of-construction or short-term shear strength.

Unfortunately, it is difficult in general to predict the magnitudes of the excess pore water pressures, especially those due to shearing, so we cannot compute the values of σ' at failure or s. Therefore, we typically analyze such problems using a total stress analysis and compute the shear strength using the total stress Mohr–Coulomb failure criterion (Equation 12.5).

If the soil is truly saturated and the loading truly undrained, then $\phi_T = 0$ (even though $\phi' > 0$) because newly applied loads are carried entirely by the pore water and do not change σ'. (See the section on the UU Test in Section 12.9 for more detailed explanation.) This is very convenient, because the second term on the right-hand side of Equation 12.5 drops out and we no longer need to compute σ. We call this a "$\phi = 0$ analysis." The corresponding shear strength is called the *undrained shear strength*, s_u, where $s_u = c_T$. Table 5.4 gives typical values of s_u.

Usually, we assign an appropriate s_u value for each saturated undrained stratum based on laboratory or field test results. In reality, s_u is probably not constant throughout a particular soil stratum, even if it appears to be homogeneous. In general, s_u increases with depth because the lower portions of the stratum have been consolidated to correspondingly greater loads, and thus have higher shear strengths. In normally consolidated clays, s_u is nearly proportional to σ'_z. The near-surface soils also have higher strengths if they have once dried out (become desiccated) and formed a crust. Finally, the natural nonuniformities in a soil stratum produce variations in s_u. We can accommodate these variations by simply taking an average value, or by dividing the stratum into smaller layers, each with its own s_u value.

Example 12.7

Revisit the proposed levee in Example 12.5, except the underlying soils are now saturated clays. The shear strength parameters at Point A are $s_u = 700\ \text{lb/ft}^2$, $c' = 300\ \text{lb/ft}^2$, $\phi' = 24°$. Compute the factor of safety against sliding at Point A.

Solution:

We solved Example 12.5 using an effective stress analysis because the underlying soils were sands, so the drained condition could be expected immediately after construction. However, if the underlying soils are clays, the undrained condition will prevail during and immediately after construction. The excess pore water pressures will be positive because the weight of the levee increases the normal stresses. Therefore, we need to evaluate this problem using the preconstruction shear strength, which is the undrained shear strength, s_u.

The factor of safety immediately after construction is then:

$$\text{Short-term } F = \frac{s}{\tau} = \frac{700\ \text{lb/ft}^2}{400\ \text{lb/ft}^2} = \mathbf{1.7}$$

As the excess pore water pressures dissipate, the factor of safety will gradually increase. Once they have completely dissipated, which may take years, the soil will reach the drained condition and the new factor of safety may be computed using the hydrostatic pore water pressures and an effective stress analysis:

$$\sigma'_z = \sum \gamma H - u$$

$$\sigma'_z = (62.4\ \text{lb/ft}^3)(22\ \text{ft}) + (125\ \text{lb/ft}^3)(9\ \text{ft}) + (119\ \text{lb/ft}^3)(15\ \text{ft}) - (62.4\ \text{lb/ft}^3)(46\ \text{ft})$$

$$\sigma'_z = 1412\ \text{lb/ft}^2$$

$$s = c' + \sigma' \tan \phi'$$

$$s = 300\ \text{lb/ft}^2 + (1412\ \text{lb/ft}^2) \tan 24°$$

$$s = 929\ \text{lb/ft}^2$$

$$\text{Long-term } F = \frac{s}{\tau} = \frac{929\ \text{lb/ft}^2}{400\ \text{lb/ft}^2} = \mathbf{2.3}$$

The use of undrained strengths in Case 2 design problems should be conservative because this method assumes none of the excess pore water pressures in the field will dissipate until well after the load is placed. In reality, some dissipation usually occurs, with corresponding increases in shear strength. Sometimes these strength increases during construction are significant, so the use of undrained strengths can be overly conservative. For example, significant strength increases might occur in soils beneath large fills that are placed slowly. In such cases, engineers sometimes perform special laboratory tests to quantify the excess pore water pressures, and then use them to perform an effective stress analysis. This methodology also includes the installation of piezometers in the field to monitor the actual pore water pressures during construction, and thus is another example of the observational method, as discussed in Chapter 11.

Case 3: Shear Strength under the Undrained Condition with Negative Excess Pore Water Pressures When the construction causes the mean normal stress in a saturated clay to decrease, negative excess pore water pressures develop. The most common example is an excavation. This negative u_e gradually dissipates, but now it causes a corresponding loss in shear strength with time as shown in Figure 12.16(b).

In Case 2, the factor of safety increased with time, which is a desirable characteristic. If a failure does occur, it probably will happen during or soon after construction. However, in Case 3, F decreases with time, which is potentially much more dangerous. The most likely time for a failure is long after construction, and probably after the site has been developed and occupied.

The lowest factor of safety now occurs under the drained condition after the excess pore water pressures have dissipated, so this condition needs to be evaluated using an effective stress analysis with the hydrostatic pore water pressures and the postconstruction effective stresses. This approach addresses the long-term stability, which should be the most critical for this case.

Example 12.8

A cut slope is to be made in a clayey soil to permit construction of a new highway, as shown in Figure 12.20. A slope stability analysis is to be performed along the potential shear surface shown in this figure. The soils are silty clays with $c' = 18$ kPa, $\phi' = 20°$, and $s_u = 100$ kPa. If the shear stress at Point A is 60 kPa, compute the short-term and long-term factors of safety at this point.

Solution:

Short-term stability

These soils are clayey, so negative excess pore water pressures will be present at Point A immediately after construction. Therefore, the short-term stability should be assessed using an undrained total stress analysis.

$$\text{Short-term } F = \frac{s}{\tau} = \frac{100 \text{ kPa}}{60 \text{ kPa}} = \mathbf{1.7}$$

Long-term stability

Eventually, the excess pore water pressures will dissipate, and the soils will attain their drained strength under the new stress conditions. The long-term stability analysis should be based on an effective stress analysis using the drained strengths.

$$\sigma'_z = \sum \gamma H - u$$

$$\sigma'_z = (18.7 \text{ kN/m}^3)(4.9 \text{ m}) + (19.2 \text{ kN/m}^3)(5.2 \text{ m}) - (9.8 \text{ kN/m}^3)(5.2 \text{ m})$$

$$\sigma'_z = 140 \text{ kPa}$$

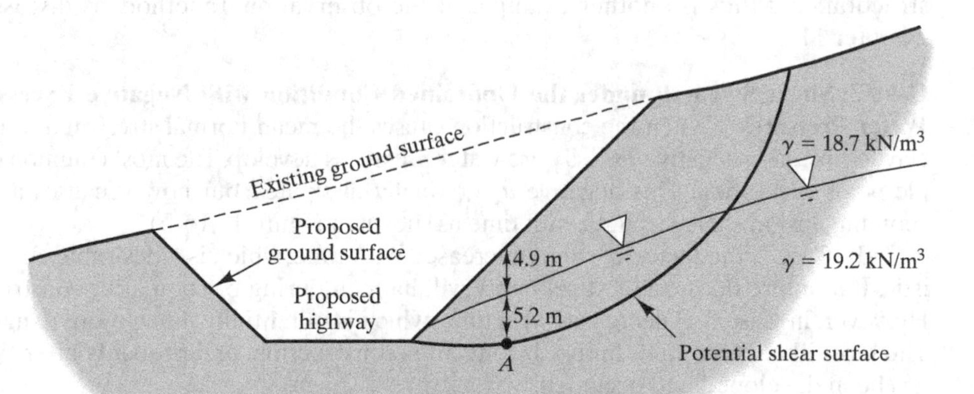

FIGURE 12.20 Proposed highway cut for Example 12.8.

$$s = c' + \sigma' \tan \phi'$$
$$s = 18 \text{ kPa} + (140 \text{ kPa}) \tan 20°$$
$$s = 69 \text{ kPa}$$

$$\text{Long-term } F = \frac{s}{\tau} = \frac{69 \text{ kPa}}{60 \text{ kPa}} = \mathbf{1.1}$$

Comments: Immediately after construction, the factor of safety at Point A is 1.7, which would usually be acceptable. However, once the negative excess pore water pressures have dissipated, F drops to only 1.1, which would generally not be acceptable. If the groundwater table rose, or if the actual c' and ϕ' values are slightly different than we think, failure could occur.

The factors of safety in this example only represent the conditions at Point A, and are intended only to illustrate the effects of negative pore water pressure dissipation. Actual slope stability analyses require assessment of the factor of safety along the entire shear surface, as discussed in Chapter 13.

Sensitivity

Some clays have a curious property called *sensitivity*, which means their strength in a remolded or highly disturbed condition is less than that in an undisturbed condition at the same moisture content. Sometimes this strength loss is very large, as shown in Figure 12.21. These highly sensitive clays, called *quick clays*, are found in certain areas

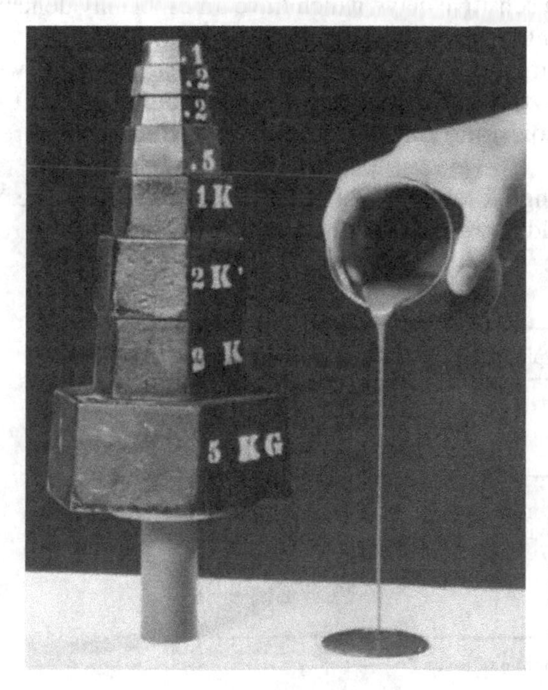

FIGURE 12.21 Undisturbed and remolded samples of Leda clay from Ottawa, Ontario. Both samples are at the same moisture content; the only difference is the remolding. This is an extreme example of a sensitive clay, with an S_t of about 1500. (National Research Council of Canada.)

of Eastern Canada, parts of Scandinavia, and elsewhere. This behavior occurs because these clays have a very delicate structure that is disturbed when they are remolded.

The degree of sensitivity is defined by the parameter S_t:

$$S_t = \frac{s_{undisturbed}}{s_{remolded}} \tag{12.7}$$

Table 12.1 presents two systems of classifying sensitivity based on S_t. Note the difference between the criteria commonly used in the United States, where highly sensitive clays are rare, and that used in Sweden, where they are common.

Shear failures in highly sensitive clays can be very dramatic because the strength loss makes the failure propagate over a wide area. This sometimes produces large flow slides, such as the one shown in Figure 2.17.

Sensitive clays also can recover from these strength losses through a process called *thixotropic hardening*.

Residual Strength

In Figure 12.5, we saw the difference in shear stress–strain curves between ductile and brittle soils and the resulting difference between peak strength and residual strength. Although this distinction is not important with sands or gravels because they have curves that are either ductile or very mildly brittle, it can be very important in clays.

Most normally consolidated clays are slightly ductile, and thus have residual strengths that are slightly less than the peak strength. This strain softening (loss of strength and stiffness with increasing strain) is largely due to particle reorientation and a breakdown of the soil fabric. In sensitive clays, which have an especially delicate fabric, the residual strength can be much less than the peak strength.

Overconsolidated clays nearly always have a brittle stress–strain curve, with the residual strength significantly less than the peak strength. This is due to the factors just described, plus an increase in void ratio during shear and the resulting increase in moisture content.

Brittle soils have two "strengths," so the data points used to develop the Mohr–Coulomb strength envelope could be based on either the peak values or the residual

TABLE 12.1 Typical Classification of Sensitivity[a]

	Sensitivity, S_t	
Classification	United States	Sweden
Low sensitivity	2–4	< 10
Medium sensitivity	4–8	10–30
High sensitivity	8–16	30–50
Quick	> 16	50–100
Extra quick	—	> 100

[a]Adapted from Holtz and Kovacs, 1981.

values, as shown in Figure 12.22. Usually we use the peak strength, so it requires no special notation. However, if the data has been assessed using residual strengths, we use a subscript "r", and express the results as c'_r and ϕ'_r. Residual strength is purely frictional (i.e., there is no cohesive strength), but the envelope is typically nonlinear. Thus, any c'_r value is solely the product of fitting a straight line to a curved envelope, as discussed earlier.

Residual strength data is especially important when evaluating shear surfaces in the field produced by landslides. The landslide movement produces a very smooth surface, known as a *slickenside*, as shown in Figure 12.23. The shear strength along such surfaces has been reduced to the residual value, which is less than the shear strength of the surrounding undisturbed soils. A geotechnical engineer would need to know this shear strength when assessing the stability of existing landslides and when designing stabilization measures.

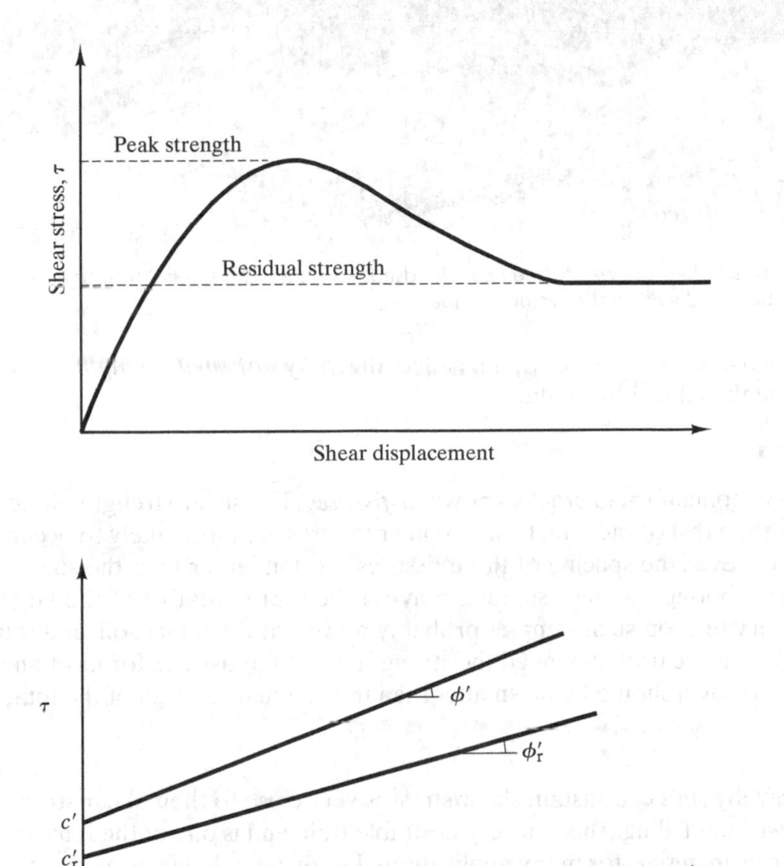

FIGURE 12.22 The peak strength envelope is obtained from the shear strengths at the highest points of the stress-stain curves, while the residual strength envelope is obtained from the strengths at large strain.

FIGURE 12.23 This slickenside was formed by a landslide in a clay. The shear strength along this surface has been reduced to the residual value.

Some analyses use a third value, called the *fully softened strength*, which lies between the peak and residual values.

Fissured Clays

Many stiff clays contain small cracks known as *fissures*. The shear strength along these fissures is less than that of the intact soil, so shear failures are more likely to occur along the fissures. However, the spacing of these fissures is often larger than the soil samples obtained from a boring, so these samples may not be representative of the larger soil mass. Laboratory tests on such samples probably represent the intact soil, and thus can be very misleading. We probably need the strength along the fissures for most analyses. However, this strength should be no smaller than the residual strength of the intact soil.

Creep

Sandy and gravelly soils can sustain shear stresses very close to their shear strength for long periods without failing. This is a very desirable trait, and is one of the reasons these soils are superior materials for many applications. Unfortunately, clayey soils are not so well behaved. When the shear stress in a clay exceeds about 70% of the shear strength, slow shear movements called *creep* begin to occur. These movements can be the source of many problems. For example, the upper soils on sloping ground sometimes exhibit creep, which causes them to slowly move downslope. Some clays exhibit significant creep at stresses as low as 50% of their shear strength. This creep behavior is one of the reasons we typically require higher factors of safety for problems involving clayey soils.

Alexandre Collin

Image courtesy of Thomas Telford, Ltd.

After graduating from the École Polytechnique in Paris, Alexandre Collin (1808–1890) worked on several canal and dam projects in France. Some of these projects experienced landslides during and after construction, and Collin had the opportunity to study them. In the process of doing so, he became the first to develop analytical methods for evaluating landslides, and the first to recognize that soil has both frictional and cohesive strengths. He also developed the first laboratory equipment to measure soil strength.

Collin attempted to publish his findings in 1840 by submitting a paper to a technical journal, but it was rejected due to the "specialty of the subject matter" (Skempton, 1949). He eventually self-published his findings in 1846, but his work was not widely circulated and was soon forgotten. Collin's contributions were finally recognized only after these principles were independently rediscovered in the early twentieth century.

12.7 SHEAR STRENGTH OF SATURATED INTERMEDIATE SOILS

The discussions in Sections 12.5 and 12.6 divided soils into two distinct categories. The sands and gravels of Section 12.5 do not develop excess pore water pressures during static loading, and thus may be evaluated assuming the drained condition using effective stress analyses and hydrostatic pore water pressures. Conversely, the clays of Section 12.6 do develop excess pore water pressures, and thus require more careful analyses, assuming the undrained condition in the short term and the drained condition in the long term. Clays also may have problems with sensitivity and creep. Although many "real-world" soils neatly fit into one of these two categories, others behave in ways that are intermediate between these two extremes. Their field behavior is typically somewhere between being drained and undrained (i.e., they develop some excess pore water pressures, but not as much as would occur in a clay).

Although there are no clear-cut boundaries, these intermediate soils typically include silts, clayey and silty soils with unified soil classifications SC, GC, SC-SM, or GC-GM, as well as some SM and GM soils. Proper shear strength evaluations for engineering analyses involving these intermediate soils require much more engineering judgment, which is guided by a thorough understanding of soil strength principles. When in doubt, it is usually conservative to evaluate these soils using the techniques described for clays in Section 12.6.

Silts are soils that have a wide range of behavior. Nonplastic silts can behave like fine sands, whereas plastic silts can behave like clays. In evaluating the shear strength of a silt, we may first categorize the silt as "sand-like" or "clay-like" and then use the methods for sands or clays, respectively. If it is difficult to categorize, both methods for sands and clays should be used.

For a "dirty" sand or gravel that has a significant amount of fines (fines content $> 5\%-12\%$), the fines will greatly influence its shearing behavior and will probably dominate its shearing behavior if the fines content is larger than about 25–35%.

12.8 SHEAR STRENGTH OF UNSATURATED SOILS

Thus far, we have only considered soils that are saturated ($S = 100\%$). The strength of unsaturated soils ($S < 100\%$) is generally greater, but more difficult to evaluate. Nevertheless, many engineering projects encounter these soils, so geotechnical engineers need to have methods of evaluating them. This has been a topic of ongoing research (Fredlund and Rahardjo, 1993), and standards of practice are not as well established as those for saturated soils.

Some of the additional strength in unsaturated soils is due to negative pore water pressures, as shown in Figure 9.25. These negative pore water pressures increase the effective stress, and thus increase the shear strength. However, this additional strength is very tenuous and is easily lost if the soil becomes saturated.

Geotechnical engineers usually base designs on the assumption that unsaturated soils could become saturated in the future. This saturation could come from a rising groundwater table, irrigation, poor surface drainage, broken pipelines, or other causes. Therefore, we usually saturate (or at least "soak") soil specimens in the laboratory before performing strength tests. This is intended to remove the apparent cohesion and thus simulate the "worst case" field conditions. We then determine the highest likely elevation for the groundwater table, which may be significantly higher than its present location, and compute positive pore water pressures accordingly. Finally, we assume $u = 0$ in soils above the groundwater table.

12.9 SHEAR STRENGTH EVALUATION

On the basis of the shear strength concepts discussed in the previous sections, we summarize the considerations in the evaluation of the shear strength of a soil in a typical static stability analysis in geotechnical practice. For such a static stability analysis, Table 12.2 gives the typical scenarios to be analyzed, and for different types of soils in the subsurface, the appropriate drainage conditions, analysis types, strength parameters, and the laboratory testing methods. It can be seen from Table 12.2 that under static loading, sands and gravels can usually be assumed to be under the drained condition at all times, in which case effective stress analyses and effective stress parameters will be appropriate. For clays, the undrained condition and total stress analyses will be appropriate for short-term analyses, and the drained condition and effective stress analyses for long-term analyses. It should be noted that even in the same analysis, different soils involved may have different drainage conditions and therefore different types of strength parameters that are applicable. An important task of the geotechnical

TABLE 12.2 Shear Strength Evaluation for Static Stability Analysis

Scenario	Soil Type	Drainage Condition	Analysis Type	Shear Strength Parameters	Lab Test Method
Short-term (End-of-construction)	Sands and gravels	Drained	Effective stress	c' and ϕ'	• Drained direct shear • CU with pore pressure measurements • CD
	Clays	Undrained	Total stress	$c_T = s_u$ $\phi_T = 0$	• Undrained direct shear • UU • CU
Long-term	Sands and gravels	Drained	Effective stress	c' and ϕ'	• Drained direct shear • CU with pore pressure measurements • CD
	Clays	Drained	Effective stress	c' and ϕ'	• Drained direct shear • CU with pore pressure measurements • CD

engineer is to evaluate the soils to come up with the appropriate strength parameters to be used in the analysis and to estimate the values of these parameters.

Geotechnical engineers can estimate the values of the shear strength parameters using empirical data or correlations. Figures 12.12 and 12.18 and Table 5.4 mentioned earlier in this chapter are examples of empirical data or correlations for ϕ' of cohesionless soils, ϕ' of normally consolidated clays and silts, and s_u of fine-grained soils, respectively.

In addition, geotechnical engineers can measure shear strength using both ex situ and in situ methods. As discussed in Chapter 3, ex situ methods involve obtaining undisturbed samples from an exploratory boring, bringing them to a soil mechanics laboratory, and testing them there. In situ methods use special equipment brought to the field and test the soils in place.

Ex Situ Methods

Most shear strength measurements are performed in laboratories using ex situ methods. Common laboratory soil strength tests include the *direct shear test*, *ring shear test*, *unconfined compression* (UC) *test*, and the *triaxial compression test.*

Direct Shear Test The earliest measurements of soil strength were probably those performed by the French engineer Alexandre Collin in 1846 (Head, 1982). His test equipment was similar to the modern direct shear machine. The direct shear test as we now know it (ASTM D3080) was perfected by several individuals during the first half of the twentieth century.

The test apparatus, shown in Figure 12.24, typically accepts a cylindrical soil specimen 60–75 mm (2.5–3.0 in.) in diameter and subjects it to a vertical load, P. The vertical total stress, σ_z, is thus equal to P/A, where A is the horizontal cross-sectional area of the specimen. The specimen is contained inside a water bath to keep it saturated, but the hydrostatic pore water pressure is very small, so we can assume σ'_z also equals P/A.

(a)

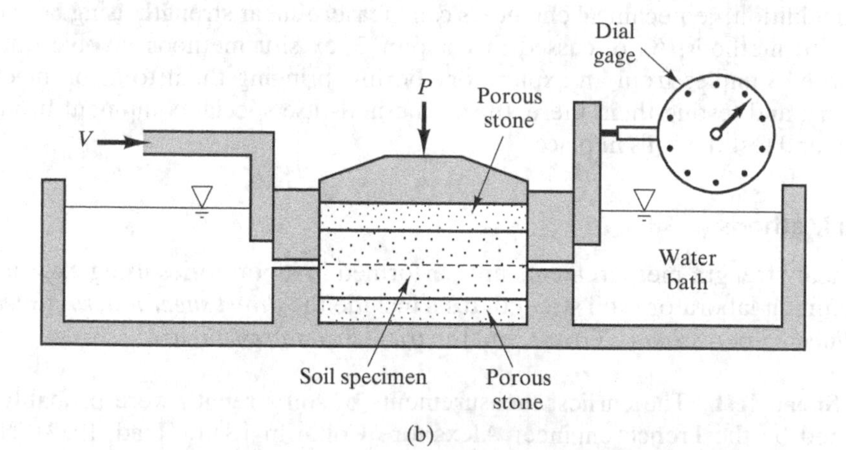

(b)

FIGURE 12.24 (a) A direct shear machine. The specimen is inside the shear box, directly below the upper dial gage; (b) cross-section through shear box showing the specimen and shearing action.

It is important to select P values such that σ'_z is close to the field stresses. The specimen is allowed to consolidate under this load.

Once the specimen has fully consolidated, a shear force, V, is gradually applied. This shear force induces a shear stress $\tau = V/A$. If drained or effective stress parameters are required, V is applied slowly enough to maintain the drained condition. In sands, the required rate of loading, usually corresponding to a constant rate of shear displacement, is such that failure occurs in a few minutes. However, clays must be loaded much more slowly, possibly requiring a time to failure of several hours.

Typical data obtained from a direct shear test are presented as a plot of the shear stress against shear displacement and a plot of the vertical displacement against shear displacement, as shown in Figure 12.25. This procedure is then repeated two more times on "identical" new specimens using different magnitudes of P.

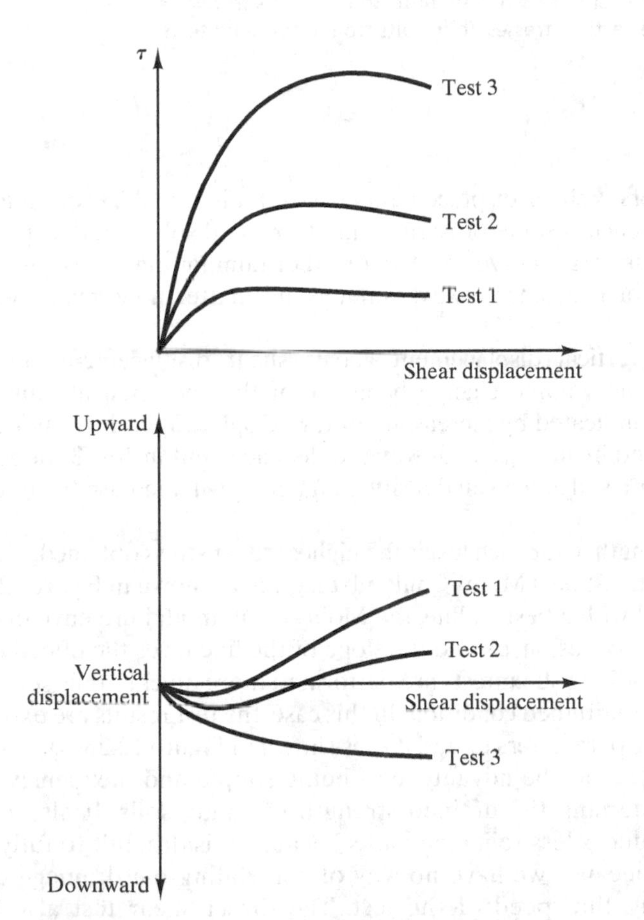

FIGURE 12.25 Plots of typical test data from three drained direct shear tests on identical sand specimens, with σ'_z in the test increasing in the following order: Test 1, Test 2, and Test 3.

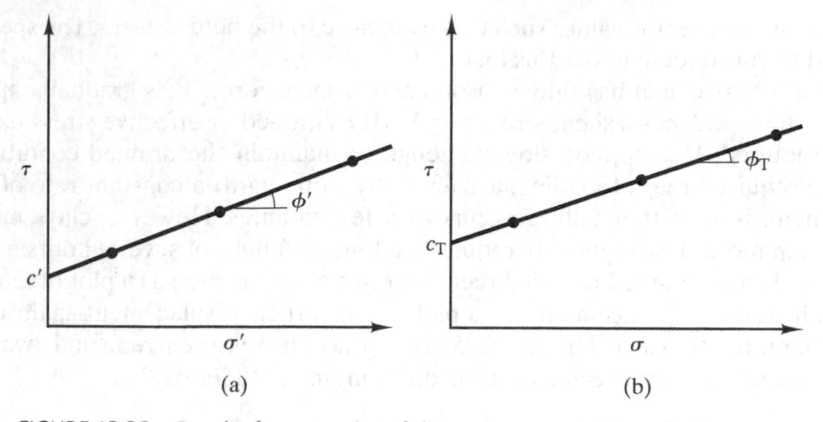

FIGURE 12.26 Results from a series of direct shear tests: (a) results from drained tests plotted using effective stresses; (b) results from undrained tests plotted using total stresses.

The shear stress versus shear displacement curves in Figure 12.25 continue until the shear displacement reaches some prescribed limit or the displacement capacity of the machine. Unlike stress–strain curves of steel or other familiar materials, there is no rupture point. Some shear resistance always remains, no matter how much displacement occurs.

Furthermore, the vertical displacement versus shear displacement curves in Figure 12.25 show that the volume change behavior of the specimen at failure can change from dilatant (as indicated by increasing upward displacement in Tests 1 and 2) to contractive (as indicated by increasing downward displacement in Test 3) depending on σ'_z used in the test. For a soil at a given density, a higher σ'_z will increase the tendency of the soil to contract.

The peak shear strength, s, for each test is the highest shear stress obtained. It can be plotted against σ'_z for each test on a Mohr–Coulomb diagram as shown in Figure 12.26(a) and the data points fitted with a best-fit line, the Mohr–Coulomb failure envelope. The τ-intercept is the effective cohesion, c', and the slope of the line gives the effective friction angle, ϕ'. Sometimes direct shear tests are performed more quickly, thus simulating a partially drained or the undrained condition. In this case, the test results are expressed in terms of the total stress parameters c_T and ϕ_T, as shown in Figure 12.26(b).

The direct shear test has the advantage of being simple and inexpensive. It is especially useful for obtaining the drained strength of sandy soils. It also can be used with clays, but produces less reliable results because it is difficult to fully saturate the specimen and because we have no way of controlling the drainage conditions other than varying the speed of the test. The direct shear test also has a disadvantage of forcing the shearing to occur along a specific plane instead of allowing the soil to fail through the weakest zone. Another disadvantage is that it produces non-uniform strains in the specimen, which can produce erroneous results in strain-softening soils.

Example 12.9

A series of three direct shear tests has been conducted on a certain saturated soil. Each test was performed on a 2.375-in. diameter, 1.00 in. tall specimen. The test has been performed slowly enough to produce the drained condition. The results of these tests are as follows:

Test Number	Normal Load (lb)	Shear Load at Failure (lb)
1	75	51
2	150	110
3	225	141

Determine c' and ϕ'.

Solution:

$$A = \frac{\pi D^2}{4} = \frac{\pi (2.375 \text{ in.})^2}{4}\left(\frac{1 \text{ ft}^2}{144 \text{ in.}^2}\right) = 0.0308 \text{ ft}^2$$

On the basis of this area and the measured forces:

Test Number	σ' (lb/ft²)	s (lb/ft²)
1	2,438	1,665
2	4,876	3,576
3	7,314	4,545

This data is plotted in Figure 12.27. It does not produce a straight line envelope. This may be due to experimental errors, slight differences in the three specimens, true nonlinearity, or other factors. We have drawn a best-fit line through these three points to obtain c' = **400 lb/ft²** and ϕ' = **31°**.

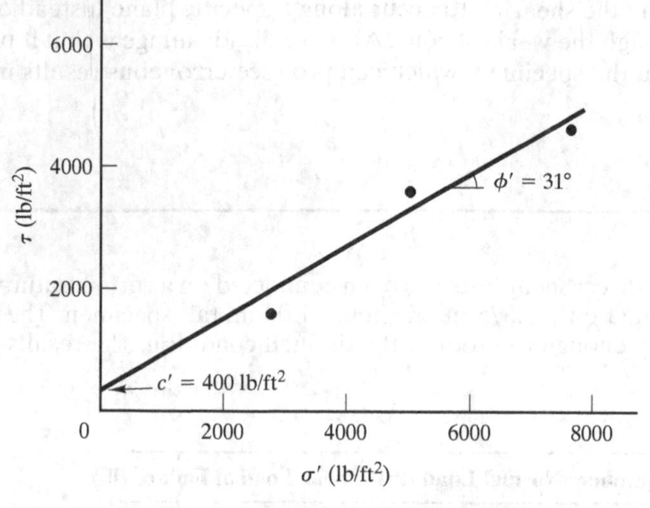

FIGURE 12.27 Direct shear test results for Example 12.9.

Ring Shear Test The direct shear test does not provide enough shear displacement to reach the residual strength in clays. In trying to measure the residual strength, sometimes engineers attempt to overcome this mechanical deficiency by shearing the specimen back and forth, or by other techniques. However, another device is available to measure the residual strength directly: the ring shear test apparatus, as shown in Figure 12.28. This test uses an annular-shaped soil specimen that is subjected to a known normal load and rotated as shown. The shear stress in the specimen may be computed from the torque required to rotate it. In theory, this test has an unlimited strain capacity, but in practice, the residual strength is normally achieved after shear displacements of less than 1 m.

Unconfined Compression Test The UC test uses a cylindrical soil specimen with no lateral confinement, as shown in Figures 12.29 and 12.30. An axial compressive load is gradually applied to the specimen until it fails (i.e., the load reaches its peak value). The load is applied fairly rapidly (typically about 1 min to failure), thus producing the undrained condition.

The major principal stress acts vertically and is equal to the compressive load P divided by the cross-sectional area A. As the specimen compresses, the center part bulges, so the area A, which is measured on a horizontal plane, increases. Under the undrained condition, the volume of the specimen remains constant, and the cross-sectional area at failure can be shown to be given by this equation:

$$A_f = \frac{A_0}{1 - \varepsilon_f} \qquad (12.8)$$

where:

A_f = cross-sectional area at failure

A_0 = initial cross-sectional area = $\pi d^2/4$

d = initial specimen diameter

ε_f = axial strain at failure

There is no lateral confinement, so $\sigma_3 = 0$ and the total stress Mohr circle intersects the origin. Thus, the Mohr circle at failure is as shown in Figure 12.31.

FIGURE 12.28 Ring shear machine. The soil specimen is placed inside the annular space shown in the photograph on the top. Then the load cap is installed and the normal load is applied. Once the soil has consolidated, a torque is applied from below, and this torque is resisted by the rods shown in the photograph on the bottom.

FIGURE 12.29 Conducting an unconfined compression test.

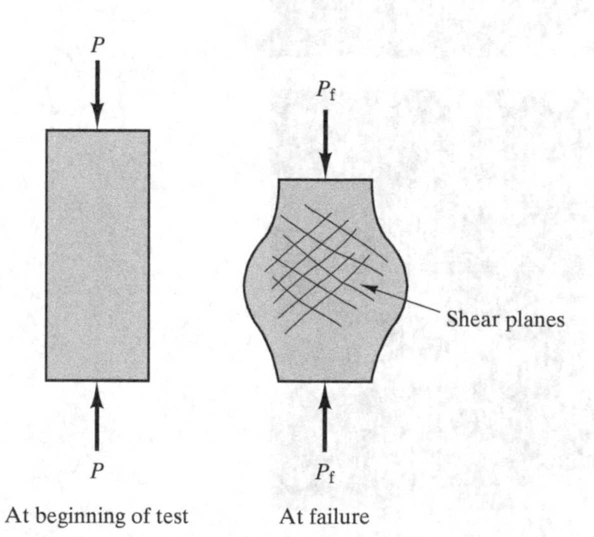

FIGURE 12.30 Loading and failure mode in an unconfined compression test.

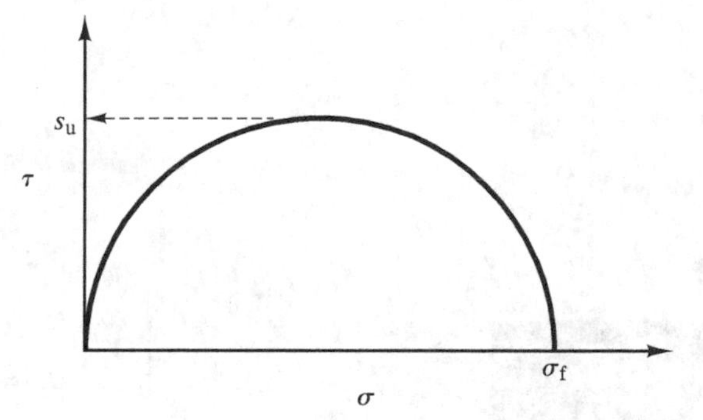

FIGURE 12.31 Mohr circle at failure in an unconfined compression test.

This test appears to measure compressive strength, and the results are often expressed that way, as follows:

$$q_u = \frac{P_f}{A_f} \tag{12.9}$$

where:

q_u = unconfined compressive strength
P_f = normal load at failure
A_f = cross-sectional area at failure

However, the soil really fails in shear along diagonal planes as shown in Figure 12.30. If the soil is soft, it fails along multiple diagonal planes and bulges in the middle as shown. If the soil is stiff, it is more likely to fail on a single distinct diagonal plane. As explained later under the section on the UU Test, the undrained shear strength, s_u, is one-half of q_u:

$$s_u = \frac{P_f}{2A_f} \tag{12.10}$$

The UC test has the advantage of being simple and inexpensive, so we can use it to obtain a large number of s_u values at a low cost. However, σ_3 in the field is actually greater than 0, so the test tends to underestimate s_u.

Triaxial Compression Test The triaxial compression test is similar to the UC test except the specimen is surrounded by a waterproof membrane and is installed in a pressure chamber known as a *cell*. Figure 12.32 shows the setup of a typical triaxial testing system, and Figure 12.33 shows a schematic of the triaxial cell. The cell is filled with water that is pressurized to provide an all-around confinement to the specimen under a specified cell pressure σ_3. A drain is attached to the specimen, and the drainage condition can be controlled by either closing or opening the drain valve. Pore pressures within the specimen can be measured by a pressure gage attached to the drain line.

FIGURE 12.32 Photograph of a typical triaxial test system.
(Courtesy of Geocomp Corp.)

To fail the specimen, a vertical load, P, is slowly applied through a rod extending through the top of the cell. This load divided by the cross-sectional area (which for undrained cases can be calculated using Equation 12.8) is the deviator stress, $\sigma_d = \sigma_1 - \sigma_3$:

$$\sigma_d = \frac{P}{A} \tag{12.11}$$

The vertical stress in the specimen, which also is the major principal stress, is the sum of the cell pressure and the deviator stress:

$$\sigma_1 = \sigma_3 + \sigma_d \tag{12.12}$$

This vertical major principal stress and the horizontal cell pressure σ_3, which also is the minor principal stress in the specimen, define a state of stress of the specimen in terms of total stresses. If the pore water pressure inside the soil specimen is measured during the test, the effective principal stresses, σ_1' and σ_3', can also be computed using Equation 9.45, defining a state of stress of the specimen in terms of effective stresses.

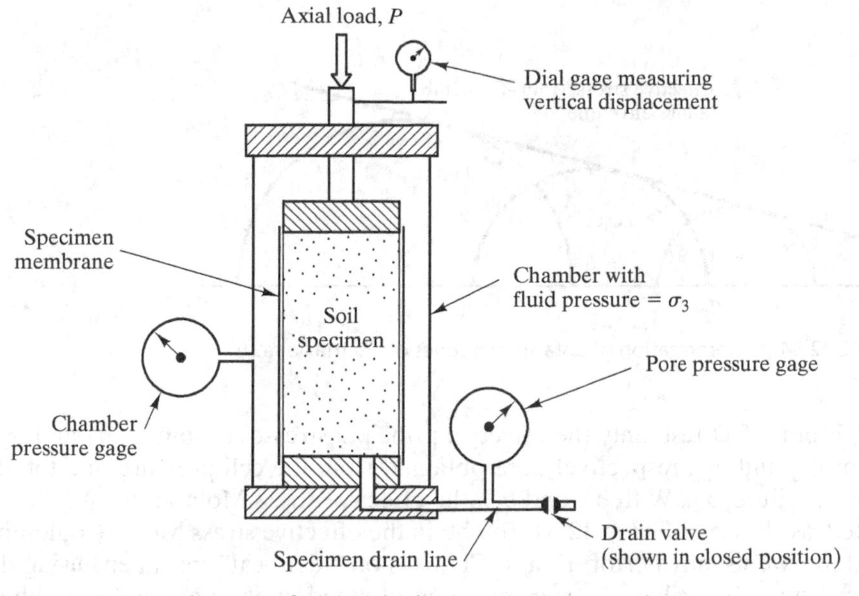

FIGURE 12.33 Schematic of a triaxial cell.

The triaxial test also has provisions to saturate specimens and to verify that the saturation process was successful. This along with the testing flexibility and other advantages makes it the standard by which other tests are compared.

There are three common triaxial test procedures: (1) the *consolidated drained* or CD test (also known as the *S* or *slow* test), (2) the *unconsolidated undrained* or UU test (also known as the *Q* or *quick* test), and (3) the *consolidated undrained* or CU test (also known as the *R* or *rapid* test). Each test consists of two main stages: first, the consolidation stage followed by the shearing stage. The first descriptor in the name of the test refers to the consolidation stage and describes whether the specimen is allowed to consolidate under the cell pressure, with the descriptor "consolidated" meaning the specimen is allowed to consolidate and "unconsolidated" meaning the opposite. The second descriptor in the name refers to the shearing stage and to the condition under which the specimen is sheared to failure by an increase in the axial load. We describe these tests in more detail in the following along with how the data obtained can be interpreted to obtain the values of appropriate shear strength parameters.

The CD Test In the consolidated drained or CD test, the specimen is consolidated during the consolidation stage and then sheared under the drained condition during the shearing stage. To allow consolidation, the drain valve is open before the cell pressure is increased to the desired value. After consolidation of the specimen is complete, that is, when all the excess pore water pressures have dissipated, the drain valve is kept open and then the specimen loaded to failure very slowly under the drained condition with no excess pore water pressures.

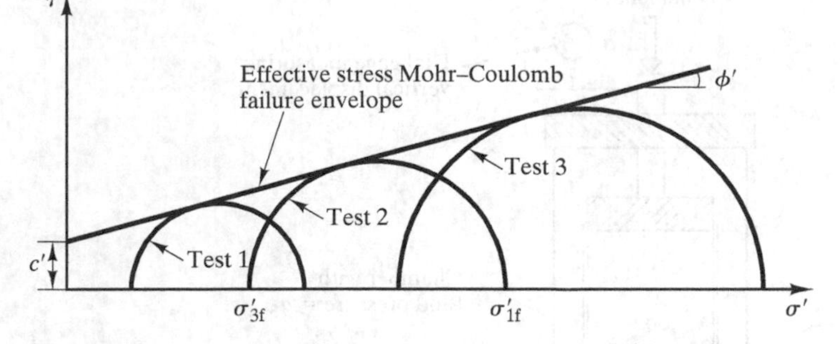

FIGURE 12.34 Interpretation of data from a series of CD triaxial tests.

From a CD test, only the effective principal stresses at failure, σ'_1 and σ'_3 at failure, or σ'_{1f} and σ'_{3f}, respectively, are obtained from the cell pressure and the deviator stress at failure, σ_{df}. With σ'_{1f} and σ'_{3f}, the effective stress Mohr circle at failure can be plotted, as shown in Figure 12.34. To obtain the effective stress Mohr–Coulomb failure envelope, we usually perform three CD tests on "identical" specimens using different cell pressures (σ_3 values). A specimen consolidated under a higher σ_3 in the CD test should be stronger than one consolidated under a lower σ_3 and should give a bigger Mohr circle at failure, as shown in Figure 12.34. Using the effective stress Mohr circles at failure, we can draw the effective stress Mohr–Coulomb failure envelope, as shown in Figure 12.34, from which we can determine the effective cohesion c' and the effective friction angle ϕ'. In practice, if there is not a failure envelope that is tangent to all three circles, we take the best-fit line to be the failure envelope.

The UU Test The simplest triaxial test is the unconsolidated undrained or UU test. In this test, the specimen is unconsolidated during the consolidation stage and then sheared under the undrained condition during the shearing stage. To prevent consolidation, the drain valve is closed before the cell pressure is increased to the desired value. The specimen is then loaded to failure under the undrained condition, achieved by having the drain valve closed throughout the test.

Because the pore water pressure in the specimen is usually not measured in this test, only the total principal stresses at failure, σ_1 and σ_3 at failure, or σ_{1f} and σ_{3f}, respectively, are obtained. With σ_{1f} and σ_{3f}, the total stress Mohr circle at failure can be plotted, as shown in Figure 12.35. If additional UU tests are performed on "identical" specimens using different cell pressures (σ_3 values), the Mohr circles at failure from these additional tests should theoretically have the same size as that from the first test, as shown in Figure 12.35. This is because, with no drainage allowed during the consolidation stage, all the specimens, regardless of what σ_3 they are under, should be under the same effective stress state and have the same density at the end of the consolidation stage. In other words, under the undrained condition, the all-around cell pressure is carried entirely by the pore water, with no changes in the effective stresses in the specimen. Under the same effective stress state, the identical specimens in the different UU tests should have the same strength. Because the strength of a specimen

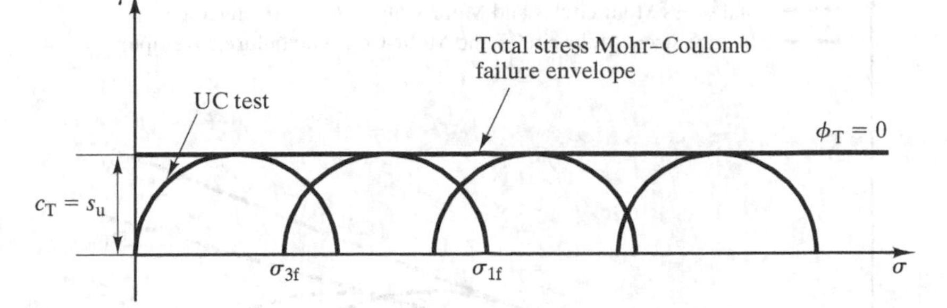

FIGURE 12.35 Interpretation of data from a series of UU triaxial tests.

in a triaxial test is directly proportional to the deviator stress at failure, σ_{df}, or the diameter of the Mohr circle at failure, the total stress Mohr circles at failure from the different UU tests using different σ_3 values should be of the same size. With the total stress Mohr circles at failure being the same size, the total stress Mohr–Coulomb failure envelope should be a horizontal line, as shown in Figure 12.35, giving a total friction angle ϕ_T of zero and a total cohesion c_T called the *undrained shear strength*, s_u. In practice, if more than one UU test is performed, all the total stress Mohr circles at failure should be used to obtain the best-fit horizontal line, and hence the s_u value. If only one UU test is performed, the s_u is equal to the radius of the lone Mohr circle at failure:

$$s_u = \left(\frac{\sigma_{1f} - \sigma_{3f}}{2}\right) = \left(\frac{\sigma_{df}}{2}\right) \tag{12.13}$$

The UC test can be considered a special case of the UU test, a UU test using a σ_3 of 0. The measured values of s_u from the UU triaxial test are more reliable than those from the UC test because the presence of $\sigma_3 > 0$ more accurately simulates the field conditions and because the membrane assures that the undrained condition is satisfied.

The CU Test In the consolidated undrained or CU test, the specimen is consolidated during the consolidation stage and then sheared under the undrained condition during the shearing stage. To allow consolidation, the drain valve is open before the cell pressure is increased to the desired value. After consolidation of the specimen is complete, that is, when all the excess pore water pressures have dissipated, the drain valve is closed and then the specimen loaded to failure under the undrained condition.

From a CU test, the total principal stresses at failure, σ_{1f} and σ_{3f}, are obtained. If the pore water pressure in the specimen is measured, the effective principal stresses at failure, σ'_{1f} and σ'_{3f}, can also be calculated from the pore water pressure at failure, u_f. With σ_{1f} and σ_{3f}, the total stress Mohr circle at failure can be plotted, as shown in Figure 12.36. With σ'_{1f} and σ'_{3f}, the effective stress Mohr circle at failure can also be plotted, also shown in Figure 12.36. Note that for a given test, the effective stress Mohr circle is shifted horizontally from the total stress Mohr circle by the amount of u_f, with the effective stress circle to the left of the total stress circle if u_f is positive or to the

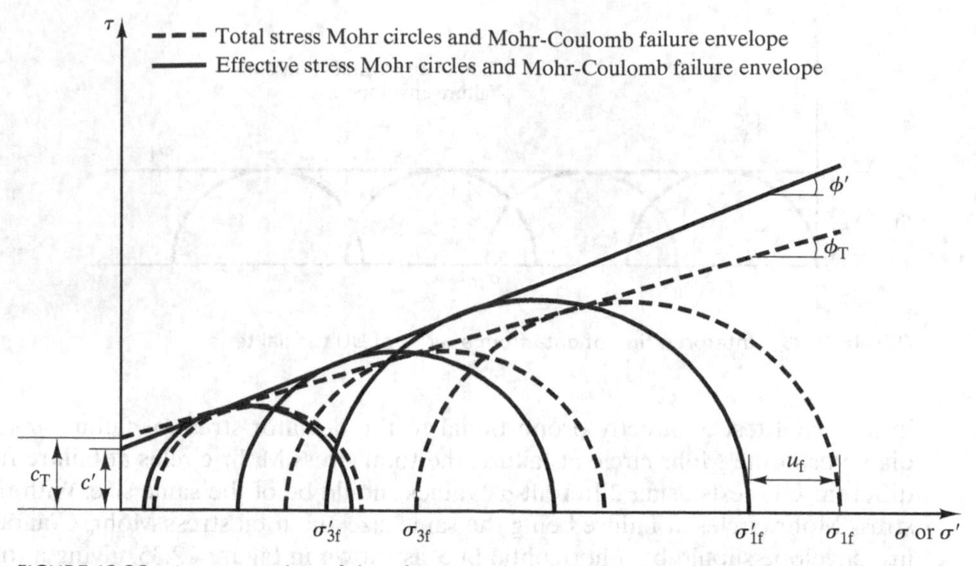

FIGURE 12.36 Interpretation of data from a series of CU triaxial tests with pore pressure measurements.

right if u_f is negative. To obtain the Mohr–Coulomb failure envelopes, we usually perform three CU tests on "identical" specimens using different cell pressures (σ_3 values). Because a specimen is allowed to consolidate in the CU test, a specimen consolidated under a higher σ_3 should be stronger than one consolidated under a lower σ_3 and should give a bigger Mohr circle at failure, as shown in Figure 12.36. Using the total stress Mohr circles at failure, we can draw the total stress Mohr–Coulomb failure envelope, as shown in Figure 12.36, from which we can determine the total cohesion c_T and the total friction angle ϕ_T. Using the effective stress Mohr circles at failure, we can draw the effective stress Mohr–Coulomb failure envelope, as shown in Figure 12.36, from which we can determine the effective cohesion c' and the effective friction angle ϕ'. In practice, if there is not a failure envelope that is tangent to all three circles, we take the best-fit line to be the failure envelope.

Because σ'_{1f} and σ'_{3f} give a Mohr circle that just touches the effective stress Mohr–Coulomb failure envelope, it can be shown that σ'_{1f} and σ'_{3f} are related by the following equation:

$$\sigma'_{1f} = \sigma'_{3f} \tan^2\left(45° + \frac{\phi'}{2}\right) + 2c' \tan\left(45° + \frac{\phi'}{2}\right) \tag{12.14}$$

Similarly, it can be shown that σ_{1f} and σ_{3f} are related by the following equation:

$$\sigma_{1f} = \sigma_{3f} \tan^2\left(45° + \frac{\phi_T}{2}\right) + 2c_T \tan\left(45° + \frac{\phi_T}{2}\right) \tag{12.15}$$

Selecting Appropriate Laboratory Tests for Strength Parameters As summarized in Table 12.2, effective stress parameters c' and ϕ' are required for effective stress analyses. We can obtain the values of c' and ϕ' from drained direct shear tests, CU tests with pore pressure measurements, or CD tests.

For total stress analyses, we may use the total stress parameters $c_T = s_u$ and $\phi_T = 0$. UU tests on specimens from identical samples taken from the field give directly the in situ s_u value of the sampled soil. It is important to note, however, that this s_u value is for the sampled soil only and does not apply to other soils in situ. CU tests, however, give data that can be interpreted to give the s_u as a function of the effective consolidation stress of the soil, and therefore, the s_u values of the same soil consolidated under different effective stresses at different depths. This interpretation of the CU test data is somewhat involved and will not be covered in this book.

Example 12.10

A series of three CU triaxial compression tests have been performed on a set of "identical" clay specimens. Each specimen had an initial diameter of 50 mm and an initial height of 120 mm. The conditions at failure are presented in the following table:

	Conditions at Failure			
Test No.	**P_f (N)**	**ε_f (%)**	**σ_{3f} (kPa)**	**u_f (kPa)**
1	89	5.0	75	42
2	180	6.1	150	69
3	220	5.8	225	109

Reduce the data and determine the effective stress parameters c' and ϕ'.

Solution:

Using Equations 9.46, 12.8, 12.11, and 12.12:

Test No.	A_f (mm^2)	σ_{df} (kPa)	σ'_{3f} (kPa)	σ'_{1f} (kPa)
1	2,060	43	33	76
2	2,090	86	81	167
3	2,080	106	116	222

The effective stress Mohr circles are plotted in Figure 12.37.

The Mohr–Coulomb failure envelope is the best-fit line that is tangent to these Mohr circles, as shown. Thus, the effective stress strength parameters are as follows:

$$c' = \mathbf{8\ kPa}$$
$$\phi' = \mathbf{16^\circ}$$

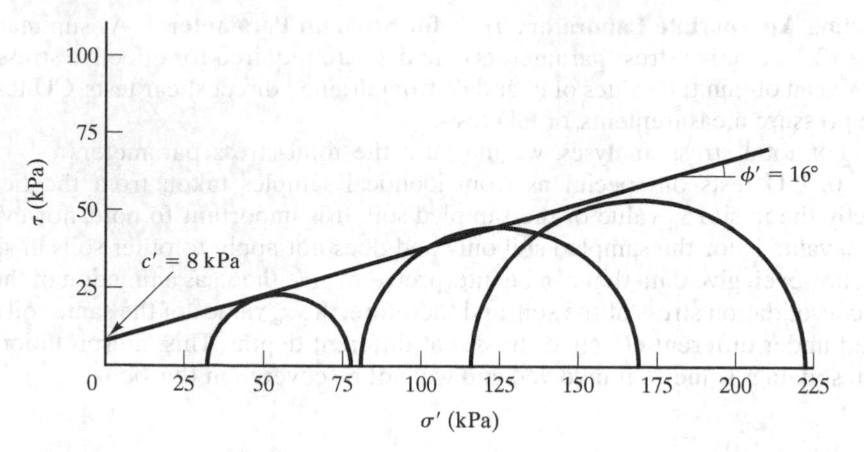

FIGURE 12.37 Effective stress Mohr circles at failure for Example 12.10.

In Situ Methods

Some in situ test methods measure shear strength directly, whereas others simply develop some index, such as the Standard Penetration Test (SPT) *N*-value, which may be combined with empirical correlations to estimate the shear strength.

Vane Shear Test The *vane shear test* (VST) consists of a four-bladed vane that is inserted into the ground as shown in Figure 12.38. A steadily increasing torque is applied until the soil fails in shear, then the undrained shear strength, s_u, is computed from this torque. This test is only usable in very soft to soft clays and silts, and usually is performed at the bottom of a boring. In very soft soils, the vane can be pressed to large depths without a boring.

The shear surface has a cylindrical shape, and the data analysis neglects any shear resistance along the top and bottom of this cylinder. Usually the vane height-to-diameter ratio is 2, which, when combined with the applied torque, produces the following theoretical formula:

$$s_u = \frac{6\,T_f}{7\pi d^3} \tag{12.16}$$

where:

s_u = undrained shear strength

T_f = torque at failure

d = diameter of vane

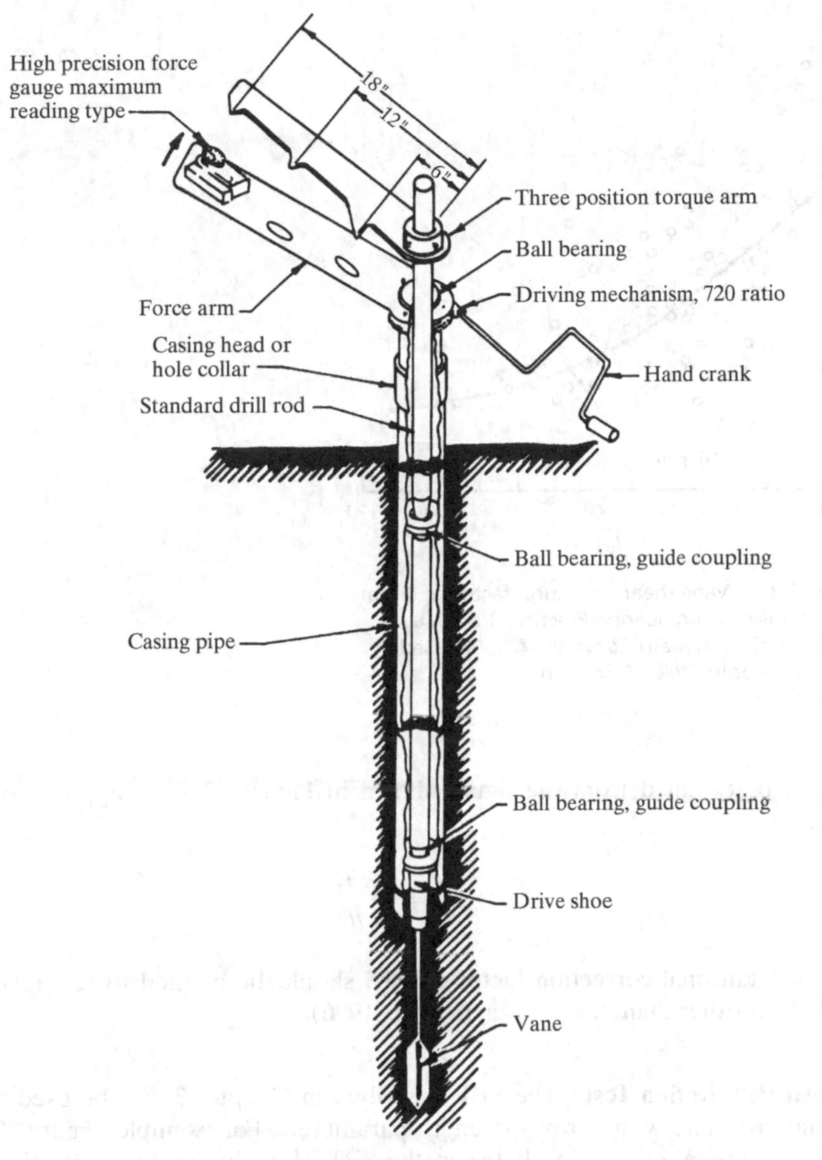

FIGURE 12.38 Vane shear test. (U.S. Navy, 1982.)

However, several researchers have analyzed failures of embankments, footings, and excavations using VST (assuming that the factor of safety was close to 1.0) and found that Equation 12.16 often overestimates s_u. Therefore, an empirical correction

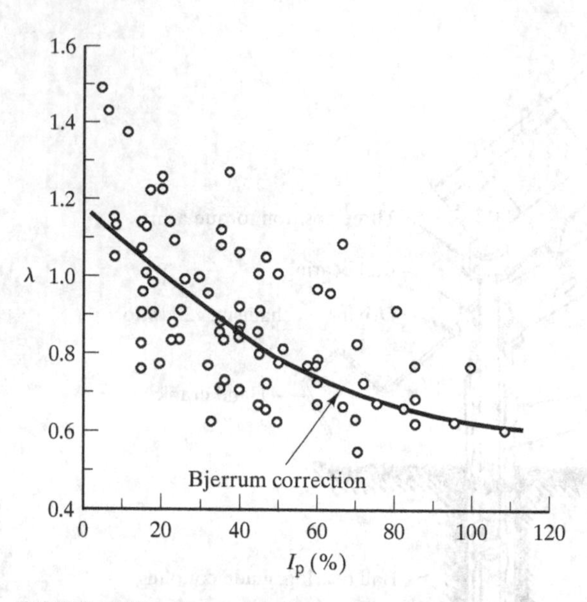

FIGURE 12.39 Vane shear correction factor, λ. (From Soil Mechanics in Engineering Practice. 3rd ed. by Terzaghi, Peck, and Mesri Copyright © 1996. Used by permission of John Wiley & Sons, Inc.)

factor, λ, determined from the figure shown in Figure 12.39, is applied to the test results:

$$s_u = \frac{6\,\lambda\,T_f}{7\pi d^3} \tag{12.17}$$

An additional correction factor of 0.85 should be applied to test results from organic soils other than peat (Terzaghi et al., 1996).

Standard Penetration Test The SPT, described in Chapter 3, can be used to obtain data that correlate with various strength parameters. For example, Figure 12.40 presents an empirical correlation between the SPT N_{60}-value and the effective friction angle, ϕ', in uncemented sands, and Table 5.4 gives an empirical correlation between the SPT N_{60}-value and the undrained shear strength, s_u.

Cone Penetration Test The cone penetration test, described in Chapter 3, also may be used to obtain data that correlate with various strength parameters. For example, Figure 12.41 presents an empirical correlation between the cone bearing, q_c, and ϕ', and Robertson and Robertson (2006) present a method to calculate s_u from cone penetration test data.

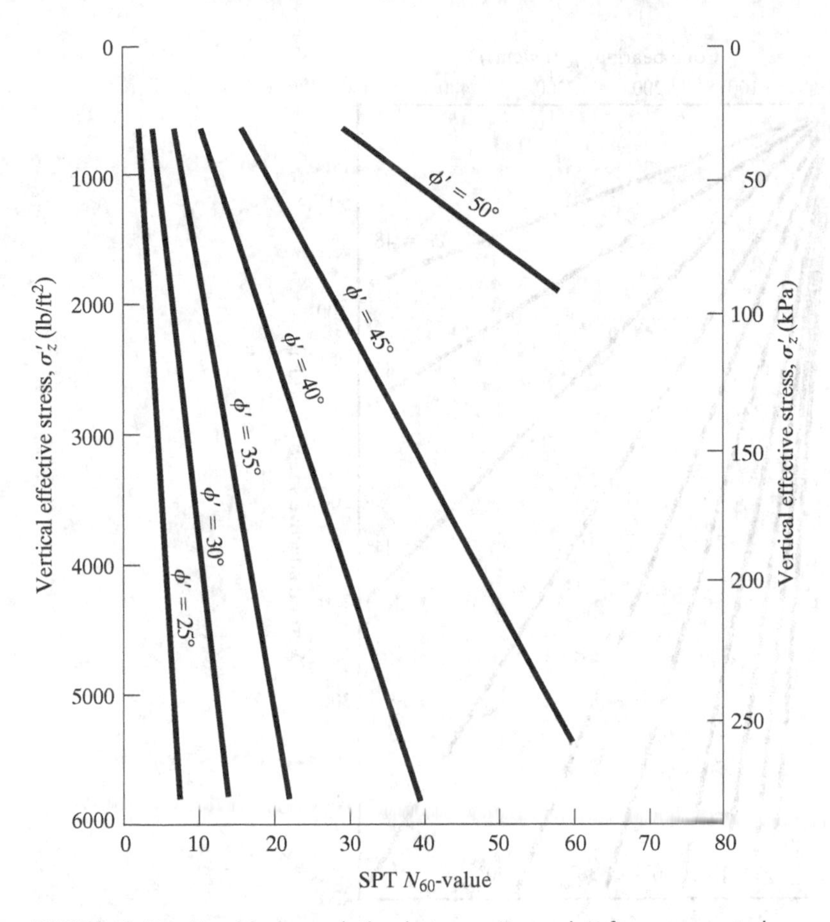

FIGURE 12.40 Empirical correlation between N_{60} and ϕ' for uncemented sands. (Adapted from DeMello, 1971.)

12.10 SHEAR STRENGTH AT INTERFACES BETWEEN SOIL AND OTHER MATERIALS

Many geotechnical construction projects use geosynthetic materials embedded into the soil for various purposes (Koerner, 1998). For example, geogrids can be used to reinforce the soil, geotextiles can be used for filtration, and geomembranes can be used to provide impervious barriers. Sometimes the shear strength along the interface between these materials and the adjacent soil becomes important and needs to be evaluated. This strength may be measured in the laboratory using a device similar to a direct shear machine.

The interface strength between geogrids and soil is generally very good because these materials are specifically designed to anchor into the ground. However, the interface strength for geomembranes is much lower and has been a source of problems. For example, in 1987, a series of high-density polyethylene (HDPE) geomembranes was installed at the Kettleman Hills landfill in California to serve as seepage barriers, thus protecting the groundwater from contamination. Unfortunately, about one year after

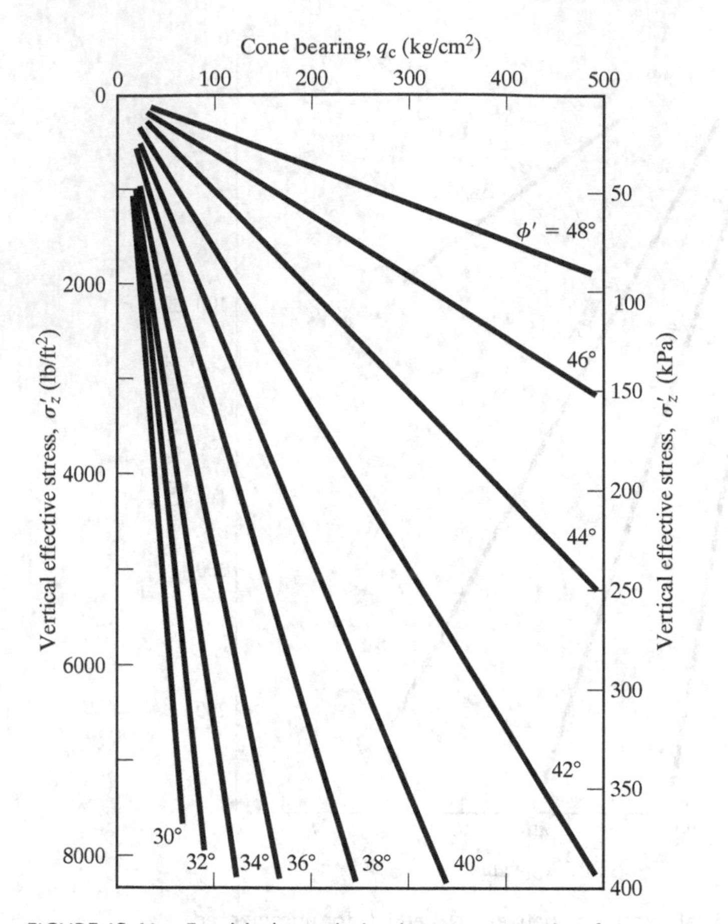

FIGURE 12.41 Empirical correlation between q_c and ϕ' for uncemented, normally consolidated quartz sands. (Adapted from Robertson and Campanella, 1983.)

refuse began to be placed over this geomembrane, a landslide occurred at the landfill. A subsequent investigation (Mitchell et al., 1990) revealed the landslide occurred along the interface between the geomembrane and the adjacent soils. Because HDPE is so smooth, the interface friction angle is as low as 8°, or even less. Thus, the geomembrane solved one problem (seepage), but introduced a new problem (shear failure).

12.11 UNCERTAINTIES IN SHEAR STRENGTH ASSESSMENTS

Shear strength assessments are subject to errors from a wide range of sources, so it is important to understand how much (or how little!) faith to place in strength values obtained from laboratory or in situ tests. The most common mistake, especially among inexperienced engineers, is to place far too much credibility in these numbers.

Even the most carefully performed laboratory tests are no better than the soil specimens tested. Are they truly representative of the soil mass? How much sample disturbance has occurred, and what effect does it have on the test results? The test methods also introduce error, because they do not simulate exactly the stress states in the field and do not shear the soils in the same way as occurs in the field. In addition, tests on unsaturated specimens may not properly account for future changes in the field moisture content.

Because of these and other factors, even carefully performed shear strength assessments can give strength values that contain errors of 25% or more. Although we typically report test results to two or three significant figures, their true accuracy is much less. Geotechnical engineers attempt to compensate for these uncertainties by using conservative interpretations of test data and applying appropriate factors of safety.

SUMMARY

Major Points

1. Empirical evidence shows that soils usually fail in shear in the field along a failure surface.
2. Shear strength is one of the most important engineering properties of soils. We use it to design structural foundations, earth slopes, retaining walls, and many other engineering projects.
3. Soil strength comes from the interaction between soil particles and has two sources: frictional strength and cohesive strength. Frictional strength is due to the sliding and rolling of the particles past each other, while cohesive strength is due to interparticle bonds, such as cementation.
4. There are two kinds of cohesive strength: true cohesion and apparent cohesion. True cohesion is strength that is truly the result of bonding between soil particles, whereas apparent cohesion is really frictional strength in disguise. The most common source of apparent cohesion is negative pore water pressures.
5. Some soils have ductile shear stress–strain curves, where the shear strength is defined as the peak shear stress. Other soils have brittle curves, which give two kinds of strengths: peak strength and residual strength.
6. We evaluate the shear strengths of saturated soils assuming two limiting conditions: the drained and undrained conditions. The drained condition is defined as the condition under which there is no excess pore water pressure in the soil. The undrained condition is the condition under which there is no pore water movement and excess pore water pressures are generated.
7. Excess pore water pressures are generated by the volume change tendencies of soils induced by changes in the mean normal stress and deviator stress.
8. Nearly all soil strength analyses in engineering practice use the Mohr–Coulomb failure criterion, which defines the shear strength in terms of the parameters cohesion and friction angle. The criterion may be expressed in terms of the total stress or effective stress.

9. The shear strength of a soil is controlled by the effective stress; therefore, effective stress analyses are fundamentally more direct descriptions of soil behavior. However, effective stress analyses become difficult to carry out when unknown excess pore water pressures are present. In such cases, it becomes necessary to resort to total stress analyses.
10. Saturated sands and gravels are almost always evaluated using effective stress analyses assuming that the drained condition applies because the excess pore water pressures are minimal.
11. Depending on the rate of loading, saturated clays can be assumed to be under the drained or undrained condition. For normal loading rates, the undrained condition prevails during and immediately after construction in the short term, but reaches the drained condition in the long term after all the excess pore water pressures have dissipated.
12. When the construction changes the mean normal stress in saturated clays, short-term (end-of-construction) strength analyses are usually total stress analyses using the undrained strength and total stress parameters. However, special effective stress analyses are possible when excess pore water pressures are measured in the laboratory or monitored in the field using piezometers.
13. Some clays lose a significant portion of their strengths when remolded. This loss is quantified by the sensitivity of the clay.
14. The strength of a soil that has undergone a large shear displacement is called the residual strength. It is useful in evaluating existing failure surfaces of landslides.
15. A soil has a higher shear strength when it is unsaturated than saturated because of the presence of apparent cohesion, but this additional strength may be lost if the soil becomes wetted in the future. Engineers usually assume a worst-case condition for unsaturated soils.
16. Several laboratory and in situ methods are available to measure the shear strength. The appropriate method depends on the required parameters, the type of soil, and cost.
17. Because of uncertainties and other factors, shear strength values obtained in practice may contain errors of 25% or more.

Vocabulary

aggregate base material
apparent cohesion
apparent mechanical forces
brittle
cementation
cohesive strength
consolidated drained test
consolidated undrained test
contraction
creep
deviator stress
dilation
direct shear test
drained condition
ductile
effective cohesion
effective friction angle
effective stress analysis
electrostatic and electromagnetic attractions
excess pore water pressure
factor of safety
failure envelope
fissured clay
frictional strength

fully softened strength
mean normal stress
micaceous sands
Mohr–Coulomb failure criterion
negative pore water pressure
peak strength
primary valence bonding
quartz sands
quick clays
quicksand
residual strength
ring shear test
sensitivity
slickenside
soil liquefaction strength
thixotropic hardening
total cohesion
total friction angle
total stress analysis
triaxial compression test
true cohesion
ultimate strength
unconfined compression test
unconsolidated undrained test
undrained condition
undrained shear strength
vane shear test

QUESTIONS AND PRACTICE PROBLEMS

Section 12.1 Strength Analyses in Geotechnical Engineering

12.1 Explain why the shear strength, not the tensile or compressive strength, is used in failure analyses of soils.

Section 12.2 Shear Failure in Soils

12.2 Describe the difference between the frictional strength and cohesive strength.

12.3 List the factors that affect the frictional strength.

12.4 Describe the difference between true cohesion and apparent cohesion.

12.5 Describe two ways in which the measured cohesion of a sand can be nonzero.

Section 12.3 The Drained and Undrained Conditions

12.6 Define the drained condition and the undrained condition.

12.7 Describe a situation in which a sand can be assumed to be under the drained condition.

12.8 Describe a situation in which a sand can be assumed to be under the undrained condition.

12.9 Describe a situation in which a clay can be assumed to be under the drained condition.

12.10 Describe a situation in which a clay can be assumed to be under the undrained condition.

12.11 Describe how changes in the mean normal stress affect the sign and magnitude of the excess pore water pressure in a soil.

12.12 Describe how changes in the deviator stress affect the sign and magnitude of the excess pore water pressure in a soil.

12.13 Describe how the density of a sand is related to its volume change during shear.

12.14 Describe the conditions under which a sand is susceptible to earthquake-induced liquefaction.

Section 12.4 Mohr–Coulomb Failure Criterion

12.15 The effective normal stress on a certain plane at a given point in a soil is 120 kPa. The effective cohesion and effective friction angle of the soil are 10 kPa and 31°, respectively. A foundation to be built nearby will induce a shear stress of 50 kPa on this plane. Using an effective stress analysis, compute the long-term factor of safety against shear failure on this plane.

12.16 A site is underlain by a soil that has a unit weight of 118 lb/ft^3. From laboratory shear strength tests that closely simulated the field conditions, the total stress parameters were measured to be c_T = 250 lb/ft^2 and ϕ_T = 29°. Estimate the shear strength on a horizontal plane at a depth of 12 ft below the ground surface at this site.

Section 12.5 Shear Strength of Saturated Sands and Gravels

12.17 A certain well-graded sand deposit has an in situ relative density of about 50%. A laboratory strength test on a specimen of this soil produced an effective friction angle of 31°. Does this test result seem reasonable? Explain the basis for your answer.

12.18 The vertical effective stress at a certain point in a loose sand is 1000 lb/ft^2. If an earthquake were to occur, how much excess pore water pressure would need to develop at this point for liquefaction to occur? Show a numerical rationale for your answer.

12.19 A temporary excavation similar to the one shown in Figure 8.7 is to be built. The soil is a clean sand with γ = 118 lb/ft^3, c' = 0, and ϕ' = 34°. According to a flow net analysis, the groundwater flow in the soil immediately below the excavation will be upward and have a hydraulic gradient of 0.76. Compute the shear strength on a horizontal plane at a depth of 3 ft below the bottom of the excavation. Discuss the significance of your answer.

Section 12.6 Shear Strength of Saturated Clays

12.20 A new building is to be built on a series of spread footing foundations that will be underlain by a saturated clay. Undisturbed soil samples have been obtained from this site and are ready to be tested. Should the laboratory test program focus on producing values of c' and ϕ', or s_u? Explain.

12.21 A steep excavation has been made in a saturated clay without the benefit of a slope stability analysis. It was completed one week ago, and thus far has not shown any signs of instability. Several people working on this project believe this is adequate demonstration of its stability, and feel it is safe. Do you agree? Why or why not?

12.22 A 5 m thick fill has recently been placed over clayey wetlands to support a new highway. The groundwater table was at or near the natural ground surface. Soon after the fill was completed, but before the paving began, a small landslide occurred in the fill and the underlying soils. Unfortunately, a sudden budget crisis stopped all work on the project and nothing has been done for ten years. At present, a new source of funding will permit construction to resume. The fill slope, at the time of its failure, can be assumed to have a factor of safety of 1.0. Is the factor of safety still equal to 1.0 at present? Will remedial construction definitely be necessary to increase the factor of safety? What should be done to evaluate this situation? Explain.

12.23 Pile foundations consist of long poles driven into the ground. They transmit structural loads into the ground through end bearing (compression between the bottom of the pile and the soil below) and through skin friction (sliding friction along the sides of the pile). Both of these depend on the shear strength of the surrounding soil.

When piles are driven into saturated clays, they push the soil aside, causing it to compress and generating excess pore water pressures. After construction, these pressures eventually dissipate.

(a) Would you expect these excess pore water pressures to be positive or negative? Why?

(b) Would you expect the load carrying capacity of the pile to increase, decrease, or remain constant with time? Why?

12.24 Soil can stand in vertical cuts only if it has cohesive strength. Even so, anyone can build a sand castle at the beach using clean fine-to-medium sand, and these castles can have vertical cuts. This appears to be a contradiction.

(a) Explain why sand castles can be built in this way.
(b) If no waves, thieves, rain, or wind disturb the castle, will the vertical cuts stand for a long time? Explain why or why not.

Section 12.9 Shear Strength Evaluation

12.25 The soils in Figure 10.26 have the following strength parameters:

Silty sand	$c' = 0$	$\phi' = 31°$
Soft clay	$c_T = 20$ kPa	$\phi_T = 0°$
Medium clay	$c_T = 45$ kPa	$\phi_T = 0°$
Glacial till	$c' = 15$ kPa	$\phi' = 40°$

In addition, the glacial till has a unit weight of 22.0 kN/m^3. Develop a plot of the shear strength applicable to short-term analyses on a horizontal plane versus depth, for a depth of 0–20 m. Keep in mind that the shear strength at a point depends on the cohesion and friction angle at that point, so it can suddenly change at strata interfaces.

12.26 Pile foundations consist of prefabricated poles, usually made of steel, wood, or concrete, that are driven into the ground with a pile hammer. The number of hammer blows per 0.1 m of pile penetration (known as the *blow count*) depends on the strength of the soil around the pile tip (along with other factors).

A series of piles is to be driven at the site described in Problem 12.25. The geotechnical engineer requires them to be driven until the tip is embedded 0.2 m into the glacial till. Could the field engineer use the blow count to determine when this penetration has been achieved? Explain.

12.27 Which laboratory and in situ tests would be appropriate for measuring ϕ' of a sand?

12.28 Which laboratory tests would be most appropriate for measuring c' and ϕ' of a clay?

12.29 Which laboratory and in situ tests would be appropriate for measuring s_u of a clay?

12.30 A series of direct shear tests has been performed on a dense well-graded sand. All tests were performed on 3.00-in. diameter, 1.25-in. tall cylindrical specimens, and were run slowly enough to produce the drained condition. The results of these tests are summarized in the following table:

Test Number	Normal Load (lb)	Shear Load at Failure (lb)
1	100	84
2	200	159
3	400	319

Assuming that the shear area remains constant during the test, determine the effective cohesion and effective friction angle from these test results. What values of these parameters would you expect? Are the test results consistent with your expectations? What values would you use for design?

12.31 An unconfined compression test has been performed on a 30 mm diameter, 75 mm long specimen of clay. The axial load and axial strain at failure were 120 N and 8.1%, respectively. Compute the undrained shear strength.

12.32 A series of UU triaxial compression tests have been performed on "identical" clay specimens. The test results were as follows:

Test Number	σ_{3f} (kPa)	σ_{1f} (kPa)
1	50	152
2	100	196
3	200	305

Plot the total stress Mohr circles at failure and draw the total stress Mohr–Coulomb failure envelope. Estimate the undrained shear strength.

12.33 A series of CU triaxial compression tests have been performed with pore pressure measurements on "identical" 2.50-in. diameter specimens of a clay. All of the specimens had an initial height of 6.00 in. The test results were as follows:

	Conditions at Failure			
Test No.	P_f (lb)	ε_f (%)	σ_{3f} (lb/in.^{2})	u_f (lb/in.^{2})
1	41.7	5.5	10.3	4.3
2	59.9	6.9	18.5	5.6
3	97.1	6.8	27.3	7.1

Plot the total stress and effective stress Mohr circles at failure and draw the total stress and effective stress Mohr–Coulomb failure envelopes. Determine c' and ϕ', and c_T and ϕ_T. Express c' and c_T in lb/ft^2.

12.34 Derive Equation 12.14 or 12.15.

12.35 If an additional CU test with pore pressure measurement is to be performed using a σ_3 of 22 lb/in^2 on a specimen identical to those in Problem 12.33, estimate the P_f and u_f for this additional test. What is theoretically the orientation of the failure plane in the specimen?

12.36 A series of vane shear tests has been performed in a stratum of inorganic clay that has a plasticity index of 50. The vane had a diameter of 50 mm and a height of 100 mm. The test results were as follows:

Depth (m)	Torque at Failure (N-m)
3.4	12.7
4.1	18.1
5.0	15.8
6.6	20.1

Compute the undrained shear strength, s_u, for each test, then combine this data to determine a single s_u value for this stratum.

Note: Geotechnical engineers frequently perform multiple tests on a single stratum, and then combine these results into one value for design. The process of doing so is somewhat subjective and requires the use of engineering judgement. Values significantly larger than the mean are typically discarded, and then a design value is typically chosen somewhere between the mean and the minimum values.

Comprehensive

12.37 When subjected to typical rates of loading in the field, sands are usually considered to be under the drained condition. Why?

12.38 A certain soil has a unit weight of 121 lb/ft^3 above the groundwater table and 128 lb/ft^3 below. It has an effective cohesion of 200 lb/ft^2, an effective friction angle of 31°, and extends from the ground surface down to a great depth. The groundwater table is at a depth of 18 ft below the ground surface, and $K = 0.78$. Compute the shear strength of this soil on both vertical and horizontal planes at depths of 15 and 30 ft below the ground surface.

12.39 A certain soil has $c' = 12$ kPa and $\phi' = 32°$. The major and minor total principal stresses at a point in this soil are 348 and 160 kPa, respectively, and the pore water pressure at this point is 96 kPa. Draw the failure envelope and the Mohr circle and determine if a shear failure will occur at this point in the soil. If so, determine the angle between the failure plane and the plane on which the major principal stress acts.

12.40 The rock outcrop shown in Figure 12.42 contains an inclined fracture. The fracture is inclined at an angle of 26° from the horizontal.

(a) Assuming the effective cohesion along the fracture to be zero, compute the lowest possible value of the effective friction angle along the fracture. Do this computation by assuming that the factor of safety against sliding is equal to 1.0.

FIGURE 12.42 Rock outcrop for Problem 12.40.

Hint: Set the weight of the rock above the fracture equal to W and the area of the fracture equal to A. Then compute the vector component of W that acts parallel to the fracture and determine what ϕ' would be required to resist this force.

(b) If the effective cohesion and friction angle along the fracture are 0 and 38°, respectively, compute the factor of safety against sliding.

12.41 A grain silo, which is a very heavy structure, was recently built on a saturated clay. Because the harvest season was fairly short and intense, the silo was completely loaded with grain fairly quickly (i.e., within a couple of weeks). This is the first time the silo has been loaded. The grain weighs about twice as much as the empty silo. The combined weight of this grain and the silo has induced both compressive and shear stresses in the soil below.

Suddenly, someone has become concerned that the soil may be about to fail in shear under the weight of the silo and the grain. This is a legitimate concern, because such failures have occurred before. Discuss the soil mechanics aspects of this situation and determine whether the risk of failure in the soil is increasing, decreasing, or remaining constant with time.

12.42 Hollywood movies sometimes show people "drowning" in quicksand and sinking to the bottom. Do such scenes accurately depict reality? What would happen in real life to a person who accidently ventured into quicksand? Explain the reasoning behind your answer.

Hint: Compare the unit weight of a human with the unit weight of the quicksand.

CHAPTER 13

Stability of Earth Slopes

Wait a minute, I think the town is getting flooded!

The mayor of Armero, Colombia, speaking on a ham radio as the town was being demolished by a massive debris flow (Voight, 1990)

Many civil engineering projects are located on or near sloping ground, and thus are potentially subject to various kinds of slope instability such as slides, flows, and falls. Slope failures often produce extensive property damage, and occasionally result in loss of life. Therefore, geotechnical engineers and engineering geologists frequently need to evaluate existing and proposed slopes to assess their stability.

Schuster (1996) estimated the cost of slope-instability-related damage in the United States alone at $1.8 billion per year, or about $7 per capita per year. Some areas are especially prone to trouble. For example, Hamilton County, Ohio (which includes the City of Cincinnati), suffers from $12.4 million in slope-instability-related damage per year (about $14 per capita per year).

Individual slope failures also can be very disastrous and expensive. For example, the 1983 Thistle debris slide near Thistle, Utah, created a huge dam across a canyon, forming a new lake as shown in Figure 13.1 (Kaliser and Fleming, 1986; Shuirman and Slosson, 1992). It caused $200 million in direct damages, including the destruction of two major highways and the main line of the Denver and Rio Grande Western Railroad. This new lake had no outlet, so it would have eventually overtopped the slide and quickly eroded it, causing a massive flood. Therefore, the lake was drained, first with pumps and later with a permanent tunnel.

Although most slope failures cause only property damage, some are fatal as well. About 25 to 50 deaths per year occur as a result of slope failures in the United States, and about 5 per year in Canada (Schuster, 1996). The worst single event on

FIGURE 13.1 The 1983 Thistle debris slide near Thistle, Utah. The lake was drained after this photograph was taken. (Utah Geological Survey.)

record is a 1786 slide in China's Sichuan Province that dammed a river in a manner similar to the Thistle slide. The river soon overtopped the new dam and rapidly eroded it, causing extensive flooding downstream. The flooding occurred very suddenly, and drowned about 100,000 people.

Sometimes civil engineering projects are built close to slopes that are naturally unstable, and eventually succumb to damage from failures that would have occurred whether or not the construction had taken place. More often, carelessly designed and constructed projects decrease the stability of nearby slopes and thus induce instability. For example, the construction of a road in hilly terrain might include creating a cut slope that undermines fractures or bedding planes in the rock, and thus induces a slide. Figure 13.2 is an example of such a project, and the disastrous results.

Because of these potential problems, it is important to retain the services of engineering geologists and geotechnical engineers when building near or on slopes. With proper evaluation, analysis, design, and construction, slope stability problems can usually be avoided. It is even possible to stabilize ground that would otherwise be unacceptable.

FIGURE 13.2 These houses were built near the top of a marginally stable slope that had been steepened as part of an earlier road construction project. The home-builders and owners also contributed to the instability by adding to the groundwater through years of landscape irrigation and poor surface drainage. The slope finally failed during a very wet winter, and the failure extended beneath the houses.

13.1 TERMINOLOGY

Civil engineers use several special terms when describing earth slopes. As shown in Figure 13.3, slopes can be categorized into three different types.

- *Natural slopes* are, as the name implies, part of the natural terrain.
- *Cut slopes* are those made by an excavation. They expose natural ground that was once buried.
- *Fill slopes* are those made by placing a fill.

In a two-dimensional section through a typical slope as shown in Figure 13.4, we can also define other terms.

- The *top* (or *crest) of slope* and *toe of slope* are the points where the slope intersects flatter ground.
- The *slope face* is the ground surface between the top of slope and toe of slope.

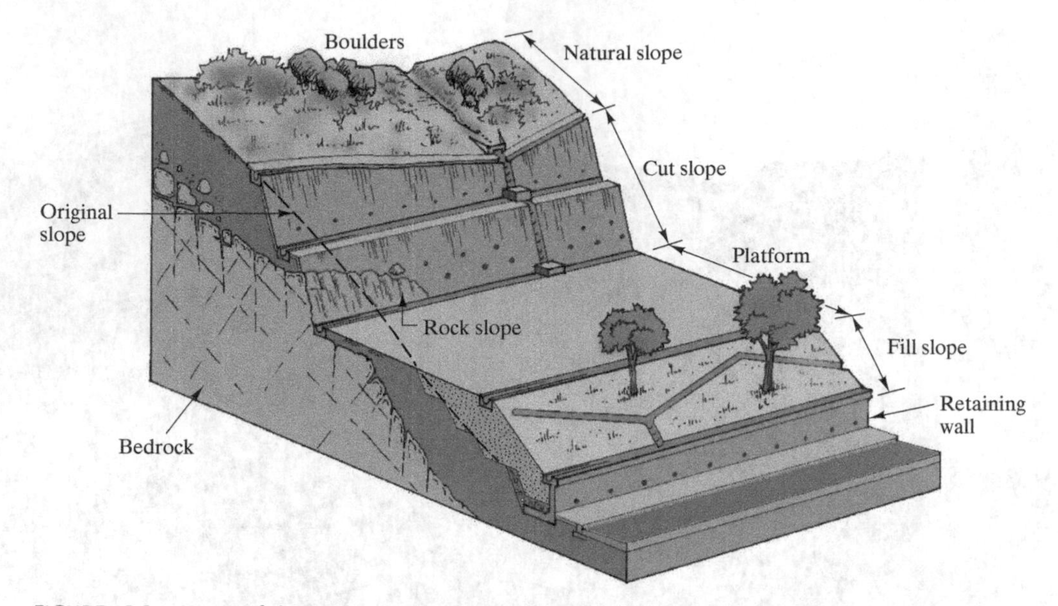

FIGURE 13.3 Types of slopes. (Used with permission of the Head of the Geotechnical Engineering Office and Director of the Civil Engineering and Development Department, the Government of the Hong Kong Special Administrative Region.)

- A *terrace*, sometimes called a *bench* or a *berm*, is a narrow level area created in cut and fill slopes to accommodate surface drainage facilities.
- The *slope ratio* describes the steepness of the slope, and is always expressed as horizontal–to–vertical. For example, a "three-to-one" (3V:1H or simply 3:1) slope is inclined at three horizontal to one vertical. Slopes steeper than 1:1 are described using fractions, such as $\frac{1}{2}$:1. This notation can be confusing to engineers used to working with roofs, which are customarily described in the opposite manner (vertical–to–horizontal).
- The *slope height*, *H*, is the difference in elevation between the top of slope and toe of slope (i.e., measured vertically, not along the face of the slope).

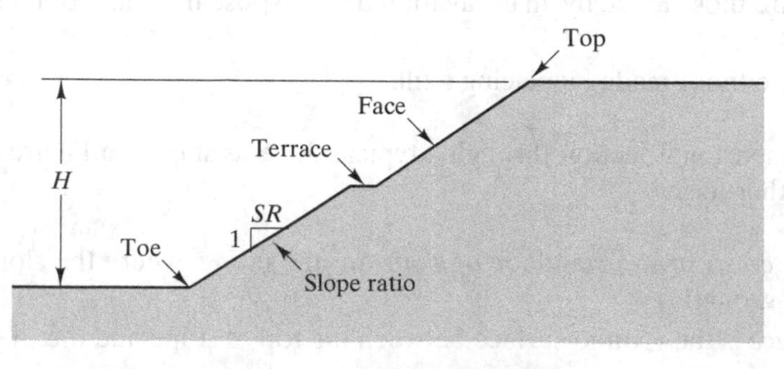

FIGURE 13.4 Terminology used to describe slopes.

13.2 MODES OF SLOPE INSTABILITY

Slopes can fail in many different ways, and several methods have been developed to classify these modes of failure. We will use the one proposed by Varnes (Varnes, 1958; Varnes, 1978; Cruden and Varnes, 1996).

The Varnes system divides slope failures into five types: *falls*, *topples*, *slides*, *spreads*, and *flows*. We will discuss each of them separately. In addition, these terms are preceded by *rock*, *debris*, or *earth* to indicate the type of material that has failed, where "rock" means bedrock or rocks, "debris" means predominantly coarse soils, and "earth" means predominantly fine soils, for example, rock slide, earth slide, rock fall, etc. Other descriptive terms also may be added to describe the rate of movement, history of movement, and other characteristics.

Falls

Falls are slope failures consisting of soil or rock fragments that drop rapidly down a slope, bouncing, rolling, and even becoming airborne along the way. Figure 13.5 shows the results of repeated falls from a rock slope, and Figures 13.6 and 13.7 show large boulders that fell down steep slopes. Falls most often occur on steep rock slopes, and are usually triggered when rock fragments are undermined by erosion, split apart by tree roots or ice, pushed out by water pressure, or shaken by an earthquake.

FIGURE 13.5 Repeated rock falls from the outcrop on this steep slope have created a fan of debris called *talus*. The road at the bottom of this slope must be cleared of debris following heavy rains, earthquakes, or other events. This slope is near Forest Falls, California (a place named for waterfalls, not rock falls!).

FIGURE 13.6 The 1992 Landers and Big Bear Earthquakes in California (magnitudes 7.5 and 6.6, respectively) generated many rock falls, including this one that landed on State Highway 38. (Photograph by Jeff Knott.)

FIGURE 13.7 This large boulder fell down a steep slope and struck the back of this house in Colorado. (Colorado Geological Survey.)

Falls usually occur very suddenly and rapidly, and thus have been responsible for many deaths. The "watch for falling rock" signs on many mountain roads warn motorists of the danger from potential rock falls.

Topples

A *topple* is similar to a fall, except that it begins with a mass of rock or stiff clay rotating away from a vertical or near-vertical joint or fissure. This mode of failure occurs only in steep slopes, as shown in Figures 13.8 and 13.9. It is especially important in schist and slate (Goodman, 1993), but also can occur in other types of rocks.

Slides

Although many people use the term *slide* or *landslide* to describe any mode of slope instability, Varnes' system uses it only to describe failures that involve one or more well-defined blocks of earth that move downslope by shearing along well-defined surfaces or thin shear zones. Slides may be described by their geometry, as shown in Figures 13.10–13.12. Common types include the following:

- *Rotational slides* move along curved shear surfaces that are concave upward. These most often occur in homogeneous materials, such as fills.
- *Translational slides* move along more planar shear surfaces. These usually reflect weak zones or bedding planes, and their thickness-to-length ratios are usually less than 0.1. When the moving blocks remain relatively intact, translational slides are sometimes called *block-glide slides*.
- *Compound slides* have a shape between those of rotational and translational slides.
- *Complex* and *composite slides* have characteristics of slides and some other mode of slope instability, such as flows.

Special terms used to describe slides are illustrated in Figure 13.12 and defined below:

- *Crown*—The nearly undisturbed ground above the main scarp
- *Main scarp*—The steep natural ground formed above the slide when it moved downhill

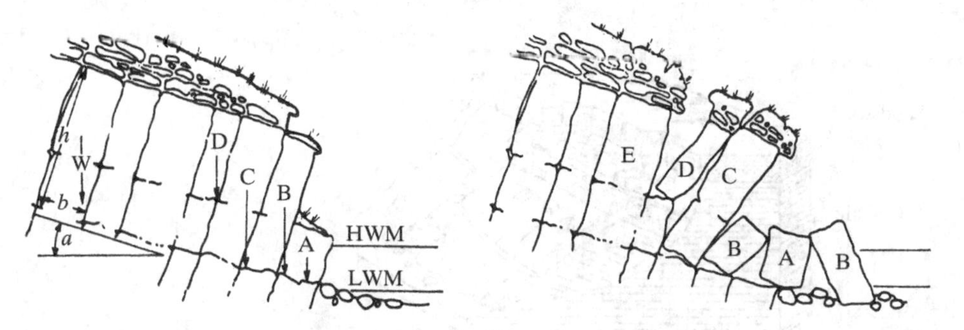

FIGURE 13.8 Topple instability, before and after failure (Varnes, 1978).

FIGURE 13.9 This rock slope in a columnar basalt contains a series of near-vertical joints and has been subjected to a series of toppling failures. The debris in the foreground is the result of these topples. (Devils Postpile National Monument, California.)

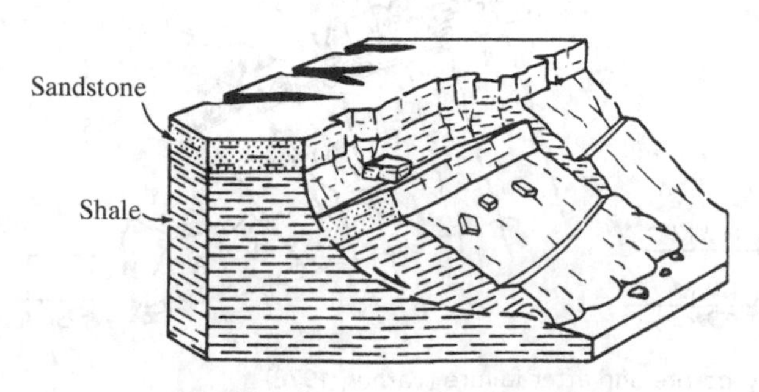

FIGURE 13.10 Rotational slide. (Varnes, 1978.)

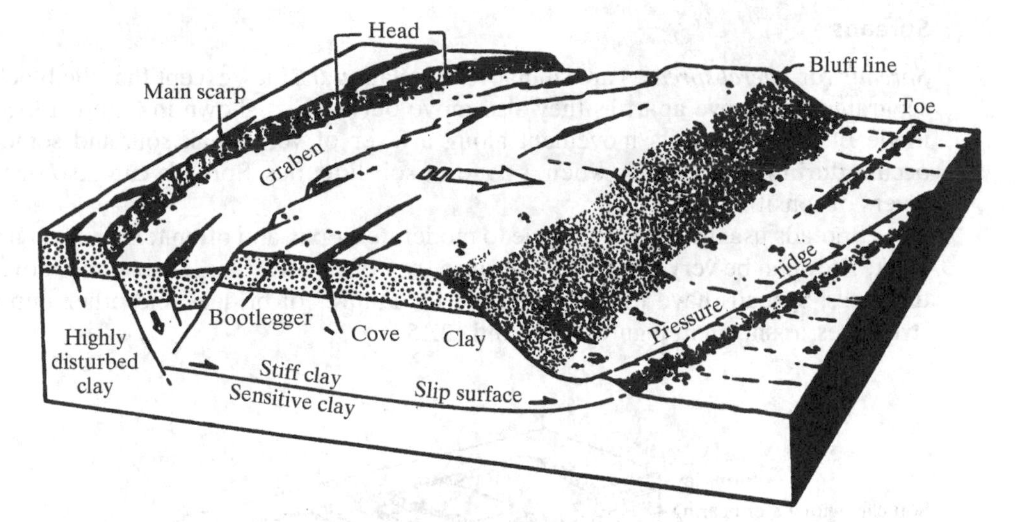

FIGURE 13.11 Block-glide translational slide. (Varnes, 1978.)

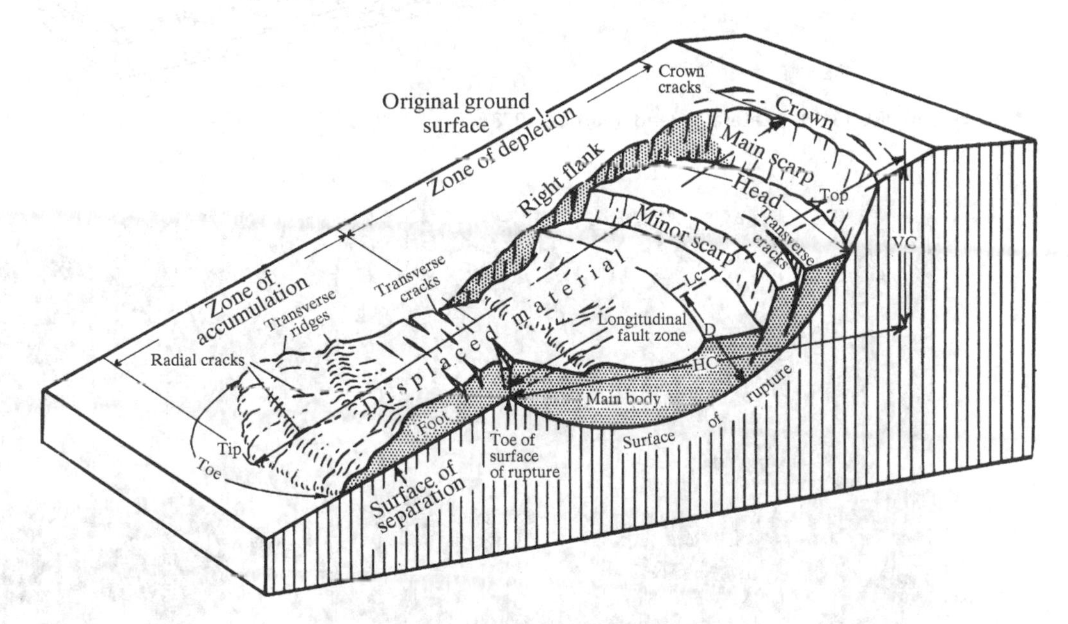

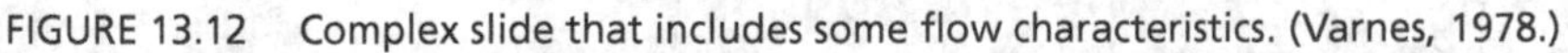

FIGURE 13.12 Complex slide that includes some flow characteristics. (Varnes, 1978.)

- *Minor scarp*—A secondary scarp created within the main body of the slide as a result of secondary failures
- *Body*—The displaced soil or rock
- *Flank*—The borders along the left and right sides of the body where it meets the relatively undisturbed ground
- *Tension cracks*—Cracks that often appear in the crown. They are roughly parallel to the top of the slope and are caused by tensile stresses in the ground.

Spreads

Spreads (or *lateral spreads*) are similar to translational slides, except that the blocks get separated and move apart as they also move outward, as shown in Figure 13.13. This mode of failure reflects movement along a layer of very weak soil, and sometimes occurs during earthquakes when a zone of soil liquefies. Spreads can also occur on layers of sensitive clay.

Spreads usually occur on gentle to moderate slopes, and often terminate at a river-bank. They can be very destructive because they often affect large areas and move long distances. Spreads have been responsible for failures of bridges and other important structures, as shown in Figures 13.14 and 13.15.

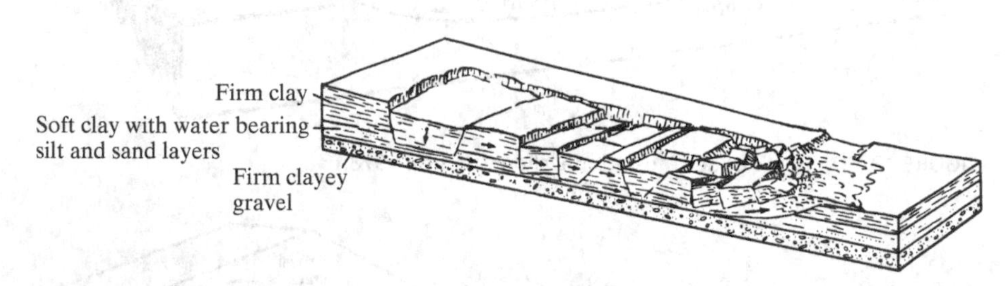

FIGURE 13.13 Lateral spread. (Varnes, 1978.)

FIGURE 13.14 Marine Research Facility, Moss Landing, California. Liquefaction during the 1989 Loma Prieta Earthquake caused a lateral spread beneath the left part of this building, which "stretched" it more than 1.5 m (5 ft). (Earthquake Engineering Research Center Library, University of California, Berkeley, Steinbrugge Collection.)

FIGURE 13.15 The Showa Bridge in Niigata, Japan, collapsed due to the formation of lateral spreads in the underlying soils during the 1964 earthquake. The lateral spreads moved the piers out of position, thus removing support from the simply supported deck. (Earthquake Engineering Research Center Library, University of California, Berkeley, Steinbrugge Collection.)

Flows

Flows are downslope movements of earth that resemble the movement of a viscous fluid. They differ from slides in that there are no well-defined blocks moving along shear surfaces. Instead, the mass flows downhill, with shear strains present everywhere. After the flow ceases, its products have a clearly fluidized appearance, as shown in Figures 2.17 and 13.16.

Flows often contain other objects, such as boulders and logs that move with the fluidized earth. These are called *debris flows* and can be very destructive. Buildings, cars, and other objects that might survive the moving mud are often destroyed by the debris contained in the mud.

Because of their high speed and ability to travel long distances, flows are the most dangerous and destructive mode of slope instability. Two of the most dramatic examples occurred in the Peruvian Andes in 1962 and 1970 (Plafker and Ericksen, 1978). The 1962 event began as an avalanche on the steep Nevados Huascarán mountain. The falling snow and ice gathered boulders and mud as it continued down the mountain, producing a large mudflow. It attained an estimated velocity of 105 mi/hr (170 km/hr) and rapidly buried nine towns, killing approximately 4000 people. The 1970 event was even larger, and began with an avalanche on the same mountain, this time triggered by an earthquake. It also gathered mud and rocks, including one boulder with a mass of about 8.2×10^6 kg, and reached an estimated velocity of 170 mi/hr (270 km/hr). This event quickly buried the city

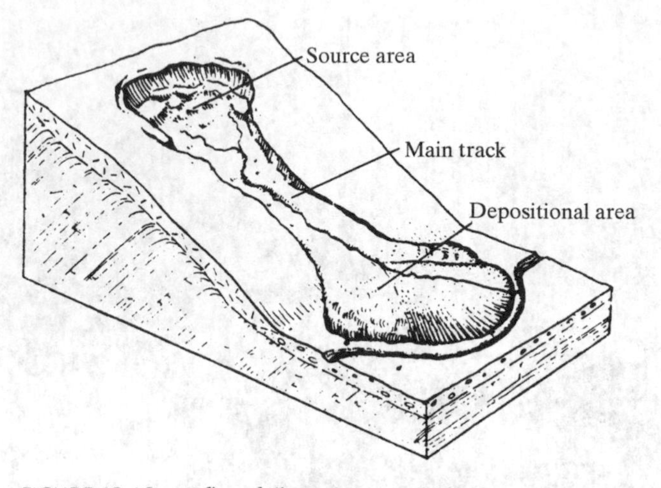

FIGURE 13.16 A flow failure. (Varnes, 1978.)

of Yungay, killing its 18,000 inhabitants. The quotation at the beginning of this chapter illustrates the suddenness of such events in areas with steep terrain.

Less dramatic examples occur throughout the world, often threatening people and property. Because flows are usually triggered by rain or snowmelt, they are often accompanied by flooding. For example, a 1934 flood and debris flow in La Cañada, California, caused over $5 million in property damage, 40 deaths, and the loss of 400 houses (Troxell and Peterson, 1937).

13.3 ANALYSIS OF SLOPE STABILITY

Geotechnical engineers and engineering geologists use both qualitative and quantitative methods to analyze slope instability problems. Some semi-quantitative methods are also very useful. These analyses often require the skills of both professions working in harmony, and need to consider both the present conditions and potential future conditions.

Analyses of potential falls and topples use both qualitative and quantitative methods. They involve geologic mapping, evaluations of past performance, and so on, and are usually performed by engineering geologists and geotechnical engineers using various rock mechanics techniques.

Flows and spreads are most amenable to semi-quantitative analysis. For flows, some engineers have attempted to use more quantitative analyses based on parameters such as shear strength and unit weight (Johnson and Rodine, 1984; Brunsden, 1984), but these methods do not appear to be widely used. Infinite slope analyses, discussed later in this chapter, may also give some insight on the potential for flows. For the analyses of spreads, especially those induced by earthquakes, some empirical or semi-empirical methods have been developed.

In contrast, slides are very amenable to quantitative analysis. Geotechnical engineers have developed methods of evaluating the potential for failure, and can express it as a factor of safety. These methods have proven to be reliable, and are used routinely in geotechnical engineering practice.

The following discussions focus primarily on the quantitative analysis methods for slides because geotechnical engineers use these methods extensively. However, this emphasis does not mean that slides are necessarily more important than other modes of failure, nor does it mean that qualitative analyses are not useful. Proper evaluation of slope stability problems requires a wide variety of techniques, so it is important not to become overly enamored with quantitative analyses at the expense of other methods.

13.4 QUANTITATIVE ANALYSIS OF SLIDES

The French engineer Alexandre Collin was probably the first to conduct quantitative analyses of landslides (Collin, 1846). Unfortunately, his work was not widely recognized at the time, and was largely forgotten. Therefore, the origin of modern slope stability analyses is traceable to another group of engineers working in Sweden during the 1920s. Apparently unaware of Collin's work, they had to begin anew and developed methods that soon became the basis for modern slope stability analyses.

Much of Scandinavia, especially Sweden and Norway, is underlain by sensitive marine clays with undrained shear strengths on the order of 300 lb/ft^2 (15 kPa). Because of this very low strength and high sensitivity, slopes in these soils are very prone to failure. These problems became especially troublesome during the late nineteenth and early twentieth centuries because of the cutting and filling associated with port construction (Petterson, 1955) and railroad construction. Both caused slides and flows. Following an especially costly failure in 1913, the Swedish State Railways formed a "Geotechnical Commission" (Statens Järnvägars, 1922) to study the problem and develop solutions (see discussion in Chapter 1). The commission's final report was issued in 1922, and is recognized as one of the early milestones in geotechnical engineering. The report presented a method of analysis we now call the *Swedish slip circle method*, discussed later in this section. It later became the basis for other methods of analysis.

Limit Equilibrium Concept and Factor of Safety

In current geotechnical engineering practice, most quantitative analyses of slides or potential slides are *limit equilibrium analyses*. In a limit equilibrium analysis, we evaluate the slope as if it were about to fail by sliding, with a well-defined body of the slide at limiting equilibrium, and determine the resulting shear stresses along the well-defined *failure surface*. Then, these equilibrium shear stresses are compared to their corresponding shear strengths to determine the *factor of safety*, *F*:

$$F = \frac{s}{\tau_e} \tag{13.1}$$

where:

F = factor of safety
s = shear strength
τ_e = equilibrium shear stress

The equilibrium shear stress is also called the *required shear strength* or *mobilized shear strength* because it is the part of the shear strength that is mobilized to keep the slope in equilibrium. The remaining shear strength is left in reserve, and contributes to the factor of safety, as illustrated by Equation 13.1. This definition of the factor of safety is the same as the one given by Equation 12.4, except that in Equation 13.1, the shear stress is specifically defined as the equilibrium shear stress.

The factor of safety typically varies from point to point along the failure surface. Some sections may have "failed" (i.e., the equilibrium shear stress equals the shear strength), while others may have a large reserve of excess shear strength (i.e., a large F). However, limit equilibrium analyses usually do not attempt to account for this variation and simply assume that the factor of safety is constant for the entire failure surface.

In theory, a factor of safety of 1 indicates incipient failure, so any slope with $F > 1$ will supposedly be stable. For design, however, we need a margin of safety to account for the consequence of failure and the many uncertainties in our analyses (including soil profile, shear strength, groundwater conditions, etc.). The most common design criterion requires a factor of safety of at least 1.5, although slightly lower values (perhaps about 1.3) may be acceptable for some roadway projects in remote areas where no structures are nearby and where a failure would only require cleaning debris from the roadway.

Most slope stability analyses quantify stability in terms of the factor of safety. In these cases, the analysis is called a *deterministic analysis*. However, it is also possible to express stability as a *probability of failure*. For example, after considering various uncertainties, we might determine that a certain slope has a probability of failure of 10^{-3} (i.e., one chance in 1000), which would then be compared to some acceptable level of risk (Wu et al., 1996; Wolff, 1996). Such an analysis is called a *probabilistic analysis*.

13.5 GENERAL PROCEDURES IN A LIMIT EQUILIBRIUM ANALYSIS OF A SLIDE

The objective of a deterministic limit equilibrium analysis of a slide is to compute the factor of safety F as defined by Equation 13.1. To compute F using this equation, we need to compute s, the shear strength of the soil along the failure surface, and the corresponding equilibrium shear stress, τ_e.

The shear strength is an inherent property of the soil that depends on the soil type and the design scenario under consideration, as discussed in detail in Chapter 12. Table 12.2 gives the appropriate analysis type (total stress or effective stress) and strength parameters that should be used for the short-term (end-of-construction) and long-term scenarios. For natural slopes that have been in existence for a long time, the drained condition prevails and, therefore, only the long-term scenario needs to be considered. For engineered cut and fill slopes that involve short-term changes to the loading of the soils in the slope, both the short- and long-term scenarios should be considered.

The equilibrium shear stress can be computed using statics, by considering equilibrium of the slide body. Different limit equilibrium analysis methods make different assumptions and have different ways to set up the slide body as a free body or free bodies to solve for F. Before presenting the details of specific methods, we will list and discuss the general steps involved in arriving at the F:

1. Obtain the three-dimensional geometry of the slope and subsurface conditions including groundwater conditions. To do this, we must usually carry out a site exploration and characterization program, as discussed in Chapter 3.
2. Select a representative two-dimensional section of the slope to analyze. Although three-dimensional methods are available, they are not often performed in practice because of their higher complexity and associated costs. In a two-dimensional analysis, we consider a slice of the slope from a representative section with unit length measured perpendicularly to the section.
3. Assume a failure surface shape. Experience and empirical observations should inform the selection of the shape of the failure surface to be analyzed. In practice, usually planar and circular surfaces are used. Circular surfaces are versatile in that they can simulate many different curved failure surfaces observed in the field, and even planar surfaces when the radii are very large. However, non-circular failure surfaces may need to be considered when the subsurface conditions are heterogeneous such as when a weak clay seam exists in the slope.
4. Select a trial failure surface and consider equilibrium of the body of the slide. If the shear strength along the entire failure surface is assumed to be uniform (e.g., as expressed by a constant undrained shear strength, s_u), draw a free body diagram of the entire slide body (as in the *planar failure analysis* and Swedish slip circle methods discussed in Sections 13.6 and 13.8, respectively) or a representative element of the slide body (as in the *infinite slope analysis* method discussed in Section 13.7). If the shear strength varies along the failure surface (as will be the case if $\phi' > 0$ or $\phi_T > 0$, or if the failure surface transcends more than one soil stratum), discretize the failure surface (to better account for the variation) and hence the slide body into several free bodies and draw a free body diagram of each (as in the *method of slices* discussed in Section 13.9). The free body of soil is taken to consist of all three phases of the soil. Therefore, the weight of the free body is from the total weight of the soil, and at any boundary of the free body, the forces transmitted include those by all three phases, i.e., the solids (from the effective stress) and water (from the pore water pressure). An important external force on each free body diagram is the shear force acting along the failure surface. Because the slide body is assumed to be at limiting equilibrium, this shear force comes from the equilibrium shear stress τ_e in Equation 13.1.
5. Write equilibrium equations. For equilibrium, three equilibrium equations can be written for each free body to solve for the unknowns, including F.
6. Count the number of equilibrium equations and the number of unknowns. If the number of equations is equal to the number of unknowns, solve the equations for F. If the number of equations is smaller than the number of unknowns, make additional assumptions to make them equal and solve for F.
7. Search for the *critical failure surface*. This is done by a trial-and-error process, repeating Steps 4 through 6 for many other trial failure surfaces to compute the

F values for these surfaces. Theoretically, the trial failure surface that gives the lowest factor of safety is the critical failure surface (i.e., the one on which sliding is most likely). The lowest F is taken to be the factor of safety for the slope. It is important to recognize that some judgment is involved in selecting the appropriate trial failure surfaces to be analyzed and that the ultimate goal is to locate the critical failure surface that corresponds to the *global* minimum factor of safety.

Using these general procedures, we can derive some specific methods in Sections 13.6 through 13.9 to calculate the factor of safety against slides in a slope.

13.6 PLANAR FAILURE ANALYSIS

A slide on a single planar failure surface, as shown in Figure 13.17, may be analyzed by considering equilibrium of the entire slide body as a single free body. This failure mechanism is common in rock masses consisting of planar joints or bedding planes.

In this two-dimensional analysis, we assume that the slope extends for an infinite distance perpendicular to the cross-section. Mathematically, we will consider a slice of length "b" from this infinitely long slope, where b is measured perpendicularly to the cross-section and is equal to 1 ft or 1 m, depending on the units of measurement. Thus, in the free body diagram shown in Figure 13.17, the weight of the sliding mass is expressed as W/b, perhaps using units of lb/ft, and the normal and shear forces acting on its base, which are actually resultant forces from pressures on the base, are N/b and T/b. For equilibrium of the free body, sum of forces in the direction normal to the failure surface is equal to zero:

$$\begin{aligned} N/b - (W/b)\cos\alpha &= 0 \\ N/b &= (W/b)\cos\alpha \end{aligned} \tag{13.2}$$

Sum of forces in the direction parallel to the failure surface is equal to zero:

$$\begin{aligned} T/b - (W/b)\sin\alpha &= 0 \\ T/b &= (W/b)\sin\alpha \end{aligned} \tag{13.3}$$

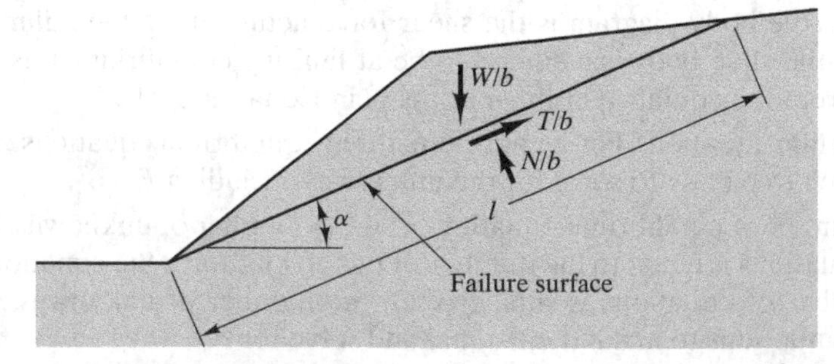

FIGURE 13.17 Slope with planar failure surface and free body diagram of the slide body.

For effective stress analyses, the shear strength s is given by the Mohr–Coulomb failure criterion and the effective stress parameters c' and ϕ'. To obtain s, we need to obtain the effective stress from the total stress, σ, which corresponds to N/b, and the pore water pressure, u. Now, the average pore water pressure, u, acting on the failure surface should be obtained from the pressure heads along the failure surface. Alternatively, the average pressure head can be approximated by the average depth of the failure surface below the groundwater table (measured vertically):

$$u = \gamma_w z_w \tag{13.4}$$

where:

γ_w = unit weight of water
z_w = the average depth of the failure surface below the groundwater table

Therefore, the average effective stress on the failure surface is given by

$$\begin{aligned} \sigma' &= \sigma - u \\ \sigma' &= \frac{N/b}{l} - u \\ \sigma' &= \frac{(W/b)\cos\alpha}{l} - u \end{aligned} \tag{13.5}$$

Using the Mohr–Coulomb failure criterion,

$$\begin{aligned} s &= c' + \sigma' \tan\phi' \\ s &= c' + \left[\frac{(W/b)\cos\alpha}{l} - u\right]\tan\phi' \end{aligned} \tag{13.6}$$

Now, the equilibrium shear stress τ_e is related to the shear force T/b:

$$\begin{aligned} \tau_e &= \frac{T/b}{l} \\ \tau_e &= \frac{(W/b)\sin\alpha}{l} \end{aligned} \tag{13.7}$$

Finally, combining Equations 13.6 and 13.7 with Equation 13.1, the factor of safety F for effective stress analyses is given by

$$\begin{aligned} F &= \frac{s}{\tau_e} \\ F &= \frac{c'l + [(W/b)\cos\alpha - ul]\tan\phi'}{(W/b)\sin\alpha} \end{aligned} \tag{13.8}$$

For the special case of $c' = 0$ and $u = 0$, Equation 13.8 reduces to

$$F = \frac{\tan \phi'}{\tan \alpha} \tag{13.9}$$

For total stress analyses and the case of $c_T = s_u$ and $\phi_T = 0$, a formula for F can be similarly derived as

$$F = \frac{s_u l}{(W/b) \sin \alpha} \tag{13.10}$$

Example 13.1

A 1.5:1 cut slope is to be made in a shale with bedding planes that dip at 16° as shown in Figure 13.18. Assume a unit weight of 20.1 kN/m^3 for the shale and strength parameters for the bedding planes of $c' = 15$ kPa and $\phi' = 20°$. Perform an effective stress analysis to compute the factor of safety against failure by sliding along the lowermost daylighting bedding plane (a bedding plane whose intersection with the slope face is exposed).

Solution:

$$W/b = \frac{1}{2}(85.0\text{ m})(11.3\text{ m})(20.1\text{ kN/m}^3) = 9650\text{ kN/m}$$

$$l = \frac{85.0\text{ m}}{\cos 16°} = 88.4\text{ m}$$

Compute the pore water pressure based on a visual estimate of z_w of 3.0 m (from a range of 0–3.2 m):

$$u = \gamma_w z_w = (9.8\text{ kN/m}^3)(3.0\text{ m}) = 29\text{ kPa}$$

$$F = \frac{c'l + [(W/b)\cos\alpha - ul]\tan\phi'}{(W/b)\sin\alpha}$$

$$F = \frac{(15\text{ kPa})(88.4\text{ m}) + [(9650\text{ kN/m})\cos 16° - (29\text{kPa})(88.4\text{ m})]\tan 20°}{(9650\text{ kN/m})\sin 16°}$$

$F = \mathbf{1.42}$

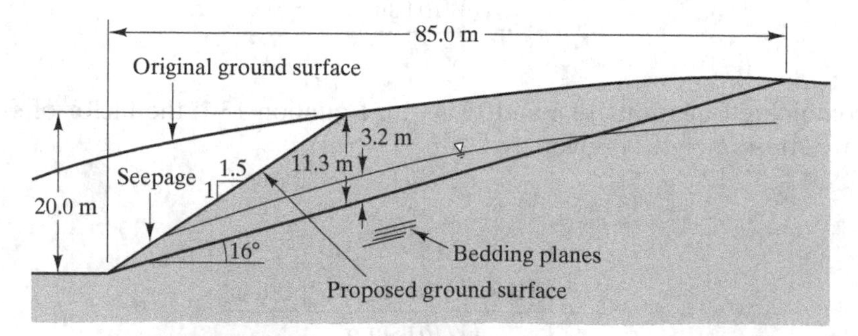

FIGURE 13.18 Cross-section for Example 13.1.

Comments: The computed factor of safety of 1.42 is slightly less than the commonly required value of 1.50. Therefore, this design is probably not acceptable.

In this example, we have computed the factor of safety for only one potential failure surface. The procedure outlined in Section 13.5 indicates that we must continue to search for other potential failure surfaces to determine the factor of safety for the slope. In this case, it is necessary to compute the factor of safety for all possible planar surfaces. The failure surface with the lowest computed factor of safety is the critical failure surface.

13.7 INFINITE SLOPE ANALYSIS

An infinite slope analysis is similar to a planar failure analysis, except that the failure surface is parallel to the slope face, and the depth to the failure surface is small compared to the height of the slope, as shown in Figure 13.19. Major assumptions of the infinite slope analysis are as follows:

1. The slope face is planar and of infinite extent.
2. The failure surface is parallel to the slope face.
3. Vertical columns of equal dimensions through the slope are identical.

The third assumption requires that any groundwater table in the slope must be parallel to the slope face and that any soil layers in the slope must have boundaries that are parallel to the slope face as well. It also allows the consideration of simply a representative element of the slide body in the analysis.

The free body diagram for this representative element, for the case of a slope in just one soil type, is shown in Figure 13.19. In this free body diagram, W/b is the weight of the element; N/b and T/b are the normal and shear forces on the base; E_l/b and S_l/b

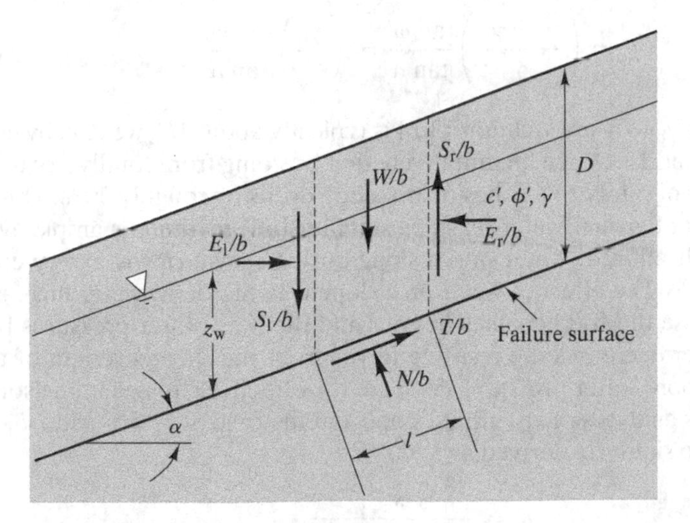

FIGURE 13.19 Infinite slope analysis method: free body diagram of a representative element of the slide body.

are normal and shear forces on the left vertical boundary; and E_r/b and S_r/b are normal and shear forces on the right vertical boundary. Because of the third assumption, the normal and shear forces on the vertical side boundaries cancel out. As a result, this free body diagram is equivalent to the one for the planar failure case shown in Figure 13.17, and the factor of safety formula for effective stress analyses, Equation 13.8, should apply to this case. To refine Equation 13.8 for the infinite slope case, we can substitute W/b and u in the formula by appropriate expressions:

$$W/b = \gamma D l \cos \alpha \tag{13.11}$$

and

$$u = \gamma_w z_w \cos^2 \alpha \tag{13.12}$$

For simplicity, Equation 13.11 assumes that the soil has the same unit weight above or below the groundwater table. Equation 13.12 can be obtained by considering that the flow of groundwater in the slope is parallel to the slope face and recognizing that the pressure head along the failure surface is equal to $z_w \cos^2 \alpha$. After substituting and simplifying, we have the factor of safety formula for the infinite slope case for effective stress analyses:

$$F = \frac{c' + (\gamma D - \gamma_w z_w) \cos^2 \alpha \tan \phi'}{\gamma D \sin \alpha \cos \alpha} \tag{13.13}$$

Equation 13.13 for the infinite slope case can be used to show the significant effect of groundwater on slope stability. Let us consider a slope in a cohesionless soil ($c' = 0$). When it is totally dry, $z_w = 0$, and Equation 13.13 reduces to Equation 13.9. When the same slope is totally saturated, the groundwater table coincides with the slope face, $z_w = D$, and the factor of safety is given by

$$F = \left(\frac{\gamma - \gamma_w}{\gamma}\right) \frac{\tan \phi'}{\tan \alpha} = \left(\frac{\gamma_b}{\gamma}\right) \frac{\tan \phi'}{\tan \alpha} \tag{13.14}$$

Because the ratio of γ_b to γ in Equation 13.14 is typically about 1/2, we can, by comparing Equations 13.9 and 13.14, see quantitatively that by going from totally dry to totally saturated, the factor of safety of a cohesionless slope drops by roughly 50%. This shows the significant effect of groundwater on slope stability. Indeed, there is ample evidence in practice that infiltration of water into a slope during rainstorms is a very common cause of slope failures. The effect of water on a slope is twofold: (a) water increases the driving forces because the soil becomes heavier and the pore water pressures become higher; and (b) water decreases the resisting forces from the shear strength of the soil because the higher pore water pressures decrease the effective stresses in the soil.

For total stress analyses of an infinite slope and the case of $c_T = s_u$ and $\phi_T = 0$, a formula for F can be similarly derived as

$$F = \frac{s_u}{\gamma D \sin \alpha \cos \alpha} \tag{13.15}$$

Note that for soils with cohesion, Equations 13.13 and 13.15 reveal that the factor of safety is a function of the depth of the failure surface and that F decreases as D increases. Therefore, the critical failure surface is theoretically located at an infinite depth. However, in practice, the depth of the critical failure surface may be controlled by actual field conditions that do not satisfy the assumptions in the infinite slope analysis. First, the assumption of a homogeneous soil in the slope may not be appropriate in practice because the shear strength usually increases with depth, making the factor of safety higher for a deeper failure surface if the higher shear strength is used in the equations. In addition, the assumed failure mechanism of sliding of a soil layer on a failure surface parallel to the slope face becomes less realistic as D increases. Therefore, a trial-and-error process using the actual field conditions may need to be carried out to arrive at the critical failure surface, by calculating the factor of safety values for different D values and keeping D below a reasonable limit. From a practical point of view, it is usually only necessary to evaluate the factor of safety at the depth where the soil strength changes significantly, for example, at the interface between a thin soil layer and a much harder underlying stratum or bedrock. In applying the infinite slope analysis to evaluate the potential for shallow flow slides, which are sometimes called *surficial slumps*, the critical failure surface is usually the interface between the surficial soil and a stronger stratum below.

In the case of a cohesionless slope, Equation 13.9 for the totally dry case and Equation 13.14 for the totally saturated case reveal that the factor of safety is not a function of the depth of the sliding surface and that all surfaces parallel to the slope have the same factor of safety. However, F is still a function of D if z_w is not equal to D or zero in Equation 13.13.

13.8 SWEDISH SLIP CIRCLE METHOD ($\phi = 0$ ANALYSIS)

Failure surfaces of slides in homogeneous or near-homogeneous soils often may be idealized in two-dimensional cross-sections as circular arcs. This *circular failure surface* geometry simplifies the mathematics because the normal forces acting on the failure surface pass through the center of the circle. It also simplifies the process of searching for the critical failure surface because each circle is defined by only three parameters: the radius, R, and the x and z coordinates of the center.

The Geotechnical Commission appointed by the Swedish State Railways developed an analysis method based on circular failure surfaces in undrained soils (Statens Järnvägars, 1922). A major assumption of this method is that the shear strength of the soils in the slope is independent of σ and is defined solely by the parameter s_u (the "$\phi = 0$ condition"). Thus, this method is only applicable to total stress analyses of saturated clays under the short-term scenario.

Let us consider the case of a slide in a slope consisting of one soil type only, as shown in Figure 13.20. We draw a free body diagram for the entire slide body as shown in Figure 13.20. The symbols in Figure 13.20 are defined as follows:

R = radius of the slip circle

s_u = undrained shear strength along the failure surface

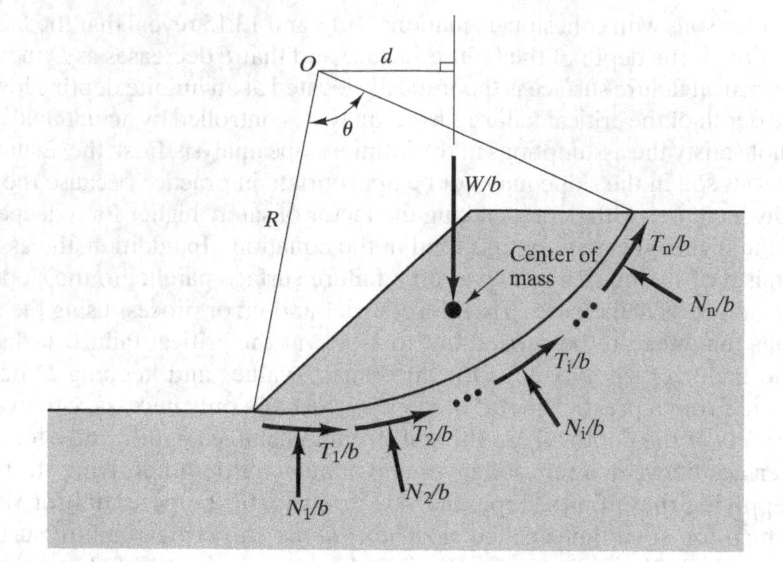

FIGURE 13.20 Swedish slip circle method: free body diagram of the slide body.

θ = central angle subtended by the circular arc that forms the failure surface (in degrees)

W/b = weight of the slide body per unit length of slope (e.g., in kN/m)

d = moment arm of the weight of the slide body about the center of the slip circle

N_i/b = i-th normal force on the failure surface per unit length of slope

T_i/b = i-th shear force on the failure surface per unit length of slope

For equilibrium, the sum of moments about point O, the center of the circle, is equal to zero:

$$(W/b)d - \sum (T_i/b)R = 0 \tag{13.16}$$

Note that the normal forces on the failure surface, N_i/b, go through the center of the circle and do not generate moments about the center because their moment arms are zero. Now, T_i/b is related to the equilibrium shear stress τ_e:

$$\sum (T_i/b) = \tau_e R\theta(\pi/180) \tag{13.17}$$

where $R\theta$ is the length of the circular arc and θ in degrees. Combining Equations 13.16 and 13.17,

$$\tau_e = \left(\frac{180}{\pi}\right)\frac{(W/b)d}{R^2\theta} \tag{13.18}$$

Therefore, the factor of safety F is given by

$$F = \frac{s}{\tau_e} = \left(\frac{\pi}{180}\right)\frac{s_u R^2 \theta}{(W/b)d} \tag{13.19}$$

For a more general case with different soil layers in the slope, as shown in Figure 13.21, the slide body can be discretized into slices by vertical planes, to better account for the different s_u values for different soil layers and to facilitate the calculation of moments about the center of the slip circle. For this case, we can similarly derive the factor of safety formula as follows:

$$F = \frac{\pi R^2}{180}\frac{\sum(s_u\theta)}{\sum[(W/b)d]} \tag{13.20}$$

where:

W/b = weight of a slice per unit length of slope

θ = central angle subtended by the circular arc that forms the base of a slice (in degrees)

d = moment arm of the weight of a slice about the center of the slip circle (note that d is positive or negative if the weight vector of the slice is to the right or left of the center, respectively, as shown in Figure 13.21)

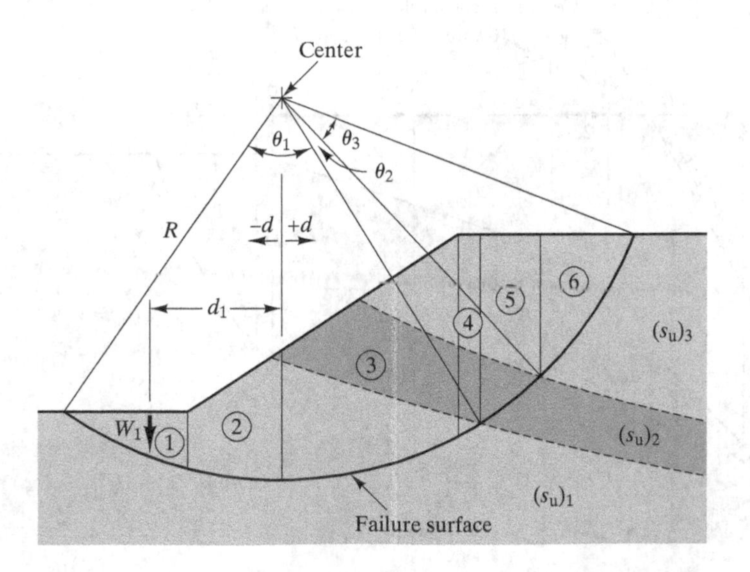

FIGURE 13.21 Swedish slip circle method: case of different soil layers.

Example 13.2

Using the Swedish slip circle method, compute the factor of safety for the trial circle shown in Figure 13.22.

Solution:

Divide the slide body into vertical slices as shown. One of the slice boundaries should, for convenience, be directly below the center of the circle (in this case, the boundary between slices 2 and 3). For convenience of computations also, draw a slice boundary wherever the slip surface intersects a new soil stratum and wherever the ground surface has a break in slope.

Then, compute the weight and moment arm for each slice using simplified computations as follows:

Weights:

$$W_1/b = 4.6\left(\frac{2.0}{2}\right)17.8 = 80 \text{ kN/m}$$

$$W_2/b = 7.0\left(\frac{2.0 + 9.8}{2}\right)17.8 = 740 \text{ kN/m}$$

$$W_3/b = 2.9\left(\frac{9.8 + 12.9}{2}\right)17.8 = 590 \text{ kN/m}$$

$$W_4/b = 7.1\left(\frac{5.0}{2}\right)17.0 + 7.1\left(\frac{12.9 + 8.0}{2}\right)17.8 = 1620 \text{ kN/m}$$

$$W_5/b = 7.2\left(\frac{5.0 + 10.3}{2}\right)17.0 + 7.2\left(\frac{8.0}{2}\right)17.8 = 1450 \text{ kN/m}$$

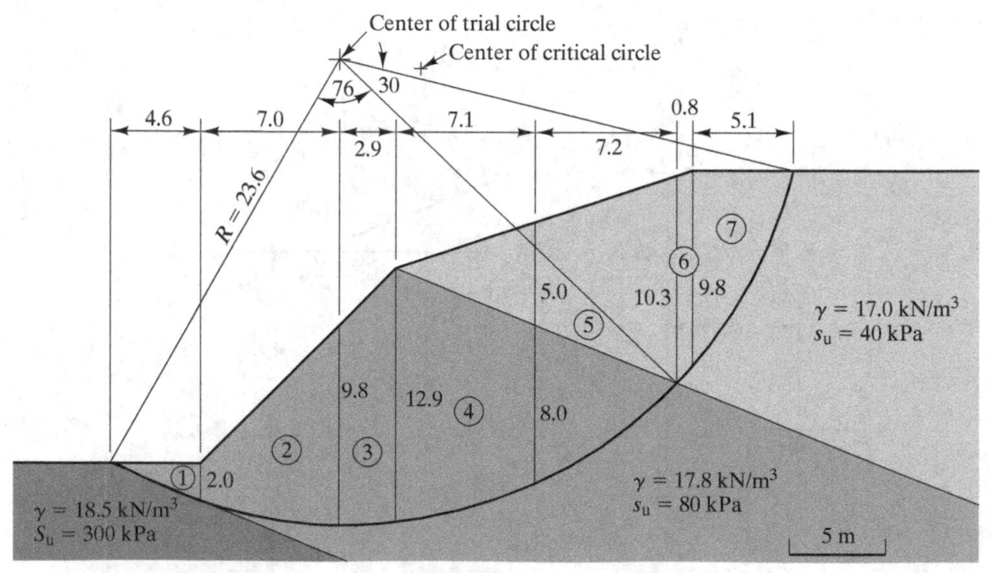

FIGURE 13.22 Cross-section for Example 13.2.

$$W_6/b = 0.8\left(\frac{10.3 + 9.8}{2}\right)17.0 = 140 \text{ kN/m}$$

$$W_7/b = 5.1\left(\frac{9.8}{2}\right)17.0 = 420 \text{ kN/m}$$

Moment arms:

$$d_1 = -7.0 - \left(\frac{4.6}{3}\right) = -8.5 \text{ m}$$

$$d_2 = -\frac{7.0}{2} = -3.5 \text{ m}$$

$$d_3 = \frac{2.9}{2} = 1.5 \text{ m}$$

$$d_4 = 2.9 + \frac{7.1}{2} = 6.5 \text{ m}$$

$$d_5 = 2.9 + 7.1 + \frac{7.2}{2} = 13.6 \text{ m}$$

$$d_6 = 2.9 + 7.1 + 7.2 + \frac{0.8}{2} = 17.6 \text{ m}$$

$$d_7 = 2.9 + 7.1 + 7.2 + 0.8 + \frac{5.1}{3} = 19.7 \text{ m}$$

Slice	s_u (kPa)	θ (Deg)	$s_u\theta$	W/b (kN/m)	d (m)	$(W/b)d$ (kN-m/m)
1				80	−8.5	−680
2				740	−3.5	−2,590
3	80	76	6,080	590	1.5	890
4				1,620	6.5	10,530
5				1,450	13.6	19,720
6	40	30	1,200	140	17.6	2,460
7				420	19.7	8,270
		Σ =	7,280		Σ =	38,600

Compute the factor of safety using Equation 13.20:

$$F = \frac{\pi R^2}{180} \frac{\sum(s_u\theta)}{\sum[(W/b)d]} = \frac{\pi(23.6)^2}{180} \frac{7280}{38{,}600} = \mathbf{1.83}$$

Comments: The computed factor of safety for this circle is 1.83. However, to find the factor of safety of the slope, we need to search for the critical circle. We use a process of controlled trial-and-error searches to find the critical circle. Figure 13.23 illustrates the

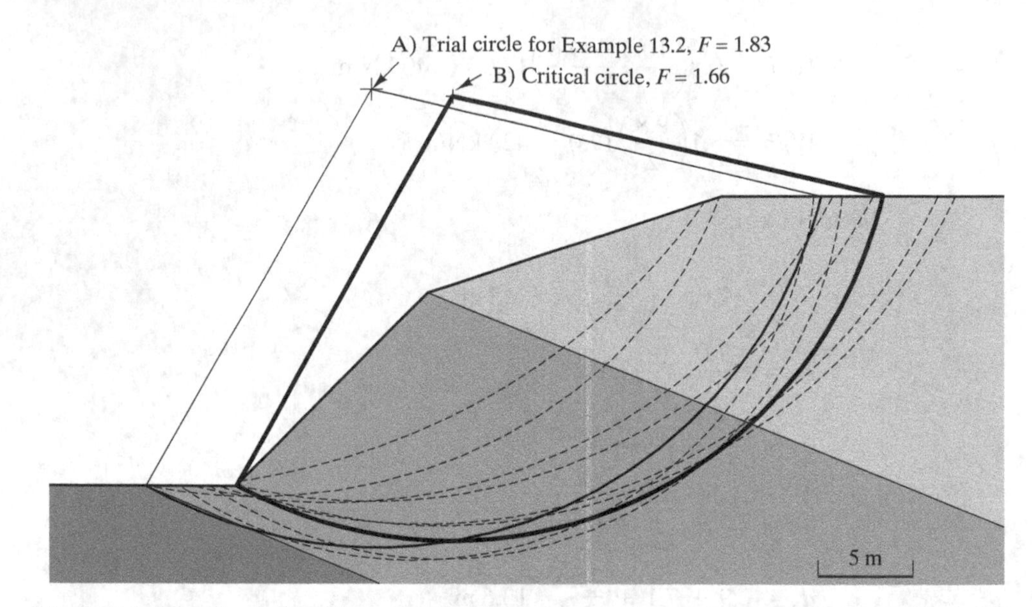

FIGURE 13.23 Illustration of the process of searching for the critical circle. Circle A is the circle used in the computations for Example 13.2. The dashed arcs represent other trial circles used in the search. Circle B is the circle with the lowest factor of safety or the critical circle.

process of finding the critical circle. The dashed arcs in this figure represent various trial circles. From this process, we find that the critical circle is centered at the location shown in Figure 13.23, has a radius of 23.4 m, and gives a factor of safety of 1.66. Thus, the computed factor of safety against a slide is 1.66, which is greater than the usual standard of 1.5, and thus is probably acceptable. In practice, we use computer programs to aid in the search for the critical circle. It is not uncommon to use hundreds or even thousands of trial circles when searching for the critical circle.

Note that the critical circle does not penetrate into the lowermost stratum, because this stratum has a shear strength significantly higher than the others.

13.9 METHOD OF SLICES

Because the Swedish slip circle method is applicable only to the $\phi = 0$ case, we need more general methods that can handle circular slides where the friction angle (either total stress or effective stress) is not equal to zero. When the friction angle is not equal to zero, the shear strength of the soil depends on the normal stress acting on the failure surface, which typically varies along the failure surface. To better account for this varying normal stress, and therefore shear strength, along the failure surface, we discretize the failure surface into small pieces, and hence the slide body into slices, as shown in Figure 13.24.

As the number of slices increases, so does the accuracy of the analysis. In practice, the slices are chosen such that the bottom of each one passes through only one type of material, and so that each slice is thin enough that its base may be approximated by the

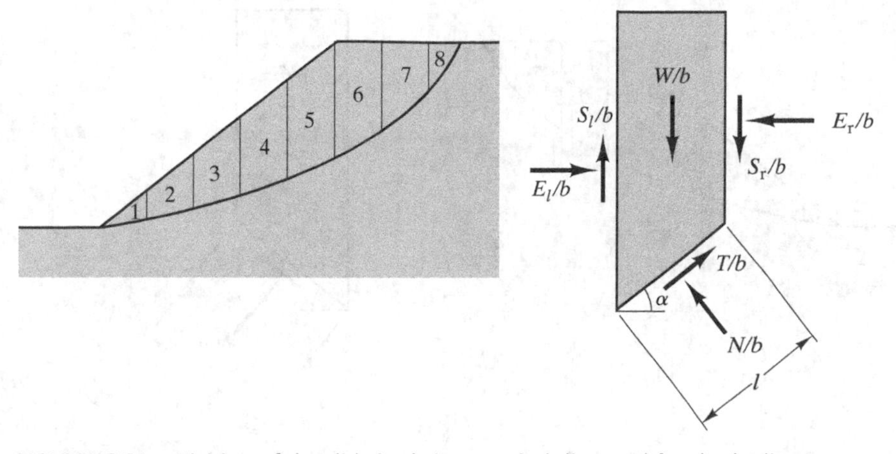

FIGURE 13.24 Division of the slide body into vertical slices and free body diagram of a typical slice.

chord that corresponds to the circular arc that forms the base. Typically 10–40 slices are sufficient for practical problems.

We can draw a free body diagram for each slice, including in each diagram both shear and normal forces acting on the base of each slice and the vertical sides between slices, as shown in Figure 13.24. These forces, as well as the factor of safety F, which is assumed to be constant for all slices, are unknowns. To solve for F, we can write three static equilibrium equations ($\Sigma F_x = 0$, $\Sigma F_z = 0$, $\Sigma M = 0$) for each free body. A more detailed inventory of all unknowns and available equations from all sources reveals that the number of unknowns is larger than the number of equations, rendering this problem a statically indeterminate one whenever the number of slices is larger than one (Duncan, 1996). In practice, most problems require tens of slices to adequately model the shear strength variation along the failure surface, and thus are statically indeterminate.

Statically indeterminate problems may be solved by increasing the number of equations, as done in indeterminate structural analyses, and/or decreasing the number of unknowns. For slope stability problems we use the latter approach, and do so by introducing simplifying assumptions. Many different analysis methods have been developed, each based on a different set of simplifying assumptions, mainly regarding the interslice or side forces (Fredlund et al., 1981). These methods are collectively called *the method of slices*.

Ordinary Method of Slices

Fellenius (1927, 1936) transformed the problem of analyzing a slide using many slices into a statically determinate problem by neglecting the effects of all interslice or side forces, the normal and shear forces acting on the two sides of each slice. This means we do not need to know their magnitudes or points of application. With this assumption, the resulting method of slices is known by different names, including the *ordinary method of slices* (OMS), the *Swedish method of slices*, and the *Fellenius method*.

Using this assumption, and the corresponding free body diagram of a typical slice shown in Figure 13.25, we can write static equilibrium equations for each free body.

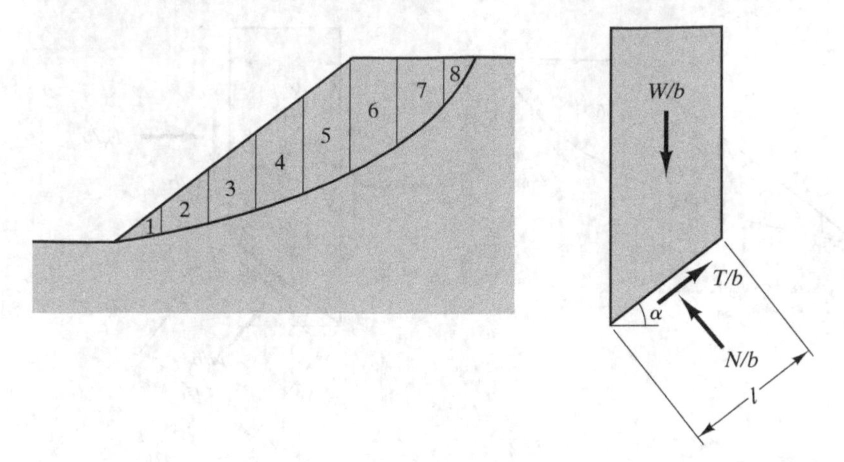

FIGURE 13.25 Ordinary method of slices: cross-section and free body diagram of a typical slice.

For equilibrium of the slice, sum of forces in the direction normal to the base of the slice is equal to zero:

$$N/b - (W/b)\cos\alpha = 0$$

$$N/b = (W/b)\cos\alpha \tag{13.21}$$

For effective stress analyses, the shear strength s is given by the Mohr–Coulomb failure criterion and the effective stress parameters c' and ϕ'. The effective stress on the base of the slice is given by

$$\sigma' = \sigma - u$$

$$\sigma' = \frac{N/b}{l} - u$$

$$\sigma' = \frac{(W/b)\cos\alpha}{l} - u \tag{13.22}$$

Using the Mohr–Coulomb failure criterion,

$$s = c' + \sigma'\tan\phi'$$

$$s = c' + \left[\frac{(W/b)\cos\alpha}{l} - u\right]\tan\phi' \tag{13.23}$$

Using the definition of F and Equation 13.23,

$$\tau_e = s/F$$

$$\tau_e = \frac{c' + \left[\dfrac{(W/b)\cos\alpha}{l} - u\right]\tan\phi'}{F} \tag{13.24}$$

Now, the equilibrium shear stress τ_e is related to the shear force T/b:

$$T/b = \tau_e l \tag{13.25}$$

Substituting τ_e given by Equation 13.24 into Equation 13.25,

$$T/b = \frac{c'l + [(W/b)\cos\alpha - ul]\tan\phi'}{F} \tag{13.26}$$

For overall moment equilibrium, the sum of moments of all forces acting on all slices about the center of the circular failure surface having a radius R is equal to zero:

$$\sum[(W/b)R\sin\alpha - (T/b)R] = 0$$

$$\sum\left\{(W/b)R\sin\alpha - \frac{c'l + [(W/b)\cos\alpha - ul]\tan\phi'}{F}R\right\} = 0 \tag{13.27}$$

Rearranging, the factor of safety for effective stress analyses from the OMS is given by

$$F = \frac{\sum\{c'l + [(W/b)\cos\alpha - ul]\tan\phi'\}}{\sum[(W/b)\sin\alpha]} \tag{13.28}$$

For total stress analyses using the OMS and the case of $c_T = s_u$ and $\phi_T = 0$, a formula for F can be similarly derived as

$$F = \frac{\sum(s_u l)}{\sum[(W/b)\sin\alpha]} \tag{13.29}$$

Example 13.3

A 30 ft tall, 1.5:1 slope is to be built as shown in Figure 13.26. The soil is homogeneous, with $c' = 400$ lb/ft^2 and $\phi' = 29°$. The unit weight is 119 lb/ft^3 above the groundwater table, and 123 lb/ft^3 below. Using the OMS, perform an effective stress analysis to compute the factor of safety for the trial circle shown in Figure 13.26.

Solution:

Weights:

$$W_1/b = 10.8\left(\frac{10.3}{2}\right)119 = 6600 \text{ lb/ft}$$

$$W_2/b = 9.4\left(\frac{10.3 + 12.5}{2}\right)119 + 9.4\left(\frac{5.2}{2}\right)123 = 15{,}800 \text{ lb/ft}$$

$$W_3/b = 12.1\left(\frac{12.5 + 14.6}{2}\right)119 + 12.1\left(\frac{5.2 + 10.0}{2}\right)123 = 30{,}800 \text{ lb/ft}$$

$$W_4/b = 12.7\left(\frac{14.6 + 16.8}{2}\right)119 + 12.7\left(\frac{10.0 + 10.7}{2}\right)123 = 39{,}900 \text{ lb/ft}$$

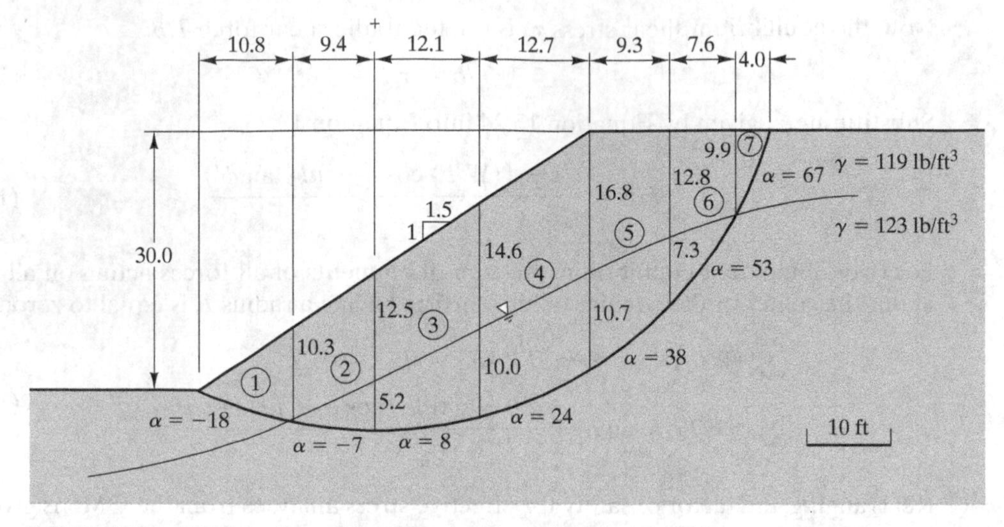

FIGURE 13.26 Cross-section of the proposed slope for Example 13.3.

$$W_5/b = 9.3\left(\frac{16.8 + 12.8}{2}\right)119 + 9.3\left(\frac{10.7 + 7.3}{2}\right)123 = 26{,}700 \text{ lb/ft}$$

$$W_6/b = 7.6\left(\frac{12.8 + 9.9}{2}\right)119 + 7.6\left(\frac{7.3}{2}\right)123 = 13{,}700 \text{ lb/ft}$$

$$W_7/b = 4.0\left(\frac{9.9}{2}\right)119 = 2400 \text{ lb/ft}$$

Average pore water pressure at base of each slice:

$$u_1 = 0$$

$$u_2 = \left(\frac{5.2}{2}\right)62.4 = 160 \text{ lb/ft}^2$$

$$u_3 = \left(\frac{5.2 + 10.0}{2}\right)62.4 = 470 \text{ lb/ft}^2$$

$$u_4 = \left(\frac{10.0 + 10.7}{2}\right)62.4 = 650 \text{ lb/ft}^2$$

$$u_5 = \left(\frac{10.7 + 7.3}{2}\right)62.4 = 560 \text{ lb/ft}^2$$

$$u_6 = \left(\frac{7.3}{2}\right)62.4 = 230 \text{ lb/ft}^2$$

$$u_7 = 0$$

Slice	W/b (lb/ft)	α (Deg)	c' (lb/ft^2)	ϕ' (Deg)	u (lb/ft^2)	l (ft)	$c'l + [(W/b) \cos \alpha - ul] \tan \phi'$	$(W/b) \sin \alpha$
1	6,600	−18	400	29	0	11.4	8,000	−2,000
2	15,800	−7	400	29	160	9.5	11,700	−1,900
3	30,800	8	400	29	470	12.2	18,600	4,300
4	39,900	24	400	29	650	13.9	20,800	16,200
5	26,700	38	400	29	560	11.8	12,600	16,300
6	13,700	53	400	29	230	12.6	8,000	10,900
7	2,400	67	400	29	0	10.2	4,600	2,200
						$\Sigma =$	84,300	46,000

Note that l of a slice is equal to the horizontal width of the slice divided by cos α. For example, for slice 1,

$$l_1 = \frac{10.8 \text{ ft}}{\cos 18°} = 11.4 \text{ ft}$$

Compute F using Equation 13.28:

$$F = \frac{84{,}300}{46{,}000} = \mathbf{1.83}$$

Note how slices 1 and 2 have negative α values because they are inclined backward.

Comments: It is still necessary to perform further trials with other circles to locate the critical circle. For this particular case, a more detailed analysis with a computer program using over 3000 trial circles and 30 slices for each circle shows that the lowest factor of safety is 1.76. Therefore, according to the OMS, the factor of safety of this slope is 1.76.

The Importance of Side Forces The simplifying assumption of no side forces in the OMS reduces the problem to one that is both statically determinate and suitable for hand computations. Nevertheless, we must ask "how valid is this assumption?"

If we consider a typical slice, as shown in Figure 13.24, the resultant of the shear and normal forces on the left side is typically larger than that on the right, and thus contributes to the normal force on the base, N/b. However, the OMS ignores this contribution and computes N based only on the weight of the slice. This produces an N value that is too low, an s value that is too low, and therefore an F value that is too low (i.e., it produces results that are conservative). Thus, the OMS can be characterized as *non-rigorous*.

This conservatism is most pronounced when α is large. For shallow circles, the computed factor of safety is generally no more than 20% less than the "correct" value, but deep, small radius circles that extend well below the groundwater table give much more error, sometimes producing computed F values as much as 50% too low (Wright, 1985).

Several researchers and engineers have developed more refined methods of analyzing circular failure surfaces based on more reasonable assumptions. Some of these methods can be characterized as *semi-rigorous*, including the *modified* (or *simplified*) *Bishop's method* and *modified* (or *simplified*) *Janbu's method*; and some *rigorous*, such as the *Spencer's method* and *Morgenstern-Price method*. We will discuss the modified Bishop's and Spencer's methods in the following sections.

Modified Bishop's Method

Bishop (1955) made a more reasonable assumption on the side forces by assuming that the shear forces on the sides of each slice are zero. Although this assumption is only an approximation to the truth, it is much better than the assumption used in the OMS. The resulting method based on this assumption is called the modified Bishop's method or the simplified Bishop's method. Careful studies have shown that it produces computed F values within a few percent of the "correct" values (Wright, 1985), and thus is sufficiently precise for virtually all circular analyses. Therefore, this is a recommended and widely used method for circular failure surfaces.

Using the assumption of no interslice shear forces and the corresponding free body diagram of a typical slice as shown in Figure 13.27, we can write equilibrium equations to derive the modified Bishop's equation for the factor of safety for effective stress analyses:

$$F = \frac{\sum\left\{\dfrac{mc' + [(W/b) - um]\tan\phi'}{\psi}\right\}}{\sum[(W/b)\sin\alpha]} \tag{13.30}$$

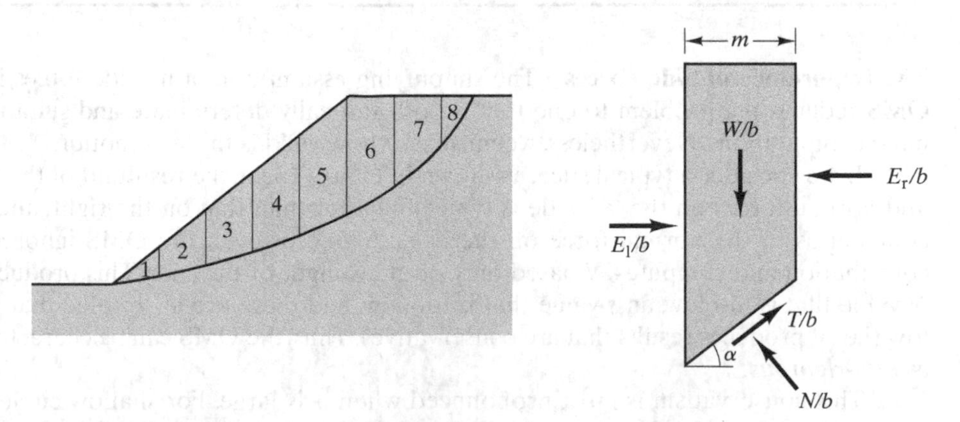

FIGURE 13.27 Modified Bishop's method: cross-section and free body diagram of a typical slice.

where for each slice,

$$\psi = \cos\alpha + \frac{\sin\alpha\tan\phi'}{F} \tag{13.31}$$

For total stress analyses and the case of $c_T = s_u$ and $\phi_T = 0$, Equations 13.30 and 13.31 combine to give

$$F = \frac{\sum\left(\dfrac{ms_u}{\cos\alpha}\right)}{\sum[(W/b)\sin\alpha]} \tag{13.32}$$

Note that, for the $\phi = 0$ case, the Swedish slip circle method, OMS, and the modified Bishop's method all give the same F.

Although a problem can be solved by hand using the modified Bishop's method, the solution for effective stress analyses is more tedious than using the OMS because the modified Bishop's equation (Equation 13.30) is not a closed-form solution, that is, the factor of safety appears on both sides of the equation. Therefore, it is necessary to use an iterative procedure to solve for F, and for each iteration: (a) estimate a value for F (which can be the F value computed using the OMS); (b) compute ψ using Equation 13.31 for each slice; and (c) compute F using Equation 13.30. The iterations continue with an updated estimate for F every time (the computed value from the previous iteration is a good choice) until the difference between the estimated and computed values is within an acceptable tolerance. Usually three iterations are sufficient. Of course, this kind of iterative procedure can be carried out much more efficiently using a spreadsheet or computer program.

Example 13.4

Solve Example 13.3 using the modified Bishop's method.

Solution:

First iteration—try $F = 1.90$

(1) = numerator in Equation 13.30

$(W/b)\sin\alpha$ = denominator in Equation 13.30

Slice	W/b (lb/ft)	α (Deg)	c' (lb/ft²)	ϕ' (Deg)	u (lb/ft²)	m (ft)	$(W/b)\sin\alpha$	Try $F = 1.90$ ψ	Try $F = 1.90$ (1)
1	6,600	−18	400	29	0	10.8	−2,000	0.861	9,300
2	15,800	−7	400	29	160	9.4	−1,900	0.957	12,200
3	30,800	8	400	29	470	12.1	4,300	1.031	18,200
4	39,900	24	400	29	650	12.7	16,200	1.032	21,900
5	26,700	38	400	29	560	9.3	16,300	0.968	16,000
6	13,700	53	400	29	230	7.6	10,900	0.835	11,600
7	2,400	67	400	29	0	4.0	2,200	0.659	4,400
							46,000		93,600

$$F = \frac{93{,}600}{46{,}000} = 2.03$$

The computed F of 2.03 is greater than the assumed value of 1.90.

Second iteration—try $F = 1.95$

① = numerator in Equation 13.30

$(W/b) \sin \alpha$ = denominator in Equation 13.30

Slice	W/b (lb/ft)	α (Deg)	c' (lb/ft²)	ϕ' (Deg)	u (lb/ft²)	m (ft)	$(W/b) \sin \alpha$	Try $F = 1.95$ ψ	①
1	6,600	−18	400	29	0	10.8	−2,000	0.863	9,200
2	15,800	−7	400	29	160	9.4	−1,900	0.958	12,200
3	30,800	8	400	29	470	12.1	4,300	1.030	18,200
4	39,900	24	400	29	650	12.7	16,200	1.029	22,000
5	26,700	38	400	29	560	9.3	16,300	0.963	16,200
6	13,700	53	400	29	230	7.6	10,900	0.829	11,700
7	2,400	67	400	29	0	4.0	2,200	0.652	4,500
							46,000		94,000

$$F = \frac{94{,}000}{46{,}000} = 2.04$$

The computed F of 2.04 is greater than the assumed value of 1.95.

Further trials will produce F = **2.05**

The computed factor of safety for this circle is 2.05, which is slightly higher than the 1.83 computed using the OMS.

Comments: Once again, a more detailed analysis with a computer program using over 3000 trial circles and 30 slices for each circle shows that the lowest factor of safety is 2.00. Therefore, according to the modified Bishop's method, the factor of safety of this slope is 2.00, which is slightly higher than the 1.76 obtained by the OMS.

Spencer's Method

Spencer's method (Spencer, 1967, 1973; Sharma and Moudud, 1992) is popular among geotechnical engineers because it combines good precision with ease of use. Although the solution still requires a computer program, the required user input is simpler than that of some other methods.

Spencer assumed that all the resultants of the normal and shear side forces are inclined at the same angle θ from the horizontal. Like the modified Bishop's method, Spencer's method requires an iterative procedure to solve for F: (a) assume an initial value for θ; (b) compute one value of F based on force equilibrium and another based on moment equilibrium; and (c) if the difference between the computed force equilibrium F value and moment equilibrium F value is within an acceptable tolerance, the solution has converged; otherwise, repeat Steps (a)–(c) with an updated value of θ. This iterative procedure is much too tedious to do by hand, but quite simple using a computer.

Spencer's assumption of inclined resultant side forces is more general than the assumptions used in non-rigorous and semi-rigorous methods. Because of this more general assumption on the side forces and the fact that the analysis satisfies both force and moment equilibrium, Spencer's method can be considered to be a rigorous method.

Besides Spencer's method, there are other similarly rigorous methods in practice, such as the Morgenstern–Price method. Many of these rigorous methods, as well as less rigorous methods, have been implemented into computer programs, available both from the public domain and from private software developers, and are routinely used in today's geotechnical engineering practice.

13.10 CHART SOLUTIONS

When the soil in the slope is homogeneous, the analysis can be substantially simplified and performed using simple charts. Several such *stability charts* have been developed (see Abramson et al., 2002), and they are useful for simple slopes that may not justify the time required to perform a computer-based analysis. Even for more complex slopes that require computer-based analyses, stability charts offer a way to check the computer-generated results. In this case, the complex slope is simplified to a slope that can be analyzed using stability charts, and the chart solution is compared with the computer solution.

Examples of stability charts are ones developed by Cousins (1978), one of which is reproduced in Figure 13.28. It is based on the geometry shown in Figure 13.29.

To use Cousins' chart, first compute $\lambda_{c\phi}$:

$$\lambda_{c\phi} = \frac{\gamma H \tan \phi}{c} \tag{13.33}$$

where $c = c'$ and $\phi = \phi'$ for effective stress analyses, or $c = c_T$ and $\phi = \phi_T$ for total stress analyses.

The chart in Figure 13.28 is valid only if all of the following conditions are met:

- The soil is homogeneous with constant unit weight and shear strength throughout
- The ground surface consists of a planar slope face with level ground above and below, as shown in Figure 13.29
- $\lambda_{c\phi} \geq 2$ and/or $\beta > 53°$ (this means the critical circle will pass through the toe of the slope)
- The groundwater table is well below the toe of the slope.

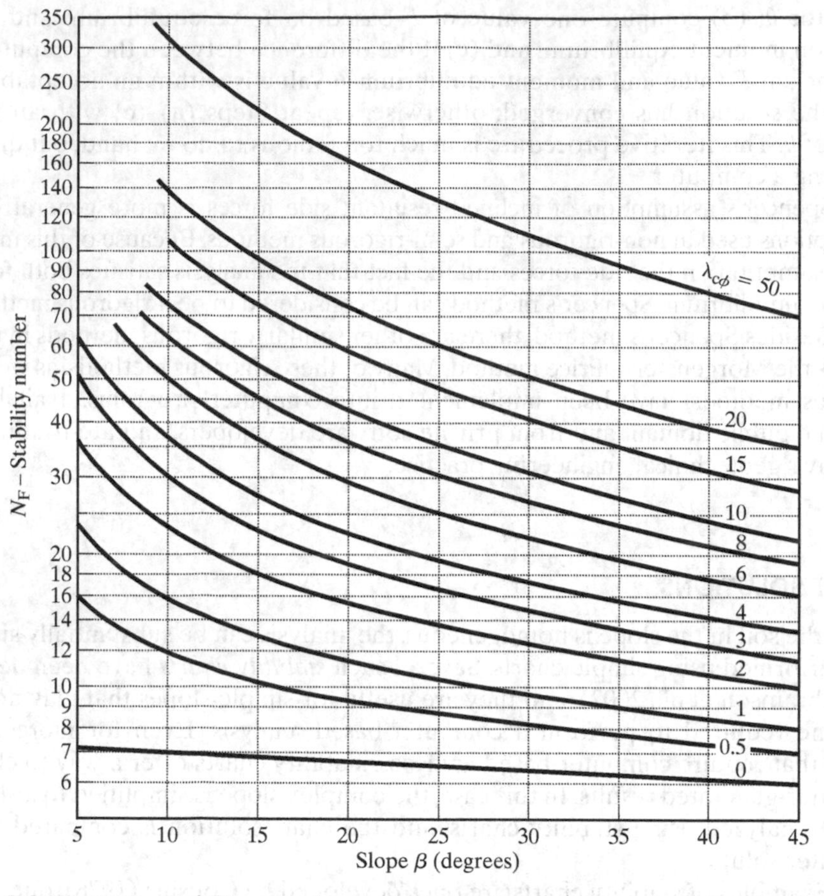

FIGURE 13.28 Stability chart for simple slopes that meet the criteria described in the text (Cousins, 1978). (Used with permission of ASCE.)

If all of these conditions are met, determine N_F from Figure 13.28, and then compute the factor of safety using

$$F = N_F \frac{c}{\gamma H} \tag{13.34}$$

where $c = c'$ or c_T as appropriate.

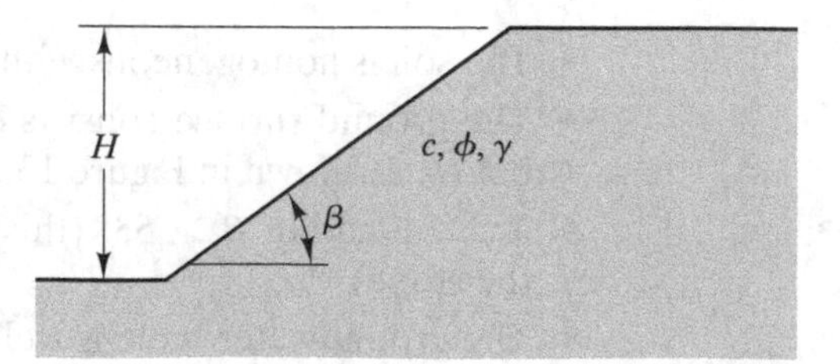

FIGURE 13.29 Slope geometry for Cousins' chart.

The chart is based on the critical slip surface, so only one iteration is required to obtain the factor of safety.

This chart is based on solutions obtained by the *friction circle method*, and generally produces computed factors of safety comparable to those obtained from the OMS. However, it is slightly less precise than a conventional solution. Cousins and others also have developed charts for slopes with more complex conditions, but such slopes are probably best analyzed using computer-based methods. However, as mentioned earlier, regardless of how complex the slope is, stability charts do offer a way to check results from computer-based methods.

Example 13.5

Using the cross-section in Figure 13.26 with a very deep groundwater table, compute the factor of safety using Cousins' chart.

Solution:

Using effective stress parameters,

$$\lambda_{c\phi} = \frac{\gamma H \tan \phi}{c} = \frac{(119\ \text{lb/ft}^3)(30\ \text{ft})\tan 29^\circ}{400\ \text{lb/ft}^2} = 4.9$$

$\lambda_{c\phi} = 4.9 > 2$, so the critical circle is a toe circle. The other relevant criteria also are satisfied, so Cousins' chart is applicable to this problem.

$$\beta - \tan^{-1}\left(\frac{1}{1.5}\right) = 34^\circ$$

From Figure 13.28, $N_F = 17.1$.

$$F = N_F \frac{c}{\gamma H} = 17.1\left[\frac{400\ \text{lb/ft}^2}{(119\ \text{lb/ft}^3)(30\ \text{ft})}\right] = \mathbf{1.92}$$

Note that this is the factor of safety for the critical circle. There is no need to search for this circle. The computed factor of safety of 1.92 is slightly higher than the 1.76 obtained by the OMS, and slightly lower than the 2.00 obtained from the modified Bishop's method. Part of this difference is because this example used a different groundwater table than was used in the previous examples.

13.11 MISCELLANEOUS ISSUES

Irregularly Shaped Failure Surfaces

Many failure surfaces are neither planar nor circular, and thus cannot be analyzed using any of the methods described thus far. Even supposedly circular surfaces are often truncated at the top due to the formation of tension cracks. Therefore, geotechnical engineers have developed additional analysis methods to accommodate irregularly shaped failure surfaces. These are sometimes called *non-circular analyses*. These methods lose the mathematical simplicity of convenient geometry, and thus are more complex and difficult to implement. Most are practical only when solved by a computer.

Randomly shaped failure surfaces also make searching routines more difficult, because the failure surface can no longer be defined by only three variables. Although some software includes searching capabilities, much more skill is required to locate the most critical failure surface.

Analysis methods have been proposed by Janbu (1957, 1973), Morgenstern and Price (1965), Spencer (1967), Sarma (1973), and others. Each method uses different simplifying assumptions to overcome the problem of static indeterminancy, and thus produces slightly different results.

Partially Submerged Slopes

Many slopes are partially submerged, and thus are subjected to external hydrostatic pressures as shown in Figure 13.30. Examples include levees and earth dams. To analyze such slopes, we simply treat the external water as if it were a "soil" with no shear strength ($c' = c_T = 0$ and $\phi' = \phi_T = 0$), and $\gamma = \gamma_w$.

In addition to the typical modes of failure, partially submerged slopes are potentially subject to a special mode of failure called *rapid drawdown failure* (Duncan et al., 1990; Borja and Kishnani, 1992). This type of failure occurs when the exterior water level has been at a certain elevation for a long time, and then quickly drops to a lower elevation as shown in Figure 13.31. If the soils have low hydraulic conductivities, the groundwater table inside the slope cannot drop nearly as rapidly as the water outside, so high pore water pressures inside the slope remain unchanged, even though the stabilizing effect of the exterior hydrostatic pressures rapidly disappears. This unfortunate situation can occur in earth dams when the reservoir is lowered too rapidly, or in levees when river water levels go down rapidly. This combination of factors may produce a landslide.

Rapid drawdown failures have occurred on occasion. For example, the Walter Bouldin Dam in Alabama experienced rapid drawdown-induced landslide in 1975 when the reservoir level was lowered 10 m in 5.5 hours (Duncan et al., 1990).

Back-Calculated Strength

Most slope stability analyses begin with measured soil strength data and compute the factor of safety. However, when working with landslides that have already occurred, it is often useful to do the reverse analysis: assume $F = 1$ (the slope was at limiting

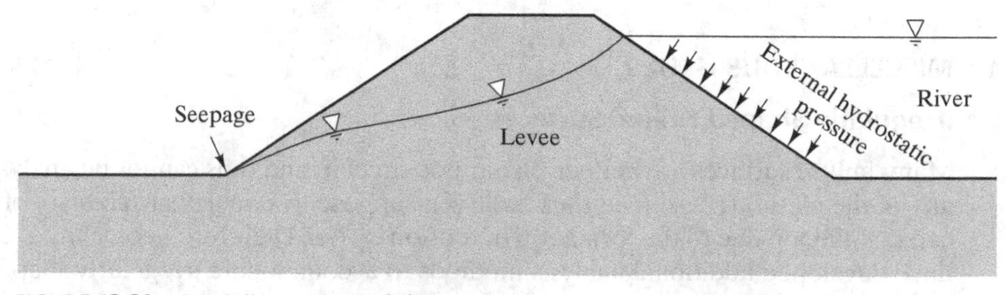

FIGURE 13.30 Partially submerged slope.

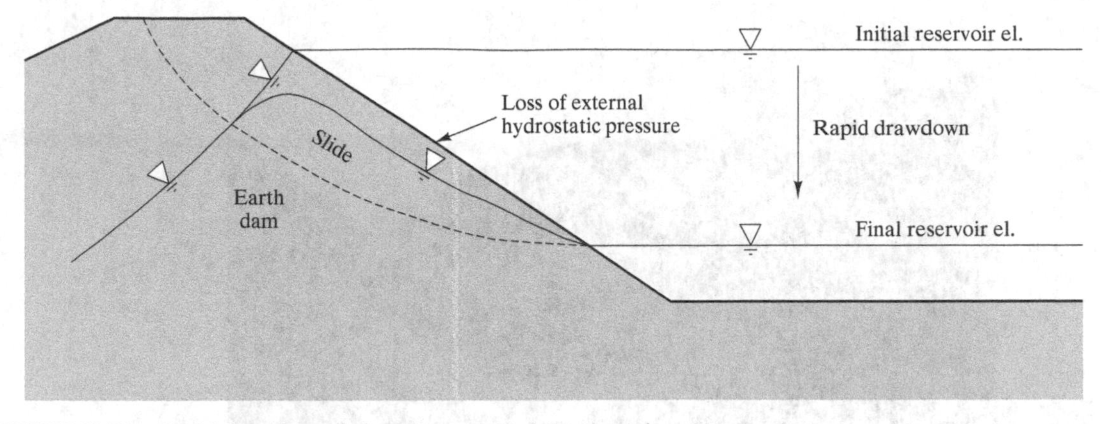

FIGURE 13.31 Rapid drawdown failure in an earth dam. el. = elevation.

equilibrium when it failed) and back-calculate the soil strength (Duncan and Stark, 1992). Soil strengths obtained in this way are generally very reliable, because they are based on the full shear surface, not on small samples. This *back-calculated strength* can then be used for subsequent analyses of proposed remediation measures.

The soil strength is usually defined by two parameters, the cohesion and friction angle, so the solution to a back-calculated strength analysis cannot produce a unique combination of values of the two parameters. Instead, we obtain a plot of the various combinations of cohesion and friction angle values that produce $F = 1$, and select one of these combinations for subsequent analyses.

13.12 SEISMIC STABILITY

A large number of slope failures have occurred during earthquakes, so geotechnical engineers working in seismically active regions routinely evaluate the seismic stability of earth slopes. This is a part of the broader discipline of *geotechnical earthquake engineering.*

Some earthquake-induced failures are very large. For example, the 1959 Hebgen Lake Earthquake triggered a massive slide in Madison Canyon, Montana, as shown in Figure 13.32. This slide had a volume of about 25 million yd^3 (20 million m^3) and traveled at an estimated velocity of 110 mi/hr (180 km/hr), creating a 220 ft (67 m) tall dam across the canyon (Sowers, 1992). This dam formed a new lake similar to the one formed by the Thistle landslide described earlier. It also killed 28 campers who were in the canyon to enjoy its world-class fishing.

Most seismic failures are much smaller than the one at Madison Canyon, and often consist of significant slope distortions without fully developing a true landslide. Nevertheless, these distortions can be large enough to cause significant property damage. For example, a large number of slope distortions occurred during the 1994 Northridge Earthquake in California (Stewart et al., 1995). Most of them involved displacements of less than 8 cm, but they produced about $100 million in property damage.

The physical mechanisms of seismically induced slope movements are very complicated, and include all of the complexities of static slope stability plus those associated

FIGURE 13.32 Madison canyon landslide in Montana as it appeared in 1997, 38 years after the failure. The scarp from the slide, which is the light-colored area in the center of the photograph, is still clearly visible on the mountainside. This slide formed a dam that created Earthquake Lake, which is visible in the foreground.

with the propagation of seismic waves and the dynamic strength of soil and rock (Rogers, 1992a). Thus, geotechnical analyses of seismic stability can sometimes be very difficult. However, there is a great deal of active research on this topic, and it is helping us better understand the physical mechanisms and develop methods of analyzing these problems.

Liquefaction-Induced Failures

Many of the most dramatic and devastating earthquake-induced slides are the result of soil liquefaction. For example, the 1964 Turnagain Heights Landslide in Anchorage, Alaska, was the result of liquefaction of buried sand strata (Seed and Wilson, 1964). This slide covered an area of about 130 acres and resulted in the destruction of 75 houses, as shown in Figure 13.33.

Another example is the failure of the Lower San Fernando Dam caused by the 1971 Sylmar Earthquake in California. This dam was a hydraulic fill dam that was completed in 1918. It contained loose sands and was founded on loose sandy soils as well. During the magnitude 6.6 Sylmar Earthquake, extensive liquefaction occurred both inside the dam and in the underlying natural soils, which resulted in a large landslide in the upstream slope. Figure 13.34 shows the dam as it appeared immediately after the landslide. Fortunately, a small freeboard remained, so the reservoir did not overtop the dam. This failure focused attention on the susceptibility of hydraulic fills to seismically induced liquefaction, and prompted the re-evaluations of hydraulic fill dams in seismic regions. As a result, a number of these dams were replaced or modified to enhance their seismic stability. The key to preventing such *liquefaction-induced failures* is to properly

FIGURE 13.33 The 1964 Turnagain Heights landslide in Anchorage, Alaska. (Earthquake Engineering Research Center Library, University of California, Berkeley, Steinbrugge Collection.)

FIGURE 13.34 The Lower San Fernando Dam looking downstream, immediately after a large landslide in the upstream slope caused by liquefaction of the soils inside this hydraulic fill dam during the 1971 Sylmar Earthquake in California. A freeboard of only 1.5 m kept the dam from being overtopped by the reservoir. The failure of this dam prompted the reconstruction or replacement of several other dams that also had been constructed by hydraulic filling.

identify potentially liquefiable soils and understand their impact on slope stability. Engineers in the field of geotechnical earthquake engineering have developed methods to evaluate the susceptibility of soils to liquefaction, the triggering of liquefaction in soils, and the effects of liquefaction; however, these methods are outside the scope of this book.

Failures Caused Directly by Ground Shaking

Slope failures can also occur as a direct result of ground shaking, even without soil liquefaction. These most often occur in steep slopes, especially those covered with loose natural soils (usually colluvium) or poorly constructed fills.

Although this mode of failure can readily be identified after it occurs, it is sometimes difficult to recognize potentially hazardous slopes before the earthquake occurs. Often such assessments can be based on empirical comparisons of soil type and unit weight, slope ratio, groundwater conditions, and other factors. For example, if dry, natural colluvial soils of a certain thickness at a certain slope ratio were found to fail during an earthquake, similar soil conditions at another location would probably fail if subjected to a similar earthquake.

Geotechnical engineers also use quantitative analyses. However, these analyses are gross simplifications of the actual physical mechanisms, and thus may not be reliable. The most common of these are the *pseudostatic method* and *Newmark's method*, as discussed below.

Pseudostatic Method The pseudostatic method is an enhancement of conventional limit equilibrium analyses that evaluates the seismic stability of an earth slope by applying a constant horizontal acceleration to each slice. This horizontal acceleration, called the *pseudostatic acceleration*, is assumed to continue indefinitely (or at least long enough for the slope to fail), and thus is idealized as a constant horizontal static force in a direction that destabilizes the slope, as shown in Figure 13.35. This static force is equal to $(a/g)W/b$, where a is the pseudostatic acceleration, g is the acceleration due to gravity, and W/b is the weight of the slice. Sometimes a is written as kg, where k is called the *pseudostatic coefficient*, making the *pseudostatic force* kW/b.

The computations then proceed like any other limit equilibrium analysis. However, this additional force produces a lower factor of safety intended to reflect the detrimental effects of the earthquake. Normally the minimum acceptable factor of safety is also lower, typically between 1.1 and 1.2.

Although this representation of the horizontal acceleration as a constant static force greatly simplifies the computations, it is not a very accurate representation of the seismic forces in a real earth slope. The differences between this representation and reality include the following:

- The real seismic accelerations cycle back and forth in opposite directions, and continue for only a limited time.
- The wavelength of the seismic waves is smaller than the dimensions of most slopes, so part of the slope may be accelerating upslope while another part is accelerating

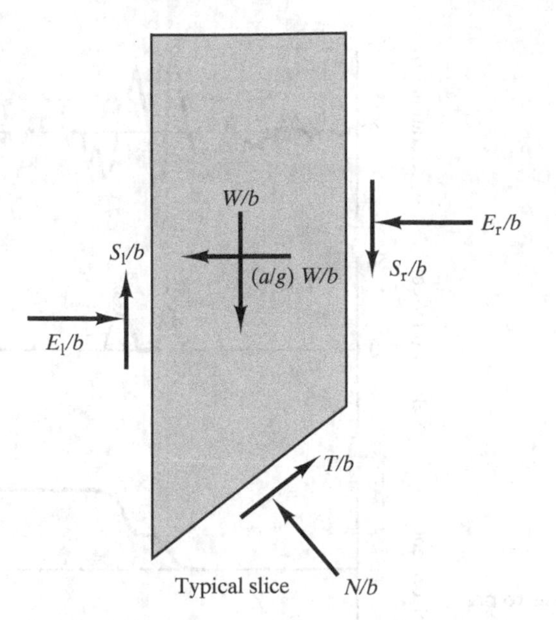

FIGURE 13.35 The pseudostatic method consists of applying a horizontal pseudostatic force to each slice.

downslope. The entire slope will not be accelerating in the same direction, even for a moment.

Because of these differences, pseudostatic analyses cannot be based on the anticipated peak ground accelerations, which are often in excess of $0.7g$. The use of such values would indicate failure in virtually all analyses. Instead, geotechnical engineers use values based on the observed behavior of slopes during earthquakes, typically $0.1\text{–}0.2g$ (Hynes-Griffin and Franklin, 1984; Kramer, 1996). Thus, the "acceleration" value in the pseudostatic analysis is really more an empirical index than a measure of the true ground accelerations.

Although the pseudostatic analysis has some value, it is only a rough approximation of the physical mechanisms acting in the field, and thus should be used only with considerable engineering judgment. In some cases, it can produce overly conservative results, and thus can dictate preventive measures that are not necessary.

When performing pseudostatic analyses, it is useful to note that slopes with static factors of safety greater than 1.70 and no liquefaction problems have never been known to fail during earthquakes (Rogers, 1992a; Hynes-Griffin and Franklin, 1984). Thus, it may be prudent to dispense with pseudostatic analyses when both these criteria have been met.

Newmark's Method Newmark (1965) developed an enhancement of the pseudostatic method that attempts to predict permanent *seismic slope displacements* during an earthquake. It does so by first establishing the *yield acceleration*, a_y, that corresponds to

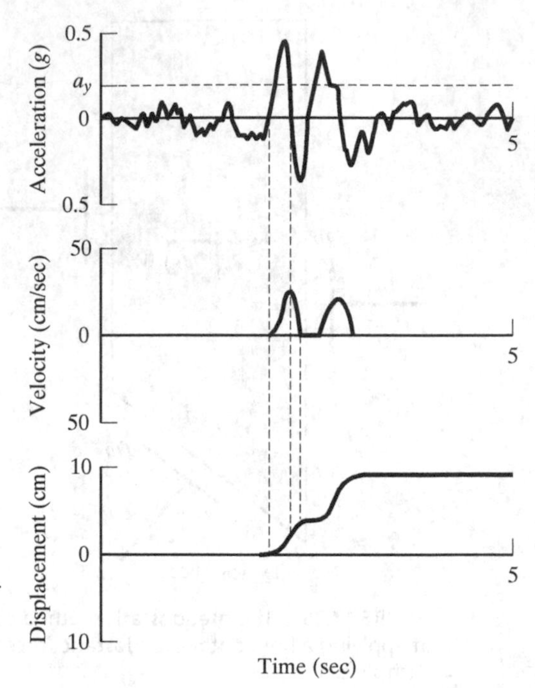

FIGURE 13.36 Use of Newmark analysis to predict permanent slope displacements from an earthquake acceleration record (Wilson and Keefer, 1985).

$F = 1$ in a conventional pseudostatic analysis. Then, the engineer obtains an acceleration versus time plot for the design earthquake, such as the one in Figure 13.36, and then identifies the time intervals where $a > a_y$. By double integrating the acceleration in these intervals, we obtain the associated permanent slope displacement, which is then compared to some maximum allowable displacement.

Although this analysis is an improvement over the pseudostatic analysis, it is still based on a simplification of the true seismic response of a real slope. It gives results that are very sensitive to the selected a_y value and other factors (Kramer, 1996), and thus requires a great deal of care to implement.

Bray (2007) reviewed Newmark's method and other simplified procedures for estimating seismic slope displacements.

13.13 STABILIZATION MEASURES

When proposed or existing slopes do not have sufficient stability, geotechnical engineers turn to various methods of slope stabilization (Hausmann, 1992; Rogers, 1992b; Abramson et al., 1996; Turner and Schuster, 1996). Many, but not all, slopes can be economically stabilized, and many methods are available. The factor of safety given by Equation 13.1 depends on both the equilibrium shear stress and the shear strength, so stabilization measures must decrease the equilibrium shear stress and/or increase the shear strength.

The selection of an appropriate stabilization plan depends on many factors, including the following:

- The subsurface conditions and potential modes of failure
- The present and required topography
- The presence of physical constraints, such as property lines or existing buildings
- The consequences of a failure (i.e., small for a rural low-traffic road, potentially catastrophic for a nuclear power plant), which determines the required reliability
- Availability of materials, equipment, and expertise (specialized methods may not be available in some areas)
- Performance history of various methods as implemented in the local area
- Aesthetics
- Time required for construction
- Cost.

Unloading

The simplest way to decrease the shear stresses in the slope is to unload it, either by reducing the slope height or by increasing the slope ratio, as shown in Figure 13.37.

For example, if the slope is associated with a proposed highway, it may be possible to decrease its height by revising the vertical alignment of the highway. Unfortunately, this solution usually results in steeper grades, which may be unacceptable. Increasing the slope ratio generally does not require a new vertical alignment, but does need a wider right-of-way and involves larger earthwork quantities. This may be quite feasible in rural areas with rolling hills, but could be prohibitive in urban areas or where the natural terrain is steep.

Another method of unloading involves construction of lightweight fills, as discussed in Chapter 6. These fills permit construction of slopes without inducing large shear stresses in the ground.

Buttressing

The short-term stability of cut slopes is generally greater than their long-term stability, so it is usually possible to make temporary construction slopes much steeper than would be permissible for the permanent slope. This is especially true when construction occurs

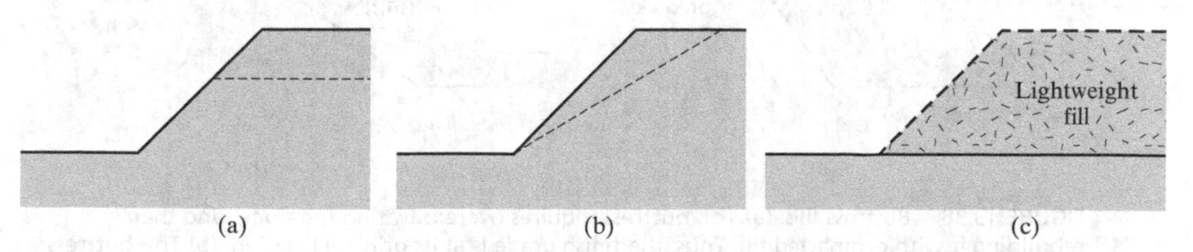

FIGURE 13.37 Slope stabilization by unloading. (a) Reducing slope height; (b) increasing slope ratio; (c) using lightweight fill.

during the dry season when the groundwater table is lower. We can use this behavior to build *buttress fills* that stabilize slopes.

The usual construction procedure is to overexcavate the proposed cut slope as shown in Figure 13.38(a), and then bring it back to the design grades using high-quality fill (i.e., one with higher shear strength than the natural soils). The size of the buttress needs to be selected so that potential failure surfaces that pass through the buttress gain enough additional strength to raise the factor of safety to an acceptable value, and that potential failure surfaces that pass below the buttress also have an acceptable factor of safety. To meet these goals, buttresses often must include downward extensions called *shear keys*, as shown in Figure 13.38.

Sometimes the buttress fill is made of crushed gravel or other very high-quality soils that have very high strength. However, the concept is also valid for normal fills, so long as they are stronger than the natural soils. For example, cut slopes that will expose daylighted bedding planes in soft sedimentary rocks often can be stabilized by making a compacted fill from the soils produced by the excavated rock. Such fills are more stable because they do not contain the weak bedding planes.

Buttresses can also be constructed without overexcavation. In this case, they become stabilization fills placed at the toe of the slope as shown in Figure 13.38(b). These can sometimes be used to stabilize existing landslides or other unstable slopes that would not tolerate steep construction excavations, or when land is not at a premium. The top of such buttresses may be level, and thus can become usable for development.

Structural Stabilization

Another option is to stabilize slopes using structural elements. These include various kinds of *earth retaining structures* and *tieback anchors*. These methods are typically very expensive, but can be cost effective in certain situations, especially in urban areas.

Earth Retaining Structures Earth retaining structures are structural systems that maintain adjacent ground surfaces at two different elevations. One example is a *retaining wall* shown in Figure 13.39. Sometimes earth-retaining structures are used in lieu of a slope,

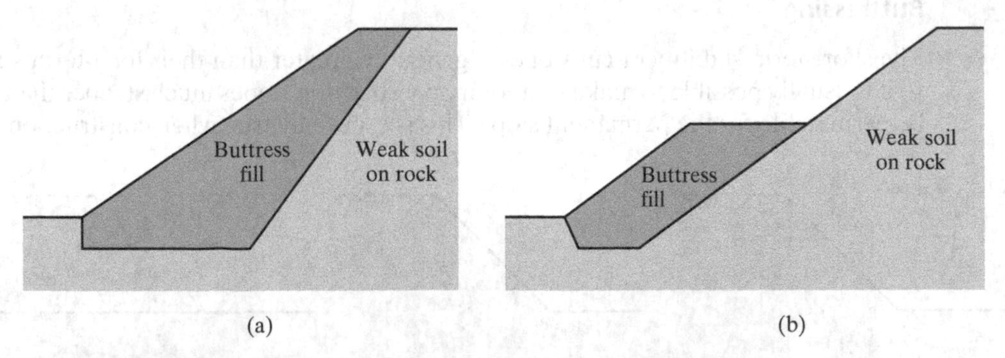

FIGURE 13.38 Buttress fills. (a) This buttress requires overexcavating the slope, and then rebuilding it with compacted fill. Thus, the finish grade is at its original location. (b) This buttress was built by adding the fill to the front of the slope without any overexcavation. In this case, the buttress is only about half the height of the slope, but any height can be built.

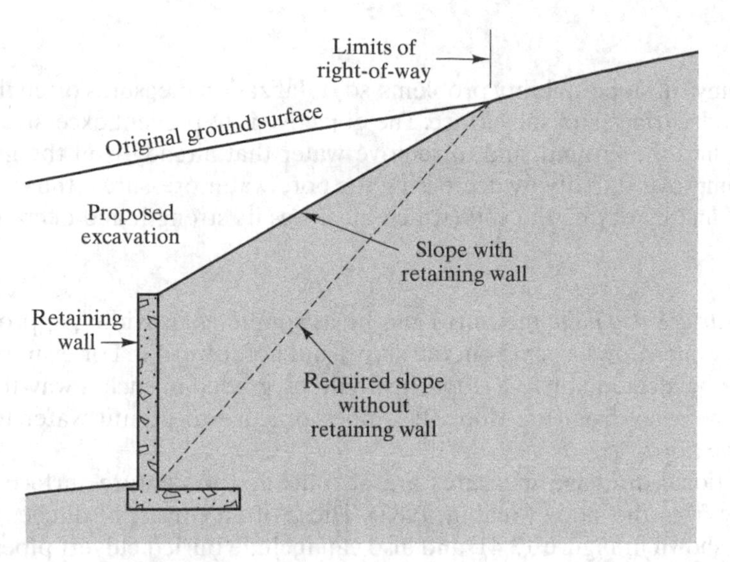

FIGURE 13.39 Use of retaining walls to stabilize slopes.

while other times they are used in conjunction with the slope to create a more stable condition. Chapter 16 discusses the various kinds of earth retaining structures.

Tieback Anchors Another structural measure is tieback anchors, or simply *tiebacks*, which are tensile members that apply stabilizing forces onto the slope as shown in Figure 13.40. Tiebacks usually consist of steel rods inside grouted holes that extend well beyond the critical failure surface. This method is generally very expensive, but may be cost effective in urban areas where space is at a premium and land is expensive.

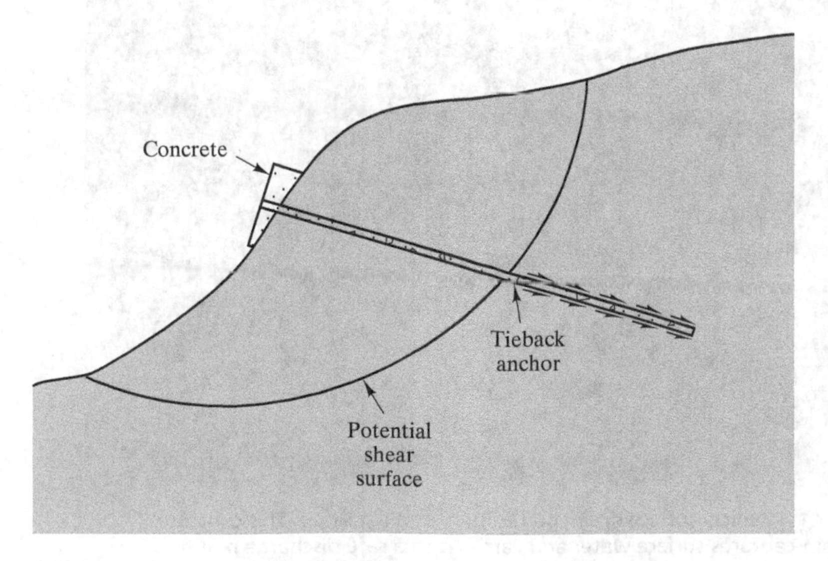

FIGURE 13.40 Use of tieback anchors to stabilize slopes.

Drainage

Water is the "enemy" in slope stability problems, so stabilization measures often involve draining water, both surface and subsurface. The objective is to prevent excessive water from percolating into the ground, and to remove water that already is in the ground. These measures improve stability by decreasing the pore water pressures (thus increasing the strength), and by drying the soil (which increases its strength and decreases its weight).

Surface Some *surface drainage* measures can be as simple as providing appropriate grades, so surface water flows away from the slope, and not toward it. For example, if a building pad is to be located above a slope, it should be graded in such a way that the surface water flows away from the slope. It is poor practice to permit water to flow over the top of the slope.

Often, additional drainage measures are also needed to capture surface water and carry it away from the slope (Scullin, 1983). These often consist of ditches paved with concrete, as shown in Figure 13.41, and also can include buried culvert pipes.

The design of surface drainage facilities is often governed by building codes, but this does not relieve the engineer of the duty to provide additional facilities when required.

In emergencies, it may be helpful to use sandbags to divert surface water away from the slope, and plastic sheets to cover the ground and reduce infiltration.

FIGURE 13.41 Typical surface drainage facilities on a cut slope. This concrete terrace drain captures surface water and carries it to a safe discharge point.

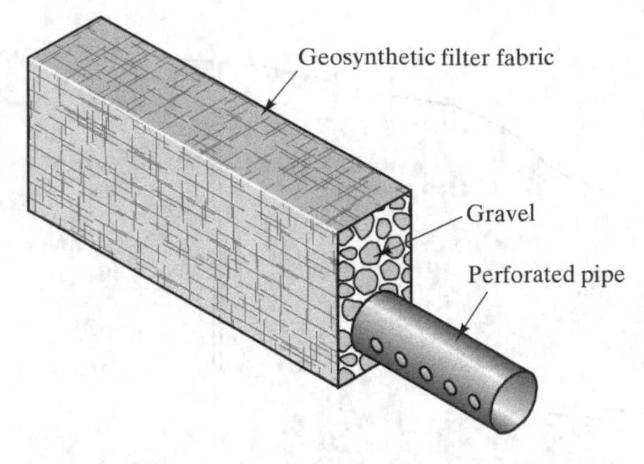

FIGURE 13.42 Perforated pipe drain.

Subsurface The objective of *subsurface drains* is to remove water that is already present in the ground. There are several types of drainage installations, including the following:

- *Perforated pipe drains* consist of special pipes with holes, buried in the ground to collect the water and carry it to a safe location. These pipes are surrounded by gravel and a *filter fabric*, to assist the entry of water and prevent finer soil from washing into the pipe and clogging it. Sometimes these drains are placed in a trench as shown in Figure 13.42 to form a *French drain*. They also may be placed below fills, behind buttress fills, and in other key locations.
- *Wells* are vertical holes drilled into the ground and equipped with pumps to remove the water as shown in Figure 13.43. Often they can double as exploratory borings. Unfortunately, the pumps are expensive to install and run, and require maintenance.
- *Horizontal drains*, shown in Figures 13.43 and 13.44, are drilled from the slope face and (in spite of their name) are inclined slightly upward. They are intended to intercept the groundwater and drain it by gravity. Horizontal drains do not require pumps, so they are less expensive to install and maintain.

Reinforcement

Structural engineers transform concrete into an efficient structural material by adding steel reinforcement at key locations. Soil fills also can be improved by incorporating synthetic reinforcement at strategic locations. These reinforcements increase strength, so they permit slopes to be built with much lower slope ratios (i.e., much steeper) than would otherwise be possible. Various materials can be used, such as steel rebars, thin steel strips, special plastic grids, and geotextiles. The plastic grids, as shown in Figures 13.45 and 13.46, are the most common tensile reinforcement material because of their durability and low cost.

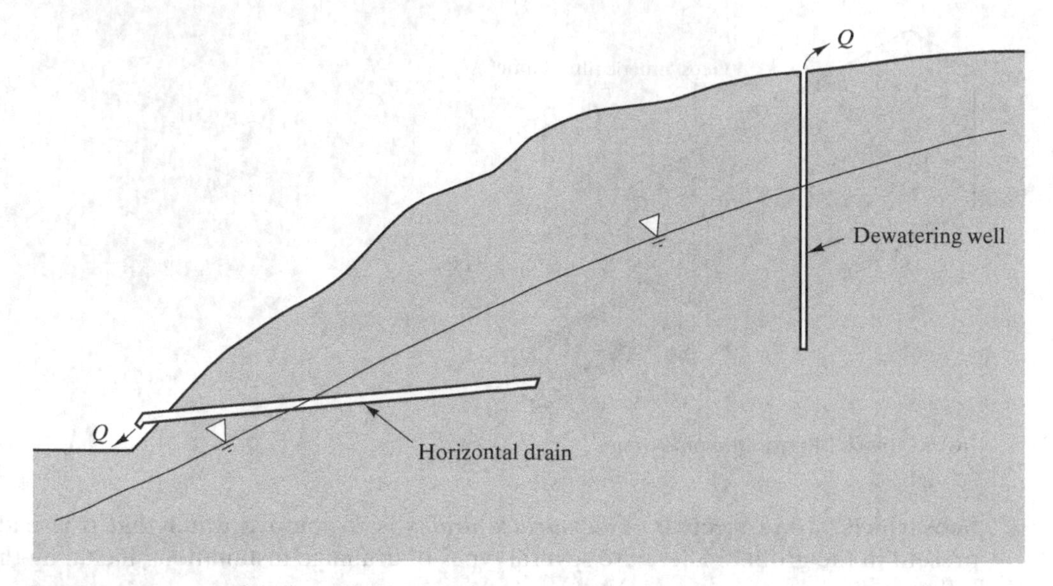

FIGURE 13.43 Use of wells and horizontal drains to remove subsurface water.

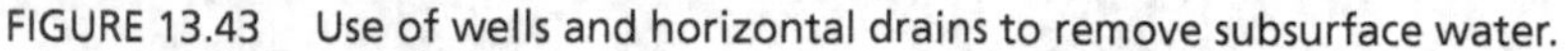

In Situ Reinforcement

In situ reinforcement methods differ from reinforced soils in that the tensile members are inserted into a soil mass rather than being embedded during placement of fill.

Soil Nailing *Soil nailing* consists of drilling near-horizontal holes into the ground, inserting steel tendons, and grouting. The face of the wall is typically covered with shotcrete, as shown in Figure 13.47.

These walls do not require a construction excavation, and thus are useful when space is limited.

Vegetation

Appropriate vegetation is an important part of most slope stabilization plans. It provides erosion protection, draws water out of the ground, provides some reinforcement of the soil, and has important aesthetic value. Although vegetation has virtually no effect on deep-seated slides, it can be very helpful in preventing shallow slides, slumps, and flows.

In arid and semi-arid areas, it is often necessary to install irrigation systems to establish and maintain the desired vegetation. These systems must be closely monitored, because excessive irrigation can introduce large quantities of water into the ground and create serious stability problems.

FIGURE 13.44 A series of horizontal drains in a cut slope along a highway. The close-up view shows one of the drains exiting the slope and discharging into a swale. This drain has a slow but continuous flow, as evidenced by an accumulation of algae at its discharge point.

13.14 INSTRUMENTATION

Geotechnical instrumentation is frequently employed in slope stability studies to help define the subsurface conditions and to monitor unstable ground. Although instrumentation is often expensive to install and monitor, it can provide valuable information that may not otherwise be available.

FIGURE 13.45 Tensar® geogrids are made in two different styles: a uniaxial geogrid, as shown on the left, is designed to resist tensile forces in one direction only; while a biaxial geogrid, as shown on the right, resists tensile forces in two perpendicular directions. (Courtesy of Tensar Earth Technologies, Inc., Atlanta, GA.)

Inclinometers

An *inclinometer* is an instrument used to measure horizontal movements in the ground as a function of depth. Inclinometers are very helpful in slope stability studies, and can be used to locate shear surfaces and monitor the rate of shear displacement in slow-moving landslides.

FIGURE 13.46 Geogrids being used to reinforce a fill for a highway. Multiple geogrids are used, each separated by a layer of soil. (Courtesy of Tensar Earth Technologies, Inc., Atlanta, Georgia, USA.)

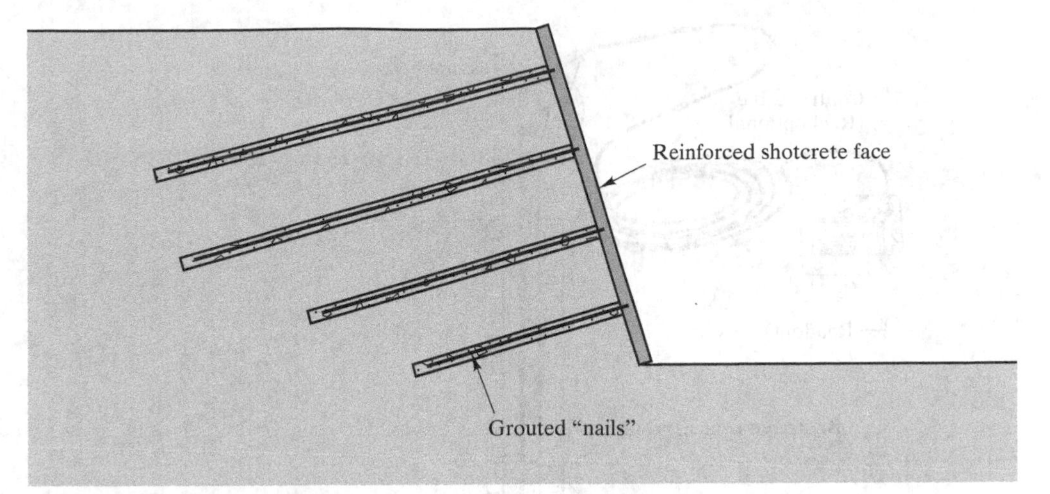

FIGURE 13.47 A soil nail wall.

To install an inclinometer, a vertical boring is drilled to a depth well below the potential zone of movement and a special plastic casing is inserted, as shown in Figure 13.48. The annular zone around the casing is then backfilled with grout to hold it firmly in place. Thus, as the ground moves horizontally, the casing deforms with it.

Once the casing is installed, an initial set of readings is obtained by lowering the *inclinometer probe* inside. This probe, shown in Figure 13.49, precisely measures the inclination of the casing in two perpendicular directions. Thus, we know the horizontal position of the upper set of wheels with respect to the lower set. We begin with the probe at the bottom of the casing and progressively raise it by intervals equal to the wheelbase, taking measurements at each interval. By summing these measurements, we know the initial horizontal position of the casing throughout its length.

Then, at some future date, we return with the probe and readout unit and obtain a second set of readings. By comparing the new horizontal configuration with the original configuration, we can determine the magnitude and direction of horizontal movements in the ground throughout the length of the casing. We continue to take additional readings at appropriate intervals as necessary. Figure 13.50 shows typical plots of horizontal movement versus depth.

Conventional Surveying

Slopes also may be monitored by installing monuments at various locations on the ground surface and measuring their positions using conventional surveying equipment, such as a *total station*. Although this approach does not provide any information on subsurface movements, it can provide extensive information on surface movements, and thus is a useful way to monitor unstable ground. In addition, each monument is far less expensive than an inclinometer, and monuments can be installed in areas difficult to access, while inclinometer installations can only take place at locations accessible by a drill rig.

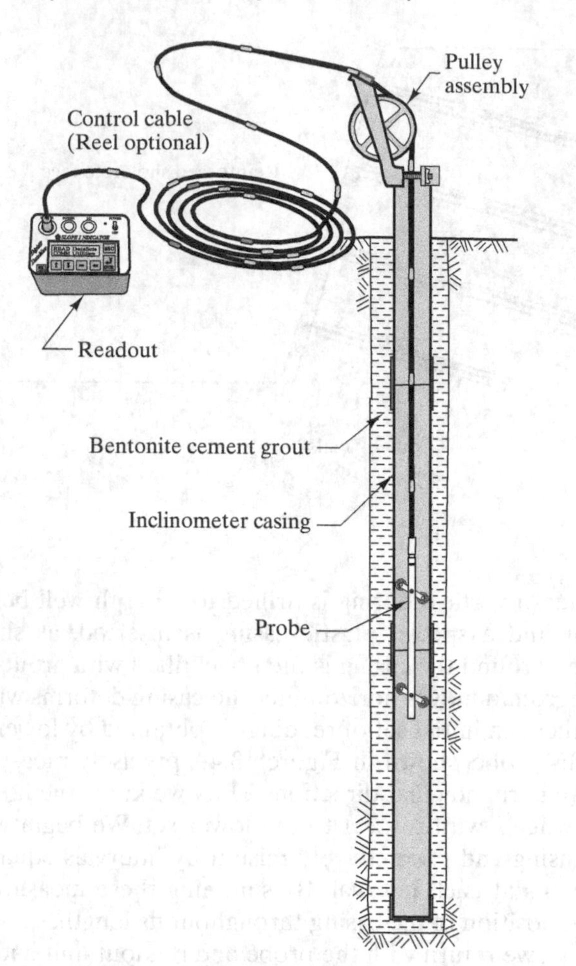

FIGURE 13.48 Cross-section of a typical inclinometer installation. (Slope Indicator Co., Bothell, WA.)

Conventional surveying methods are becoming even more attractive with the increased availability and precision of *global positioning system (GPS)* receivers. These are devices that determine position based on signals from satellites, and can achieve accuracies on the order of ±1 cm.

Groundwater Monitoring

Groundwater has a significant impact on slope stability, so information on the groundwater table position and pore water pressures is very important. Therefore, geotechnical engineers often install *observation wells* and *piezometers* as a part of slope stability studies. These devices are described in more detail in Chapters 3 and 7.

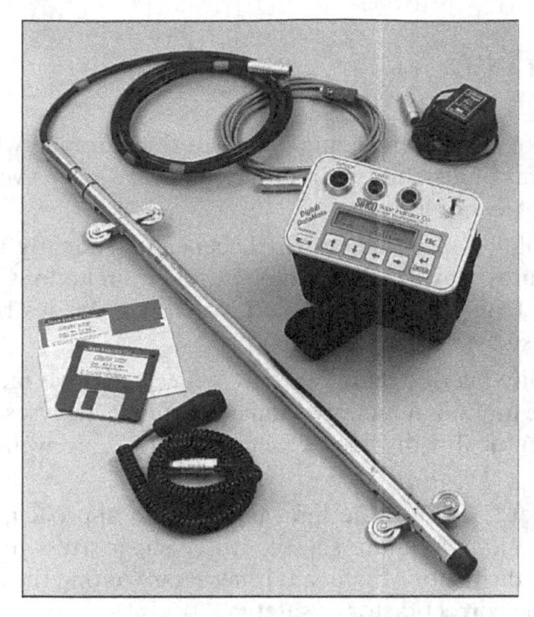

FIGURE 13.49 Inclinometer probe and readout unit. (Slope Indicator Co., Bothell, WA.)

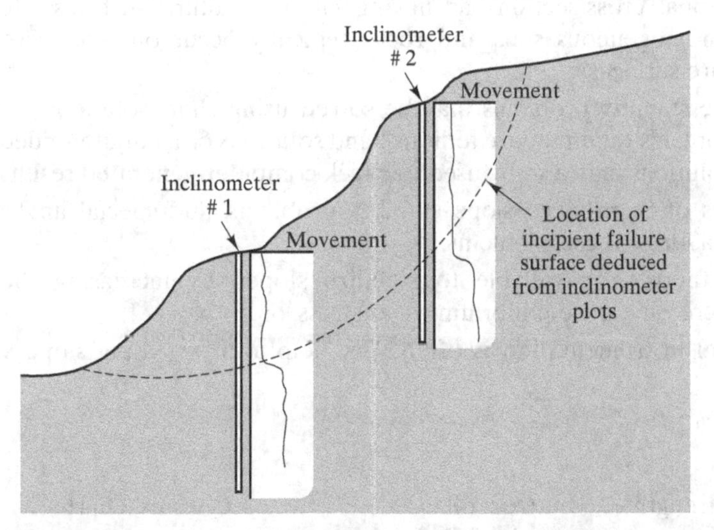

FIGURE 13.50 Plots of horizontal movement versus depth for two inclinometers. This data might be used to detect early movements along an incipient failure surface.

SUMMARY

Major Points

1. When building near or on sloping ground, engineers need to determine whether the slope is stable. Many forms of instability can occur and cause extensive property damage and occasionally loss of life.
2. The various forms of instability do not necessarily occur independently, so it is difficult to classify slope failures. Nevertheless, in spite of their limitations, classification schemes are useful. One system, developed by Varnes, divides failures into five types: falls, topples, slides, spreads, and flows.
3. Some types of slope instability may be analyzed only with qualitative methods, while other types are suitable for both qualitative and quantitative analyses. Geotechnical engineers most often deal with slides, which fortunately are well-suited for quantitative analysis.
4. Most quantitative analyses of slides use the limit equilibrium approach, which compares the shear strength along a failure surface to the shear stress required for equilibrium to determine the factor of safety. It is necessary to find the critical failure surface to compute the correct factor of safety.
5. Most slope stability problems are statically indeterminate. We overcome this difficulty by making simplifying assumptions that convert the problem into one that is statically determinate. Various assumptions have been proposed, thus producing a large number of limit equilibrium analysis methods.
6. Slides in homogeneous soils, such as compacted fills, usually can be idealized in two-dimensional cross-sections as having circular failure surfaces. However, slides in inhomogeneous soils and rocks typically occur on more irregularly shaped failure surfaces.
7. Simple slope stability problems may be solved using chart solutions, but more complex problems require more tedious hand solutions or computer-aided analyses. Chart solutions may also be used to check computer-generated results.
8. Earthquakes often produce slope stability problems, and special analyses are required to address these problems.
9. Various methods are available to stabilize slopes, by increasing the shear strength, decreasing the equilibrium shear stress, or both.
10. Geotechnical instrumentation is often very helpful in assessing slope stability problems.

Vocabulary

back-calculated strength
bench
berm
block-glide slide
body
buttress fill
circular failure surface
complex slide
composite slide
compound slide
Cousins' chart
crest of slope
critical failure surface
crown
cut slope

debris flow
deterministic analysis
earth retaining structure
effective stress analysis
equilibrium shear stress
factor of safety
failure surface
fall
Fellenius method
fill slope
filter fabric
flank
flow
French drain
friction circle method
geotechnical earthquake engineering
global positioning system (GPS)
horizontal drain
inclinometer
infinite slope analysis
in situ reinforcement
instrumentation
landslide
lateral spread
limit equilibrium analysis
liquefaction-induced failure
main scarp
minor scarp
mobilized shear strength
modified Bishop's method
modified Janbu's method
Morgenstern-Price method
natural slope
Newmark's method
non-circular analysis
non-circular failure surface
non-rigorous
observation well
ordinary method of slices (OMS)
perforated pipe drain
piezometer
planar failure analysis
probabilistic analysis
probability of failure
pseudostatic acceleration
pseudostatic coefficient
pseudostatic force
pseudostatic method
rapid drawdown failure
reinforcement
required shear strength
retaining wall
rigorous
rock fall
rotational slide
seismic slope displacement
seismic stability
semi-rigorous
shear key
simplified Bishop's method
simplified Janbu's method
slide
slope face
slope height
slope ratio
soil nailing
Spencer's method
spread
stability charts
stabilization
subsurface drain
surface drainage
Swedish method of slices
Swedish slip circle method
tension crack
terrace
tieback
tieback anchor
toe of slope
top of slope
topple
total station
total stress analysis
translational slide
undrained shear strength
well
yield acceleration

QUESTIONS AND PRACTICE PROBLEMS

Section 13.1 Terminology

13.1 Define, with the help of a sketch if appropriate, the following terms as related to slope stability:

(a) Natural slope
(b) Cut slope
(c) Fill slope
(d) Top of slope
(e) Toe of slope
(f) Slope face

(g) Terrace
(h) Slope ratio
(i) Slope height

Section 13.2 Modes of Slope Instability

13.2 Define the following types of slope instability:

(a) Falls
(b) Topples
(c) Slides
(d) Spreads
(e) Flows

13.3 Explain the difference between a flow and a slide. Which one often travels farther, and thus can be a hazard to sites far from the slope?

13.4 Cut slopes in bedded sedimentary rocks can be very problematic. What mode of failure do you think would be most common in these rocks, and how could we evaluate the potential for such a failure before construction?

Section 13.4 Quantitative Analysis of Slides

13.5 Describe the limit equilibrium analysis method.

13.6 Define the factor of safety as used in limit equilibrium-based slope stability analysis methods.

Section 13.5 General Procedures in a Limit Equilibrium Analysis of a Slide

13.7 Most limit equilibrium analysis methods include one or more simplifying assumptions. Why are these assumptions necessary? Give an example of one of the methods and its assumptions.

13.8 You are writing a computer program to perform slope stability computations. This program will consider only circular failure surfaces. What procedure might you use to locate the critical failure surface? Provide a detailed explanation.

Section 13.6 Planar Failure Analysis

13.9 State the assumptions in the planar failure analysis.

13.10 The proposed slope in Example 13.1 had an unacceptable factor of safety. We plan to remedy this situation by using a flatter slope. What slope ratio would be required to produce a factor of safety of 1.5?

Section 13.7 Infinite Slope Analysis

13.11 State the assumptions in the infinite slope analysis.

13.12 A 2.25:1 natural slope is underlain by a residual soil derived from the underlying gneiss. Compute the factor of safety for a failure surface 4 ft below the ground surface using a total stress infinite slope analysis with $s_u = 1000 \text{ lb/ft}^2$, $\phi_T = 0$, and $\gamma = 118 \text{ lb/ft}^3$.

13.13 A 2:1 natural slope is underlain by a 3-m thick (measured vertically) soil cover over bedrock. Compute the factor of safety using an effective stress infinite slope analysis with $c' = 0$ and $\phi' = 35°$ for the soil, assuming the following:

(a) The soil cover is totally dry and $\gamma = 18 \text{ kN/m}^3$.
(b) The soil cover is totally saturated and $\gamma = 20 \text{ kN/m}^3$.

Section 13.8 Swedish Slip Circle Method

13.14 State the assumptions in the Swedish slip circle method.

13.15 Prepare a spreadsheet to implement the Swedish slip circle method and use it to compute the factor of safety for the failure surface shown in Figure 13.51.

Section 13.9 Method of Slices

13.16 State the assumption on the side forces on slices in the following methods of slices:

(a) Ordinary method of slices
(b) Modified Bishop's method
(c) Spencer's method

13.17 Prepare a spreadsheet to implement the ordinary method of slices and use it to compute the factor of safety for the failure surface shown in Figure 13.52.

13.18 Prepare a spreadsheet to implement the modified Bishop's method and use it to compute the factor of safety for the failure surface shown in Figure 13.52.

Section 13.10 Chart Solutions

13.19 An 8.5-m tall, 2:1 fill slope is to be made of soil with $c' = 35$ kPa, $\phi' = 23°$, and $\gamma = 19.5$ kN/m^3. The groundwater table will be well below the toe of this slope. Using Cousins' chart, compute the factor of safety. Does this slope meet normal stability standards?

Section 13.11 Miscellaneous Issues

13.20 The slope shown in Figure 13.53 has recently failed. A geotechnical investigation indicates the failure surface was as shown. Assuming the failure occurred while undrained conditions prevailed in the slope, back-calculate the value of s_u. Use the Swedish slip circle method with the cross-section that existed immediately before it failed. Use $\gamma = 119$ lb/ft^3.

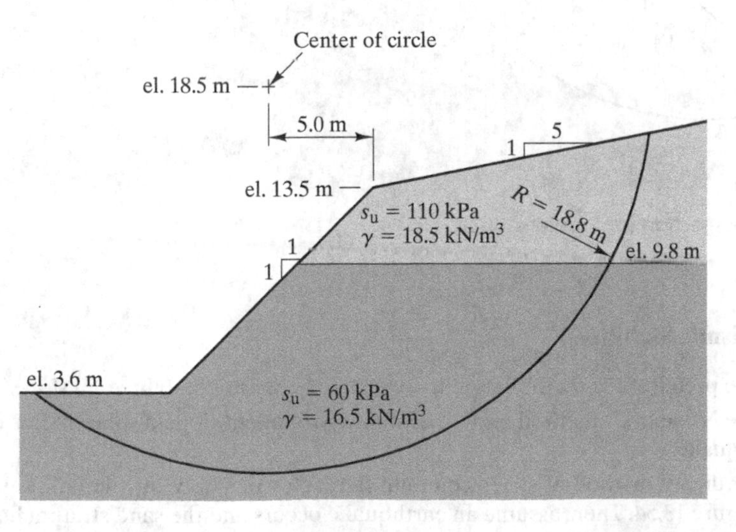

FIGURE 13.51 Cross-section for Problem 13.15. el. = elevation.

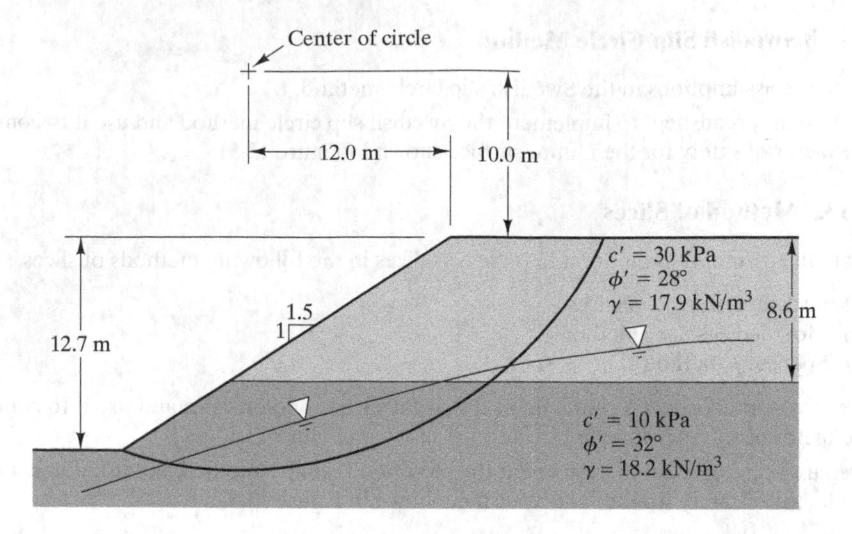

FIGURE 13.52 Cross-section for Problems 13.17 and 13.18.

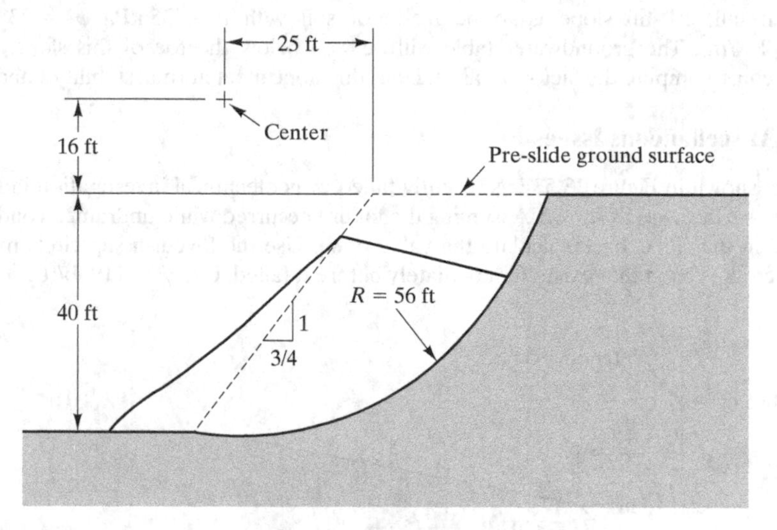

FIGURE 13.53 Cross-section for Problem 13.20.

Section 13.12 Seismic Stability

13.21 Describe the pseudostatic method used to evaluate the seismic stability of a slope.

13.22 Describe the Newmark's method used to estimate permanent slope displacements caused by an earthquake.

13.23 Using the ordinary method of slices, compute the factor of safety for the failure surface shown in Figure 13.54. Then, assume an earthquake occurs and the sand stratum liquefies and loses nearly all its strength. Assume further that the shear strength of the liquefied

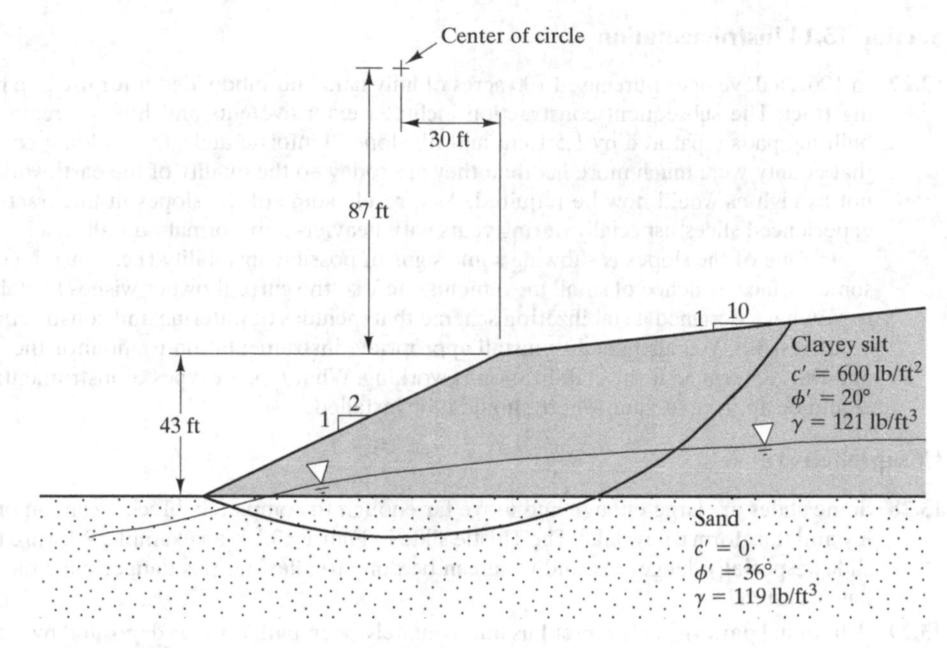

FIGURE 13.54 Cross-section for Problem 13.23.

sand is zero ($c' = 0$ and $\phi' = 0$). Compute a new factor of safety for the same failure surface. According to this analysis, will the slope survive the earthquake?

Section 13.13 Stabilization Measures

13.24 A certain slope has a factor of safety of 1.15 according to a Swedish slip circle analysis. To increase F to 1.50, you are considering the possibility of removing the upper portion of this slope, then rebuilding it to the original grades using a lightweight fill. Assuming the critical failure surface remains in the same location, how much must the weight of the potential slide body be reduced to produce the required factor of safety? Assume s_u along the failure surface remains unchanged. Express your answer as a percentage of the existing weight.

Note: In reality, the critical failure surface would probably shift to a new location, so this preliminary analysis would need to be followed by another search for the critical surface.

13.25 The hydraulic conductivity of a buttress fill is sometimes smaller than that of the adjacent natural ground. This is especially common when the natural ground is stratified, and water seeps along the more pervious strata. Could this difference in hydraulic conductivity cause any problems? Explain. If so, what might be done to remedy these problems?

13.26 The soil beneath a slope consists of alternating layers of sand and clay. These layers are nearly horizontal, but vary in thickness such that no two boring logs found these layers at the same elevations. This slope is to be stabilized by installing a series of horizontal drains that are intended to lower the groundwater table. The drains will be drilled at 20-ft intervals near the toe of the slope, and each one will be drilled at the same angle and to the same length. Would you expect the same flow rate from each drain? Why or why not?

Section 13.14 Instrumentation

13.27 In 1962, a developer purchased 100 acres of hilly land and subdivided it for use as a housing tract. The subsequent construction included extensive cuts and fills to create level building pads separated by 1.5:1 cut and fill slopes. Unfortunately, the building codes in that county were much more lax than they are today, so the quality of the earthwork was not as high as would now be required. As a result, some of the slopes in this tract have experienced slides, especially during years with heavier-than-normal rainfall.

One of the slopes is showing some signs of possible instability (i.e., tension cracks, some surface evidence of small movements, etc.), so the current owner wishes to stabilize it. You have designed a stabilization scheme that includes dewatering and construction of a buttress fill. You also need to install appropriate instrumentation to monitor the slope and thus determine if the stabilization is working. What type or types of instrumentation would be appropriate and where should it be installed?

Comprehensive

13.28 Some slides are large enough and move far enough to completely block a canyon or valley and thus form a new lake. The Thistle slide in Figure 13.1 is an example. Why are these slides especially dangerous, and what can be done to alleviate this danger once the slide has occurred?

13.29 A national park visitor's center has unfortunately been built on soils deposited by a series of earth flows. The building is located near the base of a canyon where it meets a larger valley. Ten years after construction, another earth flow occurred and deposited up to 3 ft of mud and debris around the visitor's center. Although the building was not seriously damaged, it was expensive and time consuming to clean up the mess. Everyone now recognizes that this building should have been constructed somewhere else, but there is no funding available to move it or replace it. Suggest one or two ways of protecting the building from future earth flows.

13.30 A preliminary grading plan for a proposed highway shows a 50-ft tall, 2:1 cut slope ascending from each side of the highway with level land above both slopes. Following a geotechnical study, it became necessary to change these slope ratios to 3:1. Assuming the existing right-of-way barely accommodates the 2:1 slopes, how much additional right-of-way must now be purchased because of this change?

13.31 The more rigorous limit equilibrium analysis methods, such as the modified Bishop's and Spencer's methods, produce factors of safety that are within about 5% of the "true" value. How does this error compare to the uncertainty in the soil properties (c', ϕ', and γ) and the uncertainty in the design soil profile? In light of these other sources of uncertainty, is a ±5% error tolerable? Explain.

13.32 A compacted fill slope is to be made of a soil with $c' = 200$ lb/ft^2, $\phi' = 30°$ and $\gamma = 122$ lb/ft^3. Using an infinite slope analysis and assuming a failure surface 4.0 ft below the ground surface and a groundwater table 1.0 ft below the ground surface, determine the steepest allowable slope ratio that will maintain a factor of safety of at least 1.5.

Note: This analysis considers only surficial stability. A separate analysis would need to be conducted to evaluate the potential for a deep-seated slide in the fill.

13.33 A 4-in. perforated pipe drain has been installed as part of a subsurface drainage system. The pipe has been surrounded with a poorly-graded 1.5 in. gravel. The adjacent soils are sandy silts. What is missing from this design? What mode of failure is likely to occur? What should be done to improve this design?

CHAPTER 14

Foundations

Soils never lie. People sometimes do.

Title of a lecture given by Ralph Peck, circa 1980

Geotechnical engineers are routinely involved in both the design and construction of foundations, which are the structural elements that connect buildings, bridges, and other structures to the ground. The design phase is normally performed in conjunction with a structural engineer, with the geotechnical engineer being responsible for the interaction between the foundation and the adjacent ground, and the structural engineering being responsible for structural integrity. During the construction phase, geotechnical engineers work with contractors and are responsible for comparing the soil conditions actually encountered with those anticipated in the design, providing various quality control services, and developing revised designs as needed.

Foundations must transmit the structural loads into the ground in a safe and cost-effective manner. Many kinds of foundations are available, and the proper selection depends on the magnitude and direction of the structural loads (downward, uplift, horizontal, and moments), the subsurface conditions (types and properties of soils and rocks, and groundwater), along with other factors. For convenience, we divide foundations into two broad categories: shallow foundations and deep foundations. The purpose of this chapter is to briefly introduce these two categories of foundations. Chapter 15 discusses the geotechnical design of spread footings, which are the most common type of foundation. *Foundation Design: Principles and Practices* (Coduto, 2001), the companion volume to this book, discusses foundation engineering in much more detail, and covers both geotechnical and structural design of many types of foundations.

14.1 SHALLOW FOUNDATIONS

Shallow foundations are those that transmit the structural loads to the near-surface soil or rock. There are two types: spread footings and mats, as shown in Figure 14.1. Spread footings are by far the most common type of foundation. The vast majority of one- and two-story buildings use them, and if the ground conditions are good even much larger structures can be supported on spread footings. Mats are most commonly used on moderate-size structures.

As the name implies, *spread footings* spread the structural loads across a sufficiently large soil area that the induced stresses are reduced to acceptable levels. They can be built in a wide variety of shapes and sizes to suit individual needs as shown in Figure 14.2, and are nearly always made of reinforced concrete. The most common shape is a *square footing*, which usually supports a single column. A *combined footing* is one that supports more than one column. A *continuous footing* (sometimes called a *strip footing*) supports a bearing wall. Most continuous footings are linear, but a continuous footing may wrap around in a circle to support the exterior wall of a tank, thus forming a *ring footing*. Figure 14.2 also shows the dimensions B, L, D, and T, which we use to describe the size and depth of a spread footing. The required dimensions for a specific footing depend on the magnitude of the load, the engineering properties of the underlying soils, and other factors.

The most important structural load is usually the downward load, which spread footings transfer to the ground in the form of a *bearing pressure* as shown in Figure 14.3(a). This pressure in turn induces normal and shear stresses in the underlying soil or rock. Similarly, spread footings transmit horizontal structural loads into the ground through a combination of sliding friction along the base of the foundation and bearing along the side, as shown in Figure 14.3(b). Footings have a limited ability to transfer an applied moment when combined with an axial downward load as shown in Figure 14.3(c).

An individual column is most commonly supported by a square footing. To design a footing, engineers select a width, B, such that the induced normal stress, $\Delta\sigma$, in the underlying ground is sufficiently low, and a depth D sufficient to accommodate the required footing thickness. Further increasing the depth of a footing often puts the

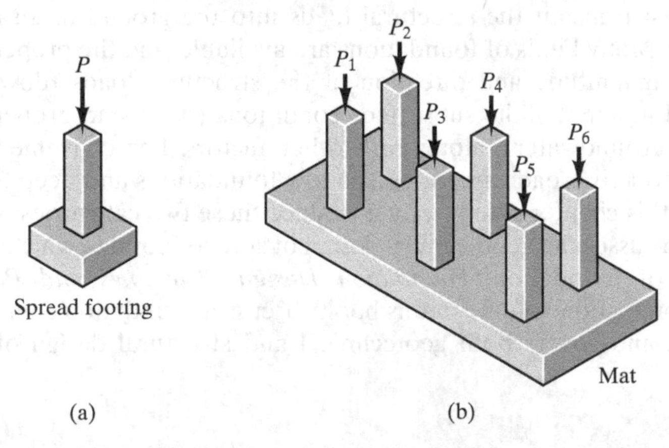

FIGURE 14.1 Shallow foundations. (a) spread footing; (b) mat.

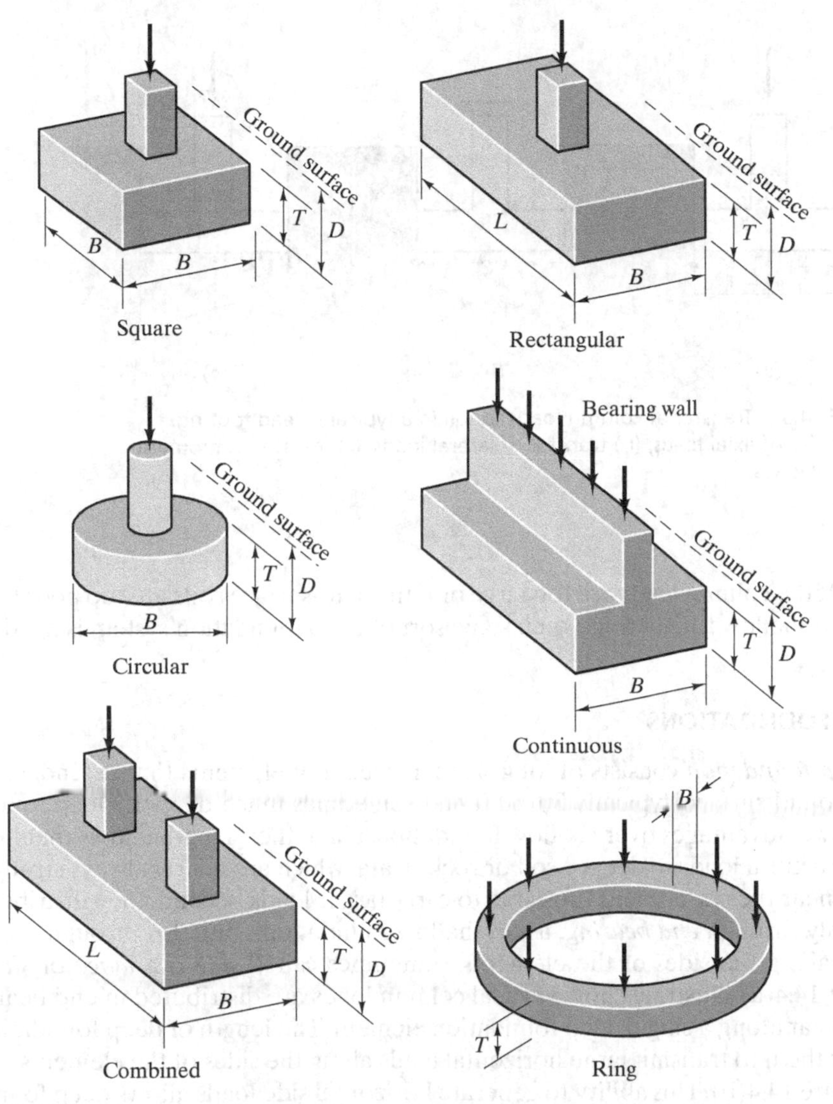

FIGURE 14.2 Spread footing shapes and dimensions.

footing in contact with stronger soils, which increases its load-bearing capacity. Therefore as column loads increase, spread footings become larger and deeper.

If the structural loads are too large or if the soil conditions are too poor, spread footings become unacceptably large so the engineer must then consider other types of foundations. One option would be to use a *mat foundation* (also known as a *raft foundation*), which is essentially one large spread footing that encompasses the entire structure. A mat distributes the weight of the structure across a larger area, thus reducing the induced stresses in the underlying soils. Mats also have the advantage of structural continuity and thus reduce the potential for differential settlements.

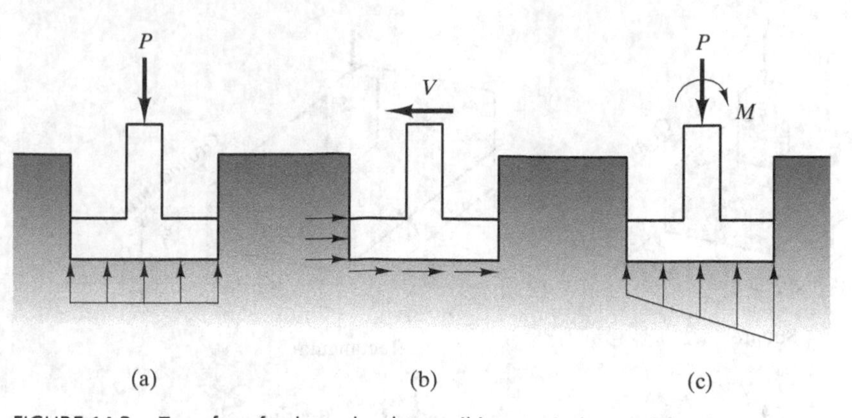

FIGURE 14.3 Transfer of column loads to soil in a typical spread footing. (a) Transfer of axial loads; (b) transfer of lateral loads; (c) transfer of moment and axial loads.

If the column loads are too large or if the near-surface soils are too poor to accommodate shallow foundations, then some sort of deep foundation system is needed.

14.2 DEEP FOUNDATIONS

A *deep foundation* consists of long slender structural elements that extend well below the ground surface, typically 30–60 ft and sometimes much deeper. Deep foundations have two advantages over shallow foundations: first, their length allows them to transmit structural loads to deeper soil or rock strata, which are nearly always stronger than those near the surface and thus able to carry larger loads; second, they distribute loads not only through *end bearing*, as do shallow foundations, but also through shear resistance along the sides of the elements, sometimes called *side resistance* or *side shear*. Figure 14.4(a) illustrates how vertical column loads are distributed in end bearing and side shear along a single deep foundation element. The length of deep foundations also allows them to transmit large horizontal loads along the sides of the elements as shown in Figure 14.4(b). This ability to generate horizontal side loads allows deep foundations to resist moments as shown in Figure 14.4(c). Thus, large and heavy structures are most often supported on deep foundations.

Deep foundation elements are often used in groups to carry large column loads. Figure 14.5 shows a four-element group supporting a single column. Another structural element, called a *cap* or *pile cap*, connects the tops of the deep foundation elements to the column.

Many different terms are used to describe the individual vertical structural elements used in deep foundations, and the terminology varies from one region to another. Some of the most common terms in use are *pile*, *pier*, *caisson*, and *drilled shafts*. We will use the term *pile* to describe this deep foundation element and add modifiers to the term *pile* to distinguish among different systems.

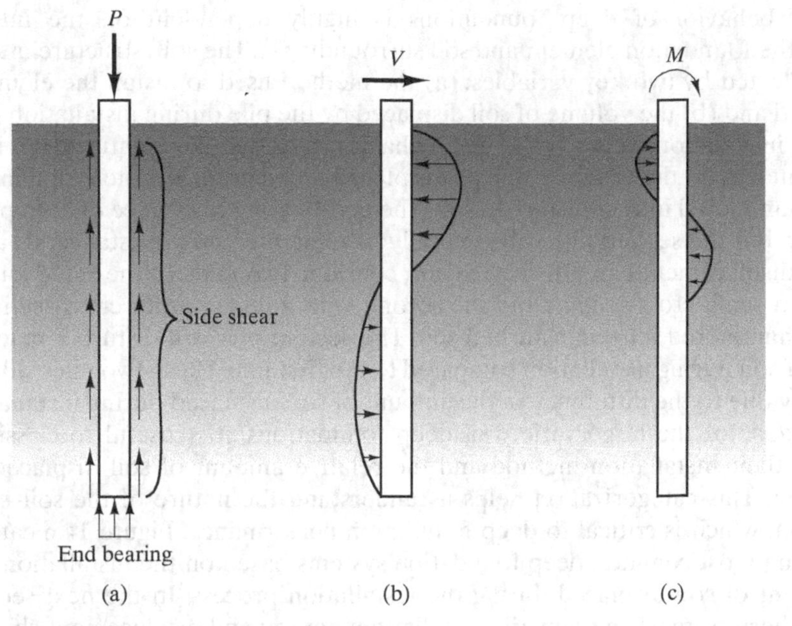

FIGURE 14.4 Transfer of column load to soil in a typical deep foundation. (a) Transfer of axial loads; (b) transfer of lateral loads; (c) transfer of moment.

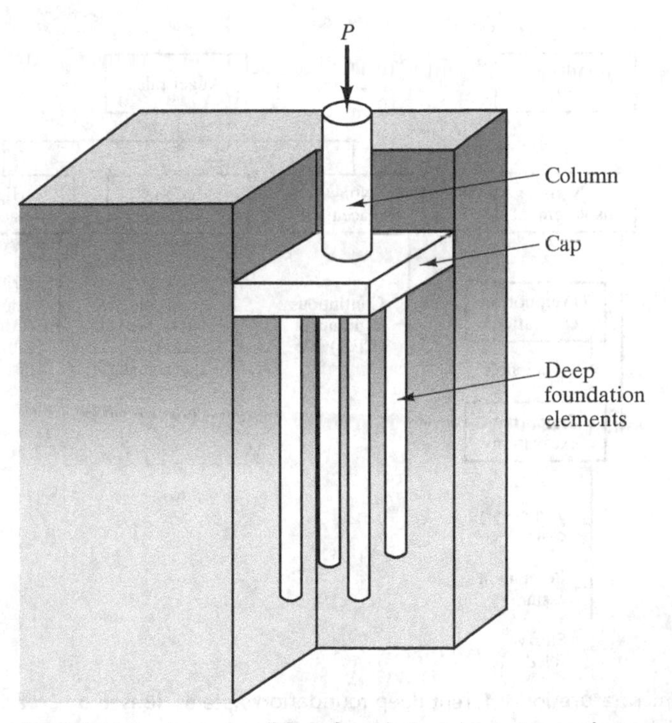

FIGURE 14.5 A group of four foundation elements carrying a single column load.

The behavior of deep foundations is highly dependent on the interaction between the foundation element and soil surrounding it. The soil–structure interaction will be affected by two key variables: (a) the method used to install the element into the ground and (b) the volume of soil displaced by the pile during installation. To illustrate how installation method can affect behavior, consider two identical piles installed in different way: in the first case, the pile is placed into a predrilled hole, and in the second it is hammered into undisturbed soil. The end-bearing resistance of both piles may be similar, but the second pile will generally have greater side resistance. To illustrate how soil displacement can affect behavior, consider two piles of the same length, the first with a small cross-section and the second with a much larger cross-section; both piles are hammered into undisturbed soil. The second pile will disturb a much larger volume of soil during installation compared to the first pile. These two piles will behave differently due to the difference in the amount of soil displaced during installation.

Because of the large variety of deep foundations, it is useful to classify them based on their installation method and the relative amount of soil displaced during installation. This categorization helps us understand the nature of the soil–structure interaction, which is critical to deep foundation performance. Figure 14.6 categorizes some of the most common deep foundation systems based on the installation method and amount of soil displaced during the installation process. In the next section, we describe these systems and how the installation method and displacement characteristics affect system performance.

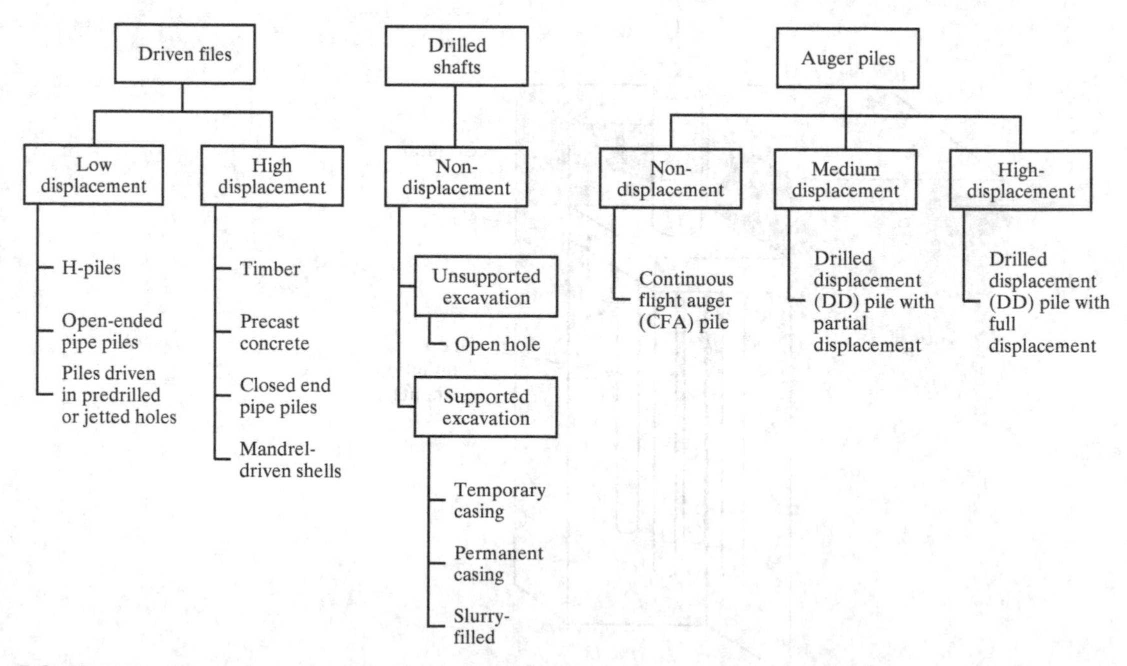

FIGURE 14.6 Categories and nomenclature for different deep foundation systems. (Based on Prezzi and Basu, 2005, and U.S. Army 1998.)

Installation Methods

Traditionally, there have been two methods for installing deep foundations: the *driven pile* method where a hammer is used to drive prefabricated piles into place, and the *drilled pile* method where a cylindrical hole is augered in the soil and then filled with reinforced concrete to form the pile. Advances in drill rig design have led to methods where concrete or grout is simultaneously pumped into the void created by the withdrawing augers. This method is called the *auger pile* method.

Driven Piles Driven piles are installed with either impact or vibratory hammers. There are four common types of impact hammers: *drop hammers, diesel hammers, hydraulic hammers*, and *air hammers*. All impact hammers operate on the same principle. A weight is raised and then falls on top of the pile to be driven. Impact hammers are rated based on the total energy the hammer is capable of generating. Hammer energy ranges from 2000 ft-lb (2.5 kN-m) for small drop hammers to over 1,000,000 ft-lb (1350 kN-m) for large diesel or hydraulic hammers.

Diesel hammers are one of the most common types of impact hammers and essentially consist of a vertically oriented single cylinder diesel engine. The piston of the engine acts as a ram that drives the pile. Figure 14.7(a) shows a diesel hammer driving a large diameter steel pipe pile. Air and hydraulic hammers operate in a manner similar to diesel hammers but use either compressed air or hydraulic fluid to raise the ram. Figure 14.7(b) shows an air hammer driving a concrete pile.

Vibratory hammers install piles by rapidly vibrating the piles up and down while placing a large weight on the top of the pile. The vibrations are generated by spinning one or more pairs of eccentrically loaded wheels in opposite directions. Figure 14.8 illustrates the vibratory mechanism used to drive the piles. Figure 14.9 shows a vibratory hammer driving an open ended pipe pile. Vibratory hammers are usually used on low-displacement piles where the vibratory action is most effective. It is difficult to drive large high-displacement piles using this technique. Because the vibratory action of these hammers is up and down, they can be used to extract piles as well as to drive them. When extracting a pile, the hammer is clamped to the top of the pile and a crane pulls up on the pile while it is vibrated.

Driven piles are typically 12–36 in. in diameter and 30–60 ft long, but much larger piles have been used on specialized projects such as offshore drilling platforms. There are four common types: *timber piles, H-piles, steel pipe piles*, and *prestressed concrete piles*, as shown in Figure 14.10. Timber and prestressed concrete piles are examples of high-displacement piles. Because these piles are solid with large cross-sectional area, they must push a large volume of soil out of the way as they are driven. H-piles are examples of low-displacement piles because they have a relatively small cross-sectional area as compared to other piles and, therefore, displace much less soil during driving. Pipe piles may be driven either as closed ended, by welding a plate on the bottom of the pile, or they may be driven open ended if no plate is used. The closed-ended piles are a high-displacement pile. The open-ended piles may be either high or low displacement depending on how the soil at the bottom of the pile behaves. If the soil at the bottom of the pile forms a plug that moves with the pile as it is driven, then the pile acts as a high-displacement pile. If no plug forms, then the pile is a low-displacement pile.

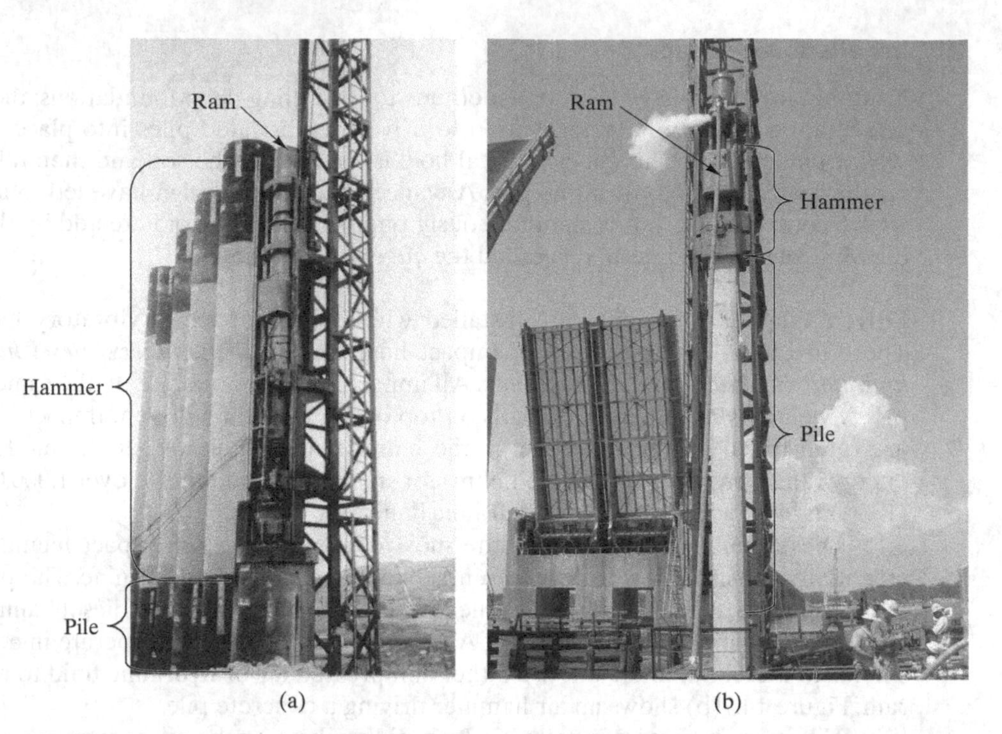

FIGURE 14.7 Piles being driving with impact hammers. (a) A diesel hammer driving a large diameter steel pipe pile. The ram is near its highest point, sticking out of the top of the hammer. It will fall from this point and drive the pile into the ground. (b) An air hammer driving a 30-inch square concrete pile. The piston at the top of the hammer raises the ram in the center of the hammer. This photo was taken after the ram has been dropped from its highest point. It is about to strike the pile. (Photo (a) courtesy of American Piledriving Equipment Inc. Photo (b) courtesy of Dan Brown and Associates.)

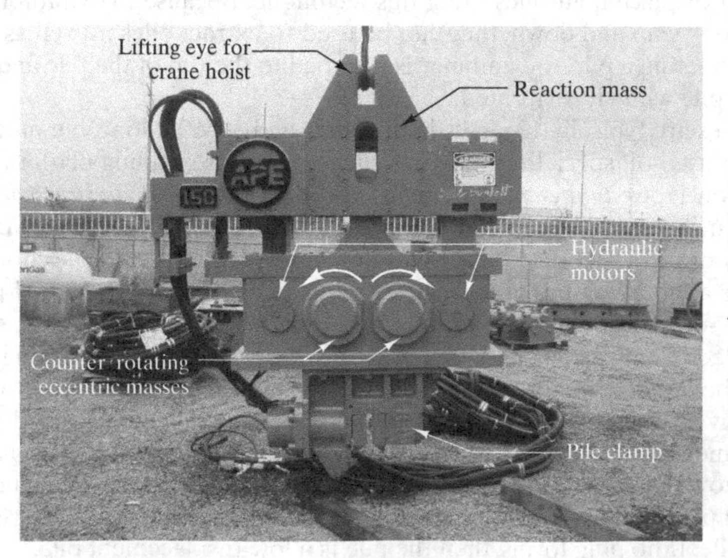

FIGURE 14.8 Photograph of vibratory hammer. The hammer is clamped on top of the pile to be driven. The hydraulic motors then turn the counter rotating eccentric masses. This causes an up and down vibratory motion. The weight of the hammer and the vibrations push the pile into the ground. (Courtesy of American Piledriving Equipment Inc.)

FIGURE 14.9 A vibratory hammer driving an open-ended pipe pile. (Courtesy of Dan Brown and Associates.)

Driven piles can cause significant changes in the soil into which they are driven: the higher the displacement of the pile, the greater the changes in the soil. In loose cohesionless soils, driving high-displacement piles will densify the soil and increase its strength. Therefore, driven piles are well suited for use in loose to medium sands and gravels where driving will densify and improve the soil properties. Driven piles are also suitable in conditions where a soft clay soil overlies a stiffer and stronger soil. The piles can be driven through the soft clay layer into the stronger soil below. However in sensitive clays or weak rock, driving piles, especially high-displacement piles, can destroy the material's structure and reduce the strength of the soil or rock. In dense sands and gravels or stiff to very stiff clays, it may not be possible to drive piles without damaging the piles themselves. In these cases, a drilled pile or auger pile may be a better choice.

(a) (b) (c) (d)

FIGURE 14.10 Typical types of driven piles. (a) Timber piles with steel toe points to protect piles during driving. (b) End of a typical H-pile with hardened steel point attached to protect pile during driving. (c) A 16-in. (405 mm) diameter steel pipe pile after being driven. (d) A stockpile of 14-in. (356-mm) prestressed concrete piles awaiting delivery to the job site. (Photos (a) and (b) courtesy of Associated Pipe and Fitting.)

Drilled Piles Drilled piles, often called *drilled shafts* or *bored piles*, are installed in a very different manner than are driven piles. The drilled pile method starts by creating a cylindrical hole in the ground and then filling the hole with concrete. An auger drill rig is used to remove the soil to create the cylindrical hole, typically 18–36 in. in diameter, but sometimes as much as 10 ft in diameter. A steel-reinforcing cage is placed in the hole to provide tensile and flexural strength; then the hole is filled with concrete. Figure 14.11 illustrates this construction sequence.

Because all the soil is removed before placing the concrete, drilled piles are a nondisplacement foundation. When constructed in competent soils that will maintain an open hole, drilled piles are normally built using the dry hole method of construction, as

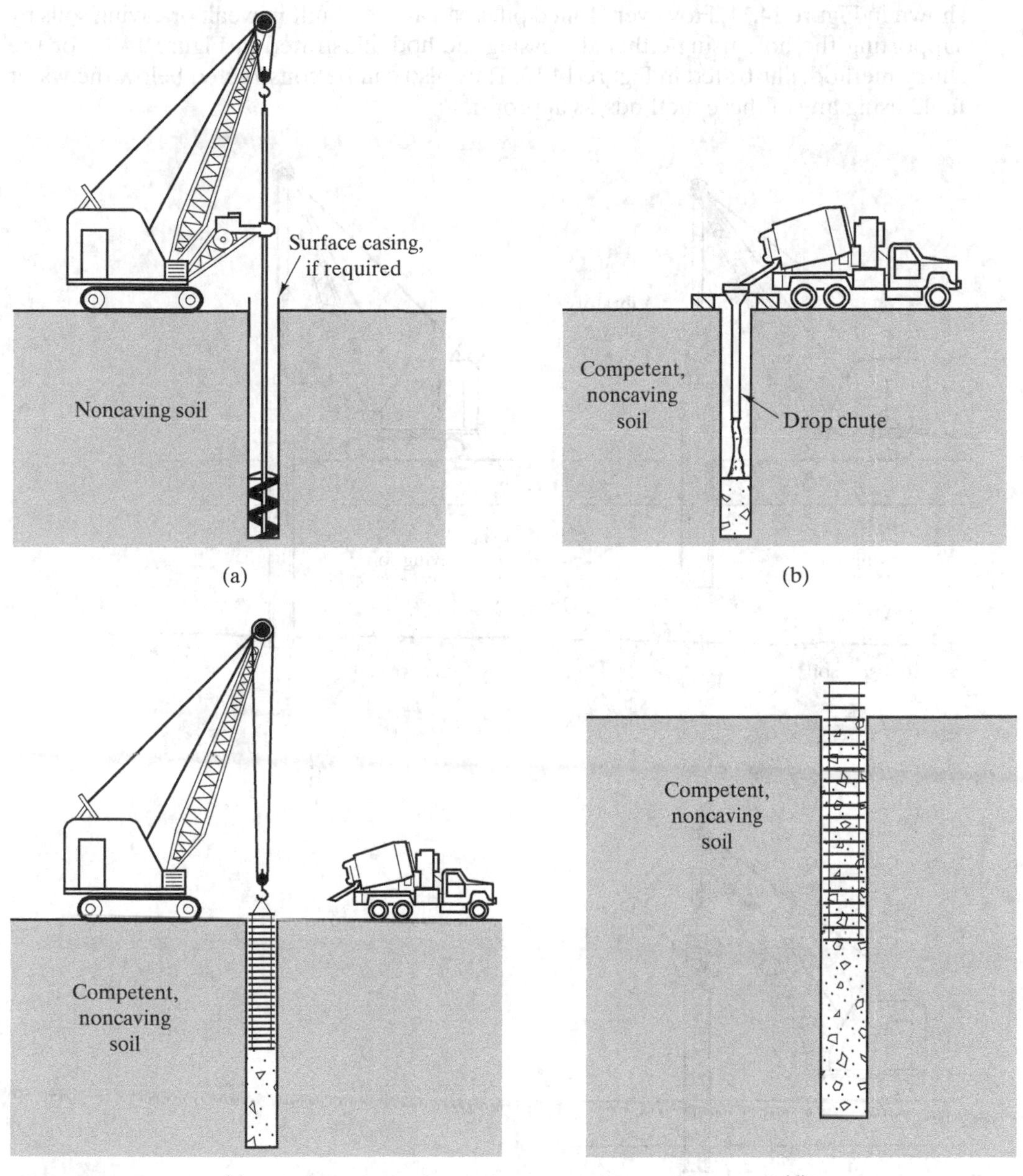

FIGURE 14.11 Drilled shaft construction in competent soils using the dry hole method. (a) Drilling the shaft; (b) starting to place the concrete; (c) placing the steel reinforcing cage; (d) the completed foundation. (Reese and O'Neil, 1988.)

shown in Figure 14.11. However, drilled piles can also be built in weak or caving soils by supporting the hole using either the casing method, illustrated in Figure 14.12, or the slurry method, illustrated in Figure 14.13. They also can be constructed below the water table using any of these methods, as appropriate.

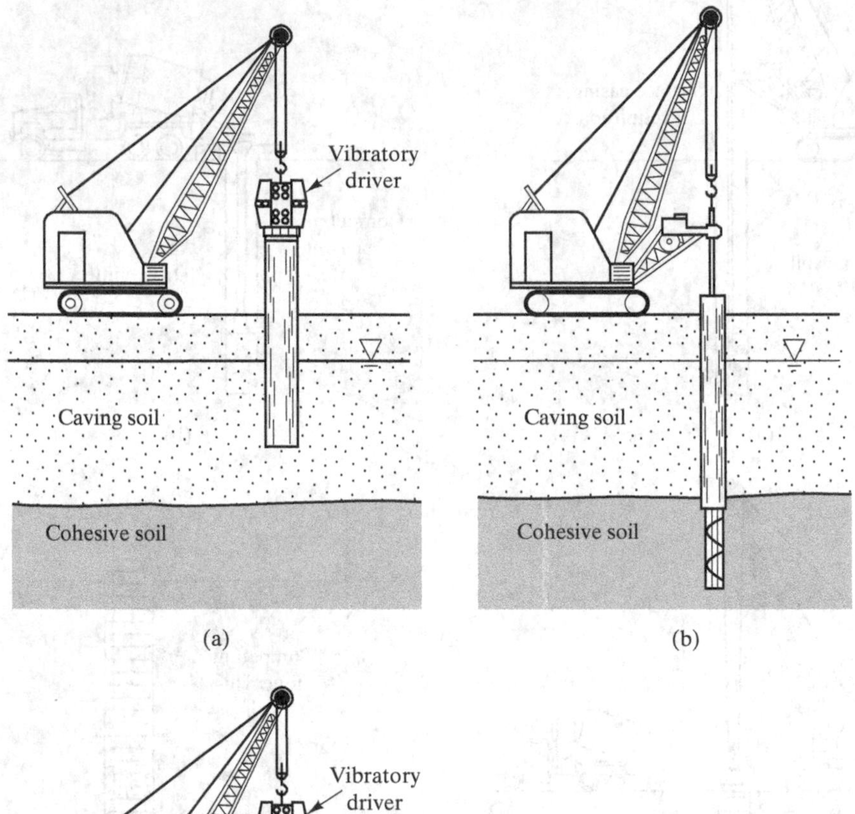

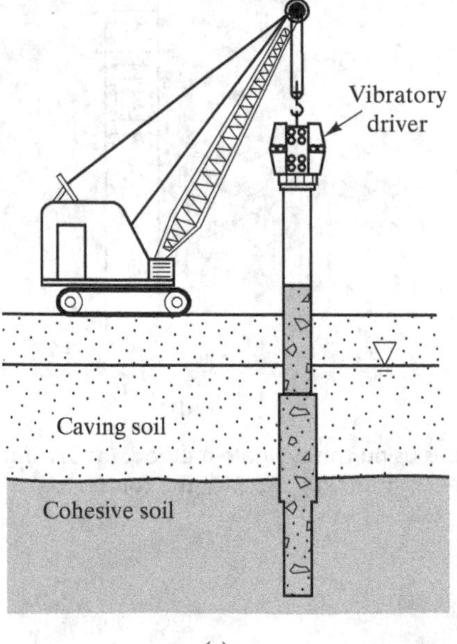

FIGURE 14.12 Installing a drilled shaft in caving or squeezing soils using the casing method. (a) Installing the casing. (b) Drilling through the casing into competent soil. (c) Removing the casing after concrete and reinforcing steel have been place in the cased hole. (Reese and O'Neil, 1988.)

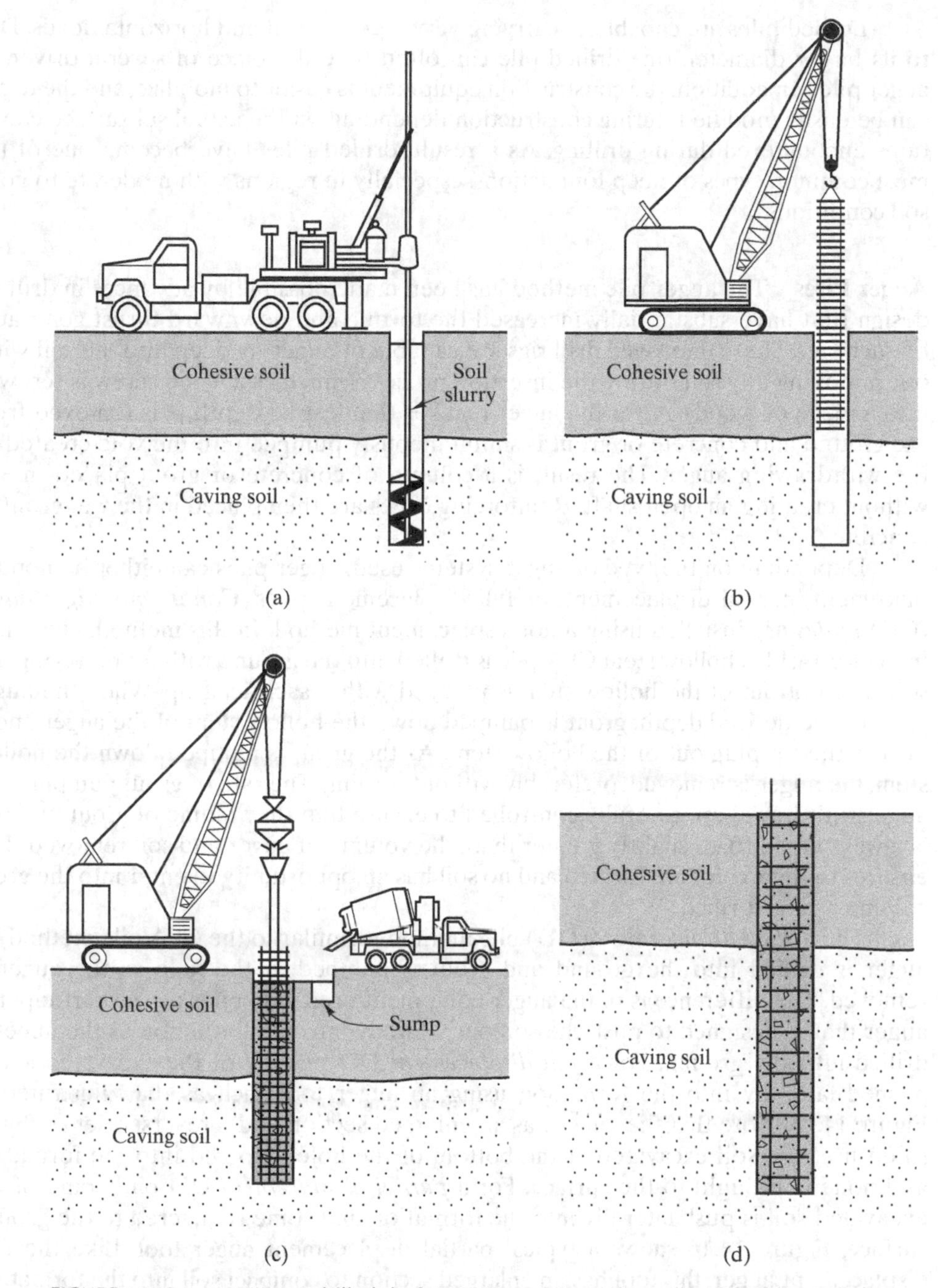

FIGURE 14.13 Installing a drilled shaft in caving or squeezing soils using the wet hole or slurry method. (a) Advancing the hole using slurry; (b) placing the reinforcement cage through the slurry; (c) placing the concrete from the bottom of the hole up using a tremie pipe. The excess slurry is collected in the sump at the ground surface; (d) the completed foundation. (Reese and O'Neil, 1988.)

Drilled piles are capable of carrying very high vertical and horizontal loads. Due to its larger diameter, one drilled pile can often take the place of several driven or auger piles. In addition, the construction equipment is easier to mobilize, and the depth can be easily modified during construction depending on the actual subsurface conditions encountered during drilling. As a result, drilled piles have become one of the most common types of deep foundations, especially in regions with moderate to good soil conditions.

Auger Piles The auger pile method has been made possible by advances in drill rig design that have substantially increased the torque and downward thrust generated by such rigs. These improved drill rigs are capable of augering deep into the soil without removing any soil during the insertion process (much as a wood screw is screwed into a piece of wood). After the auger reaches the desired depth, it is removed from the ground and concrete or grout is simultaneously pumped into the void created by the withdrawing auger. The result is a column of concrete or grout placed in situ without creating an open shaft. Reinforcing cages are then placed in the wet grout or concrete.

Depending on the type of auger systems used, auger piles can either be nondisplacement, partial displacement, or full displacement piles. *Continuous flight auger* (CFA) *piles* are installed using a nondisplacement method. In this method, illustrated in Figure 14.14, a hollow stem CFA pile is drilled into the ground without removing any soil. The bottom of the hollow stem is plugged with a sacrificial tip. When the auger reaches the desired depth, grout is pumped down the hollow stem of the auger and it pushes the tip plug out of the hollow stem. As the grout is pumped down the hollow stem, the auger is removed, preferably without turning. The rate of grout pumping and auger withdrawal are carefully controlled to ensure that the volume of grout pumped is always equal to, or slightly greater than, the volume of auger and soil removed. This ensures that no voids are formed and no soil has an opportunity to enter into the grout column as it is formed.

The *drilled displacement* (DD) pile method is similar to the CFA pile method; an auger is drilled into the ground and grout is pumped in the hole as the auger is removed. The difference is in the auger equipment used. DD piles use a short tapered auger that is designed to push the soil out laterally into the formation as the auger is drilled into the ground. For a *full-displacement* DD pile, all of the excavated soil is pushed laterally into the formation using an auger tool such as that illustrated in Figure 14.15. Note that the auger has an enlarged section and a reversed set of flights to ensure that soil excavated at the bottom of the hole is forced into the formation and none is brought to the surface. For a *partial displacement* DD pile, some of the excavated soil is push laterally into the formation and some is augered to the ground surface. Figure 14.16 shows a typical partial displacement auger tool. Like the full-displacement auger, this tool has an enlarged section to compact soil into the formation, but it also includes a smaller auger to lift some soil to the surface.

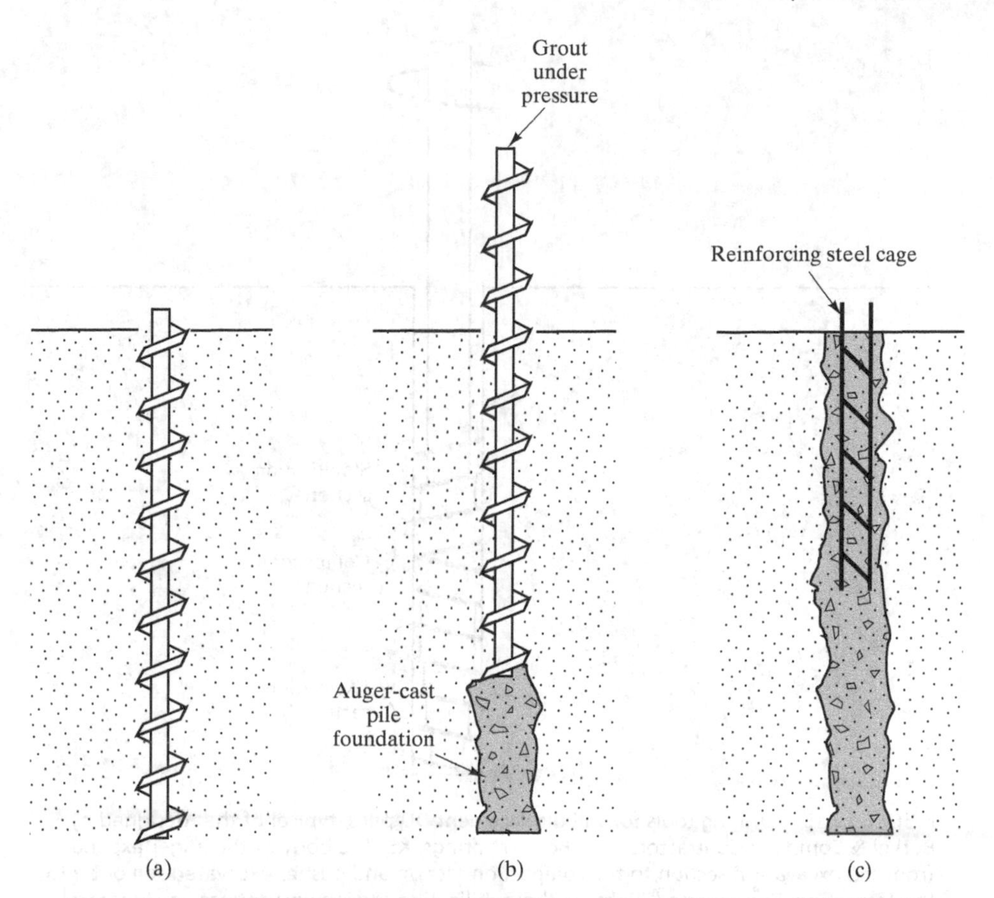

FIGURE 14.14 Construction of a continuous flight auger pile. (a) Auger is drilled to the desired depth without removing soil; (b) grout is pumped down the hollow stem of the auger completely filling the hole as the auger is withdrawn; (c) reinforcing cage is installed in the fresh grout.

Like driven piles, auger piles are well suited for use in loose or soft soils where it is difficult to maintain an open hole, as required in the drilled pile method. Partial or full-displacement auger piles can improve loose cohesionless soils by compacting soil adjacent to the pile. Because of the large torque and downward thrust needed to drill auger piles, they are limited to a diameter of approximately 2 ft (0.6 m), and diameters of 12 in. (0.3 m) to 18 in. (0.45 m) are more common. Therefore auger piles generally carry lower loads than drilled piles and often need to be used in groups to carry large loads.

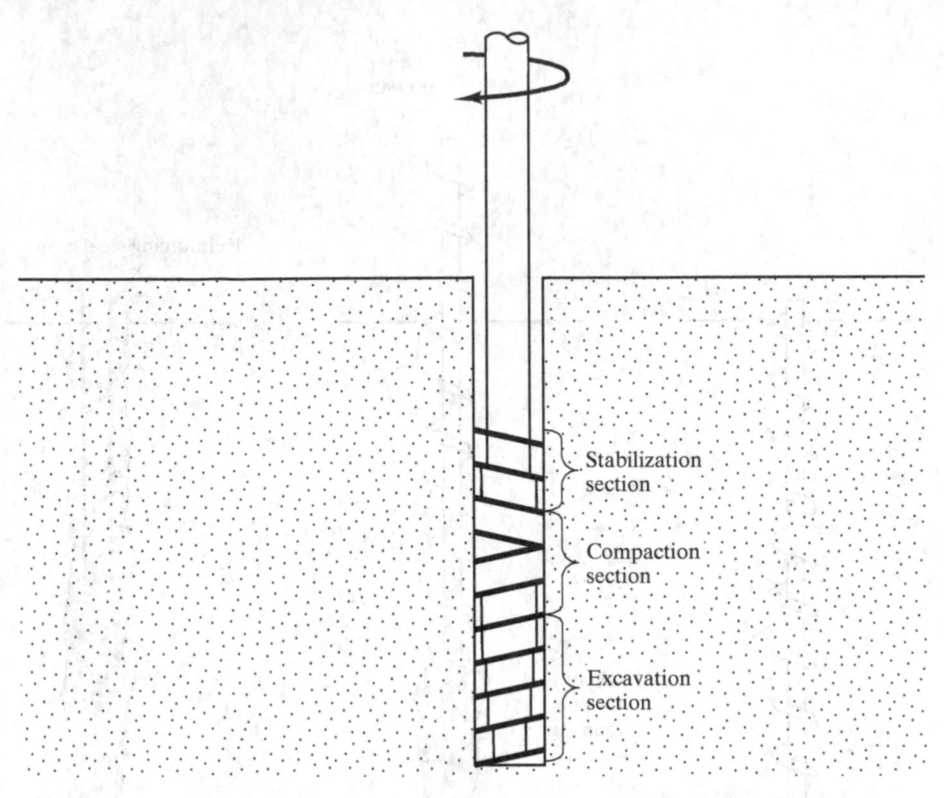

FIGURE 14.15 Drilling tools for full-displacement DD piles, typical of that designed by Berkel & Company Contractors, Inc., Bonner Springs, KS. The body of the auger expands from the excavation section to the compaction section and pushes excavated soil out into the formation. The reversed flights in the stabilization section ensure that any loose soil encountered during withdrawal of the auger is pulled back down to the compaction section and pushed out into the formation. The smooth stem above the drilling/compacting tool does not carry any soil to the surface.

SUMMARY

Major Points

1. Structural foundations are used to transmit structural loads into the ground. There are two broad categories: shallow foundations transmit the loads to the near-surface soils, while deep foundations transmit some or all of the loads to deeper soils.
2. Spread footings are one type of shallow foundation. They are the most common type, and transfers loads through end bearing.
3. Deep foundations transfer loads to the soil through both end bearing and side friction.

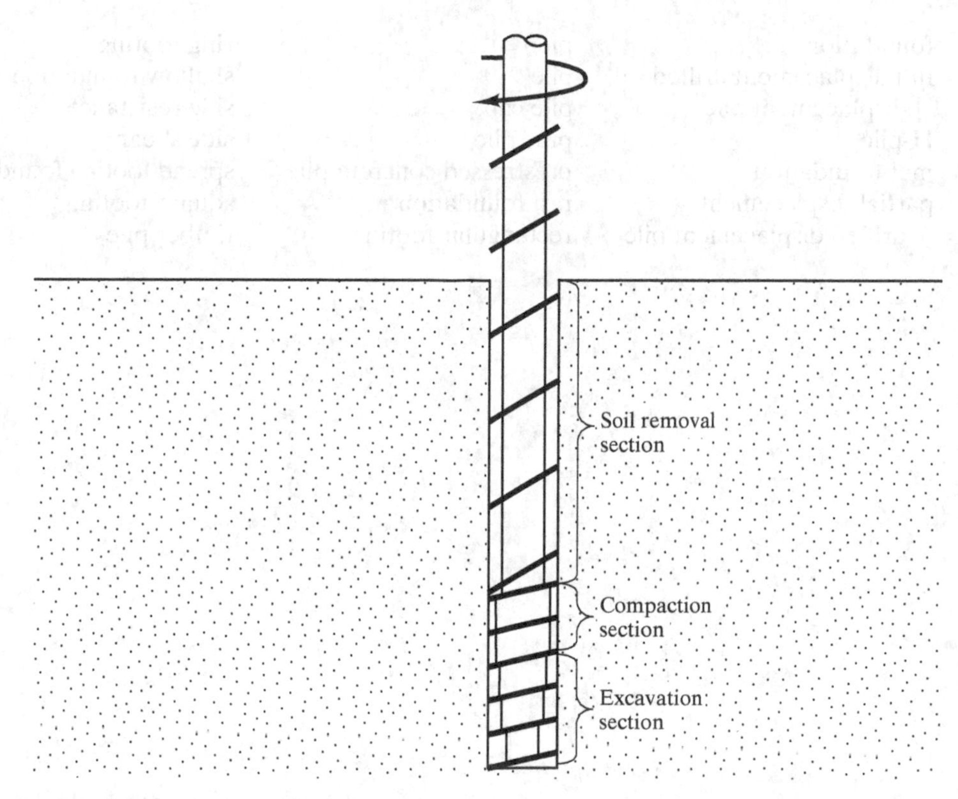

FIGURE 14.16 Drilling tools for partial displacement DD piles, typical of that designed by Berkel & Company Contractors, Inc., Bonner Springs, KS. Note that the augers in the soil removal section are very shallow and spaced far apart. This ensures that only part of the excavated soil will be removed from the hole. The remaining excavated soil will be pushed out into the formation by the compaction section of the auger.

4. Deep foundations can carry much higher loads than shallow foundations because they transfer load deep below the surface where the soil or rock is stronger than it is at the surface.
5. Deep foundations are categorized by the installation method. The installation method must be properly matched to the type of soil or rock at a particular site. The installation process can significantly change the properties of the soil or rock near the deep foundation; in some cases it improves the soil properties, in other it degrades them.

Vocabulary

auger pile
bored pile
caisson
circular footing
combined footing
continuous flight auger pile
continuous footing
deep foundation
drilled pile
drilled shaft
driven pile
end bearing

foundation
full displacement drilled displacement pile
H-pile
mat foundation
partial displacement drilled displacement pile
pier
pile
pile cap
pipe pile
prestressed concrete pile
raft foundation
rectangular footing
ring footing
shallow foundation
side resistance
side shear
spread footing foundation
square footing
timber pile

CHAPTER 15

Spread Footing Design

Therefore O students, study mathematics and do not build without foundations.

Leonardo Da Vinci

As discussed in Chapter 14, spread footings are the most common type of foundation, and they can be built in a wide range of shapes and dimensions. This chapter covers the geotechnical analysis and design of the two most common types of spread footings: a square or circular footing supporting a single column and a continuous footing supporting a load-bearing wall.

15.1 BEARING PRESSURE

During the late nineteenth and early twentieth centuries, engineers realized that the design of spread footings could be based on the contact pressure between the footing and the underlying ground. This important parameter is called the *bearing pressure* (or *gross bearing pressure*), q, and is still used in modern foundation design:

$$q = \frac{P + W_f}{A} - u \tag{15.1}$$

where:

q = bearing pressure

P = vertical column load

W_f = weight of foundation

A = base area of foundation (B^2 for square footings or $B \times L$ for rectangular footings)

u = pore water pressure at bottom of footing (i.e., at a depth D below the ground surface)

The pore water pressure term accounts for uplift pressures (buoyancy forces) that would be present if the groundwater table is (or could be) at a depth less than D. If the groundwater table is at a depth greater than D, then set $u = 0$.

The weight of the foundation, W_f, may be expressed as:

$$\frac{W_f}{A} = \gamma_c D \tag{15.2}$$

where:

γ_c = unit weight of concrete = 150 lb/ft^3 = 23.6 kN/m^3

D = depth of footing

Combining Equations 15.1 and 15.2 gives the bearing pressure equation for square, rectangular, and circular footings.

$$q = \frac{P}{A} + \gamma_c D - u \tag{15.3}$$

For a continuous footing, we express the applied load as a force per unit length, such as 2000 kN/m. For ease of computation, we identify this unit length as b, which is usually 1 m or 1 ft, as shown in Figure 15.1. Thus, the load is expressed using the notation P/b.

The bearing pressure for a continuous footing is then

$$q = \frac{P/b}{B} + \gamma_c D - u \tag{15.4}$$

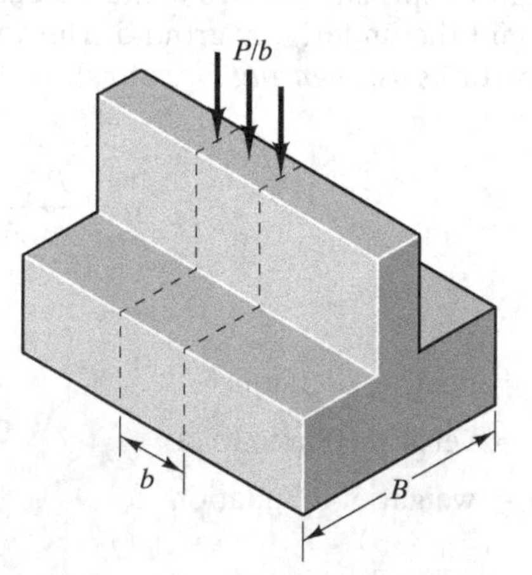

FIGURE 15.1 Definitions for loads on continuous footings.

If the column or wall load is vertical and acts through the centroid of the footing, and no moment loads are present, then we assume that the bearing pressure is uniform across the bottom of the footing. While footings can carry limited moments and horizontal loads, this chapter will deal with only downward vertical loads applied at the footing centroid.

The two most important geotechnical design requirements for spread footings are bearing capacity and settlement, and we analyze both in terms of the bearing pressure.

15.2 BEARING CAPACITY

A *bearing capacity failure* occurs when the soil fails in shear beneath the footing, as shown in Figure 15.2. This is a catastrophic foundation failure that clearly must be avoided. For example, Figures 15.3 and 15.4 show the results of a bearing capacity failure of a cement silo. This need to prevent a bearing capacity failure is called a *strength requirement* and is similar to structural engineers' requirements for strength of structural members.

Bearing Capacity Analysis

The *ultimate bearing capacity*, q_{ult}, is the bearing pressure required to produce a bearing capacity failure. In 1943, Karl Terzaghi developed the first widely accepted formulas for computing the ultimate bearing capacity. Terzaghi's bearing capacity formulas provided a rational means of designing foundations in a way that avoids bearing capacity failures. Terzaghi's original solution was for a continuous footing because the bearing capacity failure of such a footing can be simplified to a two-dimensional problem, and thus is the simplest case. For his analysis, Terzaghi assumed the symmetric geometry shown in Figure 15.5. In this figure, the soil in zone A forms a wedge under the footing and moves downward with the footing. The soil in zones B and C are in a state of general shear failure and move up and away from the footing as it moves down into the soil. Figure 15.6 shows a two-dimensional representation of the assumed failure surfaces shown in the previous figure. Note that the size of the failure surface will be controlled by the footing width, B, and the effective friction angle of the soil, ϕ'. The forces resisting bearing

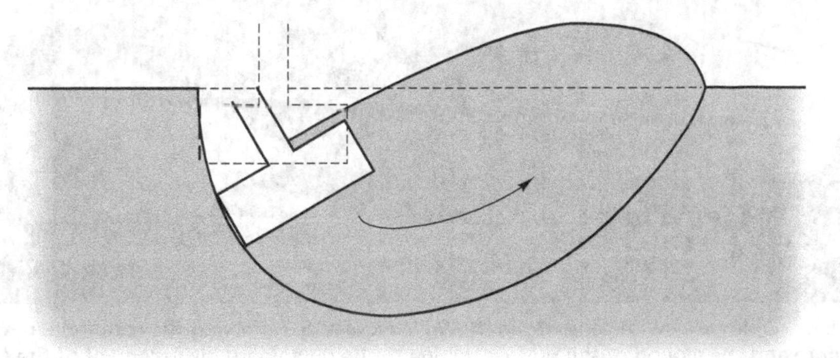

FIGURE 15.2 Generation of a failure surface and relative movement of a footing and surrounding soil during a bearing capacity failure.

FIGURE 15.3 Elevation view of a bearing capacity failure of a cement silo founded on soft clay. Note how the foundation has move down and rotated, thereby pushing a mass of soil up on the right side of the foundation. (Photograph by Lazarus White 1940, provided courtesy of Dr. Adel S. Saada, Case Western Reserve University.)

FIGURE 15.4 Oblique view of cement silo shown in Figure 15.3 showing the rotational nature of the failure and the mound of soil pushed up due to the downward displacement of the foundation. (Photograph by Lazarus White 1940, provided courtesy of Dr. Adel S. Saada, Case Western Reserve University.)

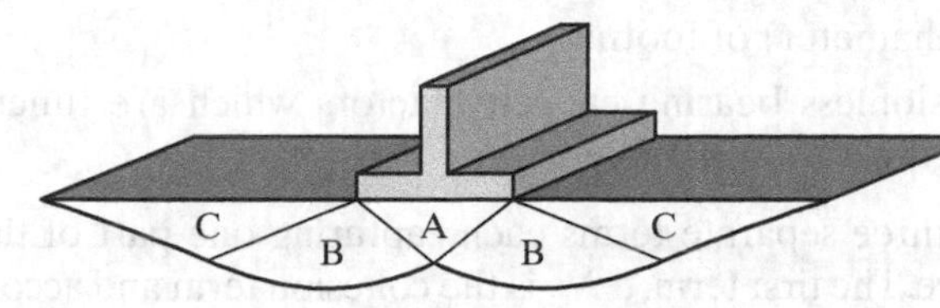

FIGURE 15.5 Footing model used by Terzaghi to determine the ultimate bearing capacity.

capacity failure include (a) the soil's effective cohesion, c', acting along the failure surface; (b) the resultant of the normal and frictional stresses along the failure surface, R, acting at an angle ϕ' to the normal to the failure surface; and (c) the vertical effective stress applied by the soil above the base of the footing, σ'_D. Terzaghi ignored the shear resistance from the part of the failure surface between the ground surface and a depth D. Using this assumed failure geometry and the Mohr–Coulomb failure model for soil strength, Terzaghi derived the following equation for the bearing capacity of continuous footings:

$$q_{ult} = c'N_c + \sigma'_D N_q + 0.5\gamma' B N_\gamma \tag{15.5}$$

where:

q_{ult} = ultimate bearing capacity

c' = effective cohesion

σ'_D = vertical effective stress at depth D below the ground surface ($\sigma'_D = \gamma D$ if depth to groundwater table is greater than D)

γ' = effective unit weight of the soil ($\gamma' = \gamma$ if the groundwater table is very deep; see discussion later in this section for shallow groundwater conditions)

D = depth of footing below ground surface

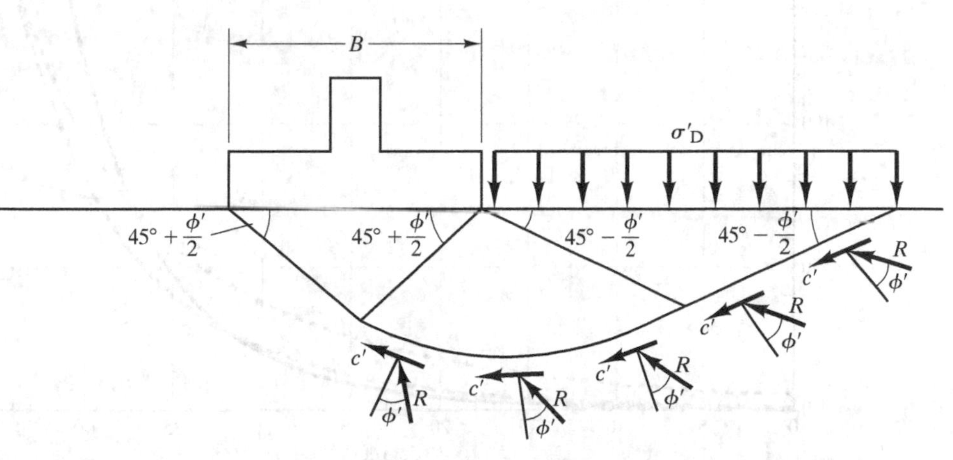

FIGURE 15.6 Two-dimensional representation of the failure surface used by Terzaghi to determine the ultimate bearing capacity.

B = width (or diameter) of footing

N_c, N_q, N_γ = are dimensionless bearing capacity factors which are functions of ϕ' (see Figure 15.7 and Table 15.1)

Equation 15.5 contains three separate terms each capturing one part of the resistance to bearing capacity failure. The first term, $c'N_c$, is the cohesion term and accounts for the cohesive resistance along the failure surface. The second term, $\sigma'_D N_q$, is the surcharge term and accounts for the resistance supplied by the mass of soil above the base of the

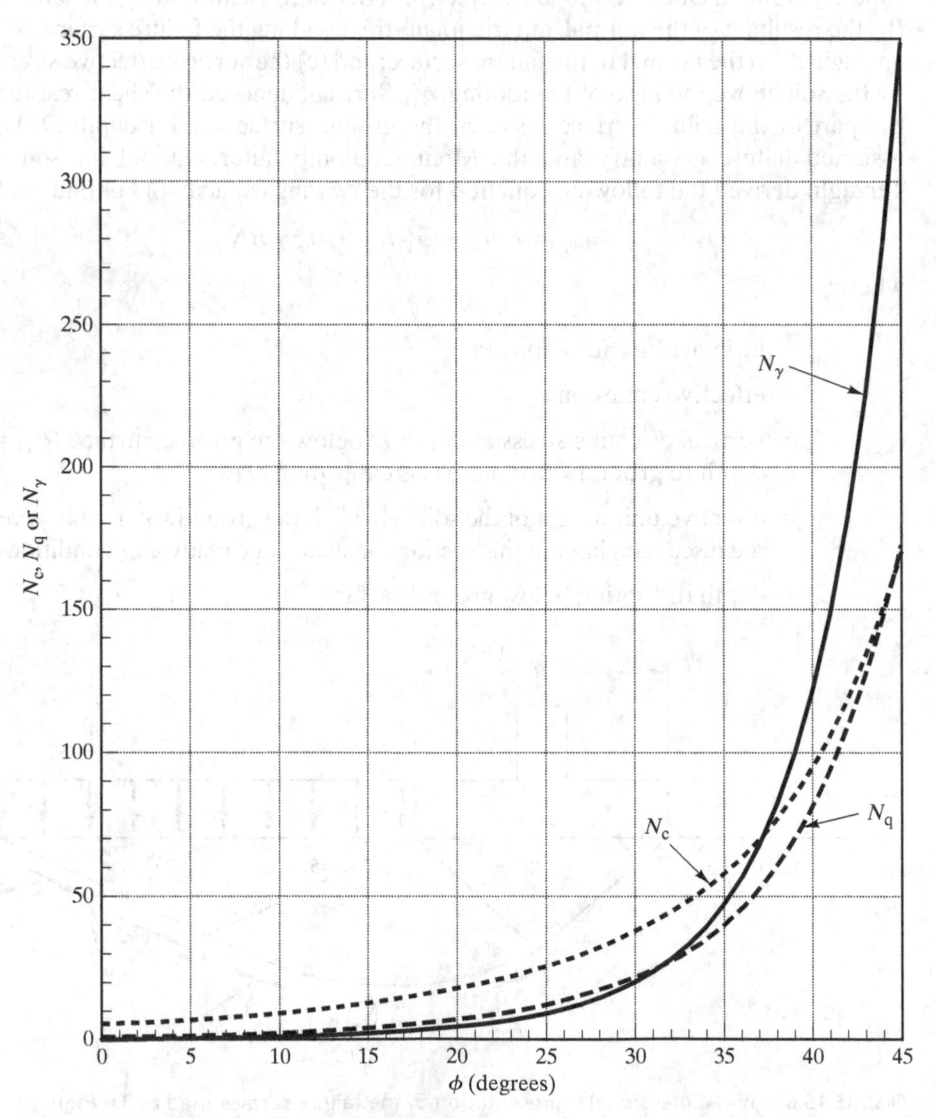

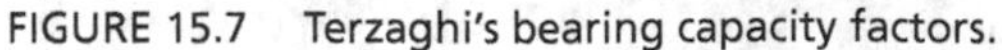
FIGURE 15.7 Terzaghi's bearing capacity factors.

TABLE 15.1 Bearing Capacity Factors for Terzaghi's Equations

ϕ' (deg)	N_c	N_q	N_γ	ϕ' (deg)	N_c	N_q	N_γ
0	5.7	1.0	0.0	21	18.9	8.3	5.1
1	6.0	1.1	0.1	22	20.3	9.2	5.9
2	6.3	1.2	0.1	23	21.7	10.2	6.8
3	6.6	1.3	0.2	24	23.4	11.4	7.9
4	7.0	1.5	0.3	25	25.1	12.7	9.2
5	7.3	1.6	0.4	26	27.1	14.2	10.7
6	7.7	1.8	0.5	27	29.2	15.9	12.5
7	8.2	2.0	0.6	28	31.6	17.8	14.6
8	8.6	2.2	0.7	29	34.2	20.0	17.1
9	9.1	2.4	0.9	30	37.2	22.5	20.1
10	9.6	2.7	1.0	31	40.4	25.3	23.7
11	10.2	3.0	1.2	32	44.0	28.5	28.0
12	10.8	3.3	1.4	33	48.1	32.2	33.3
13	11.4	3.6	1.6	34	52.6	36.5	39.6
14	12.1	4.0	1.9	35	57.8	41.4	47.3
15	12.9	4.4	2.2	36	63.5	47.2	56.7
16	13.7	4.9	2.5	37	70.1	53.8	68.1
17	14.6	5.5	2.9	38	77.5	61.5	82.3
18	15.5	6.0	3.3	39	86.0	70.6	99.8
19	16.6	6.7	3.8	40	95.7	81.3	121.5
20	17.7	7.4	4.4	41	106.8	93.8	148.5

footing. The third term, $0.5\gamma' B N_\gamma$, is the self-weight term and accounts for the frictional resistance generated along the failure surface. The self-weight term is a function of the footing width, B, because increasing the footing width increases the volume of soil in zones B and C, thereby increasing the normal forces acting on the failure surface. This increase of normal forces on the failure surface in turn increases the frictional resistance along the failure surface. Each of the bearing capacity factors, N_c, N_q, and N_γ, increases with the effective friction angle of the soil, ϕ', as shown in Figure 15.7. Note that each bearing capacity factor increases at a different rate. Terzaghi's bearing capacity factors are also presented in Table 15.1.

Since Terzaghi derived his bearing capacity equation for the two-dimensional case of a continuous footing, it is not strictly applicable to square, circular, or other finite sized footings. In the case of rectangular and circular footings, the ultimate bearing capacity, q_{ult}, should be greater than the value computed by Equation 15.5 because in these cases there is soil resisting failure on all four sides of the footing, not just two sides as assumed in Terzaghi's original analysis. Based on experimental data, Equation 15.5 has been extended as follows:

For square footings:

$$q_{ult} = 1.3c' N_c + \sigma'_D N_q + 0.4\gamma' B N_\gamma \tag{15.6}$$

For circular footings:

$$q_{ult} = 1.3c'N_c + \sigma'_D N_q + 0.3\gamma' B N_\gamma \tag{15.7}$$

Once we know q_{ult}, we can design the footing so that the actual bearing pressure is sufficiently smaller in order to provide an adequate factor of safety against a bearing capacity failure.

Terzaghi's equations also may be used in a total stress analysis. In that case, substitute c_T, ϕ_T, and σ_D for c', ϕ', and σ'_D. If the soil is saturated and the undrained condition exists, we may conduct a total stress analysis with the shear strength defined as $c_T = s_u$ and $\phi_T = 0$. In this case, $N_c = 5.7$, $N_q = 1.0$, and $N_\gamma = 0.0$.

Although subsequent work has produced formulas that are more versatile and slightly more precise, Terzaghi's formulas are still widely used, and are adequate for many practical design problems.

Groundwater Effects

Apparent Cohesion Sometimes soil samples obtained from the exploratory borings are not saturated, especially if the site is in an arid or semiarid area. These soils have additional shear strength due to the presence of apparent cohesion, as discussed in Chapter 12. However, this additional strength will disappear if the moisture content increases. Water may come from landscape irrigation, rainwater infiltration, leaking pipes, rising groundwater, or other sources. Therefore, we do not rely on the strength due to apparent cohesion.

In order to remove the apparent cohesion effects and simulate the "worst case" condition, geotechnical engineers usually wet the samples in the lab prior to testing. This may be done by simply soaking the sample, or, in the case of the triaxial test, by backpressure saturation. However, even with these precautions, the cohesion measured in the laboratory test may still include some apparent cohesion. Therefore, we often perform bearing capacity computations using a cohesion value less than that measured in the laboratory. The reduction factor applied to the laboratory values is largely a matter of engineering judgment.

Pore Water Pressure If there is enough water in the soil to develop a groundwater table, and this groundwater table is within the potential shear zone, then pore water pressures will be present, the effective stress and shear strength along the failure surface will be smaller, and the ultimate bearing capacity will be reduced (Meyerhof, 1955). We must consider this effect when conducting bearing capacity computations.

When exploring the subsurface conditions, we determine the current location of the groundwater table and worst-case (highest) location that might reasonably be expected during the life of the proposed structure. We then determine which of the following three cases describes the worst-case field conditions:

- Case 1: groundwater table is at or above base of footing ($D_w \le D$)
- Case 2: groundwater table is below the base of footing, but still within the potential shear zone, assumed to have a thickness of B, below the footing ($D < D_w < D + B$)
- Case 3: groundwater table is below the potential shear zone below the footing ($D + B \le D_w$).

All three cases are shown in Figure 15.8.

We account for the decreased effective stresses along the failure surface by adjusting the effective unit weight, γ', in the third term of Equations 15.5–15.7 (Vesić, 1973). The effective unit weight is the value that, when multiplied by the appropriate soil thickness, will give the vertical effective stress. It varies between the buoyant unit weight, γ_b, and the unit weight, γ, depending on the position of the groundwater table. We compute γ' as follows:

For Case 1 where the groundwater table is at or above the base of footing ($D_w \leq D$), we simply compute γ' as the buoyant unit weight:

$$\gamma' = \gamma_b = \gamma - \gamma_w \tag{15.8}$$

For Case 2 where the groundwater table is below the base of footing, but still within the potential shear zone below the footing ($D < D_w < D + B$), we interpolate γ' between buoyant unit weight and the unit weight using:

$$\gamma' = \gamma - \gamma_w\left[1 - \left(\frac{D_w - D}{B}\right)\right] \tag{15.9}$$

For Case 3 where the groundwater table is below the potential shear zone below the footing ($D + B \leq D_w$), no groundwater correction is necessary:

$$\gamma' = \gamma \tag{15.10}$$

In Case 1, the second term in the bearing capacity formulas is also affected, but the appropriate correction is implicit in the computation of σ'_D.

When performing a total stress analysis, do not apply any groundwater correction because the groundwater effects are supposedly implicit within the values of c_T and ϕ_T. In this case, simply use $\gamma' = \gamma$ in the bearing capacity equations, regardless of the groundwater table position.

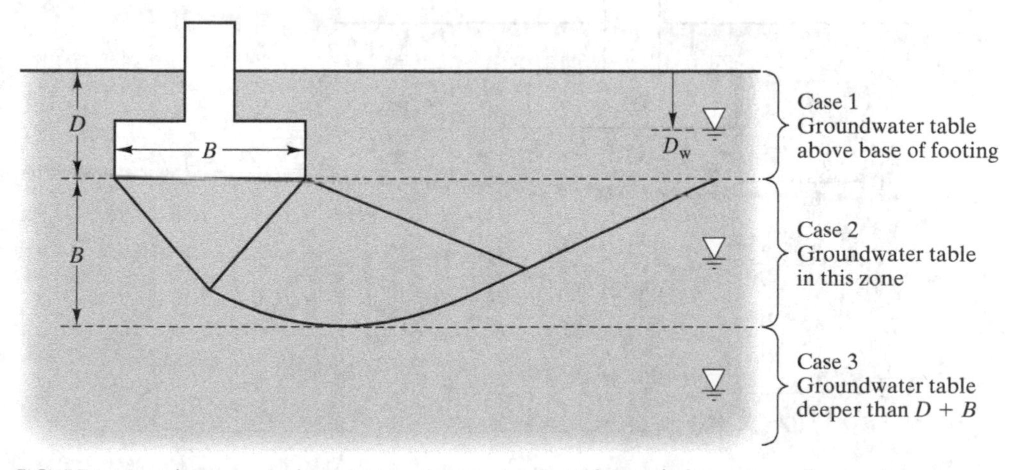

FIGURE 15.8 Three groundwater cases for bearing capacity analysis.

Allowable Bearing Capacity

We compute the *allowable bearing capacity*, q_a, using:

$$q_a = \frac{q_{ult}}{F} \tag{15.11}$$

The required factor of safety, F, depends on the type of structure, the type of soil, and other factors, and is typically between 2.0 and 3.5. Low factors of safety might be used for noncritical structures on sandy soils with extensive site characterization, while high factors of safety would more often be used for critical structures on clayey soils with minimal site characterization.

We then satisfy bearing capacity requirements by designing the footing such that $q \leq q_a$. Typically, P, c, ϕ, γ, and the groundwater conditions are fixed, so the only parameters we can vary are the footing dimensions B and D. If the soil is homogeneous, increasing D generally has very little impact, so we usually satisfy bearing capacity requirements by specifying a minimum required footing width, B. For smaller footings we normally round both B and D to multiples of 100 mm or 3 in. For larger footings, we normally use multiples of 200 mm or 6 in.

Example 15.1

Compute the factor of safety against a bearing capacity failure for the square spread footing shown in Figure 15.9 with the groundwater table at Position A.

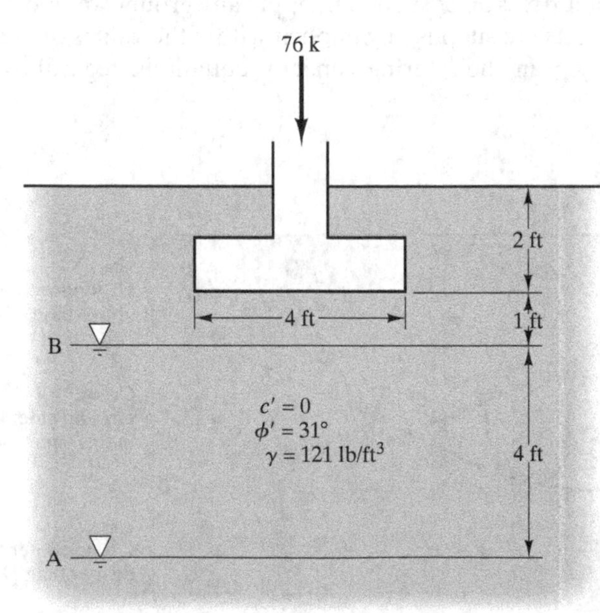

FIGURE 15.9 Proposed spread footing for Example 15.1.

Solution:

$$D = 2 \text{ ft}$$
$$D_w = 7 \text{ ft}$$
$$D + B = 6 \text{ ft}$$
$$D + B \leq D_w, \text{ so groundwater Case 3 applies} - \gamma' = \gamma$$
$$\sigma'_D = \gamma D - u = (121 \text{ lb/ft}^3)(2 \text{ ft}) - 0 = 242 \text{ lb/ft}^2$$

Per Table 15.1—$N_c = 40.4$, $N_q = 25.3$, $N_\gamma = 23.7$ when $\phi' = 31°$

$$q = 1.3c'N_c + \sigma'_D N_q + 0.4\gamma' B N_\gamma$$
$$q = 0 + (242 \text{ lb/ft}^2)(25.3) + 0.4(121 \text{ lb/ft}^3)(4 \text{ ft})(23.7)$$
$$q = 10{,}710 \text{ lb/ft}^2$$
$$q = \frac{P}{A} + \gamma_c D - u = \frac{76{,}000 \text{ lb}}{(4 \text{ ft})^2} + (150 \text{ lb/ft}^3)(2 \text{ ft}) - 0 = 5050 \text{ lb/ft}^2$$

Because we are computing the factor of safety for a given bearing pressure, we rewrite Equation 15.11 in terms of F instead of q_a, and $q_a = q$

$$F = \frac{q_{ult}}{q} = \frac{10{,}710 \text{ lb/ft}^2}{5050 \text{ lb/ft}^2} = \mathbf{2.1}$$

Comments: For clean sands, a factor of safety of 2.1 would probably be marginally acceptable.

Example 15.2

Sometime after construction, the groundwater table in Example 15.1 rose to Position B. Compute the new factor of safety against a bearing capacity failure.

Solution:

$$D = 2 \text{ ft}$$
$$D_w = 3 \text{ ft}$$
$$D + B = 6 \text{ ft}$$
$$D < D_w < D + B, \text{ so groundwater Case 2 applies}$$

$$\gamma' = \gamma - \gamma_w\left[1 - \left(\frac{D_w - D}{B}\right)\right]$$

$$\gamma' = 121 \text{ lb/ft}^2 - 62.4 \text{ lb/ft}^2\left[1 - \left(\frac{3 \text{ ft} - 2 \text{ ft}}{4 \text{ ft}}\right)\right]$$

$$\gamma' = 74.2 \text{ lb/ft}^3$$

$$q_{ult} = 1.3c'N_c + \sigma'_D N_q + 0.4\gamma' B N_\gamma$$
$$q_{ult} = 0 + (242\ \text{lb/ft}^2)(25.3) + 0.4(74.2\ \text{lb/ft}^3)(4\ \text{ft})(23.7)$$
$$q_{ult} = 8936\ \text{lb/ft}^2$$

$$F = \frac{q_{ult}}{q} = \frac{8936\ \text{lb/ft}^2}{5050\ \text{lb/ft}^2} = \mathbf{1.8}$$

Comments: The rising groundwater table has dropped the factor of safety to 1.8. Although this is still above the theoretical failure value of 1.0, it is less than the minimum acceptable F.

Example 15.3

A 1350-kN column load is to be supported on a square spread footing founded in a clay with $s_u = 150$ kPa. The depth of embedment, D, will be 500 mm, and the soil has a unit weight of 18.5 kN/m^3. The groundwater table is at a depth below the bottom of the footing. Using a factor of safety of 3.0, determine the required footing width.

Solution:

This is a case of undrained loading on clay. Since positive excess pore pressures will be generated by this loading, short-term undrained loading conditions apply and a $\phi = 0$ total stress analysis is appropriate (see Case 2 in Section 12.6).

Per Table 15.1, $N_c = 5.7$, $N_q = 1.0$, $N_\gamma = 0.0$ for $\phi = 0$

$$\sigma_D = \gamma D = (18.5\ \text{kN/m}^3)(0.5\ \text{m}) = 9\ \text{kPa}$$
$$q_{ult} = 1.3c_T N_c + \sigma_D N_q + 0.4\gamma B N_\gamma$$
$$q_{ult} = 1.3(150\ \text{kPa})(5.7) + (9\ \text{kPa})(1.0) + 0$$
$$q_{ult} = 1121\ \text{kPa}$$

$$q_a = \frac{q_{ult}}{F} = \frac{1121\ \text{kPa}}{3.0} = 374\ \text{kPa}$$

$$q_a = q = \frac{P}{B^2} + \gamma_c D - u$$

$$374\ \text{kPa} = \frac{1350\ \text{kN}}{B^2} + (23.6\ \text{kN/m}^3)(0.5\ \text{m}) - 0$$

$$B = 1.93\ \text{m}$$

Round up to B = **2.0 m.**

Example 15.4

A bearing wall for a proposed building is to be supported on a 24-in deep continuous footing founded in an unsaturated clayey sand (SC). The load from this wall will be 4.0 k/ft, and the soil has $c' = 100$ lb/ft^2, $\phi' = 28°$, and $\gamma = 119$ lb/ft^3. The groundwater table is at a very great depth. Compute the required footing width to maintain a factor of safety of 3.0 against a bearing capacity failure.

Solution:

Per Table 15.1, $N_c = 31.6$, $N_q = 17.8$, $N_\gamma = 14.6$ for $\phi' = 28°$

$$\sigma'_D = \gamma D - u = (119 \text{ lb/ft}^3)(2 \text{ ft}) - 0 = 238 \text{ lb/ft}^2$$

$$q_{ult} = c'N_c + \sigma'_D N_q + 0.5\gamma' B N_\gamma$$

$$q_{ult} = (100 \text{ lb/ft}^2)(31.6) + (238 \text{ lb/ft}^2)(17.8) + 0.5(119 \text{ lb/ft}^3)B(14.6)$$

$$q_{ult} = 7396 \text{ lb/ft}^2 + 869B$$

$$q_a = \frac{q_{ult}}{F} = \frac{7396 \text{ lb/ft}^2 + 869B}{3.0} = 2465 \text{ lb/ft}^2 + 290B$$

$$q = \frac{P/b}{B} + \gamma_c D - u = \frac{4000 \text{ lb/ft}}{B} + (150 \text{lb/ft}^3)(2 \text{ ft}) - 0$$

Setting $q = q_a$ and solving for B gives $B =$ **1 ft 9 in.**

15.3 SETTLEMENT

The structural loads applied to spread footings increase the vertical effective stress in the underlying soils, thus causing the soil to consolidate and the footings to settle. Footings must be designed so this settlement does not exceed the tolerable settlement, thus protecting the structure from excessive movement. This criterion is called a *serviceability requirement* because it is controlled by the ability of the structure to perform adequately, not by a threat of a catastrophic failure.

There are two settlement requirements for structural foundations.

Total settlement, δ, is the change in footing elevation from the original unloaded position to the final loaded position. For buildings, the allowable total settlement, δ_a, depends on the need for providing smooth pedestrian and vehicle access, avoidance of utility line shearing, maintenance of proper surface drainage, aesthetics, and other considerations.

Differential settlement, δ_D, is the difference in total settlement between two foundations or between two points on a single foundation. These differences are due to nonuniformities in the soil, differences in the structural loads, construction tolerances, and other factors. For buildings, the allowable differential settlement, δ_{Da}, depends on the ability of doors, windows, and elevators to operate if the building becomes distorted, the potential for cracks in the structure, aesthetics, and other similar concerns.

The structural engineer usually determines the maximum allowable total and differential settlements, then the geotechnical engineer performs settlement analyses to determine how to design the footings so the actual settlements do not exceed the allowable settlements ($\delta \leq \delta_a$ and $\delta_D \leq \delta_{Da}$). These analysis methods predict the total settlement, δ, based on the loads, the soil properties, and the footing geometry. The differential settlement expected to occur in the field is taken as some percentage of the total settlement or, in the case of erratic soil profiles, as the difference in results between two total settlement analyses.

If the settlement of a proposed footing is excessive, the design must be modified accordingly, usually by increasing B (thus decreasing q). Settlement requirements often dictate a larger B than needed to satisfy bearing capacity requirements, and the final design must use the larger of the B values obtained from these two independent analyses.

Footings on Clays and Silts

Settlement analyses for footings on clays and silts are similar to settlement analyses due to the placement of large areal fills as described in Chapter 10. However, there are two additional issues that need to be considered: the computation of $\Delta\sigma_z$ and the flexural rigidity of the footing.

Computation of $\Delta\sigma_z$ for Footing Loads For simplicity, we compute the vertical effective stresses beneath the center of the footing. The final vertical effective stress, σ'_{zf}, in the soil due to the application of any surface load is

$$\sigma'_{zf} = \sigma'_{z0} + \Delta\sigma_z \tag{15.12}$$

where:

σ'_{z0} = initial vertical effective stress beneath the center of footing
σ'_{zf} = final vertical effective stress beneath the center of footing
$\Delta\sigma_z$ = induced vertical stress due to the surface load

In the case of large areal fills, the induced vertical stress, $\Delta\sigma_z$, at any point in the soil is equal to the applied stress at the surface.

$$\Delta\sigma_z = \gamma_{\text{fill}} H_{\text{fill}} \tag{15.13}$$

In the case of footings, we must apply the stress distribution concepts presented in Section 9.5 to compute the induced stress in the soil under the footing. In this case, induced vertical stress, $\Delta\sigma_z$, at any location below the footing is computed as

$$\Delta\sigma_z = I_\sigma(q - \sigma'_D) \tag{15.14}$$

where:

I_σ = stress influence factor as defined in Chapter 9

Note that the induced vertical stress must be computed using the difference between the bearing pressure, q, and the vertical effective stress at the embedment depth D, σ'_D. We generally compute I_σ under the center of the footing using either Figures 9.13–9.16 or Equations 9.34, 9.35, or 9.38–9.41 as appropriate.

Example 15.5

Compute the induced vertical stress at points A and B beneath the center of the continuous footing shown in Figure 15.10.

Solution:

$$q' = q - \gamma D$$

$$q' = \frac{20{,}000\ \text{lb/ft}}{5\ \text{ft}} + (150\ \text{lb/ft}^3)(3\ \text{ft}) - (120\ \text{lb/ft}^3)(3\ \text{ft})$$

$$q' = 4000 + 450 - 360 = 4090\ \text{lb/ft}^2$$

For point A,

$z_f/B = 5/5 = 1$
from Figure 9.15 at $z_f/B = 1$
$I_\sigma = 0.55$ and
$\Delta\sigma_z = 0.55\ (4090) =$ **2250 psf**

For point B,

$z_f/B = 10/5 = 2$
from Figure 9.15 at $z_f/B = 2$
$I_\sigma = 0.31$ and
$\Delta\sigma_z = 0.31\ (4090) =$ **1268 lb/ft²**

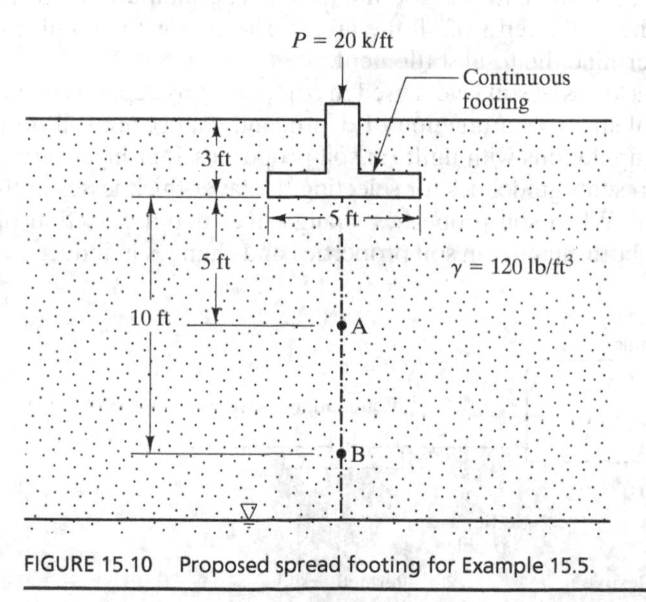

FIGURE 15.10 Proposed spread footing for Example 15.5.

Flexural Rigidity At a given depth, z_f, below the bottom of the footing, the value of $\Delta\sigma_z$ is always greater beneath the center of the footing than beneath the edges. This difference is illustrated by the stress bulbs in Figures 9.13–9.15. Because the consolidation settlement depends on $\Delta\sigma_z$, there will be more consolidation settlement below the center than below the edges. However, this conclusion is based on the assumption that the loaded area is perfectly flexible, and thus is able to settle more at the center than at the edges. A steel tank is an example of such a load. However, a spread footing is far from being flexible. Its structural rigidity is such that the settlement at the center will be essentially equal to that at the edge. Therefore, the actual settlement of a spread footing will be greater than that beneath the edge of a perfectly flexible load, but less than that beneath the center, as shown in Figure 15.11.

We account for this behavior by computing the settlement beneath the center of a perfectly flexible load, then applying a rigidity factor, r, of 0.85. In other words, the settlement of a footing is about 85% of the settlement at the center of a perfectly flexible loaded area that has the same dimensions and the same q.

Settlement Computations Settlement computations for spread footings often consider only consolidation settlement and assume that the consolidation is one-dimensional. This is the method that will be presented here. *Foundation Design: Principles and Practices* (Coduto, 2001) presents an alternative method that considers both consolidation and distortion settlements, and three-dimensional effects. In some cases, secondary compression settlement may also be important.

Computing the footing settlement due to one-dimensional consolidation is a simple matter of applying Equations 10.23–10.25, except now the final vertical effective stress, σ'_{zf}, is a function of depth and we must include the rigidity factor of 0.85. The procedure is to divide the soil beneath the footing into layers, compute the settlement of each layer, and sum the settlements of all the layers. The sum is then multiplied by the rigidity factor to determine the total settlement.

The number and thickness of soil layers used in computations depends on both the change in applied vertical stress, as a function of depth, and changes in soil properties over depth. For hand computations with uniform soil properties, four layers are usually sufficient. Figure 15.12 presents guidelines for selecting the layer thicknesses in the case where the soil is uniform. When soil properties change over depth, selection of layer thickness will depend on both changes in soil properties and changes in induced vertical

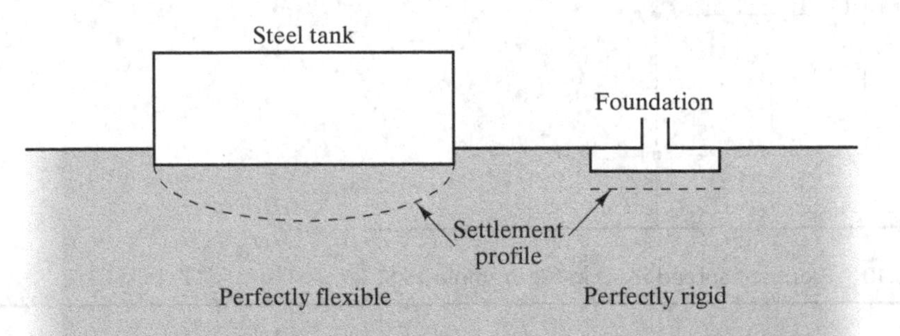

FIGURE 15.11 Influence of footing rigidity on settlement. The steel tank on the left is very flexible, so the center settles more than the edge. Conversely, the reinforced concrete footing on the right is very rigid, and thus settles uniformly.

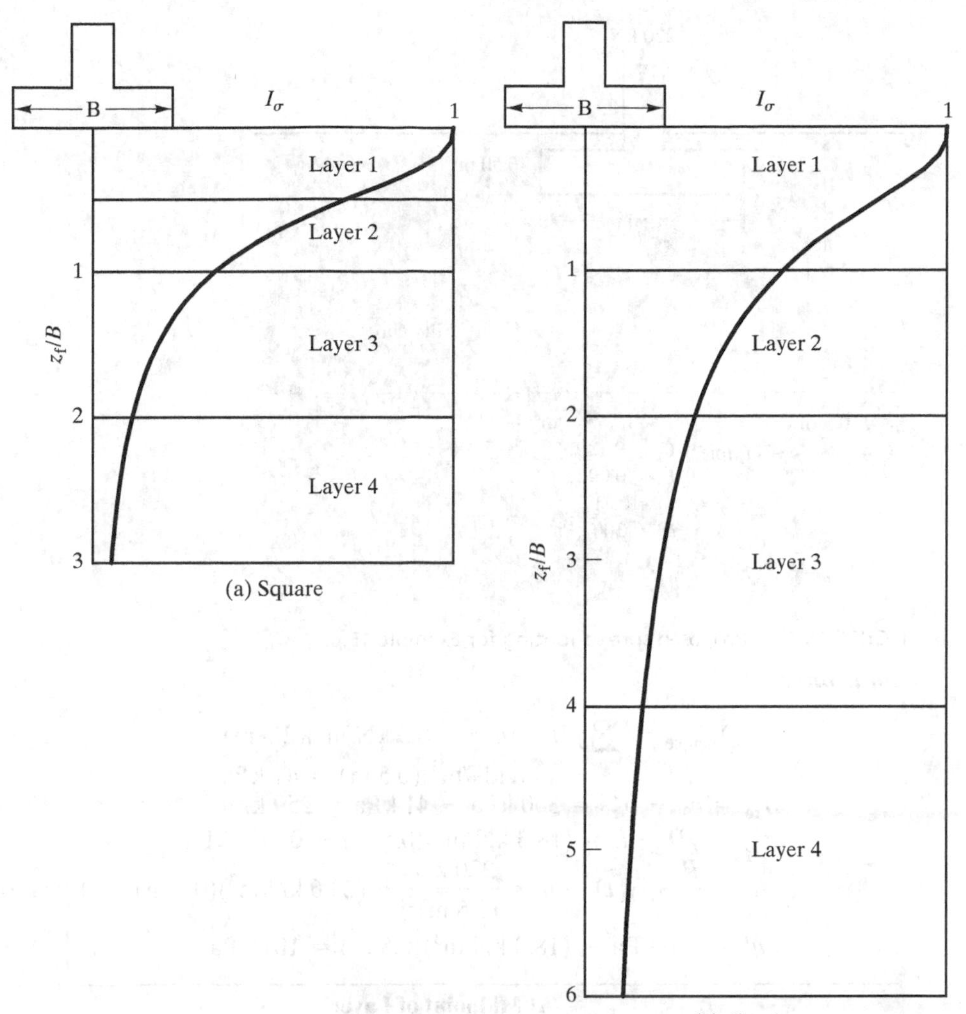

FIGURE 15.12 Guidelines for selecting layer thicknesses in hand settlement computations for uniform soils. The curved line in each figure represents the change in induced vertical stress under the center of the footing. Note that for a continuous footing in (b), the induced stress penetrates to a much greater depth than it does for a square footing in (a).

stress. If more than a few soil layers are needed, it is a relatively simple exercise to program the appropriate computations in a spreadsheet using a large number of layers, thus increasing both the speed and accuracy of the settlement computations.

Example 15.6

Compute the settlement of the spread footing in Figure 15.13. The structural engineer has determined that the allowable total settlement, δ_a, is 20 mm.

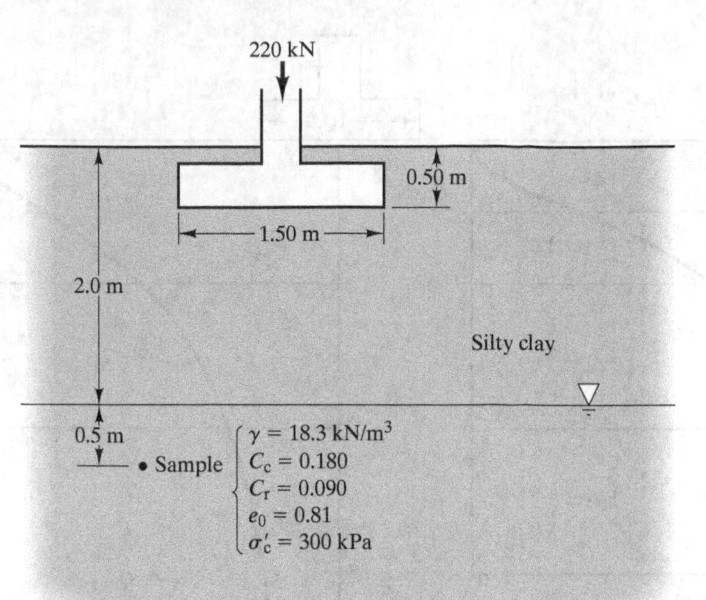

FIGURE 15.13 Proposed spread footing for Example 15.6.

Solution:

$$(\sigma'_{z0})_{\text{sample}} = \sum \gamma H - u = (18.3\ \text{kN/m}^3)(2.5\ \text{m}) - (9.8\ \text{kN/m}^3)(0.5\ \text{m}) = 41\ \text{kPa}$$

$$\sigma'_{\text{m}} = \sigma'_{\text{c}} - \sigma'_{z0} = 300\ \text{kPa} - 41\ \text{kPa} = 259\ \text{kPa}$$

$$\sigma'_{\text{D}} = \gamma D - u = (18.3\ \text{kN/m}^3)(0.5\ \text{m}) - 0 = 9\ \text{kPa}$$

$$q = \frac{P}{A} + \gamma_c D - u = \frac{220\ \text{kN}}{(1.5\ \text{m})^2} + (23.6\ \text{kN/m}^3)(0.5\ \text{m}) - 0 = 110\ \text{kPa}$$

$$q' = 110\ \text{kPa} - (18.3\ \text{kN/m}^3)(0.5\ \text{m}) = 101\ \text{kPa}$$

		At Midpoint of Layer								
Layer	**H (m)**	**z_f (m)**	**σ'_{z0} (kPa)**	**σ'_c (kPa)**	**$\Delta\sigma_z$ (kPa)**	**σ'_{zf} (kPa)**	**$\frac{C_c}{1+e_0}$**	**$\frac{C_r}{1+e_0}$**	**Case**	**$\delta_{c,ult}$ (mm)**
1	0.75	0.38	16.0	275.0	95.1	111.1	0.10	0.05	OC-I	31.5
2	0.75	1.13	29.7	288.7	48.1	77.9	0.10	0.05	OC-I	15.7
3	1.5	2.25	43.0	302.0	17.1	60.1	0.10	0.05	OC-I	10.9
4	1.5	3.75	55.7	314.7	6.7	62.5	0.10	0.05	OC-I	3.7
									Σ =	61.8
									Rigidity factor	0.85
									Settlement	52.6

Basis for computations:

The layer thicknesses are approximately equal to those in Figure 15.12(a). They have been adjusted slightly for computational convenience.

σ'_{z0}—Equation 9.47
I_σ—Equation 9.39

$\Delta\sigma_z$—Equation 15.14
σ'_{zf}—Equation 15.12
σ'_c—Equation 10.17
δ—Equation 10.24

The computed settlement is **53 mm.**

Comments: The computed settlement of 53 mm is greater than the allowable total settlement of 20 mm. Therefore, the footing width B must be increased until the settlement criterion is met.

Example 15.7

Compute the settlement of the continuous spread footing in Figure 15.14. The allowable total settlement is 1.0 in.

Solution:

Sample A

$$(\sigma'_{z0})_{\text{Sample A}} = \sum \gamma H - u = (125 \text{ lb/ft}^3)(2.5 \text{ ft}) = 312.5 \text{ lb/ft}^2$$

$$\sigma'_m = \sigma'_c - \sigma'_{z0} = 5000 \text{ lb/ft}^2 - 312.5 \text{ lb/ft}^2 = 4687.5 \text{ lb/ft}^2$$

Sample B

$$(\sigma'_{z0})_{\text{Sample B}} = \sum \gamma H - u = (125 \text{ lb/ft}^3)(5 \text{ ft}) + (121 \text{ lb/ft}^3)(5 \text{ ft}) + (121 \text{ lb/ft}^3 - 62.4 \text{ lb/ft}^3)(3 \text{ ft})$$

$$(\sigma'_{z0})_{\text{Sample B}} = 1405 \text{ lb/ft}^2$$

$$\sigma'_m = \sigma'_c - \sigma'_{z0} = 3000 \text{ lb/ft}^2 - 1405 \text{ lb/ft}^2 = 1595 \text{ lb/ft}^2$$

$$\sigma'_D = \gamma D - u = (125 \text{ lb/ft}^3)(3 \text{ ft}) - 0 = 375 \text{ lb/ft}^2$$

$$q = \frac{P/b}{B} + \gamma_c D - u = \frac{10{,}000 \text{ lb/ft}}{4 \text{ ft}} + (150 \text{ lb/ft}^3)(3 \text{ ft}) - 0 = 2950 \text{ lb/ft}^2$$

$$q' = 2950 \text{ lb/ft}^2 - (125 \text{ lb/ft}^3)(3 \text{ ft}) = 2575 \text{ lb/ft}^2$$

Note: In this case, the computed value of q is slightly conservative because the excavation has been partially filled with concrete ($\gamma = 150 \text{ lb/ft}^3$) and partly with soil ($\gamma = 121 \text{ lb/ft}^3$).

			At Midpoint of Layer							
Layer	H (ft)	z_f (ft)	σ'_{z0} (lb/ft²)	σ'_c (lb/ft²)	$\Delta\sigma_z$ (lb/ft²)	σ'_{zf} (lb/ft²)	$\frac{C_c}{1+e_0}$	$\frac{C_r}{1+e_0}$	Case	$\delta_{c,ult}$ (in.)
1	2.0	1.0	500.0	5187.5	2483.1	2983.1	0.056	0.013	OC-I	0.24
2	2.0	3.0	746.0	2341.0	1780.0	2526.0	0.069	0.020	OC-II	0.29
3	4.0	6.0	1109.0	2704.0	1038.1	2147.1	0.069	0.020	OC-I	0.28
4	8.0	12.0	1523.0	3118.0	489.0	2012.0	0.069	0.020	OC-I	0.23
5	8.0	20.0	1991.8	3586.8	259.4	2251.2	0.069	0.020	OC-I	0.10
									Σ =	1.14
								Rigidity factor		0.85
								Settlement		0.97

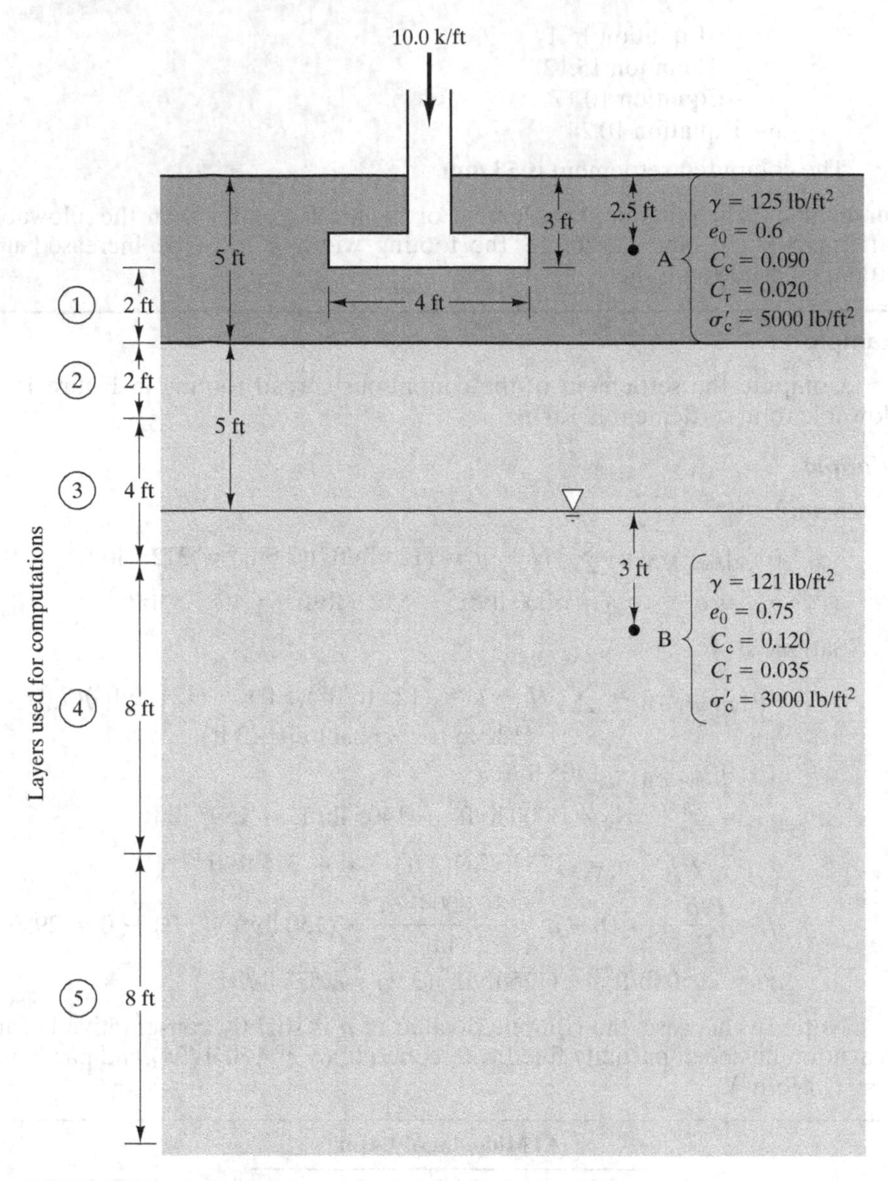

FIGURE 15.14 Proposed spread footing for Example 15.7.

The computed settlement is **1.0 in.**

Basis for computations: The layer thicknesses follow Figure 15.12(b) with the first layer split into two to account for the two different soils in the layer.

σ'_{z0}—Equation 9.47
I_σ—Equation 9.40
$\Delta\sigma_z$—Equation 15.14
σ'_{zf}—Equation 15.12

σ'_c—Equation 10.17
δ—Equations 10.24 and 10.25

Comments: The computed settlement of 1.0 in. is equal to the allowable total settlement of 1.0 in., so the current design is acceptable.

Footings on Sands and Gravels

In theory, the methods used to predict settlement of spread footings on clays and silts also could be used for sands and gravels. However, to use these methods, we would need to evaluate C_c and C_r in these soils, which would be very difficult or impossible because of the difficulties in obtaining undisturbed samples. Because of this limitation, we will use a different approach for computing settlements on sands and gravels.

Settlements in sands and gravel are generally much smaller than those in soft or medium clays. Therefore, we can apply elastic theory to compute settlements in these soils if we use appropriate empirical adjustments. Using this approach, the settlement is

$$\delta = \int C \frac{\Delta\sigma_z}{E} dz \tag{15.15}$$

where:

C = an empirical correction factor to account for nonelastic behavior (often more than one correction factor is used)
E = soil modulus
$\Delta\sigma_z$ = change in vertical stress

It is usually sufficient to replace the integral in Equation 15.15 with a finite summation giving

$$\delta = \sum_{i=1}^{n} C_i \frac{(\Delta\sigma_z)_i}{E_i} H_i \tag{15.16}$$

where:

C_i = empirical correction factor for the i-th soil layer
E = soil modulus for the i-th soil layer
$\Delta\sigma_z$ = change in vertical stress for the i-th soil layer
H_i = thickness of the i-th soil layer

A number of methods based on Equation 15.16 have been developed. The most commonly used method is that developed by Schmertmann (1978). A simplified version of that method is presented here. For a more thorough treatment of this topic, see Chapter 7 of *Foundation Design: Principles and Practices* (Coduto, 2001). Based on finite element modeling, field, and laboratory model tests, Schmertmann developed the following equation for computing settlement of footings on sands and gravels.

$$\delta = C_1 C_2 C_3 (q - \sigma'_D) \sum \frac{I_\varepsilon H}{E_s} \tag{15.17}$$

where:

δ = settlement of footing
C_1 = correction factor for footing depth
C_2 = correction factor for creep
C_3 = correction factor for footing shape
I_ε = strain influence factor
H = thickness of soil layer
E_s = equivalent modulus of soil layer

The correction factors have the following forms

$$C_1 = 1 - 0.5\left(\frac{\sigma'_D}{q - \sigma'_D}\right) \tag{15.18}$$

$$C_2 = 1 \text{ for } t < 1 \tag{15.19a}$$

$$C_2 = 1 + 0.2 \log\left(\frac{t}{0.1}\right) \text{ for } t \geq 1 \tag{15.19b}$$

$$C_3 = 1 \text{ for square footings and } 0.73 \text{ for continuous footings} \tag{15.20}$$

where:

t = time since application of load, in years

Schmertmann's strain influence factor I_ε has the forms shown in Figure 15.15. The strain influence factor accounts for both the distribution of stresses below the footing and nonlinear soil behavior immediately below the footing. The peak value of the strain influence factor, $I_{\varepsilon p}$, is

$$I_{\varepsilon p} = 0.5 + 0.1\sqrt{\frac{q - \sigma'_D}{\sigma'_{zp}}} \tag{15.21}$$

where:

σ'_{zp} = vertical effective stress at depth of the peak strain influence factor (for square footings compute σ'_{zp} at a depth of $D + B/2$; for continuous footing compute at $D + B$)

The value of the equivalent soil modulus, E_s, can be determined either using standard penetration test (SPT) N-values or cone penetration test (CPT) tip resistance, q_c. It is preferable to use CPT data rather than SPT data when possible. CPT data are more

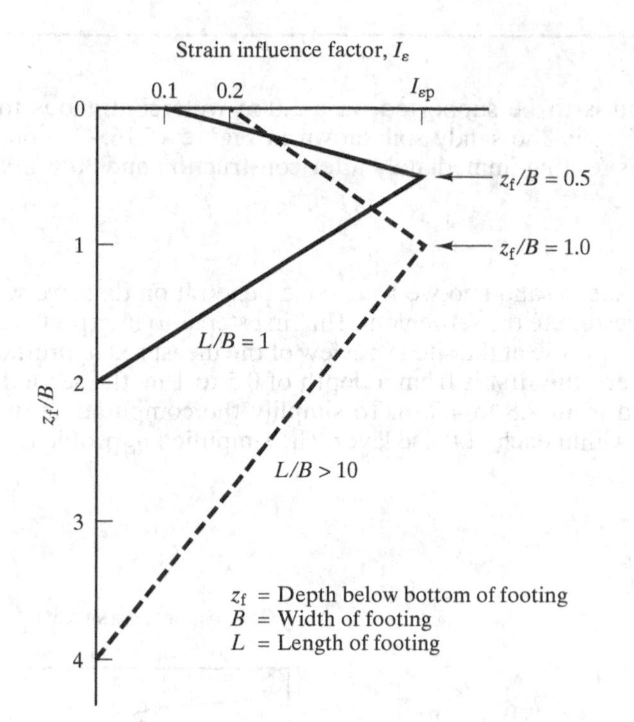

FIGURE 15.15 Distribution of strain influence factor with depth for square and continuous footings. (Adapted from Schmertmann et al., 1978, used with permission of ASCE.)

reliable and precise than SPT *N*-values. When using CPT data, E_s can conservatively be taken as

$$\begin{aligned} E_s &= 2.5q_c \text{ for clean sands} \\ &= 1.5q_c \text{ for clayey or silty sands} \end{aligned} \tag{15.22}$$

When using SPT data, the following empirical correlation between modulus and *N*-value can be used (Kulhawy and Mayne, 1990)

$$\begin{aligned} E_s &= 10N_{60}p_a \text{ for clean sands} \\ &= 5N_{60}p_a \text{ for clayey or silty sands} \end{aligned} \tag{15.23}$$

where:

N_{60} = SPT *N*-value corrected for field procedures
p_a = atmospheric pressure (2116 lb/ft^2 or 101 kPa)

For other types of soils, the reader is referred to Chapter 7 of *Foundation Design: Principles and Practices* (Coduto, 2001).

Example 15.8

A 75 kN/m wall load is to be supported on a 1.0-m wide continuous footing founded at a depth of 0.5 m in the sandy soil shown in Figure 15.16. Compute the expected settlement of this footing immediately after construction and 30 years after construction.

Solution:

Since the soil at this site is sand and we have cone penetration data, we will use Schmertmann's method to estimate the settlement. The first step is to interpret and simplify the cone penetrometer profile at the site. A review of the measured q_c profile indicate three distinct soil layers: the first is from a depth of 0.5 to 1 m, the second from 1.0 to 2.8 m, and the third from 2.8 to 4.5 m. To simplify the computations, we will assume that q_c is constant within each of these layer. The simplified q_c profile is shown in Figure 15.16.

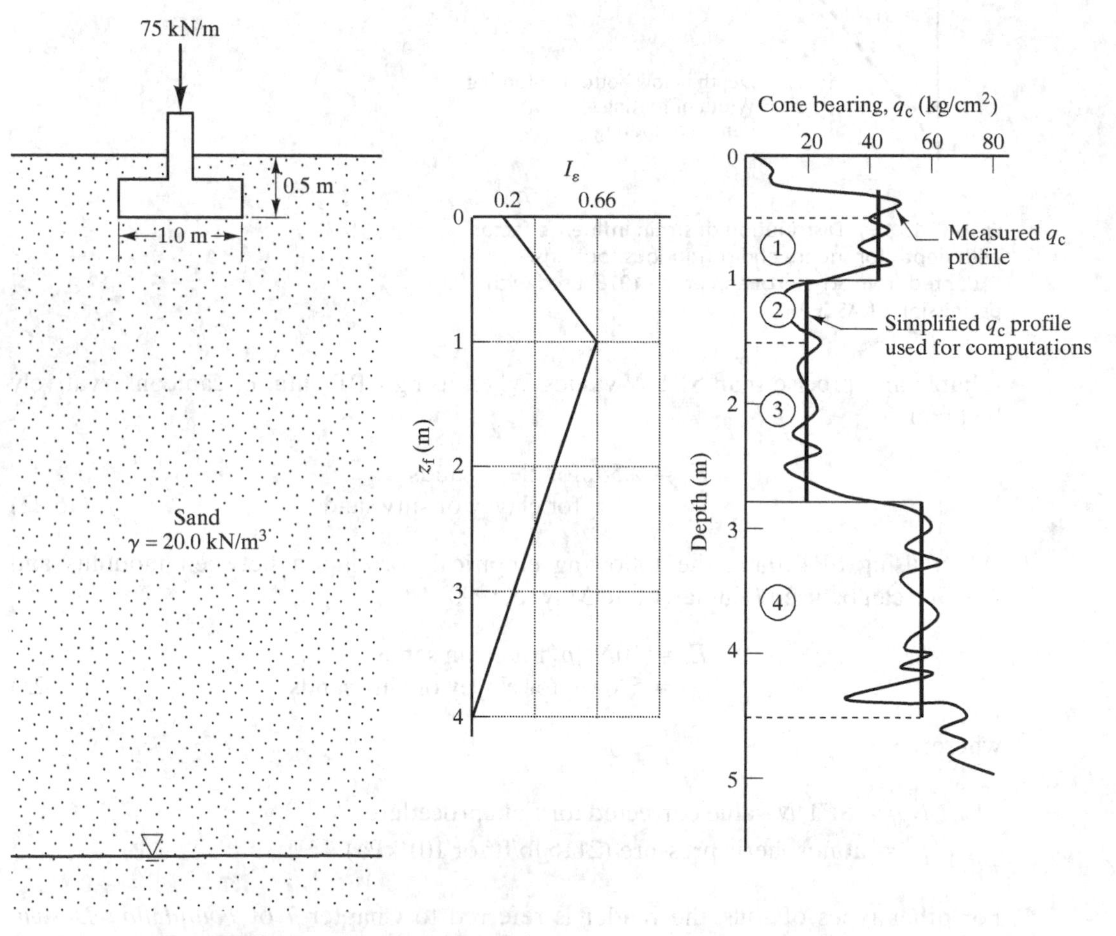

FIGURE 15.16 Proposed spread footing for Example 15.8.

Next we develop our influence diagram. Since this is a continuous footing, the depth of influence is 4B or 4 m below the base of the footing, and the peak of the influence diagram occurs at 1B or 1 m below the footing. To complete the influence diagram, we must compute $I_{\varepsilon p}$ using Equation 15.21.

$$I_{\varepsilon p} = 0.5 + 0.1\sqrt{\frac{\left(\frac{75 \text{ kN/M}}{1 \text{ m}} + (23.6 \text{ kN/m}^3)(0.5 \text{ m})\right) - (20 \text{ kN/m}^3 \times 0.5 \text{ m})}{20 \text{ kN/m}^3 \times 1.5 \text{ m}}}$$

$$= 0.5 + 0.1\sqrt{\frac{86.8 - 10}{30}} = 0.66$$

We can now plot our influence diagram as shown in Figure 15.16.

We must now break the soil profile up into layers to complete the computations. We will break the profile into a new layer wherever the simplified q_c profile changes value or the influence diagram changes slope. In this case, that will create four layers as shown in Figure 15.16. Since this is a sand profile, we will compute the modulus for each layer using

$$E_s = 2.5q_c \text{ (Equation 15.22 for clean sand)}$$
$$1 \text{ kPa} = 0.0102 \text{ kg/cm}^2$$

The soil data from the layers are then

Layer No.	Depth (m)	q_c(kg/cm^2)	E_s (kPa)
1	0.5–1.0	42	10,300
2	1.0–1.5	20	4,900
3	1.5–2.8	20	4,900
4	2.8–4.5	57	14,000

We now can compute the quantity $\sum \frac{I_\varepsilon H}{E_s}$ from Equation 15.17 using the following table. The values of z_f and I_ε are computed at the mid-height of each layer.

Layer No.	E_s (kPa)	z_f (m)	I_ε	H (m)	$I_\varepsilon H/E_s$
1	10,300	0.25	0.315	0.5	15.3×10^{-06}
2	4,900	0.75	0.545	0.5	55.6×10^{-06}
3	4,900	1.65	0.517	1.3	137.2×10^{-06}
4	14,000	3.15	0.187	1.7	22.7×10^{-06}
					$\Sigma = 230.8 \times 10^{-06}$

Compute correction factors, using Equations 15.18–15.20:

$$C_1 = 1 - 0.5\left(\frac{10}{86.8 - 10}\right) = 1 - 0.5(0.130) = 0.935$$

$$C_3 = 0.73 \text{ (for continuous footings)}$$

At the end of construction,

$$C_2 = 1$$

Using Equation 15.17

$$\delta = (0.935)(1)(0.73)(86.8 - 10)(230.8 \times 10^{-6}) = \mathbf{0.012\ m} = \mathbf{12\ mm}$$

Thirty years after construction,

$$C_2 = 1 + 0.2 \log\left(\frac{30}{0.1}\right) = 1.50$$

Using Equation 15.17,

$$\delta = (0.935)(1.50)(0.73)(86.8 - 10)(230.8 \times 10^{-6}) = \mathbf{0.018\ m} = \mathbf{18\ mm}$$

15.4 SPREAD FOOTINGS—SUMMARY AND DESIGN CONCERNS

Spread footings must be designed to satisfy both bearing capacity and settlement criteria. Given a column load and the required soil properties, the geotechnical engineer can use the techniques described in this chapter to determine the footing width required to provide an adequate factor of safety against bearing capacity failure and a different footing width required to limit settlements to the maximum allowable settlement. The greater of these two widths then becomes the design value of B.

Other Geotechnical Concerns

This chapter is only an introduction to spread footing design, so our discussions have been limited to bearing capacity and settlement. However, other issues also may be important. These include the following:

- *Frost heave*—in areas with cold climates, the upper soils may heave due to the formation of ice lenses. To avoid heave-induced damage, spread footings in such soils are typically founded below the frost depth.
- *Expansive* or *collapsing soils*—some soils are subject to additional heave or collapse when wetted, and these motions can cause substantial damage to spread footings. Special designs are often required to minimize the potential for such damage.
- *Footings* on or near slopes—if the ground surface is not level, special designs are necessary to maintain adequate stability.

- *Highly stratified soils*—in some cases, it may be more effective to satisfy bearing capacity and settlement criteria by deepening the footing to reach a stronger and less compressible soil.

Recognizing the Need for More Extensive Foundations

Although spread footings provide suitable support for many structures, and are the most common type of foundation, there are times when they are not suitable. These include the following:

- Structures with loads so high and/or soil properties so poor that the footings would be very large. Usually some other types of foundation needs to be considered if the total footing area is more than 50% of the building footprint area.
- Locations where a footing might be undermined, such as from riverbed scour adjacent to bridge foundations or from future excavations near the foundation.
- A foundation that must penetrate through water, such as for a bridge pier.
- Requirements for a large uplift capacity (the uplift capacity of spread footings is limited to their dead weight).
- Requirements for a large lateral load capacity.

In these cases, we need to consider either a mat foundation or some type of deep foundation as described in Chapter 14.

Example 15.9

A proposed warehouse will have design column loads of 150 k. Each column is to be supported on a square spread footing foundation underlain by sandy soils with the following engineering properties: $c' = 0$, $\phi' = 35°$, $\gamma = 118\ \text{lb/ft}^3$. A cone penetration test for the site is shown in Figure 15.17. The minimum acceptable factor of safety against a bearing capacity failure is 2.0 and the maximum allowable settlement is 1.0 in. immediately after construction. The groundwater table is at a depth of 50 ft and all of the footings will be located 2 ft below the ground surface. What footing width should be used for the columns?

Solution:

Bearing capacity analysis:

Per Table 15.1: $N_q = 41.4$, $N_\gamma = 47.3$

$$\sigma'_D = \gamma D - u = (118\ \text{lb/ft}^3)(2\ \text{ft}) - 0 = 236\ \text{lb/ft}^2$$

$$q_{ult} = 1.3c'N_c + \sigma'_D N_q + 0.4\gamma B N_\gamma$$
$$q_{ult} = 0 + (236\ \text{lb/ft}^2)(41.4) + 0.4(118\ \text{lb/ft}^2)B(47.3)$$
$$q_{ult} = 9770 + 2233B$$

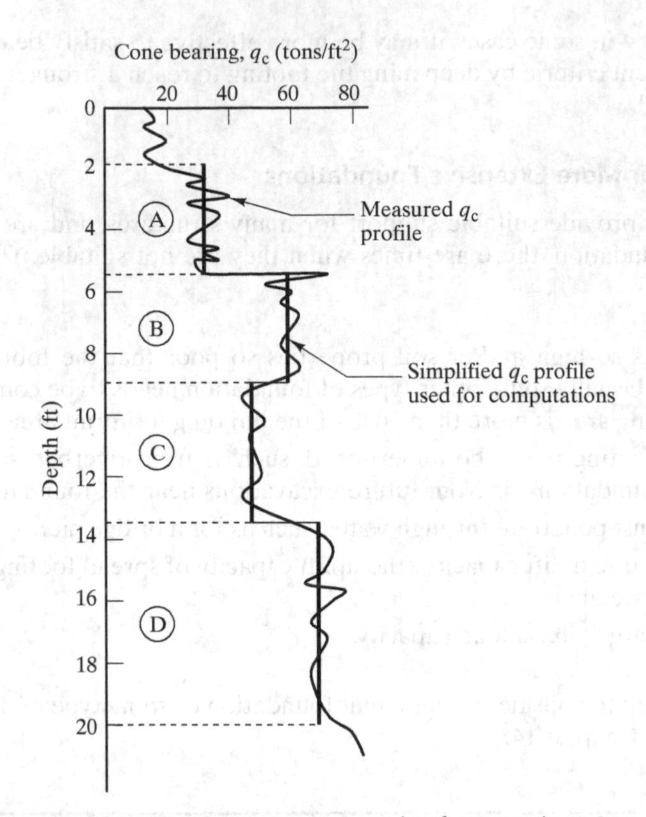

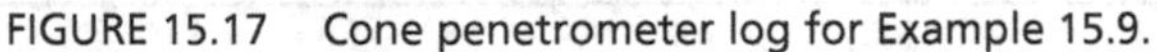

FIGURE 15.17 Cone penetrometer log for Example 15.9.

$$q_a = \frac{q_{ult}}{F} = \frac{9770 + 2233B}{2} = 4885 + 1120B$$

$$q = \frac{P}{A} + \gamma_c D - u$$

$$q = \frac{150{,}000 \text{ lb}}{B^2} + (150 \text{ lb/ft}^3)(2 \text{ ft}) - 0$$

$$q = \frac{150{,}000 \text{ lb}}{B^2} + 300 \text{ lb/ft}^2$$

Setting $q_a = q$ and solving for B gives $B = 4.05$ ft

$$q_a = 4885 + 1120B = 4885 + 1120(4.05) = \mathbf{9420\ lb/ft^2}$$

Settlement analysis:

First we create our simplified q_c profile and divide the profile into layers as shown in Figure 15.17. From this simplified profile and using

$$E_s = 2.5\, q_c \qquad \text{(Equation 15.22 for clean sand)}$$

We have layers with the following properties.

Layer	Depth (ft)	q_c (tons/ft²)	E_s (lb/ft²)
A	2.0–5.2	32	160,000
B	5.2–8.8	59	295,000
C	8.8–13.3	47	235,000
D	13.3–20.0	69	345,000

To perform the settlement analysis, we must assume a footing width. We will start with the footing width of 4.5 ft based on our bearing capacity analysis.

$$\sigma'_D = 236 \text{ lb/ft}^2$$

$$\sigma'_{zp} = (118 \text{ lb/ft}^3)(2 + 4.5/2 \text{ ft}) = 502 \text{ lb/ft}^2$$

$$q = (150{,}000 \text{ lb}/(4.5 \text{ ft})^2) + 300 \text{ lb/ft}^2 = 7700 \text{ lb/ft}^2$$

$$I_{\varepsilon p} = 0.5 + 0.1\sqrt{\frac{7700 - 236}{502}} = 0.89$$

The strain influence distribution for this footing is shown in Figure 15.18. Using this distribution, we can compute the settlement.

Four layers are required for this computation as shown.

Layer No.	E_s (kPa)	z_f (m)	I_ε	H (m)	$I_\varepsilon H/E_s$
1	160,000	1.13	0.495	2.25	6.96×10^{-06}
2	160,000	2.73	0.827	0.95	4.91×10^{-06}
3	295,000	5.00	0.527	3.60	6.44×10^{-06}
4	235,000	7.90	0.145	2.20	2.20×10^{-06}
					$\Sigma = 19.7 \times 10^{-06}$

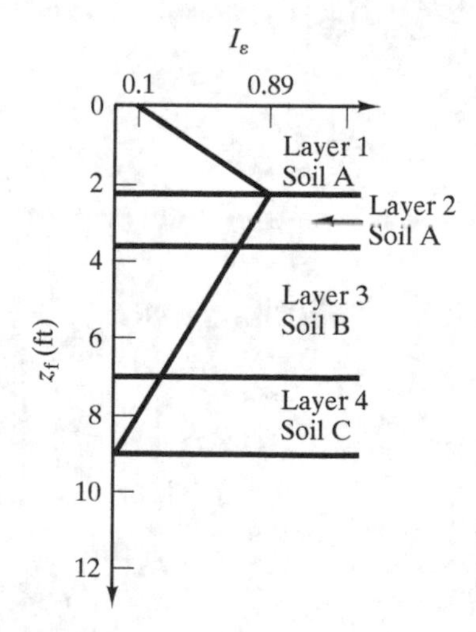

FIGURE 15.18 Strain influence distribution profile for a 4.5-ft square footing for Example 15.9.

Compute correction factors

$$C_1 = 1 - 0.5\left(\frac{236}{7700 - 236}\right) = 1 - 0.5\left(0.032\right) = 0.984$$

$C_2 = 1.0$ (immediately after construction)

$C_3 = 1.0$ (for square footings)

Using Equation 15.17

$$\delta = (0.984)(1.0)(1.0)(7700 - 236)(19.6 \times 10^{-6}) = 0.14 \text{ ft} = 1.7 \text{ in.}$$

Settlement is too large. Try a wider footing. Let $B = 6$ ft

$$\sigma'_{zp} = (118 \text{ lb/ft}^2)(2 + 6/2 \text{ ft}) = 590 \text{ lb/ft}^2$$

$$q = (150{,}000 \text{ lb}/(6 \text{ ft})^2) + 300 \text{ lb/ft}^2 = 4470 \text{ lb/ft}^2$$

$$I_{\varepsilon p} = 0.5 + 0.1\sqrt{\frac{4470 - 236}{590}} = 0.77$$

The strain influence distribution for this footing is shown in Figure 15.19. Using this distribution, we can compute the settlement.

Three layers are required for this computation as shown.

Layer No.	E_s (kPa)	z_f (M)	I_ε	H (M)	$I_\varepsilon H/E_s$
1	160,000	1.60	0.511	3.20	10.2×10^{-06}
2	295,000	5.00	0.613	3.60	7.47×10^{-06}
3	235,000	9.40	0.228	5.20	5.03×10^{-06}
					$\Sigma = 22.7 \times 10^{-06}$

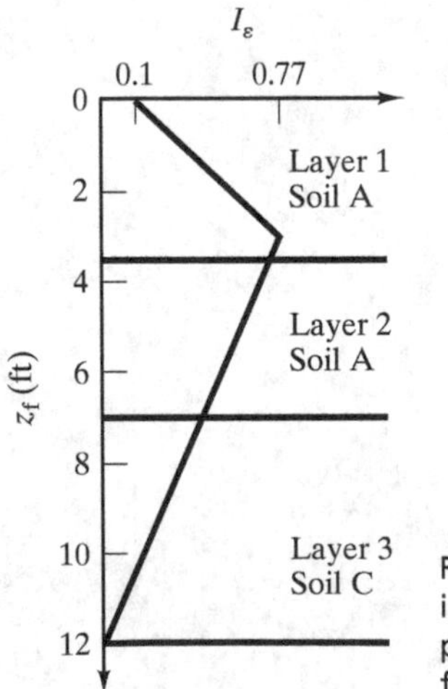

FIGURE 15.19 Strain influence distribution profile for 6-ft square footing for Example 15.9.

Recompute C_1, other correction factors are the same

$$C_1 = 1 - 0.5\left(\frac{236}{4470 - 236}\right) = 1 - 0.5(0.056) = 0.972$$

Using Equation 15.17

$$\delta = (0.972)(1.0)(1.0)(4470 - 236)(22.7 \times 10^{-6}) = 0.093 \text{ ft} = 1.1 \text{ in.}$$

Settlement is still too large. If we did another iteration with B = 6 ft 6 in., we would find the settlement would be just less than 1 in.

Therefore required $\boldsymbol{B}$ = **6 ft 6 in.**

Commentary: In this case, the footing design was controlled by settlement rather than bearing capacity. This is normally the case for footings on sand.

SUMMARY

Major Points

1. Spread footings are one type of shallow foundation. It is the most common type, and the only type considered in this chapter.
2. The bearing pressure is the contact pressure between the bottom of the footing and the soil.
3. The primary geotechnical strength requirement for spread footings is called *bearing capacity*. It addresses the potential for shear failure in the soil.
4. Terzaghi's bearing capacity theory can be used to determine the ultimate bearing capacity. In this theory there are three sources of resistance to bearing capacity failure:
   - **a.** Cohesion resistance represented by $c'N_c$
   - **b.** Surcharge resistance represented by $\sigma'_D N_q$
   - **c.** Frictional resistance represented by $0.5\gamma' B N_\gamma$
5. The allowable bearing capacity is the bearing pressure required to produce a bearing capacity failure divided by a factor of safety.
6. The primary serviceability requirement for spread footings is settlement. The allowable settlement depends on the tolerance of the structure to movements. The actual settlement depends on the loads, footing geometry, and soil conditions.
7. Settlement of spread footings founded on clay or silt soils is computed in a manner similar to the computation of settlement of large fills (Chapter 10) except two additional factors must be considered:
   - **a.** The decrease in induced vertical stress as a function of depth due to the finite size of spread footing
   - **b.** The rigidity of the footing.

8. Settlement of spread footings founded on sand or gravel soils is computed using pseudoelastic methods. The most widely used method is that of Schmertmann (1978).
9. The design footing width is the larger of the width required to resist bearing capacity failure and the width required to limit settlement.

Vocabulary

allowable bearing capacity
allowable differential settlement
allowable total settlement
bearing capacity factor
bearing capacity failure
bearing pressure
continuous footing
differential settlement
effective unit weight
equivalent soil modulus
gross bearing pressure
rectangular footing
serviceability requirement
shallow foundation
spread footing foundation
square footing
strain influence factor
strength requirement
total settlement
ultimate bearing capacity

QUESTIONS AND PRACTICE PROBLEMS

Section 15.1 Bearing Pressure

15.1 A 5-ft square, 3-ft deep footing supports a column load of 110 k. The groundwater table is at a depth greater than 3 ft. Compute the bearing pressure.

15.2 An 800-mm wide, 400-mm deep continuous footing supports a wall load of 120 kN/m. The groundwater table is at a great depth. Compute the bearing pressure.

15.3 A rectangular footing 8 ft by 12 ft founded at a depth of 5 ft supports two columns each bearing a load of 135 k. The groundwater is currently at a depth of 7 ft, but the historic high water table is at a depth of 2 ft. Compute the bearing pressure under the current groundwater conditions and if the groundwater reaches its historic high level.

Section 15.2 Bearing Capacity

15.4 The bearing wall of a structure is to be supported on a 3-ft wide continuous footing founded at a depth of 2.5 ft in a clayey sand with $c' = 100 \text{ lb/ft}^2$, $\phi' = 27°$, and $\gamma = 118 \text{ lb/ft}^3$. Compute the allowable wall load if a factor of safety of 2.5 is required.

15.5 A proposed column is to be supported by a 1.5-m wide, 0.5-m deep square footing. The soil beneath this footing is a silty sand with $c' = 0$, $\phi' = 29°$, and $\gamma = 18.0 \text{ kN/m}^3$. The groundwater table is at a depth of 10 m below the ground surface. The factor of safety against a bearing capacity failure must be at least 2.75. Compute the maximum allowable column load.

15.6 You are reviewing the footing design for a hospital founded on clay soils. The controlling column load is 130 k. The proposed footing is square with a width of 7 ft. It is founded at a depth of 3 ft in a medium to stiff clay with $s_u = 1100 \text{ lb/ft}^2$ and $\gamma = 127 \text{ lb/ft}^3$. The groundwater table is at a depth of 2 ft. Is this design acceptable? Explain.

15.7 Write a spreadsheet program that will compute the ultimate and allowable bearing capacity for both continuous and square footings using Terzaghi's bearing capacity theory. The program must allow users to input column or wall load, footing dimensions, footing depth,

depth to groundwater table, factor of safety, and the cohesion, friction angle, and unit weight of the soil. It should accept either SI or English units.

15.8 A 39-in. wide, 24-in. deep continuous footing supports a wall load of 12 k/ft. This footing is underlain by a fine-to-medium sand with $c' = 0, \phi' = 31°$, and $\gamma = 122 \text{ lb/ft}^3$. The groundwater table is currently at a depth of 10 ft below the ground surface, but could rise to 4 ft below the ground surface during the life of the project. The factor of safety against a bearing capacity failure must be at least 3.0. Is the design acceptable? Provide computations to justify your answer. Comment on any special considerations.

15.9 A 949-kN column load is to be supported on a square spread footing that will be underlain by a clayey silt with $s_u = 125$ kPa. The bottom of this footing will be 1.0 m below the ground surface, and the groundwater table is more than 30 m below the ground surface. Using a factor of safety of 3.0, compute the required footing width.

15.10 You are designing a continuous footing to support the walls of a tilt-up structure. The wall loads will be 25 k/ft. The footing will be founded at a depth of 3 ft on a medium dense sand with $c' = 0, \phi' = 33°$, and $\gamma = 124 \text{ lb/ft}^3$. The groundwater is at a depth of 4 ft. Using a factor of safety of 2.75, compute the required footing width.

15.11 A proposed cylindrical steel water tank is to be built on a medium clay that has an undrained shear strength, s_u, of 31 kPa. The tank diameter will be 35.0 m, and it will contain 10.0 m of water. Its empty mass will be 253,000 kg. Assuming both the weight of the empty tank and that of the water are spread evenly along the bottom, compute the factor of safety against a bearing capacity failure. Is this factor of safety acceptable? If not, how could the design be modified to provide an acceptable F?

Note: Although a ring footing would be present along the perimeter of this tank to support the weight of the walls, the live load (i.e., the weight of the water) is spread evenly across the bottom of the tank. This live load is a large fraction of the total load, so the bearing capacity analysis should be based on a circular load with a diameter equal to the diameter of the tank.

Section 15.3 Settlement

15.12 The footing described in Example 15.6 has been redesigned so B now equals 2.50 m. The column load and depth of embedment remain the same. Compute the new settlement, δ. Does this new design satisfy the allowable settlement criteria described in Example 15.6?

15.13 The proposed footing shown in Figure 15.20 has an allowable total settlement of 1.0 in. Compute the settlement and determine if it meets this criterion.

15.14 Write a spreadsheet program that will compute the settlement of square or continuous footings founded on clay soil. The program must allow users to input column or wall load, footing dimensions, footing depth, C_c, C_r, σ'_c, and unit weight of the soil. It should divide the soil into at least 30 layers and should accept either SI or English units.

15.15 A proposed 3-ft 6-in. wide continuous footing is to be built at the site described in Problem 15.13. The depth, D, the bearing pressure, q, and the allowable settlement, δ_a, are the same as before. Compute the predicted settlement and determine if it meets the settlement criteria.

15.16 Compute the settlement of the proposed footing shown in Figure 15.21 using the CPT log shown.

15.17 Write a spreadsheet program that will compute the settlement of square or continuous footings founded on sand using Schmertmann's method. The program must allow users to

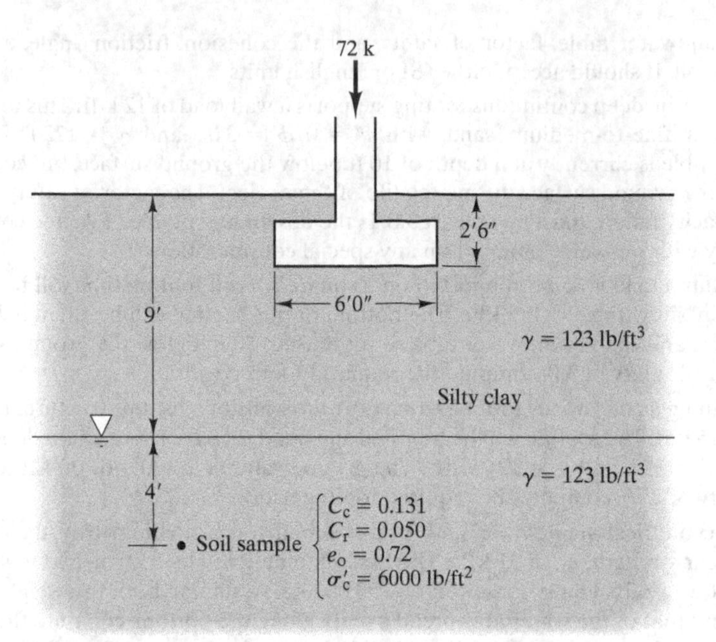

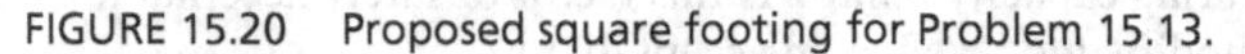
FIGURE 15.20 Proposed square footing for Problem 15.13.

input column or wall load, footing dimensions, footing depth, E_s and unit weight of the soil and time after construction. It should divide the soil into at least 30 layers and should accept either SI or English units.

15.18 According to the structural engineer, the footing described in Problem 15.16 must not settle more than 25 mm. Determine the required B that meets this criterion and produces the most economical design.

Comprehensive

15.19 A proposed 1200-mm wide, 400-mm deep square footing will be built on a sandy soil with $c' = 0$, $\phi' = 34°$, $\gamma = 20.1\ \text{kN/m}^3$ and $N_{60} = 30$. The groundwater table is at a depth of 2 m below the ground surface. Determine the maximum allowable column load that may be placed on this footing while maintaining a factor of safety of at least 2.5 against a bearing capacity failure and a total settlement of no more than 15 mm. Which controls this design, bearing capacity or settlement?

15.20 A 103-k column load is to be supported on a square footing embedded 2.5 ft into the ground. The underlying soil is a silty clay with $C_c/(1 + e_0) = 0.11$, $C_r/(1 + e_0) = 0.03$, $\sigma'_m = 5000\ \text{lb/ft}^2$, $\gamma = 119\ \text{lb/ft}^3$, and $s_u = 2000\ \text{lb/ft}^2$. The groundwater table is at a depth of 40 ft below the ground surface. The factor of safety against a bearing capacity failure must be at least 3, and the total settlement must not exceed 1 in. Find the design footing width, B.

15.21 A proposed building is to be built on the soil profile shown in Figure 15.21. The column loads are 1500 kN, and will be supported on a 0.8-m deep square footings. The allowable

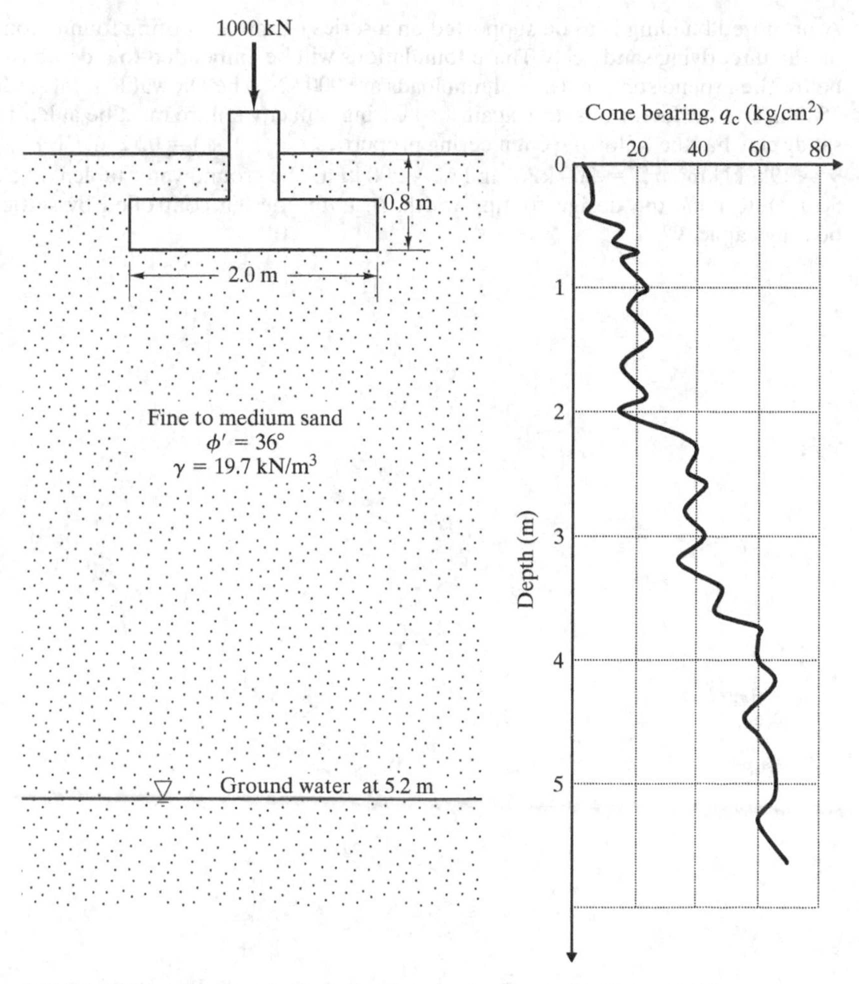

FIGURE 15.21 Proposed square footing for Problems 15.16 and 15.21.

total settlement is 25 mm, and the factor of safety against a bearing capacity failure must be at least 2.0. Determine the design footing width, *B*. Is the design controlled by settlement or bearing capacity?

15.22 A proposed bent for a bridge will impart a vertical load of 3100 k onto a spread footing foundation that will be embedded 6 ft into the ground. The underlying soils are dense well-graded sands with $c' = 0$, $\phi' = 37°$, $\gamma = 128$ lb/ft^3, and $N_{60} = 36$. The groundwater table is at a depth of 12 ft below the ground surface. This footing must have a factor of safety of at least 2.75 against a bearing capacity failure and a total settlement of no more than 1.5 in. Determine the design footing width, *B*. Is the design controlled by settlement or bearing capacity?

15.23 A proposed building is to be supported on a series of spread footing foundations resting on the underlying sandy clay. These foundations will be embedded to a depth of 500 mm below the ground surface. The column loads are 900 kN. The allowable total settlement is 25 mm, and the factor of safety against a bearing capacity failure must be at least 3.0. The sandy clay has the following engineering properties: $C_c/(1 + e_0) = 0.12$, $C_r/(1 + e_0) = 0.03$, $\gamma = 19.5 \text{ kN/m}^3$, $\sigma'_m = 300 \text{ kPa}$, and $s_u = 200 \text{ kPa}$. The groundwater table is at a depth of 5 m. Determine the design footing width, B. Is the design controlled by settlement or bearing capacity?

CHAPTER 16

Earth Retaining Structures

There are two modes of acquiring knowledge, namely by reasoning and by experience. Reasoning draws a conclusion and makes us grant the conclusion, but it does not make the conclusion certain; nor does it remove doubt so that the mind may rest on the intuition of truth, unless the mind discovers it by the path of experience.

Roger Bacon (1220–1292)

Geotechnical engineers often participate in the design of *earth retaining structures* (also known as *retaining walls*), which are civil engineering works that maintain adjacent ground surfaces at two different elevations. These are vertical or near-vertical structures that retain soil or rock. Figure 16.1 shows typical uses of earth retaining structures in civil engineering projects.

Many kinds of retaining structures are available, each best suited for particular applications, and *Foundation Design: Principles and Practices* (Coduto, 2001) discusses the selection, analysis, and design of retaining walls in detail. This chapter gives a brief introduction to the topic.

O'Rourke and Jones (1990) classified earth retaining structures into two broad categories: *externally stabilized systems* and *internally stabilized systems*, as shown in Figure 16.2. Some hybrid methods combine features from both categories.

16.1 EXTERNALLY STABILIZED SYSTEMS

Externally stabilized systems are those that resist the applied earth loads by virtue of their weight and stiffness. This was the only type of retaining structure available before 1960, and they are still very common. O'Rourke and Jones subdivided these structures into two categories: *gravity walls* and *in situ walls*.

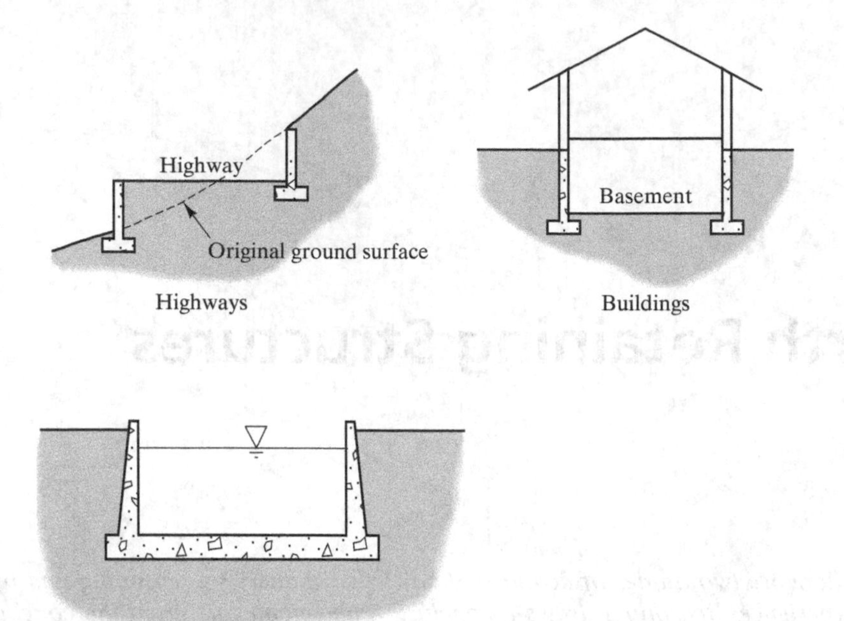

FIGURE 16.1 Typical applications of earth retaining structures.

Gravity Walls

Massive Gravity Walls The earliest retaining structures were *massive gravity walls*, such as the one shown in Figure 16.3. They were often made of mortared stones, masonry, or unreinforced concrete and resisted the lateral forces from the backfill by virtue of their sheer weights. The construction of these walls is very labor intensive and requires large quantities of materials, so they are rarely used today except for very short walls.

Cantilever Gravity Walls The *cantilever gravity wall*, shown in Figure 16.4, is a refinement of the massive wall. These walls have a much smaller cross-section and thus require much less material. However, these walls have large flexural stresses, so they are typically made of reinforced concrete or reinforced masonry.

Crib Walls A *crib wall*, shown in Figure 16.5, is another type of gravity retaining structure. It consists of precast concrete members linked together to form a crib. These members resemble a child's Lincoln Log toy. The zone between the members is filled with compacted soil.

In Situ Walls

In situ walls differ from gravity walls in that they rely primarily on their flexural strengths, not their weights.

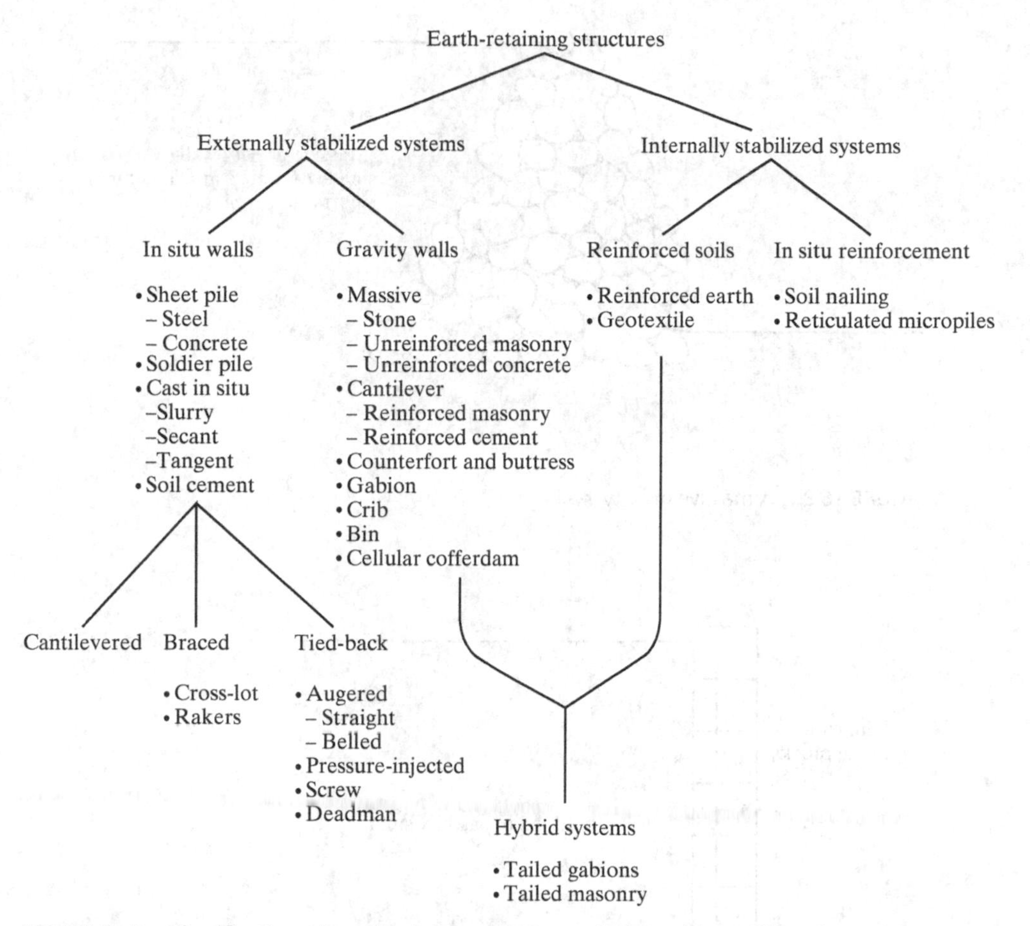

FIGURE 16.2 Classification of earth retaining structures. (Adapted from O'Rourke and Jones, 1990; used with permission of ASCE.)

Sheet Pile Walls A *sheet pile* is a thin, wide pile driven into the ground using a pile hammer. A series of interlocking sheet piles in a row form a *sheet pile wall*, as shown in Figure 16.6. Most sheet piles are made of steel, but some are made of reinforced concrete.

It may be possible to simply cantilever short sheet piles out of the ground, as shown in Figure 16.7(a). However, taller sheet pile walls usually need lateral support at one or more levels above the ground. This may be accomplished in either of two ways: by *internal braces,* as shown in Figure 16.7(b), or by *tieback anchors,* as shown in Figure 16.7(c).

Internal braces are horizontal or diagonal compression members that support the wall, as shown in Figure 16.7. Tieback anchors are tension members drilled into the ground behind the wall. The most common type is a grouted anchor with a steel tendon.

Soldier Pile Walls *Soldier pile walls* consist of vertical wide flange steel members with horizontal timber lagging, as shown in Figure 16.8. They are often used as temporary retaining structures for construction excavations.

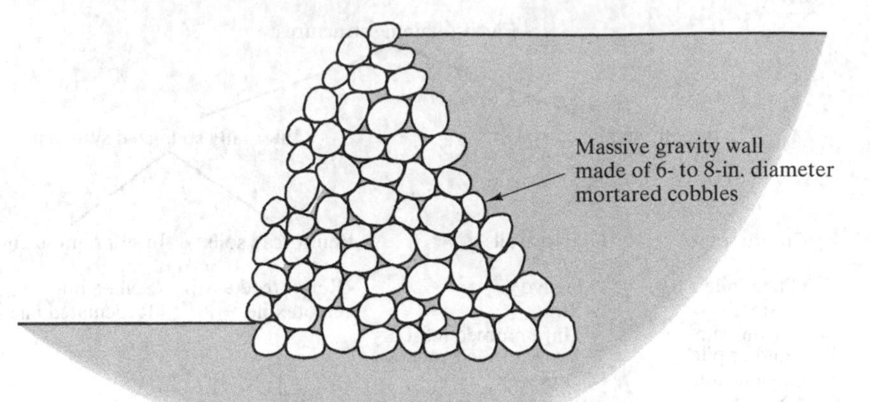

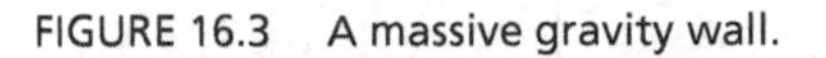
FIGURE 16.3 A massive gravity wall.

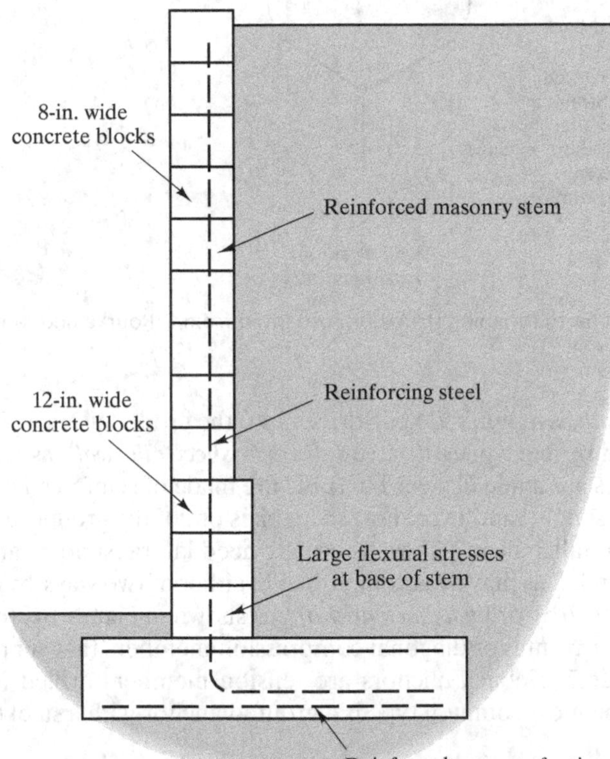

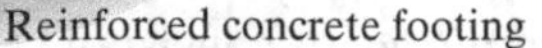

FIGURE 16.4 A cantilever gravity wall with a concrete block stem.

FIGURE 16.5 A crib wall.

FIGURE 16.6 A sheet pile wall.

Slurry Walls *Slurry walls* or *diaphragm walls* are cast-in-place concrete walls built using bentonite slurry. The contractor digs a trench along the proposed wall alignment and keeps it open using the slurry. Then, the reinforcing steel cage is inserted and the concrete is placed using tremie pipes or pumps. As the concrete fills the trench, it displaces the slurry out of it.

Slurry walls have been used as basement walls in large urban construction, and often eliminate the need for temporary walls.

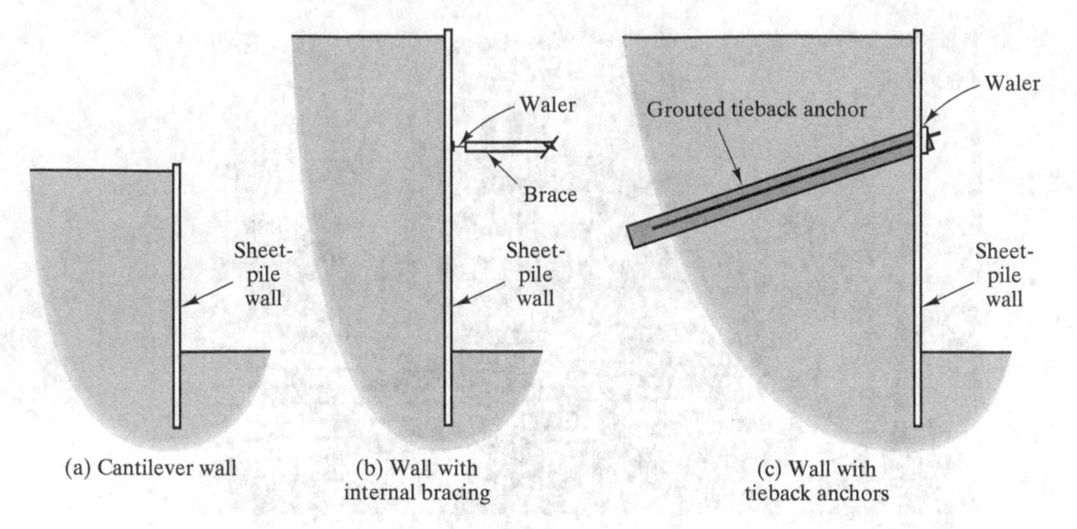

FIGURE 16.7 Short sheet pile walls often can cantilever, but taller walls usually require bracing or tieback anchors.

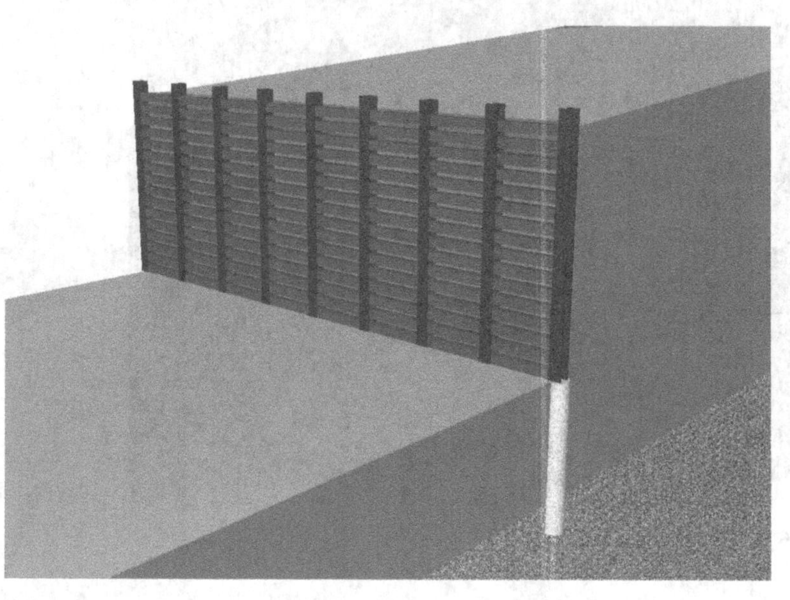

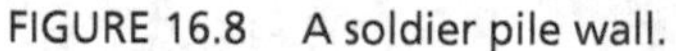

FIGURE 16.8 A soldier pile wall.

16.2 INTERNALLY STABILIZED SYSTEMS

Internally stabilized systems reinforce the soil to provide the necessary stability. Various schemes are available, all of which have been developed since 1960. They can be subdivided into two categories: *reinforced soils* and *in situ reinforcement.*

Reinforced Soils

Soil is strong in compression, but has virtually no tensile strength. Therefore, the inclusion of tensile reinforcing members in a soil can significantly increase its strength and load-bearing capacity, much the same way that placing rebars in concrete increases its strength. The resulting reinforced soil is sometimes called *mechanically stabilized earth (MSE)*.

Often MSE is used so that slopes may be made steeper than would otherwise be possible. Thus, this method forms an intermediate alternative between earth slopes and retaining walls. MSE also may be used with vertical or near-vertical faces, thus forming a type of retaining wall. In this case, it becomes necessary to place some type of facing panels on the vertical surface, even though the primary soil support comes from the reinforcement, not the panels. Such structures are called *MSE walls*. The earliest MSE walls were developed by Henri Vidal in the early 1960, using the trade name *reinforced earth*. This design uses strips of galvanized steel for the reinforcement and precast concrete panels for the facing, as shown in Figures 16.9 and 16.10.

Many other similar methods also are used to build MSE walls. The reinforcement can consist of steel strips, polymer geogrids, wire mesh, geosynthetic fabric, or other materials. The facing can consist of precast concrete panels, precast concrete blocks, rock filled cages called *gabions*, or other materials. Sometimes the reinforcement is simply curved around to form a type of facing. Figure 16.11 shows an MSE wall being built using wire mesh reinforcement and gabion facing.

MSE walls are becoming very popular for many applications, especially for highway projects. Their advantages include low cost and tolerance of differential settlements.

FIGURE 16.9 Reinforced Earth® walls consist of precast concrete facing panels and steel or polymer reinforcing strips that extend into the retained soil. This wall is under construction. (The Reinforced Earth Company.)

FIGURE 16.10 A completed Reinforced Earth® wall. (The Reinforced Earth Company.)

FIGURE 16.11 A mechanically stabilized earth (MSE) wall under construction using galvanized wire mesh as the tensile reinforcement and rock-filled cages called *gabions* for the facing. (Federal Highway Administration.)

In Situ Reinforcement

In situ reinforcement methods was mentioned in Section 13.13 as being different from reinforced soils in that the tensile members are inserted into a soil mass rather than being embedded during placement of fill.

A common in situ reinforcement method is soil nailing, which consists of installing grouted steel rebars in near-horizontal holes, as shown in Figure 13.47. The face of the wall is typically covered with shotcrete. These walls do not require a temporary construction excavation, and thus are useful when space is limited.

16.3 DESIGN OF EARTH RETAINING STRUCTURES

The stresses and deformations in an earth retaining structure depend on the applied loads from the retained soil. Likewise, the loads imparted by the soil depend on the deformation of the structure. This inter-dependency is called *soil-structure interaction.* In addition, the construction methods used to build the structure also impact its behavior. Many of these factors are only partially understood, so the analysis and design of earth retaining structures is often based on empirical or semi-empirical methods. A detailed discussion of these methods is beyond the scope of this book.

The design process for a simple cantilever gravity wall illustrates the general design process. As shown in Figure 16.12, we can use the free body diagram shown in this figure to analyze the wall with respect to various design considerations. These design considerations relate to the various potential failure mechanisms and include the following:

- *Sliding stability*—the wall must have adequate factor of safety against sliding on the base of the footing of the wall, as shown in Figure 16.13(a).

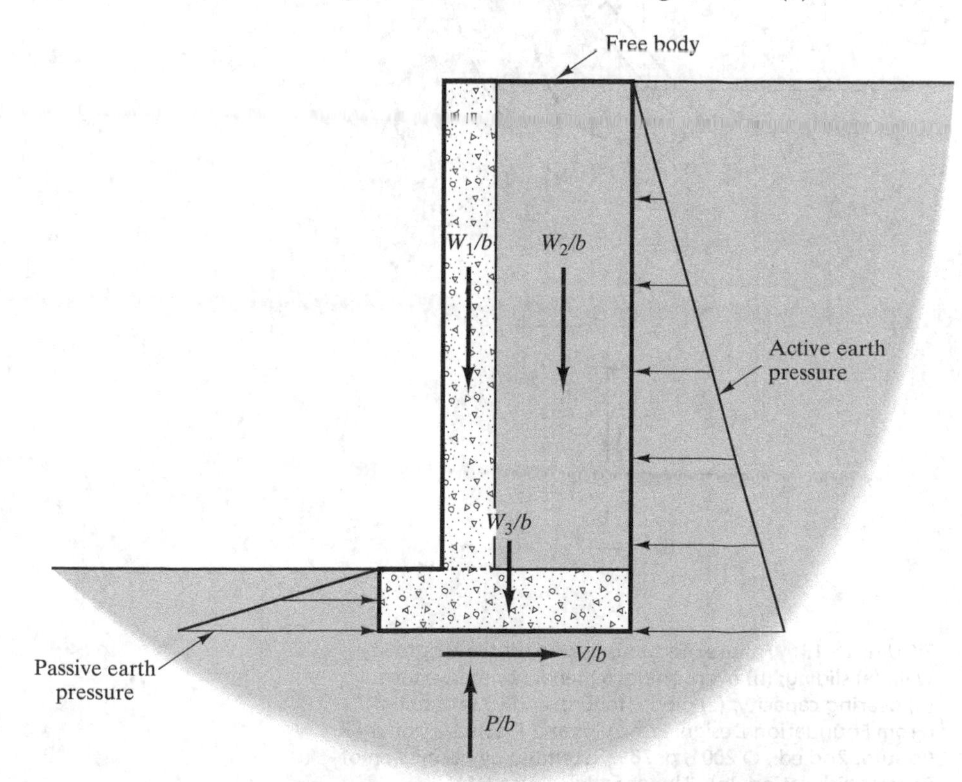

FIGURE 16.12 A free body diagram for the design of a cantilever gravity wall.

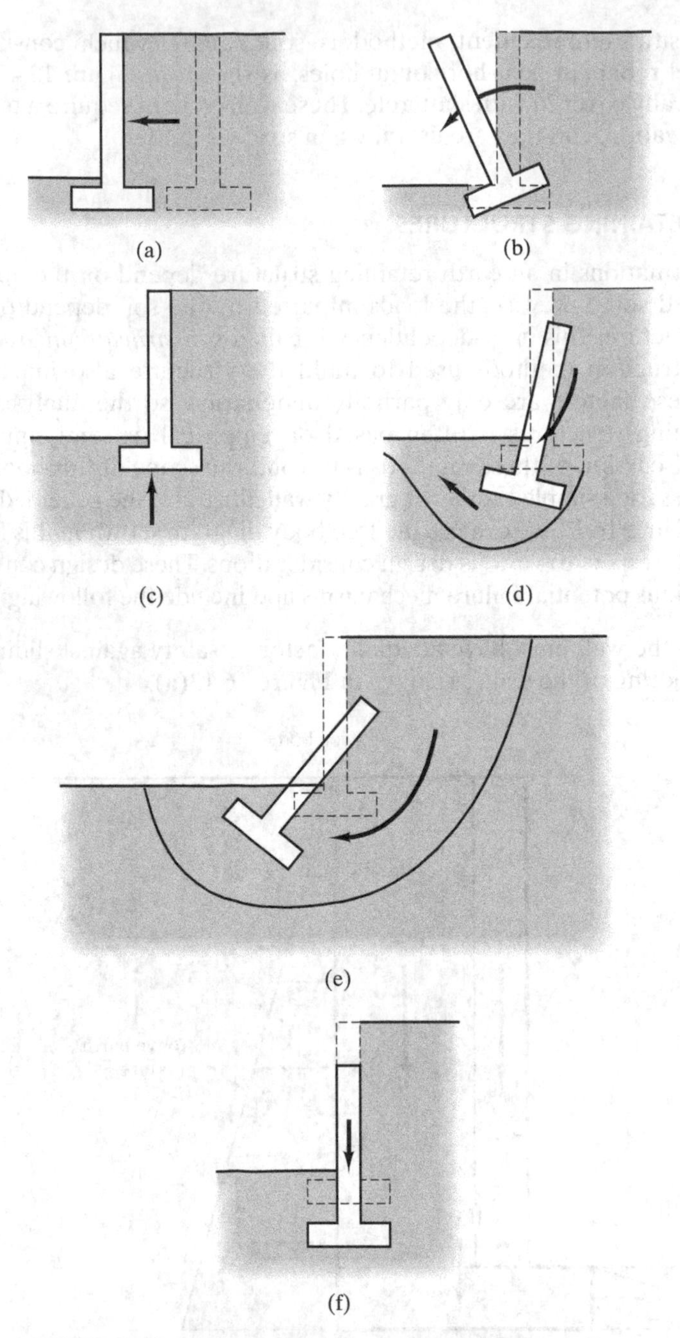

FIGURE 16.13 Failure mechanisms of a cantilever gravity wall: (a) sliding; (b) overturning; (c) tendency to overturn; (d) bearing capacity; (e) global failure; and (f) settlement. (From Foundation Design: Principles and Practices, Donald P. Coduto, 2nd ed., © 2001, p. 789. Reprinted by permission of Pearson Education, Inc., Upper Saddle River, NJ.)

- *Overturning stability*—the wall must have adequate factor of safety against overturning, as shown in Figure 16.13(b).
- *Tendency to overturn*—this requires that the resultant normal force on the base of the footing be within the middle third of the base, as shown in Figure 16.13(c).
- *Bearing capacity*—the wall must have adequate factor of safety against bearing capacity failure of the footing of the wall, as shown in Figure 16.13(d).
- *Global stability*—the wall must have adequate factor of safety against global slope failures in which the entire wall is embedded in the slide body, as shown in Figure 16.13(e).
- *Settlement*—the settlement of the wall, as shown in Figure 16.13(f), must be acceptable.

It can be seen in Figure 16.12 that some important external forces on the free body diagram are from lateral earth pressures. Chapter 17 presents fundamental concepts on lateral earth pressures and common methods used in practice to compute these pressures.

SUMMARY

Major Points

1. Retaining walls are vertical or near-vertical structures designed to retain soil or rock. Many different types are available. Externally stabilized walls resist the applied loads by virtue of their weights and stiffnesses, whereas internally stabilized walls rely on reinforcements within the ground.
2. Externally stabilized walls include gravity walls and in situ walls.
3. Internally stabilized walls include reinforced soils (also known as mechanically stabilized earth or MSE) and in situ reinforced walls.
4. The design of earth retaining structures can be considered a soil-structure interaction problem that often requires the use of empirical or semi-empirical methods.

Vocabulary

cantilever gravity wall
cantilever wall
crib wall
diaphragm wall
earth retaining structure
externally stabilized system
flexible wall
gabion
gravity wall
in situ reinforcement
in situ wall
internal braces
internally stabilized system
massive gravity wall
mechanically stabilized earth (MSE)
reinforced earth wall
reinforced soil
retaining wall
sheet pile wall
slurry wall
soil nailing
soil-structure interaction
soldier pile wall
tieback anchor

CHAPTER 17

Lateral Earth Pressures

Things should be made as simple as possible, but not one bit simpler.

Albert Einstein

In Chapter 16, we described various types of earth retaining structures and mentioned that the design of a retaining structure requires an understanding of the interaction between the structure and the soils it retains. A major result of this interaction is the development of *lateral earth pressures* between the soils and these near-vertical structures. These pressures may include both normal and shear stresses, as shown in Figure 17.1. They are a major input into the analysis and design of many types of earth retaining structures or retaining walls. Their evaluation, using common methods, illustrates how soil mechanics material from previous chapters can be applied to solve a practical geotechnical engineering problem.

We will first present fundamental concepts on lateral earth pressures and then discuss common methods used in practice to compute these pressures.

17.1 LATERAL EARTH PRESSURES AND WALL MOVEMENT

In this chapter, we will compute stresses using the same x, y, z coordinate system as used in Chapter 9. For convenience, we will align the axes such that the x axis is oriented perpendicular to the wall face.

Lateral earth pressures are the direct result of horizontal stresses in the soil. In Chapter 9, we defined the ratio of the horizontal effective stress to the vertical effective stress at any point in a soil as the *coefficient of lateral earth pressure*, K:

$$K = \frac{\sigma'_x}{\sigma'_z} \tag{17.1}$$

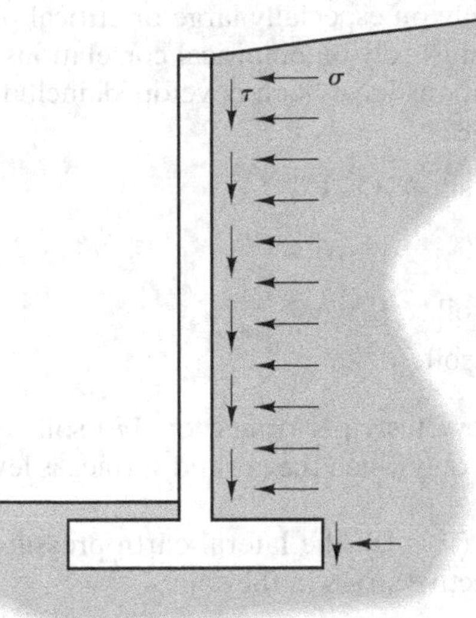

FIGURE 17.1 Lateral earth pressures imparted by a soil onto a vertical or near-vertical structure.

where:

K = coefficient of lateral earth pressure
σ'_x = horizontal effective stress
σ'_z = vertical effective stress

In the context of this chapter, K is important because it is an indicator of the lateral earth pressures acting on a retaining wall.

For purposes of describing lateral earth pressures, geotechnical engineers have defined three important soil conditions: *the at-rest condition*, *the active condition*, and *the passive condition*.

The At-Rest Condition

Let us assume that a certain retaining wall is both *rigid* and *unyielding*. In this context, a rigid wall is one that does not experience any significant flexural movements. The opposite would be a *flexible* wall—one that has no resistance to flexure. The term *unyielding* means the wall does not translate or rotate, as compared with a *yielding* wall that can do either or both. Let us also assume this wall is built so that no lateral strains occur in the ground. Therefore, the lateral stresses in the ground are the same as they were in its natural undisturbed state.

The K in this case is K_0, the *coefficient of lateral earth pressure at rest.* The most reliable method of assessing K_0 is to use in situ tests such as the dilatometer test (DMT) or pressuremeter test (PMT) discussed in Chapter 3. It also may be measured using special laboratory tests on undisturbed samples. However, because of cost constraints, engineers generally use these methods only on especially large or critical projects. For the vast majority of projects, we usually must rely on empirical correlations to develop design values of K_0. Several such correlations have been developed, including the following one from Mayne and Kulhawy (1982):

$$K_0 = (1 - \sin \phi')\text{OCR}^{\sin \phi'} \tag{17.2}$$

where:

ϕ' = effective friction angle of soil
OCR = overconsolidation ratio of soil

Equation 17.2 is based on laboratory tests performed on 170 soil samples that ranged from clay to gravel. It is applicable only when the ground surface is level. Usually K_0 is between 0.3 and 1.4.

If no groundwater table is present ($u = 0$), the lateral earth pressure, σ, acting on the wall is equal to the horizontal effective stress in the soil:

$$\sigma = \sigma'_x = \sigma'_z K_0 \tag{17.3}$$

Lateral earth pressures below the groundwater table are discussed later in this chapter.

In the at-rest case, we assume the shear stress, τ, acting between the soil and the wall is zero.

In a homogeneous soil above the groundwater table, K_0 is constant and σ'_z varies linearly with depth. Therefore, in theory, σ'_x also varies linearly with depth, forming a triangular pressure distribution, as shown in Figure 17.2. Thus, if the at-rest condition is

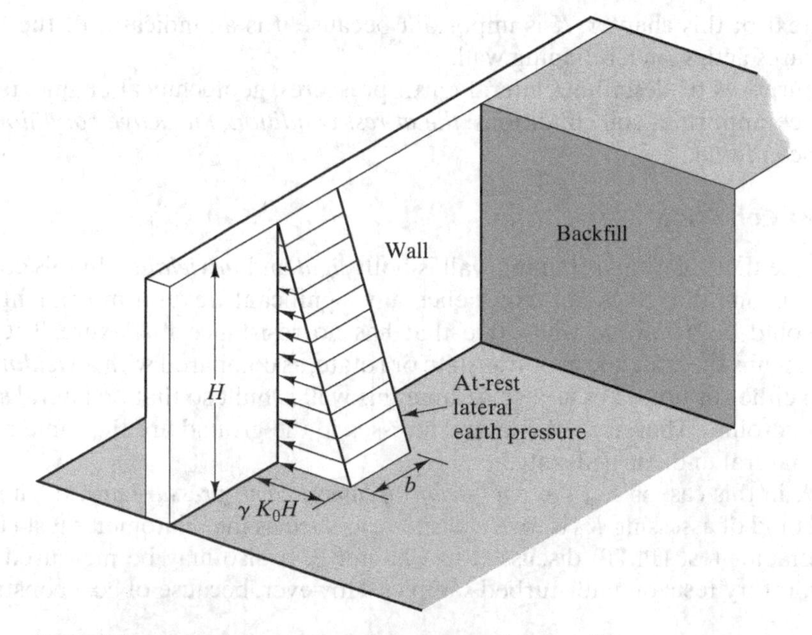

FIGURE 17.2 At-rest pressure acting on a retaining wall.

present, the resultant horizontal force acting on a unit length of a vertical wall is the area of this triangle:

$$P_0/b = \frac{\gamma H^2 K_0}{2} \tag{17.4}$$

where:

P_0/b = resultant horizontal force between wall and soil under the at-rest condition per unit length of wall
b = unit length of wall (usually 1 ft or 1 m)
γ = unit weight of soil
H = height of wall

This resultant horizontal force acts through the centroid of the triangle and therefore acts at a point on the wall that is at a distance $H/3$ above the base of the wall.

Example 17.1

An 8-ft tall basement wall retains a soil that has the following properties: $c' = 0$, $\phi' = 35°$, $\gamma = 127\ \text{lb/ft}^3$, OCR = 2. The ground surface is horizontal and level with the top of the wall. The groundwater table is well below the bottom of the wall. Consider the soil to be in the at-rest condition and compute the force that acts between the wall and the soil.

Solution:

$$K_0 = (1 - \sin\phi')\text{OCR}^{\sin\phi'}$$
$$K_0 = (1 - \sin 35°)2^{\sin 35°}$$
$$K_0 = 0.635$$

$$P_0/b = \frac{\gamma\, H^2\, K_0}{2}$$

$$P_0/b = \frac{(127\ \text{lb/ft}^2)(8\ \text{ft})^2(0.635)}{2}$$

$$P_0/b = \mathbf{2580\ lb/ft}$$

Because the theoretical pressure distribution is triangular, this resultant force acts at the lower third point on the wall.

The Active Condition

The at-rest condition applies only if the wall does not move. Although this may seem to be a criterion that all walls should meet, even very small movements alter the lateral earth pressures.

Now, we examine what happens to the lateral earth pressures if we permit the wall to move a short distance, either away from the backfill or toward the backfill. First, we look at the case of the wall moving away from the backfill. This movement may be either translational or rotational about the bottom of the wall. It relieves some of the horizontal stress by mobilizing the shear strength of the soil behind the wall to stabilize the vertical slope. As the amount of outward movement increases, more and more of the shear strength of the soil is mobilized and the lateral earth pressures decrease with the amount of movement, as shown in Figure 17.3. If we continue to move the wall outward, all the shear strength of the soil will eventually be mobilized and the soil will fail in shear. A soil that has reached this state is said to be in the active condition, meaning that the soil appears to be actively pushing the wall outward to reach this condition. The K for a soil in the active condition is known as K_a, the *coefficient of active earth pressure.*

Once the soil attains the active condition, the horizontal stress in the soil (and thus the pressure acting on the wall) will have reached its lower bound, as shown in Figure 17.3. The amount of movement required to reach the active condition depends on the soil type and the wall height, as given in Table 17.1. For example, in a loose

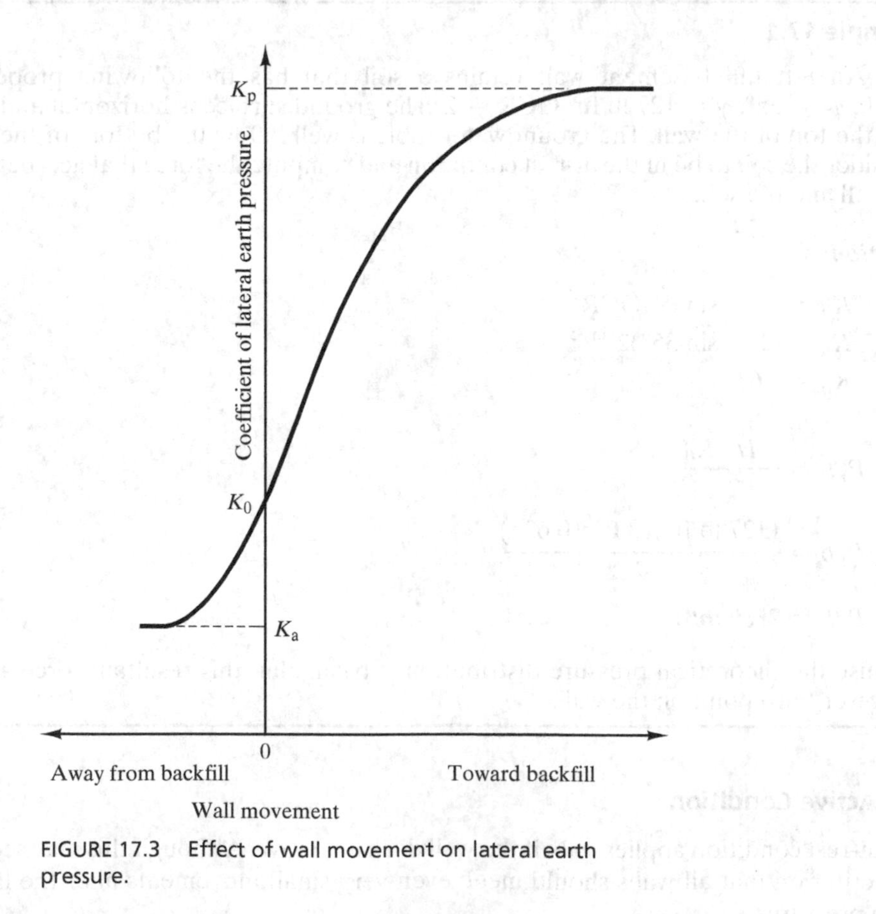

FIGURE 17.3 Effect of wall movement on lateral earth pressure.

TABLE 17.1 Wall Movement Required to Reach the Active Condition[a]

Soil Type	Horizontal Movement Required to Reach the Active Condition
Dense cohesionless	0.001 H
Loose cohesionless	0.004 H
Stiff cohesive	0.010 H
Soft cohesive	0.020 H

H = Wall height.
Cohesionless soils include sands and gravels.
Cohesive soils are those with significant clay content.
[a]Adapted from CGS, 1992.

cohesionless soil, the active condition is reached if the wall moves outward from the backfill a distance equal to only 0.004 H (about 12 mm for a 3 m tall wall). Although basement walls, being braced at the top, cannot move even that distance, a *cantilever wall* (one in which the top is not connected to a building or other structure) could very easily move 12 mm outward, and such a movement would usually be acceptable. Thus, a basement wall may need to be designed to resist the at-rest pressure, whereas the design of a free-standing cantilever wall could use the active pressure, as shown in Figure 17.4. Because the active pressure is smaller, the design of free-standing walls will be more economical.

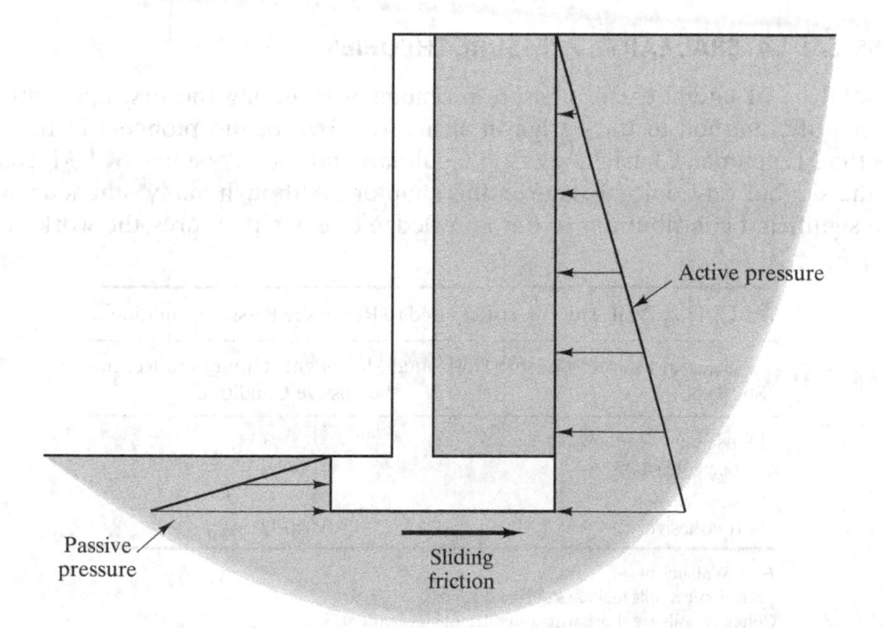

FIGURE 17.4 Active and passive pressures acting on a cantilever retaining wall.

The Passive Condition

The passive condition is the opposite of the active condition. In this case, the wall moves *into* the backfill, and the soil appears to be passively being pushed by the wall. As the wall pushes into the soil, the shear strength of the soil is again mobilized, this time to resist the push. This generates an increase in the lateral earth pressures, as shown in Figure 17.3. When all the shear strength of the soil has been mobilized, the soil fails in shear and reaches the passive condition. The K for a soil in the passive condition is known as K_p, the *coefficient of passive earth pressure*. This is the upper bound of K, as shown in Figure 17.3, and produces the upper bound of pressure that can act on the wall.

Engineers often use the passive pressure that develops along the toe of a retaining wall footing to help resist sliding, as shown in Figure 17.4. In this case, the "wall" is the side of the footing. More movement must occur to attain the passive condition than for the active condition. Typical required movements for various soils are given in Table 17.2.

Although movements on the order of those listed in Tables 17.1 and 17.2 are necessary to reach the active and passive states, respectively, much smaller movements also cause significant changes in the lateral earth pressure, as shown in Figure 17.3. While conducting a series of full-scale tests on retaining walls, Terzaghi (1934b) observed:

> ***"With compacted sand backfill, a movement of the wall over an insignificant distance (equal to one ten-thousandth of the depth of the backfill) decreases the [coefficient of lateral earth pressure] to 0.20 or increases it up to 1.00."***

This effect is not as dramatic in other soils, but even with those soils, only the most rigid and unyielding structures are truly subjected to at-rest pressures.

17.2 CLASSICAL LATERAL EARTH PRESSURE THEORIES

The solution of lateral earth pressure problems was among the first applications of the scientific method to the design of structures. Two of the pioneers in this effort were the Frenchman Charles Augustin Coulomb and the Scotsman W. J. M. Rankine (see the sidebar on Coulomb, later in this chapter). Although many others have since made significant contributions to our knowledge of earth pressures, the work of these

TABLE 17.2 Wall Movement Required to Reach the Passive Condition[a]

Soil Type	Horizontal Movement Required to Reach the Passive Condition
Dense cohesionless	0.020 H
Loose cohesionless	0.060 H
Stiff cohesive	0.020 H
Soft cohesive	0.040 H

H = Wall height.
Cohesionless soils include sands and gravels.
Cohesive soils are those with a significant clay content.

[a]Adapted from CGS, 1992.

two men was so fundamental that it still forms the basis for earth pressure calculations today. More than 50 earth pressure theories are now available; all of them have their roots in Coulomb's or Rankine's theory.

Coulomb presented his theory in 1773 and published it 3 years later (Coulomb, 1776). Rankine developed his theory more than 80 years after Coulomb (Rankine, 1857). In spite of this chronology, it is conceptually easier for us to discuss Rankine's theory first.

In this book, we will only consider lateral earth pressures in soils that are homogeneous and isotropic. Homogeneous implies that the soil properties have the same values everywhere, and isotropic implies at any given point each soil property is the same in all directions. We further assume that the soil is cohesionless. This is the simplest case. Soils that have high clay contents require special considerations that are beyond the scope of this discussion. *Foundation Design: Principles and Practices* (Coduto, 2001) discusses lateral earth pressures in layered soils, clayey soils, and in soils with cohesion.

Lateral earth pressure theories may be used with either effective stress analyses (c', ϕ') or total stress analyses (c_T, ϕ_T). However, effective stress analyses are appropriate for cohesionless soils normally preferred as backfills, and are the only type we will consider in this chapter.

Rankine's Theory for Cohesionless Soils

Assumptions Rankine approached the lateral earth pressure problem by considering the state of stress of the soil mass behind the wall and made the following assumptions:

1. The soil is homogeneous and isotropic, as defined earlier.
2. The most critical shear surface is a plane. In reality, it is slightly concave upward, but this is a reasonable assumption (especially for the active case) and it simplifies the analysis.
3. The backfill surface is planar (although it does not necessarily need to be level).
4. The wall is infinitely long so that a representative two-dimensional section of the wall may be analyzed, assuming there is no strain in the direction perpendicular to the section. We refer to this as a *plane strain condition*.
5. The wall moves sufficiently to develop the active or passive condition.
6. The resultant of the normal and shear forces that act on the back of the wall is inclined at an angle parallel to the ground surface (Coulomb's theory provides a more general model of shear forces acting on the wall).

We will first examine the special case in which the backfill surface is horizontal.

Horizontal Backfill Surface—Active Condition Suppose Mohr circle A in Figure 17.5 represents the state of stress at a point in the soil behind the wall in Figure 17.6, and that this soil is in the at-rest condition. In this case, σ'_z and σ'_x are the major and minor principal stresses, respectively, and the inclined lines are the Mohr–Coulomb failure envelopes. Because the Mohr circle does not touch the failure envelope, the shear stress, τ, is less than the shear strength, s. This is typically true because in the at-rest condition, the soil has not failed and is in equilibrium.

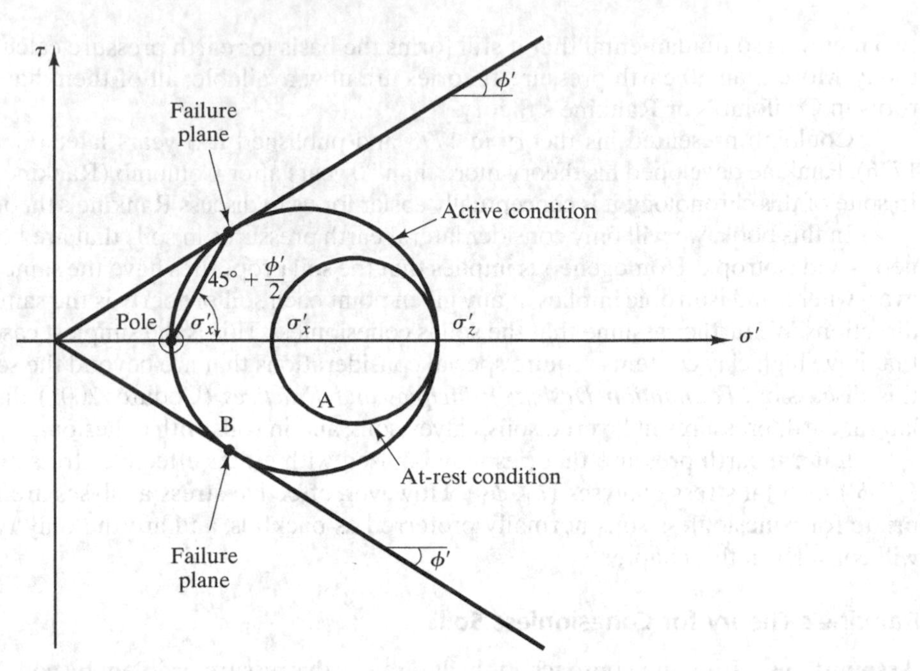

FIGURE 17.5 Changes in the stress state in a soil as it transitions from the at-rest condition to the active condition.

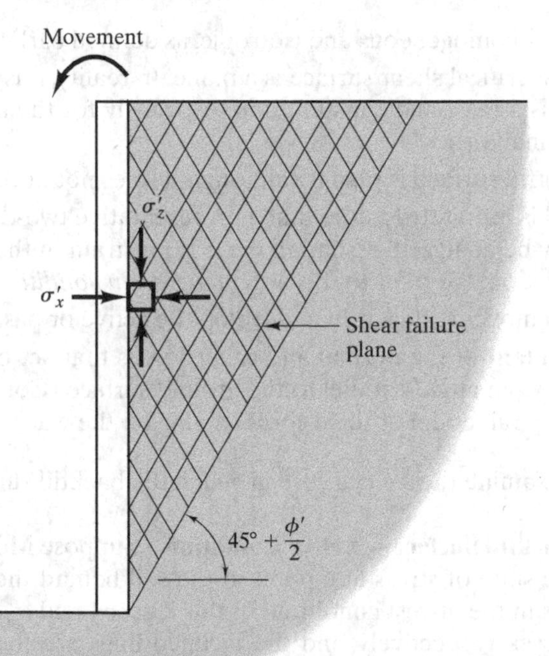

FIGURE 17.6 Development of shear failure surfaces in the soil behind a wall as it reaches the active condition.

Now, consider the active condition in which the wall is permitted to move outward a short distance. Note that in this case, because of Assumption 6, there are no shear forces acting on the back of the wall, that is, the wall is assumed to be frictionless. Therefore, σ'_z and σ'_x will remain the major and minor principal stresses, respectively, with σ'_x decreasing as the wall moves away from the soil. This causes the Mohr circle to expand to the left until the circle touches the failure envelope and the soil fails in shear (circle B). This shear failure will occur along the failure surfaces shown in Figure 17.6. The inclination of these failure surfaces can be obtained by the pole method to be $45° + \phi'/2$ with the horizontal, as shown in Figure 17.5. At this state, the soil is in the active condition. To derive a formula for K_a, we consider Mohr circle B in Figure 17.5, which depicts a state of stress at failure. Now, $\sigma'_{1f} = \sigma'_z$, and $\sigma'_{3f} = \sigma'_x = K_a\sigma'_z$. Using Equation 12.14 that relates σ'_{1f} and σ'_{3f}, we have:

$$\sigma'_z = K_a\sigma'_z \tan^2\left(45° + \frac{\phi'}{2}\right) \tag{17.5}$$

$$K_a = \frac{1}{\tan^2\left(45° + \frac{\phi'}{2}\right)} = \tan^2\left(45° - \frac{\phi'}{2}\right) \tag{17.6}$$

Horizontal Backfill Surface—Passive Condition The passive condition is the opposite of the active condition. In this case, the wall moves into the backfill, as shown in Figure 17.7, and the Mohr circle changes, as shown in Figure 17.8. Notice how the vertical stress remains constant, whereas the horizontal stress changes in response to the induced horizontal strains.

In a homogeneous soil, the failure surfaces in the passive case are inclined at an angle of $45° - \phi'/2$ with the horizontal, which can be obtained by the pole method, as shown in Figure 17.8. Note that the failure surfaces in the passive case are much flatter than in the active case. To derive a formula for K_p, we consider Mohr circle B in Figure 17.8, which depicts a state of stress at failure. Now, $\sigma'_{1f} = \sigma'_x = K_p\sigma'_z$, and $\sigma'_{3f} = \sigma'_z$. Using Equation 12.14 that relates σ'_{1f} and σ'_{3f}, we have

$$K_p\sigma'_z = \sigma'_z \tan^2\left(45° + \frac{\phi'}{2}\right) \tag{17.7}$$

$$K_p = \tan^2\left(45° + \frac{\phi'}{2}\right) \tag{17.8}$$

We now consider the more general case of an inclined backfill surface using Rankine's theory.

Inclined Backfill Surface—Active Condition For an inclined backfill surface having a constant slope angle of β, Rankine considered the state of stress of the soil behind the wall as shown in Figure 17.9(a) for the active case. The solution can be presented in terms of the *resultant* active pressure, p_a in Figure 17.9(a), as a function of the vertical stress γz, or in terms of the total *resultant* force, P_a in Figure 17.9(a). Then, the coefficient of active

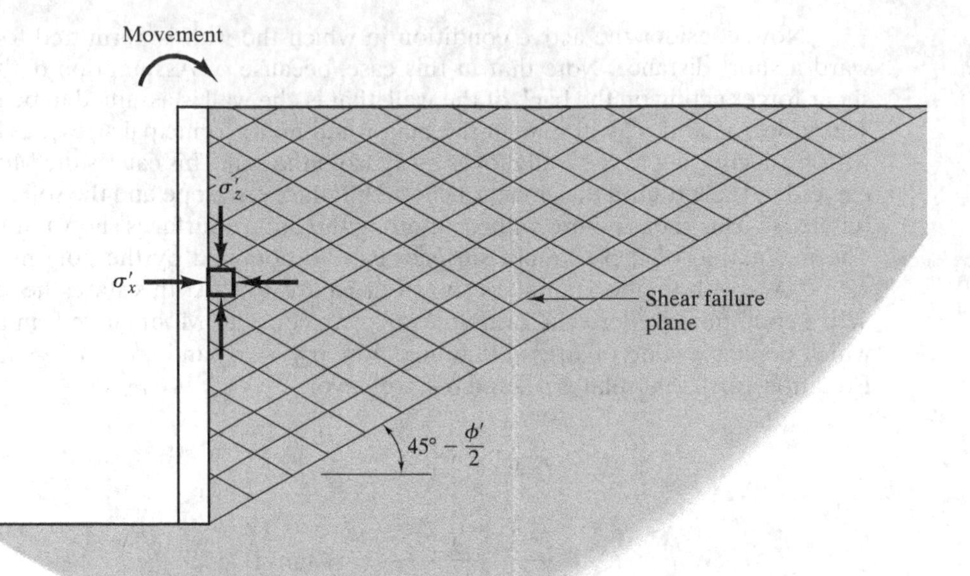

FIGURE 17.7 Development of shear failure surfaces in the soil behind a wall as it reaches the passive condition.

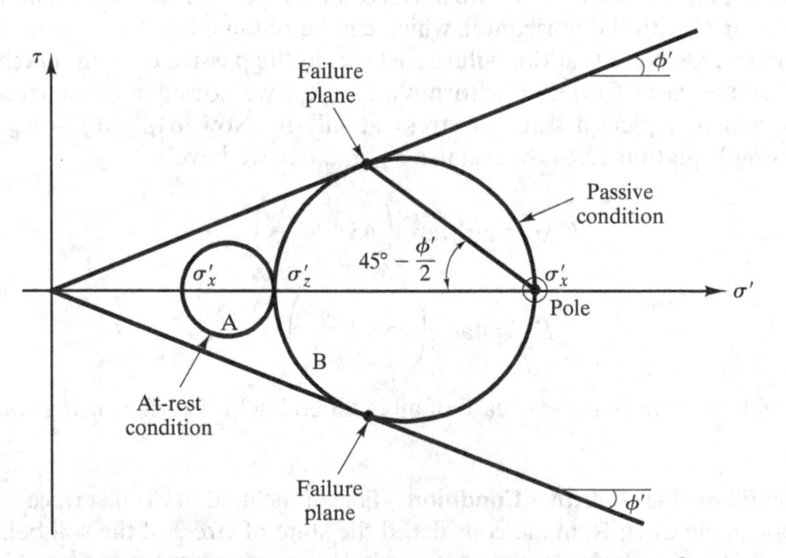

FIGURE 17.8 Changes in the stress state in a soil as it transitions from the at-rest condition to the passive condition.

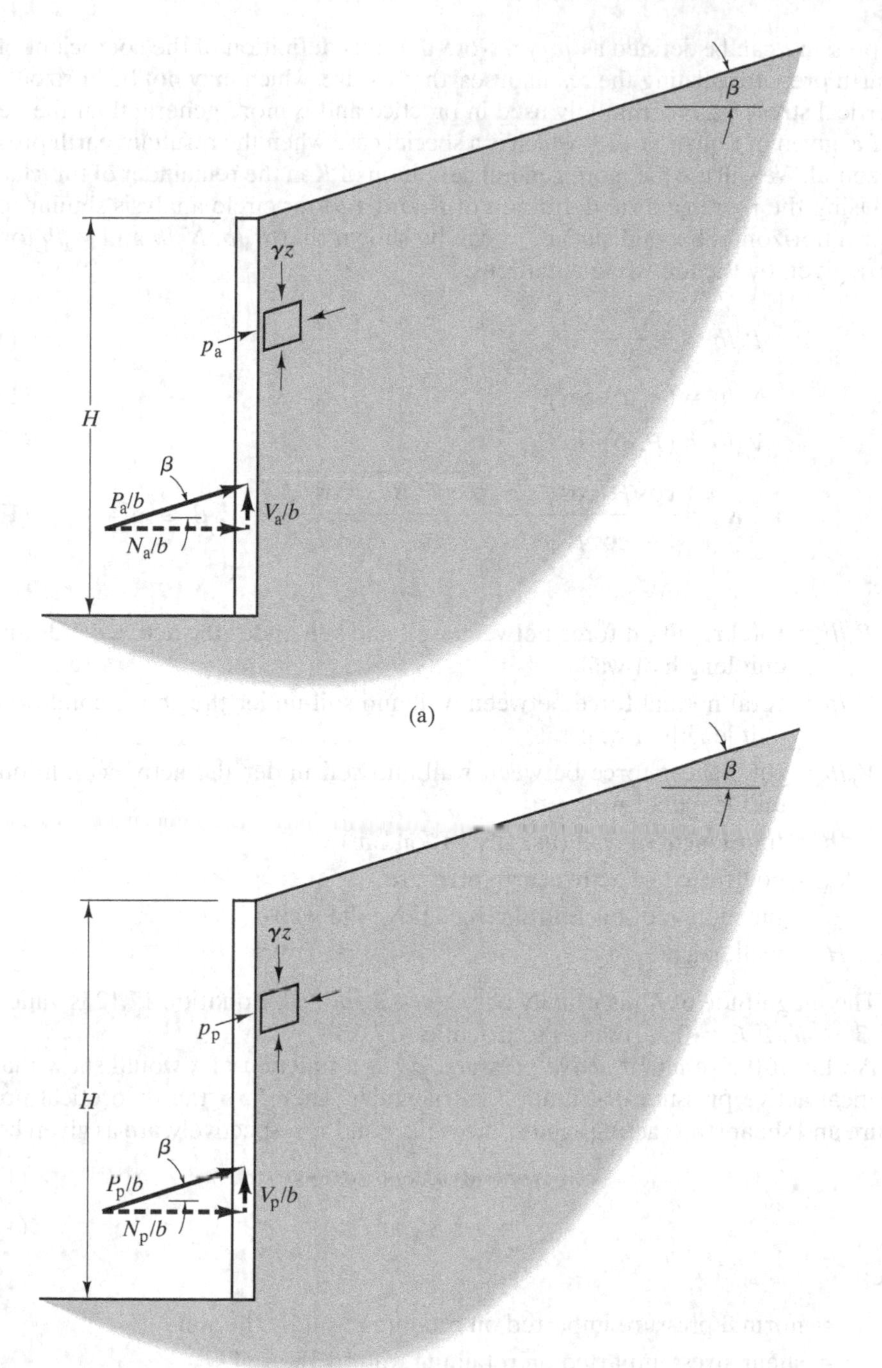

FIGURE 17.9 Rankine's theory for inclined backfill surface: (a) active case and (b) passive case.

earth pressure can be defined as $p_a/\gamma z$. Note that this definition of the coefficient of lateral earth pressure relating the resultant earth pressure, which may not be horizontal, to the vertical stress γz, is commonly used in practice and is more general than the definition of K given in Equation 17.1, which is a special case when the resultant earth pressure is horizontal. We will use the more general definition of K in the remainder of this chapter.

Using the more general definition of K and a Mohr circle analysis similar to the one for a horizontal backfill surface, it can be shown that P_a/b, N_a/b and V_a/b for this case are given by the following equations:

$$P_a/b = \frac{\gamma H^2 K_a}{2} \tag{17.9}$$

$$N_a/b = (P_a/b)\cos\beta \tag{17.10}$$

$$V_a/b = (P_a/b)\sin\beta \tag{17.11}$$

$$K_a = \frac{\cos\beta\,(\cos\beta - \sqrt{\cos^2\beta - \cos^2\phi'})}{\cos\beta + \sqrt{\cos^2\beta - \cos^2\phi'}} \qquad \beta \le \phi' \tag{17.12}$$

where:

P_a/b = total resultant force between wall and soil under the active condition per unit length of wall

N_a/b = total normal force between wall and soil under the active condition per unit length of wall

V_a/b = total shear force between wall and soil under the active condition per unit length of wall

b = unit length of wall (usually 1 ft or 1 m)

K_a = coefficient of active earth pressure

β = inclination of backfill surface above the wall

H = wall height

The magnitude of K_a is usually between 0.2 and 0.9. Equation 17.12 is valid only when $\beta \le \phi'$. If $\beta = 0$, it reduces to Equation 17.6.

A plot of the resultant active pressure, p_a, as a function of z would show that the theoretical active pressure distribution is triangular. Therefore, the theoretical normal pressure and shear stress acting against the wall, σ and τ, respectively, are as given below:

$$\sigma = \gamma z\, K_a \cos\beta \tag{17.13}$$

$$\tau = \gamma z\, K_a \sin\beta \tag{17.14}$$

where:

σ = normal pressure imparted on retaining wall by the soil

τ = shear stress imparted on retaining wall by the soil

K_a = coefficient of active earth pressure

z = depth below top of wall

β = inclination of backfill surface above the wall

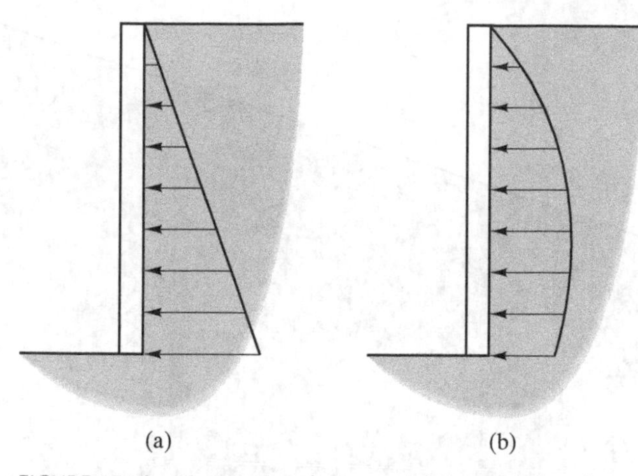

FIGURE 17.10 Comparison between (a) theoretical and (b) observed distributions of earth pressures acting behind retaining structures.

However, observations and measurements from real retaining structures indicate that the true pressure distribution, as shown in Figure 17.10, is not triangular. This difference is because of wall deflections, arching, and other factors. The magnitudes of N_a/b and V_a/b are approximately correct, but the resultant acts at about $0.40H$ from the bottom, not $0.33H$ as predicted by theory (Duncan et al., 1990).

Example 17.2

A 6-m tall cantilever wall retains a soil that has the following properties: $c' = 0$, $\phi' = 30°$, and $\gamma = 19.2 \text{ kN/m}^3$. The ground surface behind the wall is inclined at a slope of 3 horizontal to 1 vertical, and the wall has moved sufficiently to develop the active condition. Determine the total normal and shear forces acting on the back of this wall using Rankine's theory.

Solution:

$$\beta = \tan^{-1}(1/3)$$

$$\beta = 18°$$

$$K_a = \frac{\cos \beta(\cos \beta - \sqrt{\cos^2 \beta - \cos^2 \phi'})}{\cos \beta + \sqrt{\cos^2 \beta - \cos^2 \phi'}}$$

$$K_a = \frac{\cos 18°(\cos 18° - \sqrt{\cos^2 18° - \cos^2 30°})}{\cos 18° + \sqrt{\cos^2 18° - \cos^2 30°}}$$

$$K_a = 0.395$$

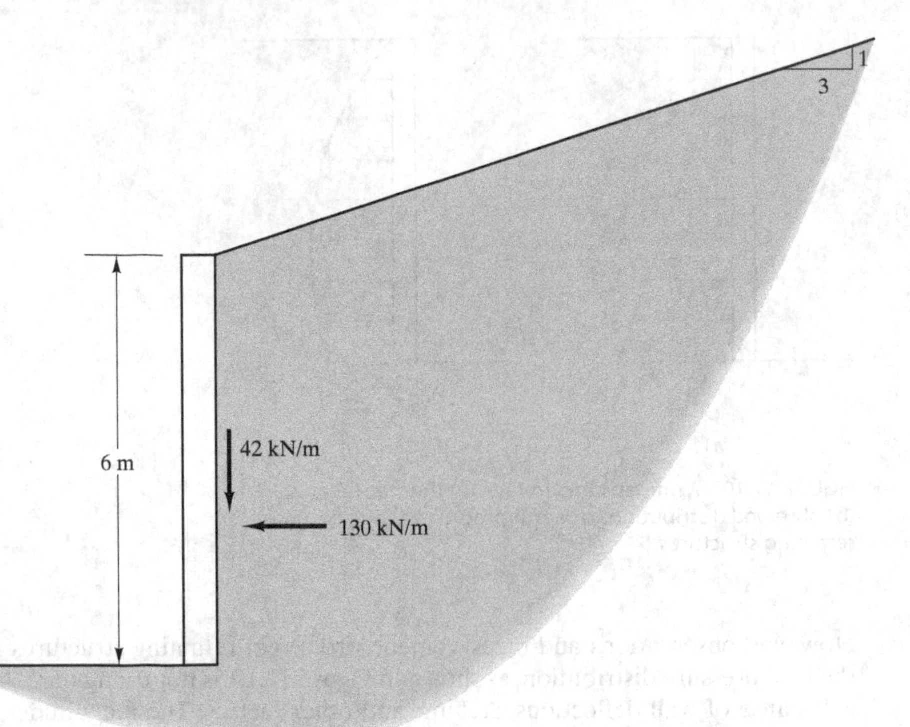

FIGURE 17.11 Results from Example 17.2.

$$N_a/b = \frac{\gamma H^2 K_a \cos \beta}{2} = \frac{(19.2\ \text{kN/m}^3)(6\ \text{m})^2(0.395)\cos 18^\circ}{2} = \mathbf{130\ kN/m}$$

$$V_a/b = \frac{\gamma H^2 K_a \sin \beta}{2} = \frac{(19.2\ \text{kN/m}^3)(6\ \text{m})^2(0.395)\sin 18^\circ}{2} = \mathbf{42\ kN/m}$$

These results are shown in Figure 17.11.

Inclined Backfill Surface—Passive Condition For the Rankine case as shown in Figure 17.9(b), the total resultant, normal and shear forces, P_p/b, N_p and V_p/b, respectively, acting on the wall in the passive case can be similarly derived as follows:

$$P_p/b = \frac{\gamma H^2 K_p}{2} \tag{17.15}$$

$$N_p/b = (P_p/b)\cos \beta \tag{17.16}$$

$$V_p/b = (P_p/b)\sin \beta \tag{17.17}$$

$$K_p = \frac{\cos\beta(\cos\beta + \sqrt{\cos^2\beta - \cos^2\phi'})}{\cos\beta - \sqrt{\cos^2\beta - \cos^2\phi'}} \qquad \beta \le \phi' \tag{17.18}$$

where:

P_p/b = total resultant force between wall and soil under the passive condition per unit length of wall
N_p/b = total normal force between wall and soil under the passive condition per unit length of wall
V_p/b = total shear force between wall and soil under the passive condition per unit length of wall
b = unit length of wall (usually 1 ft or 1 m)
K_p = coefficient of passive earth pressure
β = inclination of backfill surface above the wall
H = wall height

The magnitude of K_p is typically between 2 and 6. Equation 17.18 is valid only when $\beta \le \phi'$. If $\beta = 0$, it reduces to Equation 17.8.

The theoretical normal pressure and shear stress acting against the wall, σ and τ, respectively, are given below:

$$\sigma = \gamma z\, K_p \cos\beta \tag{17.19}$$

$$\tau = \gamma z\, K_p \sin\beta \tag{17.20}$$

where:

σ = normal pressure imparted on retaining wall by the soil
τ = shear stress imparted on retaining wall by the soil
K_p = coefficient of passive earth pressure
z = depth below top of wall
β = inclination of backfill surface above the wall

Example 17.3

A six-story building with plan dimensions of 150 ft × 150 ft has a 12-ft-deep basement. This building is subjected to horizontal wind loads, and the structural engineer wishes to transfer these loads into the ground through the basement walls. The maximum horizontal force acting on the basement wall is limited by the passive pressure in the soil. Using Rankine's theory, compute the maximum force between one of the basement walls and the adjacent soil assuming the passive condition develops, and then convert it to an allowable force using a factor of safety of 3. The soil is a silty sand with $c' = 0$, $\phi' = 30°$, and $\gamma = 119\ \text{lb/ft}^3$, and the ground surface surrounding the building is essentially level.

Solution:

$$K_p = \tan^2\left(45° + \frac{\phi'}{2}\right) = \tan^2\left(45° + \frac{30°}{2}\right) = 3.00$$

$$P_p/b = \frac{\gamma H^2 K_p}{2}$$

$$P_p/b = \frac{(119 \text{ lb/ft}^3)(12 \text{ ft})^2(3.00)}{2}$$

$$P_p/b = 25{,}700 \text{ lb/ft}$$

$$P_p = \frac{(25{,}700 \text{ lb/ft}^3)(150 \text{ ft})}{1000 \text{ lb/k}} = 3860 \text{ k}$$

The allowable passive force, $(P_p)_a$ is given below:

$$(P_p)_a = \frac{P_p}{F} = \frac{3860 \text{ k}}{3} = \mathbf{1290\ k}$$

Note: The actual design computations for this problem would be more complex because we would need to consider the active pressure acting on the opposite wall, sliding friction along the basement floor, lateral resistance in the foundations, and other factors. In addition, the horizontal displacement required to develop the full passive resistance may be excessive, so the design value may need to be reduced accordingly. Finally, to take advantage of this resistance, the wall would need to be structurally designed to accommodate this large load, which is much greater than that due to the active or at-rest pressure.

Coulomb's Theory for Cohesionless Soils

Active Condition Coulomb's theory differs from Rankine's in that the resultant of the normal and shear forces acting on the wall is inclined at an angle ϕ_w with the normal to the wall, where $\tan \phi_w$ is the coefficient of friction between the wall and the soil, as shown in Figure 17.12(a) for the active case. This is a more general and realistic model, and thus can handle more general cases. For the active case as shown in Figure 17.12(a), we treat the wedge of soil that fails behind the wall as a free body, and evaluate the problem using the principles of statics. This is similar to the slope stability analysis methods we used in Chapter 13, and is known as a *limit equilibrium analysis*, which means that we consider the conditions that would exist if the soil along the failure surface at the base of the failure wedge was about to fail in shear.

As in a slope stability analysis, the critical failure surface must in general be obtained through a trial-and-error process. In some cases, weak seams or other heterogeneity in the soil may control the inclination of the critical failure surface. Consider the

Charles Augustin Coulomb

Charles Augustin Coulomb (1736–1806) was a French physicist who is best remembered for his work in electricity and magnetism. However, he also made important contributions in other fields, including the computation of lateral earth pressures.

Coulomb graduated from the Mézières School of Military Engineers in France at the age of 26. Two years later, the young officer was sent to the Caribbean island of Martinique where he was placed in charge of building a fort to protect the harbor. In the process of finalizing the design of the fort, he became dissatisfied with the rules of thumb for sizing retaining walls because they dictated walls that were too large. Although some theoretical analyses had already been attempted, they were flawed. He later wrote (Kerisel, 1987):

> "I have often come across situations in which all the theories based on hypotheses or on small-scale experiments in a physics laboratory have proved inadequate in practice."

Therefore, he began studying the problem, and eventually developed a new theory of lateral earth pressures. This work is generally recognized as the first important quantitative contribution to what would become geotechnical engineering. Coulomb was the first to define soil strength using both cohesion and friction, the first to consider wall friction, and the first to analytically search for the orientation of the most critical failure plane (which turned out to be at an angle of $45° + \phi'/2$ for the active case, as shown in Figure 17.6). He also developed other important insights.

Coulomb published his results in 1776 as a paper titled *Essai sur une application des règles de maximis et minimis à quelques problèmes de statique relatifs à l'architecture* (Essay on an Application of the Rules of Maximum and Minimum to Some Statical Problems, Relevant to Architecture).[1] This paper also addressed other problems, including the stability of arches and the strength of beams.

He had a sense of both theory and practice. For example, his *Essai* also discussed the detrimental effects of groundwater and noted "Even though, to avoid this problem, vertical pipes are placed in practice behind retaining walls, and the drains at the feet of the same walls, so that the water can run off, these drains get blocked, either by soil carried along with the water, or by ice, and sometimes become useless."

Although Coulomb's work provided important insights into the earth pressure problem, it was difficult to apply to practical problems because nobody had the ability to measure the cohesion and friction angle of a soil. The first significant soil strength tests would not be performed until about 70 years later by another Frenchman, Alexandre Collin, and Coulomb's work was not widely recognized. As a practical matter, Coulomb's work did not reach its full potential until the twentieth century, when soil strength tests became common.

[1]Heyman (1972) provides an English translation and commentary.

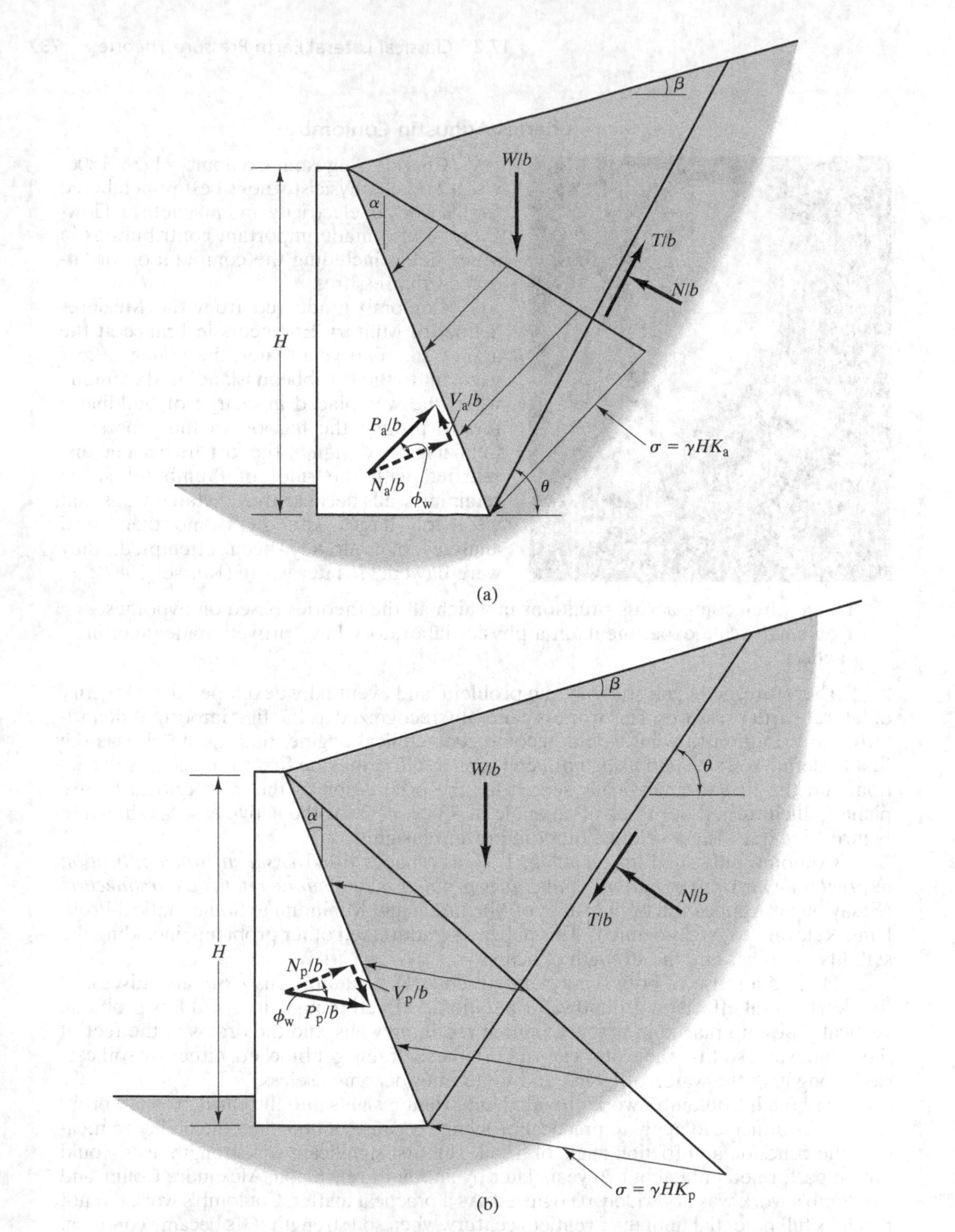

FIGURE 17.12 Free body diagram of a failure wedge behind a retaining wall using Coulomb's theory: (a) active case and (b) passive case.

free body of the failure wedge bounded by a trial planar failure surface inclined at an angle θ with the horizontal, as shown in Figure 17.12(a). Note that the resultant force of N/b and T/b is inclined at an angle ϕ' with respect to the normal to the failure surface, and that the resultant of N_a/b and V_a/b is inclined at ϕ_w with respect to the normal to the inclined back surface of the wall. Therefore, these two resultant forces are the only two unknowns that can be computed using statics by drawing a force polygon or writing two force equilibrium equations in two orthogonal directions. For the active case as shown in Figure 17.12(a), the critical failure surface that gives the largest P_a/b can be obtained analytically.

Coulomb presented his earth pressure formula in a difficult form, so others have rewritten it in a more convenient fashion, as follows (Müller Breslau, 1906; Tschebotarioff, 1951):

$$P_a/b = \frac{\gamma H^2 K_a}{2} \tag{17.21}$$

$$N_a/b = (P_a/b)\cos\phi_w \tag{17.22}$$

$$V_a/b = (P_a/b)\sin\phi_w \tag{17.23}$$

$$K_a = \frac{\cos^2(\phi' - \alpha)}{\cos^2\alpha\cos(\alpha + \phi_w)\left[1 + \sqrt{\dfrac{\sin(\phi' + \phi_w)\sin(\phi' - \beta)}{\cos(\alpha + \phi_w)\cos(\alpha - \beta)}}\right]^2} \quad \beta \le \phi' \tag{17.24}$$

where:

P_a/b = total resultant force between wall and soil under the active condition per unit length of wall

N_a/b = total normal force between wall and soil under the active condition per unit length of wall

V_a/b = total shear force between wall and soil under the active condition per unit length of wall

b = unit length of wall (usually 1 ft or 1 m)

K_a = coefficient of active earth pressure

α = inclination of back surface of wall from vertical

β = inclination of backfill surface above the wall

ϕ_w = wall–soil interface friction angle

Equation 17.21 is valid only for $\beta \le \phi'$. When designing concrete or masonry walls, it is common practice to use $\phi_w = (2/3)\phi'$. Steel walls have less sliding friction, with $\phi_w = (1/3)\phi'$. Note again that K_a relates the resultant earth pressure, which may not be horizontal, to the vertical stress γz, as shown in Figure 17.12(a). This definition is more general than the definition of the earth pressure coefficient, used earlier in the book, which relates the horizontal stress to vertical stress.

Example 17.4

Using Coulomb's method, compute the total active normal and shear forces acting on the reinforced concrete retaining wall shown in Figure 17.13.

Solution:

$$\beta = \tan^{-1}\left(\frac{1}{2}\right) = 27°$$

$$\phi_w = (2/3)\phi' = (2/3)(32°) = 21°$$

$$K_a = \frac{\cos^2(\phi' - \alpha)}{\cos^2\alpha\cos(\alpha + \phi_w)\left[1 + \sqrt{\dfrac{\sin(\phi' + \phi_w)\sin(\phi' - \beta)}{\cos(\alpha + \phi_w)\cos(\alpha - \beta)}}\right]^2}$$

$$= \frac{\cos^2(32° - 2°)}{\cos^2 2°\cos(2° + 21°)\left[1 + \sqrt{\dfrac{\sin(32° + 21°)\sin(32° - 27°)}{\cos(2° + 21°)\cos(2° - 27°)}}\right]^2}$$

$$= 0.491$$

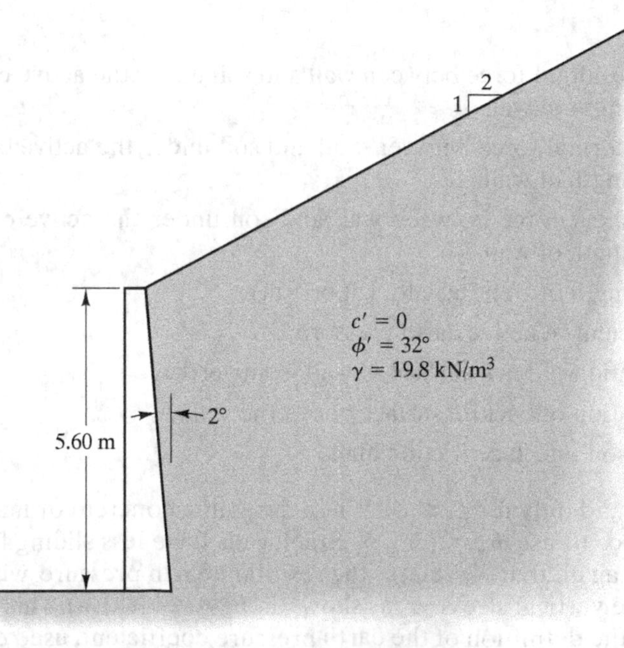

FIGURE 17.13 Retaining wall for Example 17.4.

$$N_a/b = \frac{\gamma H^2 K_a \cos\phi_w}{2}$$

$$N_a/b = \frac{(19.8\ \text{kN/m}^3)(5.60\ \text{m})^2(0.491)\cos 21°}{2}$$

$$N_a/b = \mathbf{142\ kN/m}$$

$$V_a/b = \frac{\gamma H^2 K_a \sin\phi_w}{2}$$

$$V_a/b = \frac{(19.8\ \text{kN/m}^3)(5.60\ \text{m})^2(0.491)\sin 21°}{2}$$

$$V_a/b = \mathbf{55\ kN/m}$$

Passive Condition Coulomb did not develop a formula for the passive earth pressure, although others have used his theory to do so. The Coulomb analysis of the passive condition is similar to that of the active condition except that in the free body diagram as shown in Figure 17.12(b), each of the shear forces acting along the base of the wedge and along the interface between the wall and the soil now acts in the opposite direction (it always opposes the movement of the wedge). For the passive case as shown in Figure 17.12(b), the formulas based on Coulomb's theory are as follows:

$$P_p/b = \frac{\gamma H^2 K_p}{2} \tag{17.25}$$

$$N_p/b = (P_p/b)\cos\phi_w \tag{17.26}$$

$$V_p/b = (P_p/b)\sin\phi_w \tag{17.27}$$

$$K_p = \frac{\cos^2(\phi' + \alpha)}{\cos^2\alpha\cos(\alpha - \phi_w)\left[1 - \sqrt{\frac{\sin(\phi' + \phi_w)\sin(\phi' + \beta)}{\cos(\alpha - \phi_w)\cos(\alpha - \beta)}}\right]^2} \tag{17.28}$$

where:

P_p/b = total resultant force between wall and soil under the passive condition per unit length of wall

N_p/b = total normal force between soil and wall under the passive condition per unit length of wall

V_p/b = total shear force between soil and wall under the passive condition per unit length of wall

b = unit length of wall (usually 1 ft or 1 m)

K_p = coefficient of passive earth pressure

α = inclination of back surface of wall from vertical

β = inclination of backfill surface above the wall
ϕ_w = wall–soil interface friction angle

One limitation of the Coulomb method is that when the wall friction is high ($\phi_w > (1/3)\phi'$), the failure surface in the passive case deviates greatly from the planar surface assumption, leading to computed passive pressures that are too high and therefore unconservative (Dunn et al., 1980). Therefore, for such cases, we use the log spiral method that assumes a more realistic failure surface that consists of a piece of the log spiral.

Log Spiral Method for Cohesionless Soils

The log spiral method is similar to the method derived from Coulomb's theory, except that the failure surface is now assumed to be formed partly by a piece of the log spiral, as shown in Figure 17.14 (U.S. Navy, 1982). This figure also shows the chart for the active and passive earth pressures obtained by the log spiral method. Figure 17.14 is an example of chart solutions that are available in practice for estimating earth pressures, similar to the slope stability chart solutions discussed in Section 13.10 for estimating factors of safety of slopes.

17.3 EQUIVALENT FLUID PRESSURE

As discussed earlier, the theoretical distribution of lateral earth pressure acting on a wall is triangular. This is the same shape as the pressure distribution that would be imposed if the wall was backfilled with a fluid instead of with soil. Further, if this fluid had the proper unit weight, the magnitude of the lateral earth pressure also would be equal to that from the soil, that is, the *equivalent fluid pressure* distribution would be the same as the lateral earth pressure distribution.

Engineers often use this similarity when expressing lateral earth pressures for design purposes. Instead of quoting K values, we define the lateral earth pressure using the *equivalent fluid density*, G_h. This is the unit weight of a fictitious fluid that would impose the same horizontal pressures on the wall as the soil. We give this value to a civil engineer or structural engineer who wishes to design a wall, and they proceed using the principles of fluid statics.

This method is popular because it reduces the potential for confusion and mistakes. All engineers understand the principles of fluid statics (at least they should!), so G_h is easy to apply.

For homogeneous, isotropic soils with $c' = 0$, the equivalent fluid density is

$$G_h = \gamma K \cos \phi_w \tag{17.29}$$

And the total normal force between the soil and the wall per unit length of the wall is

$$N/b = \frac{G_h H^2}{2} \tag{17.30}$$

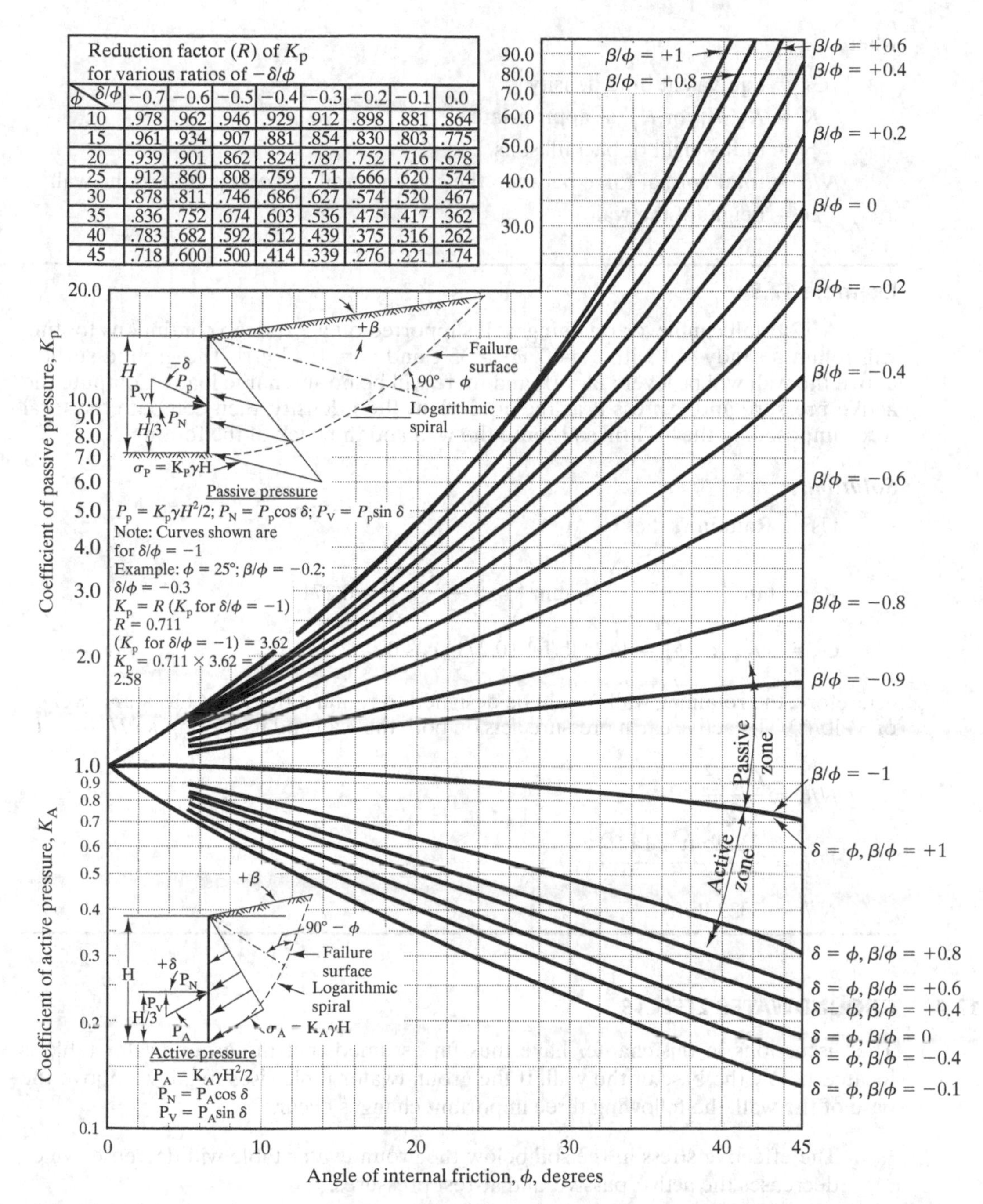

ϕ \ δ/ϕ	−0.7	−0.6	−0.5	−0.4	−0.3	−0.2	−0.1	0.0
10	.978	.962	.946	.929	.912	.898	.881	.864
15	.961	.934	.907	.881	.854	.830	.803	.775
20	.939	.901	.862	.824	.787	.752	.716	.678
25	.912	.860	.808	.759	.711	.666	.620	.574
30	.878	.811	.746	.686	.627	.574	.520	.467
35	.836	.752	.674	.603	.536	.475	.417	.362
40	.783	.682	.592	.512	.439	.375	.316	.262
45	.718	.600	.500	.414	.339	.276	.221	.174

FIGURE 17.14 Chart for the log spiral method. (U.S. Navy, 1982.)

where:

G_h = equivalent fluid density
K = K_a, K_0, or K_p, as appropriate
γ = unit weight of backfill soils
N/b = total normal force between the soil and wall per unit length of the wall
H = height of the wall

Example 17.5

A 12 ft tall cantilever retaining wall supported on a 2 ft deep continuous footing will retain a sandy soil with $c' = 0$, $\phi' = 35°$, and $\gamma = 124\ \text{lb/ft}^3$. The ground surface above the wall will be level ($\beta = 0$) and there will be no surcharge loads. Compute the active pressure and express it as the equivalent fluid density, then compute the total force imposed by the backfill soils onto the wall and the back of the footing.

Solution:

Using Rankine's theory,

$$K_a = \tan^2\left(45° - \frac{\phi'}{2}\right) = \tan^2\left(45° - \frac{35°}{2}\right) = 0.271$$

$$G_h = \gamma K_a \cos\phi_w = (124\ \text{lb/ft}^3)(0.271)\cos 0° = 34\ \text{lb/ft}^3$$

Therefore, the retaining wall should be designed to retain a fluid that has a unit weight of 34 lb/ft^3. The active earth pressure acts on both the wall and its footing, so $H = 14$ ft.

$$N/b = \frac{G_h H^2}{2}$$

$$N/b = \frac{(34\ \text{lb/ft}^3)(14\ \text{ft})^2}{2}$$

$$N/b = 3332\ \text{lb/ft}$$

17.4 GROUNDWATER EFFECTS

The discussions in this chapter have thus far assumed that the groundwater table is located below the base of the wall. If the groundwater table rises to a level above the base of the wall, the following three important changes occur:

1. The effective stress in the soil below the groundwater table will decrease, which decreases the active, passive, and at-rest pressures.
2. Horizontal hydrostatic pressures will develop against the wall and must be superimposed onto the lateral earth pressures.
3. The effective stress between the bottom of the footing and the soil becomes smaller, so there is less sliding friction.

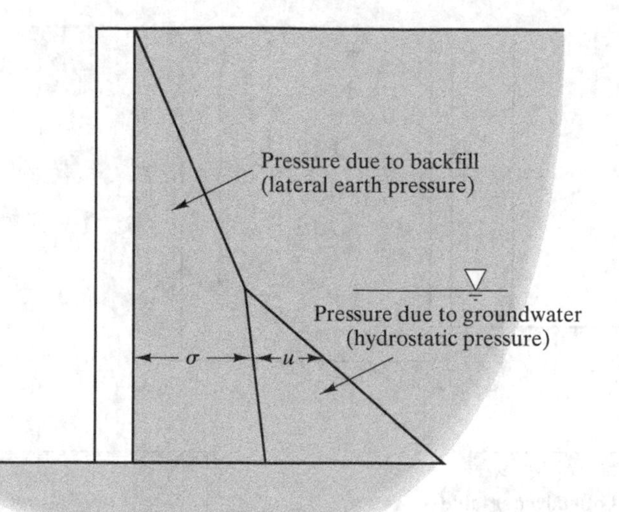

FIGURE 17.15 Theoretical lateral earth pressure distribution with shallow groundwater table.

The net effect of the first two changes is a large increase in the total horizontal pressure acting on the wall (i.e., the increase in hydrostatic pressure more than offsets the decrease in lateral earth pressure). The resulting pressure diagram is shown in Figure 17.15.

Example 17.6

Compute the lateral pressure distribution acting on the cantilever wall with the groundwater table at locations **a** and **b**, as shown in Figure 17.16. Assume that this cantilever wall has moved sufficiently to create the active condition.

The soil properties are $c' = 0$, $\phi' = 30°$, $\gamma = 20.4\ \text{kN/m}^3$, and $\gamma_{sat} = 22.0\ \text{kN/m}^3$.

Solution:

Use Rankine's method.

$$K_a = \tan^2\left(45° - \frac{\phi'}{2}\right) = \tan^2\left(45° - \frac{30°}{2}\right) = 0.333$$

With the groundwater table at **a**:

$$\sigma = \sigma'_z K_a \cos\beta$$
$$\sigma = \gamma z K_a \cos\beta$$
$$\sigma = 20.4z(0.333)\cos 0$$
$$\sigma = 6.79z$$

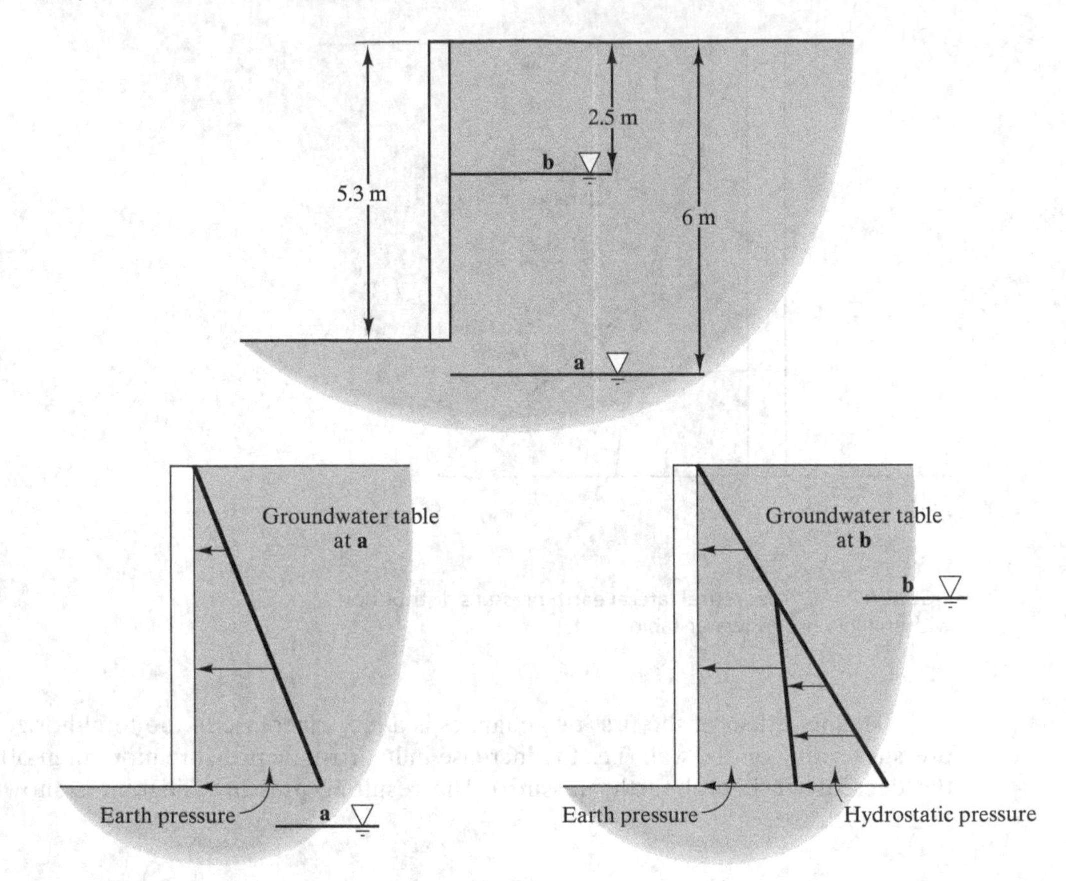

FIGURE 17.16 Retaining wall for Example 17.6.

where:

z = depth below the top of the wall
$\beta = 0 \therefore V_a = 0$ (per Rankine)

With the groundwater table at **b**:

$$\sigma \text{ @ } z \geq 2.5 \text{ m}$$

$$\sigma = \sigma'_z K_a \cos \beta$$

$$\sigma = (\sum \gamma H - u) K_a \cos \beta$$

$$\sigma = [(20.4 \text{ kN/m}^3)(2.5 \text{ m}) + (22.0 \text{ kN/m}^3)(z - 2.5 \text{ m}) - u](0.333) \cos 0$$

$$\sigma = 7.33z - 0.33u - 1.33$$

$$u = 9.80 \text{ kN/m}^3 (z - 2.5 \text{ m}) \geq 0$$

Total horizontal pressure on wall $= \sigma + u$

	Groundwater table at a		Groundwater table at b	
z (m)	σ (kPa)	u (kPa)	σ (kPa)	Total pressure (kPa)
0.0	0.00	0.00	0.00	0.00
0.5	3.40	0.00	3.40	3.40
1.0	6.79	0.00	6.79	6.79
1.5	10.19	0.00	10.19	10.19
2.0	13.58	0.00	13.58	13.58
2.5	16.98	0.00	16.98	16.98
3.0	20.37	4.90	19.04	23.94
3.5	23.77	9.80	21.09	30.89
4.0	27.16	14.70	23.14	37.84
4.5	30.56	19.60	25.19	44.79
5.0	33.95	24.50	27.24	51.74
5.3	35.99	27.44	28.46	55.90

Example 17.6 demonstrates the profound impact of groundwater on retaining walls. If the groundwater table rises from **a** to **b**, the total horizontal force acting on the wall increases by about 30%. Therefore, the factor of safety against sliding and overturning could drop from 1.5 to about 1.0, and the flexural stresses in the stem would be about 30% larger than anticipated. There are two ways to avoid these problems:

1. Design the wall for the highest probable groundwater table. This can be very expensive, but it may be the only available option.
2. Install drains to prevent the groundwater from rising above a certain level. These could consist of *weep holes* drilled in the face of the wall or a *perforated pipe drain* installed behind the wall. Drains such as these are the most common method of designing for groundwater.

Further problems can occur if the groundwater becomes frozen and ice lenses form. The same processes that cause frost heave at the ground surface also produce large horizontal pressures on retaining walls. This is another good reason to provide good surface and subsurface drainage around retaining walls.

SUMMARY

Major Points

1. Lateral earth pressures are those imparted by soil onto vertical or near-vertical structures. These pressures include both compression and shear.
2. The coefficient of lateral earth pressure, K, is the ratio of the horizontal to vertical effective stresses. In undisturbed soil, it is equal to the coefficient of lateral

earth pressure at rest, K_0. However, if the structure (usually a retaining wall) moves a sufficient distance out from the backfill, the soil reaches the active condition, and K becomes K_a, which is smaller than K_0. Conversely, if the structure moves a sufficient distance toward the backfill, the soil reaches the passive condition, and K becomes K_p, which is larger than K_0.

3. The active and passive earth pressures may be computed using classical earth pressure theories. Rankine's theory and Coulomb's theory are the most commonly used.
4. For the passive case and high wall friction ($\phi_w > (1/3)\phi'$), the log spiral method is more appropriate than the Rankine or Coulomb method.
5. If the groundwater table is located above the base of the wall, the resulting hydrostatic pressures will significantly increase the total force acting on the wall. In some cases, these hydrostatic pressures may be greater than the lateral earth pressures from the soil. Therefore, it is very important for walls to have good drainage.

Vocabulary

active condition
at-rest condition
cantilever wall
coefficient of active earth pressure
coefficient of lateral earth pressure
coefficient of lateral earth pressure at rest
coefficient of passive earth pressure
Coulomb's theory
earth retaining structure
equivalent fluid density
equivalent fluid pressure
lateral earth pressure
limit equilibrium analysis
log spiral method
passive condition
perforated pipe drain
plane strain condition
Rankine's theory
rigid wall
unyielding wall
weep hole
yielding wall

QUESTIONS AND PRACTICE PROBLEMS

Section 17.1 Lateral Earth Pressures and Wall Movement

17.1 A massive gravity wall is to be built on a hard bedrock, then backfilled with a very loose uncompacted cohesionless soil. Which should be used for the design earth pressure acting on the back of this wall, the at-rest pressure, the active pressure, or the passive pressure? Why?

17.2 Explain the difference between the active, at-rest, and the passive earth pressure conditions.

17.3 Which of the three earth pressure conditions should be used to design a rigid basement wall? Why?

17.4 A basement is to be built using 2.5-m tall masonry walls. These walls will be backfilled with a silty sand that has $c' = 0$, $\phi' = 35°$, and $\gamma = 19.7\ \text{kN/m}^3$. Assuming the at-rest condition will exist and using an overconsolidation ratio of 2, compute the normal force per meter acting on the back of this wall. Also, draw a pressure diagram and indicate the lateral earth pressure acting at the bottom of the wall.

17.5 A 4-m tall cantilever wall is to be backfilled with a dense silty sand. How far must this wall move to attain the active condition in the soil behind it? Is it appropriate to use the active pressure for design? Explain.

Section 17.2 Classical Lateral Earth Pressure Theories

17.6 State the assumptions in the Rankine's method of calculating lateral earth pressures.

17.7 State the assumptions in the Coulomb's method of calculating lateral earth pressures.

17.8 A 10-ft tall concrete wall with a vertical back is to be backfilled with a silty sand that has a unit weight of 122 lb/ft^3, an effective cohesion of 0, and an effective friction angle of 32°. The ground behind the wall will be level. Using Rankine's method, compute the total normal force per foot acting on the back of the wall. Assume the wall moves sufficiently to develop the active condition in the soil. Also, draw a pressure diagram and indicate the lateral earth pressure acting at the bottom of the wall.

17.9 The wall described in Problem 17.8 has a foundation that extends from the ground surface to a depth of 2 ft. As the wall moves slightly away from the backfill soils to create the active condition, the footing moves into the soils below the wall, creating the passive condition as shown in Figure 17.4. Using Rankine's method, compute the total normal force per foot acting on the front of the foundation. Draw the distribution of passive pressure on the front of the foundation. Determine the magnitude of the total resultant passive force and its theoretical point of application.

17.10 A 12-ft tall concrete wall with a vertical back is to be backfilled with a clean sand that has a unit weight of 126 lb/ft^3, an effective cohesion of 0, and an effective friction angle of 36°. The ground behind the wall will be inclined at a slope of 2 horizontal to 1 vertical. Using Rankine's method, compute the total normal and shear forces per foot acting on the back of the wall. Assume the wall moves sufficiently to develop the active condition in the soil.

17.11 Repeat Problem 17.8 using Coulomb's method.

17.12 Repeat Problem 17.9 using Coulomb's method.

17.13 Repeat Problem 17.10 using Coulomb's method.

17.14 A proposed concrete retaining wall is to be built as shown in Figure 17.17. Using Rankine's method, compute the horizontal component of the active earth pressure acting on the 14.3-ft tall dashed line and the horizontal passive earth pressure acting on the front of the footing. Present your results as pressure diagrams. Then compute the resultant of the active earth pressure distribution and the resultant of the passive earth pressure distribution and show them as horizontal point loads.

Note: Another important force has not been considered in this analysis: The sliding friction force along the bottom of the footing. In a properly designed wall, the combination of this force and the resultant of the horizontal passive pressure is greater than the resultant of the horizontal active pressure with an appropriate factor of safety.

17.15 Repeat Problem 17.14 using Coulomb's method. Compare your answers with the ones obtained in Problem 17.14.

Section 17.3 Equivalent Fluid Pressure

17.16 A 3-m tall cantilever retaining wall with a vertical back is to be backfilled with a soil that has an equivalent fluid density of 6.0 kN/m^3. Compute the total lateral force per meter acting on the back of this wall.

Section 17.4 Groundwater Effects

17.17 Using a groundwater table at Level A and Rankine's method, compute the lateral earth pressure acting on the back of the concrete wall in Figure 17.18. Present your results in the form of a pressure diagram, and then compute the total normal force acting on the wall and the bending moment at the bottom of the stem.

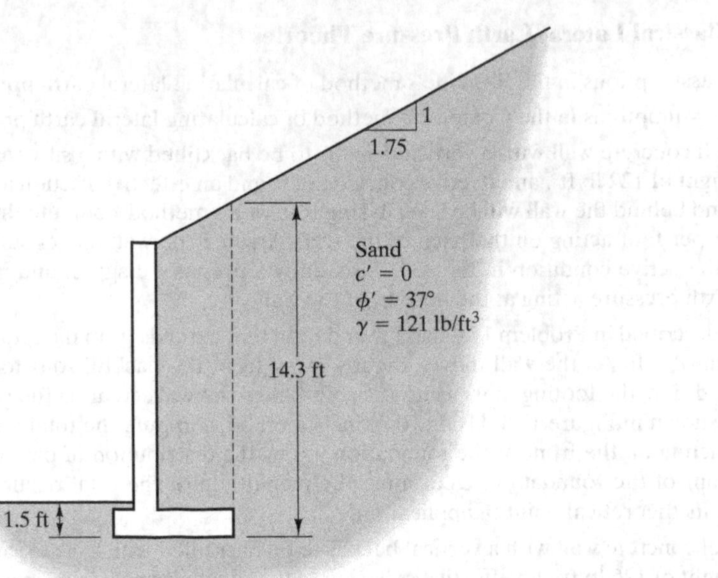

FIGURE 17.17 Proposed retaining wall for Problems 17.14 and 17.15.

17.18 Using the information from Problem 17.17 and a groundwater table at Level B, recompute the lateral earth pressures and compute the hydrostatic pressures acting on the back of the wall. Present your results in the form of a pressure diagram, and then compute the total normal force acting on the wall and the bending moment at the bottom of the stem. Compare the results with those obtained in Problem 17.17.

17.19 Repeat Problem 17.17 using Coulomb's method.

17.20 Repeat Problem 17.18 using Coulomb's method. Compare the results with those obtained in Problems 17.17, 17.18, and 17.19.

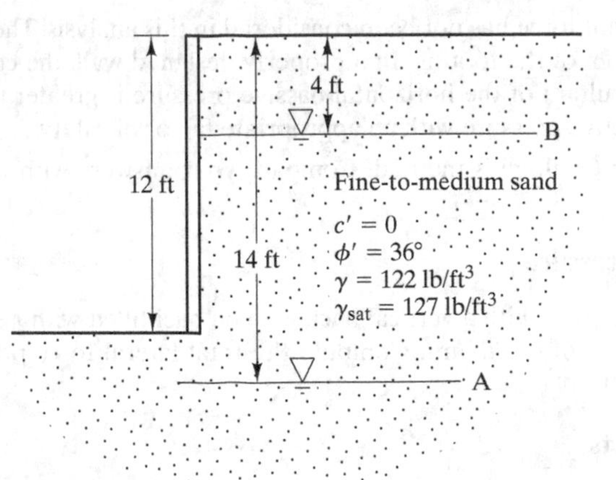

FIGURE 17.18 Proposed retaining wall for Problems 17.17 through 17.20.

APPENDIX A

Finite Difference Solutions to Flow Problems

As discussed in Section 8.3, two-dimensional flow problems that are too complex to be solved using hand-drawn flow nets can be solved using numerical methods. This appendix presents one of the simplest and most straight forward numerical methods, the *Finite Difference Method.*

A.1 FINITE DIFFERENCE FORMULATION

To understand the finite difference method, it is simplest to start with a one-dimensional example. Consider a function, $f(x)$, as shown in Figure A.1. If we know the value of the function at a point, x, we can determine the value of the function at a point Δx away from x by using the Taylor series expansion.

$$f(x + \Delta x) = f(x) + \frac{\mathrm{d}f}{\mathrm{d}x}(x)\Delta x + \frac{1}{2!}\frac{\mathrm{d}^2f}{\mathrm{d}x^2}(x)\Delta x^2 + \frac{1}{3!}\frac{\mathrm{d}^3f}{\mathrm{d}x^3}(x)\Delta x^3 + \ldots \quad \text{(A.1)}$$

Using the notation $f(x_i) = f_i$, $f(x_i + \Delta x) = f_{i+1}$ and $f(x_i - \Delta x) = f_{i-1}$ as shown in Figure A.1, we can use Equation A.1 to expand f_i in the forward direction and get

$$f_{i+1} = f_i + \left.\frac{\mathrm{d}f}{\mathrm{d}x}\right|_i \Delta x + \left.\frac{1}{2!}\frac{\mathrm{d}^2f}{\mathrm{d}x^2}\right|_i \Delta x^2 + \left.\frac{1}{3!}\frac{\mathrm{d}^3f}{\mathrm{d}x^3}\right|_i \Delta x^3 + \ldots \quad \text{(A.2)}$$

Similarly, we can expand f_i in the backward direction and get

$$f_{i-1} = f_i - \left.\frac{\mathrm{d}f}{\mathrm{d}x}\right|_i \Delta x + \left.\frac{1}{2!}\frac{\mathrm{d}^2f}{\mathrm{d}x^2}\right|_i \Delta x^2 - \left.\frac{1}{3!}\frac{\mathrm{d}^3f}{\mathrm{d}x^3}\right|_i \Delta x^3 + \ldots \quad \text{(A.3)}$$

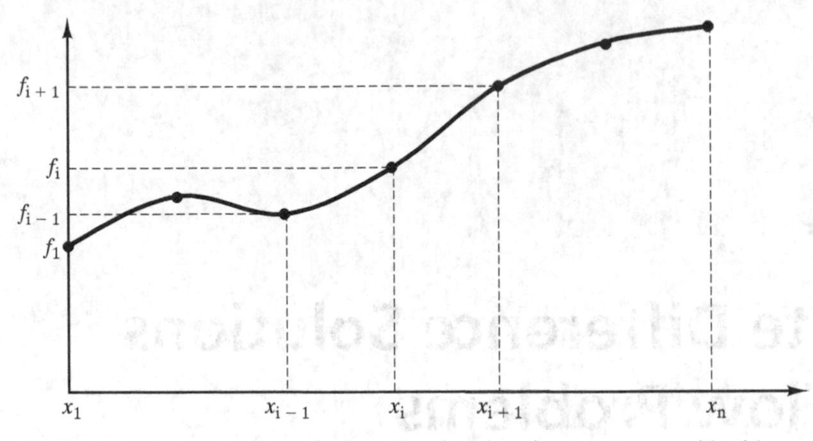

FIGURE A.1 Discretization of a function showing the notation used in this appendix.

If we subtract Equation A.3 from Equation A.2 the odd terms in the series cancel out and we have

$$f_{i+1} - f_{i-1} = 0 + 2\left.\frac{df}{dx}\right|_i \Delta x + 0 + \left.\frac{1}{3!}\frac{d^3f}{dx^3}\right|_i \Delta x^3 + \ldots \tag{A.4}$$

If we keep only the first three terms of Equation A.4, we have the following approximation for the first derivative of $f(x)$ at $x = x_i$:

$$\left.\frac{df}{dx}\right|_i \cong \frac{f_{i+1} - f_{i-1}}{2\Delta x} \tag{A.5}$$

Similarly if we add Equations A.3 and A.4, the even terms of the series cancel out and we have the following approximation for the second derivative of $f(x)$ at $x = x_i$ as

$$\left.\frac{d^2f}{dx^2}\right|_i \cong \frac{f_{i+1} + f_{i-1} - 2f_i}{\Delta x^2} \tag{A.6}$$

Equations A.5 and A.6 are second-order approximations to the first and second derivatives of $f(x)$ even though they contain only one term from the series. This is possible because when adding or subtracting Equations A.2 and A.3 every other term drops out; therefore, the second-order term for the sum is zero as shown in Equation A.4.

The finite difference method discretizes a function at a fixed number of points, thus the word *finite*, and then uses equations A.5 and A.6 to approximate the derivatives of the function, thus the word *difference*, because the first derivative is computed by subtracting the forward and backward Taylor's series expansions of $f(x)$.

Finite Differences in Two Dimensions

The finite difference formulation is easily extended into two dimensions. Consider a function $g(x, y)$, which is discretized at a series of nodes as shown in Figure A.2. Using the notation $g(x_i, y_j) = g_{i,j}$, the second-order approximations of first-partial differentials become

$$\left.\frac{\partial g}{\partial x}\right|_i \cong \frac{g_{i+1,j} - g_{i-1,j}}{2\Delta x} \tag{A.7}$$

$$\left.\frac{\partial g}{\partial y}\right|_i \cong \frac{g_{i,j+1} - g_{1,j-1}}{2\Delta y} \tag{A.8}$$

And the second-order approximations of second-partial differentials become

$$\left.\frac{\partial^2 g}{\partial x^2}\right|_i \cong \frac{g_{i+1,j} + g_{i-1,j} - 2g_{i,j}}{\Delta x^2} \tag{A.9}$$

$$\left.\frac{\partial^2 g}{\partial y^2}\right|_i \cong \frac{g_{i,j+1} + g_{i,j-1} - 2g_{i,j}}{\Delta y^2} \tag{A.10}$$

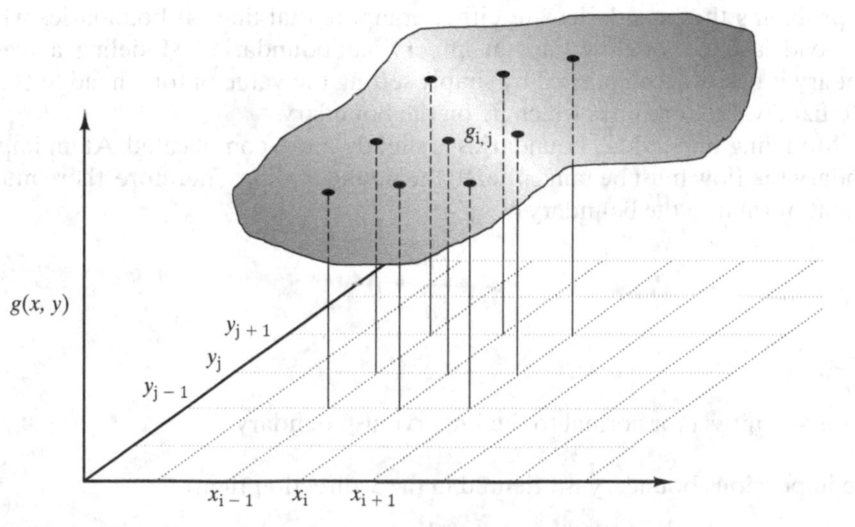

FIGURE A.2 Discretization of a two-dimensional function.

A.2 APPLICATION TO TWO-DIMENSIONAL FLOW

Discretization of the LaPlace Equation

In Section 8.1, we showed that the LaPlace equation (Equation 8.8) governs flow in two dimensions,

$$\frac{\partial^2 h}{\partial x^2} + \frac{\partial^2 h}{\partial z^2} = 0 \tag{8.8}$$

where $h(x,z)$ is the total head or potential function, in our normal x-z coordinate system.

By substituting Equations A.9 and A.10 in Equation 8.8 we can derive the following second-order approximation to the LaPlace equation

$$\frac{h_{i+1,j} + h_{i-1,j} - 2h_{i,j}}{\Delta x^2} + \frac{h_{i,j+1} + h_{i,j-1} - 2h_{i,j}}{\Delta z^2} = 0 \tag{A.11}$$

If we let $\Delta x = \Delta z$ and solve Equation A.10 for $h_{i,j}$ we have

$$h_{i,j} = \frac{h_{i+1,j} + h_{i-1,j} + h_{i,j+1} + h_{i,j-1}}{4} \tag{A.12}$$

In other words, the total head at any given node, $h_{i,j}$, is equal to the average of the head at the four adjacent nodes.

Boundary Conditions

To solve a specific two-dimensional flow problem using finite difference method, we must properly model the boundary conditions. As discussed in Chapter 8, for confined flow problems the boundaries are either equipotential lines at boundaries where the total head is fixed, or flow lines at impervious boundaries. Modeling a fixed head boundary is easily accomplished by simply setting the value of total head to the appropriate fixed value for nodes which lie on the boundary.

Modeling impervious boundaries is slightly more complicated. At an impervious boundary the flow must be tangential to the boundary line. Therefore, there must be no gradient normal to the boundary or

$$\frac{\partial h}{\partial \vec{n}} = 0 \tag{A.13}$$

where:

$\vec{n}$ = unit vector normal to the impervious boundary

If the impervious boundary is oriented in the x-direction then

$$\frac{\partial h}{\partial z} = 0 \tag{A.14}$$

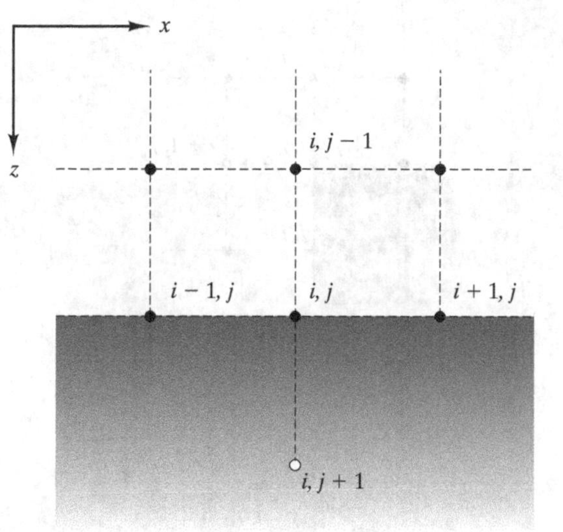

FIGURE A.3 Use of a virtual node to model an impervious boundary.

Figure A.3 illustrates an impervious boundary oriented in the x-direction. We can ensure there is no flow in the y-direction by introducing a virtual node outside the flow as shown in Figure A.3. With this virtual node introduced the body condition can be represented as

$$\frac{\partial h}{\partial z} \cong h_{i,j+1} - h_{i,j-1} = 0 \tag{A.15}$$

or

$$h_{i,j+1} = h_{i,j-1} \tag{A.16}$$

Combining Equations A.16 and A.12, we can compute $h_{i,j}$ as along the boundary as

$$h_{i,j} = \frac{h_{i+1,j} + h_{i-1,j} + 2h_{i,j-1}}{4} \tag{A.17}$$

Notice that the potential at the virtual node does not appear in Equation A.17. The equation is a function of only the potential at interior nodes.

Using the same technique, we can compute the equations for other geometries of impervious boundaries. Figure A.4 shows the equations for various impervious boundaries. The proof of these equations is left to the reader.

Assembling a Problem

To solve a specific two-dimensional flow problem, we must first assemble the series of equations describing that problem. We start by discretizing our x, z space into a series

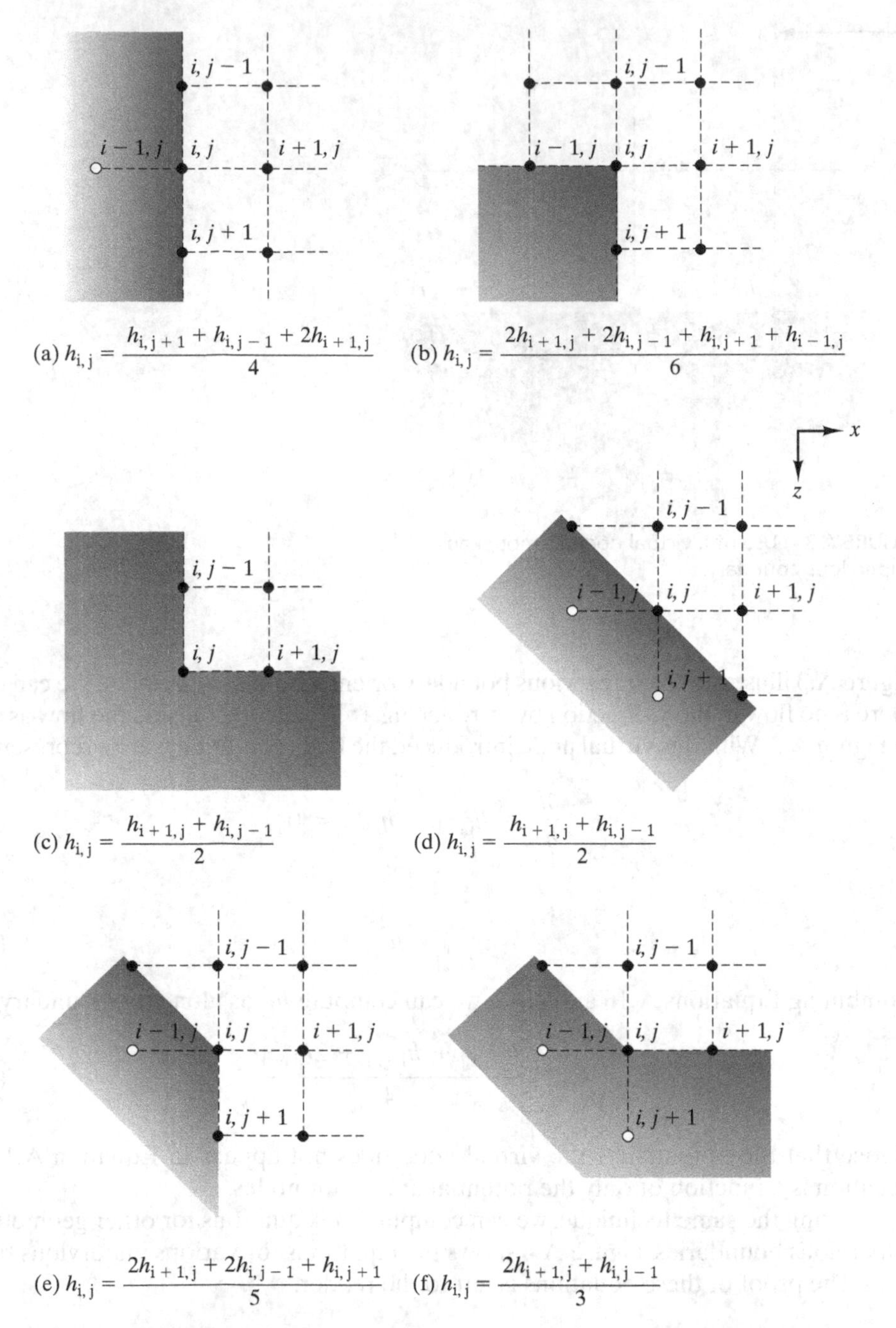

FIGURE A.4 Finite difference formulas for various different geometries of impervious boundaries.

of uniformly spaced nodes: (1,1), (1,2) ... (n, m). We then apply the boundary conditions and write the potential function for each node.

$$\begin{aligned}
b_{1,1} &= a_{1,1}h_{1,1} + a_{1,2}h_{1,2} \cdots + a_{1,n}h_{1,n} \\
b_{1,2} &= a_{2,1}h_{1,1} + a_{2,2}h_{1,2} \cdots + a_{2,n}h_{2,n} \\
&\vdots \\
b_{i,j} &= a_{i,1}h_{1,j} + a_{i,2}h_{2,j} \cdots + a_{i,m}h_{n,j} \\
&\vdots \\
b_{n,m} &= a_{n,1}h_{1,m} + a_{n,2}h_{2,m} \cdots + a_{n,m}h_{n,m}
\end{aligned}$$

This process will create a series of $n \times m$ linear equations which we can write in matrix form as

$$\begin{bmatrix} a_{1,1} & \cdots & a_{1,m} \\ \vdots & \ddots & \vdots \\ a_{n,1} & \cdots & a_{n,m} \end{bmatrix} \begin{bmatrix} h_{1,1} & \cdots & h_{1,m} \\ \vdots & \ddots & \vdots \\ h_{n,1} & \cdots & h_{n,m} \end{bmatrix} = \begin{bmatrix} b_{1,1} & \cdots & b_{1,m} \\ \vdots & \ddots & \vdots \\ b_{n,1} & \cdots & b_{n,m} \end{bmatrix} \quad \text{(A.18)}$$

or in matrix notation as

$$\mathbf{A} \times \mathbf{H} = \mathbf{B} \quad \text{(A.19)}$$

The continuity matrix, **A**, describes which nodes are adjacent to one another, the **H** matrix contains the potential values at each node, and the **B** matrix contains constants controlled by the boundary conditions.

A.3 SOLVING FINITE DIFFERENCE PROBLEMS

The system of linear equations described by Equation A.19 can directly by inverting the continuity matrix, **A**, and solving for the potential matrix, **H**.

$$\mathbf{H} = \mathbf{A}^{-1} \times \mathbf{B} \quad \text{(A.20)}$$

While the direct method is straight forward, it can be very difficult and time consuming to invert the continuity matrix.

An alternative approach is to use an iterative solution method that does not require inverting the continuity matrix. We will be using a class of iterative methods collectively known as the *relaxation methods* because they are well suited to spreadsheet calculations. Specifically, we will use the *Jacobi method* named after Carl Gustav Jacob Jacobi (1804–1851). The Jacobi method is the simplest of the relaxation methods and is adequate for our purposes.

The Jacobi method essentially consists of estimating the values of the potential function at each node and then solving the system of equations one at a time using the estimated values of the potential function. This will generate a new set of values for the potential function, which will be different than the initial estimate. The new values are then used to generate another set of values for the potential function. This process continues until successive computed values of the potential function are within an accepted tolerance.

A.4 APPLICATIONS USING SPREADSHEETS

It is relatively easy to implement the Jacobi method in a spreadsheet. This section will describe how to implement the Jacobi solution to two-dimensional flow problems using Microsoft Excel®. To illustrate the process we will use Excel to solve the problem shown in Figure A.5. This is a simple flow problem using a relatively coarse 5 by 4 grid

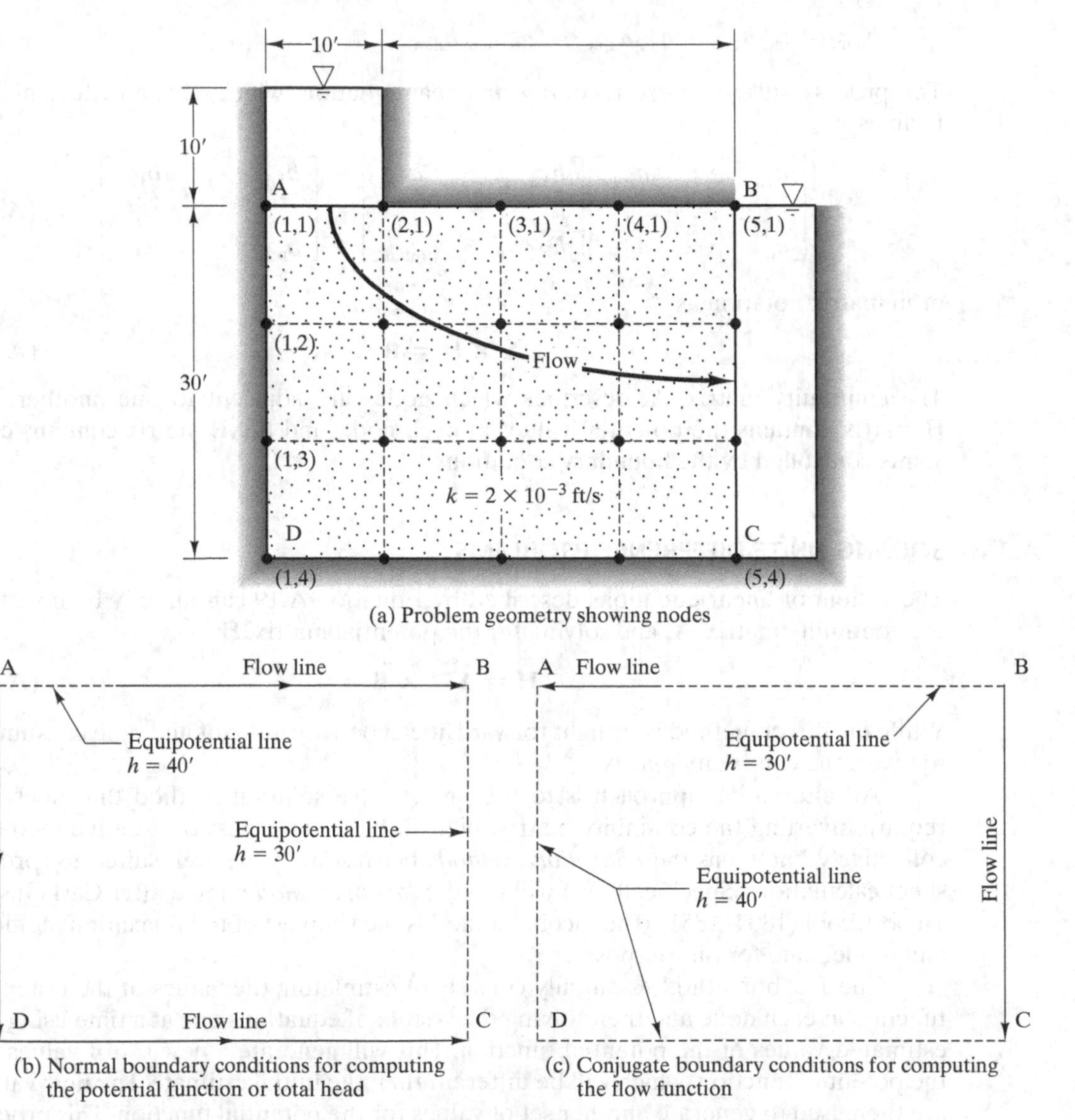

FIGURE A.5 Discretization and boundary conditions for a two-dimensional flow problem.

of 20 nodes, as shown in Figure A.5. To implement this solution follow the procedure given below.

Computing Potential

1. **Set the spreadsheet to allow iterative calculations.** First you must display the window that allows us to set the calculation mode. In Excel 2007 click the *Microsoft Office Button*, click *Excel Options*, and then click the *Formula* category. (You can also use the key strokes "<alt> F I" in Excel 2007) You should now see a window that allows you to select between manual and automatic calculations and allows you to enable iterative calculations. Set the calculation to manual and check the box which enables iterative calculations. Set the maximum number of iterations to 1 and the maximum change to 0.001.
2. **Enter the appropriate finite difference equations.** In the spreadsheet enter the appropriate finite difference equations for the head in each cell. Cell A-1 represents node (1,1), point A in Figure A.5, and cell E-4 represents node (5,4), point C in Figure A.5. Figure A.6 shows the appropriate equations for each cell. Note that it is necessary to enter the appropriate boundary condition equations for the nodes at the edges (see Figure A.4 for the appropriate equations) and the general equation, Equation A.17, for the central nodes.
3. **Perform the calculations.** We are now ready to perform the calculations. In step 1 we set the spreadsheet calculation controls to perform only one iteration at a time. This will allow us to watch the solution converging. Press the F9 key one time and the spreadsheet will perform one iteration. Figure A.7(a) shows the results for this problem after one iteration. Continue to press the F9 key to perform additional iterations. You should be able to see the values in the spreadsheet converging. Figure A.7 shows the results of the computations for various numbers of iterations. For practical problems it is easier to set the maximum number of iterations to 1000 rather than pressing the F9 key that many times.

Upper equipotential boundary

Upper flow line boundary

	A	B	C	D	E
1	40	=A1	=(B1+D1+2*C2)/4	=(C1+E1+2*D2)/4	30
2	=(A1+A3+2*B2)/4	=(B1+B3+A2+C2)/4	=(C1+C3+B2+D2)/4	=(D1+D3+C2+E2)/4	=E1
3	=(A2+A4+2*B3)/4	=(B2+B4+A3+C3)/4	=(C2+C4+B3+D3)/4	=(D2+D4+C3+E3)/4	=E2
4	=(A3+B4)/2	=(A4+C4+2*B3)/4	=(B4+D4+2*C3)/4	=(C4+E4+2*D3)/4	=E3

Flow line boundary on left and lower side

Right side equipotential boundary

FIGURE A.6 Nodal equations, for example, shown in Figure A.5.

	A	B	C	D	E
1	40	40	10.000	10.000	30
2	10.000	12.500	5.625	11.406	30
3	2.500	3.750	2.344	10.938	30
4	1.250	2.188	1.719	13.398	30

(a) 1 iteration

	A	B	C	D	E
1	40	40	34.964	32.172	30
2	36.653	36.190	34.016	31.941	30
3	34.782	34.448	33.214	31.661	30
4	34.277	33.997	32.979	31.575	30

(b) 20 iterations

	A	B	C	D	E
1	40	40	35.465	32.524	30
2	37.445	36.899	34.669	32.316	30
3	35.984	35.485	33.995	32.072	30
4	35.523	35.062	33.757	31.975	30

(c) 50 iterations

FIGURE A.7 Values of the potential or head function after various numbers of iterations.

Once the solution has converged with satisfactory accuracy, we can plot the results using the Excel contour plotting options. Figure A.8 shows the contour plot of the equipotential lines generated from this model. The contours are rather rough due to the coarseness of the grid used. Figure A.9 shows the results of a finite difference computation using a grid twice as dense as that shown in Figure A.5. More accuracy can be achieved by increasing the number of nodes even further.

Computing Flow Lines

Recall from Chapter 8 that the potential, Φ, and the flow function, Ψ, are conjugate functions, and equipotential lines and flow line intersect at right angles. We can use this property to easily compute values of the flow function instead of values of the potential

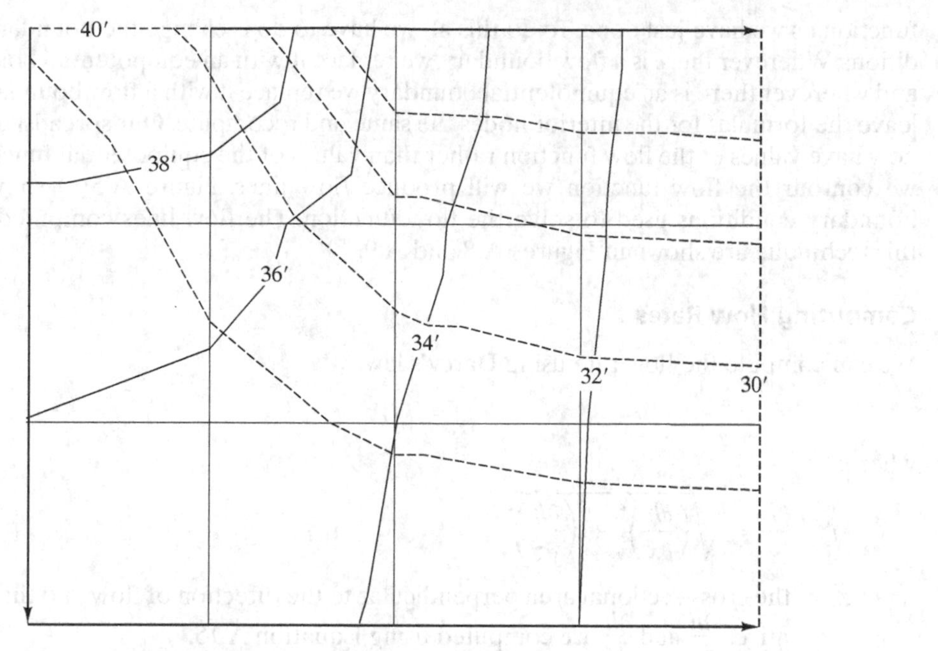

FIGURE A.8 Solution generated by the model shown in Figure A.5 showing equipotential lines (solid) and flow lines (dashed).

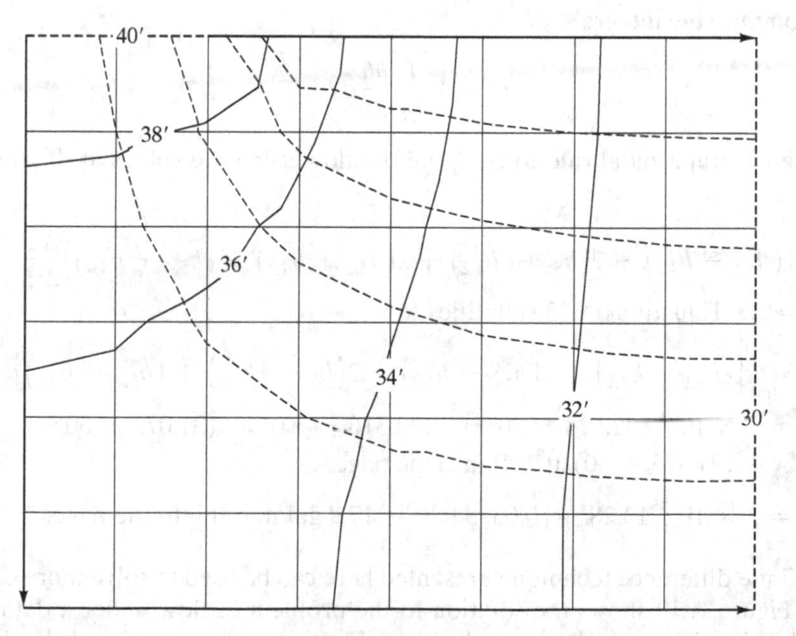

FIGURE A.9 A more accurate solution using a finite difference grid of twice the density of that shown in Figure A.5.

function as we have just done. To do this all we have to do is change the boundary conditions. Wherever there is a flow boundary, we replace it with an equipotential boundary and wherever there is an equipotential boundary we replace it with a flow boundary. We leave the formulas for the interior nodes the same and recompute. Our spreadsheet will now have values of the flow function rather than values of the equipotential function. If we contour the flow function we will produce flow lines. Figure A.5(c) shows the boundary conditions used to solve the flow function. The flow lines computed using this technique are shown in Figures A.8 and A.9.

Computing Flow Rates

We can compute the flow rate using Darcy's law

$$Q = kiA$$

where:

$$i = \frac{\partial h}{\partial n} = \sqrt{\left(\frac{\partial h}{\partial x}\right)^2 + \left(\frac{\partial h}{\partial z}\right)^2} \tag{A.21}$$

A = the cross-sectional area perpendicular to the direction of flow and the derivatives $\frac{\partial h}{\partial x}$ and $\frac{\partial h}{\partial z}$ are computed using Equation A.15.

The flow can be computed across any area that encloses the entire flow area. It is simplest to pick a location where the flow is in the x- or z-direction. In this case we can compute the flow across the area from points B to C in Figure A.5. To get the total flow we must compute the integral

$$Q = k\int_B^C \frac{\partial h}{\partial x} dz \tag{A.22}$$

We can use the trapezoidal rule to compute the flow using the values at the nodes. In this case

$$Q = k\left[\left(h_{5,1} - h_{4,1}\right) + 2\left(h_{5,2} - h_{4,2}\right) + 2\left(h_{5,3} - h_{4,3}\right) + \left(h_{5,4} - h_{4,4}\right)\right]\frac{\Delta z}{\Delta x} \tag{A.23}$$

Since $\Delta x = \Delta z$, Equations A.23 simplifies to

$$Q = k\left[\left(h_{5,1} - h_{4,1}\right) + 2\left(h_{5,2} - h_{4,2}\right) + 2\left(h_{5,3} - h_{4,3}\right) + \left(h_{5,4} - h_{4,4}\right)\right]$$

$$Q = 3 \times 10^{-3}[(32.524 - 30) + 2(32.316 - 30) + 2(31.072 - 30) + (31.975 - 30)]\text{ft}^3\text{/s/ft into the page}$$

$$Q = 3 \times 10^{-3}[13.28] = 0.0398\ \text{ft}^3\text{/s} = 17.9\ \text{gal/min/ft into the page.} \tag{A.24}$$

The finite difference technique presented here can be used to solve more complex problems. Figure A.10 shows the solution to the problem of flow under a dam with a cutoff wall. This solution was generated with an Excel spreadsheet using the techniques presented in this chapter. This technique can be used to solve many such problems

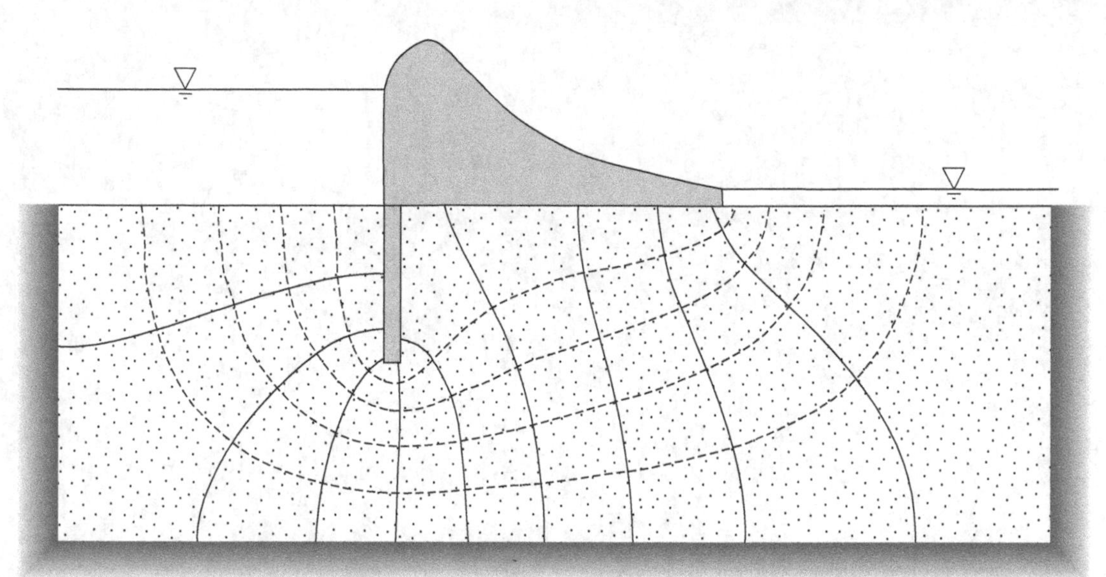

FIGURE A.10 Solution to a more complex flow problem of flow under a dam with a cutoff wall. This solution was developed using techniques presented in this appendix.

including problems with layers of soils and with anisotropic hydraulic conductivity. For more information on the finite difference technique the reader is referred to Section 4.5 of *Experimental Soil Mechanics*, Bardet (1997).

Limitations

The chief limitation of the finite difference technique is the requirement to use a uniform grid throughout the entire problem area. This makes it difficult to model irregular boundaries or to increase the density of the grid in an area where the flow is changing rapidly. Some techniques have been developed to allow for changes in grid spacing but these methods are quite complicated. For problems that require modeling complex boundaries or extremely fine meshes, the finite element technique is more appropriate, but beyond the scope of this appendix.

References

AASHTO (1993), "Recommended Practice for the Classification of Soils and Soil–Aggregate Mixtures for Highway Construction Purposes," AASHTO designation M 145–91, *Standard Specifications for Transportation Materials and Methods of Sampling and Testing*, American Association of State Highway and Transportation Officials, Washington, DC.

ABRAMSON, LEE W.; LEE, THOMAS S.; SHARMA, SUNIL; AND BOYCE, GLENN M. (2002), *Slope Stability and Stabilization Methods*, 2nd ed., John Wiley, New York.

ACHARYA, PRASANNA KUMAR (1980), *Architecture of Manasara*, Translated from the original Sanskrit, 2nd ed., Oriental Books, New Delhi.

AMERICAN GEOLOGICAL INSTITUTE (1976), *Dictionary of Geological Terms*, Revised ed., Anchor Books.

ASTM (2010), *Annual Book of ASTM Standards*, Volume 04.08—Soil and Rock, American Society for Testing and Materials, Philadelphia.

ATTERBERG, A. (1911), "Über die Physicalische Bodenuntersuchung und über die Plastizität der Tone," *Internationale Mitteilungen fur Bodenkunde*, Vol. 1 (in Swedish).

BAGUELIN, F.; JÉZÉQUEL, J.F.; AND SHIELDS, D.H. (1978), *The Pressuremeter and Foundation Engineering*, Trans Tech, Clausthal, Germany.

BARDET, JEAN–PIERRE (1997), *Experimental Soil Mechanics*, Prentice Hall, Upper Saddle River, NJ.

BERKEY, CHARLES P. (1939), "Geology in Engineering," *Frontiers in Geology*, p. 31–34, Geological Society of America.

BISHOP, ALAN W. (1955), "The Use of the Slip Circle in the Stability Analysis of Slopes," *Géotechnique*, Vol. 5, p. 7–17.

BISWAS, ASIT K. (1970), *History of Hydrology*, North Holland Publishing Company, Amsterdam.

BJERRUM, LAURITS AND FLODIN, NILS (1960), "The Development of Soil Mechanics in Sweden: 1900–1925," *Géotechnique*, Vol. 10, No. 1, p. 1–18.

BLACKALL, T.E. (1952), "A.M. Atterberg 1846–1916," *Géotechnique*, Vol. 3, p. 17–19.

BORJA, RONALDO I. AND KISHNANI, SUNIL (1992), "Movement of Slopes During Rapid and Slow Drawdown," *Stability and Performance of Slopes and Embankments–II*, Geotechnical Special Publication No. 31, Vol. I, p. 404–413, Raymond B. Seed and Ross W. Boulanger, Eds., ASCE.

BOUSSINESQ, J. (1885), *Application des Potentiels à L'Étude de L'Équilibre et du Mouvement des Solides Élastiques*, Gauthier–Villars, Paris (in French).

BRAY, JONATHAN D. (2007), "Simplified Seismic Slope Displacement Procedures," *Earthquake Geotechnical Engineering*, Chapter 14, p. 327–353. K.D. Pitilakis, Ed., Springer.

BRIAUD, JEAN–LOUIS (1992), *The Pressuremeter*, A.A. Balkema, Rotterdam.

BRIAUD, JEAN–LOUIS AND MIRAN, JEROME (1991), *The Cone Penetration Test*, Report No. FHWA–TA–91–004, Federal Highway Administration, McLean, VA.

BRUMUND, WILLIAM F. (1995), "Environmental Geotechnology—Examples of Where We Are and How We Got Here," *Geoenvironment 2000*, p. 1622–1629, Yalcin B. Acar and David E. Daniel, Eds., ASCE.

BRUMUND, WILLIAM F.; JONAS, ERNEST; AND LADD, CHARLES C. (1976), "Estimating In–Situ Maximum Past (Preconsolidation) Pressure of Saturated Clays From Results of Laboratory Consolidometer Tests," *Estimation of Consolidation Settlement*, Special Report 163, p. 4–12, Transportation Research Board, Washington, DC.

BRUNSDEN, D. (1984), "Mudslides" Chapter 9 in *Slope Instability*, D. Brunsden and D.B. Prior, Eds., John Wiley, New York.

BRUNSDEN, DENYS, AND PRIOR, DAVID B., Eds. (1984), *Slope Instability*, John Wiley, New York.

BURLAND, J.; JAMIOLKOWSKI, M.; AND VIGGIANI, C. (2003), "The Stabilisation of the Leaning Tower of Pisa," *Soils and Foundations*, Vol. 43, No. 5, pp. 63–80.

BURLAND, J.B.; JAMIOLKOWSKI, M.; AND VIGGIANI, C. (2009), "Leaning Tower of Pisa: Behaviour after Stabilization Operations," *International Journal of Geoengineering Case Histories*, Vol. 1, No. 3, pp. 156.

BURMISTER, D.M. (1962), "Physical, Stress–Strain, and Strength Responses of Granular Soils," *Symposium on Field Testing of Soils*, STP 322, p. 67–97, ASTM.

CARMAN, P.C. (1956), *Flow of Gases Through Porous Media*, Butterworths Scientific Publications, London.

CARRIER III, W. (2003), "Goodbye, Hazen; Hello, Kozeny-Carman," *Journal of Geotechnical and Geoenvironmental Engineering,* Vol. 129, No. 11, p. 1054–1056.

CARROLL, R.G., JR. (1983), "Geotextile Filter Criteria," *Engineering Fabrics in Transportation Construction*, Transportation Research Record 916, p. 46–53.

CASAGRANDE, A. (1936), "The Determination of the Pre–Consolidation Load and Its Practical Significance," Discussion D–34, *Proceedings of the First International Conference on Soil Mechanics and Foundation Engineering*, Vol. III, p. 60–64.

CASAGRANDE, ARTHUR M. (1948), "Classification and Identification of Soils," *ASCE Transactions*, Vol. 113, p. 901–991.

CASAGRANDE, A. AND FADUM, R.E. (1940), Notes *on Soil Testing for Engineering Purposes.* Harvard Soil Mechanics Series: Harvard University, Graduate School of Engineering, Cambridge, MA, p. 74.

CATERPILLAR (1993), *Caterpillar Performance Handbook*, 24th ed., Caterpillar, Inc., Peoria, IL.

CEDERGREN, HARRY R. (1989), *Seepage, Drainage, and Flow Nets,* 3rd ed., John Wiley, New York.

CGS (1992), *Canadian Foundation Engineering Manual*, 3rd ed., Canadian Geotechnical Society, BiTech, Vancouver, BC.

CHUNG, C.K. AND FINNO, R.J. (1992), "Influence of Depositional Processes on the Geotechnical Parameters of Chicago Glacial Clays," *Engineering Geology*, Vol. 32, p. 225–242.

CHURCH, HORACE K. (1981), *Excavation Handbook*, McGraw Hill, New York.

CLAYTON, C.R.I. (1990), "SPT Energy Transmission: Theory, Measurement, and Significance," *Ground Engineering*, Vol. 23, No. 10, p. 35–43.

CLAYTON, C.R.I.; MÜLLER STEINHAGEN, H.; AND POWRIE, W. (1995), "Terzaghi's Theory of Consolidation and the Discovery of Effective Stress," *Proceedings of the Institution of Civil Engineers, Geotechnical Engineering*, Vol. 113, p. 191–205. Includes an English translation of a portion of Terzaghi (1925a).

CLOUGH, G. WAYNE; SITAR, NICHOLAS; BACHUS, ROBERT C.; AND RAD, NADER SHAFII (1981), "Cemented Sands Under Static Loading," *Journal of the Geotechnical Engineering Division*, Vol. 107, No. GT6, p. 799–817, ASCE.

CODUTO, DONALD P. (2001), *Foundation Design: Principles and Practices*, 2nd ed., Prentice Hall, Upper Saddle River, NJ.

CODUTO, DONALD P. AND HUITRIC, RAYMOND (1990), "Monitoring Landfill Movements Using Precise Instruments," *Geotechnics of Waste Fills: Theory and Practice*, STP 1070, p. 358–370, ASTM.

COLLIN, A. (1846), *Recherches Expérimentales sur les Glissements Spontanés des Terrains Argileux, accompagnées de Considerations sur Quelques Principes de la Méchanique Terrestre,* Carilian–Goery and Dalmont, Paris (in French). Translated to English by W.R. Schriever as *Landslides in Clays by Alexandre Collin, 1846*. University of Toronto Press, 1956.

COULOMB, C.A. (1776), "Essai sur une application des règles de maximis et minimis à quelques problèmes de statique relatifs à l'architecture," *Mémoires de mathématique et de physique présentés à l'Académie Royale des Sciences*, Paris, Vol. 7, p. 343–382 (in French). Translated to English by Heyman, 1972.

COUSINS, B.F. (1978), "Stability Charts for Simple Earth Slopes," *Journal of the Geotechnical Engineering Division*, Vol. 104, No. GT2, p. 267–282, ASCE.

CRUDEN, DAVID M. AND VARNES, DAVID J. (1996), "Landslide Types and Processes," Chapter 3 in *Landslides: Investigation and Mitigation*, A. Keith Turner and Robert L. Schuster, Eds., National Academy Press, Washington, DC.

DANIEL, DAVID E. AND BENSON, CRAIG H. (1990), "Water Content-Density Criteria for Compacted Soil Liners," *Journal of Geotechnical Engineering,* Vol. 116, No. 12, p. 1811–1830, ASCE.

DARCY, H. (1856), *Les Fontaines Publiques de la Ville de Dijon* (The Water Supply of the City of Dijon), Dalmont, Paris (in French). Translated into English by Patricia Bobeck and published by Kendall-Hunt, 2004

DE MELLO, V. (1971), "The Standard Penetration Test—A State-of-the-Art Report," *Fourth Pan–American Conference on Soil Mechanics and Foundation Engineering*, Vol. 1, p. 1–86.

DE RUITER, J. (1981), "Current Penetrometer Practice," *Cone Penetration Testing and Experience*, p. 1–48, ASCE.

DIBIAGIO, E. AND FLAATE, K. (2000), *Ralph B. Peck: Engineer, Educator, A Man of Judgement.* Norwegian Geotechnical Institute, Oslo.

DOBRIN, MILTON B. (1988), *Introduction to Geophysical Prospecting*, McGraw Hill, New York.

DOWDING, CHARLES H. (1979), *Site Characterization & Exploration*, ASCE.

DOWDING, CHARLES H. (1996), *Construction Vibrations*, Prentice Hall, Upper Saddle River, NJ.

DRISCOLL, FLETCHER G. (1986), *Groundwater and Wells*, 2nd ed., Johnson Filtration Systems, Inc., St. Paul, MN.

DUNCAN, J. MICHAEL (1996), "Soil Slope Stability Analysis," Chapter 13 in *Landslides: Investigation and Mitigation*, Special Report 247, A. Keith Turner and Robert L. Schuster, Eds., Transportation Research Board, Washington, DC.

DUNCAN, J. MICHAEL; CLOUGH, G. WAYNE; AND EBELING, ROBERT M. (1990), "Behavior and Design of Gravity Earth Retaining Structures," *Design and Performance of Earth Retaining Structures*, Philip C. Lambe and Lawrence A. Hansen, Eds., p. 251–277, ASCE.

DUNCAN, J. MICHAEL AND STARK, TIMOTHY D. (1992), "Soil Strengths from Back Analysis of Slope Failures," *Stability and Performance of Slopes and Embankments–II*, Geotechnical Special Publication No. 31, Vol I, p. 890–904, R.B. Seed and R.W. Boulanger, Eds., ASCE.

DUNCAN, JAMES M.; WRIGHT, STEPHEN G; AND WONG, KAI S. (1990), "Slope Stability During Rapid Drawdown," Chapter 12 in *H. Bolton Seed Memorial Symposium Proceedings*, J. Michael Duncan, Ed., Vol. 2, BiTech, Vancouver, BC.

DUNN, I.S.; ANDERSON, L.R.; AND KIEFER, F.W. (1980), *Fundamentals of Geotechnical Analysis*, John Wiley, New York.

FAIR, G. AND HATCH, L. (1933), "Fundamental Factors Governing the Stream-line Flow of Water Through Sand," *Journal of the American Water Works Association*, Vol. 25, p. 1551–1565.

FELLENIUS, WOLMAR (1927), *Erdstatische Berechnungen mit Reibung and Kohaesion*, Ernst, Berlin.

FELLENIUS, WOLMAR (1936), "Calculation of Stability of Earth Dams," *Transactions, Second Congress on Large Dams*, Washington, Vol. 4, p. 445–462.

FETTER, C.W. (1993), *Contaminant Hydrogeology*, MacMillan, New York.

FLODIN, NILS AND BROMS, BENGT (1981), "Historical Development of Civil Engineering in Soft Clay," Chapter 1 in Soft Clay Engineering, E.W Brand and R.P. Brenner, Eds., Elsevier, Amsterdam.

FORCHHEIMER, PHILIP (1914), Hydraulik, p. 26, 494–495, Leipzig (in German).

FORCHHEIMER, P. (1917), "Zur Grundwasserbewegung nach isothermischen Kurvenscharen" (Concerning Groundwater Movement in Accordance with Isothermal Families of Curves), *Sitzungsberichte/Akademie der Wissenschaften in Wien, IIa*, Vol. 126, p. 409–440 (in German).

FOX, PATRICK J. (1995), "Consolidation and Settlement Analysis," Chapter 18 in *The Civil Engineering Handbook*, W.F. Chen, Ed., CRC Press.

FREDLUND, D.G.; KRAHN, J.; AND PUFAHL, D.E. (1981), "The Relationship Between Limit Equilibrium Slope Stability Methods," *Proceedings, International Conference on Soil Mechanics and Foundation Engineering*, p. 409–416.

FREDLUND, D.G. AND RAHARDJO, H. (1993), *Soil Mechanics for Unsaturated Soils*, John Wiley, New York.

FREEZE, R. ALLAN AND CHERRY, JOHN A. (1979), *Groundwater*, Prentice Hall, Inc., Englewood,Cliffs, NJ.

FRONTARD, J. (1914), "Notice sur l'accident de la digue de Charmes," *Annales Ponts et Chaussées*, 9th Series, Vol. 23, p. 173–280 (in French).

GOODMAN, RICHARD E. (1990), "Soils Versus Rocks as Engineering Materials," *H. Bolton Seed Memorial Symposium Proceedings*, J. Michael Duncan, Ed., Vol. 2, p. 111–133, BiTech, Vancouver, BC.

GOODMAN, RICHARD E. (1993), Engineering Geology: *Rock in Engineering Construction*, John Wiley, New York.

HAMBLIN, W. KENNETH AND HOWARD, JAMES D. (1975), *Exercises in Physical Geology*, 4th ed., Burgess Publishing Co., Minneapolis.

HAMILTON, DOUGLAS H. AND MEEHAN, RICHARD L. (1992), "Cause of the 1985 Ross Store Explosion and Other Gas Ventings, Fairfax District, Los Angeles," *Engineering Geology Practice in Southern California*, Special Publication No. 4, Association of Engineering Geologists.

HANDY, RICHARD L. (1980), "Realism in Site Exploration: Past, Present, Future, and Then Some—All Inclusive," *Site Exploration on Soft Ground Using In-Situ Techniques*, p. 239–248, Report No. FHWA–TS–80–202, Federal Highway Administration, Washington, DC.

HARDER, LESLIE F. AND SEED, H. BOLTON (1986), *Determination of Penetration Resistance for Coarse-Grained Soils Using the Becker Hammer Drill*, Report No. UCB/EERC–86/06, Earthquake Engineering Research Center, Richmond, CA.

HARR, MILTON E. (1962), *Groundwater and Seepage*, McGraw Hill, New York.

HAUSMANN, MANFRED R. (1992), "Slope Remediation," *Stability and Performance of Slopes and Embankments–II*, Geotechnical Special Publication No. 31, Vol II, p. 1274–1317, R.B. Seed and R.W. Boulanger, Eds., ASCE.

HAZEN, A. (1911), "Discussion of 'Dams on Sand Foundations' by A.C. Koenig," *ASCE Transactions*, Vol. 73, p. 199.

HEAD, K.H. (1982), *Manual of Soil Laboratory Testing*, Vol. 2, Pentech, London.

HEYMAN, JACQUES (1972), *Coulomb's Memoir on Statics*, Cambridge University Press.

HILF, JACK W. (1991), "Compacted Fill," Chapter 8 in *Foundation Engineering Handbook*, 2nd ed., Hsai–Yang Fang, Ed., Van Nostrand Reinhold, New York.

HIRIART, FERNANDO AND MARSAL, RAUL J. (1969), "The Subsidence of Mexico City," *Nabor Carrillo; El Hundimiento de la Ciudad de Mexico y Proyecto Texcoco* (Nabor Carrillo; The Subsidence of Mexico City and Texcoco Project), p. 109–147, in English and Spanish.

HOLTZ, R.D. AND BROMS, B.B. (1972), "Long-Term Loading Tests at Skå-Edeby, Sweden," *Proceedings of the ASCE Specialty Conference on Performance of Earth and Earth-Supported Structures*, Vol. I, Part 1, p. 435–464, ASCE.

HOLTZ, ROBERT D. AND KOVACS, WILLIAM D. (1981), *An Introduction to Geotechnical Engineering*, Prentice Hall, Upper Saddle River, NJ.

HORVATH, JOHN S. (1995), *Geofoam Geosynthetic*, Horvath Engineering, Scarsdale, NY.

HOUGH, B.K. (1969), *Basic Soils Engineering*, 2nd ed., The Ronald Press, New York.

HVORSLEV, M. JUUL (1949), Subsurface Exploration and Sampling of Soils for Civil Engineering Purposes, ASCE.

HYNES–GRIFFIN, MARY ELLEN AND FRANKLIN, ARLEY G. (1984), *Rationalizing the Seismic Coefficient Method*, Miscellaneous paper GL–84–13, Waterways Experiment Station, U.S. Army Corps of Engineers.

IFAI (1997), *Geotechnical Fabrics Report Specifier's Guide*, Industrial Fabrics Association, International, St. Paul, MN (published annually).

INDEPENDENT PANEL TO REVIEW CAUSE OF TETON DAM FAILURE, UNITED STATES, AND IDAHO (1976), *Report to U.S. Department of the Interior and State of Idaho on Failure of Teton Dam,* The Panel, Idaho Falls, ID, p. 578.

JANBU, N. (1957), "Earth Pressure and Bearing Capacity Calculations by Generalized Procedure of Slices," *Proceedings, Fourth International Conference on Soil Mechanics and Foundation Engineering*, London, Vol. 2, p. 17–26.

JANBU, N. (1973), "Slope Stability Computations," *Embankment Dam Engineering—Casagrande Volume*, p. 47–86, John Wiley, New York.

JOHNSON, A.M. AND RODINE, J.R. (1984), "Debris Flow," Chapter 8 in *Slope Instability*, D. Brunsden and D.B. Prior, Eds., John Wiley, New York.

JOHNSON, A.W. AND SALLBERG, J.R. (1960), *Factors that Influence Field Compaction of Soils*, Bulletin 272, Highway Research Board, Washington, DC.

KALISER, BRUCE N. AND FLEMING, ROBERT W. (1986), "The 1983 Landslide Dam at Thistle, Utah," *Landslide Dams: Processes, Risk, and Mitigation*, Geotechnical Special Publication No. 3, p. 59–83, Robert L. Schuster, Ed., ASCE.

KASHEF, ABDEL–AZIZ (1986), *Groundwater Engineering*, McGraw Hill, New York.

KERISEL, JEAN (1987), *Down to Earth; Foundations Past and Present: The Invisible Art of the Builder*, A.A. Balkema, Rotterdam.

KIM, JAE Y.; EDIL, TUNCER B.; AND PARK, JAE K. (1997), "Effective Porosity and Seepage Velocity in Column Tests on Compacted Clay," *Journal of Geotechnical and Geoenvironmental Engineering*, Vol. 123, No. 12, p. 1135–1142.

KITCH, WILLIAM A. (2009), *Geotechnical Engineering Lab Manual,* Bent Tree Press, Reno, NV.

KOERNER, ROBERT M. (1998), *Designing with Geosynthetics*, 4th ed., Prentice Hall, Upper Saddle River, NJ.

KOVACS, W.D.; SALOMONE, L.A.; AND YODEL, F.Y. (1981), *Energy Measurements in the Standard Penetration Test*, Building Science Series 135, National Bureau of Standards, Washington, DC.

KOZENY, J. (1927), "Ueber kapillare Leitung des Wassers im Boden," *Akademie der Wissenschaften in Wien,* Vol. 136, No. 2a, pp. 271–306 (in German).

KRAMER, STEVEN L. (1996), *Geotechnical Earthquake Engineering*, Prentice Hall, Upper Saddle River, NJ.

KULHAWY, F.H. AND MAYNE, P.W. (1990), *Manual on Estimating Soil Properties for Foundation Design*, Report EL–6800, Electric Power Research Institute, Palo Alto, CA.

KULHAWY, FRED H.; ROTH, MARY JOEL S.; AND GRIGORIU, MIRCEA D. (1991), "Some Statistical Evaluations of Geotechnical Properties," *Proceedings of ICASP6,*

Sixth International Conference on Applications of Statistics and Probability in Civil Engineering, Vol. 2, p. 707–712, L. Esteva and S.E. Ruiz, Eds.

KULHAWY, F.H.; TRAUTMANN, C.H.; BEECH, J.F.; O'ROURKE, T.D.; MCGUIRE, W.; WOOD, W.A.; AND CAPANO, C. (1983), *Transmission Line Structure Foundations for Uplift-Compression Loading*, Report No. EL–2870, Electric Power Research Institute, Palo Alto, CA.

LADD, C.C. AND LUSCHER, U. (1965), "Engineering Properties of Soils Underlying the MIT Campus," *Research Report R65–68, Soils Publication 185*, Department of Civil Engineering, Massachusetts Institute of Technology.

LAMBE, T.W. (1958), "The Engineering Behavior of Compacted Clay," *Journal of the Soil Mechanics and Foundations Division*, Vol. 84, No. SM2, p. 16551–165535, ASCE.

LAMBE, T. WILLIAM AND WHITMAN, ROBERT V. (1969), *Soil Mechanics*, John Wiley, New York.

LEDESMA, JOSÉ (1936), "The National Theater Building and Efforts Made to Prevent Its Further Sinking," *Proceedings, International Conference on Soil Mechanics and Foundation Engineering*, Vol. I, p. 119–123.

LEE, KENNETH L. (1965), Triaxial Compressive Strength of Saturated Sands Under Seismic Loading Conditions, PhD Dissertation, University of California, Berkeley.

LEGGET, ROBERT F. AND HATHEWAY, ALLEN W. (1988), *Geology and Engineering*, 3rd ed., McGraw Hill, New York.

LEONARDS, G.A. AND GIRAULT, P. (1961), "A Study of the One-Dimensional Consolidation Test," *Proceedings of the Fifth International Conference on Soil Mechanics and Foundation Engineering*, Vol. I, p. 116–130.

LIAO, S.S.C. AND WHITMAN, R.V. (1986), "Overburden Correction Factors for SPT in Sand," *Journal of Geotechnical Engineering*, Vol. 112, No. 3, p. 373–377, ASCE.

LOUDON, A. (1952), "The Computation of Permeability from Simple Soil Tests," *Géotechnique,* Vol. 3, No. 4, p. 165–183.

LOWE, J.; ZACCHEO, P.F.; AND FELDMAN, H.S. (1964), "Consolidation Testing with Back Pressure," *Journal of the Soil Mechanics and Foundations Division*, Vol. 90, No. SM5, p. 69–86, ASCE.

MARCHETTI, SILVANO (1980), "In-Situ Tests by Flat Dilatometer," *Journal of the Geotechnical Engineering Division*, Vol. 106, No. GT3, p. 299–321 (also see discussions, Vol. 107, No. GT8, p. 831–837), ASCE.

MAYNE, PAUL W. AND KULHAWY, FRED H. (1982), "K_0-OCR Relationships in Soil," *Journal of the Geotechnical Engineering Division*, Vol. 108, No. SM5, p. 63–91, ASCE.

MEANS, R.E. AND PARCHER, J.V. (1963), *Physical Properties of Soils*, Charles E. Merrill Books, Inc.

MEIGH, A.C. (1987), *Cone Penetration Testing: Methods and Interpretation*, Butterworths, London.

MESRI, G. AND CHOI, Y.K. (1985), "Settlement Analysis of Embankments on Soft Clays," *Journal of Geotechnical Engineering,* Vol. 111, No. 4, p. 441–464.

MESRI, GHOLAMREZA; LO, DOMINIC O. KWAN; AND FENG, TAO–WEI (1994), "Settlement of Embankments on Soft Clays," *Vertical and Horizontal Deformations of Foundations and Embankments*, Vol. 1, p. 8–56, ASCE.

MEYERHOF, G.G. (1955), "Influence of Roughness of Base and Ground-Water Conditions on the Ultimate Bearing Capacity of Foundations," *Géotechnique*, Vol. 5, p. 227–242 (reprinted in Meyerhof, 1982).

MITCHELL, JAMES K. (1978), "In-Situ Techniques for Site Characterization," *Site Characterization and Exploration*, p. 107–129, C.H. Dowding, Ed., ASCE.

MITCHELL, JAMES K. AND HOOPER, D.R. (1965), "Permeability of Compacted Clay," *Journal of Soil Mechanics and Foundation Engineering*, Vol. 94, No. 4, p. 41–65, ASCE.

MITCHELL, JAMES K.; SEED, RAYMOND B.; AND SEED, H. BOLTON (1990), "Stability Considerations in the Design and Construction of Lined Waste Depositories," *Geotechnics of Waste Fills: Theory and Practice*, p. 209–224, ASTM.

MITCHELL, JAMES K. AND SOGA, KENICHI (2005), *Fundamentals of Soil Behavior*, 3rd ed., John Wiley, New York.

MORGENSTERN, N.R. AND PRICE, V.E. (1965), "The Analysis of the Stability of General Slip Surfaces," *Géotechnique*, Vol. 15, No. 1, p. 79–93.

MÜLLER–BRESLAU, H. (1906), "Erddruk auf Stützmauern," Alfred Kröner Verlag, Stuttgart (in German).

NEGUSSEY, DAWIT (1997), *Properties and Applications of Geofoam*, Society of the Plastics Industry.

NEWMARK, NATHAN M. (1935), *Simplified Computation of Vertical Pressures in Elastic Foundations*, Engineering Experiment Station Circular No. 24, University of Illinois, Urbana.

NEWMARK, N. (1965), "Effects of Earthquakes on Dams and Embankments," *Géotechnique*, Vol. 15, No. 2, p. 139–160.

NIV (1984), *The Holy Bible, New International Version*, International Bible Society, Colorado Springs, CO.

NIXON, IVAN K. (1982), "Standard Penetration Test State-of-the-Art Report," *Second European Symposium on Penetration Testing* (ESOPT II), Vol. 1, p. 3–24, A. Verruijt, F.L. Beringen, and E.H. de Leeuw, Eds. Amsterdam.

NOORANY, IRAJ; SWEET, JOEL A.; AND SMITH, IAN M. (1992), "Deformation of Fill Slopes Caused by Wetting," *Stability and Performance of Slopes and Embankments II*, R.B. Seed and R.W. Boulanger, Eds., p. 1244–1257, ASCE.

NOORANY, IRAJ AND STANLEY, JEFFREY (1994), "Settlement of Compacted Fills Caused by Wetting," *Vertical and Horizontal Deformations of Foundations and Embankments*, p. 1516–1530, A.T. Yeung and G.Y. Felio, Eds., ASCE.

OLSON, R. (1977), *"Consolidation Under Time-Dependent Loading,"* Journal of the Geotechnical Engineering Division, Vol. 103, No. 1, p. 55–60.

O'NEIL, M. AND REESE, L. (1999), *Drilled Shafts: Construction Procedures and Design Methods*, FHWA, Washington, DC, p. 758.

O'ROURKE, T.D. AND JONES, C.J.F.P. (1990), "Overview of Earth Retention Systems: 1970–1990," *Design and Performance of Earth Retaining Structures*, Geotechnical Special Publication No. 25, p. 22–51, P.C. Lambe and L.A. Hansen, Eds., ASCE Assessment.

PECK, RALPH B. (1969), "Advantages and Limitations of the Observational Method in Applied Soil Mechanics," *Géotechnique*, Vol. 19, p. 171–187 (reprinted in

Dunnicliff and Deere (1984), *Judgement in Geotechnical Engineering*, p. 122–127).

PETTERSON, KNUT E. (1955), "The Early History of Circular Sliding Surfaces," *Géotechnique*, Vol. 5, p. 275–296.

PLAFKER, G. AND ERICKSEN, G.E. (1978), "Nevados Huascarán Avalanches, Peru," *Rockslides and Avalanches, 1: Natural Phenomena*, B. Voight, Ed., p. 277–314, Elsevier, Amsterdam.

POULOS, H.G. AND DAVIS, E.H. (1974), *Elastic Solutions for Soil and Rock Mechanics*, John Wiley, New York.

POWERS, J. PATRICK (1992), *Construction Dewatering: New Methods and Applications*, 2nd ed., John Wiley, New York.

PREZZI, M. AND BASU, P. (2005), "Overview of Construction and Design of Auger Cast-in-Place and Drilled Displacement Piles," *Proceedings of the 30th Annual Conference on Deep Foundations*, Chicago, IL, p. 497–512.

PRIEST, STEPHEN D. (1993), *Discontinuity Analysis for Rock Engineering*, Chapman and Hall, London.

PROCTOR, R.R. (1933), "Fundamental Principles of Soil Compaction" p. 245–248; "Description of Field and Laboratory Procedures," p. 286–289; "Field and Laboratory Verification of Soil Suitability," p. 348–351; and "New Principles Applied to Actual Dam-Building," p. 372–376; *Engineering News Record*, Vol. 111, No. 9.

RANKINE, W.J.M. (1857), "On the Stability of Loose Earth," *Philosophical Transactions of the Royal Society*, Vol. 147, London.

REDDI, LAKSHMI N. AND BONALA, MOHAN V.S. (1997), "Analytical Solution for Fine Particle Accumulation in Soil Filters," *Journal of Geotechnical and Geoenvironmental Engineering*, Vol. 123, No. 12, p. 1143–1152.

ROBERTSON, P.K. AND CAMPANELLA, R.G. (1983), "Interpretation of Cone Penetration Tests: Parts 1 and 2," *Canadian Geotechnical Journal*, Vol. 20, p. 718–745.

ROBERTSON, P.K. AND CAMPANELLA, R.G. (1989), *Guidelines for Geotechnical Design Using the Cone Penetrometer Test and CPT With Pore Pressure Measurement,* 4th ed., Hogentogler & Co., Columbia, MD.

ROBERTS, DON V. AND DARRAGH, ROBERT D. (1962), "Areal Fill Settlements and Building Foundation Behavior at the San Francisco Airport," *Field Testing of Soils*, Special Publication 322, p. 211–230, ASTM.

ROBERTSON, P.K. AND ROBERTSON, K.L. (2006), *Guide to Cone Penetration Testing and its Application to Geotechnical Engineering,* Gregg Drilling & Testing, Inc., p. 108.

ROGERS, J. DAVID (1992a), "Mechanisms of Seismically-Induced Slope Movements," Unpublished manuscript presented at Transportation Research Board meeting.

ROGERS, J. DAVID (1992b), "Recent Developments in Landslide Mitigation Techniques," Chapter 10 in *Landslides/Landslide Mitigation*, Reviews in Engineering Geology, Vol. IX, Geological Society of America.

ROGERS, J. DAVID (1992c), "Long Term Behavior of Urban Fill Embankments," *Stability and Performance of Slopes and Embankments II*, R.B. Seed and R.W. Boulanger, Eds., p. 1258–1273, ASCE.

ROGERS, J. DAVID (1995), "A Man, A Dam, and A Disaster: Mulholland and the St. Francis Dam," *The St. Francis Dam Disaster Revisited*, Doyce B. Nunnis, Jr., Ed., p. 1–109, Historical Society of Southern California, Los Angeles.

SALGADO, R. (2008), *The Engineering of Foundations.* McGraw Hill, Boston, p. 882.

SANTAMARINA, J. CARLOS (1997), "Cohesive Soil: A Dangerous Oxymoron," *Electronic Journal of Geotechnical Engineering*, published at web site geotech.civen.okstate.edu/magazine/oxymoron/dangeoxi.htm.

SARMA, S.K. (1973), "Stability Analysis of Embankments and Slopes," *Géotechnique*, Vol. 23, No. 3, p. 423–433.

SCHMERTMANN, JOHN H. (1955), "The Undisturbed Consolidation Behavior of Clay," *Transactions of the American Society of Civil Engineers*, Vol. 120, p. 1201–1233, ASCE.

SCHMERTMANN, JOHN H. (1978), *Guidelines for Cone Penetration Test: Performance and Design*, Report FHWA–TS–78–209, Federal Highway Administration, Washington, DC.

SCHMERTMANN, J.H. (1986a), "Suggested Method for Performing the Flat Dilatometer Test," *Geotechnical Testing Journal*, Vol. 9, No. 2, p. 93–101.

SCHMERTMANN, JOHN H. (1986b), "Dilatometer to Compute Foundation Settlement," *Use of In-Situ Tests in Geotechnical Engineering*, p. 303–319, Samuel P. Clemence, Ed., ASCE.

SCHMERTMANN, JOHN H. (1988a), "Dilatometers Settle In," *Civil Engineering*, Vol. 58, No. 3, p. 68–70, March 1988, ASCE.

SCHMERTMANN, JOHN H. (1988b), *Guidelines for Using the CPT, CPTU and Marchetti DMT for Geotechnical Design*, Vol. I–IV, Federal Highway Administration, Washington, DC.

SCHUSTER, ROBERT L. (1996), "Socioeconomic Significance of Landslides," Chapter 2 in *Landslides: Investigation and Mitigation*, A. Keith Turner and Robert L. Schuster, Eds., Transportation Research Board.

SCHUYLER, JAMES DIX (1905), *Reservoirs for Irrigation, Water-Power, and Domestic Water-Supply,* John Wiley, New York.

SCS (1986), *Guide for Determining the Gradation of Sand and Gravel Filters*, Soil Mechanics Note No. 1, 210–VI, U.S. Dept. of Agriculture, Soil Conservation Service, Lincoln, NE.

SCULLIN, C. MICHAEL (1983), *Excavation and Grading Code Administration, Inspection, and Enforcement*, Prentice Hall, Upper Saddle River, NJ.

SEED, H. BOLTON AND CHAN, C.K. (1959), "The Structure and Strength Characteristics of Compacted Clays," *Journal of the Soil Mechanics and Foundations Division,* Vol. 85, No. SM5, p. 87–128, ASCE.

SEED, H. BOLTON; TOKIMATSU, K.; HARDER, L.F.; AND CHUNG, RILEY M. (1985), "Influence of SPT Procedures in Soil Liquefaction Resistance Evaluations," ASCE *Journal of Geotechnical Engineering*, Vol. 111, No. 12, p. 1425–1445.

SEED, H. BOLTON AND WILSON, STANLEY D. (1964), *The Turnagain Heights Landslide in Anchorage, Alaska*, Department of Civil Engineering, University of California, Berkeley.

SHARMA, SUNIL AND MOUDUD, ABDUL (1992), "Interactive Slope Analysis Using Spencer's Method," *Stability and Performance of Slopes and Embankments–II*,

Geotechnical Special Publication No. 31, p. 506–520, R.B. Seed and R.W. Boulanger, Eds., ASCE.

SHERARD, J.L.; DUNNIGAN, L.P.; AND TALBOT, J.R. (1984a), Basic Properties of Sand and Gravel Filters," *Journal of Geotechnical Engineering*, Vol. 110, No. 6, p. 684–700, ASCE.

SHERARD, J.L.; DUNNIGAN, L.P.; AND TALBOT, J.R. (1984b), Filters for Silts and Clays," *Journal of Geotechnical Engineering*, Vol. 110, No. 6, p. 701–718, ASCE.

SHERARD, J.L. AND DUNNIGAN, L.P. (1985), "Filters and Leakage Control in Embankment Dams," *Seepage and Leakage Control in Embankment Dams*, p. 1–29, ASCE.

SHERARD, J.L. AND DUNNIGAN, L.P. (1989), "Critical Filters for Impervious Soils," *Journal of Geotechnical Engineering*, Vol. 115, No. 7, p. 927–947, ASCE.

SHUIRMAN, GERARD AND SLOSSON, JAMES E. (1992), *Forensic Engineering: Environmental Case Histories for Civil Engineers and Geologists,* Academic Press, San Diego.

SICHART, W. AND KYRIELEIS, W. (1930), *Grundwasser Absekungen bei Fundierungsarbeiten* (in German), as quoted in Powers (1992).

SIMMONS, C. (2006), "Henry Darcy (1803–1858): Immortalised by His Scientific Legacy," *International Symposium: Aquifers Systems Management*, Dijon, France, #119.

SKEMPTON, A.W. (1949), "Alexandre Collin, A Note on His Pioneer Work in Soil Mechanics," *Géotechnique*, Vol. 1, No. 4, p. 216–221.

SKEMPTON, A.W. (1954), "The Pore-Pressure Coefficients *A* and *B*," *Géotechnique*, Vol. IV, p. 143–147.

SKEMPTON, A.W. (1960), "Significance of Terzaghi's Concept of Effective Stress," *From Theory to Practice in Soil Mechanics*, p. 42–53.

SKEMPTON, A.W. (1986), "Standard Penetration Test Procedures and the Effects in Sands of Overburden Pressure, Relative Density, Particle Size, Aging, and Overconsolidation," *Géotechnique*, Vol. 36, No. 3, p. 425–447.

SMALLEY, I. (1992). "The Teton Dam: Rhyolite Foundation + Loess Core = Disaster," *Geology Today*, Vol. 8, No. 1, 19–22.

SMITH, RONALD E., Ed. (1987), *Foundations and Excavations in Decomposed Rock of the Piedmont Province*, Geotechnical Special Publication No. 9, ASCE.

SOIL SURVEY STAFF (1975), *Soil Taxonomy*, Agriculture Handbook No, 436, Soil Conservation Service, U.S. Department of Agriculture, Washington, DC.

SOLAVA, S. AND DELATTE, N. (2003), "Lessons from the Failure of the Teton Dam," *ASCE Conference Proceedings*, ASCE, San Diego, pp. 178–189.

SOOY SMITH, W. (1892), "The Building Problem in Chicago From an Engineering Standpoint," *The Technograph*, No. 6, p. 9–19, University of Illinois.

SOUTHWEST BUILDER AND CONTRACTOR (1936), "Tamping Feet of a Flock of Sheep Gave Idea for Sheepsfoot Roller," Aug 7, 1936, p. 13.

SOWERS, GEORGE F. (1979), *Introductory Soil Mechanics and Foundations: Geotechnical Engineering*, 4th ed., Macmillan, New York.

SOWERS, GEORGE F. (1992), "Natural Landslides," *Stability and Performance of Slopes and Embankments–II*, Geotechnical Special Publication No. 31, Vol. I, p. 804–833, R.B. Seed and R.W. Boulanger, Eds., ASCE.

SPANN, STEVE W. (1986), "Available Compaction Equipment," *Earthmoving and Heavy Equipment*, Garold D. Oberlender, Ed., p. 10–13, ASCE.

SPENCER, E. (1967), "A Method of Analysis of the Stability of Embankments Assuming Parallel Interslice Forces," *Géotechnique*, Vol. 17, No. 1, p. 11–26.

SPENCER, E. (1973), "The Thrust Line Criterion in Embankment Stability Anaysis," *Géotechnique*, Vol. 23, No. 1, p. 85–100.

STAMATOPOULOS, ARIS C. AND KOTZIAS, PANAGHIOTIS C. (1985), *Soil Improvement by Preloading*, John Wiley, New York.

STARK, TIMOTHY D. AND DUNCAN, J. MICHAEL (1991), "Mechanisms of Strength Loss in Stiff Clays," *Journal of Geotechnical Engineering*, Vol. 117, No. 1, p. 139–154, ASCE.

STATENS JÄRNVÄGARS GEOTEKNISKA KOMMISSION: SLUTBETÄNKANDE (1922), (The State Railways Geotechnical Commission: Final Report), Stockholm (in Swedish).

STEWART, JONATHAN P.; BRAY, JONATHAN D.; MCMAHON, DAVID J.; AND KROPP, ALAN L. (1995), "Seismic Performance of Hillside Fills," *Landslides Under Static and Dynamic Conditions—Analysis, Monitoring, and Mitigation*, Geotechnical Special Publication No. 52, p. 76–95, David K. Keefer and Carlton L. Ho, Eds., ASCE.

TAYLOR, DONALD W. (1948), *Fundamentals of Soil Mechanics*, John Wiley, New York.

TELFORD, THOMAS (1830), "Inland Navigation," in *Edinburgh Encyclopedia*, Vol. 15, p. 209–315.

TERZAGHI, KARL (1920), "New Facts About Surface-Friction," *Physical Review*, Vol. 16, No. 1, p. 54–61 (reprinted in *From Theory to Practice in Soil Mechanics*, John Wiley, New York, 1960).

TERZAGHI, KARL (1921), "Die physikalischen Grundlagen der technischgeologischen Gutachtens," *Österreichischer Ingenieur und Architekten-Verein Zeitschrift*, Vol. 73, No. 36/37, p. 237–241 (in German).

TERZAGHI, KARL (1923a), "Die Beziehungen zwischen Elastizität und Innendruck," *Akademic der Wissenschaften in Wien. Sitzungsberichte.* Mathematisch-naturwissenschaftliche Klasse, Part IIa, Vol. 132, No. 3⁄4, p. 105–124 (in German).

TERZAGHI, KARL (1923b), "Die Berechnung der Durchlässigkeitsziffer des Tones aus dem Verlauf der hydrodynamischen Spannungserscheinungen" (A Method of Calculating the Coefficient of Permeability of Clay from the Variation of Hydrodynamic Stress with Time), *Akademic der Wissenschaften in Wien. Sitzungsberichte.* Mathematisch-naturwissenshaftliche Klasse, Part IIa, Vol. 132, No. 3⁄4, p. 125–138 (in German) (reprinted in *From Theory to Practice in Soil Mechanics*, John Wiley, New York, 1960, p. 133–146); Translated into English by Clayton et al. (1995).

TERZAGHI, KARL (1924), "Die Theorie der hydrodynamischen Spannungerscheinungen und ihr erdbautechnisches Anwendungsgebiet," *Proceedings International Congress on Applied Mechanics*, Delft, p. 288–294 (in German).

TERZAGHI, KARL (1925a), *Erdbaumechanik auf bodenphysikalischer Grundlage*, Deuticke, Vienna (in German) (Forward translated into English in *From Theory to Practice in Soil Mechanics*, John Wiley, New York, 1960, p. 58–61). Also see Clayton et al. (1995).

TERZAGHI, KARL (1925b), "Principles of Soil Mechanics," *Engineering News Record*, Vol. 95, No. 19–23 and 25–27, p. 742–746, 796–800, 832–836, 874–878, 912–915, 987–990, 1026–1029, 1064–1068.

TERZAGHI, KARL (1925c), "Modern Conceptions Concerning Foundation Engineering," *Journal of the Boston Society of Civil Engineers*, Vol. 12, No. 10, p. 397–439.

TERZAGHI, KARL (1929), "The Mechanics of Shear Failures on Clay Slopes and the Creep of Retaining Walls," *Public Roads*, Vol. 10, No. 10, p. 177–192.

TERZAGHI, KARL (1934a), "Die Ursachen der Schiefstellung des Turmes von Pisa," *Der Bauingenieur*, Vol. 15, No. 1/2, p. 1–4 (reprinted in *From Theory to Practice in Soil Mechanics*, L. Bjerrum, A. Casagrande, R.B. Peck, and A.W. Skempton, Eds, p. 198–201, John Wiley, New York, 1960) (in German).

TERZAGHI, KARL (1934b), "Large Retaining Wall Tests," a series of articles in *Engineering News-Record*, Vol. 112; 2/1/34, 2/22/34, 3/8/34, 3/29/34, 4/19/34, and 5/17/34.

TERZAGHI, KARL (1936), "Discussion on Instruction in Soil Mechanics," *Proceedings, International Conference on Soil Mechanics and Foundation Engineering*, Vol. III, p. 261–263.

TERZAGHI, KARL (1939), "Soil Mechanics—A New Chapter in Engineering Science," *Journal of the Institution of Civil Engineers*, Vol. 12, p. 106–141.

TERZAGHI, KARL (1943), *Theoretical Soil Mechanics*, John Wiley, New York.

TERZAGHI, KARL AND PECK, RALPH B. (1967), *Soil Mechanics in Engineering Practice*, 2nd ed., John Wiley, New York.

TERZAGHI, KARL; PECK, RALPH B.; AND MESRI, GHOLAMREZA (1996), *Soil Mechanics in Engineering Practice*, 3rd ed., John Wiley, New York.

THORNLEY, J.H.; SPENCER, C.B.; AND ALBIN, PEDRO (1955), "Mexico's Palace of Fine Arts Settles 10 ft," *Civil Engineering*, p. 356–360, 576, 616, and 707, ASCE.

THORP, JAMES (1936), *Geography of the Soils of China*, National Geological Survey of China, Nanking.

TROXELL, HAROLD C. AND PETERSON, JOHN Q. (1937), *Flood in La Cañada Valley, California*, U.S. Geological Survey Water Supply Paper 796C.

TSCHEBOTARIOFF, GREGORY P. (1951), *Soil Mechanics, Foundations, and Earth Structures*, McGraw Hill, New York.

TURNER, A. KEITH AND SCHUSTER, ROBERT L., Eds. (1996), *Landslides—Investigation and Mitigation*, National Research Council, Transportation Research Board, Washington, DC.

U.S. ARMY (1992), Field Manual 5-410, *Military Soils Engineering.*

U.S. ARMY CORPS OF ENGINEERS (1998), *Design of Deep Foundations: TI 818-02,* Technical Instructions, DoD, p. 195.

U.S. NAVY (1982), *Soil Mechanics*, NAVFAC Design Manual 7.1, Naval Facilities Engineering Command, Arlington, VA.

VARNES, D.J. (1958), "Landslide Types and Processes," *Landslides and Engineering Practice*, Special Report 29, p. 20–47, E.B. Eckel, Ed., Highway Research Board, National Research Council.

VARNES, DAVID J. (1978), "Slope Movement Types and Processes," Chapter 2 in *Landslides: Analysis and Control*, Special Report 176, Robert L. Schuster and Raymond J. Krizek, Eds., Transportation Research Board, Highway Research Council.

VESIĆ, ALEKSANDAR S. (1973a), "Analysis of Ultimate Loads of Shallow Foundations," *Journal of the Soil Mechanics and Foundations Division*, Vol. 99, No. SM1, p. 45–73, ASCE.

VESIĆ, ALEKSANDAR S. (1973b), "Analysis of Ultimate Loads of Shallow Foundations," *Journal of the Soil Mechanics and Foundation Division*, Vol. 99, No. SM1, pp. 45–73, ASCE.

VOIGHT, BARRY (1990), "The 1985 Nevado del Ruiz Volcano Catastrophe: Anatomy and Retrospection," *Journal of Volcanology and Geothermal Research*, Vol. 42, p. 151–188.

WALLACE, G.B. AND OTTO, W.C. (1964), "Differential Settlement at Selfridge Air Force Base," *Journal of the Soil Mechanics and Foundations Division*, Vol. 90, No. SM5, p. 197–220, ASCE.

WESTERGAARD, H.M. (1938), "A Problem of Elasticity Suggested by a Problem of Soil Mechanics: Soft Material Reinforced by Numerous Strong Horizontal Sheets," *Contributions to the Mechanics of Solids*, MacMillan, New York.

WHETTON, NATHAN; WEAVER, JAMES; HUMPHREY, DANA; AND SANDFORD, THOMAS (1997), "Rubber Meets the Road in Maine," *Civil Engineering*, Vol. 67, No. 9, p. 60–63, ASCE.

WHITE, L. (1940), *Report . . . on the Collapse . . . Silos*, (unpublished) as cited by Tschebotarioff, 1951.

WILSON, R.C. AND KEEFER, D.K. (1985), "Predicting Areal Limits of Earthquake-Induced Landsliding," *Evaluating Earthquake Hazards in the Los Angeles Region*, USGS Professional Paper 1360, p. 317–345, J.I. Ziony, Ed., USGS.

WOLFF, THOMAS F. (1996), "Probabilistic Slope Stability in Theory and Practice," *Uncertainty in the Geologic Environment*, Vol. 1, p. 419–433.

WOOD, STUART (1977), *Heavy Construction: Equipment and Methods*, Prentice Hall, Upper Saddle River, NJ.

WRIGHT, STEPHEN G. (1985), "Limit Equilibrium Slope Analysis Procedures," *Design of Non-Impounding Waste Dumps*, p. 63–77, American Institute of Mining Engineers.

WU, TIEN H.; TANG, WILSON H.; AND EINSTEIN, HERBERT H. (1996), "Landslide Hazard and Risk Assessment," Chapter 6 in *Landslides: Investigation and Mitigation*, Special Report 247, Transportation Research Board, National Research Council, A.K. Turner and R.L. Schuster, Eds.

YOUD, T.L. (1973), "Factors Controlling Maximum and Minimum Densities of Sands," *Evaluation of Relative Density and its Role in Geotechnical Projects Involving Cohesionless Soils*, STP 523, p. 98–112, ASTM.

Subject Index

A

B

F

G

H

I

Q

R

T

U

V